Integers - Division

$\frac{+28}{+7} = +4$ $\frac{+28}{-7} = -4$

$\frac{-28}{-7} = +4$ $\frac{-28}{+7} = -4$

Integers - Multiply

$-6 \times -6 = +36$ ← same sign

$-6 \times +6 = -36$ ← different sign

HOLT
INTRODUCTORY
ALGEBRA 1

Russell F. Jacobs

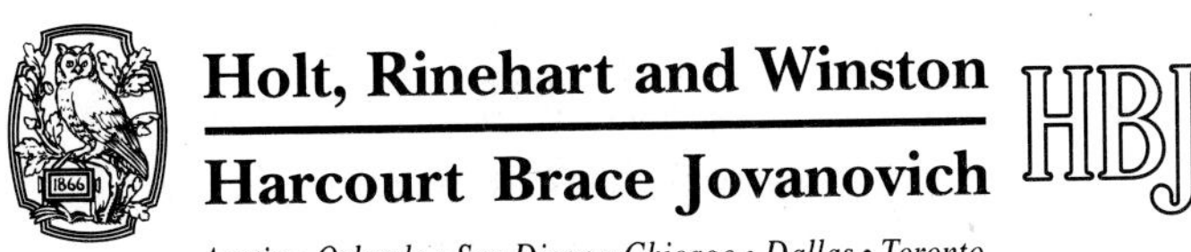
Holt, Rinehart and Winston
Harcourt Brace Jovanovich
Austin • Orlando • San Diego • Chicago • Dallas • Toronto

ABOUT THE AUTHOR

RUSSELL F. JACOBS

Formerly Mathematics Supervisor for the Phoenix Union High School System Phoenix, Arizona

EDITORIAL ADVISORS

Ronald Fisher
Supervisor of Mathematics
North Plainfield High School
North Plainfield, New Jersey

Elizabeth Anne Kloman
Mathematics Teacher
McPherson Junior High School
Orange, California

Tim Stephens
Mathematics Teacher
Canyon High School
Anaheim, California

Robert Tindall
Chairperson, Department of Mathematics
Shelby County High School
Shelbyville, Kentucky

Richard J. Wyllie
Chairperson, Department of Mathematics
Downers Grove South High School
Downers Grove, Illinois

Printed in the United States of America

ISBN 0-03-076979-5

7 8 9 040 99 98 97

Contents

Expressions and Formulas

CHAPTER 1

To store a number in a computer, you assign the number to a letter (**variable**), such as X. Once numbers are stored in **memory,** they can be added, subtracted, multiplied, and so on. The information is often stored on a **magnetic tape** or **diskette.**

1.1 Numerical Expressions

A ***numerical expression*** includes at least one of the operations of addition, subtraction, multiplication, or division. These are examples of numerical expressions.

$$15 \div 5 \qquad 12\tfrac{1}{2} + 3\tfrac{1}{3} \qquad 86 - 2.4 \qquad 7 \times 8$$

To find the ***value*** of a numerical expression, perform the indicated operation or operations.

Expression	$15 \div 5$	$12\frac{1}{2} + 3\frac{1}{3}$	$86 - 2.4$	7×8
Value	3	$15\frac{5}{6}$	83.6	56

An expression with the multiplication sign, such as 7×8, can also be written in the following ways.

$$7 \cdot 8, \quad \text{or } 7(8), \quad \text{or } (7)(8)$$

To find the value of an expression when there is more than one operation, follow these rules.

Order of Operations

1. When the only operations are addition and subtraction, perform the operations in order from left to right.
2. When the only operations are multiplication and division, perform the operations in order from left to right.
3. When multiplication or division is involved along with addition or subtraction, perform the multiplication and division before the addition and subtraction.

The Examples in the table illustrate the rules for order of operations.

Examples	Order	Computation
1a. $12 + 6 - 3$	[1] Add.	$\mathbf{12 + 6} - 3 = \mathbf{18} - 3$
	[2] Subtract.	$= 15$
b. $20 - 12 + 6$	[1] Subtract.	$\mathbf{20 - 12} + 6 = \mathbf{8} + 6$
	[2] Add.	$= 14$

Examples	Order	Computation
2a. $24 \div 6 \cdot 4$	[1] Divide. [2] Multiply.	$\mathbf{24 \div 6} \cdot 4 = \mathbf{4} \cdot 4$ $= 16$
b. $18 \cdot 2 \div 4$	[1] Multiply. [2] Divide.	$\mathbf{18 \cdot 2} \div 4 = \mathbf{36} \div 4$ $= 9$
3a. $12 \cdot 3 - 4 \div 2$	[1] Multiply. [2] Divide. [3] Subtract.	$\mathbf{12 \cdot 3} - 4 \div 2 = \mathbf{36} - 4 \div 2$ $36 - \mathbf{4 \div 2} = 36 - \mathbf{2}$ $= 34$
b. $24 - 18 \div 3 \cdot 4$	[1] Divide. [2] Multiply. [3] Subtract.	$24 - \mathbf{18 \div 3} \cdot 4 = 24 - \mathbf{6} \cdot 4$ $24 - \mathbf{6 \cdot 4} = 24 - \mathbf{24}$ $= 0$
c. $3 + 8 \cdot 9 \div 4$	[1] Multiply. [2] Divide. [3] Add.	$3 + \mathbf{8 \cdot 9} \div 4 = 3 + \mathbf{72} \div 4$ $3 + \mathbf{72 \div 4} = 3 + \mathbf{18}$ $= 21$

P–1 **Find the value of each expression.**

a. $6 - 4 + 7$ **b.** $12 + 5 - 9$ **c.** $18 \div 9 \cdot 7$

d. $8 \cdot 6 \div 3$ **e.** $11 - 6 \div 3 \cdot 5$ **f.** $3 + 5 \cdot 6 \div 10$

CLASSROOM EXERCISES

For Exercises 1–24, name the operation that should be performed first. (Examples 1–2)

1. $6 + 4 - 5$ **2.** $10 - 3 + 5$ **3.** $12 - 3 + 4$ **4.** $16 + 5 - 7$

5. $4 \cdot 6 \div 2$ **6.** $12 \div 4 \cdot 6$ **7.** $36 \div 6 \cdot 3$ **8.** $2 \cdot 10 \div 5$

9. $4 + 7 - 9$ **10.** $24 \div 3 \cdot 2$ **11.** $3 \cdot 8 \div 4$ **12.** $8 - 8 + 4$

(Example 3)

13. $6 + 9 \div 3$ **14.** $12 - 8 \div 2$ **15.** $20 - 2 \cdot 3 + 4$ **16.** $12 \div 3 + 5 \cdot 2$

17. $24 + 3 \cdot 4 \div 2$ **18.** $36 - 16 \div 2 \cdot 4$ **19.** $3 \cdot 8 - 4 \div 2$ **20.** $4 - 1 + 10 \div 2$

21. $15 - 3 \cdot 4$ **22.** $2 \cdot 4 + 3 \cdot 6$ **23.** $9 \div 3 + 12 \div 6$ **24.** $6 + 2 \cdot 5$

Complete. (Examples 1–3)

25. $6 + 3 - 4 = \underline{\ ?\ } - 4$
$= \underline{\ ?\ }$

26. $12 \div 3 \cdot 2 = \underline{\ ?\ } \cdot 2$
$= \underline{\ ?\ }$

27. $8 + 12 \div 4 \cdot 3 = 8 + 3 \cdot 3$
$= 8 + \underline{\ ?\ }$
$= \underline{\ ?\ }$

28. $9 - 3 + 18 \div 6 = 9 - 3 + 3$
$= \underline{\ ?\ } + 3$
$= \underline{\ ?\ }$

29. $24 \div 6 + 5 \cdot 3 = \underline{\ ?\ } + 5 \cdot 3$
$= \underline{\ ?\ } + \underline{\ ?\ }$
$= \underline{\ ?\ }$

30. $4 + 16 \cdot 2 \cdot 3 = 4 + \underline{\ ?\ } \cdot 3$
$= 4 + \underline{\ ?\ }$
$= \underline{\ ?\ }$

WRITTEN EXERCISES

Goal: To find the value of a numerical expression
Sample Problem: $13 - 10 \div 2 + 1$ **Answer:** 9

For Exercises 1–52, find the value of each expression.
(Example 1)

1. $5 + 6 - 3$ **2.** $15 - 3 + 6$ **3.** $29 + 6 - 8$ **4.** $23 - 9 + 14$
5. $19 - 7 + 16$ **6.** $8 + 12 - 15$ **7.** $16 + 22 - 30$ **8.** $45 - 41 + 36$

(Example 2)

9. $12 \div 3 \cdot 6$ **10.** $4 \cdot 6 \div 12$ **11.** $24 \div 4 \cdot 9$ **12.** $4 \cdot 15 \div 3$
13. $28 \div 7 \cdot 25$ **14.** $5 \cdot 6 \div 15$ **15.** $7 \cdot 8 \div 14$ **16.** $32 \div 4 \cdot 7$

(Example 3)

17. $12 - 4 \cdot 2$ **18.** $15 \div 3 + 7$ **19.** $8 \cdot 5 + 7$ **20.** $16 - 20 \div 5$
21. $45 - 2 + 51 \div 3$ **22.** $29 - 42 \div 2 + 15$ **23.** $33 - 9 + 21 \div 7$ **24.** $23 + 18 \div 6 + 52$

MIXED PRACTICE

25. $4 + 8 \cdot 3$ **26.** $5 + 3 \cdot 2 - 8$ **27.** $27 - 3 \cdot 4$ **28.** $42 - 3 \cdot 4 + 6$
29. $56 - 4 \div 4 - 10$ **30.** $19 + 5 \div 5 \cdot 4$ **31.** $14 + 3 \cdot 12$ **32.** $33 \div 3 \cdot 5 - 11$
33. $15 \div 3 \cdot 7$ **34.** $13 - 4 + 12$ **35.** $6 \cdot 8 \div 16$ **36.** $3 \div 3 \cdot 9 - 5$
37. $30 - 9 \cdot 3 + 16$ **38.** $3 + 9 - 8$ **39.** $54 - 8 \cdot 2 \div 4$ **40.** $12 \cdot 5 + 6 \div 6$
41. $16 - 9 + 11$ **42.** $16 \div 4 \cdot 3 - 6$ **43.** $32 \div 4 \cdot 7$ **44.** $16 - 8 - 4 + 12$
45. $3 \cdot 4 \cdot 2 \div 6$ **46.** $8 + 9 - 11$ **47.** $9 + 7 \cdot 8 + 25$ **48.** $3 + 9 \cdot 12$
49. $6 + 7 + 2 \cdot 4$ **50.** $14 \div 2 \cdot 3 - 5$ **51.** $36 \div 3 \div 6 \div 2$ **52.** $15 \cdot 4 - 18 \div 2$

1.2 Algebraic Expressions

A ***variable*** is a letter representing one or more numbers. The variable n represents the length of each side of the square at the right. The sum of the lengths of the sides is called the ***perimeter.*** The perimeter of the square can be represented in the following ways.

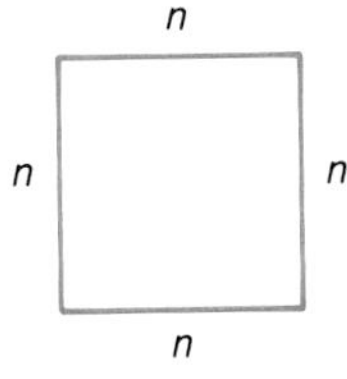

$$n + n + n + n \qquad \text{or} \qquad 4 \cdot n \qquad \text{or} \qquad 4(n)$$

Each of these expressions for perimeter is an algebraic expression. An ***algebraic expression*** contains at least one variable.

P–1 **Which of these are algebraic expressions?**

a. $5 \cdot y$ **b.** $15 \div 3$ **c.** $5.8 - 3$ **d.** $3.1m + 2$

An expression such as $4 \cdot n$ or $4(n)$ can also be written as $4n$.

$4n$ means "4 times the value of n."
$13.2t$ means "13.2 times the value of t."
$3xy$ means "3 times the value of x times the value of y."

To find the value of an algebraic expression, follow these steps. ***Evaluate*** means to find the value.

Steps for Evaluating an Algebraic Expression

1. Replace each variable with its value.
2. Perform the operations according to the rules for the order of operations.

EXAMPLE 1 Evaluate $36 + 9w$ when $w = 3$.

Solution:
[1] Replace the variable. ⟶ $36 + 9w = 36 + 9(3)$
[2] Multiply. ⟶ $= 36 + 27$
[3] Add. ⟶ $= 63$

P–2 **Evaluate each expression when the variable is equal to 4.**

a. $4k - 6$ **b.** $12s + 3s$ **c.** $12 \div p + 2 \cdot p$

An algebraic expression may contain more than one variable.

EXAMPLE 2 Evaluate $5x - 3y$ when $x = 7$ and $y = 8$.

Solution: [1] Replace the variables. ⟶ $5x - 3y = 5(7) - 3(8)$

[2] Multiply. ⟶ $= 35 - 24$

[3] Subtract. ⟶ $= 11$

P–3 **Evaluate each expression when $r = 9$ and $t = 6$.**

a. $5r - 2t$ **b.** $9t + 3r$ **c.** $27 \div r + 144 \div 2t$

CLASSROOM EXERCISES

Write the expression after each variable is replaced. (Step 1, Examples 1 and 2.)

1. $15p + 32$ when $p = 5$
2. $29 - 3k$ when $k = 8$
3. $24q + 15r$ when $q = 4$ and $r = 7$
4. $36a - 27b$ when $a = 13$ and $b = 10$
5. $24c + 18d$ when $c = 19$ and $d = 15$
6. $27m + 14 - 11g$ when $m = 15$ and $g = 10$
7. $2xy - y \cdot y$ when $x = 3$ and $y = 4$
8. $x \cdot x + 5xy + y \cdot y$ when $x = 5$ and $y = 6$

Evaluate each expression when the variable is equal to 2. (Example 1)

9. $8 - 3n$
10. $x + x$
11. $4y \div 2$
12. $7 - y + 2$
13. $3a + a$
14. $4 + 2y$
15. $6n \div 3$
16. $8 - t \div 2$

Evaluate each expression when $x = 4$ and $y = 10$. (Example 2)

17. $x + 2y$
18. $3x - y$
19. $5x - 2y$
20. $9x + 5y$
21. $12 \div x + 90 \div y$
22. $7x + 10y$
23. $15y - 5y$
24. $y - x + xy$

WRITTEN EXERCISES

Goal: To find the value of an algebraic expression for given replacements of the variable or variables

Sample Problem: Evaluate $8 + 4a - 6$ when $a = 3$. **Answer:** 14

Evaluate each expression when the variable is equal to 3. (Example 1)

1. $2a - 1$
2. $4n + 2$
3. $2 + 2y$
4. $1 + 4c$
5. $2 + g - 1$
6. $5 - k - 1$
7. $2 + 6 \div x$
8. $12 \div p \cdot 4$

For Exercises 9–32, evaluate each expression.
(Example 2)

9. $p + 8q$ when $p = 12$ and $q = 3$
10. $r + 12s$ when $r = 3$ and $s = 6$
11. $12c - 3d$ when $c = 5$ and $d = 3$
12. $15x - 8y$ when $x = 8$ and $y = 15$
13. $3j - 12 + 3k$ when $j = 6$ and $k = 8$
14. $16 \div p - 6 \cdot r$ when $p = 2$ and $r = 1$
15. $2v + 3 \cdot x$ when $v = 5$ and $x = 12$
16. $h - 7 \cdot 6 \div k$ when $h = 23$ and $k = 14$

MIXED PRACTICE

17. $8p - 19$ when $p = 6$
18. $16x + 3y$ when $x = 2$ and $y = 13$
19. $16 + 12s$ when $s = 7$
20. $12h - 25m$ when $h = 12$ and $m = 3$
21. $x \cdot x + 3x - 12$ when $x = 6$
22. $32r - 92$ when $r = 4$
23. $15 \div a + 10 \div b$ when $a = 3$ and $b = 2$
24. $16 - 5 \cdot d$ when $d = 2$
25. $16 + 78 \div q$ when $q = 3$
26. $14m - mp$ when $m = 5$ and $p = 10$
27. $3 \cdot s \div 5 - t$ when $s = 15$ and $t = 2$
28. $15 + 8z$ when $z = 12$
29. $3f - 8g$ when $f = 21$ and $g = 5$
30. $q \div 7 + b \cdot 9$ when $q = 119$ and $b = 8$
31. $3h - 5$ when $h = 12$
32. $8t + 2u$ when $t = 4$ and $u = 5$

MORE CHALLENGING EXERCISES

33. If $n = 25$, what is $n \cdot n$?
34. If $r = 10$, what is $r \cdot r - 1$?
35. If $a + b = 17$, what is the value of $a + b + 12$?
36. If $p - q = 29$, what is the value of $p - q - 10$?
37. If $r - s + 29 = 100$, what is the value of $r - s$?
38. If $t + u - 29 = 21$, what is the value of $t + u$?

REVIEW CAPSULE FOR SECTION 1.3

Round each number to the nearest tenth.

EXAMPLES: **a.** 7.05 **b.** 1.33

SOLUTIONS: **a.** Since the number in the hundredths place (5) is 5 or more, round <u>up</u> to 7.1.

b. Since the number in the hundredths place (3) is less than 5, round <u>down</u> to 1.3.

1. 2.07	**2.** 4.55	**3.** 6.00	**4.** 2.98	**5.** 1.05	**6.** 3.34
7. 9.25	**8.** 7.12	**9.** 5.93	**10.** 1.45	**11.** 3.99	**12.** 2.64

1.3 Using Formulas: Unit Price

You can save money if you use unit price to compare the costs of products. ***Unit price*** is the price per gram, per ounce, and so on.

DESCRIPTION	UPC ITEM
PAPER TOWELS 100 SHEETS	PT1004 302-IADO-1
UNIT PRICE $0.0095/SHEET	$0.95

The following word rule tells you how to find unit price.

Word Rule: Unit price equals the price of an item divided by the number of units.

A ***formula*** is a shorthand way of writing a word rule. In a formula, variables and symbols are used to represent words.

Formula: $U = p \div n$, or $U = \frac{p}{n}$

U = unit price
p = price of an item
n = number of units

EXAMPLE 1

Find the unit price of this can of soup. Write your answer to the nearest tenth of a cent.

Solution:

1. Write the formula. ⟶ $U = p \div n$
2. Identify known variables. ⟶ $p = 59¢$; $n = 16$
3. Replace the variables. ⟶ $U = 59 \div 16$
4. Divide to hundredths. ⟶ $U = 3.68¢$

The unit price is 3.7¢ per ounce.

Rounded to the nearest tenth.

P–1 **Replace the variables in the formula $U = p \div n$ with the given values.**

a. a 12–ounce can of peaches for 89¢
b. a 100–gram jar of oregano for 65¢

Definition

When you compare the unit prices of two or more sizes of a product, the size with the lower unit price is the ***better buy.*** This assumes that the items are of the same quality.

When the price of an item is \$1.00 or more, change the price to cents <u>before</u> replacing the variables.

EXAMPLE 2 **a.** Find the unit price of this can of soup to the nearest tenth of a cent.

b. Compare this with the unit price in Example 1 to determine the better buy.

Solutions: **a.** [1] $U = p \div n$

[2] $p = \$1.48$, or 148¢; $n = 48$

[3] $U = 148 \div 48$

[4] $U = 3.08¢$

To the nearest tenth, the unit price is 3.1¢ per ounce.

b. From Example 1, the unit price of the 16-ounce can is 3.7¢ per ounce. Since 3.1¢ is less than 3.7¢, the 48-ounce can is the better buy.

CLASSROOM EXERCISES

Subtract.

1. $27.3 - 9.5$ **2.** $0.282 - 0.105$ **3.** $1.673 - 0.938$ **4.** $286.4 - 89.9$ **5.** $17.35 - 3.855$ **6.** $2.000 - 1.763$

7. $8.27 - 4.18$ **8.** $15.23 - 9.67$ **9.** $112.62 - 78.301$ **10.** $2765.2 - 1823.7$

Divide.

11. $2.4 \div 4$ **12.** $0.27 \div 0.9$ **13.** $37 \div 1.8$ **14.** $0.7 \div 0.25$

15. $0.17\overline{)0.816}$ **16.** $0.08\overline{)1.84}$ **17.** $0.03\overline{)742}$ **18.** $30\overline{)0.095}$

Replace the variables in $U = p \div n$ with the known values.
(Step 3, Examples 1 and 2)

19. Price: 65¢ Size: 8 g

20. Price: 79¢ Size: 100 mL

21. Price: 98¢ Size: 3.5 oz

22. Price: 55¢ Size: 2 lb

23. Price: \$1.19 Size: 10 fl oz

24. Price: \$1.40 Size: 3 kg

25. Price: \$2.95 Size: 3 L

26. Price: \$3.49 Size: 24 oz

WRITTEN EXERCISES

Goal: To find and compare the unit prices of two sizes of a product to determine the better buy

Sample Problem: 23–ounce size: $1.59; 35–ounce size: $1.98

Answer: The unit price of the 23–ounce size is 6.9¢. The unit price of the 35–ounce size is 5.7¢. Since 5.7¢ is less than 6.9¢, the 35–ounce size is the better buy.

Find the unit price of each item to the nearest tenth of a cent. (Example 1)

	Price of Item	Number of Units	Unit Price
1.	76¢	20 oz	?
2.	$1.19	58 g	?
3.	59¢	13 fl oz	?
4.	$2.35	80 mL	?
5.	65¢	128 g	?
6.	99¢	32 oz	?
7.	$1.38	21 fl oz	?
8.	$2.95	205 g	?

9. The price of a 4–fluid ounce bottle of perfume is $29.95. Find the unit price of the perfume to the nearest tenth of a cent.

10. The price of a 7–ounce tube of toothpaste is $1.69. Find the unit price of the toothpaste to the nearest tenth of a cent.

Find the unit price of each pair of items to the nearest tenth of a cent. Compare to find which is the better buy. (Example 2)

	Price/Size	Unit Price	Price/Size	Unit Price	Better Buy
11.	$1.59/250 oz	?	$2.15/450 oz	?	?
12.	$3.49/2 kg	?	$6.89/5 kg	?	?
13.	79¢/16 oz	?	$1.19/24 oz	?	?
14.	$1.23/135 g	?	$1.58/195 g	?	?
15.	$1.15/3 lb	?	$2.09/5 lb	?	?
16.	$2.19/60 mL	?	$5.69/150 mL	?	?

17. The price of a 900–milliliter carton of juice is 89¢, and the price of the 2000–milliliter carton is $1.75. Find which size is the better buy.

18. The price of a 500–gram can of coffee is $3.49, and the price of the 1000–gram can is $6.95. Find which size is the better buy.

19. The price of an 18–ounce box of cereal is $1.59, and the price of a 24–ounce box is $2.09. Which size is the better buy?

20. The price of a 2–pound bag of flour is 49¢, and the price of the 5–pound bag is $1.19. Which size is the better buy?

MID-CHAPTER REVIEW

Find the value of each expression. (Section 1.1)

1. $8 - 6 + 4$
2. $7 - 5 + 12$
3. $5 \cdot 4 \div 2$
4. $8 \cdot 12 \div 6$
5. $3 + 8 \cdot 5$
6. $15 \div 3 + 4$
7. $25 \div 5 \cdot 9 - 3$
8. $7 - 5 \cdot 6 \div 10$

Evaluate each expression. (Section 1.2)

9. $12 - 4c$ when $c = 2$
10. $3r + 15$ when $r = 9$
11. $16y - 5z$ when $y = 3$ and $z = 7$
12. $17 - p + 3q$ when $p = 11$ and $q = 16$
13. $k + 3 - 4m$ when $k = 21$ and $m = 6$
14. $a \cdot a - b \div 3$ when $a = 7$ and $b = 12$

Solve each problem. (Section 1.3)

15. The price of a 12–ounce can of pears is 78¢. Find the unit price of the pears to the nearest tenth of a cent.

16. The price of a 16–ounce carton of milk is 49¢ and the price of a 64–ounce carton of milk is $1.58. Find which size is the better buy.

REVIEW CAPSULE FOR SECTION 1.4

Round each amount to the nearest whole dollar.

EXAMPLES: **a.** $1.04 **b.** $3.95

SOLUTIONS: **a.** Since the number in the tenths place (0) is less than 5, round down to $1.
b. Since the number in the tenths place (9) is 5 or more, round up to $4.

1. $5.05
2. $8.49
3. $6.90
4. $10.08
5. $18.25
6. $43.80
7. $52.55
8. $61.48
9. $36.85
10. $49.51
11. $38.05
12. $86.50
13. $24.09
14. $32.99
15. $74.52
16. $18.47
17. $54.08
18. $99.52

1.4 Using Formulas: Heating/Cooling Costs

One way to conserve fuel and to save heating costs is to set the thermostat at a lower temperature.

The following word rule and formula will help you find the estimated savings that result from doing this.

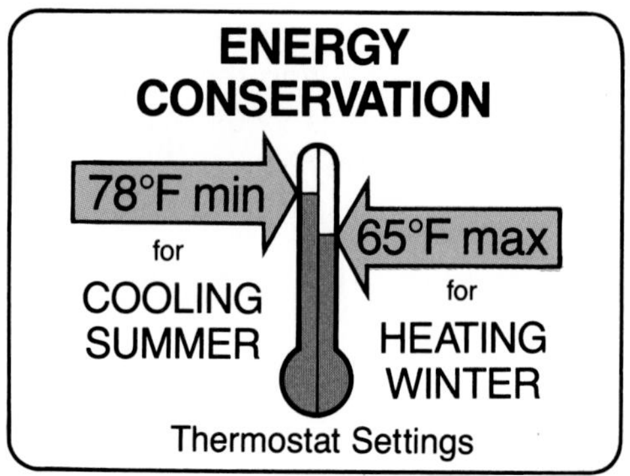

ENERGY CONSERVATION
78°F min
for
COOLING SUMMER
65°F max
for
HEATING WINTER
Thermostat Settings

Word Rule: The savings equals the heating zone factor times the yearly heating cost times the number of degrees Fahrenheit (F) that the thermostat is lowered.

Formula: $S = k \cdot c \cdot t$, or $S = kct$

***S* = savings**
***k* = heating zone factor**
***c* = yearly heating cost**
***t* = number of degrees lowered**

The heating zone factor, *k*, depends on where you live. The map at the left below shows the various heating zones. If you live in a heating zone where $k = 0$, you don't have to worry about heating your home.

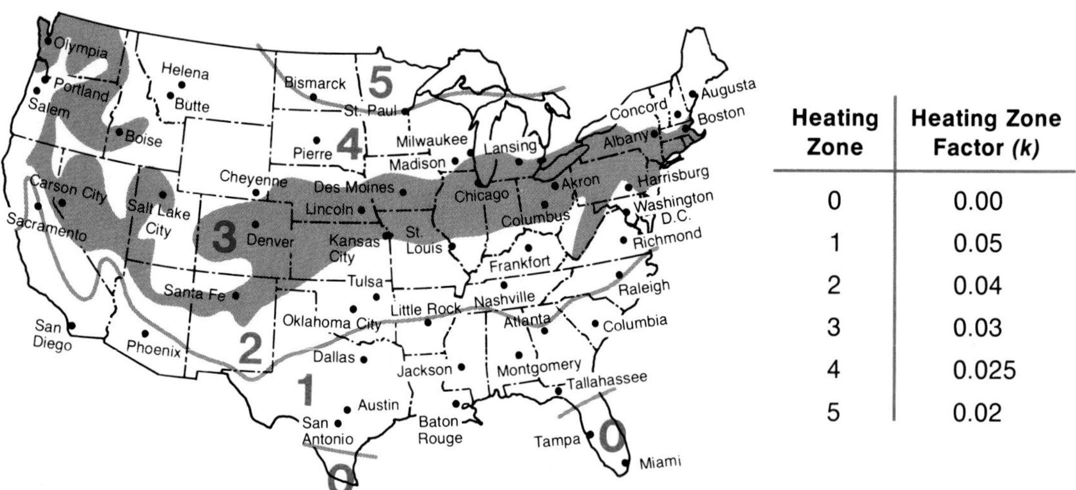

Heating Zone	Heating Zone Factor (*k*)
0	0.00
1	0.05
2	0.04
3	0.03
4	0.025
5	0.02

P–1 **Find the heating zone and the heating zone factor for each city.**

a. Tampa, Florida **b.** Chicago, Illinois **c.** Boise, Idaho

EXAMPLE 1

The Stewart family lives in Akron, Ohio. Last year they spent \$950 to heat their home. This year they will set the thermostat at 3°F lower. Find their estimated savings for this year. Write your answer to the nearest whole dollar.

Solution: From the map, Akron is in heating zone 3.

1. Write the formula. ⟶ $S = kct$
2. Identify known values. ⟶ $k = 0.03$, $c = \$950$, $t = 3°$ ◀ ***From the table for heating zone 3.***
3. Replace the variables and multiply. ⟶ $S = (0.03)(950)(3)$
   $S = \$85.50$

They will save about \$86.00. ◀ ***Rounded to the nearest dollar.***

You can conserve electricity and save money by setting the thermostat on the air conditioner at a higher temperature.

The following word rule and formula tell you how to compute the estimated savings.

Word Rule: The savings equals 0.02 times the yearly cooling costs times the number of degrees Fahrenheit that the thermostat is raised.

Formula: $S = 0.02ct$

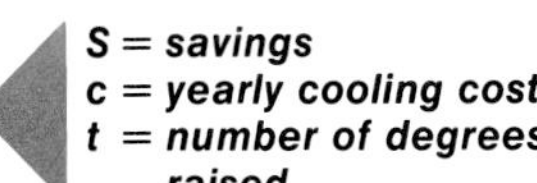

S = savings
c = yearly cooling costs
t = number of degrees raised

EXAMPLE 2

The Ruiz family spent \$426 last year to keep their house cool. This year they will set the thermostat on the air conditioner 3° higher. Find their estimated savings for this year. Write your answer to the nearest whole dollar.

Solution:

1. Write the formula. ⟶ $S = 0.02ct$
2. Identify known values. ⟶ $c = \$426$; $t = 3°$
3. Replace the variables and multiply. ⟶ $S = (0.02)(426)(3)$
   $S = \$25.56$

They will save about \$26.00.

Rounded to the nearest dollar.

CLASSROOM EXERCISES

Find **a.** *the heating zone and* **b.** *the heating zone factor for each city.* (Map, Table)

1. Dallas, Texas
2. Milwaukee, Wisconsin
3. Boston, Massachusetts
4. Tulsa, Oklahoma
5. Kansas City, Kansas
6. San Diego, California

Replace the variables in $S = kct$ with the known values. (Step 3, Example 1)

7. Heating zone: 4
Yearly heating cost: $725
Thermostat setting: lowered 2°

8. Heating zone: 2
Yearly heating cost: $690
Thermostat setting: lowered 3°

9. Heating zone: 1
Yearly heating cost: $585
Thermostat setting: lowered 4°

10. Heating zone: 5
Yearly heating cost: $780
Thermostat setting: lowered 5°

Replace the variables in $S = 0.02ct$ with the known values. (Step 3, Example 2)

11. Yearly cooling cost: $230
Thermostat setting: raised 2°

12. Yearly cooling cost: $390
Thermostat setting: raised 5°

13. Yearly cooling cost: $550
Thermostat setting: raised 3.5°

14. Yearly cooling cost: $465
Thermostat setting: raised 3°

WRITTEN EXERCISES

Goal: To use the formulas $S = kct$ and $S = 0.02ct$ to compute estimated savings on heating and cooling costs

Sample Problem: Find the estimated savings on yearly cooling costs of $280 if the thermostat is set 2.5° higher.

Answer: $14

Find the estimated savings on the yearly heating costs for each exercise. Round your answers to the nearest dollar. (Example 1)

	City	Yearly Heating Costs	Thermostat Setting	Estimated Savings
1.	Des Moines, Iowa	$845	lowered 2°	?
2.	Portland, Oregon	$730	lowered 3°	?
3.	Frankfort, Kentucky	$793	lowered 3°	?
4.	St. Louis, Missouri	$660	lowered 4°	?
5.	San Antonio, Texas	$690	lowered 3.5°	?

6. Find the estimated savings on yearly heating costs of $812 in Richmond, Virginia, if the thermostat is set 4° lower.

7. Find the estimated savings on yearly heating costs of $952 in Chicago, Illinois, if the thermostat is set 2° lower.

8. Find the estimated savings on yearly heating costs of $412 in Atlanta, Georgia, if the thermostat is set 4° lower.

9. Find the estimated savings on yearly heating costs of $1045 in Butte, Montana, if the thermostat is set 3° lower.

Find the estimated savings on the yearly cooling costs for each exercise. Round your answers to the nearest dollar. (Example 2)

	Yearly Cooling Costs	Thermostat Setting	Estimated Savings
10.	$375	raised 2°	?
11.	$280	raised 1°	?
12.	$420	raised 1.5°	?
13.	$515	raised 2.5°	?
14.	$495	raised 4°	?

15. Find the estimated savings on yearly cooling costs of $398 if the thermostat is set 3° higher.

16. Find the estimated savings on yearly cooling costs of $212 if the thermostat is set 5° higher.

MORE CHALLENGING EXERCISES

17. a. Evaluate $S = kct$ for $k = 0.02$, $c =$ $100, and $t = 3°$. Then evaluate S for $k = 0.04$, $c =$ $100, and $t = 3°$.
    b. How does S change when k changes from 0.02 to 0.04?
    c. If k is doubled while c and t remain the same, what is the effect on S?

18. a. Evaluate $S = 0.02ct$ for $c =$ $200 and $t = 2°$. Then evaluate S for $c =$ $400 and $t = 4°$.
    b. How does S change when c changes from $200 to $400 and t changes from 2° to 4°?
    c. If c and t are both doubled, what is the effect on S?

NON-ROUTINE PROBLEMS

19. There are 12 dollars in one dozen. How many dimes are there in one dozen?

20. If five days ago was the day after Saturday, what was the day before yesterday?

21. If 5 persons can pack 5 boxes of flowers in 5 minutes, how many persons are needed to pack 50 boxes in 50 minutes?

22. Complete the pattern.

    F ⅂ ꟻ T ? ⊥ N ? ?

PROBLEM SOLVING AND APPLICATIONS

1.5 Patterns and Formulas

You can write a word rule to state how two quantities are related. Then you can use the word rule to write a formula.

EXAMPLE 1

Several package prices are available for student pictures. A processing fee of $3 is added to each order. The table below shows how the package price and the total cost are related.

Package price: *p*	$10	$15	$20	$25	$30	$35	$40	$45
Total cost: *t*	$13	$18	$23	$28	$33	$38	$43	$48

a. Write a word rule for the pattern in the table.
b. Write a formula for the word rule.

Solutions: Look for a sum, a difference, a product, or a quotient that is common to each pair (package price, total cost).

Pattern: $10 + \mathbf{3} = 13$ $\quad 15 + \mathbf{3} = 18 \quad$ $20 + \mathbf{3} = 23$

$25 + \mathbf{3} = 28 \quad$ $30 + \mathbf{3} = 33 \quad$ $35 + \mathbf{3} = 38$, and so on

a. Word Rule: **Total cost equals package price added to 3.**

$$t \quad = \quad p \quad + \quad 3$$

b. Formula: $t = p + 3$

EXAMPLE 2

Eight students at Wilkins High School received the results of their first biology test. Their teacher made a table showing each number of correct answers and the test score for each number.

Number of correct answers: c	25	24	23	22	21	20	19	18
Test score: *s*	100	96	92	88	84	80	76	72

a. Write a word rule for the pattern in the table.
b. Write a formula for the word rule.

Solutions: Look for an operation that is common to each of these pairs, (correct answers, test score), in the table.

Pattern: $4 \times 25 = 100$ $\quad 4 \times 24 = 96$ $\quad 4 \times 23 = 92$

$4 \times 22 = 88$ $\quad 4 \times 21 = 84$ $\quad 4 \times 20 = 80$, and so on

a. Word Rule: **Test score equals 4 multiplied by number of correct answers.**

$s \quad = \quad 4 \quad \cdot \quad c$

b. Formula: $s = 4 \cdot c$, or $s = 4c$

CLASSROOM EXERCISES

For Exercises 1–5:

a. *State the operation that is common to each given pair of numbers (x,y) in the table.*

b. *Complete the table.*

1.

x	$350	$400	$450	$500	$550	$600	$650	$700	$750	$800
y	$50	$100	$150	$200	$250	?	?	?	?	?

2.

x	1	2	3	4	5	6	7	8	9	10
y	37	38	39	40	41	?	?	?	?	?

3.

x	3	4	5	6	7	8	9	10	11	12
y	$7.50	$10.00	$12.50	$15.00	$17.50	?	?	?	?	?

4.

x	12	14	16	18	20	22	24	26	28	30
y	2	4	6	8	?	?	?	?	?	?

5.

x	6	9	12	15	18	21	27	33	36	42
y	2	3	4	?	6	?	?	?	12	?

WRITTEN EXERCISES

Goal: To use patterns to write word rules and formulas

Sample Problem: Write a word rule and a formula for the pattern in the table.

Hours worked: *h*	1	2	3	4	5
Wages: *w*	\$6.50	\$13.00	\$19.50	\$26.00	\$32.50

Answers: Word Rule: The amount of wages, w, equals 6.50 multiplied by the hours worked, h.

Formula: $w = 6.50 \cdot h$, or $w = 6.50h$

*For Exercises 1–4, choose the word rule that corresponds to the pattern in the table. Write **a, b,** or **c.***

1.

Quiz score: *q*	100	90	80	70	60
Total score: *t*	105	95	85	75	65

a. The total score, T, equals 5 added to the quiz score, q.
b. The total score, T, equals 5 subtracted from the quiz score, q.
c. The quiz score, q, equals 5 multiplied by the total score, T.

2.

Distance traveled: *d*	50	100	150	200	250
Number of hours: *h*	1	2	3	4	5

a. The number of hours, h, equals 50 divided by the distance traveled, d.
b. The number of hours, h, equals the distance traveled, d, divided by 50.
c. The distance traveled, d, multiplied by 50 equals the number of hours, h.

3.

Total sales: *t*	\$250	\$300	\$350	\$400	\$450	\$500
Profit: *p*	\$25	\$75	\$125	\$175	\$225	\$275

a. The profit, p, multiplied by 225 equals the total sales, t.
b. The profit, p, equals 225 added to the total sales, t.
c. The profit, p, equals 225 subtracted from the total sales, t.

4.

Number of rows: *r*	4	5	6	10	12	15
Marchers in each row: *m*	15	12	10	6	5	4

a. The number of rows, r, equals the number of marchers per row, m, divided by 60.
b. The number of marchers per row, m, multiplied by 60 equals the number of rows, r.
c. The number of rows, r, multiplied by the number of marchers per row, m, equals 60.

Match each word rule in Exercises 5–8 with its corresponding formula in a–h.

5. The number of seniors, n, in the chorus equals the number of sophomores, s, divided by 2.
6. The present checking account balance, n, equals 2 subtracted from the previous balance, s.
7. The number of new club members, n, equals 2 added to the number of current members, s.
8. The total sales, s, equals 2 multiplied by the number of tickets sold, n.

a. $s = \frac{n}{2}$

b. $s = n - 2$

c. $n = s - 2$

d. $n = \frac{s}{2}$

e. $n = 2s$

f. $s = n + 2$

g. $n = s + 2$

h. $s = 2n$

For Exercises 9–12, complete the table. Then write a word rule and a formula.

9. This table shows how the ages of two brothers are related.

Reggie's age: *r*	16	18	20	22	24	26	28	30	32	34
Dexter's age: *d*	9	11	13	15	17	?	?	?	?	?

Word Rule: ? Formula: ?

10. This table shows how the lengths of two pieces of string are related.

Length of red string: *r*	12	16	20	24	28	32	36	40	44	48
Length of yellow string: *y*	24	32	40	48	56	?	?	?	?	?

Word Rule: ? Formula: ?

11. Mr. Yin's math class is learning how to find an average. This table shows how the total of their test scores and their average test score are related.

Total score: *T*	320	340	360	380	400	420	440	460	480	500
Average test score: *a*	64	68	72	76	80	?	?	?	?	?

Word Rule: ? Formula: ?

12. The owner of the Thrift Mart store is pricing merchandise. This table shows how the cost for an item and its price are related.

Owner's cost: *c*	\$8	\$10	\$12	\$14	\$16	\$20	\$24	\$28	\$34	\$40
Price: *p*	\$13	\$15	\$17	?	\$21	?	?	?	\$39	?

Word Rule: ? Formula: ?

Using Statistics

Counting Strategies

There are four students running for president of the Student Council: Rosa, Frank, Hiromi, and Madge. There are two candidates for vice-president: Caitlin and Ralph. The **tree diagram** at the right shows the eight possible pairings for president and vice-president.

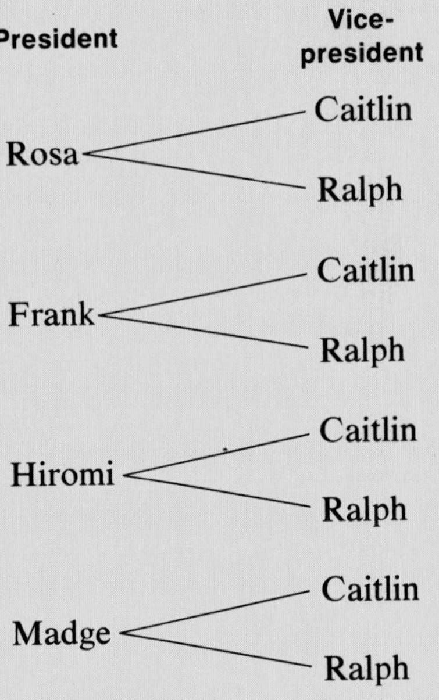

You can also use the **Counting Principle** to determine the number of pairings. Since each person running for president can be paired with each person running for vice-president, there are

p ways ⟶ 4×2, or **8** possible pairings.
q ways ⟶ (points to 2)

> **Counting Principle** If there are p ways that a first choice can be made, and q ways that a second choice can be made, then there are $p \times q$ ways of pairing the choices in that order.

EXERCISES *Solve. Use the method you prefer.*

1. Tim has three sports coats and five pairs of slacks. All are color coordinated. In how many ways can Tim choose an outfit?

2. Ann is choosing a telephone from seven colors, four styles, and either dial or pushbutton. How many different choices does she have?

Sandra plans to buy a Model Q car. She can select from the choices shown in the table.

3. How many choices does she have in all?
4. How many choices does she have if she decides on a cream interior?
5. How many choices does she have if she decides on a beige car with an automatic transmission?
6. How many choices does Sandra have if she decides on a white car with a brown interior?

MODEL Q CARS

Exterior	*Interior*	*Transmission*
Beige Black Blue Cream Red White	Cream Blue Brown	Automatic Manual

CHAPTER SUMMARY

IMPORTANT TERMS	Numerical expression *(p. 2)* Value *(p. 2)* Variable *(p. 5)* Perimeter *(p. 5)* Algebraic expression *(p. 5)*	Evaluate *(p. 5)* Unit price *(p. 8)* Formula *(p. 8)* Better buy *(p. 8)* Heating zone factor *(p. 12)*

IMPORTANT IDEAS

1. Rules for Order of Operations

a. When the only operations are addition and subtraction, perform the operations in order from left to right.

b. When the only operations are multiplication and division, perform the operations in order from left to right.

c. When multiplication or division is involved along with addition or subtraction, perform the multiplication and division before the addition and subtraction.

2. Steps for Evaluating an Algebraic Expression

a. Replace each variable with its value.

b. Perform the operations according to the rules for the order of operations.

3. Steps for Solving Problems with Formulas

a. Write the formula.

b. Identify known values.

c. Replace the variables.

d. Perform the indicated operations.

CHAPTER REVIEW

SECTION 1.1

Evaluate each expression.

1. $13 + 5 - 6$

2. $32 - 4 + 16$

3. $15 \div 3 \cdot 4$

4. $6 \cdot 9 \div 3$

5. $32 + 6 \cdot 5$

6. $20 \div 4 + 6$

7. $42 - 18 \div 6 \cdot 8$

8. $16 \div 2 + 2 \cdot 3$

SECTION 1.2

Evaluate each expression.

9. $19 + 3y$ when $y = 4$

10. $2t - 17$ when $t = 16$

11. $12a - 13b$ when $a = 5$ and $b = 3$

12. $3p - q + 2p$ when $p = 15$ and $q = 24$

13. $h - 3 \cdot k \div 4$ when $h = 7$ and $k = 8$

14. $16r \cdot r + 13r \cdot s$ when $r = 3$ and $s = 6$

Solve each problem.

SECTION 1.3

15. The price of a 100–milliliter bottle of cologne is $8.95, and the price of the 250–milliliter bottle is $21.95. Find which size is the better buy.

16. The price of a 480–gram box of detergent is $1.25. The price of the 1200–gram box is $2.29. Find which is the better buy.

SECTION 1.4

For Exercises 17–20, round your answer to the nearest dollar.

17. Find the estimated savings on yearly cooling costs of $512 if the thermostat is set 3° higher.

18. Find the estimated savings on yearly heating costs of $960 in Denver, Colorado, if the thermostat is set 4° lower.

19. Find the estimated savings on yearly heating costs of $690 in Columbia, South Carolina, if the thermostat is set 2° lower.

20. Find the estimated savings on yearly cooling costs of $345 if the thermostat is set 4° higher.

SECTION 1.5

Complete the table. Then write a word rule and a formula.

21.

Hours worked: *h*	1	2	3	4	5	6	7	8	9	10
Wages: *w*	$5	$10	$15	$20	$25	?	?	?	?	?

Word Rule: ? Formula: ?

22.

List price: *p*	$30	$40	$50	$60	$70	$80	$90	$100
Sale price: *s*	$15	$25	$35	$45	?	?	?	?

Word Rule: ? Formula: ?

CHAPTER 2
Properties of Numbers

The photograph below of Stonehenge, which is in England, shows a few of the 56 small white circular regions called *Aubrey Holes*. Using six stones spaced at 9, 9, 10, 9, 9, 10 holes apart and moving each stone one hole counterclockwise each year, the eclipses could be predicted by this "computer."

2.1 The Special Numbers 0 and 1

You know that the sum of 0 and any number is the number. This is the Addition Property of Zero.

You also know that the product of 0 and any number is 0. This is the Multiplication Property of Zero.

The number one also has a special property, the Multiplication Property of One. The product of 1 and any number is the number.

Addition Property of Zero

For any number a, $\quad 4\frac{1}{3} + 0 = 4\frac{1}{3}$

$$\mathbf{a + 0 = 0 + a = a} \qquad 2m = 0 + 2m$$

Multiplication Property of Zero

For any number a, $\quad (7)(0) = 0$

$$\mathbf{a \cdot 0 = 0 \cdot a = 0} \qquad 0 = (0)(0.6p)$$

Multiplication Property of One

For any number a, $\quad (17)(1) = 17$

$$\mathbf{a \cdot 1 = 1 \cdot a = a} \qquad 3y = 3y(1)$$

$a + 0 = a$, $0 + 4\frac{1}{3} = 4\frac{1}{3}$, and $3y = 3y(1)$ are all mathematical ***sentences.***

EXAMPLE 1 Find the value of the variable that makes each sentence true. Name the property that gives the reason for your choice.

a. $x + 8 = 8$ **b.** $p = (35.7)(0)$
c. $3.45 = 3.45 + n$ **d.** $t = (\frac{5}{8})(1)$

Solutions:

	Sentences	Value	Property
a.	$x + 8 = 8$	$x = 0$	Addition Property of Zero
b.	$p = (3.57)(0)$	$p = 0$	Multiplication Property of Zero
c.	$3.45 = 3.45 + n$	$n = 0$	Addition Property of Zero
d.	$t = (\frac{5}{8})(1)$	$t = \frac{5}{8}$	Multiplication Property of One

The Multiplication Property of One suggests that a number divided by itself equals 1. This is true for all numbers except 0.

$$5 \div 5 = 1 \qquad \frac{7.3}{7.3} = 1 \qquad 4\frac{1}{2} \div 4\frac{1}{2} = 1 \qquad \frac{n}{n} = 1,\ n \neq 0$$

EXAMPLE 2 Find the value of the variable that makes each sentence true.

a. $2(\frac{5}{5}) = t$ **b.** $\frac{13}{13}(k) = 45$ **c.** $13(\frac{r}{24}) = 13$

Solutions: **a.** Since $2(\frac{5}{5}) = t$ and $\frac{5}{5} = 1$, then $2(1) = t$. Thus, $t = 2$.

b. Since $\frac{13}{13}(k) = 45$ and $\frac{13}{13} = 1$, then $1(k) = 45$. Thus, $k = 45$.

c. Since $13(\frac{r}{24}) = 13$ and $13(1) = 13$, then $\frac{r}{24} = 1$. Thus, $r = 24$.

P–1 **Find the value of the variable that makes each sentence true.**

a. $r = (3\frac{1}{2})(0)$ **b.** $(1.5)(y) = 1.5$ **c.** $102 = 102 + b$ **d.** $q = 3(\frac{20}{20})$

CLASSROOM EXERCISES

Find the value of the variable that makes each sentence true. (Example 1)

1. $4 + n = 4$ **2.** $13 = t + 13$ **3.** $v = 18 + 0$ **4.** $x + 0 = 57$

5. $(1\frac{3}{4})(0) = k$ **6.** $0 = v\,(\frac{11}{12})$ **7.** $0.75 = (1)(c)$ **8.** $(1.2)(z) = 1.2$

(Example 2)

9. $\frac{5}{5}(m) = 24$ **10.** $\frac{18}{18}(11) = x$ **11.** $(\frac{s}{7})(5) = 5$ **12.** $16 = 16(\frac{91}{w})$

13. $(\frac{3}{3})0.3 = k$ **14.** $q = 1.5(\frac{8}{8})$ **15.** $(2.5)(\frac{31}{31}) = f$ **16.** $g = (\frac{5}{5})(0.2)$

WRITTEN EXERCISES

Goal: To use the Addition Property of Zero, the Multiplication Property of Zero and the Multiplication Property of One

Sample Problems: **a.** $x + 0 = 4$ **b.** $\frac{3}{3}(a) = 0$ **c.** $1 \cdot 6 = y$

Answers: **a.** $x = 4$ **b.** $a = 0$ **c.** $y = 6$

Find the value of the variable that makes each sentence true. (Example 1)

1. $6 + n = 6$ **2.** $t + 11 = 11$ **3.** $0.45 = 0 + x$ **4.** $0 + y = 0.8$

5. $12y = 0$ **6.** $0 = 3c$ **7.** $(m)(3.9) = 0$ **8.** $0 = (n)(14.1)$

9. $225n = 225$ **10.** $46 = 46t$ **11.** $4\frac{1}{2}k = 4\frac{1}{2}$ **12.** $2\frac{3}{8} = 2\frac{3}{8}m$

Find the value of the variable that makes each sentence true. (Example 2)

13. $x(\frac{3}{3}) = 13$
14. $15 = y(\frac{10}{10})$
15. $z(\frac{7}{7}) = 19$
16. $28 = k(\frac{15}{15})$
17. $9(\frac{5}{5}) = n$
18. $y = 34(\frac{11}{11})$
19. $k = (\frac{8}{8})23$
20. $(\frac{2}{2})39 = t$
21. $16 = 16(\frac{3}{c})$
22. $(\frac{z}{10})21 = 21$
23. $21.3 = 21.3(\frac{4}{m})$
24. $(\frac{f}{11})0.8 = 0.8$

MIXED PRACTICE

25. $17n = 17$
26. $23 = (t)(23)$
27. $75 + r = 75$
28. $16(\frac{8}{8}) = n$
29. $25 = x + 25$
30. $256k = 0$
31. $46 = (\frac{g}{8})(46)$
32. $0 = (s)(93)$
33. $(\frac{4}{4})w = 25$
34. $18 + 0 = v$
35. $(\frac{2}{2})10.83 = p$
36. $0.09c = 0.09$
37. $(\frac{3}{5})0 = x$
38. $\frac{5}{9} + g = \frac{5}{9}$
39. $4.9 = 4.9(\frac{g}{4})$
40. $(\frac{19}{19})r = 7.5$

MORE CHALLENGING EXERCISES

Write Any Number or None to show which values of the variable will make each sentence true.

41. $(n)(0) = 0$
42. $(\frac{3}{3})y = y$
43. $x + 1 = x$
44. $(\frac{n}{n})3 = 3$
45. $(0)(k) = 1.7$
46. $n = n + (\frac{6}{6})$
47. $(\frac{1.5}{1.5})t = t$
48. $v = 0 + v$

Write the value of the variable that makes each sentence true.

49. $12 + n(\frac{3}{3}) - 42(0) = 18$
50. $5 \div b + b(0) = 1$
51. $(11 - 2 \cdot 5)(3x + x \cdot 0) = 12$
52. $(5 \cdot z + 5 \cdot 0)(4 + z \cdot 0) = 20$
53. $(14 - 3 \cdot 4)(\frac{5}{5} \cdot x) = 24$
54. $(\frac{r}{36} \cdot 14) + 39(0) = 14$
55. $45 \div 3a + (\frac{3}{3})(0) = 3$
56. $(18 \cdot k - 18 \cdot 0)(5 + k \cdot 0) = 90$

REVIEW CAPSULE FOR SECTION 2.2

Multiply. (Pages 2–4)

1. a. 6×7
   b. 7×6
2. a. 3×9
   b. 9×3
3. a. 11×2
   b. 2×11
4. a. $10 \times 8 \times 1$
   b. $10 \times 1 \times 8$
5. a. $4 \times 3 \times 2$
   b. $3 \times 2 \times 4$
6. a. $6 \times 9 \times 7$
   b. $9 \times 7 \times 6$
7. a. $8 \times 10 \times 12$
   b. $12 \times 10 \times 8$
8. a. $5 \times 7 \times 5$
   b. $5 \times 5 \times 7$
9. a. $11 \times 10 \times 5$
   b. $5 \times 10 \times 11$
10. a. $6 \times 2 \times 3 \times 4$
    b. $3 \times 4 \times 6 \times 2$
11. a. $5 \times 9 \times 2 \times 2$
    b. $5 \times 2 \times 2 \times 9$
12. a. $3 \times 4 \times 5 \times 6$
    b. $4 \times 5 \times 3 \times 6$

2.2 Properties of Multiplication

You know that you can multiply two numbers in either order. This is the Commutative Property of Multiplication.

When you multiply three or more numbers, you may group them in any way. This is the Associative Property of Multiplication.

Commutative Property of Multiplication

Any two numbers can be multiplied in either order.

$4(\frac{1}{2}) = \frac{1}{2}(4)$

$3(y) = y(3)$

$$\mathbf{a \cdot b = b \cdot a}$$

Associative Property of Multiplication

The way numbers are grouped for multiplication does not affect the product.

$16(2 \cdot \frac{1}{3}) = (16 \cdot 2)\frac{1}{3}$

$(2g)g = 2(g \cdot g)$

$$\mathbf{(a \cdot b) \cdot c = a \cdot (b \cdot c)}$$

EXAMPLE 1 Find the value of the variable that makes each sentence true. Name the property or properties that give the reason for your choice.

a. $3 \cdot 4 = 4(d)$ **b.** $(12 \cdot 9)5 = 12(m \cdot 5)$ **c.** $(a \cdot 7)9 = 7 \cdot (9 \cdot 6)$

Solutions:

	Sentences	Value	Property
a.	$3 \cdot 4 = 4(d)$	$d = 3$	Commutative Property of Multiplication
b.	$(12 \cdot 9)5 = 12(m \cdot 5)$	$m = 9$	Associative Property of Multiplication
c.	$(a \cdot 7)9 = 7 \cdot (9 \cdot 6)$	$a = 6$	Commutative and Associative Properties of Multiplication

P–1 **Find the value of the variable that makes each sentence true. Which property or properties did you use?**

a. $(c \cdot 4)13 = 8(4 \cdot 13)$ **b.** $17p = 6(17)$ **c.** $3(5 \cdot 7) = (3 \cdot 7)h$

The Commutative and Associative Properties can be used to simplify algebraic expressions that involve multiplication.

EXAMPLE 2 Multiply: $(6x)(7y)$

Solution: $(6x)(7y)$ means $6 \cdot x \cdot 7 \cdot y$

[1] Change the order. ⟶ $6 \cdot x \cdot 7 \cdot y = 6 \cdot 7 \cdot x \cdot y$

[2] Group the factors and multiply. ⟶ $= (6 \cdot 7)(x \cdot y)$

$= 42xy$

Commutative and Associative Properties

P–2 **Multiply.**

a. $3(10y)$ **b.** $(6k)(8m)$ **c.** $(9t)(4s)$ **d.** $(11z)(3c)$

When a number or variable is multiplied by itself two or more times, the number is raised to a ***power.*** The second power of 3 is written as 3^2. The raised two is called an ***exponent.*** The exponent indicates how many times a number is multiplied by itself.

3^2 ← exponent (2), base (3)

$3^2 = 3 \cdot 3 = 9$

$n^2 = n \cdot n$

3^2 is read "3 squared;" n^2 is read "n squared"

$c^3 = c \cdot c \cdot c$

c^3 is read "c cubed"

EXAMPLE 3 Multiply: $(3a)(2ab)(5b^2)$

Solution: $(3a)(2ab)(5b^2) = 3 \cdot a \cdot 2 \cdot a \cdot b \cdot 5 \cdot b \cdot b$

$= \underline{3 \cdot 2 \cdot 5} \cdot \underline{a \cdot a} \cdot \underline{b \cdot b \cdot b}$

$= 30 \cdot a^2 \cdot b^3$

$= 30a^2b^3$

P–3 **Multiply.**

a. $(10c)(4cd)$ **b.** $(8m)(4p)(2mp)$ **c.** $(5x)(2yz)(3x^2)$

CLASSROOM EXERCISES

Find the value of the variable that makes each sentence true. (Example 1)

1. $g(8 \cdot 3) = 5(3 \cdot 8)$
2. $(7 \cdot 10)b = 7(10 \cdot 2)$
3. $10(20 \cdot 30) = (10 \cdot m)30$
4. $(17 \cdot 3)8 = 8(x \cdot 17)$
5. $(2 \cdot 16)30 = 2(16 \cdot y)$
6. $20(t \cdot 5) = 5(15 \cdot 20)$
7. $3 \cdot 4 \cdot 5 \cdot 2.3 = (3 \cdot s)(5 \cdot 2.3)$
8. $(0.5)(1.2y) = (0.5 \cdot 1.2)7$

Multiply. (Example 2)

9. $10(5y)$
10. $4(12t)$
11. $20(10t)$
12. $(11)(5k)$
13. $(4k)7$
14. $(9r)9$
15. $(6x)(3y)$
16. $(2a)(4b)$
17. $(14m)(0.5)$
18. $(12t)(0.3)$
19. $(100r)(0.3s)$
20. $(2r)(4.1s)$

(Example 3)

21. $(8a)(7a)$
22. $(2x)(3xy)$
23. $(3m^2)(8n)(2mn)$
24. $(3pq)(qr)(7p)$
25. $(2b^2)(3bc)(4c)$
26. $(6f^2g)(5gh)(3f)$
27. $(2rt)(3s^2)(rst)$
28. $(7n^2)(k^2n)(7jkn)$

WRITTEN EXERCISES

Goal: To simplify algebraic expressions that involve multiplication
Sample Problem: $(3x^2y)2xy$ **Answer:** $6x^3y^2$

Find the value of the variable that makes each sentence true. (Example 1)

1. $7 \cdot 6 = 6(a)$
2. $x(9) = 9 \cdot 4$
3. $(11 \cdot 2)7 = 11(n \cdot 7)$
4. $13(r \cdot 6) = (13 \cdot 4)6$
5. $(9 \cdot k)5 = (5 \cdot 9)4$
6. $3(t \cdot 7) = 7(15 \cdot 3)$
7. $4 \cdot 2 \cdot 7 \cdot 10 = (4 \cdot x)(7 \cdot 10)$
8. $3 \cdot 2 \cdot 7 \cdot 4 = (7 \cdot f)(4 \cdot 3)$
9. $(\frac{1}{3} \cdot b)\frac{1}{4} = (\frac{1}{4} \cdot \frac{1}{3})\frac{1}{2}$
10. $(0.2)(1.3y) = (0.2 \cdot 1.3)10$

For Exercises 11–50, multiply. (Example 2)

11. $5(7x)$
12. $6(4y)$
13. $7(6a)$
14. $9(10b)$
15. $(8t)7$
16. $(12r)8$
17. $(1.2n)7$
18. $(2.3k)4$
19. $(9a)(7b)$
20. $(8s)(12t)$
21. $(13x)(3y)$
22. $(6p)(14q)$
23. $(5r)(20s)$
24. $(32a)(6b)$
25. $(5.6p)(q)(2r)$
26. $(1.9r)(3s)(t)$

(Example 3)

27. $(13x)(3x)$ **28.** $(6y)(14y)$ **29.** $(12mn)(3n)$ **30.** $(3k)(17kt)$

31. $(9x)(4xy)$ **32.** $(7m)(6m^2n)$ **33.** $(10p)(0.2np)$ **34.** $(15st)(0.4t^2)$

35. $(3x^2)(8xy)$ **36.** $(20mn^2)(4mn)$ **37.** $(5k)(3gt^2)(8kt)$ **38.** $(9dp)(3a^2p)(3d)$

MIXED PRACTICE

39. $(2x)(9y)$ **40.** $(6m)(12n)$ **41.** $7(12t)$ **42.** $11(27r)$

43. $(0.5x^2)(18xy)$ **44.** $(12m)(0.5mn^2)$ **45.** $(15w)0.5$ **46.** $(28a)0.2$

47. $(2r)(3s)(5t)$ **48.** $(4a)(3b)(2c)$ **49.** $(r)(rs)(2st)(3t^2)$ **50.** $(2mn)(m^2p)(3p)(4n)$

MORE CHALLENGING EXERCISES

*Let * mean: "Square the first number. Then multiply by the second number." Compute each value.*

EXAMPLE: $3 * 2$ **SOLUTION:**

$$
\begin{aligned}
3 * 2 &= 3^2 \cdot 2 \\
&= 9 \cdot 2 \\
&= 18
\end{aligned}
$$

51. $2 * 3$ **52.** $3 * 5$ **53.** $5 * 3$ **54.** $6 * 2$

55. Compare Exercises 52 and 53. Is * a commutative operation? Explain.

Let ☆ mean: "Cube the first number and square the second number. Then multiply."

Compute each value.

EXAMPLE: $2 \star x$ when $x = 3$ **SOLUTION:**

$$
\begin{aligned}
2 \star x &= 2^3 \cdot x^2 \\
&= 8 \cdot 3^2 \\
&= 8 \cdot 9, \text{ or } 72
\end{aligned}
$$

56. $3 \star y$ when $y = 2$ **57.** $x \star 6$ when $x = 4$ **58.** $p \star 2$ when $p = 5$

59. $2 \star p$ when $p = 5$ **60.** $r \star 1$ when $r = 6$ **61.** $4 \star t$ when $t = 3$

62. Compare Exercises 58 and 59. Is ☆ a commutative operation? Explain.

REVIEW CAPSULE FOR SECTION 2.3

Write without a multiplication symbol. (Pages 2–4)

1. $3 \cdot 6 + 3 \cdot 5$ **2.** $4 \cdot 7 - 2 \cdot 7$ **3.** $5 \cdot 9 + 7 \cdot 9$ **4.** $10 \cdot 4 - 10 \cdot 1$

5. $6 \cdot p - 6 \cdot q$ **6.** $5 \cdot a + 5 \cdot 2$ **7.** $2r \cdot r + 2r \cdot 5$ **8.** $3d \cdot 4 - 3d \cdot d$

9. $c \cdot 8 + d \cdot 8$ **10.** $f \cdot 12 - g \cdot 12$ **11.** $k \cdot m - j \cdot m$ **12.** $z \cdot y + z \cdot x$

2.3 Writing Products As Sums or Differences

Here are two ways to find the value of $6(29 + 24)$.

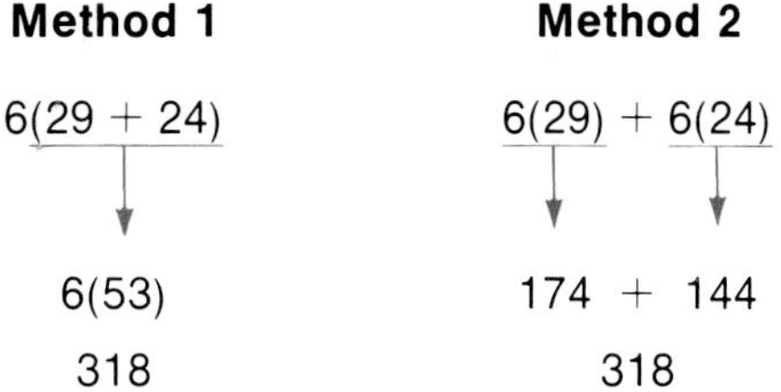

Method 1
6(29 + 24)
6(53)
318
Method 2
6(29) + 6(24)
174 + 144
318

Thus, $6(29 + 24) = 6(29) + 6(24)$. This illustrates the <u>Distributive Property of Multiplication over Addition</u>.

The expression $3(12 - 8)$ can also be evaluated in two ways.

Method 1	**Method 2**
$3(12 - 8)$	$3(12) - 3(8)$
↓	↓ ↓
$3(4)$	$36 - 24$
12	12

You can see that $3(12 - 8) = 3(12) - 3(8)$. This illustrates the <u>Distributive Property of Multiplication over Subtraction</u>.

Distributive Property of Multiplication over Addition

For any numbers a, b, and c,

1. $\mathbf{a(b + c) = ab + ac}$, and $\qquad 3(2 + w) = 3 \cdot 2 + 3w$
2. $\mathbf{(b + c)a = ba + ca}$. $\qquad (4 + 1)x = 4x + 1x$

Distributive Property of Multiplication over Subtraction

For any numbers a, b, and c,

1. $\mathbf{a(b - c) = ab - ac}$, and $\qquad 8(h - 3) = 8h - 8 \cdot 3$
2. $\mathbf{(b - c)a = ba - ca}$. $\qquad (x - 1)4 = x \cdot 4 - 1 \cdot 4$

The Distributive Properties can be used to write products as sums or differences.

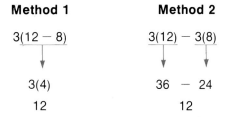

EXAMPLE 1

Write each product as a sum or difference.

a. $3(m + 3)$ **b.** $(k - 6)4$ **c.** $5(x - y)$

Solutions:	Product	Distributive Property	Sum or Difference
a.	$3(m + 3)$	$3 \cdot m + 3 \cdot 3$	$3m + 9$
b.	$(k - 6)4$	$k \cdot 4 - 6 \cdot 4$	$4k - 24$
c.	$5(x - y)$	$5 \cdot x - 5 \cdot y$	$5x - 5y$

P–1 **Write each product as a sum or difference.**

a. $11(4 + k)$ **b.** $(a - 4)3$ **c.** $(5f + g)16$ **d.** $2(3r - 5s)$

EXAMPLE 2

a. Write $2m(3p + 5m)$ as a sum of two terms.
b. Write $16v(2v - 3z)$ as a difference of two terms.

Solutions: **a.**
$$\begin{aligned} 2m(3p + 5m) &= (2m \cdot 3p) + (2m \cdot 5m) \\ &= (2 \cdot 3 \cdot m \cdot p) + (2 \cdot 5 \cdot m \cdot m) \\ &= 6mp + 10m^2 \end{aligned}$$

b.
$$\begin{aligned} 16v(2v - 3z) &= (16v \cdot 2v) - (16v \cdot 3z) \\ &= (16 \cdot 2 \cdot v \cdot v) - (16 \cdot 3 \cdot v \cdot z) \\ &= 32v^2 - 48vz \end{aligned}$$

P–2 **Write each product as a sum or difference.**

a. $(3c - 4d)6c$ **b.** $2a(a + 14)$ **c.** $(2g + f)3f$ **d.** $5rs(6s - 4r^2)$

When evaluating numerical expressions that involve parentheses, it may be simpler not to use the Distributive Property. That is, perform the operations inside the parentheses first. Then multiply. For example,

$$\begin{aligned} (45 + 32)12 &= (77)12 \\ &= 924 \end{aligned}$$

P–3 **Evaluate: a. 16(20 + 15) b. 32(24 − 18) c. (81 − 29)22**

CLASSROOM EXERCISES

For Exercises 1–28, write each product as a sum or difference. (Example 1)

1. $(5 + c)7$ **2.** $6(m - 4)$ **3.** $12(j - 8)$ **4.** $(15 + a)22$
5. $4(5 - d)$ **6.** $(2p + 5)7$ **7.** $(9 - 5w)9$ **8.** $13(2t + 10)$
9. $(7 + 2a)4$ **10.** $3(15 + 6r)$ **11.** $(19m - 9)10$ **12.** $5(42 + 13x)$

(Example 2)

13. $x(x + y)$ **14.** $(r - s)7s$ **15.** $2m(m + 2)$ **16.** $(q - 3)3q$
17. $3q(4p + 2q)$ **18.** $(4n - 6)5n$ **19.** $a(a - 1.8)$ **20.** $(20 + 4t)0.5$
21. $(5m - n)2n$ **22.** $2c(3 + 4c)$ **23.** $2k(3k - k^2)$ **24.** $(3p + 15q)p^2$
25. $2v(3m + 6v)$ **26.** $(12r - f)3f$ **27.** $(7s + k^2)5s$ **28.** $2.1p(11p - 9a)$

Evaluate. (P-3)

29. $(8 + 13)15$ **30.** $(12 + 21)16$ **31.** $25(32 - 18)$ **32.** $35(41 - 25)$
33. $52(24 + 33)$ **34.** $(55 - 20)40$ **35.** $(73 + 87)34$ **36.** $75(107 - 35)$

WRITTEN EXERCISES

Goal: To write a product as a sum or difference of two terms
Sample Problem: $(8 + y)4$ **Answer:** $32 + 4y$

For Exercises 1–40, write each product as a sum or difference. (Example 1)

1. $4(4 - d)$ **2.** $(t + 5)2$ **3.** $(3g + 15)3$ **4.** $7(9 - 5v)$
5. $(r + 6)5$ **6.** $(8 - a)8$ **7.** $7(5w - 3)$ **8.** $(18 + 6y)5$
9. $10(r + 3)$ **10.** $9(f - 3)$ **11.** $(14 - 3m)6$ **12.** $(2x + 16)7$

(Example 2)

13. $(m + t)15m$ **14.** $2q(p - q)$ **15.** $x(r + 2x)$ **16.** $(7 - k)yk$
17. $(0.5n + 8)4n$ **18.** $t(7s + 0.4t)$ **19.** $p(6p + 3q)$ **20.** $(8y + 12z)2y$
21. $(x + y)2x$ **22.** $c(3b + c)$ **23.** $3r(4r + 5v)$ **24.** $(4a + 3b)ab$

MIXED PRACTICE

25. $4k(7k + 5g)$ **26.** $4(a + 3)$ **27.** $(6t - 7s)2$ **28.** $5(3x - 9)$
29. $4(2a + b)$ **30.** $(12c - 9cx)3x$ **31.** $16(12 + 7s)$ **32.** $(s - 13)s$
33. $(5m - 2n)m$ **34.** $3(6p + r)$ **35.** $6y(4yz - z^2)$ **36.** $(2d + 9f)d^2f$
37. $(g - 5)8$ **38.** $3h(k + 4h)$ **39.** $(0.9m + q)15$ **40.** $0.5m(8m - 30)$

MORE CHALLENGING EXERCISES

Write each product as a sum of three terms.

EXAMPLE: $3c(c^2 + 5c + 9)$ **SOLUTION:** $3c^3 + 15c^2 + 27c$

41. $4(3a^2 + 9a + 7)$ **42.** $(6t^2 + 2t + 12)9$ **43.** $(2x^3 + 10x + 7)5x$

44. $8k(7k^2 + 9k + 11)$ **45.** $(3m^2 + 12mn + 9)4mn$ **46.** $3b(a^2b + 4ab + 5ab^2)$

47. $6r(9r^2 + 9r + 9)$ **48.** $(4s^3 + 11s^2 + 3)7s$ **49.** $(12y^2z + 8y^2 + 9z)2yz$

NON-ROUTINE PROBLEMS

50. Write plus signs and decimal points between the digits to make each equation true.

Example: 3 4 1 2 5 6 = 12.72

$3 + 4.12 + 5.6 = 12.72$

a. 9 1 4 6 7 3 = 164.46

b. 4 4 3 8 6 5 = 18.03

51. A building has six stories that are each the same height. It takes an elevator 6 seconds to rise to the third floor. How many additional seconds will it take the elevator to rise from the third floor to the sixth floor?

MID-CHAPTER REVIEW

Write the value of the variable that makes each sentence true. (Section 2.1)

1. $12.5 + t = 12.5$ **2.** $0.26 = r + 0.26$ **3.** $0 = 3\frac{1}{4}m$

4. $(k)(21.7) = 0$ **5.** $275 = (\frac{4}{4})(a)$ **6.** $(p)(38) = 38$

7. $14.3 = b + 0$ **8.** $0 + w = \frac{3}{5}$ **9.** $16 = (\frac{y}{5})(16)$

Multiply. (Section 2.2)

10. $8(7t)$ **11.** $(9c)(7)$ **12.** $(24w)2$ **13.** $3(18r)$

14. $(1.2p)(50q)$ **15.** $(102d)(0.5c)$ **16.** $(3x)(6x)$ **17.** $(28y)(5y)$

18. $(5m^2n)(3mn)$ **19.** $(4ab^2)(9ac)$ **20.** $(4rs)(2st)(10rt)$ **21.** $(3pq^2)(7qr)(6p^2r)$

Write each product as a sum or difference. (Section 2.3)

22. $4(7 - 4r)$ **23.** $(3c + 9)15$ **24.** $3y(7y - 8)$

25. $(9a + 7)6a$ **26.** $7r(2r - s)$ **27.** $(5t + n)8n$

28. $6k(12kt - 32)$ **29.** $(0.5rs + 1.5s)(4s)$ **30.** $8pq(2.5p^2q - 0.5q)$

2.4 Writing Sums or Differences as Products

The Distributive Property can be used to write sums or differences as products. If the terms of the sum or difference share a ***common factor,*** a product can be written with the common factor as one of the factors. This is called ***factoring.***

Sum	Product	Difference	Product
$\mathbf{a}b + \mathbf{a}c =$	$\mathbf{a}(b + c)$	$\mathbf{a}b - \mathbf{a}c =$	$\mathbf{a}(b - c)$
$\mathbf{2}x + \mathbf{2}y =$	$\mathbf{2}(x + y)$	$\mathbf{3}r - \mathbf{3} \cdot 1 =$	$\mathbf{3}(r - 1)$
$b\mathbf{a} + c\mathbf{a} =$	$(b + c)\mathbf{a}$	$b\mathbf{a} - c\mathbf{a} =$	$(b - c)\mathbf{a}$
$4\mathbf{t} + s\mathbf{t} =$	$(4 + s)\mathbf{t}$	$k\mathbf{x} - 1\mathbf{x} =$	$(k - 1)\mathbf{x}$

EXAMPLE 1 Write $4k - 4j$ as a product.

Solution:

[1] Factor each term. ⟶ $\mathbf{4}k - \mathbf{4}j = (\mathbf{4} \cdot k) - (\mathbf{4} \cdot j)$

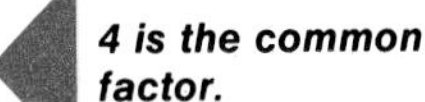

4 is the common factor.

[2] Write a product using the common factor, 4. ⟶ $= \mathbf{4}(k - j)$

By the Distributive Property

P–1 **Write each expression as a product.**

a. $4t - 4$ **b.** $12m + 12$ **c.** $8q + 8p$ **d.** $17j - j$

EXAMPLE 2 Factor: $x^2 - 5x$

Solution: $x^2 - 5x = x \cdot x - 5 \cdot x$

$= (x - 5)x$

$= x(x - 5)$

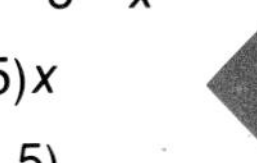

x is the common factor.

EXAMPLE 3 Factor: $5y + 10$

Solution: $5y + 10 = 5 \cdot y + 5 \cdot 2$

$= 5(y + 2)$

P–2 **Factor each of the following.**

a. $m^2 + 7m$ **b.** $r^2 - 12r$ **c.** $6k - 36$ **d.** $9t - 45$

CLASSROOM EXERCISES

For Exercises 1–24, factor each expression. (Example 1)

1. $3p + 3q$ **2.** $5r - 5s$ **3.** $12m - 12q$ **4.** $x(6) - 5(6)$

5. $2r - 2$ **6.** $10m + 10$ **7.** $xr - 3x$ **8.** $9w - hw$

(Example 2)

9. $n^2 + 2n$ **10.** $a^2 - 5a$ **11.** $t^2 - 10t$ **12.** $y^2 + 100y$

13. $5w + w^2$ **14.** $8t + t^2$ **15.** $12c - c^2$ **16.** $9m - m^2$

(Example 3)

17. $3x - 6$ **18.** $2m + 10$ **19.** $2y - 6$ **20.** $5k - 10$

21. $6b + 12$ **22.** $5y - 15$ **23.** $12 + 3r$ **24.** $20 - 4t$

WRITTEN EXERCISES

Goal: To factor an algebraic expression

Sample Problem: $32 + 4y$ **Answer:** $4(8 + y)$ or $(8 + y)4$

For Exercises 1–32, factor each expression. (Example 1)

1. $7k + 7n$ **2.** $13r - 13t$ **3.** $19p - 19q$ **4.** $15k + 15a$

5. $h(5) - 7(5)$ **6.** $8(m) + 4(8)$ **7.** $4y - y$ **8.** $5y - yz$

(Example 2)

9. $y^2 + 5y$ **10.** $p^2 - 9p$ **11.** $k^2 - 4k$ **12.** $h^2 + 3h$

13. $13t + t^2$ **14.** $15b - b^2$ **15.** $\frac{1}{4}a - a^2$ **16.** $\frac{4}{5}r + r^2$

(Example 3)

17. $2n + 4$ **18.** $7y - 14$ **19.** $2r - 6$ **20.** $2t + 18$

21. $3y + 15$ **22.** $15 + 5m$ **23.** $21 + 7c$ **24.** $18 - 9f$

MIXED PRACTICE

25. $\frac{1}{2}y + \frac{1}{2}$ **26.** $5p - 35$ **27.** $7n + 28$ **28.** $\frac{3}{4} - \frac{3}{4}y$

29. $tn + tq$ **30.** $3g^2 + g$ **31.** $na - ta$ **32.** $36 - 9h$

REVIEW CAPSULE FOR SECTION 2.5

Choose the unit you would use to measure each of the following.

1. Length of a soccer field **a.** meter **b.** centimeter **c.** millimeter

2. Width of a book **a.** kilometer **b.** meter **c.** centimeter

3. Height of a door **a.** kilometer **b.** millimeter **c.** meter

4. Length of a state park **a.** meter **b.** kilometer **c.** centimeter

2.5 Geometry Formulas: $P = 2(l + w)$; $A = s^2$

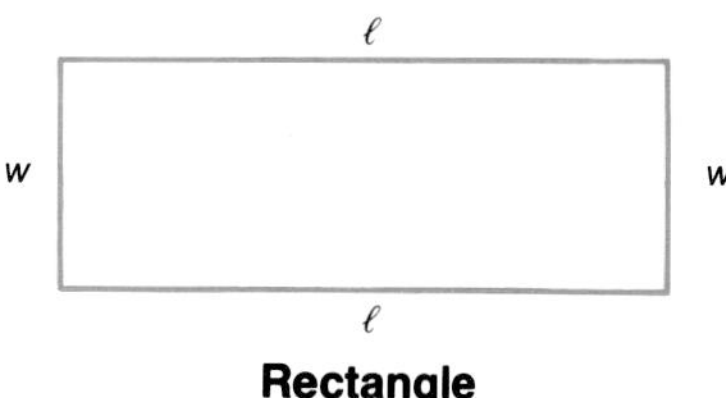

Rectangle

The ***perimeter*** of a figure is the distance around it. To find perimeter, add the lengths of the sides.

The following word rule and formula tell how to find the perimeter of any rectangle.

Word Rule: The perimeter of a rectangle equals twice the length plus twice the width.

Formula: $p = 2\ell + 2w$, or by factoring, $p = 2(\ell + w)$

p = perimeter
ℓ = length
w = width

Perimeter is expressed in linear units such as kilometers (km), meters (m), centimeters (cm), miles (mi), yards (yd), feet (ft), and so on.

When you replace a variable in the perimeter formula with a number, you are substituting for the variable.

EXAMPLE 1

The length of a rectangular park is 425 yards and its width is 200 yards. Find the perimeter.

Solution:
1. Write the formula. ——→ $p = 2(\ell + w)$
2. Identify known values. ——→ $\ell = 425$ yd; $w = 200$ yd
3. Substitute in the formula and add. ——→ $p = 2(425 + 200)$
4. Multiply. ——→ $p = 2(625)$

$p = 1250$

The perimeter is 1250 yards.

A **square** is a rectangle with four equal sides. The following word rule and formula tell you how to find the area of a square.

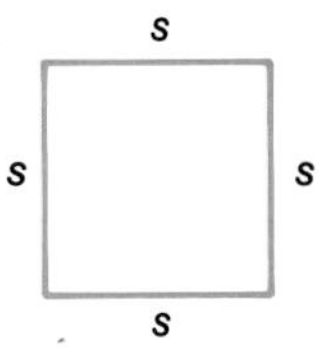

Square

Word Rule: The area of a square equals the product of the lengths of any two sides.

Formula: $A = s \cdot s$, or $A = s^2$

A = area
s = length of a side

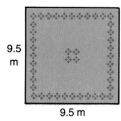

EXAMPLE 2 One side of a square rug is 9.5 meters long. Find the area of the rug.

9.5 m

9.5 m

Solution:

[1] Write the formula. ⟶ $A = s^2$

[2] Identify known values. ⟶ $s = 9.5$ m

[3] Substitute and multiply. ⟶ $A = (9.5)(9.5) = 90.25$

The area of the rug is 90.25 m². ◀ ***Remember to express area in square units.***

CLASSROOM EXERCISES

Find the perimeter of each rectangle. (Example 1)

1. $\ell = 10$ cm; $w = 5$ cm **2.** $\ell = 5$ m; $w = 3$ m **3.** $\ell = 2.5$ cm; $w = 1.5$ cm

4. $\ell = 1.2$ km; $w = 0.3$ km **5.** $\ell = 13$ ft; $w = 10$ ft **6.** $\ell = 7$ in; $w = 3$ in

7. $\ell = 4$ yd; $w = 2$ yd **8.** $\ell = 15$ mi; $w = 10$ mi **9.** $\ell = 24$ ft; $w = 20$ ft

Find the area of each square. (Example 2)

10. $s = 8$ cm **11.** $s = 12$ km **12.** $s = 0.5$ m **13.** $s = 1.2$ m

14. $s = 15$ mm **15.** $s = 12$ yd **16.** $s = 8$ ft **17.** $s = 8$ mi

WRITTEN EXERCISES

Goal: To solve word problems that involve perimeter and area
Sample Problem: Each side of a square room is 12 feet long. Find the area.
Answer: The area is 144 ft².

For Exercises 1–14, solve each problem. (Example 1)

1. A rectangular yard is 3 meters long and 2.7 meters wide. How many meters of fence are needed to enclose the yard?

2. A rectangular garden is 5 yards long and 3 yards wide. What is the perimeter of the garden?

3. Two tennis courts are laid out side by side. The total length of the courts is 21.95 meters and the width is 23.7 meters. Find the perimeter.

4. The rectangular building in which spacecraft are assembled at the Kennedy Space Center in Florida is 654 feet long and 474 feet wide. Find the perimeter.

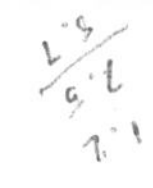

(Example 2)

5. Floor tiles are sold in squares that are 30 centimeters on a side. What is the area of each tile?

6. A city square is 40.5 meters on each side. Find the area of the square.

7. The length of each side of a square box is 14 inches. Find the area of the bottom of the box.

8. Each side of a square piece of sheet metal is 1 yard long. Find the area of the piece.

MIXED PRACTICE

9. A square slab of concrete for the foundation of a building is 16.5 meters long on each side. Find the area.

10. A swimming pool is 164 feet long and 80 feet wide. Find the perimeter of the pool.

11. The formula for the area of a rectangle is $\boldsymbol{A} = \boldsymbol{\ell w}$ where ℓ is the length and w is the width. Use this formula to find the area of the roof below.

12. The formula for the perimeter of a square is $\boldsymbol{p} = \boldsymbol{4s}$ where s is the length of a side. Use this formula to find the perimeter of the roof below.

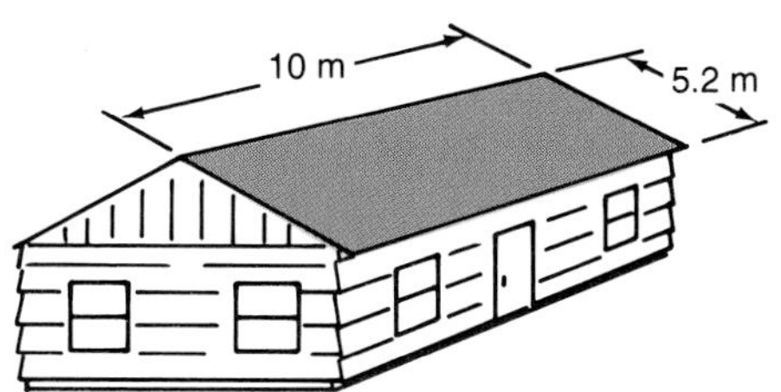

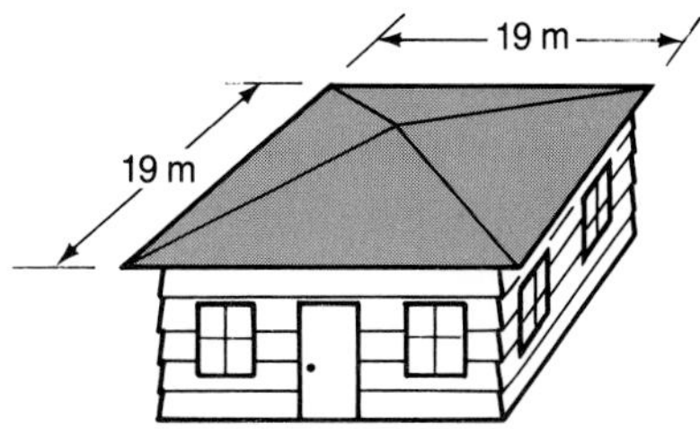

13. Each floor of a parking garage is 42 yards wide and 108 yards long. The floors are rectangular in shape. Find the area of each floor.

14. A square picture frame is 33 centimeters long on each side. Find the perimeter of the frame.

REVIEW CAPSULE FOR SECTION 2.6

Write each product as a sum or difference. (Pages 31–33)

1. $(2 + 3)x$
2. $(6 + 8)mp$
3. $r^2s(20 + 5)$
4. $ab^2c(1.8 + 4.7)$
5. $(19 - 2)t^2$
6. $pq^2(23 - 14)$
7. $(7.8 - 5.3)xy^2$
8. $a^3d(13.6 - 12.6)$
9. $(18 + 5 + 3)rs$
10. $(4.7 + 1.3 + 3.6)mn^2p$
11. $t(5\frac{3}{4} + 3\frac{1}{8} + 1\frac{1}{2})$
12. $(56 + 39 + 11)c^3d^2$

2.6 Like Terms

The ***numerical coefficient*** of $5x^2$ is 5. In x^2y the numerical coefficient is 1.

P–1 **What is the numerical coefficient of each expression?**

a. $\frac{1}{2}x^2$ **b.** $0.8x^2$ **c.** mn^2 **d.** $12x^2y^3z$

Notice that $\frac{1}{2}x^2$ and $0.8x^2$ have the same variable and the same power of the variable.

Definition

Two or more ***like terms*** have
1. the same variables, and
2. the same powers of these variables.

Like: $5x^2y$, $3x^2y$
Unlike: $5x^2y$, $3xy^2$

EXAMPLE 1 For each term in List I, select one or more like terms from List II.

List I	List II	
1. $5xy$	**a.** xy^2	**b.** $4x^2y$
2. $3rs$	**c.** $5x$	**d.** $3r$
3. $2x^2y$	**e.** $3rs^2$	**f.** $2xy$
4. $8xy^2$	**g.** $0.5rs$	**h.** $\frac{1}{2}yx^2$

Solutions: **1.** f **2.** g **3.** b, h **4.** a

To add or subtract like terms, you can use the Distributive Property.

Distributive Property: $ba + ca = (b + c)a$

Sum of Like Terms: $10x + 3x = (10 + 3)x$
$= 13x$

However, you can add or subtract like terms directly by using these steps.

Steps for Adding and Subtracting Like Terms

1. Add or subtract the numerical coefficients. $10x + 3x = 13x$
2. Use the same variable(s) and exponent(s). $6x^2 - 2x^2 = 4x^2$

When you add or subtract like terms, you are ***combining*** like terms.

EXAMPLE 2 Combine like terms: $5x^2y + x^2y$

Solution: [1] Add the numerical coefficients. ⟶ $\mathbf{5 + 1 = 6}$

[2] Use the same variables and exponents. ⟶ $\mathbf{5x^2y + x^2y = 6x^2y}$

P–2 **Combine like terms.**

a. $3n + n$ **b.** $5cd + 9cd$ **c.** $a + a$ **d.** $6v^2 + 16v^2$

EXAMPLE 3 Combine like terms: $2.8rst^2 - 1.3rst^2$

Solution:

[1] Subtract the numerical coefficients. ⟶ $\mathbf{2.8 - 1.3 = 1.5}$

[2] Use the same variables and exponents. ⟶ $2.8\mathbf{rst^2} - 1.3\mathbf{rst^2} = 1.5\mathbf{rst^2}$

P–3 **Combine like terms.**

a. $7b - 4b$ **b.** $8a - 7a$ **c.** $4r^2 - 3r^2$ **d.** $7ab - 7ab$

CLASSROOM EXERCISES

Identify any like terms in each exercise. (Example 1)

1. $3mn$; $5x^2$; $7m^2n$; $6x$; mn; $\frac{1}{2}m^2n$; $3x^2y$; $1.8x$ **2.** $5rs$; $3a^2b$; ab^2; $12r$; rst; $6.2a^2b$; $9r$; rs

Add or subtract the numerical coefficients in each exercise. (Step 1, Examples 2 and 3)

3. $7mn + 15mn$ **4.** $13rs^2 - 5rs^2$ **5.** $32t - 19t$ **6.** $80xy + 23xy$

7. $5k - 4k$ **8.** $26m^2n^2 + 17m^2n^2$ **9.** $0.9yz^2 - 0.4yz^2$ **10.** $4.8abc + 1.3abc$

For Exercises 11–26, combine like terms. (Example 2)

11. $8x + 3x$ **12.** $6ab + ab$ **13.** $15cd + 19cd$ **14.** $t + 99t$

15. $13r^2 + 8r^2$ **16.** $9m^2n + 22m^2n$ **17.** $0.3ad + 0.6ad$ **18.** $1.2bc^3 + 2.5bc^3$

(Example 3)

19. $9y - 4y$ **20.** $12xy - xy$ **21.** $40ab - 15ab$ **22.** $19f^2g - 6f^2g$

23. $4.5mn - mn$ **24.** $5.9rt^2 - 1.3rt^2$ **25.** $35r^3st - 15r^3st$ **26.** $80hjk - 79hjk$

WRITTEN EXERCISES

Goal: To combine like terms

Sample Problems: a. $1.2ab^3 + 2.6ab^3$ **b.** $16rs - 4.5rs$

Answers: a. $3.8ab^3$ **b.** $11.5rs$

For each term in Exercises 1–10, select one or more like terms from A–T. (Example 1)

1. $5x$
2. $3ab^2$
3. $2y^3$
4. x^2y
5. $7y^2$
6. $3rst$
7. $4r^2t$
8. $0.7mn^3$
9. $\frac{1}{2}rs^2t$
10. $1.8a^2b$

A. $\frac{3}{4}rst$
B. $\frac{1}{4}r^2t$
C. m^3n
D. $13rs^2t$
E. $\frac{1}{2}m^2n^2p$
F. $4y^3$
G. x
H. b^2a
I. $3tr^2$
J. mn^3
K. $12a^2b$
L. $27ab^2$
M. $3y^2$
N. $12x^2$
O. $12x^2y$
P. $3r^2st$
Q. $3m^2n^2p$
R. $0.1yx^2$
S. $14.2ba^2$
T. $0.5xy^2$

For Exercises 11–43, combine like terms. (Example 2)

11. $7a + 9a$
12. $6m + 8m$
13. $12ab + ab$
14. $rs + 9rs$
15. $35y + 4y$
16. $41n + 9n$
17. $2.4x + 1.5x$
18. $3.2y + 2.5y$
19. $6mn + 6mn$
20. $7ab^2 + 9ab^2$
21. $8x^2 + 6x^2$
22. $11xy + 4xy$

(Example 3)

23. $12s - 5s$
24. $13t - 7t$
25. $6a^2 - a^2$
26. $9x^3 - x^3$
27. $7.8rt - rt$
28. $4.9b^2 - b^2$
29. $7ab - 6ab$
30. $9rs - 2rs$
31. $3.2pq^2 - 1.9pq^2$
32. $5.7mn^3 - 2.9mn^3$
33. $6t^3 - 3t^3$
34. $14k^2 - 9k^2$

MIXED PRACTICE

35. $16np - 14np$
36. $24ab^2c - 9ab^2c$
37. $22stv + 23stv$
38. $19r^3st - 8r^3st$
39. $16abc + 11abc$
40. $6cp - 5cp$
41. $38m^5n^2 + 121m^5n^2$
42. $5.3f^2g - 2.9f^2g$
43. $122.8p^2q^2 + 23.6p^2q^2$

REVIEW CAPSULE FOR SECTION 2.7

Find each sum.

1. a. $12 + 9$
b. $9 + 12$

2. a. $4.6 + 9.8$
b. $9.8 + 4.6$

3. a. $12 + 7 + 8$
b. $7 + 12 + 8$

4. a. $12.6 + 5.9 + 7.4$
b. $7.4 + 5.9 + 12.6$

5. a. $26 + 18 + 35$
b. $18 + 26 + 35$

6. a. $9 + 13 + 17$
b. $9 + 17 + 13$

2.7 Properties of Addition

You know that you can add two numbers in either order. This is the Commutative Property of Addition.

When you add three or more numbers, you may group them in any way. This is the Associative Property of Addition.

Commutative Property of Addition

Any two numbers can be added in either order.

$$\mathbf{a + b = b + a}$$

$p + 6 = 6 + p$

$9 + \frac{1}{2} = \frac{1}{2} + 9$

Associative Property of Addition

The way numbers are grouped for addition does not affect their sum.

$$\mathbf{(a + b) + c = a + (b + c)}$$

$(7 + \frac{3}{4}) + \frac{1}{4} = 7 + (\frac{3}{4} + \frac{1}{4})$

$(2 + s) + s = 2 + (s + s)$

EXAMPLE 1 Find the value of the variable that makes each sentence true. Name the property or properties that give the reason for your choice.

a. $(6 + 4) + w = 6 + (4 + 9)$ **b.** $3.1 + 2.6 = p + 3.1$

c. $16 + (2\frac{1}{2} + 4) = (16 + b) + 2\frac{1}{2}$

Solutions:

	Sentences	Value	Property
a.	$(6 + 4) + w = 6 + (4 + 9)$	$w = 9$	Associative Property of Addition
b.	$3.1 + 2.6 = p + 3.1$	$p = 2.6$	Commutative Property of Addition
c.	$16 + (2\frac{1}{2} + 4) = (16 + b) + 2\frac{1}{2}$	$b = 4$	Commutative and Associative Properties of Addition

P–1 **Find the value of the variable that makes each sentence true. Which property or properties did you use?**

a. $19 + g = 42 + 19$ **b.** $16 + (18 + 9) = (16 + 9) + z$

c. $(12 + 14) + c = 12 + (14 + 3)$ **d.** $25 + 9 = 9 + y$

The Commutative and Associative Properties can be used to simplify algebraic expressions that involve addition of more than two terms.

EXAMPLE 2 Combine like terms: $13n + 18 + 29n$

Solution:

[1] Change the order. ⟶ $13n + 18 + 29n = 13n + 29n + 18$

[2] Group like terms and add. ⟶ $= (13n + 29n) + 18$

$= 42n + 18$

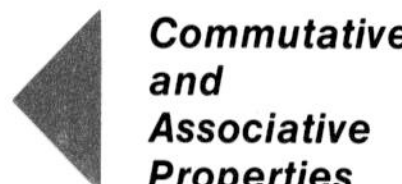

Commutative and Associative Properties

P-2 **Combine like terms.**

a. $26r^2 + 9 + 7r^2$ **b.** $12pq + 3 + 13pq$ **c.** $kr + 16 + 4kr$

EXAMPLE 3 Combine like terms: $4x^2y + xy + 3xy + 7x^2y$

Solution: [1] $4x^2y + xy + 3xy + 7x^2y = 4x^2y + 7x^2y + xy + 3xy$

[2] $= (4x^2y + 7x^2y) + (xy + 3xy)$

$= 11x^2y + 4xy$

P-3 **Combine like terms.**

a. $a + 4a + 6a$ **b.** $cd + 2c^2d + 3cd + 5c^2d$ **c.** $10p + 9q^3 + 12q^3 + 4p$

CLASSROOM EXERCISES

Find the value of the variable that makes each sentence true. (Example 1)

1. $10 + (15 + 20) = (10 + a) + 20$

2. $(3 + 4) + x = 5 + (3 + 4)$

3. $7 + (9 + 11) = (9 + 7) + r$

4. $(c + 100) + 25 = 50 + (100 + 25)$

5. $(20 + 13) + 15 = (b + 13) + 20$

6. $(12 + y) + 7 = (12 + 7) + 13$

For Exercises 7–20, combine like terms. (Example 2)

7. $8x + 3 + 3x$ **8.** $12b + 5 + 7b$ **9.** $14y + 23 + 20y$ **10.** $6t + 18 + 9t$

11. $36r + 10 + 16r$ **12.** $14st + 25 + 28st$ **13.** $7.1n^2 + 9 + 1.6n^2$ **14.** $1.2tw + 8 + 0.9tw$

(Example 3)

15. $3x + 2x + 4x + 9x$ **16.** $s^3 + 5s + 3 + 2s^3 + 9s$ **17.** $3ab + 5a^2b + ab + a^2b$

18. $8t + t^2 + 7t + 3t^2$ **19.** $5rs + 7rs + 8 + 9rs + 10$ **20.** $15yz + 9z^3 + 30z^3 + 4yz$

WRITTEN EXERCISES

Goal: To simplify algebraic expressions that involve addition of more than two terms

Sample Problem: $6a^2 + 6a + 12a^2 + 4 + 10a$ **Answer:** $18a^2 + 16a + 4$

Find the value of the variable that makes each sentence true. (Example 1)

1. $27 + 19 = 19 + r$

2. $24 + q = 35 + 24$

3. $(7 + 9) + x = 7 + (4 + 9)$

4. $(12 + 15) + 8 = 12 + (b + 15)$

5. $(41 + c) + 90 = (41 + 90) + 6$

6. $(m + 14) + 25 = 15 + (14 + 25)$

7. $(\frac{1}{2} + \frac{1}{3}) + \frac{1}{4} = (b + \frac{1}{3}) + \frac{1}{2}$

8. $(3\frac{1}{2} + 7\frac{1}{3}) + 1\frac{1}{4} = f + (1\frac{1}{4} + 3\frac{1}{2})$

For Exercises 9–28, combine like terms. (Example 2)

9. $12a + 15 + 17a$ **10.** $23t + 9 + 18t$ **11.** $0.9m + 4.2 + 3.4m$

12. $8.7r + 0.5 + 12.8r$ **13.** $21k + 7 + 34k$ **14.** $33n + 16 + 39n$

(Example 3)

15. $14t + t + 19t$ **16.** $6y^2 + 2y^2 + y^2 + 9y^2$ **17.** $0.3p + 1.8r + 0.9p + r$

18. $3.9x + y + 0.1y + 9.4x$ **19.** $5q^3 + 3q + q^3 + 12q + 7$ **20.** $9v + v^2 + 7v + 3 + 9v^2$

MIXED PRACTICE

21. $4xy^2 + 15xy^2 + xy^2$

22. $8mp + 19 + 9mp$

23. $24r^2s + 19rs^2 + r^2s + 29rs + 28rs^2$

24. $24p^3 + 9 + 15p^3$

25. $43a^2 + 6b^2 + 17a^2 + 12b^2$

26. $12ab^2c + abc + 5abc^2 + 9abc + 28ab^2c$

27. $1.3ma + ma + 1.9ma$

28. $3.1rs + 2.2r^2s + 3.7r^2s + 1.9rs$

NON-ROUTINE PROBLEM

29. On Thursday, Russell paid \$1.00 for parking and \$2.00 for admission at the county fair. He spent half the money he had left on rides and exhibits. After repeating this spending pattern on Friday and on Saturday, he had \$1.00 left. Find how much money Russell had on Thursday before he went to the fair.

Using Formulas

Electricity

The **watt, volt,** and **ampere** are units of electrical measure in the metric system that are common terms to the **electrician.** The terms are related by the following **power formula.**

$W = EI$

W = power in number of watts
E = number of volts
I = current in amperes

In planning the wiring for a home, an electrician follows these two basic conditions.

1. Allow about 32.3 watts of electrical power per square meter of floor area. As a formula:

$W = 32.3A$

A = area in m^2

2. 1 circuit is required for each 46 square meters of floor area. As a formula:

$C = A \div 46$

C = number of circuits

EXAMPLE: The floor area of a house is 225.63 square meters.

a. To the nearest hundred, how many watts are needed?

b. Find the number of circuits required.

SOLUTIONS: **a.** $W = 32.3A$

$W = (32.3)(225.63)$

$W = 7287.849$, or **7300 watts**

b. $C = A \div 46$

$C = 225.63 \div 46$

$C = 4.905$, or **5 circuits**

EXERCISES

For Exercises 1–2, find each number.

a. watts needed to the nearest hundred

b. circuits required

1. A two-story house with total floor area of 318 square meters
2. A ranch-style house with dimensions 14 meters by 11 meters
3. In the formula $W = EI$, suppose that E always equals 120. When the number of amperes, I, increases, does the power in watts, W, increase or decrease?
4. In the formula $C = A \div 46$, how does the number of circuits, C, change when the area, A, decreases?

CHAPTER SUMMARY

IMPORTANT TERMS

Sentence *(p. 24)*
Power *(p. 28)*
Exponent *(p. 28)*
Common factor *(p. 35)*
Factoring *(p. 35)*
Perimeter *(p. 37)*
Numerical coefficient *(p. 40)*
Like terms *(p. 40)*
Combining like terms *(p. 41)*

IMPORTANT IDEAS

1. *Addition Property of Zero*
$a + 0 = 0 + a = a$
2. *Multiplication Property of Zero*
$a \cdot 0 = 0 \cdot a = 0$
3. *Multiplication Property of One*
$a \cdot 1 = 1 \cdot a = a$
4. A number divided by itself equals 1.
$\frac{n}{n} = 1, n \neq 0$
5. *Commutative Property of Multiplication*
$a \cdot b = b \cdot a$
6. *Associative Property of Multiplication*
$(a \cdot b) \cdot c = a \cdot (b \cdot c)$
7. *Distributive Properties*
$a(b + c) = ab + ac \qquad (b + c)a = ba + ca$
$a(b - c) = ab - ac \qquad (b - c)a = ba - ca$
8. Steps for Adding and Subtracting Like Terms
a. Add or subtract the numerical coefficients.
b. Use the same variable(s) and exponent(s).
9. *Commutative Property of Addition*
$a + b = b + a$
10. *Associative Property of Addition*
$(a + b) + c = a + (b + c)$

CHAPTER REVIEW

SECTION 2.1

Write the value of the variable that makes each sentence true.

1. $n + 16 = 16$
2. $r + 0 = 16.3$
3. $73 = t(73)$
4. $a(18) = 0$
5. $12.1 = 12.1 + x$
6. $k(\frac{2}{2}) = 35$
7. $0 = 1.4q$
8. $3\frac{1}{2}p = 3\frac{1}{2}$
9. $\frac{15}{16} = (y)(\frac{15}{16})$

SECTION 2.2

Multiply.

10. $10(5x)$ **11.** $(12r)8$ **12.** $(4a)(9b)$ **13.** $(6m)(5n)$

14. $(7rs)(12r)$ **15.** $(13k)(9kg)$ **16.** $(15x)(5xy)$ **17.** $(0.4pq)(22qp)$

18. $(3.1z)(8z^2a)$ **19.** $(7d^2)(8df^2)$ **20.** $(2a)(3ab)(2b)(a^2)$ **21.** $(5m)(3mp)(m)(p)$

SECTION 2.3

Write each product as a sum or difference.

22. $9(a + 7)$ **23.** $(k - 4)12$ **24.** $19(p - q)$ **25.** $(c + d)25$

26. $7(2x - 5)$ **27.** $(3p - 9)6$ **28.** $3y(8y - 4t)$ **29.** $(9g + 3h)6g$

30. $12f(3f + 4)$ **31.** $(16 - 9s)3s$ **32.** $y(4y - 2yz)$ **33.** $2pq(6p + 8q)$

SECTION 2.4

Factor.

34. $16c + 16d$ **35.** $24m - 24w$ **36.** $k(9.3) - 7(9.3)$ **37.** $(\frac{1}{3})(r) + (\frac{1}{3})(5)$

38. $xk - 19k$ **39.** $23q + pq$ **40.** $20t + 20$ **41.** $39 - 39z$

42. $4.2r - r^2$ **43.** $2.2pb + pa$ **44.** $\frac{1}{6}y - \frac{1}{6}$ **45.** $c^3 - 2c$

SECTION 2.5

Solve.

46. Each side of a square ceiling tile is 24 centimeters. What is the area of the tile?

47. A rectangular scarf is 54 inches long and 18 inches wide. How many inches of lace are needed to go around the edge of the scarf?

SECTION 2.6

Combine like terms.

48. $13r + 24r$ **49.** $29t - t$ **50.** $4.2w + 3.7w$ **51.** $12.1n - 7.9n$

52. $19ab^2 + 23ab^2$ **53.** $42x^3y - 17x^3y$ **54.** $5.7m^2n^2 - m^2n^2$ **55.** $13.3a^2bc - 9.7a^2bc$

SECTION 2.7

Combine like terms.

56. $17q + 29q + 34q$ **57.** $19ab + 13ab + 47ab$

58. $5t^2 + 3t + t^2 + 9t$ **59.** $12n^3 + 9n^2 + 4n^2 + 3n^3$

60. $7xy + 3y + 21x + 8y + 7x$ **61.** $12.9ab + 3.8a^2b^2 + 5.3ab + a^2b$

62. $2x^3 + 7.3x + 1.2 + 7x^3 + 9.4x$ **63.** $7p^2 + 5p + 8p^3 + 6p^2 + 15 + 9p$

Basic Number Concepts

CHAPTER 3

Many of the facts about prime numbers were discovered by the ancient Greek mathematicians. There is evidence that the Greeks may have even tried to make a computer. The fragment shown at the right below is believed to be part of a mechanism designed to compute the position of the stars.

3.1 Factors, Multiples, and Divisibility

Definition A number is ***divisible*** by another number if the remainder is 0.

$$\begin{array}{r} 3 \\ 8\overline{)24} \\ \underline{24} \\ 0 \end{array}$$

24 is divisible by 8.

$$\begin{array}{r} 2 \\ 6\overline{)13} \\ \underline{12} \\ 1 \end{array}$$

13 is not divisible by 6.

Since 24 is divisible by 8, 8 is a ***factor*** of 24. Here are all the factors of 24.

Factors of 24: 1, 2, 3, 4, 6, 8, 12, 24

NOTE: The only factors that we will consider are counting numbers.

Counting Numbers: $\{1, 2, 3, 4, 5, \cdots\}$

These are also called natural numbers.

The number 24 is a ***multiple*** of each of its factors.

$4 \cdot 6 = 24$ $\quad$ $8 \cdot 3 = 24$ $\quad$ $12 \cdot 2 = 24$ $\quad$ $1 \cdot 24 = 24$

Thus, divisible by, factor of, and multiple of are related terms.

24 is divisible by 8.
8 is a factor of 24.
24 is a multiple of 8.

EXAMPLE 1

Write True or False for each sentence. Write a reason for each answer.

a. 36 is a multiple of 9. **b.** 19 is divisible by 8.
c. 1 is a factor of 34. **d.** 12 is a multiple of 12.

Solutions: **a.** True, since $9 \cdot 4 = 36$.

b. False, since $19 \div 8 = 2$ R3.

c. True, since 1 is a factor of every number.

d. True, since $12 \cdot 1 = 12$.

Unless otherwise indicated, consider only counting numbers when referring to factors and multiples in this text.

EXAMPLE 2 List the factors of 15.

Solution: [1] List the numbers through 15.

1, 2, 3, 4, 5, 6, 7, 8, 9, 10, 11, 12, 13, 14, 15

[2] Divide 15 by each number in the list. Reject each divisor that produces a nonzero remainder.

1, ~~2~~, 3, ~~4~~, 5, ~~6~~, ~~7~~, ~~8~~, ~~9~~, ~~10~~, ~~11~~, ~~12~~, ~~13~~, ~~14~~, 15

Answer: 1, 3, 5, 15

P–1 **What are the factors of each of these numbers?**

a. 22 **b.** 29 **c.** 34 **d.** 35 **e.** 41

EXAMPLE 3 Write 20 as the product of two numbers. Do not use 1 or 20. There may be more than one answer.

Solution: [1] List the factors of 20 except for 1 and 20. ⟶ 2, 4, 5, 10

[2] Select pairs of factors whose product is 20. ⟶ $2 \cdot 10 = 20$

or $4 \cdot 5 = 20$

P–2 **Write each number as the product of two numbers. Do not use 1 or the given number.**

a. 10 **b.** 22 **c.** 39 **d.** 33 **e.** 30

CLASSROOM EXERCISES

Write a true statement about each pair. Use the words is divisible by.

1. 35 and 7 **2.** 9 and 45 **3.** 10 and 10 **4.** 26 and 13
5. 5 and 30 **6.** 20 and 100 **7.** 18 and 1 **8.** 3 and 21

Write a true statement about each pair. Use the words is a factor of.

9. 2 and 10 **10.** 15 and 3 **11.** 35 and 7 **12.** 5 and 75
13. 19 and 1 **14.** 23 and 23 **15.** 17 and 34 **16.** 39 and 13

Write a true statement about each pair. Use the words is a multiple of.

17. 4 and 28 **18.** 18 and 6 **19.** 36 and 9 **20.** 7 and 21
21. 27 and 27 **22.** 1 and 19 **23.** 45 and 5 **24.** 6 and 30

Write True or False for each sentence. Write a reason for each answer. (Example 1)

25. 18 is a factor of 6.

26. 18 is divisible by 6.

27. 17 is divisible by 17.

28. 10 is a multiple of 60.

29. 23 is a multiple of 7.

30. 1 is a factor of 93.

31. 13 is a factor of 26.

32. 9 is divisible by 27.

33. 33 is a multiple of 13.

34. 48 is a factor of 16.

35. 19 is divisible by 19.

36. 14 is a multiple of 7.

Write the factors of each number. (Example 2)

37. 4	**38.** 6	**39.** 5	**40.** 7
41. 8	**42.** 9	**43.** 10	**44.** 14
45. 17	**46.** 26	**47.** 24	**48.** 31

Write each number as the product of two numbers. Do not use 1 or the given number. Some exercises may have more than one answer. (Example 3)

49. 6	**50.** 8	**51.** 15	**52.** 14
53. 9	**54.** 25	**55.** 21	**56.** 35
57. 38	**58.** 42	**59.** 44	**60.** 32

WRITTEN EXERCISES

Goal: To identify the factors of a number

Sample Problem: List the factors of 6. Then write 6 as the product of two counting numbers. Do not use 1 or 6.

Answer: 1, 2, 3, 6; $2 \cdot 3$

Write True or False for each sentence. Write a reason for each answer. (Example 1)

1. 21 is divisible by 7.

2. 34 is divisible by 17.

3. 6 is a factor of 43.

4. 9 is a factor of 54.

5. 55 is a multiple of 5.

6. 57 is a multiple of 7.

7. 258 is divisible by 13.

8. 168 is divisible by 14.

9. 126 is a multiple of 126.

10. 7 is a factor of 158.

11. 13 is a factor of 237.

12. 228 is a multiple of 228.

List all the factors of each number. (Example 2)

13. 12 **14.** 16 **15.** 11 **16.** 13
17. 19 **18.** 21 **19.** 20 **20.** 18
21. 23 **22.** 30 **23.** 25 **24.** 27
25. 28 **26.** 32 **27.** 48 **28.** 64

Write each number as the product of two numbers. Do not use 1 or the given number. Some exercises may have more than one answer. (Example 3)

29. 12 **30.** 16 **31.** 18 **32.** 24
33. 27 **34.** 28 **35.** 36 **36.** 30
37. 33 **38.** 34 **39.** 77 **40.** 55
41. 49 **42.** 63 **43.** 93 **44.** 81

MIXED PRACTICE

*Write Yes or No to the questions for **a, b,** and **c** in each row. Then list the factors for **d** in each row.*

n	Is *n* Divisible by This Number?		Is This Number a Factor of *n*?		Is *n* a Multiple of This Number?		Set of Factors of *n*
45. 56	**a.** 9	?	**b.** 8	?	**c.** 112	?	**d.** ?
46. 60	**a.** 12	?	**b.** 9	?	**c.** 15	?	**d.** ?
47. 77	**a.** 27	?	**b.** 11	?	**c.** 7	?	**d.** ?
48. 95	**a.** 19	?	**b.** 25	?	**c.** 25	?	**d.** ?
49. 72	**a.** 24	?	**b.** 18	?	**c.** 48	?	**d.** ?
50. 96	**a.** 18	?	**b.** 16	?	**c.** 4	?	**d.** ?
51. 120	**a.** 15	?	**b.** 25	?	**c.** 360	?	**d.** ?
52. 144	**a.** 36	?	**b.** 24	?	**c.** 36	?	**d.** ?

NON-ROUTINE PROBLEM

53. A 5-digit number contains the digits shown at the right. Use the clues below to find the number.

Clue 1: The number is less than 40,000.
Clue 2: The 4 is next to the 2.
Clue 3: The 2 is not next to the 1 or 3.
Clue 4: The 5 is not next to the 1 or 3.
Clue 5: The 4 is next to the 3 or 5.

1 2 3
4 5

3.2 Tests for Divisibility

The set of whole numbers consists of 0 and the counting numbers.

Whole Numbers: {0, 1, 2, 3, 4, 5, · · ·}

There are tests for divisibility that can help you with division. These tests involve the ***digits*** 0, 1, 2, 3, 4, 5, 6, 7, 8, and 9.

A whole number is divisible by
- 2 if it ends in 0, 2, 4, 6, or 8.
- 3 if the sum of its digits is divisible by 3.
- 4 if its last two digits name a multiple of 4.
- 5 if it ends in 0 or 5.
- 9 if the sum of its digits is divisible by 9.
- 10 if it ends in 0.

Any whole number that does not meet one of these tests is not divisible by the number shown.

Recall that the following statements are related.

456 is divisible by 2. 456 is a multiple of 2. 2 is a factor of 456.

EXAMPLE 1

Write True or False for each reason for each.

a. 438 is a multiple of 2. **b.** 722 is divisible by 4.
c. 5 is a factor of 920.

	Statement	True or False? Why?
Solutions: **a.**	438 is a multiple of 2.	True. 438 ends in 8.
b.	722 is divisible by 4.	False. 22 is not divisible by 4.
c.	5 is a factor of 920.	True. 920 ends in 0.

A number that is divisible by 2 is an ***even number.*** A number that is not divisible by 2 is an ***odd number.***

P–1 **Which of these numbers are divisible by 2? by 4? by 5? by 10?**

a. 327 **b.** 1084 **c.** 160 **d.** 1045 **e.** 1240

EXAMPLE 2

Write True or False for each statement. Write a reason for each.

a. 823 is divisible by 3. **b.** 3 is a factor of 846. **c.** 46, 845 is divisible by 9.

Solutions:

	Statement	Sum of Digits	True or False? Why?
a.	823 is divisible by 3.	$8+2+3=13$	False. 13 is not divisible by 3.
b.	3 is a factor of 846.	$8+4+6=18$	True. 18 is divisible by 3.
c.	46, 845 is divisible by 9.	$4+6+8+4+5=27$	True. 27 is divisible by 9.

P–2 **Which of these numbers are divisible by 9?**

a. 58,014 **b.** 768 **c.** 2346 **d.** 40,693

P–3 **Select a, b, c, or d to indicate which number(s) in P-2 have 3 as a factor.**

You can use the tests for divisibility to solve some word problems.

EXAMPLE 3

At a banquet for 357 people, a caterer plans to seat 9 persons per table.

a. Can the caterer do this so that there are exactly 9 persons at each table?

b. If not, how many more persons are needed to completely fill the tables?

Solutions:

[1] Determine whether 357 is divisible by 9. ⟶ $3+5+7=15$ — ***15 is not divisible by 9.***

[2] Find the nearest whole number greater than 15 and divisible by 9. ⟶ $15+3=18$ — ***18 is divisible by 9.***

a. Since 357 is not divisible by 9, 357 persons cannot be seated with exactly 9 persons per table.

b. Adding 3 more persons will completely fill the tables with exactly 9 persons per table.

CLASSROOM EXERCISES

For Exercises 1–14, write True or False for each sentence. Write a reason for each. (Example 1)

1. 958 is divisible by 2.

2. 10,381 is a multiple of 2.

3. 2 is a factor of 706.

4. 4 is a factor of 1738.

5. 756 is divisible by 5.

6. 8630 is a multiple of 5.

7. 12,690 is a multiple of 10.

8. 10 is a factor of 9864.

(Example 2)

9. 96 is divisible by 3.

10. 3 is a factor of 382.

11. 276 is a multiple of 3.

12. 4362 is divisible by 9.

13. 81,234 is a multiple of 9.

14. 9 is a factor of 14,825.

WRITTEN EXERCISES

Goal: To test a whole number for divisibility by 2, 3, 4, 5, 9, or 10

Sample Problem: Write True or False: 47,205 is divisible by 9. Write a reason.

Answer: True. $4 + 7 + 2 + 0 + 5 = 18$; 18 is divisible by 9. So 47,205 is divisible by 9.

For Exercises 1–32, write True or False for each sentence. Write a reason for each. (Example 1)

1. 546 is divisible by 2.

2. 513 is divisible by 2.

3. 2 is a factor of 811.

4. 2 is a factor of 838.

5. 814 is divisible by 4.

6. 324 is divisible by 4.

7. 1028 is a multiple of 4.

8. 938 is a multiple of 4.

9. 56 is divisible by 5.

10. 45 is divisible by 5.

(Example 2)

11. 813 is divisible by 3.

12. 747 is divisible by 3.

13. 3 is a factor of 1252.

14. 3 is a factor of 14,628.

15. 10,521,175 is a multiple of 3.

16. 15,281,923 is a multiple of 3.

17. 65,736 is divisible by 9.

18. 23,567 is divisible by 9.

19. 713,685 is a multiple of 9.

20. 1,203,403 is a multiple of 9.

MIXED PRACTICE

21. 4 is a factor of 495.
22. 6260 is a multiple of 9.
23. 6120 is divisible by 10.
24. 5 is a factor of 975.
25. 8,346,492 is a multiple of 3.
26. 83,025 is divisible by 2.
27. 222 is divisible by 2.
28. 4 is a factor of 3084.
29. 5 is a factor of 829.
30. 49,473 is a multiple of 3.
31. 873 is a multiple of 9.
32. 3,895,875 is divisible by 10.

APPLICATIONS

Solve. (Example 3)

33. A grocer has 2948 apples to pack in boxes. Each box will contain 10 apples.
    a. Will all the boxes be completely filled?
    b. If not, how many more apples are needed to fill the boxes?

34. Maria built shelves for her collection of 983 books. Each shelf holds 10 books.
    a. Will the 983 books completely fill the shelves?
    b. If not, how many more books are needed to do this?

35. Mr. Velez wishes to invest $50,709 in equal amounts for his four grand-children.
    a. Can the $50,709 be divided into four equal whole-dollar amounts?
    b. If not, how many dollars must be added so that this can be done?

36. A grocer plans to sell 2677 kilograms of potatoes in 5-kilogram bags.
    a. Will all the bags be completely filled?
    b. If not, how many kilograms of potatoes are needed to completely fill the bags?

37. A sewing club with 9 members plans to make 230 stuffed toys.
    a. Can the work be divided so that each member makes the same number of toys?
    b. If not, how many more toys are needed in order to do this?

REVIEW CAPSULE FOR SECTION 3.3

Find the value of each expression. (Pages 27–30)

1. 2^2
2. 3^2
3. 3^3
4. 2^3
5. $2^2 \cdot 5^2$
6. $3^2 \cdot 5^2$
7. $2 \cdot 3^2 \cdot 7^2$
8. $3 \cdot 5^2 \cdot 11^2$

3.3 Prime Numbers

Each number below has only two factors, 1 and the number itself. These numbers are called prime numbers.

2, 3, 5, 7, 11, 13, 17, 19, 23

Definition

A ***prime number*** is a counting number greater than 1 that has exactly two counting-number factors, 1 and the number itself.

$\mathbf{7} = 7 \cdot 1$

$\mathbf{41} = 41 \cdot 1$

Numbers such as 4, 6, 8, 15, 50, etc., are called composite numbers.

Definition

A ***composite number*** is a counting number greater than 1 that is not a prime number.

$\mathbf{22} = 2 \cdot 11$

$\mathbf{45} = 3 \cdot 3 \cdot 5$

P–1 **What factors does each of these composite numbers have other than 1 and the given number?**

a. 4 **b.** 6 **c.** 8 **d.** 15 **e.** 50 **f.** 28

EXAMPLE 1

Write the smallest prime number that is a factor of 161.

Solution:

1. Write several prime numbers in order. ⟶ 2, 3, 5, 7, 11, 13, 17, 19, 23
2. Is 161 divisible by 2? ⟶ No; 161 is an odd number.
3. Is 161 divisible by 3? ⟶ $1 + 6 + 1 = 8$. Since 8 is not divisible by 3, then 161 is not divisible by 3.
4. Is 161 divisible by 5? ⟶ No; 161 does not end in either 0 or 5.
5. Is 161 divisible by 7? ⟶

$$\begin{array}{r} 23 \\ 7\overline{)161} \\ \underline{14} \\ 21 \\ \underline{21} \\ 0 \end{array}$$

Since the remainder is 0, 161 is divisible by 7.

The smallest prime factor of 161 is 7.

P–2 **What is the smallest prime factor of each of these numbers?**

a. 452 **b.** 261 **c.** 550 **d.** 365 **e.** 209

A product of prime numbers can often be simplified by using exponents.

EXAMPLE 2

Use exponents to rewrite the following.

$3 \cdot 7 \cdot 2 \cdot 5 \cdot 2 \cdot 2 \cdot 13 \cdot 5 \cdot 3$

Solution:

[1] Write each prime number once as a factor. (Order from least to greatest.) → $2 \cdot 3 \cdot 5 \cdot 7 \cdot 13$

[2] Write an exponent for each prime number to show the number of times it is a factor. → $2^3 \cdot 3^2 \cdot 5^2 \cdot 7 \cdot 13$

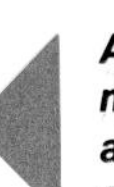

An exponent is not used when a factor appears only once.

The answer to Example 2 is read as: "two cubed times three squared times five squared times seven times thirteen."

CLASSROOM EXERCISES

Identify each number as prime or composite.

1. 29 **2.** 33 **3.** 18 **4.** 43
5. 49 **6.** 17 **7.** 15 **8.** 27

Write the smallest prime factor of each number. (Example 1)

9. 12 **10.** 10 **11.** 9 **12.** 35
13. 20 **14.** 15 **15.** 40 **16.** 49
17. 25 **18.** 358 **19.** 417 **20.** 245

Use exponents to write each product. (Example 2)

21. $2 \cdot 5 \cdot 2 \cdot 3$
22. $2 \cdot 7 \cdot 5 \cdot 5 \cdot 2$
23. $2 \cdot 11 \cdot 3 \cdot 5 \cdot 3 \cdot 2 \cdot 2$
24. $7 \cdot 5 \cdot 7 \cdot 5 \cdot 5 \cdot 5$
25. $17 \cdot 5 \cdot 5 \cdot 17 \cdot 11 \cdot 17$
26. $23 \cdot 3 \cdot 2 \cdot 23 \cdot 11 \cdot 3 \cdot 3$
27. $2 \cdot 2 \cdot 2 \cdot 3 \cdot 3 \cdot 3 \cdot 3 \cdot 7 \cdot 7$
28. $3 \cdot 3 \cdot 5 \cdot 5 \cdot 5 \cdot 5 \cdot 5 \cdot 7 \cdot 11$
29. $11 \cdot 17 \cdot 5 \cdot 7 \cdot 5 \cdot 17 \cdot 11 \cdot 7 \cdot 5 \cdot 11$
30. $9 \cdot 11 \cdot 7 \cdot 13 \cdot 13 \cdot 9 \cdot 11 \cdot 9 \cdot 11 \cdot 7$

WRITTEN EXERCISES

Goal: To rewrite a product of prime factors with exponents
Sample Problem: $2 \cdot 2 \cdot 3 \cdot 5 \cdot 2 \cdot 3$ **Answer:** $2^3 \cdot 3^2 \cdot 5$

Find the smallest prime factor of each number. (Example 1)

1. 26	**2.** 34	**3.** 38	**4.** 46	**5.** 55	**6.** 57
7. 69	**8.** 87	**9.** 65	**10.** 95	**11.** 77	**12.** 91
13. 119	**14.** 253	**15.** 451	**16.** 481	**17.** 201	**18.** 413

Use exponents to rewrite each product. (Example 2)

19. $2 \cdot 3 \cdot 3 \cdot 5$
20. $2 \cdot 2 \cdot 3 \cdot 5 \cdot 5$
21. $2 \cdot 3 \cdot 5 \cdot 5 \cdot 7 \cdot 7$
22. $2 \cdot 3 \cdot 3 \cdot 5 \cdot 7 \cdot 7$
23. $7 \cdot 2 \cdot 3 \cdot 5 \cdot 7 \cdot 3 \cdot 3$
24. $5 \cdot 3 \cdot 3 \cdot 7 \cdot 2 \cdot 5 \cdot 2 \cdot 2$
25. $3 \cdot 7 \cdot 13 \cdot 7 \cdot 7 \cdot 5 \cdot 2 \cdot 3$
26. $11 \cdot 2 \cdot 3 \cdot 7 \cdot 2 \cdot 2 \cdot 5 \cdot 13$
27. $17 \cdot 11 \cdot 3 \cdot 5 \cdot 11 \cdot 7 \cdot 3 \cdot 2 \cdot 2 \cdot 17 \cdot 5$
28. $23 \cdot 19 \cdot 2 \cdot 3 \cdot 2 \cdot 2 \cdot 5 \cdot 7 \cdot 23 \cdot 5$

APPLICATIONS

You can use a calculator to find certain prime numbers p by using the formula $\mathbf{p = 2^n - 1}$ *for some values of the number n.*

EXAMPLE Use the formula $p = 2^n - 1$ to find p when $n = 5$.

SOLUTION On some calculators, you can keep pressing the "=" key to find powers.

[2] [×] [=] [=] [=] [=] [−] [1] [=] 31.

Use the formula $p = 2^n - 1$ *to find p for each value of n given.*

29. 2	**30.** 3	**31.** 7	**32.** 13	**33.** 17	**34.** 19

MORE CHALLENGING EXERCISES

The ***prime-number sieve*** *is a method for finding all the prime numbers that are less than a given number. It is used in the example below.*

EXAMPLE: Use the prime-number sieve to write the prime numbers less than 50.

[1] Arrange the numerals from 2 to 50.

	2	3	4	5	6	7	8	9	10
11	12	13	14	15	16	17	18	19	20
21	22	23	24	25	26	27	28	29	30
31	32	33	34	35	36	37	38	39	40
41	42	43	44	45	46	47	48	49	50

[2] Cross out numerals after **2** for numbers divisible by 2. (4, 6, 8, 10, · · ·)

[3] Cross out numerals after **3,** the next prime, for numbers divisible by 3. (6, 9, 12, 15, · · ·)

[4] Cross out numerals after **5** for numbers divisible by 5. (10, 15, 20, 25, · · ·)

[5] Cross out numerals after **7** for numbers divisible by 7. (14, 21, 28, 35, · · ·)

2	3	[2] ~~4~~	5	[2] ~~6~~	7	[2] ~~8~~	[3] ~~9~~	[2] ~~10~~
11	[2] ~~12~~	13	[2] ~~14~~	[3] ~~15~~	[2] ~~16~~	17	[2] ~~18~~	19	[2] ~~20~~
[3] ~~21~~	[2] ~~22~~	23	[2] ~~24~~	[4] ~~25~~	[2] ~~26~~	[3] ~~27~~	[2] ~~28~~	29	[2] ~~30~~
31	[2] ~~32~~	[3] ~~33~~	[2] ~~34~~	[4] ~~35~~	[2] ~~36~~	37	[2] ~~38~~	[3] ~~39~~	[2] ~~40~~
41	[2] ~~42~~	43	[2] ~~44~~	[3] ~~45~~	[2] ~~46~~	47	[2] ~~48~~	[5] ~~49~~	[2] ~~50~~

[6] Look for numerals after **11** for numbers divisible by 11. Since there are none, you are finished. The numerals not crossed out represent the prime numbers less than 50: 2, 3, 5, 7, 11, 13, 17, 19, 23, 29, 31, 37, 41, 43, 47

35. Use the prime–number sieve to write the prime numbers from 2 to 75.

36. Use the prime–number sieve to write the prime numbers from 2 to 100.

MID-CHAPTER REVIEW

List all the factors of each number. Then write each number as the product of two counting numbers. Do not use 1 or the given number. (Section 3.1)

1. 22 **2.** 26 **3.** 38 **4.** 39 **5.** 40 **6.** 42

Write True or False for each statement. Write a reason for each answer. (Section 3.2)

7. 3 is a factor of 42,055.

8. 29,739 is divisible by 3.

9. 5610 is a multiple of 2.

10. 4 is a multiple of 724.

11. 4 is divisible by 1008.

12. 2 is a factor of 3825.

Solve. (Section 3.2)

13. Sheila wishes to spend $56 to buy gifts for three friends.

a. Can the $56 be divided into three equal whole-dollar amounts?

b. If not, how many more dollars are needed in order to do this?

Find the smallest prime factor of each number. (Section 3.3)

14. 381,045 **15.** 502,865 **16.** 623 **17.** 1067

Use exponents to rewrite each product. (Section 3.3)

18. $2 \cdot 3 \cdot 3 \cdot 5 \cdot 5 \cdot 5 \cdot 17 \cdot 17$

19. $2 \cdot 3 \cdot 5 \cdot 5 \cdot 5 \cdot 7 \cdot 7$

20. $2 \cdot 2 \cdot 2 \cdot 2 \cdot 5 \cdot 5 \cdot 7 \cdot 7 \cdot 7$

21. $3 \cdot 5 \cdot 7 \cdot 11 \cdot 11 \cdot 13 \cdot 13 \cdot 13$

Using Diagrams and Tables

Surface Area

Landscape gardeners apply strategies such as **drawing a diagram** and **using a geometric formula** to solve problems related to their work.

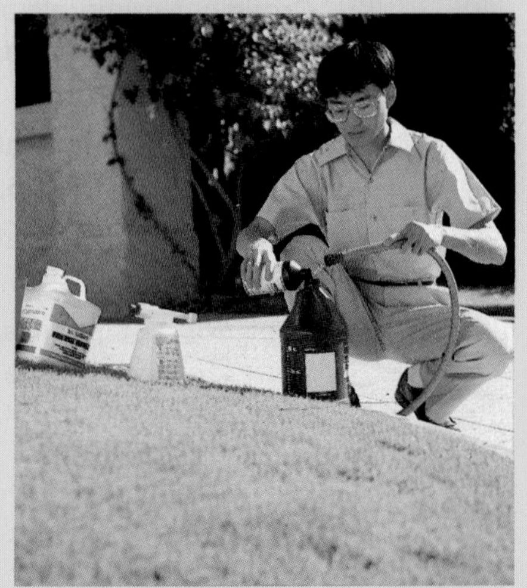

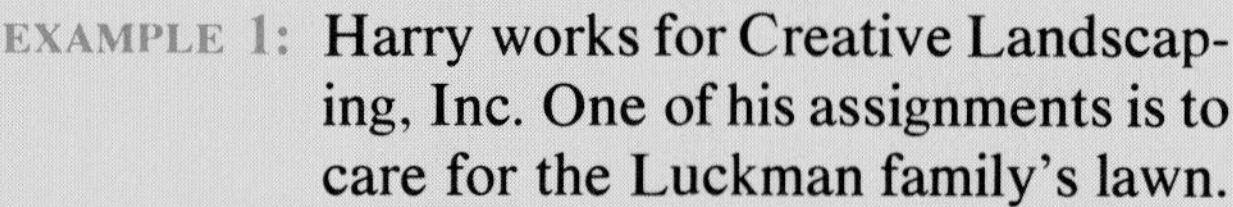

EXAMPLE 1: Harry works for Creative Landscaping, Inc. One of his assignments is to care for the Luckman family's lawn.

a. What information does Harry need to order the required amount of fertilizer for the lawn?

SOLUTION: Since the amount of fertilizer depends on the size of the lawn, Harry needs to know the area of the lawn.

b. How can Harry find the area?

SOLUTION:

1. To find the area of the lawn, Harry drew and labeled the diagram at the right.

2. Harry used the formula $A = lw$ and a calculator to compute the area.

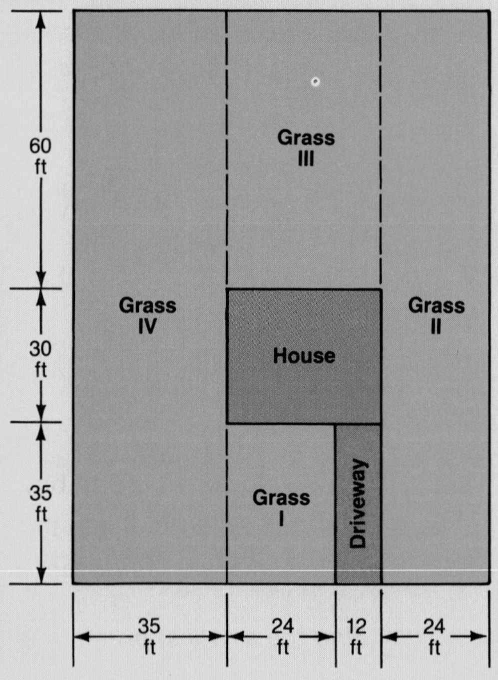

I. $A = (35)(24) = 840$ square feet
II. $A = (125)(24) = 3000$ square feet
III. $A = (60)(36) = 2160$ square feet
IV. $A = (125)(35) = \underline{4375}$ square feet
Total area: **10,375 square feet**

Harry knows that Brand A of a fertilizer requires 1 pound of fertilizer for each 250 square feet of grass.

EXAMPLE 2: **a.** How much fertilizer will be needed?

SOLUTION: Since 250 square feet require 1 pound, divide the total area by 250.

1 0 3 7 5 [÷] 2 5 0 [=] 41.5

Harry will need **42 pounds** of fertilizer.

b. If Brand A is sold in 20-pound bags at \$9.95 per bag and in 40-pound bags at \$16.95 per bag, how much will the fertilizer cost?

SOLUTION: One 40-pound bag: \$16.95
One 20-pound bag: \$ 9.95
Total: **\$26.90**

EXERCISES

Suppose Harry could obtain another brand of fertilizer, Brand B, that required 1 pound for each 100 square feet of lawn. Brand B comes in 20-pound bags at \$4.99 per bag and in 40-pound bags at \$7.99 per bag.

1. What would it cost Harry to use Brand B?
2. Which brand would you suggest that Harry use? Why?

Suppose you decide to paint a room in your house. You will need these facts for your project.

Room dimensions:	15 feet by 12 feet
Ceiling height:	8 feet
Door dimensions:	3 feet by $6\frac{1}{2}$ feet (The 3 doors will not be painted.)
Window dimensions:	3 feet by $3\frac{1}{2}$ feet (There are 2 windows.)

The walls and the ceiling will receive two coats of paint.
One gallon of paint covers 400 square feet on the first coat.
One gallon of paint covers 500 square feet on the second coat.
One gallon of paint costs \$10.97 and one quart costs \$5.79.

3. How much paint will you need for the first coat? The second coat? How much in all?
4. How much paint will you buy? Choose one of the following.
   i. 2 gallons **ii.** 3 gallons **iii.** 2 gallons plus 1 quart
   iv. 2 gallons plus 2 quarts **v.** 2 gallons plus 3 quarts

5. How much would 3 gallons cost?
6. How much would 2 gallons plus 1 quart cost?
7. Why would you not buy 2 gallons plus 2 quarts or 2 gallons plus 3 quarts?

REVIEW CAPSULE FOR SECTION 3.4

Evaluate. (Pages 2–4)

1. a. $6 \cdot 21$
b. $3 \cdot 2 \cdot 7 \cdot 3$

2. a. $14 \cdot 9$
b. $7 \cdot 2 \cdot 3 \cdot 3$

3. a. $6 \cdot 7$
b. $2 \cdot 3 \cdot 7$

4. a. $4 \cdot 17$
b. $2 \cdot 2 \cdot 17$

5. a. $6 \cdot 25$
b. $3 \cdot 2 \cdot 5 \cdot 5$

6. a. $15 \cdot 21$
b. $5 \cdot 3 \cdot 3 \cdot 7$

3.4 Prime Factorization

Every composite number can be expressed as a product of prime factors.

$$126 = 21 \cdot 6 = 3 \cdot 7 \cdot 2 \cdot 3$$

Such a product is called the ***prime factorization*** of the number.

The factors in a prime factorization are usually written in order from smallest to largest. For example, the prime factorization $3 \cdot 7 \cdot 2 \cdot 3$ is written as $2 \cdot 3 \cdot 3 \cdot 7$ or preferably as $2 \cdot 3^2 \cdot 7$.

The prime numbers less than 24 are listed below for reference in the examples that follow.

2, 3, 5, 7, 11, 13, 17, 19, 23

EXAMPLE 1 Write the prime factorization of 126.

Solution:

[1] Divide 126 by 2. Write the quotient 63.

$$\begin{array}{r|l} \mathbf{2} & 126 \\ \hline & 63 \end{array}$$

[2] Is 2 a factor of 63? No.

[3] Select the next greater prime number 3. You know that 3 is a factor of 63. Divide and write the quotient 21.

$$\begin{array}{r|l} 2 & 126 \\ \hline \mathbf{3} & 63 \\ \hline & 21 \end{array}$$

[4] Check again to see whether 3 is a factor of 21. It is. The quotient is 7. Since 7 is a prime number, the problem is finished.

$$\begin{array}{r|l} 2 & 126 \\ \hline 3 & 63 \\ \hline \mathbf{3} & 21 \\ \hline & 7 \end{array}$$

[5] Write the prime factorization of 126.

$$126 = 2 \cdot 3 \cdot 3 \cdot 7 \quad \text{or} \quad 2 \cdot 3^2 \cdot 7$$

P–1 **Write the prime factorization of 105.**

You can also use a **factor tree** to find the prime factorization of a composite number.

EXAMPLE 2 Write the prime factorization of 72.

Solution:

[1] Choose any two factors of 72. ⟶ 8 · 9

[2] Keep factoring until only prime numbers appear. ⟶ 4 · 2 · 3 · 3

```
                72
             /      \
            8   ·    9
           / \      / \
          4 · 2 · 3 · 3
         / \   |   |   |
        2 · 2 · 2 · 3 · 3
```

$72 = 2 \cdot 2 \cdot 2 \cdot 3 \cdot 3$, or $2^3 \cdot 3^2$

P–2 **Use a factor tree to write the prime factorization of 120.**

Examples 1 and 2 illustrate the following idea.

> There is only one possible selection of prime numbers in the prime factorization of a composite number.

CLASSROOM EXERCISES

Write the smallest prime factor of each number.

1. 462 **2.** 95 **3.** 87 **4.** 860 **5.** 225 **6.** 99

Write the prime factorization of each number. (Step 5, Example 1)

7. 455
```
5 | 455
7 | 91
    13
```

8. 261
```
3 | 261
3 | 87
    29
```

9. 150
```
2 | 150
3 | 75
5 | 25
    5
```

10. 204
```
2 | 204
2 | 102
3 | 51
    17
```

11. 492
```
2 | 492
2 | 246
3 | 123
    41
```

12. 315
```
3 | 315
3 | 105
5 | 35
    7
```

13. 198
```
  | 198
  | 99
  | 33
    11
```

14. 220
```
  | 220
  | 110
  | 55
    11
```

15. 261
```
  | 261
  | 87
    29
```

16. 396
```
  | 396
  | 198
  | 99
  | 33
    11
```

17. 450
```
  | 450
  | 225
  | 75
  | 25
    5
```

18. 312
```
  | 312
  | 156
  | 78
  | 39
    13
```

Write the prime factorization of each number.
(Example 1)

19. 6 **20.** 14 **21.** 15 **22.** 33
23. 35 **24.** 34 **25.** 39 **26.** 55

Use a factor tree to write the prime factorization of each number. (Example 2)

27. 8 **28.** 12 **29.** 16 **30.** 20
31. 18 **32.** 24 **33.** 27 **34.** 32
35. 36 **36.** 40 **37.** 100 **38.** 90

WRITTEN EXERCISES

Goal: To write the prime factorization of a composite number
Sample Problem: 980 **Answer:** $2^2 \cdot 5 \cdot 7^2$

For Exercises 1–32, write the prime factorization of each number. Use whichever method you prefer. (Examples 1 and 2)

1. 22 **2.** 26 **3.** 25 **4.** 49
5. 21 **6.** 38 **7.** 39 **8.** 46
9. 51 **10.** 57 **11.** 77 **12.** 65
13. 85 **14.** 91 **15.** 143 **16.** 161
17. 48 **18.** 50 **19.** 54 **20.** 56
21. 78 **22.** 105 **23.** 84 **24.** 60
25. 125 **26.** 64 **27.** 168 **28.** 128
29. 378 **30.** 364 **31.** 324 **32.** 288

NON-ROUTINE PROBLEMS

33. The odometer on a car showed 15951 miles. This number is called a **palindrome** since the same number results when you reverse the digits. Two hours later the odometer showed the next successive palindrome. How many miles had the car traveled in 2 hours?

34. At a club meeting, each person present shakes hands with each other person exactly one time. There are a total of 28 handshakes. Find the number of people at the meeting.

REVIEW CAPSULE FOR SECTION 3.5

List the first six multiples of each number. (Pages 50–53)

1. 2 **2.** 4 **3.** 3 **4.** 5 **5.** 9 **6.** 6

7. List any numbers that are in the answers to both Exercises **1** and **2**.

8. List any numbers that are in the answers to both Exercises **3** and **5**.

9. List any numbers that are in the answers to both Exercises **2** and **6**.

10. List any numbers that are in the answers to both Exercises **1** and **4**.

3.5 Least Common Multiple

The number 12 is a common multiple of 3 and 4. The number 24 is also a common multiple of 3 and 4.

Multiples of 3: 3, 6, 9, **12,** 15, 18, 21, **24,** 27, · · ·

Multiples of 4: 4, 8, **12,** 16, 20, **24,** 28, 32, · · ·

Since 12 is the smallest number that is a common multiple of 3 and 4, it is called their least common multiple.

Definition

The ***least common multiple*** (*LCM*) of two or more counting numbers is the smallest counting number that is divisible by the given numbers.

18 is the *LCM* of 6 and 9.

You can use prime factors to find the *LCM*.

EXAMPLE 1

Write the least common multiple of 12 and 18.

Solution:

[1] Write the prime factorizations.

$12 = 4 \cdot 3 = \mathbf{2^2 \cdot 3}$ $\qquad 18 = 9 \cdot 2 = \mathbf{3^2 \cdot 2}$

[2] Write a product using each prime factor only once. → $2 \cdot 3$

[3] For each factor, write the highest exponent used in any of the prime factorizations. → $2^2 \cdot 3^2$

[4] Multiply. → $4 \cdot 9 = 36$ ◀ ***LCM of 12 and 18***

Some numbers, such as 8 and 9, do not have common factors. When this is the case, the *LCM* is simply their product.

$8 \cdot 9 = 72$ ◀ ***LCM of 8 and 9***

> To find the *LCM* of two or more numbers
>
> 1. Write the prime factorization of each number.
> 2. Write a product using each prime factor only once.
> 3. For each factor, write the highest exponent used in any of the prime factorizations.
> 4. Multiply these factors.

EXAMPLE 2

Write the least common multiple of 8, 30, and 45.

Solution:

[1] $8 = \mathbf{2^3}$ $30 = \mathbf{2 \cdot 3 \cdot 5}$ $45 = \mathbf{3^2 \cdot 5}$

[2] $2 \cdot 3 \cdot 5$

[3] $2^3 \cdot 3^2 \cdot 5$

[4] 360 ◀ ***LCM of 8, 30, and 45***

You can use the least common multiple to solve some word problems.

EXAMPLE 3

Both Walter and Susan start a new job on the same day. Walter works every fourth day, and Susan works every sixth day. How many days will it be before they will again be working on the same day?

Solution: Find the least common multiple of 4 and 6.

[1] $4 = 2^2$ $6 = 2 \cdot 3$

[2] $2 \cdot 3$

[3] $2^2 \cdot 3$

[4] 12 They will both be working on the 12th day.

JULY

S	M	T	W	T	F	S
			1	2	3	4
5	6	7	8	9	10	11
12	13	14	15	16	17	18
19	20	21	22	23	24	25
26	27	28	29	30	31	

CLASSROOM EXERCISES

The prime factorizations of two or three numbers are given. Write the prime factorization of their least common multiple. (Step 3, Examples 1 and 2)

1. $2 \cdot 3$
$2^2 \cdot 5$

2. $3^2 \cdot 5 \cdot 7$
$5^2 \cdot 2$

3. $11 \cdot 13$
17

4. $19 \cdot 23$
23^2

5. $2^2 \cdot 3 \cdot 5$
$2 \cdot 3^2 \cdot 5$

6. $2 \cdot 3^2 \cdot 5$
$2^3 \cdot 3 \cdot 5^3$

7. $2 \cdot 3$
$3 \cdot 5$
$3^2 \cdot 5^2$

8. $2 \cdot 3 \cdot 5 \cdot 7$
$2 \cdot 3^2 \cdot 11$
$3 \cdot 5^2 \cdot 7$

For Exercises 9–40, write the least common multiple.
(Example 1)

9. 4 and 6	**10.** 2 and 8	**11.** 2 and 3	**12.** 3 and 5
13. 3 and 15	**14.** 4 and 12	**15.** 8 and 12	**16.** 4 and 10
17. 2 and 7	**18.** 2 and 5	**19.** 6 and 9	**20.** 6 and 15
21. 10 and 15	**22.** 12 and 9	**23.** 5 and 7	**24.** 3 and 11

(Example 2)

25. 3, 5, and 15	**26.** 4, 5, and 10	**27.** 2, 6, and 8	**28.** 6, 8, and 12
29. 2, 4, and 8	**30.** 3, 6, and 12	**31.** 2, 6, and 9	**32.** 3, 8, and 12
33. 5, 10, and 15	**34.** 6, 8, and 9	**35.** 4, 10, and 25	**36.** 2, 8, and 12
37. 6, 9, and 12	**38.** 5, 8, and 10	**39.** 3, 10, and 15	**40.** 10, 15, and 20

WRITTEN EXERCISES

Goal: To write the least common multiple of two or more numbers
Sample Problem: 9, 12, and 15 **Answer:** 180

For Exercises 1–52, write the least common multiple. (Example 1)

1. 6 and 12	**2.** 8 and 24	**3.** 10 and 28	**4.** 12 and 20
5. 5 and 13	**6.** 3 and 17	**7.** 45 and 54	**8.** 36 and 80
9. 20 and 24	**10.** 18 and 27	**11.** 7 and 8	**12.** 5 and 12
13. 25 and 30	**14.** 50 and 60	**15.** 84 and 96	**16.** 60 and 150

(Example 2)

17. 8, 12, and 15	**18.** 12, 18, and 30	**19.** 12, 16, and 20	**20.** 18, 21, and 28
21. 14, 21, and 24	**22.** 24, 36, and 40	**23.** 9, 15, and 16	**24.** 15, 25, and 35
25. 24, 30, and 42	**26.** 16, 20, and 24	**27.** 27, 35, and 63	**28.** 27, 45, and 75

MIXED PRACTICE

29. 9 and 36
30. 8 and 40
31. 3, 9, and 27
32. 5, 15, and 30
33. 14, 20, and 35
34. 12 and 30
35. 18 and 24
36. 12, 14, and 16
37. 13 and 19
38. 8, 32, and 128
39. 9, 27, and 81
40. 17 and 23
41. 10, 12, and 20
42. 18, 20, and 24
43. 60 and 126
44. 54 and 180
45. 24, 36, and 48
46. 36 and 45
47. 42 and 70
48. 18, 36, and 210
49. 17 and 18
50. 33, 57, and 187
51. 65, 70, and 119
52. 19 and 20

APPLICATIONS

Solve. (Example 3)

53. Margaret and Alberto have part-time jobs. Margaret works every third day, and Alberto works every fourth day. If both work on Monday, how many days will pass before they both will again be working on the same day?

54. Rita can wash a car in 12 minutes. It takes Sam 9 minutes to wash a car. They begin at the same time. How much time will they spend washing cars if they finish at the same time?

55. In a grocery store, produce is delivered every third business day, milk every fourth business day, and paper goods every twentieth business day. All three items are delivered on Monday. How many business days will pass before all three are again delivered on the same day?

56. Three salespeople take business trips. The first salesperson takes a trip every fifth business day, the second salesperson every tenth business day, and the third every fifteenth business day. They all take a trip on Tuesday. How many business days will pass before all three are again taking a trip on the same day?

You can use the formula $\mathbf{p = n^2 - n + 41}$ *to find a prime number, p. The formula works for the counting numbers from 1 to 40.*

EXAMPLE: Use the formula $p = n^2 - n + 41$ to find p when $n = 23$.

SOLUTION:
$$\begin{aligned} p &= n^2 - n + 41 \\ &= (23)^2 - 23 + 41 \\ &= 529 - 23 + 41 \\ &= 547 \end{aligned}$$

Use a calculator.

Use the formula $p = n^2 - n + 41$ *to find p for each value of n.*

57. 12
58. 18
59. 21
60. 29
61. 35
62. 40

Focus on Reasoning: Venn Diagrams

Using **Venn diagrams** can help you to determine whether a statement is true or false.

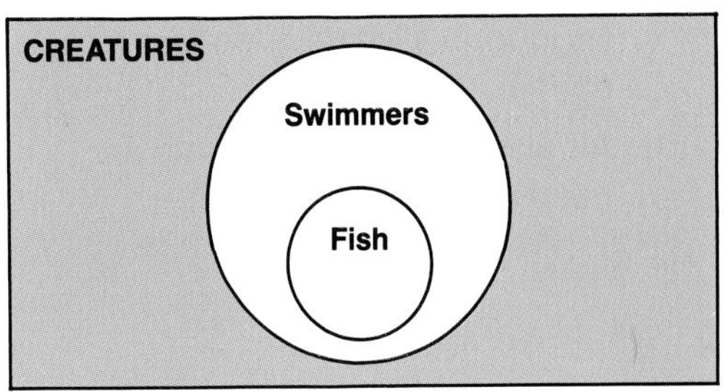

EXAMPLE 1 Use the Venn diagram to determine whether each statement is true or false. Give a reason for each answer.

Statements	**Reasons**
a. All fish can swim.	**a. True.** The circle for "Fish" is entirely contained in the circle for "Swimmers."
b. No creature can swim.	**b. False.** The rectangle for "Creatures" contains some creatures that are "Swimmers."
c. All creatures that swim are fish.	**c. False.** The circle for "Fish" does not completely fill the circle for "Swimmers." Thus, some creatures that swim are not fish.

EXERCISES

In each exercise, use the Venn diagram to determine whether each statement is true or false. Give a reason for each answer.

1. Statements:

a. All students work hard.

b. Some students work hard.

c. No students like holidays.

d. No student who works hard likes holidays.

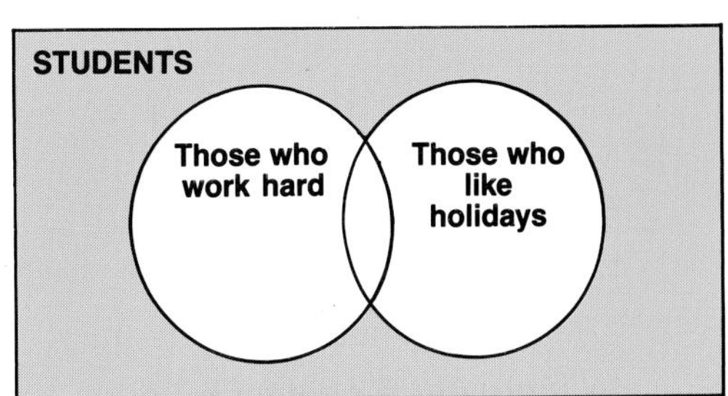

2. Statements:

a. Some cats are not mammals.

b. All dogs and cats are mammals.

c. No dog is a warm-blooded animal.

d. All warm-blooded animals are mammals.

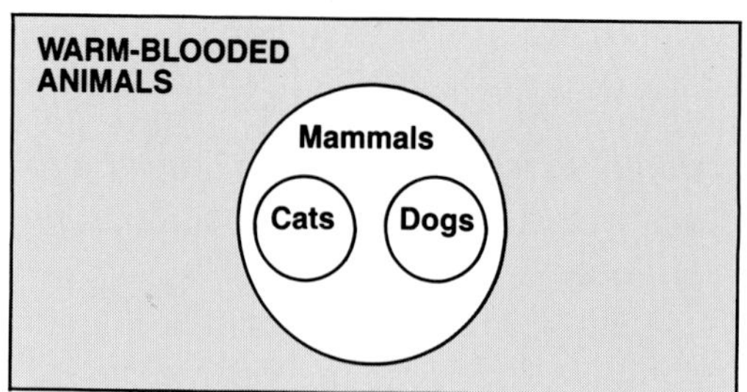

3. Statements:

a. All people who eat yogurt are healthy.

b. All healthy people eat yogurt.

c. Some people who eat yogurt are not healthy.

d. Some healthy people eat yogurt.

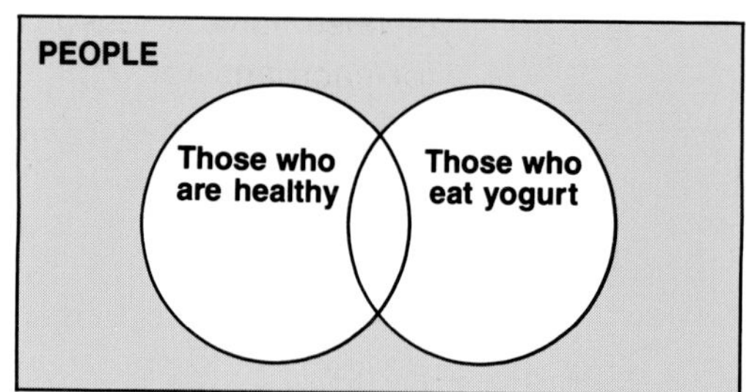

4. Statements:

a. Spinach is a green vegetable.

b. Some vegetables are not green.

c. All vegetables are green or yellow.

d. No green vegetable is a yellow vegetable.

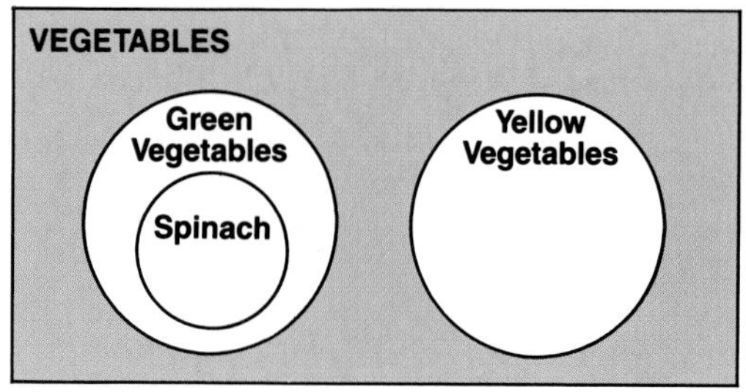

5. Statements:

a. All whole numbers are divisible by 5.

b. All whole numbers divisible by 5 are also divisible by 10.

c. Some whole numbers divisible by 10 are not divisible by 5.

d. Some whole numbers divisible by 5 are not divisible by 10.

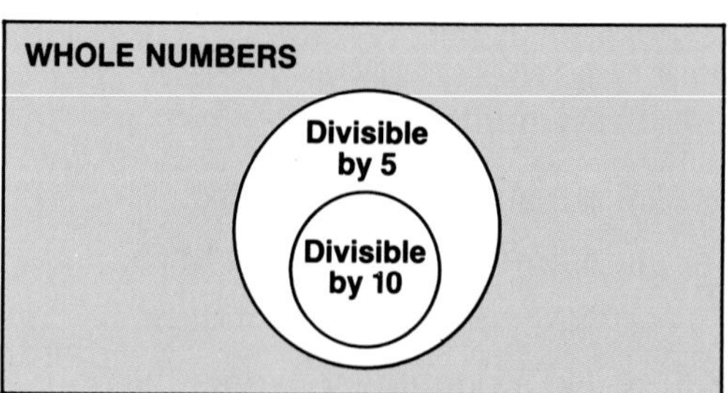

CHAPTER SUMMARY

IMPORTANT TERMS	Divisible by *(p. 50)*	Even number *(p. 54)*
	Factor *(p. 50)*	Odd number *(p. 54)*
	Counting number *(p. 50)*	Prime number *(p. 58)*
	Multiple *(p. 50)*	Composite number *(p. 58)*
	Whole number *(p. 54)*	Prime factorization *(p. 64)*
	Digit *(p. 54)*	Least common multiple *(p. 67)*

IMPORTANT IDEAS

1. The terms divisible by, factor of, and multiple of are related.

2. A number is divisible by
- 2 if it ends in 0, 2, 4, 6, or 8.
- 3 if the sum of its digits is divisible by 3.
- 4 if its last two digits name a multiple of 4.
- 5 if it ends in 0 or 5.
- 9 if the sum of its digits is divisible by 9.
- 10 if it ends in 0.

Any whole number that does not meet one of these tests is not divisible by the number shown.

3. There is only one possible selection of prime numbers in the prime factorization of a composite number.

4. To find the least common multiple of two or more numbers, write each prime factor the greatest number of times it is a factor of any one of the given numbers. Then multiply these factors.

5. The least common multiple of two numbers that have no common factor is the product of the two numbers.

CHAPTER REVIEW

SECTION 3.1

List the factors of each number. Then write each number as a product of two counting numbers. Do not use 1 or the given number.

1. 27 **2.** 40 **3.** 50 **4.** 56 **5.** 44

6. 62 **7.** 82 **8.** 85 **9.** 90 **10.** 98

SECTION 3.2

Write True or False for each sentence. Give a reason for each answer.

11. 4725 is divisible by 3.

12. 90,286 is divisible by 3.

13. 4 is a factor of 1802.

14. 4 is a factor of 12,328.

15. 43,902 is a multiple of 9.

16. 715,824 is a multiple of 9.

17. 73,195 is divisible by 5 and 10.

18. 9760 is a multiple of 5 and 10.

19. 2 is a factor of 26,703.

20. 2 is a factor of 8,906.

Solve.

21. A grocer plans to sell 3424 pounds of onions in 5–pound bags.
    a. Will all the bags be completely filled?
    b. If not, how many pounds of onions are needed to do this?

22. At a reception for 135 people, six persons will be seated at each table.
    a. Will there be exactly six persons at each table?
    b. If not, how many more persons are needed to do this?

SECTION 3.3

Use exponents to rewrite each product.

23. $19 \cdot 5 \cdot 5 \cdot 7 \cdot 3 \cdot 2 \cdot 2 \cdot 7 \cdot 2$

24. $37 \cdot 17 \cdot 2 \cdot 5 \cdot 2 \cdot 3 \cdot 5 \cdot 2 \cdot 3 \cdot 5$

25. $5 \cdot 3 \cdot 5 \cdot 3 \cdot 3 \cdot 2 \cdot 2 \cdot 5 \cdot 5 \cdot 2 \cdot 3$

26. $29 \cdot 23 \cdot 7 \cdot 7 \cdot 2 \cdot 5 \cdot 3 \cdot 5 \cdot 3 \cdot 3 \cdot 3$

SECTION 3.4

Write the prime factorization of each number.

27. 12
28. 16
29. 28
30. 44
31. 68
32. 74
33. 92
34. 98

SECTION 3.5

The prime factorizations of two or three numbers are given in each exercise. Write the prime factorization of their least common multiple.

35. $2 \cdot 3^3 \cdot 5^2 \cdot 13$, $2 \cdot 3 \cdot 11 \cdot 13^2$

36. $2^5 \cdot 3 \cdot 7^2 \cdot 11$, $2^3 \cdot 5^2 \cdot 11^3$

37. $2 \cdot 3 \cdot 5^2$, $3^4 \cdot 5 \cdot 7$, $2^3 \cdot 5 \cdot 11$

38. $2^3 \cdot 3 \cdot 17$, $2 \cdot 3^4 \cdot 11$, $2^2 \cdot 5^2 \cdot 7^3$

Write the least common multiple of the given numbers.

39. 8 and 20
40. 14 and 10
41. 12 and 18
42. 24 and 30
43. 7, 5, and 11
44. 3, 11, and 13
45. 8, 12, and 20
46. 9, 21, and 49
47. 25, 35, and 21
48. 18, 24, and 32
49. 8, 18, and 20
50. 15, 18, and 45

51. Tim and Sabrina have part–time jobs. Tim works every third day and Sabrina works every fifth day. If they both work on Tuesday, how many days will pass before they both will again be working on the same day?

52. Cartons 36 inches tall, 40 inches tall, and 24 inches tall are being stacked next to each other in three separate piles. What is the shortest height possible so that all three piles will have the same height?

CUMULATIVE REVIEW: CHAPTERS 1–3

Evaluate each expression. (Section 1.1)

1. $3 + 7 \cdot 8$

2. $12 - 11 + 1$

3. $12 \div 4 \cdot 3$

4. $14 + 28 \div 7 - 2$

Evaluate each expression. (Section 1.2)

5. $25 - 4w$ when $w = 3$

6. $15a + 13b$ when $a = 3$ and $b = 8$

7. $y + 24 \div x \cdot 2$ when $x = 4$ and $y = 5$

8. $p - 14 \div q + 11$ when $p = 21$ and $q = 7$

Use the formula $U = p \div n$ to find the unit price of each item to the nearest tenth of a cent. Compare to find which is the better buy. (Section 1.3)

9. The price of a 25–pound bag of dog food is $9.69, and the price of a 10-pound bag of dog food is $4.95.

10. The price of an 11–ounce bottle of shampoo is $2.79, and the price of a 16-ounce bottle is $4.18.

Solve each problem. (Section 1.4)

11. In Ames, Iowa, the heating zone factor is 0.025. Use the formula $S = kct$ to find the estimated savings on yearly heating costs of $775 if the thermostat is set 2° lower.

12. Use the formula $S = 0.02ct$ to find the estimated savings on yearly cooling costs of $490 if the thermostat is set 3° higher.

For Exercises 13–14, refer to the table below. (Section 1.5)

Original price: *p*	$18	$20	$22	$24	$26	$28	$30
Sale price: *s*	$13	$15	$17	$19	$21	$23	$27

13. Write a word rule for the pattern in the table.

14. Write a formula for the word rule.

Find the value of the variable that makes each sentence true. (Section 2.1)

15. $x + 7.6 = 7.6$

16. $0 = n(3\frac{1}{2})$

17. $(\frac{6}{6})y = 17$

Multiply. (Section 2.2)

18. $3(7b)$

19. $(8h)(0.5k)$

20. $(6m)(0.2m^2p)$

21. $(4gt^2)(3gt)(g)$

Write each product as a sum or difference. (Section 2.3)

22. $7(8x + 3)$

23. $(1 - 4p)12$

24. $(a - 3b)5a$

25. $3x(6x + y)$

Factor each expression. (Section 2.4)

26. $11a + 11b$ **27.** $2x + 12$ **28.** $a^2 - 7a$ **29.** $6x^2 + x$

Solve each problem. (Section 2.5)

30. A volleyball court is 69 feet long and 30 feet wide. Find the perimeter of the court.

31. A square rug is 5 meters on each side. What is the area of the rug?

Combine like terms. (Sections 2.6 and 2.7)

32. $14x^2 + x^2$ **33.** $16ab + 7ab$ **34.** $1.9f^2g - 1.2f^2g$

35. $1 + 9y + 3y$ **36.** $6c + d + 5d + 8c$ **37.** $1.6x^2y + 3xy^2 + 5.4x^2y$

List all the factors of each number. (Section 3.1)

38. 28 **39.** 34 **40.** 72 **41.** 100

Answer yes or no. Give a reason for each answer. (Section 3.2)

42. Is 5 a factor of 9735?

43. Is 23,568 divisible by 9?

44. Is 12,034 a multiple of 4?

45. Is 436 a factor of 2?

Use exponents to rewrite each product. (Section 3.3)

46. $11 \cdot 3 \cdot 3 \cdot 2 \cdot 11$

47. $3 \cdot 7 \cdot 5 \cdot 2 \cdot 2 \cdot 7 \cdot 7$

48. $7 \cdot 5 \cdot 13 \cdot 2 \cdot 5 \cdot 7 \cdot 2 \cdot 5$

49. $17 \cdot 2 \cdot 5 \cdot 5 \cdot 3 \cdot 2 \cdot 5 \cdot 3 \cdot 2$

Write the prime factorization of each number. (Section 3.4)

50. 92 **51.** 90 **52.** 132 **53.** 200 **54.** 315

Write the least common multiple. (Section 3.5)

55. 12 and 60 **56.** 28 and 21 **57.** 5, 10, and 18

Solve. (Section 3.5)

58. Martin can jog once around a track in 6 minutes. Eloise can jog once around the track in 8 minutes. They begin at the same time. How much time will pass before they are again at the same point on the track?

59. Edward mows his lawn once every 7 days. Grace mows her lawn once every 5 days. If they both mow on Saturday, how many days will pass before they again mow on the same day?

CHAPTER 4

Fractions

Atomic clocks are used to regulate regular timepieces. They are also used in navigation systems to determine the position of a ship, plane, or satellite. An atomic clock is accurate to $3 \times \frac{1}{100,000,000,000}$ second. Some are reliable to about 1 second every 30,000 years.

4.1 Writing Fractions in Lowest Terms

Numerals such as $\frac{3}{5}$, $\frac{12}{4}$, and $\frac{2}{6}$ are called ***fractions.***

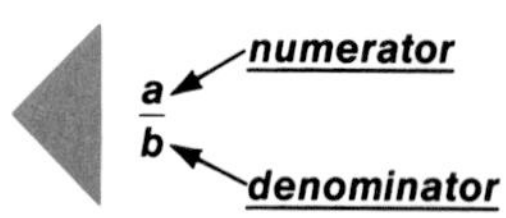

Since you cannot divide by zero, the denominator of a fraction can never equal zero. However, a numerator can equal zero.

Any fraction with the same nonzero numerator and denominator equals 1.

$\frac{a}{a} = 1$ ***(a is not zero.)***

$$\frac{35}{35} = 1$$

$$\frac{2 \cdot 2 \cdot 3}{2 \cdot 2 \cdot 3} = 1$$

Any fraction with a denominator of 1 names the same number as its numerator.

$\frac{a}{1} = a$

$$\frac{35}{1} = 35$$

$$\frac{0}{1} = 0$$

P–1 **What whole number does each fraction name?**

a. $\frac{10}{1}$ **b.** $\frac{0}{12}$ **c.** $\frac{5}{5}$ **d.** $\frac{100}{100}$ **e.** $\frac{250}{250}$

Product Rule for Fractions

1. Multiply the numerators.
2. Multiply the denominators.

$\frac{a}{b} \cdot \frac{c}{d} = \frac{a \cdot c}{b \cdot d}$ ***(b and d are not zero.)***

$$\frac{2}{3} \cdot \frac{5}{7} = \frac{2 \cdot 5}{3 \cdot 7} = \frac{10}{21}$$

Definition

A fraction is in ***lowest terms*** when the numerator and denominator have no common prime factor.

$$\frac{6}{35} = \frac{2 \cdot 3}{5 \cdot 7}$$

P–2 **In each of these fractions, what common prime factor do the numerator and denominator share?**

a. $\frac{3}{6}$ **b.** $\frac{2}{4}$ **c.** $\frac{1}{2}$ **d.** $\frac{5}{10}$ **e.** $\frac{9}{18}$

The first step in writing a fraction in lowest terms is to write the numerator and denominator as a product of prime factors.

EXAMPLE 1 Write in lowest terms: $\frac{7}{42}$

Solution:

[1] Write the prime factorization. ⟶ $\frac{7}{42} = \frac{7}{2 \cdot 3 \cdot 7}$

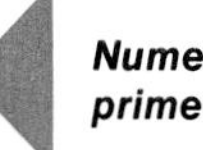

Numerator is a prime number.

[2] Write as a product of two fractions (with one fraction having equal terms). ⟶ $= \frac{7}{7} \cdot \frac{1}{2 \cdot 3}$

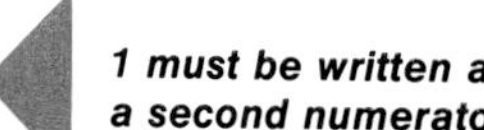
1 must be written as a second numerator.

[3] Use the Multiplication Property of One. ⟶ $= 1 \cdot \frac{1}{6}$

$= \frac{1}{6}$

P–3 **Write each of the following in lowest terms.**

a. $\frac{5}{20}$ **b.** $\frac{11}{33}$ **c.** $\frac{3}{9}$ **d.** $\frac{2}{22}$ **e.** $\frac{6}{18}$

Example 2 illustrates a short method for writing a fraction in lowest terms.

EXAMPLE 2 Write in lowest terms: $\frac{12}{30}$

Solution: [1] Write the prime factorization. ⟶ $\frac{12}{30} = \frac{2 \cdot 2 \cdot 3}{2 \cdot 3 \cdot 5}$

[2] Divide the numerator and denominator by their common factors. ⟶ $= \frac{\overset{1}{\cancel{2}} \cdot 2 \cdot \overset{1}{\cancel{3}}}{\underset{1}{\cancel{2}} \cdot \underset{1}{\cancel{3}} \cdot 5}$

[3] Write in lowest terms. ⟶ $= \frac{2}{5}$

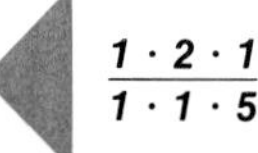
$\frac{1 \cdot 2 \cdot 1}{1 \cdot 1 \cdot 5}$

P–4 **Write each of the following in lowest terms.**

a. $\frac{10}{18}$ **b.** $\frac{9}{15}$ **c.** $\frac{25}{35}$ **d.** $\frac{16}{36}$ **e.** $\frac{21}{49}$

To compute the **odds** in favor of an event occurring or the odds against an event occurring, you write fractions in lowest terms.

EXAMPLE 3

In a class of 24 girls and 18 boys, a student will be chosen as the student council member for the class. The selection will be made by drawing one name from a box.

a. What are the odds in favor of a girl's being chosen?
b. What are the odds against a girl's being chosen?

Solutions:

a. Odds in favor of an event $= \dfrac{\text{number of ways an event can occur}}{\text{number of ways an event cannot occur}}$

Odds in favor of choosing a girl $= \dfrac{24}{18}$ ← Number of ways of choosing a girl / ← Number of ways of not choosing a girl

$= \dfrac{4}{3}$ ◀ ***Lowest Terms***

The odds in favor of a girl's name being chosen are 4 to 3.

b. Odds against an event $= \dfrac{\text{number of ways an event cannot occur}}{\text{number of ways an event can occur}}$

Odds against choosing a girl $= \dfrac{18}{24}$ ← Number of ways of not choosing a girl / ← Number of ways of choosing a girl

$= \dfrac{3}{4}$ ◀ ***Lowest terms***

The odds against a girl's name being chosen are 3 to 4.

Each fraction below equals $\frac{2}{3}$.

$$\frac{4}{6} = \frac{\overset{1}{\cancel{2}} \cdot 2}{\underset{1}{\cancel{2}} \cdot 3} = \frac{2}{3} \qquad \frac{6}{9} = \frac{2 \cdot \overset{1}{\cancel{3}}}{\underset{1}{\cancel{3}} \cdot 3} = \frac{2}{3} \qquad \frac{10}{15} = \frac{2 \cdot \overset{1}{\cancel{5}}}{3 \cdot \underset{1}{\cancel{5}}} = \frac{2}{3}$$

Definition

Equivalent fractions name the same number. $\frac{4}{6} = \frac{6}{9} = \frac{10}{15}$

P–5 **Which of these fractions are equivalent?**

a. $\frac{10}{14}$ b. $\frac{3}{4}$ c. $\frac{21}{15}$ d. $\frac{5}{7}$ e. $\frac{12}{16}$

CLASSROOM EXERCISES

For Exercises 1–30, write each fraction in lowest terms.
(Steps 2 and 3, Example 1)

1. $\frac{2}{2 \cdot 5}$
2. $\frac{3}{3 \cdot 11}$
3. $\frac{2}{5 \cdot 2}$
4. $\frac{3}{3 \cdot 7}$
5. $\frac{2 \cdot 2}{2 \cdot 3}$
6. $\frac{3 \cdot 5}{3 \cdot 7}$
7. $\frac{3 \cdot 2}{3 \cdot 2}$
8. $\frac{5 \cdot 3}{5 \cdot 11}$
9. $\frac{4 \cdot 7}{9 \cdot 4}$
10. $\frac{5 \cdot 3}{3 \cdot 8}$
11. $\frac{7 \cdot 2}{5 \cdot 7}$
12. $\frac{9 \cdot 4}{11 \cdot 9}$

(Steps 2 and 3, Example 2)

13. $\frac{2 \cdot 2 \cdot 5}{2 \cdot 2 \cdot 2 \cdot 3}$
14. $\frac{2 \cdot 3 \cdot 3 \cdot 5}{3 \cdot 5 \cdot 7}$
15. $\frac{2 \cdot 2 \cdot 3}{2 \cdot 2 \cdot 2 \cdot 2 \cdot 3 \cdot 3}$
16. $\frac{2 \cdot 3 \cdot 5 \cdot 7 \cdot 7}{2 \cdot 3 \cdot 7 \cdot 11}$
17. $\frac{2 \cdot 3 \cdot 3 \cdot 3}{2 \cdot 2 \cdot 2 \cdot 3 \cdot 3}$
18. $\frac{2 \cdot 7 \cdot 11 \cdot 19}{3 \cdot 7 \cdot 11 \cdot 11}$

(Example 1)

19. $\frac{2}{16}$
20. $\frac{3}{12}$
21. $\frac{3}{15}$
22. $\frac{2}{12}$
23. $\frac{4}{14}$
24. $\frac{6}{15}$

(Example 2)

25. $\frac{8}{12}$
26. $\frac{12}{30}$
27. $\frac{20}{28}$
28. $\frac{12}{28}$
29. $\frac{15}{27}$
30. $\frac{10}{45}$

WRITTEN EXERCISES

Goal: To write a fraction in lowest terms

Sample Problems: a. $\frac{6}{24}$ **b.** $\frac{12}{40}$ **Answers: a.** $\frac{1}{4}$ **b.** $\frac{3}{10}$

For Exercises 1–54, write each fraction in lowest terms.
(Example 1)

1. $\frac{7}{21}$
2. $\frac{4}{28}$
3. $\frac{2}{18}$
4. $\frac{10}{40}$
5. $\frac{8}{24}$
6. $\frac{15}{30}$
7. $\frac{14}{28}$
8. $\frac{4}{24}$
9. $\frac{8}{16}$
10. $\frac{4}{16}$
11. $\frac{14}{42}$
12. $\frac{16}{48}$

(Example 2)

13. $\frac{12}{18}$ 14. $\frac{12}{20}$ 15. $\frac{24}{30}$ 16. $\frac{20}{32}$ 17. $\frac{12}{64}$ 18. $\frac{24}{28}$

19. $\frac{27}{36}$ 20. $\frac{14}{42}$ 21. $\frac{60}{72}$ 22. $\frac{24}{64}$ 23. $\frac{12}{32}$ 24. $\frac{36}{45}$

MIXED PRACTICE

25. $\frac{16}{32}$ 26. $\frac{45}{80}$ 27. $\frac{15}{21}$ 28. $\frac{12}{21}$ 29. $\frac{2}{28}$ 30. $\frac{20}{36}$

31. $\frac{65}{80}$ 32. $\frac{70}{154}$ 33. $\frac{16}{96}$ 34. $\frac{39}{51}$ 35. $\frac{20}{42}$ 36. $\frac{33}{55}$

37. $\frac{40}{154}$ 38. $\frac{28}{42}$ 39. $\frac{120}{132}$ 40. $\frac{165}{210}$ 41. $\frac{36}{48}$ 42. $\frac{15}{45}$

43. $\frac{39}{48}$ 44. $\frac{35}{65}$ 45. $\frac{19}{114}$ 46. $\frac{13}{169}$ 47. $\frac{78}{114}$ 48. $\frac{105}{225}$

49. $\frac{49}{56}$ 50. $\frac{80}{105}$ 51. $\frac{96}{112}$ 52. $\frac{16}{256}$ 53. $\frac{25}{205}$ 54. $\frac{29}{203}$

APPLICATIONS

Write the odds as a fraction in lowest terms. (Example 3)

55. A drawer contains 8 blue pens and 6 red pens. Without looking, Margo reaches into the drawer. What are the odds against Margo's choosing a blue pen?

56. A committee consists of 10 Republicans and 12 Democrats. A spokesperson for the committee is chosen at random. What are the odds against choosing a Democrat as the spokesperson?

57. Laura has two 1980 dimes, four 1982 dimes, and six 1985 dimes in her purse. She reaches in and picks out one dime. What are the odds that he chooses a 1982 dime?

58. Brian will choose a topic for a speech from slips of paper in a box. Of the ten remaining topics in the box, he likes four of them. What are the odds he chooses a topic he likes?

REVIEW CAPSULE FOR SECTION 4.2

Write each expression as a product. (Pages 27–30)

EXAMPLE: t^4 **ANSWER:** $t \cdot t \cdot t \cdot t$

1. x^3 2. m^2 3. t^6 4. a^5

5. r^2s^3 6. m^3n 7. p^2qr^2 8. a^3b^2c

9. $3n^3$ 10. $5b^4$ 11. $7c^3d^5$ 12. $13r^4t^3$

4.2 Algebraic Fractions

Fractions such as $\frac{3}{x}$, $\frac{a+4}{5}$, and $\frac{2x}{y}$ are <u>algebraic fractions.</u>

Definition In an ***algebraic fraction,*** a variable appears in either the numerator or denominator or both. $\frac{2}{x-3}$

P–1 **Which of the following are algebraic fractions?**

a. $\frac{x}{5}$ **b.** $\frac{5}{12}$ **c.** $\frac{13}{19-5}$ **d.** $\frac{x+3}{12}$ **e.** $\frac{x}{3x}$

Algebraic fractions can be expressed in lowest terms using the same methods as for arithmetic fractions.

NOTE: In the Examples and Exercises of this text, it is understood that a variable cannot be replaced by a number that will make a denominator equal 0.

EXAMPLE 1

Write in lowest terms: $\frac{4x}{10x}$

Solution: [1] Write the prime factorization. ⟶ $\frac{4x}{10x} = \frac{2 \cdot 2 \cdot x}{2 \cdot 5 \cdot x}$

[2] Write as a product of fractions. ⟶ $= \frac{2}{2} \cdot \frac{x}{x} \cdot \frac{2}{5}$

[3] Use the Multiplication Property of One. ⟶ $= 1 \cdot 1 \cdot \frac{2}{5}$

$= \frac{2}{5}$

P–2 **Write each fraction in lowest terms.**

a. $\frac{6d}{8d}$ **b.** $\frac{2n}{6n}$ **c.** $\frac{9w}{15w}$ **d.** $\frac{12t}{36t}$ **e.** $\frac{48r}{54r}$

Example 2 illustrates a short method.

EXAMPLE 2 Write in lowest terms: $\frac{6ab^2c}{9abc^3}$

Solution:

1 Write the prime factorization. → $\frac{6ab^2c}{9abc^3} = \frac{2 \cdot 3 \cdot a \cdot b \cdot b \cdot c}{3 \cdot 3 \cdot a \cdot b \cdot c \cdot c \cdot c}$

2 Divide the numerator and denominator by their common factors. → $= \frac{2 \cdot \overset{1}{\cancel{3}} \cdot \overset{1}{\cancel{a}} \cdot \overset{1}{\cancel{b}} \cdot b \cdot \overset{1}{\cancel{c}}}{\underset{1}{\cancel{3}} \cdot 3 \cdot \underset{1}{\cancel{a}} \cdot \underset{1}{\cancel{b}} \cdot \underset{1}{\cancel{c}} \cdot c \cdot c}$

3 Write in lowest terms. → $= \frac{2b}{3c^2}$

P–3 Write each fraction in lowest terms.

a. $\frac{5mn}{8mn}$ b. $\frac{4tp}{10t^2p}$ c. $\frac{9x^2y}{15xy^2}$ d. $\frac{6j^3k}{10jk^2}$ e. $\frac{7b^3x}{14bx}$

CLASSROOM EXERCISES

For Exercises 1–12, find the prime factors of each numerator and denominator. (Step 1, Example 1)

1. $\frac{4n}{6m}$ 2. $\frac{12x}{14c}$ 3. $\frac{9rt}{15s}$ 4. $\frac{10p}{25qy}$ 5. $\frac{10zx}{18t}$ 6. $\frac{28z}{35bx}$

(Step 1, Example 2)

7. $\frac{2x^2y}{18xy^3}$ 8. $\frac{8mn^2}{16m^2n^3}$ 9. $\frac{24abc}{26b^2c^3}$ 10. $\frac{27rs^4}{36s^2t^3}$ 11. $\frac{5x^2y^2}{20xy^3}$ 12. $\frac{40r^2m}{48mt^2}$

For Exercises 13–32, express each fraction in lowest terms. (Steps 2 and 3, Example 1)

13. $\frac{2 \cdot a}{3 \cdot a}$ 14. $\frac{x \cdot y}{5 \cdot y}$ 15. $\frac{3 \cdot r \cdot s}{3 \cdot s \cdot t}$ 16. $\frac{2 \cdot 2 \cdot t}{2 \cdot 2 \cdot 3 \cdot t}$

(Steps 2 and 3, Example 2)

17. $\frac{a \cdot b \cdot b \cdot c}{5 \cdot a \cdot a \cdot b}$ 18. $\frac{2 \cdot 3 \cdot r}{2 \cdot 3 \cdot r \cdot r}$ 19. $\frac{5 \cdot m \cdot m}{5 \cdot 7 \cdot m \cdot n}$ 20. $\frac{3 \cdot p \cdot q \cdot q}{2 \cdot 3 \cdot p \cdot p \cdot q}$

(Example 1)

21. $\frac{4x}{6x}$ 22. $\frac{x}{5x}$ 23. $\frac{3r}{12r}$ 24. $\frac{6rs}{14st}$ 25. $\frac{7tw}{49yt}$ 26. $\frac{25mn}{35mq}$

(Example 2)

27. $\frac{4xy^2}{12xy}$ 28. $\frac{a}{a^2}$ 29. $\frac{10r^2s^2}{35r^3s}$ 30. $\frac{8x^2y}{12xy^2}$ 31. $\frac{16a^2b}{18b^2}$ 32. $\frac{24xy^3}{32x^3y}$

WRITTEN EXERCISES

Goal: To write an algebraic fraction in lowest terms

Sample Problem: $\frac{5x^2y}{15xy^4}$ **Answer:** $\frac{x}{3y^3}$

For Exercises 1–36, write each fraction in lowest terms.
(Example 1)

1. $\frac{5r}{11r}$ 2. $\frac{2m}{7m}$ 3. $\frac{3y}{xy}$ 4. $\frac{st}{11t}$ 5. $\frac{6p}{8p}$ 6. $\frac{8t}{12t}$

(Example 2)

7. $\frac{3pn}{5p^2}$ 8. $\frac{4rt}{7t^2}$ 9. $\frac{2y^2}{3y^3}$ 10. $\frac{4w}{15w^3}$ 11. $\frac{n}{n^2}$ 12. $\frac{k}{2k^2}$

13. $\frac{8m^2n}{18mnp}$ 14. $\frac{6rs}{16r^2t}$ 15. $\frac{15ra^2}{24ab}$ 16. $\frac{12s^2t}{15rt}$ 17. $\frac{10r^2y^2}{12r^2y^3t}$ 18. $\frac{16rst}{12r^2s^2t}$

MIXED PRACTICE

19. $\frac{4t}{30t}$ 20. $\frac{20qr}{15qr^2s}$ 21. $\frac{9acd^2}{6bc^2d}$ 22. $\frac{6r}{15r^2}$ 23. $\frac{30ab}{63bc}$ 24. $\frac{27cde}{60def}$

25. $\frac{3rs}{3s^2}$ 26. $\frac{2n^2k}{5nk^2}$ 27. $\frac{14p^2qr}{70p^2q^2r^3}$ 28. $\frac{24rst}{21r^2s^2t}$ 29. $\frac{24xz}{16x^3yz^2}$ 30. $\frac{12mn^2q}{72m^2n^3q^2}$

31. $\frac{15x^4y^2}{63xy^3z^2}$ 32. $\frac{26abc}{65bcd}$ 33. $\frac{21rs}{35stp}$ 34. $\frac{18a^2b^3}{42ab^4c^2}$ 35. $\frac{36x^2y^4z}{80x^3y^2z^3}$ 36. $\frac{17rs^2}{51rs^3t}$

MORE CHALLENGING EXERCISES

37. If the fraction $\frac{3x^2}{15x}$ has a value greater than 1, why must the value of x be greater than 5?

38. If the fraction $\frac{4m}{2m^2}$ has a value less than 1, why must the value of m be greater than 2?

39. For what value of n is the following true?

$$\frac{5+n}{8+n}=\frac{3}{4}$$

40. If the numerator of a fraction is a multiple of 3 and the denominator is a multiple of 4, is the fraction in lowest terms? Explain. Use examples to illustrate your answer.

.3 Multiplication

You can use the Product Rule to write a product of fractions as a single fraction.

$$\frac{3}{16} \cdot 5 = \frac{3}{16} \cdot \frac{5}{1}$$ **Note:** $5 = \frac{5}{1}$

$$= \frac{3 \cdot 5}{16 \cdot 1}$$ ***Single fraction***

$$= \frac{15}{16}$$

The following steps for multiplying with fractions will always give you a product in lowest terms.

> ***Steps for Multiplying with Fractions***
>
> 1. Factor the numerators and denominators completely.
> 2. Divide the numerator and denominator by any common factors.
> 3. Multiply the factors remaining in the numerator and denominator.

In Step 1 above, factor completely means to write the prime factors.

P–1 **What are the prime factors of 9? of 14? of 30?**

EXAMPLE 1 Multiply: $\frac{9}{14} \cdot \frac{7}{30}$

Solution:

[1] Factor the numerators and denominators completely. ⟶ $\frac{9}{14} \cdot \frac{7}{30} = \frac{3 \cdot 3}{2 \cdot 7} \cdot \frac{7}{2 \cdot 3 \cdot 5}$

[2] Divide the numerator and denominator of the single fraction by any common factors. ⟶ $= \frac{\overset{1}{\cancel{3}} \cdot 3 \cdot \overset{1}{\cancel{7}}}{2 \cdot \underset{1}{\cancel{7}} \cdot 2 \cdot \underset{1}{\cancel{3}} \cdot 5}$

[3] Multiply the factors remaining in the numerator and denominator. ⟶ $= \frac{3}{20}$

P–2 **Multiply.**

a. $\frac{6}{7} \cdot \frac{14}{15}$ **b.** $\frac{10}{11} \cdot \frac{33}{35}$ **c.** $\frac{9}{10} \cdot \frac{20}{21}$ **d.** $\frac{5}{9} \cdot \frac{18}{25}$

EXAMPLE 2

Multiply: $\frac{3}{xy} \cdot 2y$

Solution:

$$\frac{3}{xy} \cdot 2y = \frac{3}{xy} \cdot \frac{2y}{1}$$

$2y = \frac{2y}{1}$

$$= \frac{3 \cdot 2 \cdot \overset{1}{\cancel{y}}}{x \cdot \underset{1}{\cancel{y}} \cdot 1}$$

$$= \frac{6}{x}$$

P–3 **Multiply.**

a. $\frac{4}{5} \cdot 10n$ **b.** $16p \cdot \frac{7}{8s}$ **c.** $3r^2 \cdot \frac{2y}{5r}$ **d.** $\frac{8p}{9q} \cdot 21pt$

EXAMPLE 3

Multiply: $\frac{2x^2}{9y^2} \cdot \frac{3y}{10x}$

Solution:

$$\frac{2x^2}{9y^2} \cdot \frac{3y}{10x} = \frac{(2 \cdot x \cdot x) \cdot (3 \cdot y)}{(3 \cdot 3 \cdot y \cdot y) \cdot (2 \cdot 5 \cdot x)}$$

$$= \frac{\overset{1}{\cancel{2}} \cdot \overset{1}{\cancel{x}} \cdot x \cdot \overset{1}{\cancel{3}} \cdot \overset{1}{\cancel{y}}}{\underset{1}{\cancel{3}} \cdot 3 \cdot \underset{1}{\cancel{y}} \cdot y \cdot \underset{1}{\cancel{2}} \cdot 5 \cdot \underset{1}{\cancel{x}}} = \frac{x}{15y}$$

P–4 **Multiply.**

a. $\frac{2b}{3c} \cdot \frac{9}{10b}$ **b.** $\frac{1}{f} \cdot \frac{3g}{7}$ **c.** $\frac{14}{8r} \cdot \frac{r}{21}$ **d.** $\frac{12}{25f^2} \cdot \frac{10fg}{9h^2}$

CLASSROOM EXERCISES

Multiply. Be sure that each answer is in lowest terms. (Example 1)

1. $\frac{2}{3} \cdot 4$ **2.** $\frac{3}{7} \cdot 8$ **3.** $6 \cdot \frac{5}{12}$ **4.** $8 \cdot \frac{7}{24}$ **5.** $\frac{1}{2} \cdot \frac{1}{7}$ **6.** $\frac{1}{4} \cdot \frac{1}{5}$

7. $\frac{2}{5} \cdot \frac{3}{2}$ **8.** $\frac{5}{12} \cdot \frac{3}{5}$ **9.** $\frac{4}{3} \cdot \frac{9}{16}$ **10.** $\frac{5}{2} \cdot \frac{4}{25}$ **11.** $\frac{5}{6} \cdot \frac{3}{10}$ **12.** $\frac{3}{8} \cdot \frac{4}{15}$

For Exercises 13–24, multiply. Write each fraction in lowest terms.
(Example 2)

13. $\frac{2}{3} \cdot x$ **14.** $\frac{3}{5} \cdot 2a$ **15.** $\frac{1}{3} \cdot 6x$ **16.** $\frac{3}{4} \cdot 4y$ **17.** $\frac{5}{6b} \cdot 12b$ **18.** $\frac{9}{2xz} \cdot 7z$

(Example 3)

19. $\frac{2}{x} \cdot \frac{3}{y}$ **20.** $\frac{a}{3} \cdot \frac{2}{b}$ **21.** $\frac{2r}{s} \cdot \frac{3}{t}$ **22.** $\frac{2x}{3y} \cdot \frac{2x}{5}$ **23.** $\frac{a^2}{5} \cdot \frac{3}{a}$ **24.** $\frac{3}{2x} \cdot \frac{x^2}{3y}$

WRITTEN EXERCISES

Goal: To multiply with fractions

Sample Problem: $\frac{12x^2}{15y^3} \cdot \frac{5xy}{2}$ **Answer:** $\frac{2x^3}{y^2}$

For Exercises 1–36, multiply. Be sure that each answer is in lowest terms.
(Example 1)

1. $\frac{1}{2} \cdot 7$ **2.** $\frac{1}{4} \cdot 11$ **3.** $\frac{2}{3} \cdot 5$ **4.** $\frac{3}{4} \cdot 9$ **5.** $24 \cdot \frac{5}{9}$ **6.** $18 \cdot \frac{3}{14}$

7. $36 \cdot \frac{11}{45}$ **8.** $27 \cdot \frac{18}{63}$ **9.** $\frac{2}{3} \cdot \frac{5}{2}$ **10.** $\frac{3}{7} \cdot \frac{2}{3}$ **11.** $\frac{7}{8} \cdot \frac{8}{9}$ **12.** $\frac{4}{5} \cdot \frac{5}{11}$

(Example 2)

13. $\frac{3}{4} \cdot x$ **14.** $\frac{5}{6} \cdot y$ **15.** $\frac{1}{3} \cdot 2r$ **16.** $\frac{1}{8} \cdot 5t$ **17.** $3a \cdot \frac{7}{ab}$ **18.** $xy \cdot \frac{8}{x}$

(Example 3)

19. $\frac{4a^2}{7} \cdot \frac{1}{ab}$ **20.** $\frac{2m}{6} \cdot \frac{5}{3m^2}$ **21.** $\frac{3}{x^2} \cdot \frac{x}{7}$ **22.** $\frac{a^3}{5} \cdot \frac{3}{a^2}$ **23.** $\frac{3st^2}{5} \cdot \frac{5}{2t}$ **24.** $\frac{11y}{4x^2} \cdot \frac{7x}{y^2}$

MIXED PRACTICE

25. $\frac{4}{5} \cdot \frac{5}{8}$ **26.** $\frac{6}{7} \cdot \frac{7}{12}$ **27.** $30n \cdot \frac{11}{35n^2}$ **28.** $28t \cdot \frac{17}{21t^2}$

29. $\frac{5}{ab^2} \cdot \frac{bc}{10}$ **30.** $\frac{4xy}{3z} \cdot \frac{15z}{16y^2}$ **31.** $\frac{7}{8} \cdot \frac{12}{35}$ **32.** $\frac{5}{27} \cdot \frac{18}{55}$

33. $\frac{3a}{13} \cdot \frac{17}{5ab}$ **34.** $\frac{11}{19m^2} \cdot \frac{7mn}{2}$ **35.** $\frac{9}{35t^2} \cdot 56t$ **36.** $\frac{5}{77rt} \cdot 84r^2$

APPLICATIONS

The **circumference,** C, of a circle is the distance around the circle. You can compute the approximate circumference of a circle by using the formula $C = \frac{22}{7} \cdot d$ or $C = \frac{22d}{7}$, where d is the circle's diameter (distance across).

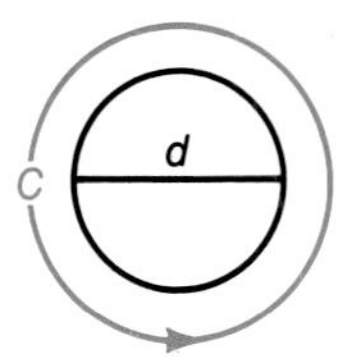

EXAMPLE Find the circumference of a circle with a 161-inch diameter.

SOLUTION $C = \frac{22d}{7}$

$C = 22 \cdot 161 \div 7$

$C = 506$

Approximate circumference: 506 inches

Use the formula $C = \frac{22d}{7}$ *to compute C for each value of d given.*

37. 259 feet **38.** 364 centimeters **39.** 686 inches **40.** 1001 millimeters

MID-CHAPTER REVIEW

Write each fraction in lowest terms. (Section 4.1)

1. $\frac{15}{24}$ **2.** $\frac{20}{24}$ **3.** $\frac{18}{32}$ **4.** $\frac{24}{36}$ **5.** $\frac{45}{48}$ **6.** $\frac{54}{72}$

7. $\frac{30}{36}$ **8.** $\frac{18}{48}$ **9.** $\frac{14}{28}$ **10.** $\frac{36}{81}$ **11.** $\frac{35}{50}$ **12.** $\frac{72}{84}$

Write the odds as a fraction in lowest terms.

13. A drawer contains 5 blue socks and 3 black socks. Without looking, Carl reaches into the drawer. What are the odds against Carl's choosing a black sock?

14. A booth at a carnival uses a grab bag to award prizes. The grab bag contains 4 red watches and 6 black watches. What are the odds in favor of choosing a black watch?

Write each fraction in lowest terms. (Section 4.2)

15. $\frac{3}{12m}$ **16.** $\frac{5}{25w}$ **17.** $\frac{4k^2}{18k}$ **18.** $\frac{15t}{24t^2}$ **19.** $\frac{21mn}{49np}$ **20.** $\frac{20ab}{35bc}$

21. $\frac{12x^2y}{20xy^3}$ **22.** $\frac{27rs^3}{45r^2s^2}$ **23.** $\frac{32pq^2}{12rp^2}$ **24.** $\frac{48c^2d}{40d^3e}$ **25.** $\frac{80m^4n}{96m^2n^3}$ **26.** $\frac{60xyz^2}{72x^3yz}$

Multiply. Be sure that each answer is in lowest terms. (Section 4.3)

27. $20 \cdot \frac{3}{8}$ **28.** $\frac{3}{4} \cdot \frac{10}{21}$ **29.** $\frac{5}{12} \cdot 3t$ **30.** $\frac{7}{17a} \cdot \frac{3a}{14}$

31. $\frac{28rs}{15xy} \cdot \frac{25xz}{21rw}$ **32.** $\frac{18fg}{35de} \cdot \frac{25cd}{24gh}$ **33.** $\frac{4x^2y}{9pq} \cdot \frac{3p^2}{20xy^3}$ **34.** $\frac{14ab^2}{27cd} \cdot \frac{45c^3d}{77a^2b^2}$

Using Methods of Computation

Making a Choice

In everyday situations, you often have to consider whether you need an **exact answer** to a problem or whether an **estimate** will give an answer that is close enough. Then you have to decide whether to use **mental computation,** a **calculator, paper and pencil,** or some combination of these to solve the problem efficiently.

EXERCISES

1. Raoul stopped at the supermarket on his way home from work. He placed the items on the list at the right in a shopping cart before discovering that he had exactly $15.10 in his wallet. Raoul thought: "I'll have to find the total cost to be sure I have enough money to pay for everything."

a. Does Raoul need to find the exact total or will an estimate be close enough? Explain.

b. Which would be the more efficient way to estimate the sum, using a calculator or using mental computation? Explain.

Grapes $1.85
Meat $4.90
Milk $1.49
Bread $1.10
Detergent $2.05

Raoul used **front-end estimation** in this way.

[1] Add the dollars first. $1 + $4 + $1 + $1 + $2 = $9

[2] Estimate the cents.

0.85 ← About $1
0.90 ← About $1
0.49, 0.10, 0.05 ← About 65¢

Estimate: $9 + $2 + 65¢ = **$11.65**

c. How far was Raoul's estimate from the actual cost?

d. Did Raoul have enough money to purchase all the items?

2. The manager of the Barlow High School softball team updates a list of the batting averages of the team members after each game. She uses this formula to find the batting average. Then she writes each average as a decimal rounded to the nearest thousandth.

$$\text{Batting Average} = \frac{\text{Number of hits}}{\text{Times at bat}}$$

Player	Number of Hits	Times at Bat	Average
Pam	15	60	?
Mary	19	61	?
Janine	17	59	?
Audrey	20	60	?
Carlotta	14	57	?
Sandy	17	49	?
Hannah	21	65	?
Corliss	18	59	?
Helen	16	55	?

a. Should the manager compute an exact average or use an estimate? Why?

b. To find the average before rounding, would it be more efficient to use a calculator or to use paper and pencil? Why?

c. To round the quotients to the nearest thousandth, would it be more efficient to use paper and pencil or to use mental computation? Why?

d. Find the average for each player.

e. Rank the players according to batting average, from highest to lowest.

f. What computation method did you use to arrange the rankings? Why?

3. Claudia plans to make gingersnaps for her younger brother's birthday party. In order to have enough, she triples the recipe.

Ingredient	Single Batch	Triple Batch
Flour	$3\frac{3}{4}$ cups	?
Sugar	2 cups	?
Eggs	2	?
Molasses	$\frac{1}{2}$ cup	?
Oleo	$\frac{3}{4}$ cup	?
Soda	1 tsp	?
Cinnamon	$\frac{1}{2}$ tsp	?
Cloves	$\frac{1}{4}$ tsp	?

a. Should Claudia estimate the amount of each ingredient to triple the recipe? Why or why not?

b. What computation method or methods would you use to triple the recipe? Give reasons for your answer.

c. Find the amount needed for each ingredient when the recipe is tripled.

4.4 Division

Definition

Two numbers having 1 as their product are called ***reciprocals.***

$\frac{2}{3}$ and $\frac{3}{2}$

$4x$ and $\frac{1}{4x}$

Recall that no denominator can equal zero.

Number	Equivalent Fraction	Reciprocal
$\frac{5}{16}$	$\frac{5}{16}$	$\frac{16}{5}$
12	$\frac{12}{1}$	$\frac{1}{12}$
$3\frac{1}{2}$	$\frac{7}{2}$	$\frac{2}{7}$
0.7	$\frac{7}{10}$	$\frac{10}{7}$
$\frac{c}{d}$	$\frac{c}{d}$	$\frac{d}{c}$
$7n$	$\frac{7n}{1}$	$\frac{1}{7n}$

P–1 **What is the reciprocal of each number?**

a. $\frac{8}{13}$ **b.** 9 **c.** $\frac{x}{y}$ **d.** $6g$ **e.** 0.2

Note that dividing by a number is the same as multiplying by its reciprocal.

$$14 \div 2 = 7 \qquad 14 \cdot \frac{1}{2} = 7$$

Quotient Rule for Fractions

To divide by a number, multiply by its reciprocal.

$\frac{a}{b} \div \frac{c}{d} = \frac{a}{b} \cdot \frac{d}{c}$ ***(b, c, d are not zero.)***

$14 \div 2 = 14 \cdot \frac{1}{2}$

$\frac{x}{2} \div \frac{x}{4} = \frac{x}{2} \cdot \frac{4}{x}$

EXAMPLE 1 Divide: $\frac{2}{3} \div \frac{x}{7}$ Write your answer in lowest terms.

Solution: [1] Use the Quotient Rule for Fractions. ⟶ $\frac{2}{3} \div \frac{x}{7} = \frac{2}{3} \cdot \frac{7}{x}$

[2] Use the Product Rule for Fractions. ⟶ $= \frac{14}{3x}$

P–2 **Divide.**

a. $\frac{3}{5} \div \frac{3}{k}$ **b.** $\frac{1}{3} \div \frac{t}{6}$ **c.** $\frac{3}{4} \div \frac{3}{4t}$ **d.** $\frac{2}{w} \div \frac{6}{7}$

When the steps for multiplying with fractions are followed, the answer to a division problem will always be in lowest terms.

EXAMPLE 2 Divide: $\frac{a^2}{3} \div a$

Solution:

[1] Use the Quotient Rule for Fractions. ⟶ $\frac{a^2}{3} \div a = \frac{a^2}{3} \cdot \frac{1}{a}$

The reciprocal of a is $\frac{1}{a}$.

[2] Use the Product Rule for Fractions. ⟶ $= \frac{a \cdot a \cdot 1}{3 \cdot a}$

[3] Divide the numerator and denominator by any common factors. ⟶ $= \frac{a \cdot \overset{1}{\cancel{a}} \cdot 1}{3 \cdot \underset{1}{\cancel{a}}}$

$= \frac{a}{3}$

P–3 **Divide.**

a. $\frac{1}{6} \div 3p$ **b.** $\frac{2}{3} \div 4n$ **c.** $\frac{4y}{7} \div 2y$ **d.** $\frac{3}{5g} \div 9h$

EXAMPLE 3 Divide: $\frac{12st^2}{35r^2} \div \frac{18st}{49r}$

Solution: $\frac{12st^2}{35r^2} \div \frac{18st}{49r} = \frac{12st^2}{35r^2} \cdot \frac{49r}{18st}$

$= \frac{\overset{1}{\cancel{2}} \cdot 2 \cdot \overset{1}{\cancel{3}} \cdot \overset{1}{\cancel{s}} \cdot \overset{1}{\cancel{t}} \cdot t \cdot \overset{1}{\cancel{7}} \cdot 7 \cdot \overset{1}{\cancel{r}}}{5 \cdot \underset{1}{\cancel{7}} \cdot \underset{1}{\cancel{r}} \cdot r \cdot \underset{1}{\cancel{2}} \cdot \underset{1}{\cancel{3}} \cdot 3 \cdot \underset{1}{\cancel{s}} \cdot \underset{1}{\cancel{t}}}$

$= \frac{14t}{15r}$

P–4 **Divide.**

a. $\frac{6y}{7z} \div \frac{3}{10z}$ b. $\frac{m}{2} \div \frac{m}{8}$ c. $\frac{5bc}{18d} \div \frac{15c^2}{16bd}$ d. $\frac{9mn}{10k} \div \frac{3m^2}{40k^2}$

CLASSROOM EXERCISES

For Exercises 1–12, express each quotient as a product. (Step 1, Examples 1, 2, and 3)

1. $\frac{3}{4} \div \frac{y}{3}$
2. $\frac{1}{5} \div \frac{2}{n}$
3. $\frac{4y}{3} \div \frac{1}{8}$
4. $\frac{2a}{5} \div \frac{4}{b}$
5. $\frac{3}{8} \div y$
6. $\frac{7}{4} \div 3n$
7. $\frac{4a^2}{2b} \div 5b^2$
8. $\frac{3mn}{7t} \div 3t^2$
9. $\frac{7a^2}{8} \div \frac{4}{3ab}$
10. $\frac{5}{2n^2} \div \frac{8mn}{7}$
11. $\frac{r}{s} \div \frac{1}{t}$
12. $\frac{k}{a} \div \frac{w}{q}$

For Exercises 13–28, divide. Write each answer in lowest terms. (Example 1)

13. $\frac{2}{5} \div \frac{n}{5}$
14. $\frac{1}{4} \div \frac{3t}{4}$
15. $\frac{3x}{4} \div \frac{1}{2}$
16. $\frac{r}{4} \div \frac{3}{2}$

(Example 2)

17. $\frac{3}{4} \div 2y$
18. $\frac{4}{5} \div 3m$
19. $\frac{2s}{11} \div s$
20. $\frac{3y}{x} \div 2y$

(Example 3)

21. $\frac{r}{4} \div \frac{r}{5}$
22. $\frac{x}{w} \div \frac{y}{w}$
23. $\frac{x^2}{3} \div \frac{x}{6}$
24. $\frac{4}{7p^2} \div \frac{10q}{21p}$
25. $\frac{f^2}{2g} \div \frac{4f^2}{3g^2}$
26. $\frac{4c^2d^2}{9b} \div \frac{2cb}{15d^2}$
27. $\frac{8xz^2}{3y} \div \frac{5x^2y^2}{6z}$
28. $\frac{jk^2}{15h} \div \frac{4j^2h}{3k}$

WRITTEN EXERCISES

Goal: To divide with fractions

Sample Problem: $\frac{6x^2}{10y} \div \frac{2x}{3}$ **Answer:** $\frac{9x}{10y}$

For Exercises 1–36, divide. Write each answer in lowest terms. (Example 1)

1. $\frac{x}{2} \div \frac{5}{3}$
2. $\frac{3}{4} \div \frac{7}{n}$
3. $\frac{2}{3} \div \frac{k}{3}$
4. $\frac{2}{7} \div \frac{r}{7}$
5. $\frac{m}{4} \div \frac{3}{10}$
6. $\frac{y}{6} \div \frac{4}{15}$
7. $\frac{5}{12} \div \frac{p}{18}$
8. $\frac{2}{15} \div \frac{c}{21}$

(Example 2)

9. $\frac{2t}{3} \div t$ **10.** $\frac{3r}{4} \div 2r$ **11.** $4b \div \frac{b}{2c}$ **12.** $10m \div \frac{5m}{p}$

13. $\frac{15x^2}{13} \div 21x$ **14.** $\frac{14pr}{17} \div 26r^2$ **15.** $24c^2d \div \frac{60cde}{19}$ **16.** $26rst \div \frac{65s^2t}{23}$

(Example 3)

17. $\frac{r}{s} \div \frac{rt}{s^2}$ **18.** $\frac{b^2}{a^2c} \div \frac{b}{a}$ **19.** $\frac{4p}{3r} \div \frac{2q^2}{9r}$ **20.** $\frac{6a}{c} \div \frac{9b}{c^2}$

21. $\frac{rs^2}{t^3} \div \frac{rs}{t}$ **22.** $\frac{a^2b^2}{cd^2} \div \frac{a^3b}{c^3d}$ **23.** $\frac{21kd}{20p} \div \frac{14d^2t}{18p^2}$ **24.** $\frac{9nc}{14dh^2} \div \frac{30n^2d}{77h}$

MIXED PRACTICE

25. $\frac{9mn}{16} \div 3n$ **26.** $\frac{4p^2q}{27} \div 2pq$ **27.** $\frac{r}{21} \div \frac{q}{14}$ **28.** $\frac{12}{m} \div \frac{30}{a}$

29. $\frac{98y}{39} \div \frac{105}{26}$ **30.** $24xy^2z \div \frac{44xyz}{29}$ **31.** $36acd^2 \div \frac{42b^2cd}{11}$ **32.** $\frac{84}{85} \div \frac{154}{51q}$

33. $\frac{2w}{5x} \div \frac{14wz}{15xy}$ **34.** $\frac{4rs}{9tu} \div \frac{11s^2t}{13ru^3}$ **35.** $\frac{a^3b^2c^4}{rst} \div \frac{abc^3}{r^2s^2t^3}$ **36.** $\frac{c^2de^4}{fg^3h} \div \frac{cd^2e}{f^4g^3h^2}$

NON-ROUTINE PROBLEMS

37. A cardboard strip is one inch wide and 48 inches long. It is cut with scissors at one-inch intervals, making 48 square inches. If each cut takes one second, how long will it take to make all the cuts?

38. An ant climbs up a pole five feet during the day and slides back four feet during the night. The pole is 30 feet high. How many days will it take the ant to climb from ground level to the top of the pole?

REVIEW CAPSULE FOR SECTION 4.5

Add or subtract like terms. (Pages 40–42)

1. $8t - 5t$ **2.** $3r + 9r$ **3.** $13m + m$ **4.** $19k - k$

5. $3a^2b + 4a^2b$ **6.** $15rs^3 - 9rs^3$ **7.** $1.6pq + pq$ **8.** $4.3x^2y - x^2y$

Simplify each expression. (Pages 40–45)

9. $10x - x + 2y$ **10.** $w + 7w - 4r$ **11.** $5a + 9a - 5$

12. $13t - 9t + 5v$ **13.** $19x^2 + 14x^2 - 2x^2$ **14.** $16rs^2 - r^2s + rs^2$

4.5 Addition and Subtraction

Like fractions have a common denominator.

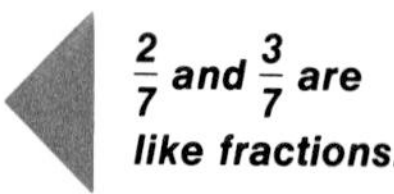
$\frac{2}{7}$ and $\frac{3}{7}$ are
like fractions.

Steps for Adding with Like Fractions

1. Add the numerators.
2. Write the sum over the common denominator.

$$\frac{2}{7}+\frac{3}{7}=\frac{2+3}{7}$$
$$=\frac{5}{7}$$

$\frac{a}{c}+\frac{b}{c}=\frac{a+b}{c}$ ***(c is not zero.)***

P–1 **What is the missing numerator?**

a. $\frac{4}{13}+\frac{6}{13}=\frac{?}{13}$ **b.** $\frac{5}{x}+\frac{8}{x}=\frac{?}{x}$ **c.** $\frac{1}{2y}+\frac{3}{2y}=\frac{?}{2y}$ **d.** $\frac{7}{mn}+\frac{9}{mn}=\frac{?}{mn}$

Steps for Subtracting with Like Fractions

1. Subtract the numerators.
2. Write the difference over the common denominator.

$$\frac{8}{y}-\frac{2}{y}=\frac{8-2}{y}$$
$$=\frac{6}{y}$$

$\frac{a}{c}-\frac{b}{c}=\frac{a-b}{c}$ ***(c is not zero.)***

P–2 **What is the missing numerator?**

a. $\frac{12}{15}-\frac{4}{15}=\frac{?}{15}$ **b.** $\frac{7}{5x}-\frac{3}{5x}=\frac{?}{5x}$ **c.** $\frac{11}{9t}-\frac{5}{9t}=\frac{?}{9t}$ **d.** $\frac{14}{rs}-\frac{8}{rs}=\frac{?}{rs}$

EXAMPLE 1 Add or subtract.

a. $\frac{3m}{7}+\frac{5m}{7}$ **b.** $\frac{5a}{3b}-\frac{a}{3b}$

Solutions:

	a.	**b.**
1 Add or subtract the numerators. →	$\frac{3m}{7}+\frac{5m}{7}=\frac{3m+5m}{7}$	$\frac{5a}{3b}-\frac{a}{3b}=\frac{5a-a}{3b}$
2 Write the result over the common denominator. →	$=\frac{8m}{7}$	$=\frac{4a}{3b}$

P–3 **Add or subtract.**

a. $\frac{4k}{11} + \frac{3k}{11}$ **b.** $\frac{5b}{9} - \frac{4b}{9}$ **c.** $\frac{6c}{7r} + \frac{3c}{7r}$ **d.** $\frac{8r}{4s} - \frac{3r}{4s}$

Be careful when the terms in the numerator are not like terms.

EXAMPLE 2 Add or subtract.

a. $\frac{r}{5t} + \frac{2s}{5t}$ **b.** $\frac{3p}{16} - \frac{q}{16}$

Solutions:

a.

$$\frac{r}{5t} + \frac{2s}{5t} = \frac{r + 2s}{5t}$$

b.

$$\frac{3p}{16} - \frac{q}{16} = \frac{3p - q}{16}$$

P–4 **Add or subtract.**

a. $\frac{t}{6} + \frac{u}{6}$ **b.** $\frac{3m}{5} + \frac{4n}{5}$ **c.** $\frac{7p}{8r} - \frac{5q}{8r}$ **d.** $\frac{19j}{20z} - \frac{11k}{20z}$

The answer should always be written in lowest terms.

EXAMPLE 3 Add or subtract. Write in lowest terms.

a. $\frac{19x}{12y} - \frac{11x}{12y}$ **b.** $\frac{2a^2}{3bc} + \frac{7a^2}{3bc}$

Solutions:

a.

$$\frac{19x}{12y} - \frac{11x}{12y} = \frac{19x - 11x}{12y}$$

$$= \frac{8x}{12y}$$

$$= \frac{\overset{1}{\cancel{2}} \cdot \overset{1}{\cancel{2}} \cdot 2 \cdot x}{\underset{1}{\cancel{2}} \cdot \underset{1}{\cancel{2}} \cdot 3 \cdot y}$$

$$= \frac{2x}{3y}$$

b.

$$\frac{2a^2}{3bc} + \frac{7a^2}{3bc} = \frac{2a^2 + 7a^2}{3bc}$$

$$= \frac{9a^2}{3bc}$$

$$= \frac{\overset{1}{\cancel{3}} \cdot 3 \cdot a \cdot a}{\underset{1}{\cancel{3}} \cdot b \cdot c}$$

$$= \frac{3a^2}{bc}$$

P–5 **Add or subtract. Write each answer in lowest terms.**

a. $\frac{19g}{20} - \frac{7g}{20}$ **b.** $\frac{15s}{6t} + \frac{5s}{6t}$ **c.** $\frac{3}{5y^2} + \frac{2}{5y^2}$ **d.** $\frac{16a^2b}{20c} - \frac{8a^2b}{20c}$

CLASSROOM EXERCISES

For Exercises 1–30, add or subtract. Write each answer in lowest terms. (Example 1)

1. $\frac{1}{3}+\frac{1}{3}$
2. $\frac{1}{4}+\frac{2}{4}$
3. $\frac{2}{x}+\frac{3}{x}$
4. $\frac{x}{5}+\frac{x}{5}$
5. $\frac{8}{n}-\frac{5}{n}$
6. $\frac{7t}{13}-\frac{3t}{13}$
7. $\frac{8s}{17}+\frac{s}{17}$
8. $\frac{5}{3c}-\frac{3}{3c}$
9. $\frac{a}{4}+\frac{2a}{4}$
10. $\frac{7m}{n}-\frac{2m}{n}$

(Example 2)

11. $\frac{x}{3}+\frac{y}{3}$
12. $\frac{m}{8}+\frac{5}{8}$
13. $\frac{4}{9}-\frac{2r}{9}$
14. $\frac{3a}{7}-\frac{2b}{7}$
15. $\frac{12}{n}-\frac{m}{n}$
16. $\frac{r}{t}-\frac{s}{t}$
17. $\frac{6p}{5}-\frac{3q}{5}$
18. $\frac{9}{2b}+\frac{3c}{2b}$
19. $\frac{5}{6z}+\frac{5a}{6z}$
20. $\frac{8x}{30}-\frac{7y}{30}$

(Example 3)

21. $\frac{10y}{2x}+\frac{y}{2x}$
22. $\frac{3r}{4z}+\frac{6r}{4z}$
23. $\frac{4y}{8d}-\frac{y}{8d}$
24. $\frac{13t}{10h}-\frac{7t}{10h}$
25. $\frac{5f}{14w}+\frac{17f}{14w}$
26. $\frac{7x^2}{7n}-\frac{3x^2}{7n}$
27. $\frac{19m}{5jk}-\frac{2m}{5jk}$
28. $\frac{b^2}{3a}+\frac{4b^2}{3a}$
29. $\frac{11c}{s^2}+\frac{3c}{s^2}$
30. $\frac{23k}{bc}-\frac{14k}{bc}$

WRITTEN EXERCISES

Goal: To add or subtract with fractions that have common denominators

Sample Problem: **a.** $\frac{2x}{5y^2}+\frac{3x}{5y^2}$ **b.** $\frac{4y}{7z}-\frac{3k}{7z}$

Answer: **a.** $\frac{x}{y^2}$ **b.** $\frac{4y-3k}{7z}$

For Exercises 1–50, add or subtract. Write each answer in lowest terms. (Example 1)

1. $\frac{3}{11}+\frac{5}{11}$
2. $\frac{4}{17}+\frac{11}{17}$
3. $\frac{7}{y}+\frac{12}{y}$
4. $\frac{6}{r}+\frac{11}{r}$
5. $\frac{r}{9}+\frac{r}{9}$
6. $\frac{k}{5}+\frac{k}{5}$
7. $\frac{14}{16}-\frac{1}{16}$
8. $\frac{18}{10}-\frac{9}{10}$
9. $\frac{25}{3x}-\frac{20}{3x}$
10. $\frac{19}{4b}-\frac{18}{4b}$

(Example 2)

11. $\frac{a}{5}+\frac{b}{5}$
12. $\frac{c}{4}-\frac{d}{4}$
13. $\frac{2y}{3}-\frac{x}{3}$
14. $\frac{3r}{10}+\frac{s}{10}$
15. $\frac{4b}{15}-\frac{k}{15}$
16. $\frac{3x}{z}-\frac{5y}{z}$
17. $\frac{4p}{n}+\frac{5q}{n}$
18. $\frac{7a}{3c}+\frac{5b}{3c}$
19. $\frac{6s}{7w}-\frac{5t}{7w}$
20. $\frac{9m}{5h}+\frac{9j}{5h}$

(Example 3)

21. $\frac{9b}{16} - \frac{5b}{16}$ 22. $\frac{3n}{12} + \frac{7n}{12}$ 23. $\frac{3x}{10} + \frac{5x}{10}$ 24. $\frac{8a^2}{15} + \frac{2a^2}{15}$ 25. $\frac{7}{12x} - \frac{5}{12x}$

26. $\frac{17}{18k^2} - \frac{15}{18k^2}$ 27. $\frac{7r}{16s} + \frac{5r}{16s}$ 28. $\frac{3a}{10b} + \frac{3a}{10b}$ 29. $\frac{13k}{12m^2} - \frac{5k}{12m^2}$ 30. $\frac{22x^2}{24y} - \frac{x^2}{24y}$

MIXED PRACTICE

31. $\frac{9h}{11} - \frac{h}{11}$ 32. $\frac{17t^2}{5} - \frac{t^2}{5}$ 33. $\frac{3m}{5} + \frac{2n}{5}$ 34. $\frac{15a}{4p} - \frac{3a}{4p}$ 35. $\frac{19r^2}{3g} - \frac{4r^2}{3g}$

36. $\frac{b}{4} + \frac{3c}{4}$ 37. $\frac{2w}{13} + \frac{6w}{13}$ 38. $\frac{5a}{12b} - \frac{a}{12b}$ 39. $\frac{11m}{18n} - \frac{m}{18n}$ 40. $\frac{7q}{9} + \frac{13q}{9}$

41. $\frac{k}{9} + \frac{7}{9}$ 42. $\frac{5}{7} + \frac{y}{7}$ 43. $\frac{14t}{24r^2} + \frac{t}{24r^2}$ 44. $\frac{45ab}{4c} - \frac{9ab}{4c}$ 45. $\frac{29mn^2}{8p} - \frac{5mn^2}{8p}$

46. $\frac{g}{14h} + \frac{5g}{14h}$ 47. $\frac{3a}{n} - \frac{2a}{n}$ 48. $\frac{3p^2}{2q} + \frac{5p^2}{2q}$ 49. $\frac{4r^2}{3s} + \frac{5r^2}{3s}$ 50. $\frac{7m}{x} - \frac{6m}{x}$

MORE CHALLENGING EXERCISES

51. If the value of $\frac{a}{7} + \frac{b}{7}$ is less than 1, how does the value of $a + b$ compare with 7? Explain.

52. If the value of $\frac{p}{5} - \frac{q}{5}$ equals zero, how does the value of p compare with the value of q? Explain.

53. If $\frac{2a}{3b} = 1$, give three possible values of a and b. How many values of a and b are possible?

54. If $\frac{a}{c} + \frac{b}{c} = \frac{4a}{c}$, how does the value of b compare with the value of a? Explain.

55. If $\frac{4r^2}{3s} + \frac{5r^2}{3s} = 27$ when $s = 1$, what is the value of r?

56. If $\frac{5}{12} + \frac{3}{12} = \frac{8}{9} - \frac{x}{9}$, what is the value of x?

REVIEW CAPSULE FOR SECTION 4.6

Write the least common multiple of the given numbers. (Pages 67–70)

1. 6 and 18 **2.** 9 and 63 **3.** 4 and 10 **4.** 5 and 15

5. 35 and 80 **6.** 20 and 9 **7.** 12 and 15 **8.** 8 and 14

9. 2, 3, and 4 **10.** 4, 5, and 6 **11.** 9, 12, and 15 **12.** 7, 9, and 21

13. 5, 7, and 10 **14.** 3, 8, and 12 **15.** 12, 15, and 20 **16.** 9, 15, and 30

4.6 Unequal Denominators

To add with fractions having unequal denominators, use equivalent fractions that have the same denominator.

Steps for Adding or Subtracting with Fractions

1. Find the least common multiple of the denominators.
2. Write each fraction as an equivalent fraction with the least common multiple as the denominator.
3. Add or subtract the numerators.
4. Write in lowest terms.

EXAMPLE 1 Add: $\frac{1}{4} + \frac{7}{16}$

Solution: 1 Find the least common multiple of 4 and 16. → $\left.\begin{array}{r} 4 = 2^2 \\ 16 = 2^4 \end{array}\right\}$ $LCM = 2^4 = 16$

2 Write an equivalent fraction for $\frac{1}{4}$ with a denominator of 16. → $\frac{1}{4} + \frac{7}{16} = \left(\frac{1}{4} \cdot \frac{4}{4}\right) + \frac{7}{16}$

$= \frac{4}{16} + \frac{7}{16}$

3 Add the numerators. → $= \frac{4+7}{16}$

$= \frac{11}{16}$ 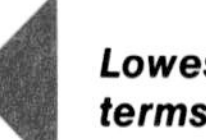 ***Lowest terms***

P–1 **Add or subtract.**

a. $\frac{2}{3} + \frac{5}{6}$ **b.** $\frac{3}{8} - \frac{1}{4}$ **c.** $\frac{4}{5} - \frac{7}{15}$ **d.** $\frac{3}{4} + \frac{9}{10}$

P–2 **What is the missing fraction?**

a. $\frac{x}{3} \cdot \frac{?}{?} = \frac{4x}{12}$ **b.** $\frac{x}{4} \cdot \frac{?}{?} = \frac{3x}{12}$ **c.** $\frac{x}{6} \cdot \frac{?}{?} = \frac{2x}{12}$ **d.** $\frac{x}{2} \cdot \frac{?}{?} = \frac{6x}{12}$

EXAMPLE 2 Subtract: $\frac{x}{3} - \frac{x}{4}$

Solution: [1] Find the least common multiple of 3 and 4. ⟶ $\left.\begin{array}{l} 3 = 3 \\ 4 = 2^2 \end{array}\right\}$ $LCM = 3 \cdot 2^2 = 12$

[2] Write equivalent fractions. ⟶ $\frac{x}{3} - \frac{x}{4} = \left(\frac{x}{3} \cdot \frac{4}{4}\right) - \left(\frac{x}{4} \cdot \frac{3}{3}\right)$

$= \frac{4x}{12} - \frac{3x}{12}$

[3] Subtract the numerators. ⟶ $= \frac{4x - 3x}{12}$

$= \frac{x}{12}$

P–3 **Add or subtract.**

a. $\frac{3p}{10} - \frac{p}{5}$ **b.** $\frac{b}{2} + \frac{b}{3}$ **c.** $\frac{6h}{13} + \frac{h}{26}$ **d.** $\frac{m}{8} - \frac{m}{10}$

EXAMPLE 3 Add: $\frac{13}{21x} + \frac{3}{14x}$ Write the answer in lowest terms.

Solution:

[1] Find the least common multiple of $21x$ and $14x$. ⟶ $\left.\begin{array}{l} 21x = 3 \cdot 7 \cdot x \\ 14x = 2 \cdot 7 \cdot x \end{array}\right\}$ $LCM = 2 \cdot 3 \cdot 7 \cdot x$ $= 42x$

[2] Write equivalent fractions. ⟶ $\frac{13}{21x} + \frac{3}{14x} = \left(\frac{13}{21x} \cdot \frac{2}{2}\right) + \left(\frac{3}{14x} \cdot \frac{3}{3}\right)$

$= \frac{26}{42x} + \frac{9}{42x}$

[3] Add the numerators. ⟶ $= \frac{26 + 9}{42x}$

$= \frac{35}{42x}$

[4] Write in lowest terms. ⟶ $= \frac{5 \cdot \overset{1}{\cancel{7}}}{2 \cdot 3 \cdot \underset{1}{\cancel{7}} \cdot x}$

$= \frac{5}{6x}$

P–4 **Add or subtract.**

a. $\frac{1}{6y}+\frac{1}{3y}$ b. $\frac{5}{2j}-\frac{2}{3j}$ c. $\frac{7}{10x}-\frac{6}{15x}$ d. $\frac{1}{4v}+\frac{2}{3v}$

CLASSROOM EXERCISES

For Exercises 1–10, find the least common multiple of the denominators. (Step 1, Examples 1, 2, and 3)

1. $\frac{1}{4}+\frac{3}{8}$ 2. $\frac{5}{12}+\frac{7}{24}$ 3. $\frac{1}{3}-\frac{1}{6}$ 4. $\frac{4}{7}-\frac{5}{14}$ 5. $\frac{4}{9}+\frac{3}{18}$

6. $\frac{5x}{6}-\frac{x}{4}$ 7. $\frac{n}{8}+\frac{n}{12}$ 8. $\frac{7}{8t}-\frac{5}{6t}$ 9. $\frac{3}{10r}+\frac{1}{6r}$ 10. $\frac{7}{15m}-\frac{2}{5m}$

Find the missing fractions. (Step 2, Examples 1, 2, and 3)

11. $\frac{2}{3}\cdot\frac{?}{?}=\frac{4}{6}$ 12. $\frac{5}{6}\cdot\frac{?}{?}=\frac{15}{18}$ 13. $\frac{7}{8}\cdot\frac{?}{?}=\frac{28}{32}$ 14. $\frac{3}{5}\cdot\frac{?}{?}=\frac{18}{30}$

15. $\frac{x}{6}\cdot\frac{?}{?}=\frac{4x}{24}$ 16. $\frac{3x}{7}\cdot\frac{?}{?}=\frac{9x}{21}$ 17. $\frac{5}{9a}\cdot\frac{?}{?}=\frac{15}{27a}$ 18. $\frac{3}{4y}\cdot\frac{?}{?}=\frac{9}{12y}$

For Exercises 19–33, add or subtract. Write each answer in lowest terms. (Example 1)

19. $\frac{1}{2}+\frac{1}{4}$ 20. $\frac{5}{8}-\frac{1}{4}$ 21. $\frac{11}{12}-\frac{3}{4}$ 22. $\frac{3}{4}+\frac{1}{12}$ 23. $\frac{7}{9}-\frac{2}{3}$

(Example 2)

24. $\frac{a}{3}+\frac{a}{6}$ 25. $\frac{3x}{8}-\frac{x}{4}$ 26. $\frac{3a}{2}-\frac{a}{4}$ 27. $\frac{y}{4}+\frac{2y}{3}$ 28. $\frac{2b}{6}-\frac{2b}{7}$

(Example 3)

29. $\frac{1}{3x}+\frac{2}{x}$ 30. $\frac{2}{n}+\frac{3}{2n}$ 31. $\frac{5}{4y}-\frac{1}{8y}$ 32. $\frac{6}{5t}-\frac{2}{3t}$ 33. $\frac{9}{7a}+\frac{2}{4a}$

WRITTEN EXERCISES

Goal: To add or subtract with fractions that have unequal denominators

Sample Problem: $\frac{7y}{10x}+\frac{2y}{15x}$ **Answer:** $\frac{5y}{6x}$

Write the missing numerals. (Examples 1 and 2)

1. $\frac{2}{3}+\frac{4}{5}=\left(\frac{2}{3}\cdot\frac{?}{5}\right)+\left(\frac{4}{5}\cdot\frac{?}{3}\right)$ 2. $\frac{4}{7}+\frac{1}{3}=\left(\frac{4}{7}\cdot\frac{?}{3}\right)+\left(\frac{1}{3}\cdot\frac{?}{7}\right)$ 3. $\frac{x}{3}+\frac{3x}{10}=\left(\frac{x}{3}\cdot\frac{?}{10}\right)+\left(\frac{3x}{10}\cdot\frac{?}{3}\right)$

For Exercises 4–48, add or subtract. Write each answer in lowest terms. (Example 1)

4. $\frac{3}{8}+\frac{1}{4}$ **5.** $\frac{5}{12}+\frac{1}{6}$ **6.** $\frac{9}{10}-\frac{3}{5}$ **7.** $\frac{13}{16}-\frac{3}{8}$ **8.** $\frac{2}{3}+\frac{11}{12}$

(Example 2)

9. $\frac{x}{3}+\frac{x}{5}$ **10.** $\frac{r}{7}+\frac{r}{2}$ **11.** $\frac{3d}{2}-\frac{d}{5}$ **12.** $\frac{5x}{4}-\frac{x}{3}$ **13.** $\frac{4m}{5}-\frac{2m}{7}$

14. $\frac{w}{6}+\frac{w}{8}$ **15.** $\frac{t}{6}+\frac{t}{4}$ **16.** $\frac{3k}{4}-\frac{2k}{5}$ **17.** $\frac{5n}{6}-\frac{n}{9}$ **18.** $\frac{7x}{8}+\frac{6x}{7}$

(Example 3)

19. $\frac{3}{2n}+\frac{1}{n}$ **20.** $\frac{2}{3t}+\frac{3}{t}$ **21.** $\frac{2}{x}-\frac{3}{5x}$ **22.** $\frac{3}{r}-\frac{2}{3r}$ **23.** $\frac{1}{s}-\frac{3}{4s}$

24. $\frac{5}{2s}+\frac{1}{6s}$ **25.** $\frac{7}{12y}+\frac{1}{6y}$ **26.** $\frac{2}{3x}-\frac{1}{9x}$ **27.** $\frac{4}{5q}-\frac{1}{10q}$ **28.** $\frac{3}{11k}+\frac{5}{k}$

MIXED PRACTICE

29. $\frac{11}{12}+\frac{5}{3}$ **30.** $\frac{7}{8}-\frac{7}{24}$ **31.** $\frac{3}{2r}+\frac{1}{3r}$ **32.** $\frac{5}{6y}+\frac{1}{4y}$ **33.** $\frac{8z}{9}-\frac{6z}{7}$

34. $\frac{7}{2m}-\frac{3}{3m}$ **35.** $\frac{8}{2w}-\frac{4}{3w}$ **36.** $\frac{2r}{5}-\frac{r}{7}$ **37.** $\frac{3n}{4}-\frac{n}{5}$ **38.** $\frac{3k}{5}+\frac{4k}{7}$

39. $\frac{2}{3q}-\frac{1}{5q}$ **40.** $\frac{3}{2x}-\frac{1}{7x}$ **41.** $\frac{5p}{6}+\frac{p}{15}$ **42.** $\frac{3t}{20}+\frac{t}{12}$ **43.** $\frac{4t}{9}+\frac{11t}{15}$

44. $\frac{12w}{7}-\frac{3w}{5}$ **45.** $\frac{7}{2s}+\frac{12}{5s}$ **46.** $\frac{8}{3c}+\frac{4}{11c}$ **47.** $\frac{4}{5b}-\frac{3}{4b}$ **48.** $\frac{9}{10a}-\frac{2}{3a}$

NON-ROUTINE PROBLEMS

49. The number of single-cell organisms in a jar doubles every minute. The jar is full in one hour. When was the jar half-full?

50. How many rectangles are there in the figure below?

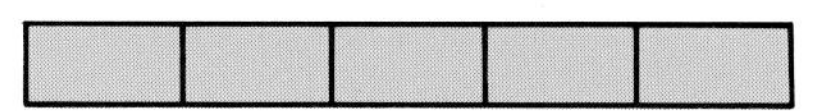

REVIEW CAPSULE FOR SECTION 4.7

Write a fraction for each mixed numeral.

1. $2\frac{3}{4}$ **2.** $7\frac{1}{2}$ **3.** $2\frac{5}{16}$ **4.** $3\frac{11}{12}$ **5.** $12\frac{1}{5}$

Multiply. Write your answer as a mixed numeral or a whole number.

6. $3\frac{1}{5}\times 25$ **7.** $4\frac{1}{6}\times 12$ **8.** $11\frac{1}{3}\times 27$ **9.** $19\frac{1}{2}\times 10$ **10.** $12\frac{1}{4}\times 100$

PROBLEM SOLVING AND APPLICATIONS

4.7 Using Triangle Formulas: Perimeter/Area

Recall that perimeter is the distance around a figure.

This word rule tells how to find the perimeter of any triangle.

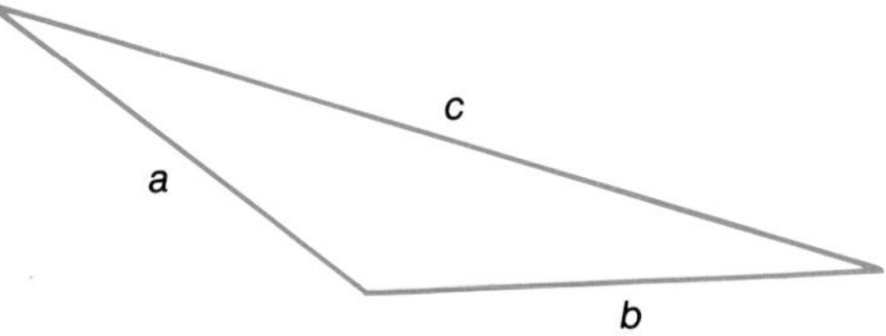

Word Rule: The perimeter of a triangle equals the sum of the lengths of the sides.

The following formula uses variables and symbols to represent the words in the word rule.

Formula: $\mathbf{p = a + b + c}$ ◀ ***p = perimeter*** ***a, b, c = lengths of the sides***

Recall that perimeter is expressed in linear units such as miles (mi), feet (ft), yards (yd), kilometers (km), meters (m), and so on.

EXAMPLE 1

The course for a boat race is triangular in shape. One leg of the race is $50\frac{1}{2}$ miles long. The second leg is $45\frac{1}{4}$ miles long, and the third leg is $60\frac{3}{4}$ miles. Find the total length of the race.

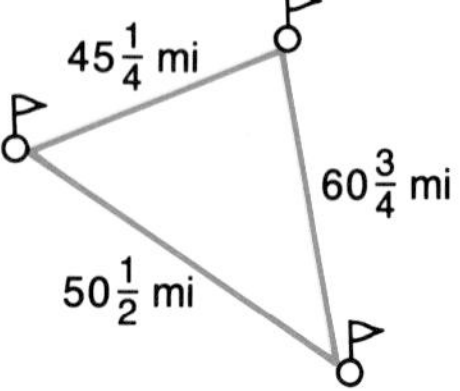

Solution: Since the course has the shape of a triangle, use the formula for the perimeter of a triangle.

[1] Write the formula. ⟶ $p = a + b + c$

[2] Identify known values. ⟶ $a = 50\frac{1}{2}$ mi; $b = 45\frac{1}{4}$ mi; $c = 60\frac{3}{4}$ mi

[3] Substitute and add. ⟶ $p = 50\frac{1}{2} + 45\frac{1}{4} + 60\frac{3}{4}$

$p = 156\frac{1}{2}$

The total length of the race is $156\frac{1}{2}$ miles.

The rule for the area of a triangle is given in terms of the base and height of the triangle.

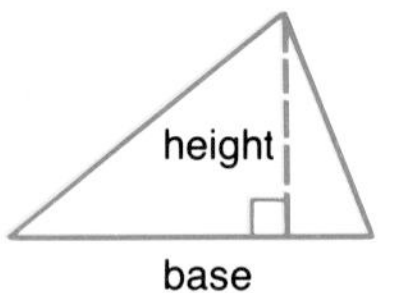

◀ ***The height of a triangle is measured along a line perpendicular to the base.***

Word Rule: The area of a triangle equals one-half times the length of the base times the height.

This formula corresponds to the word rule.

Formula: $A = \frac{1}{2}bh$ A = area, b = base, h = height

Area is expressed in square units, such as square miles (mi^2), square yards (yd^2), square meters (m^2), and so on.

EXAMPLE 2 A scarf having the shape of a triangle is shown at the right. Find the number of square inches of cloth in the scarf.

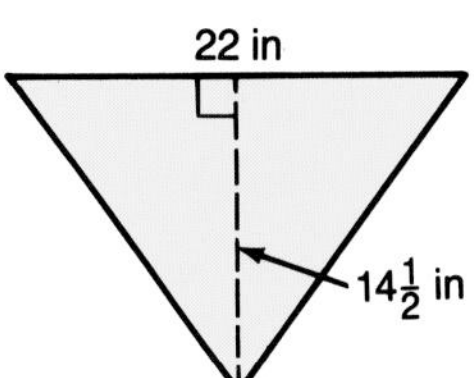

Solution: Find the area of the scarf. Write a fraction for $14\frac{1}{2}$ before multiplying.

[1] Write the formula. ⟶ $A = \frac{1}{2}bh$

[2] Identify known values. ⟶ $b = 22$ in; $h = 14\frac{1}{2}$ in

[3] Substitute. Write a fraction for the mixed numeral. ⟶ $A = \frac{1}{2} \times 22 \times \frac{29}{2}$

[4] Multiply. ⟶ $A = \frac{1}{\cancel{2}_1} \times \frac{\cancel{22}^{11}}{1} \times \frac{29}{2}$

$A = \frac{319}{2} = 159\frac{1}{2}$

There are $159\frac{1}{2}$ square inches of cloth in the scarf.

CLASSROOM EXERCISES

Find the perimeter of each triangle. (Example 1)

1. $a = 5$ in; $b = 4\frac{1}{2}$ in; $c = 8\frac{3}{4}$ in

2. $a = 3\frac{1}{5}$ mi; $b = 2\frac{2}{5}$ mi; $c = 4\frac{3}{5}$ mi

3. $a = 6\frac{1}{3}$ yd; $b = 4\frac{2}{3}$ yd; $c = 3$ yd

4. $a = 2\frac{1}{10}$ ft; $b = 2\frac{4}{5}$ ft; $c = 1\frac{9}{10}$ ft

Find the area of each triangle. (Example 2)

5. $b = 9$ in; $h = 4\frac{1}{2}$ in

6. $b = 13$ mi; $h = 5\frac{1}{2}$ mi

7. $b = 20\frac{1}{3}$ yd; $h = 18$ yd

8. $b = 9\frac{1}{4}$ ft; $h = 8$ ft

WRITTEN EXERCISES

Goal: To solve word problems that involve perimeter and area

Sample Problem: The sides of a triangular bicycle path are 3 kilometers, 4 kilometers, and 5 kilometers. Find the length of the path.

Answer: The length of the bicycle path is 12 kilometers.

For Exercises 1–12, solve each problem. (Example 1)

1. The sides of a triangular wedge of cheese are $5\frac{1}{2}$ inches, $4\frac{3}{4}$ inches, and 6 inches long. Find the perimeter of the wedge.

2. A corner shelf in the shape of a triangle has equal sides. Each side is $1\frac{1}{8}$ feet long. Find the perimeter of the shelf.

3. A triangular flower bed has sides that are 3 feet, $2\frac{1}{6}$ feet, and $3\frac{5}{12}$ feet long. Find the perimeter of the bed.

4. The sides of a triangular lot are $44\frac{1}{2}$ yards, $40\frac{3}{4}$ yards and $34\frac{7}{12}$ yards long. Find the perimeter of the lot.

(Example 2)

5. The mainsail of a boat is triangular in shape. The base of the sail is 7 feet long and the height is 8 feet. Find the area of the mainsail.

6. The jib sail on a boat has the shape of a triangle. The base of the sail is $1\frac{1}{6}$ yards long and the height is $2\frac{2}{3}$ yards. Find the area of the jib sail.

7. One side of a roof is a triangle with a base 20 feet long. The height of the triangle is $10\frac{5}{6}$ feet. Find the area of the side of the roof.

8. The red pennants flown for gale warnings are triangular in shape. The base of the triangle is $7\frac{1}{2}$ feet and the height is 15 feet. Find the area of a pennant.

MIXED PRACTICE

9. This triangle shows a path laid out for joggers. How many miles does a jogger run before retracing the route?

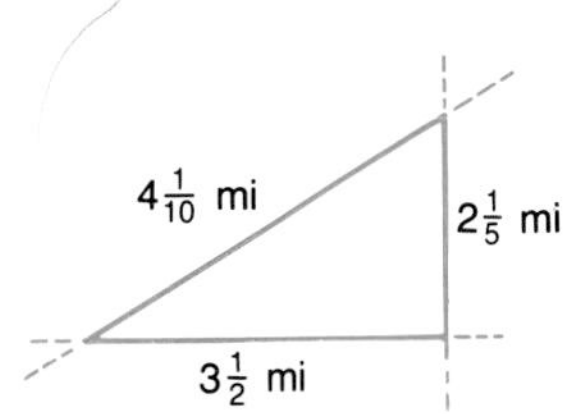

10. A safety zone at an intersection is shown below. Find the area of the safety zone.

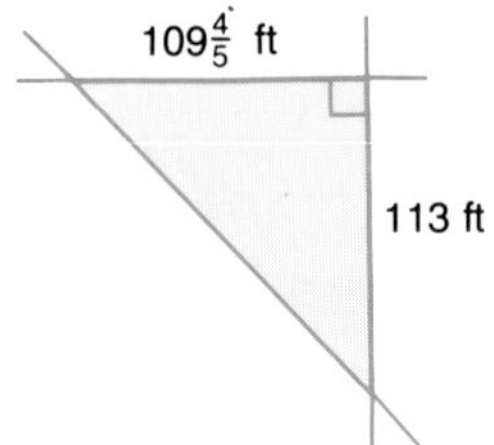

11. A triangular football pennant has a 7 inch base and a 15 inch height. Find the area of the pennant.

12. The sides of a triangular bicycle path are 4 miles, $5\frac{1}{2}$ miles, and $2\frac{3}{4}$ miles long. Find the perimeter.

CHAPTER SUMMARY

IMPORTANT TERMS	Fraction *(p. 78)* Numerator *(p. 78)* Denominator *(p. 78)* Lowest terms *(p. 78)* Odds *(p. 80)*	Equivalent fractions *(p. 80)* Algebraic fraction *(p. 83)* Reciprocals *(p. 92)* Like fractions *(p. 96)* Common denominator *(p. 96)*

IMPORTANT IDEAS

1. The denominator of a fraction can never equal 0.
2. Special Fractions
   a. Any fraction with the same nonzero numerator and denominator equals 1. $\frac{a}{a} = 1$ (a is not zero.)
   b. Any fraction with a denominator of 1 names the same number as its numerator. $\frac{a}{1} = a$
3. A fraction is in lowest terms when the numerator and denominator have no common prime factor.
4. Steps for Multiplying with Fractions
   See page 86.
5. Quotient Rule for Fractions
   $\frac{a}{b} \div \frac{c}{d} = \frac{a}{b} \cdot \frac{d}{c}$ (b, c, and d are not zero.)
6. Steps for Adding with Like Fractions
   See page 96.
7. Steps for Subtracting with Like Fractions
   See page 96.
8. Steps for Adding or Subtracting with Fractions
   See page 100.

CHAPTER REVIEW

SECTION 4.1

Write each fraction in lowest terms.

1. $\frac{2\cdot 3\cdot 3\cdot 5}{3\cdot 5\cdot 5}$ **2.** $\frac{2\cdot 5\cdot 5\cdot 7}{2\cdot 3\cdot 5\cdot 7}$ **3.** $\frac{8}{14}$ **4.** $\frac{8}{12}$ **5.** $\frac{14}{18}$ **6.** $\frac{20}{24}$

Write the odds as a fraction in lowest terms.

7. Bill entered his name 16 times to win a door prize. If the winner's name is drawn from a box of 240 entries, what are the odds in favor of Bill's name being drawn?

8. A committee consists of 3 women and 2 men. A committee representative is chosen at random. What are the odds against choosing a woman as the representative?

SECTION 4.2

Write each fraction in lowest terms.

9. $\frac{10r}{8rs}$ 10. $\frac{3pq}{15pr}$ 11. $\frac{x}{5x^2}$ 12. $\frac{2t}{3t^2}$ 13. $\frac{14a^2b}{24ab^2}$ 14. $\frac{27mn^2}{15m^3n}$

SECTION 4.3

Multiply. Be sure that each answer is in lowest terms.

15. $\frac{4n}{5} \cdot \frac{7}{12n}$ 16. $\frac{14}{5a^2} \cdot \frac{3a}{10}$ 17. $\frac{3}{4} \cdot 12r$ 18. $\frac{2}{3} \cdot 39x$

19. $\frac{4a}{3m} \cdot \frac{15m^2}{18a^2}$ 20. $\frac{6rs}{5t} \cdot \frac{15rt}{18st}$ 21. $\frac{5a}{2b^2} \cdot \frac{2b}{3a}$ 22. $\frac{m}{3k} \cdot \frac{9k}{4m^2}$

SECTION 4.4

Divide. Be sure that each answer is in lowest terms.

23. $\frac{3}{10} \div \frac{6}{5}$ 24. $\frac{15}{12} \div \frac{35}{8}$ 25. $\frac{3}{8b^2} \div \frac{5}{2b}$ 26. $\frac{6a}{7} \div \frac{9a^2}{5}$

27. $\frac{6ab}{25c} \div \frac{9b}{5ac^2}$ 28. $\frac{12q^2}{7pq} \div \frac{30q}{21p}$ 29. $\frac{r^2t}{2s} \div \frac{r}{3st}$ 30. $\frac{3h}{10m^2} \div \frac{6hp}{5m}$

SECTION 4.5

Add or subtract. Write each answer in lowest terms.

31. $\frac{2}{17} + \frac{13}{17}$ 32. $\frac{3}{a} + \frac{5}{a}$ 33. $\frac{3x}{11} + \frac{6x}{11}$ 34. $\frac{13x}{4} - \frac{3x}{4}$ 35. $\frac{13}{12y} - \frac{4}{12y}$ 36. $\frac{17t}{8} - \frac{5t}{8}$

SECTION 4.6

Add or subtract. Write each answer in lowest terms.

37. $\frac{x}{5} + \frac{3x}{10}$ 38. $\frac{5y}{4} - \frac{y}{3}$ 39. $\frac{5}{3t} + \frac{2}{t}$ 40. $\frac{3}{r} + \frac{6}{5r}$ 41. $\frac{2}{3n} - \frac{3}{5n}$ 42. $\frac{7}{6y} - \frac{3}{8y}$

SECTION 4.7

Solve each problem.

43. The sides of the Transamerica Pyramid building in San Francisco are triangular in shape. The base of the triangle is $38\frac{1}{2}$ yards and the height is 286 yards. Find the area of a side of the building.

44. A triangular coat hanger uses $16\frac{1}{3}$ inches of wire for the base of the triangle and $10\frac{1}{4}$ inches for each side. The hook takes $3\frac{1}{3}$ inches of wire. How much wire is needed to make one hanger?

CHAPTER 5

Solving Equations

The Wright brothers' first flight covered 120 feet and took 12 seconds. What is the rate? The first crossing of the Atlantic by balloon covered 3233 miles at an average rate of 24 mi/hr. How many hours did it take? The Voyager flight took 216 hours at an average rate of 120 mi/hr. How far did it travel?

5.1 Subtraction Property for Equations

The following table shows that subtraction is the opposite or ***inverse operation*** of addition.

Start with a number.	Add any number.	Subtract the same number.	Get back the original number.
10	$10 + 3$	$10 + 3 - 3$	10
$5\frac{1}{2}$	$5\frac{1}{2} + \frac{3}{4}$	$5\frac{1}{2} + \frac{3}{4} - \frac{3}{4}$	$5\frac{1}{2}$
6.2	$6.2 + 3.1$	$6.2 + 3.1 - 3.1$	6.2
n	$n + 5$	$n + 5 - 5$	n

Thus, subtraction reverses the addition step. This idea can be applied to solving equations. An ***equation*** is a sentence that contains the equality symbol "=."

To ***solve an equation*** means to find the number or numbers that will make the equation true.

P–1 **If *n* is replaced by 7, which of these equations are true?**

a. $n + 5 = 12$ **b.** $n + 5 - 5 = 12 - 5$ **c.** $n = 7$

The number 7 is the ***solution*** to all three equations. Equations that have the same solution are ***equivalent equations.*** Thus, equations **a, b,** and **c** are equivalent. Note that the only difference between equations **a** and **b** is that 5 is subtracted from each side of the equation.

This suggests the following property.

Subtraction Property for Equations

Subtracting the same number from each side of an equation forms an equivalent equation.

$$n + 5 = 12$$
$$n + 5 - \mathbf{5} = 12 - \mathbf{5}$$
$$n = 7$$

The goal in solving an equation is to get the variable alone on one side of the equation. The Subtraction Property for Equations is a means of doing this, as is demonstrated in the following examples.

EXAMPLE 1 Solve and check: $n+8=20$ Show all steps.

Solution: To get n alone, subtract 8 from each side.

$n+8=20$

[1] $n+8-\mathbf{8}=20-\mathbf{8}$ ◀ $8-8=0$

[2] $n=12$

Check: $n+8=20$

$12+8$

20

As Example 1 shows, you check the solution by replacing the variable with the solution to see whether it makes the equation true.

P–2 **Solve and check:** $\mathbf{14=x+5}$

EXAMPLE 2 Solve and check: $8.2=x+1.3$ Show all steps.

Solution: $8.2=x+1.3$

[1] $8.2-\mathbf{1.3}=x+1.3-\mathbf{1.3}$

[2] $6.9=x$

Check: $8.2=x+1.3$

$6.9+1.3$

8.2

P–3 **Solve and check:** $\mathbf{n+9.6=17.8}$

In order to make a profit and to cover expenses, an amount called the ***markup*** is added to a dealer's cost. The following word rule and formula relate the selling price, the cost, and the markup.

Word Rule: The selling price equals the sum of the cost and the markup.

Formula: $\mathbf{p=c+m}$ ◀ ***p* = selling price**, ***c* = cost**, ***m* = markup**

EXAMPLE 3 A music store sells a popular record for \$4.95. The markup is \$3.15. Find the store's cost for the record.

Solution: You are given $p=\$4.95$ and $m=\$3.15$. Find c.

[1] Write the formula. ⟶ $p=c+m$

[2] Replace the variables. ⟶ $4.95=c+3.15$

[3] Solve the equation. ⟶ $4.95-\mathbf{3.15}=c+3.15-\mathbf{3.15}$

$1.80=c$

The store's cost is \$1.80.

CLASSROOM EXERCISES

Simplify each expression. (Table)

1. $x + 7 - 7$
2. $r + 30 - 30$
3. $b + 0.5 - 0.5$
4. $t + \frac{1}{2} - \frac{1}{2}$
5. $n + 19 - 19$
6. $a + \frac{3}{4} - \frac{3}{4}$
7. $k + 12.5 - 12.5$
8. $p + 2\frac{1}{2} - 2\frac{1}{2}$
9. $12 + n - n$
10. $3.6 + x - x$
11. $112 + r - r$
12. $x + n - n$

Write the equivalent equation formed as the first step in solving each equation. (Step 1, Examples 1 and 2)

13. $x + 5 = 19$
14. $a + 12 = 14$
15. $105 = m + 92$
16. $87 = r + 36$
17. $t + 1.8 = 2.3$
18. $8.1 = w + 2.9$
19. $184 = c + 102$
20. $z + 56 = 91$
21. $x + 12.5 = 21.3$
22. $14.6 = p + 9.7$
23. $b + \frac{1}{2} = 19$
24. $6\frac{7}{8} = y + 2\frac{3}{4}$

WRITTEN EXERCISES

Goal: To solve an equation using the Subtraction Property for Equations
Sample Problem: $x + 2.7 = 5.3$
Answer: 2.6

For Exercises 1–36, solve and check. Show all steps. (Example 1)

1. $n + 14 = 21$
2. $x + 18 = 35$
3. $46 = y + 29$
4. $42 = r + 23$
5. $t + 19 = 43$
6. $s + 34 = 51$
7. $a + 39 = 56$
8. $b + 45 = 71$
9. $64 = q + 37$
10. $120 = t + 90$
11. $h + 48 = 97$
12. $n + 17 = 71$

(Example 2)

13. $k + 5.8 = 7.0$
14. $x + 3.4 = 7.6$
15. $5.8 = y + 2.3$
16. $6.9 = n + 4.5$
17. $t + 6.8 = 9.9$
18. $b + 12.8 = 15.1$
19. $r + 23.7 = 35.2$
20. $s + 0.43 = 0.9$
21. $0.85 = y + 0.47$
22. $0.74 = x + 0.24$
23. $m + 1.05 = 22.79$
24. $p + 4.31 = 18.19$

MIXED PRACTICE

25. $51 = a + 38$
26. $t + 29 = 48$
27. $y + 1906 = 4037$
28. $7.03 = g + 4.28$
29. $x + 97 = 103$
30. $9378 = n + 8417$
31. $x + 0.39 = 1.46$
32. $39 = s + 27$
33. $9.6 = b + 1.45$
34. $w + 5.9 = 8.3$
35. $m + \frac{1}{6} = \frac{2}{3}$
36. $2\frac{1}{4} = p + \frac{1}{2}$

APPLICATIONS

Solve each problem. Use the formula $p = c + m$. (Example 3)

37. A school bookstore sells calculators for \$8. The markup is \$3. Find the bookstore's cost for each calculator.

38. A hardware store sells hammers for \$7.95 each. The store's cost for a hammer is \$4.80. Find the markup.

39. A supply store sells a package of typing paper for \$7.83. The store's cost for the paper is \$4.97. Find the markup.

40. The selling price for a pair of tennis shoes is \$25.50. The markup on the shoes is \$10.75. What is the store's cost?

41. A popular music store sells record albums for \$8.98 each. The markup on each album is \$3.53. Find the store's cost.

42. A fruit and vegetable stand sells a small bag of apples for \$1.34. The stand's cost for the bag of apples is \$0.78. Find the markup.

43. Each year, the Athletic Booster Club sells school megaphones for \$2.50 each. The club's cost for each megaphone is \$1.62. Find the markup.

44. A sporting goods store sells soccer balls for \$12.95. The markup on the balls is \$7.13. Find the store's cost for each ball.

NON-ROUTINE PROBLEMS

45. Robert has \$15 more than Marlene. Marlene has \$10 more than Albert. Together, they have a total of \$56. How much does Robert have?

46. A building has 6 doors. Find the number of ways a person can enter the building by one door and leave by another door.

REVIEW CAPSULE FOR SECTION 5.2

Simplify. (Pages 2–4)

1. $23 - 7 + 7$

2. $8.3 - 4.7 + 4.7$

3. $98 - 16 + 16$

4. $107 - 92 + 92$

5. $10 - 5.6 + 5.6$

6. $45 - 19.2 + 19.2$

7. $18 - 3\frac{1}{3} + 3\frac{1}{3}$

8. $42 - 10\frac{7}{8} + 10\frac{7}{8}$

9. $102 - 86 + 86$

5.2 Addition Property for Equations

The following table shows that addition is the opposite or inverse operation of subtraction.

Start with a number.	Subtract any number.	Add the same number.	Get back the original number.
13	$13 - 8$	$13 - 8 + 8$	13
9.7	$9.7 - 3.2$	$9.7 - 3.2 + 3.2$	9.7
n	$n - 19$	$n - 19 + 19$	n

P–1 **What is the value of $5 - 18 + 18$?**

You can find the answer to **P-1** even though you have not yet learned how to solve problems such as $5 - 18$. Just write $5 - 18 + 18$ as $5 + 18 - 18$. You solved this type of problem in the last section.

P–2 **If x is replaced by 8, which of these equations are true?**

a. $x - 3 = 5$ **b.** $x - 3 + 3 = 5 + 3$ **c.** $x = 8$

Equations **a, b,** and **c** are equivalent. Note that 3 was added to each side of Equation **a** to get Equation **b** and Equation **c.** This suggests the following property.

Addition Property for Equations

Adding the same number to each side of an equation forms an equivalent equation.

$$x - 3 = 5$$
$$x - 3 + \mathbf{3} = 5 + \mathbf{3}$$
$$x = 8$$

EXAMPLE 1 Solve and check: $x - 12 = 9$. Show all steps.

Solution: To get x alone, add 12 to each side.

$$x - 12 = 9$$

[1] $x - 12 + \mathbf{12} = 9 + \mathbf{12}$

[2] $x = 21$

Check: $x - 12 = 9$

$21 - 12$

9

P–3 **Solve and check: $x - 5 = 13$**

EXAMPLE 2 Solve and check: $9.8 = t - 17.3$. Show all steps.

Solution:

$$9.8 = t - 17.3$$

[1] $9.8 + \mathbf{17.3} = t - 17.3 + \mathbf{17.3}$

[2] $27.1 = t$

Check: $9.8 = t - 17.3$

$27.1 - 17.3$

9.8

P–4 **Solve and check: $x - 3.1 = 4.7$**

The regular price of an article is the ***list price.***
The following word rule and formula relate the sale price, the list price, and the discount.

Word Rule: The sale price equals the list price less the discount.

Formula: $\mathbf{s = \ell - d}$

***s* = sale price**
ℓ = list price
***d* = discount**

EXAMPLE 3 A women's clothing store is having a sale on sweaters for \$49 each. The discount on each sweater is \$16. Find the list price.

Solution: You are given $s = \$49$ and $d = \$16$. Find ℓ.

[1] Write the formula. ⟶ $s = \ell - d$

[2] Replace the variables. ⟶ $49 = \ell - 16$

[3] Solve the equation. ⟶ $49 + \mathbf{16} = \ell - 16 + \mathbf{16}$

$65 = \ell$

The list price of a sweater is \$65.

CLASSROOM EXERCISES

Simplify each expression. (Table)

1. $r - 12 + 12$ **2.** $s - 37 + 37$ **3.** $x - 2.8 + 2.8$

4. $a - \frac{1}{4} + \frac{1}{4}$ **5.** $n - 0.8 + 0.8$ **6.** $p - 6\frac{1}{2} + 6\frac{1}{2}$

7. $b - 19.2 + 19.2$ **8.** $72 - s + s$ **9.** $3\frac{1}{4} - x + x$

Write the equivalent equation formed as the first step in solving each equation. (Examples 1–3)

10. $x - 7 = 19$
11. $a - 12 = 5$
12. $23 = s - 27$
13. $47 = t - 83$
14. $q - 2.7 = 3.9$
15. $12.9 = b - 5.6$
16. $r - 56 = 23$
17. $w - 3 = 21$
18. $y - 13.2 = 17.9$
19. $8.3 = t - 15.9$
20. $4\frac{1}{4} = n - 1\frac{7}{8}$
21. $q - \frac{3}{4} = 6\frac{1}{2}$

Decide whether the value shown is the solution for the given equation. Answer Yes *or* No. (Check of Examples 1–2)

22. $m - 23 = 56$; $m = 33$
23. $16.7 = p - 4.6$; $p = 21.3$
24. $x - \frac{2}{3} = 8\frac{1}{6}$; $x = 8\frac{5}{6}$

WRITTEN EXERCISES

Goal: To solve an equation using the Addition Property for Equations
Sample Problem: $n - 3.2 = 6.7$
Answer: $n = 9.9$

For Exercises 1–39, solve and check. Show all steps. (Example 1)

1. $x - 32 = 26$
2. $r - 23 = 36$
3. $t - 48 = 53$
4. $p - 59 = 47$
5. $75 = n - 128$
6. $87 = b - 136$
7. $185 = a - 207$
8. $95 = p - 235$
9. $m - 169 = 169$
10. $378 = k - 378$
11. $y - 4021 = 86$
12. $x - 17 = 8013$

(Example 2)

13. $s - 4.2 = 5.9$
14. $q - 7.9 = 6.7$
15. $0.86 = w - 0.27$
16. $0.54 = n - 0.89$
17. $19.7 = g - 28.3$
18. $28.5 = h - 42.7$
19. $r - 9.08 = 15.29$
20. $x - 12.37 = 9.09$
21. $c - 10.8 = 9.7$
22. $d - 90.6 = 7.9$
23. $6.08 = p - 104.1$
24. $113.2 = t - 8.4$

MIXED PRACTICE

25. $49 = r - 78$
26. $g - 24.9 = 13.9$
27. $24 = n - 19$
28. $57 = t - 96$
29. $k - 268 = 105$
30. $5.9 = p - 2.7$
31. $t - 276 = 307$
32. $m - 135.8 = 94.7$
33. $2.97 = y - 0.08$
34. $3.3 = d - 6.8$
35. $b - 1005 = 105$
36. $1356 = f - 867$
37. $567 = q - 1228$
38. $y - \frac{3}{4} = \frac{7}{8}$
39. $1\frac{7}{8} = h - 3\frac{1}{2}$

APPLICATIONS

Solve each problem. Use the formula $s = \ell - d$. (Example 3)

40. A sports store has a sale price of \$16 on tennis shirts. The discount is \$11. Find the list price.

41. The sale price of a certain clock is \$39.50. The discount is \$14.50. Find the list price.

42. A hobby shop has a sale price of \$10.49 on a model plane. The discount is \$9.46. What is the list price?

43. The sale price of a certain washing machine is \$410. The discount is \$75. What is the list price?

MORE CHALLENGING EXERCISES

EXAMPLE: In the formula, $s = \ell - d$, s represents the sale price, ℓ represents the list price, and d represents the discount. Solve the formula for ℓ.

SOLUTION: To get ℓ alone, add d to each side.

$$s = \ell - d$$
$$s + \mathbf{d} = \ell - d + \mathbf{d}$$
$$s + d = \ell$$

In the formula, $p = c + m$, *p represents the selling price, c represents the cost, and m represents the markup.*

44. Solve the formula for c.

45. Solve the formula for m.

In the formula, $p = a + b + c$, *p represents perimeter, and a, b, and c represent the lengths of the three sides of a triangle.*

46. Solve the formula for a.

47. Solve the formula for b.

REVIEW CAPSULE FOR SECTION 5.3

Simplify. (Pages 78–85)

1. $\frac{12 \cdot 8}{12}$

2. $\frac{3 \cdot 15}{3}$

3. $\frac{19 \cdot 17}{19}$

4. $\frac{27 \cdot 27}{27}$

5. $\frac{15 \cdot 4}{15}$

6. $\frac{30 \cdot 1}{30}$

7. $\frac{5x}{5}$

8. $\frac{12y}{12}$

9. $\frac{100a}{100}$

10. $\frac{56n}{56}$

11. $\frac{99r}{99}$

12. $\frac{76p}{76}$

5.3 Division Property for Equations

The following table shows that division is the inverse operation of multiplication.

Start with a number.	Multiply by any number.	Divide by the same number.	Get back the original number.
8	$8 \cdot 3$	$8 \cdot 3 \div 3$	8
27	$27 \cdot 5$	$\frac{27 \cdot 5}{5}$	27
n	$9n$	$\frac{9n}{9}$	n

P–1 **If n is replaced by 3, which of these equations are true?**

a. $4n = 12$ **b.** $\frac{4n}{4} = \frac{12}{4}$ **c.** $n = 3$

Equations **a, b,** and **c** are equivalent. Note that each side of Equation **a** was divided by 4 to get Equation **b** and Equation **c.**

This leads to the following property.

Division Property For Equations

Dividing each side of an equation by the same nonzero number forms an equivalent equation.

$$4n = 12$$
$$\frac{4n}{\mathbf{4}} = \frac{12}{\mathbf{4}}$$
$$n = 3$$

EXAMPLE 1 Solve and check: $3n = 48$. Show all steps.

Solution: To get n alone, divide each side by 3.

$$3n = 48$$

[1] $\frac{3n}{\mathbf{3}} = \frac{48}{\mathbf{3}}$ ◀ $\frac{3}{3} = 1$

[2] $n = 16$

Check: $3n = 48$

$3 \cdot 16$

48

P–2 By what number would you divide $8w$ to get w as the result?

EXAMPLE 2 Solve and check: $50 = 8w$. Show all steps.

Solution: $50 = 8w$

[1] $\frac{50}{\mathbf{8}} = \frac{8w}{\mathbf{8}}$

[2] $6\frac{1}{4} = w$ or $6.25 = w$

Check: $50 = 8\ w$

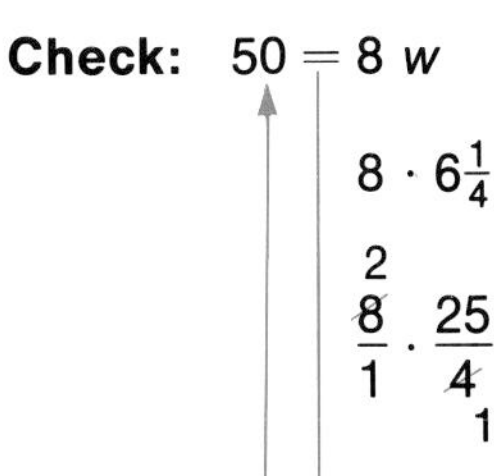

P-3 **Solve and check: a. $5n = 24$ b. $1.2 = 4n$ c. $8.8 = 2.2n$**

The following word rule and formula relate distance, rate, and time.

Word Rule: Distance equals rate multiplied by time.

Formula: $d = rt$

d = distance
r = rate
t = time

EXAMPLE 3 The Indianapolis 500 is won by the first driver who completes 500 miles. In 1986, Bobby Rahal won the race with an average speed of about 171 miles per hour. What was his time in hours? Round your answer to the nearest tenth.

Solution: You are given $d = 500$ miles and $r = 171$ miles per hour. Find t.

[1] Write the formula. ⟶ $d = rt$

[2] Replace the variables. ⟶ $500 = 171t$

[3] Solve the equation. ⟶ $\frac{500}{\mathbf{171}} = \frac{171t}{\mathbf{171}}$

$2.92 = t$ ***Round to the nearest tenth.***

Bobby Rahal's time was about 2.9 hours.

CLASSROOM EXERCISES

Simplify each expression. (Table)

1. $\frac{13x}{13}$
2. $\frac{27r}{27}$
3. $\frac{9t}{9}$
4. $4q \div 4$
5. $\frac{12.8k}{12.8}$
6. $3.5y \div 3.5$
7. $\frac{35n}{n}$, $n \neq 0$
8. $\frac{42s}{s}$, $s \neq 0$

Write the equivalent equation formed as the first step in solving each equation. (Step 1, Examples 1–2)

9. $4n = 32$ **10.** $9x = 72$ **11.** $44 = 11y$ **12.** $87 = 53t$

13. $25a = 75$ **14.** $21 = 0.7w$ **15.** $1.8 = 2.4x$ **16.** $108b = 226$

WRITTEN EXERCISES

Goal: To solve an equation using the Division Property for Equations

Sample Problem: $7x = 84$ **Answer:** 12

For Exercises 1–32, solve and check. Show all steps. (Example 1)

1. $6x = 54$ **2.** $9x = 63$ **3.** $12n = 60$ **4.** $7w = 91$

5. $8y = 256$ **6.** $7a = 161$ **7.** $16q = 80$ **8.** $18t = 72$

(Example 2)

9. $48 = 6r$ **10.** $63 = 9n$ **11.** $30 = 8k$ **12.** $47 = 6w$

13. $108 = 12m$ **14.** $144 = 16t$ **15.** $9.6 = 8s$ **16.** $7.2 = 3p$

MIXED PRACTICE

17. $168 = 14x$ **18.** $270 = 18y$ **19.** $9t = 216$ **20.** $8p = 216$

21. $375 = 3n$ **22.** $136 = 8r$ **23.** $29k = 377$ **24.** $4.8x = 456$

25. $56 = 24s$ **26.** $18g = 96$ **27.** $104 = 16a$ **28.** $114 = 15m$

29. $624 = 9.6h$ **30.** $306 = 24n$ **31.** $3x = 4\frac{1}{2}$ **32.** $2x = 3\frac{1}{3}$

APPLICATIONS

For Exercises 33–36, solve each problem. Use the formula $d = rt$. Round your answers to the nearest tenth. (Example 3)

33. In December, 1986, the airplane *Voyager* flew 26,000 miles around the world without landing or refueling. The flight took 216 hours. Find the average speed in miles per hour.

34. The *Voyager* flew a test mission of 11,600 miles up and down the coast of California. The trip took 108 hours. Find the average speed in miles per hour.

35. The *Voyager* has a range of 28,000 miles and an average speed of 110 miles per hour. How many hours would it take to fly 28,000 miles?

36. The top speed of the *Voyager* is 120 miles per hour. At that rate, how many hours would it take to fly 2,000 miles?

MID-CHAPTER REVIEW

Solve and check. Show all steps. (Section 5.1)

1. $r + 56 = 103$
2. $c + 89 = 117$
3. $k + 35 = 71$
4. $s + 15 = 36$
5. $0.86 = t + 0.39$
6. $4.02 = m + 0.98$

Solve each problem. (Section 5.1)

7. An office supply store sells pens for \$1.79 each. The store's cost is \$0.98 per pen. Use the formula $p = c + m$ to find the markup.

8. A bakery has a markup of \$0.65 on a loaf of rye bread. The bakery's selling price is \$1.55 per loaf. Use the formula $p = c + m$ to find the cost.

Solve and check. Show all steps. (Section 5.2)

9. $47 = y - 19$
10. $63 = n - 38$
11. $p - 21 = 37$
12. $d - 35 = 12$
13. $a - 43.5 = 98.7$
14. $w - 67.9 = 70.8$

Solve each problem. (Section 5.2)

15. The sale price for a tennis racket is \$72. The discount on the racket is \$26. Use the formula $s = \ell - d$ to find the list price.

16. The sale price for a pair of running shoes is \$18. The discount on the shoes is \$15. Use the formula $s = \ell - d$ to find the list price.

Solve and check. Show all steps. (Section 5.3)

17. $14q = 266$
18. $648 = 24g$
19. $48.6 = 0.09x$
20. $2.15g = 567.6$
21. $80 = 15s$
22. $12t = 160$

Solve. Use the formula $d = rt$. (Section 5.3)

23. The fastest passenger train in France can travel 690 miles in 3 hours. Find the average speed in miles per hour.

24. An elephant runs at a rate of 31 feet per second. At this rate, how many seconds will it take the elephant to run 248 feet?

5.4 Multiplication Property for Equations

The following table shows that multiplication is the inverse operation of division.

Start with a number.	Divide by any number (not zero).	Multiply by the same number.	Get back the original number.
12	$12 \div 4$	$12 \div 4 \cdot 4$	12
45	$\frac{45}{9}$	$\frac{45}{9} \cdot 9$	45
n	$\frac{n}{8}$	$\frac{n}{8} \cdot 8$	n

P–1 **If x is replaced by 12, which of these equations are true?**

a. $\frac{x}{4} = 3$ **b.** $\frac{x}{4} \cdot 4 = 3 \cdot 4$ **c.** $x = 12$

Equations **a, b,** and **c** are equivalent. Note that each side of Equation **a** was multiplied by 4 to get Equation **b** and Equation **c.**

This suggests the following property.

> ***Multiplication Property for Equations***
>
> Multiplying each side of an equation by the same nonzero number forms an equivalent equation.
>
> $\frac{x}{4} = 3$
>
> $\frac{x}{4} \cdot \mathbf{4} = 3 \cdot \mathbf{4}$
>
> $x = 12$

EXAMPLE 1 Solve and check: $\frac{x}{15} = 9$ Show all steps.

Solution: To get x alone, multiply each side by 15.

$$\frac{x}{15} = 9$$

[1] $\frac{x}{15} \cdot \mathbf{15} = 9 \cdot \mathbf{15}$

[2] $x = 135$

Check: $\frac{x}{15} = 9$

$\frac{135}{15}$

9

EXAMPLE 2 Solve and check: $12 = \frac{y}{8}$. Show all steps.

Solution: $12 = \frac{y}{8}$

[1] $12 \cdot \mathbf{8} = \frac{y}{8} \cdot \mathbf{8}$

[2] $96 = y$

Check: $12 = \frac{y}{8}$

$\frac{96}{8}$

12

P–2 **Solve and check. Show all steps.**

a. $\frac{t}{18} = 3$ **b.** $40 = \frac{m}{5}$ **c.** $\frac{s}{5} = 9$ **d.** $\frac{r}{8} = 1.2$

The ***fuel economy*** of a car refers to how many miles it can travel on one gallon of gasoline. The following word rule and formula relate the fuel economy, the distance driven, and the number of gallons of gasoline used.

Word Rule: Fuel economy equals the distance driven divided by the number of gallons of gasoline used.

Formula: $\mathbf{f = \frac{d}{g}}$

f = fuel economy
d = distance driven
g = gallons used

EXAMPLE 3 On a trip by car, the Denton family used 16 gallons of gasoline. The fuel economy of their car was 28 miles per gallon. Find the distance they traveled in miles.

Solution: You are given $g = 16$ gallons and $f = 28$ miles per gallon. Find d.

[1] Write the formula. ⟶ $f = \frac{d}{g}$

[2] Replace the variables. ⟶ $28 = \frac{d}{16}$

[3] Solve the equation. ⟶ $28 \cdot \mathbf{16} = \frac{d}{16} \cdot \mathbf{16}$

$448 = d$

The Denton family traveled 448 miles.

CLASSROOM EXERCISES

Simplify each expression. (Table)

1. $a \div 19 \cdot 19$
2. $\frac{s}{89} \cdot 89$
3. $9 \cdot \frac{r}{9}$
4. $\frac{16p}{11} \cdot 11$
5. $\frac{12h}{12}$

Write the equivalent equation formed as the first step in solving each equation. (Step 1, Examples 1–2)

6. $\frac{x}{15} = 13$
7. $7 = \frac{s}{23}$
8. $12 = \frac{y}{1.8}$
9. $\frac{w}{14} = 19$
10. $12 = \frac{r}{19}$

For Exercises 11–22, solve and check. Show all steps. (Example 1)

11. $\frac{x}{2} = 5$
12. $\frac{y}{3} = 4$
13. $\frac{n}{8} = 2$
14. $\frac{r}{6.2} = 5$

(Example 2)

15. $4 = \frac{a}{5}$
16. $2 = \frac{k}{13}$
17. $9 = \frac{m}{11}$
18. $2 = \frac{w}{3.1}$

(Examples 1 and 2)

19. $2\frac{1}{2} = \frac{p}{2}$
20. $\frac{t}{3} = 4\frac{1}{3}$
21. $\frac{c}{8} = 1\frac{1}{4}$
22. $3\frac{1}{8} = \frac{q}{8}$

WRITTEN EXERCISES

Goal: To solve an equation using the Multiplication Property For Equations

Sample Problem: $\frac{w}{15} = 42$ **Answer:** 630

For Exercises 1–32, solve and check. Show all steps. (Example 1)

1. $\frac{t}{8} = 17$
2. $\frac{r}{12} = 9$
3. $\frac{x}{23} = 12$
4. $\frac{x}{34} = 15$
5. $\frac{w}{7} = 126$
6. $\frac{k}{9} = 208$
7. $\frac{p}{12} = 4.5$
8. $\frac{a}{16} = 8.5$

(Example 2)

9. $8 = \frac{a}{9}$
10. $9 = \frac{n}{7}$
11. $83 = \frac{s}{9}$
12. $56 = \frac{t}{7}$
13. $8 = \frac{w}{3.7}$
14. $4 = \frac{p}{5.2}$
15. $23.5 = \frac{x}{4}$
16. $40.4 = \frac{y}{3}$

MIXED PRACTICE

17. $2.8 = \frac{r}{12}$
18. $\frac{p}{46} = 3$
19. $\frac{n}{26} = 9$
20. $3.6 = \frac{t}{15}$
21. $\frac{q}{54} = 5$
22. $\frac{w}{42} = 9$
23. $112 = \frac{a}{8}$
24. $5 = \frac{b}{16}$
25. $\frac{m}{9.6} = 5$
26. $202 = \frac{c}{8}$
27. $9 = \frac{d}{23}$
28. $\frac{r}{8} = 92$
29. $\frac{t}{8} = 96$
30. $\frac{d}{48} = 5$
31. $6\frac{2}{9} = \frac{w}{3}$
32. $3\frac{5}{6} = \frac{n}{2}$

APPLICATIONS

Solve. Use the formula $f = \frac{d}{g}$. (Example 3)

33. The fuel economy of a certain compact car is 34 miles per gallon. How far can the car be expected to travel on 12 gallons of gasoline?
34. The fuel economy of a certain pickup truck is 15 miles per gallon. How far can the truck be expected to travel on 40 gallons of gasoline?

Solve. Use the formula $k = \frac{d}{\ell}$ *where* k = *kilometers per liter,* d = *distance in kilometers, and* ℓ = *the number of liters of fuel used.*

35. A certain motorbike can travel about 25 kilometers per liter of fuel. Find how many kilometers the motorbike can be expected to travel on 30 liters of fuel.
36. A certain luxury car can travel about 16.5 kilometers per liter of fuel. Find how many kilometers the car can be expected to travel on 75 liters of fuel.

MORE CHALLENGING EXERCISES

37. **a.** In the formula $f = \frac{d}{g}$, solve for d when $f = 40$ and $g = 10$. Then solve for d when $f = 40$ and $g = 20$.
    b. How does d change when g changes from 10 to 20?
    c. If g is doubled while f remains the same, what is the effect on d?
38. **a.** In the formula $d = rt$, solve for t when $d = 400$ and $r = 25$. Then solve for t when $d = 400$ and $r = 50$.
    b. How does t change when r changes from 25 to 50?
    c. If r is doubled while d remains the same, what is the effect on t?

5.5 From Words to Symbols

The first step in solving word problems is to translate words into algebraic symbols. The following table shows some word expressions and the corresponding algebraic expressions.

Table

Operation	Word Expression	Algebraic Expression
Addition	The sum of a number, n, and 10	$n + 10$
	Five more than the number of apples, a	$a + 5$
	The number of meters, m, plus 1.6	$m + 1.6$
	Thirteen increased by the number of years, y	$13 + y$
	Some number, n, added to $3\frac{1}{4}$	$3\frac{1}{4} + n$
Subtraction	The difference between some number, w, and 36	$w - 36$
	The number of centimeters, c, minus 14.3	$c - 14.3$
	One hundred decreased by an unknown number, x	$100 - x$
	The number of inches, s, less $2\frac{3}{8}$	$s - 2\frac{3}{8}$
	Fifteen less than the number of records, r	$r - 15$
Multiplication	The product of 18 and some number, x	$18x$
	Nineteen times the number of tables, t	$19t$
	Twice the number of hours, h	$2h$
	The number of chairs, c, doubled	$2c$
Division	The number of kilograms, k, divided by 10	$\frac{k}{10}$
	The quotient of an unknown number, w, and 1.6	$\frac{w}{1.6}$

EXAMPLE 1 Write an algebraic expression for this word expression.

The number of marathon runners plus 28

Solution:
1. Choose a variable. ⟶ Let $r =$ the number of runners
2. Identify the operation. ⟶ addition ("plus")
3. Write an algebraic expression. ⟶ $r + 28$

Example 2 illustrates that similar word expressions may correspond to different algebraic expressions.

EXAMPLE 2

Write an algebraic expression for each word expression.

a. Six meters less than the length of a room

b. Six meters less the length of a room

Solutions:

a.

1 Let ℓ = the length

2 subtraction ("less than")

3 $\ell - 6$ — *6 meters less than ℓ*

b.

Let ℓ = the length

subtraction ("less")

$6 - \ell$ — *6 meters less ℓ*

Example 3 illustrates that a word expression may be represented by an algebraic expression with more than one operation.

EXAMPLE 3

Write an algebraic expression for this word expression.

50 dollars more than twice the cost of a trip by auto

Solution:

1 Choose a variable. → Let c = the cost in dollars of a trip by auto.

2 Identify the operations. → addition ("more than"); multiplication ("twice")

3 Write an algebraic expression. → $50 + 2c$, or $2c + 50$

CLASSROOM EXERCISES

Match each item in 1–5 with one item from a–e. (Table)

1. $3\frac{1}{2}$ increased by some number

2. The product of $3\frac{1}{2}$ and some number

3. Some number less $3\frac{1}{2}$

4. The quotient of some number and $3\frac{1}{2}$

5. Some number less than $3\frac{1}{2}$

a. $q \cdot 3\frac{1}{2}$

b. $3\frac{1}{2} - t$

c. $3\frac{1}{2} + b$

d. $\frac{n}{3\frac{1}{2}}$

e. $r - 3\frac{1}{2}$

For Exercises 6–19, write an algebraic expression. Use the variable n. (Examples 1 and 2)

6. Six more than the number of fish.

7. The number of points less 20

8. The number of meters divided by 1000

9. Five less than the number of swimmers

10. The product of 10 and the number of centimeters

11. The number of kilometers per hour increased by 5

12. Fifteen less the number of scuba divers

13. Fifteen less than the number of scuba divers

14. The number of typists increased by 3

15. Three times the number of typists

16. The total cost in dollars divided by fifteen

17. Fifteen decreased by the total cost in dollars

18. The quotient of forty-two and a number of kilowatt-hours

19. The quotient of a number of kilowatt-hours and forty-two

For Exercises 20–22, match each item with one item in a–c. (Example 3)

20. 27 less than three times the distance traveled

21. 27 minus the quotient of some number and 3

22. 3 fewer than 27 times the number of tennis rackets

a. $27 - \frac{y}{3}$

b. $27q - 3$

c. $3p - 27$

WRITTEN EXERCISES

Goal: To write an algebraic expression for a word expression

Sample Problems: **a.** 5 less than a number **b.** 5 less a number

Answers: **a.** $n - 5$ **b.** $5 - n$

For Exercises 1–32, write an algebraic expression. Use the variable n. (Examples 1 and 2)

1. The number of hours increased by 12

2. Five times the number of dollars

3. The number of cakes decreased by two

4. The product of fourteen and the number of lemons

5. Fifteen less than the number of athletes

6. Two more than the number of kilometers

7. The number of months less 120

8. The number of teachers less 31

9. The number of satellites decreased by three

10. The number of satellites divided by three

11. Twice as many computer operators

12. The number of computer operators increased by two

13. The total cost in dollars divided by five hundred

14. Five hundred divided by the total cost in dollars

15. Five meters less the length of a rectangle

16. Five meters less than the length of a rectangle

(Example 3)

17. Seven more than twice the number of doctors

18. Ten fewer than three times the number of accidents

19. Eight less than sixteen times the number of square centimeters

20. Twenty-five more than triple the number of years

21. Fifteen less than the quotient of the number of meters and one hundred

22. Forty-five plus the number of club members less ten

23. Eight-hundred minus the number of items sold increased by 425

24. The product of one hundred and the number of kilometers less 76.3

MIXED PRACTICE

25. Thirty-six more pages than in last year's directory

26. One hundred twenty-six fewer cars sold than last month

27. Twenty-five centimeters less than the width of the rectangle

28. The quotient of the number of pages printed and twenty-eight dollars

29. Twenty-five centimeters less the width of the rectangle

30. The quotient of twenty-eight dollars and the number of pages printed

31. Eight hundred fifty kilometers fewer than twice the number of kilometers driven in Car B

32. The difference between the winning team's score and the losing team's score of 96 points

MORE CHALLENGING EXERCISES

Choose the mathematical expression in the box that best corresponds to each word expression.

$7y - x$	$7(y - x)$
$\frac{y}{7} - x$	$\frac{y - x}{7}$

33. The product of 7 and the difference of y and x

34. The quotient of the difference of y and x and 7

35. The quotient of y and 7 less x

36. x less than the product of 7 and y

5.6 From Words to Equations

Translating from word sentences to equations is another important step in solving word problems. The table below shows some word sentences and their corresponding equations.

Table

Word Sentence	Equation
The sum of the number of tickets, *t*, and 16 **totals** 44.	$t + 16 = 44$
Eighteen more than the number of quarters, *q*, **equals** 102.	$q + 18 = 102$
The number of runners, *r*, decreased by 6 **is** 24.	$r - 6 = 24$
The difference between the cost, *c*, and $12.50 **is** $45.	$c - 12.50 = 45$
The product of 24 and the number of gallons, *g*, **equals** 192.	$24g = 192$
The quotient of the number of trucks, *t*, and 15 **is** 6.	$\frac{t}{15} = 6$

EXAMPLE 1 Write an equation for this word sentence.
The number of new cars sold plus 12 is 45.

Solution: [1] Choose a variable. ⟶ Let $c =$ the number of new cars sold.
[2] Identify an operation ⟶ addition ("plus")
[3] Write the equation.

Think: Number of new cars sold plus 12 is 45.

Translate: $c + 12 = 45$

EXAMPLE 2

Write an equation for this word sentence.

The amount of a phone bill decreased by $13 equals $30.

Solution:

[1] Let $b =$ the amount of the phone bill.

[2] subtraction ("decreased by")

Think: Amount of phone bill decreased by $13 equals $30.

[3] **Translate** $b - 13 = 30$

EXAMPLE 3

Write an equation for each word sentence.

a. The product of the number of weeks and 40 is 240.

b. The quotient of the cost and 6 is $22.

Solutions:

a. [1] Let $w =$ the number of weeks.

[2] multiplication ("product")

[3] $40w = 240$

b. [1] Let $c =$ the cost.

[2] division ("quotient")

[3] $c \div 6 = 22$, or $\frac{c}{6} = 22$

CLASSROOM EXERCISES

Match each word sentence in Exercises 1–8 with the corresponding equation from a–h below. (Table)

a. $p - 9 = 63$	**b.** $2w = 76$	**c.** $s + 4 = 76$	**d.** $w + 11 = 77$
e. $9k = 63$	**f.** $3n = 77$	**g.** $\frac{a}{6} = 9$	**h.** $\frac{77}{x} = 11$

1. Seventy–seven divided by some number equals eleven.

2. The product of 3 and some number is 77.

3. The sum of the number of weeks and 11 is 77.

4. Twice the number of tickets sold equals 76.

5. Four more than the number of seats is 76.

6. The product of 9 and the number of coins is 63.

7. The quotient of the number of apples and 6 is 9.

8. The number of passengers minus 9 is 63.

For Exercises 9–22, write an equation for each word sentence. (Example 1)

9. The sum of the number of nickels, n, and 75 is 108.

10. The number of kilometers driven, k, plus 16 equals 72.

11. Thirteen more than the number of miles, m, is 20.

12. The sum of 12 and the number of meeting rooms, r, is 56.

(Example 2)

13. Last Tuesday's average temperature, t, minus 7° is 76°.

14. Sixty minutes decreased by the number of minutes, m, for commercials is 52.

15. The difference between the length of the hall, ℓ, and 5 equals 14.

16. The number of passengers, p, on the bus minus 47 is 12.

(Example 3)

17. The product of 22 and the number of chairs per table, c, equals 220.

18. Twice the number of students, s, in a class is 48.

19. Six times the height, h, of a tree is 72.

20. The cost of a dinner, c, divided by 6 is \$6.25.

21. The quotient of the distance traveled, d, and 16 is 24.

22. The number of pounds, p, divided by 2000 is 4.

WRITTEN EXERCISES

Goal: To write an algebraic equation for a word sentence

Sample Problem: Three times the cost, c, of a concert ticket is \$24.

Answer: $3c = 24$

For Exercises 1–16:

a. *Choose a variable and write what it represents.*

b. *Write an equation for the word sentence.* (Example 1)

1. The number of sailboats plus four ships is 15.

2. The sum of the number of liters and 7 equals 22.

3. Two dollars more than the cost of a ticket is \$6.

4. The number of votes increased by 95 is 308.

(Example 2)

5. The number of animals in a pet store minus 14 equals 57.

6. The number of commuters decreased by 17 is 39.

7. The difference between the number of millimeters of rain and 3.2 is 24.5.

8. Fourteen less than the number of sandwiches is 25.

(Example 3)

9. The product of a number and 6 is 42.

10. Six times the number of days amounts to 126.

11. Twice the number of kilometers walked is 12.4.

12. The product of 8 and the number of quarters is 96.

13. The amount of a loan divided by 12 equals $72.

14. The quotient of some number and 6 is 24.

15. The quotient of the number of cars and 18 is 5.

16. The area of a parking lot divided by 32 is 40.

MIXED PRACTICE

Write an equation for each word sentence.

17. The product of the number of points scored, p, and 6 is 54.

18. The cost of a baseball glove, c, increased by $6 totals $28.

19. The number of miles, m, divided by 12 equals 18.

20. The number of football players, f, decreased by 3 is 16.

21. The number of words typed, w, minus 16 equals 33.

22. The sum of the number of musicians, m, and 7 is 52.

NON-ROUTINE PROBLEM

23. Copy the figure at the right. Then find the greatest number of boxes that can be shaded without shading more than 2 boxes in a straight line of 5 boxes.

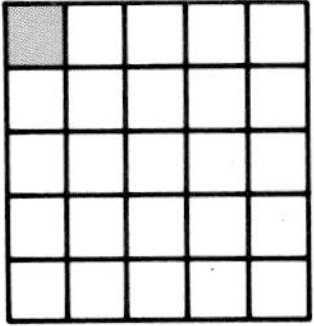

REVIEW CAPSULE FOR SECTION 5.7

Solve and check. (Pages 110–125)

1. $g + 13 = 48$

2. $\frac{r}{3} = 42$

3. $\frac{f}{3.1} = 8.6$

4. $19n = 95$

5. $y - 15 = 82$

6. $k - 4\frac{1}{2} = 16\frac{1}{2}$

7. $103 = v + 28$

8. $3\frac{1}{2}h = 24$

5.7 One Unknown

Thus far in this text, you have solved word problems by using formulas and replacing variables in the formulas, by using number concepts (Chapter 3), and by calculating odds (Chapter 4). You are now at the point in your study of algebra when you will learn how to write equations to solve word problems. The following steps for doing this are illustrated in the examples.

1. **Understand** the problem: Read the problem carefully to determine what information is given and what you are asked to find. Use this information to choose a variable to represent the unknown(s).
2. Make a **plan:** In algebra, the plan usually involves writing an equation.
3. **Solve** the problem.
4. **Check** your results with the statements in the problem. Write the answer in a complete sentence.

EXAMPLE 1 Tom worked a total of 90 hours during the past two weeks. Last week he worked 44 hours. How many hours has he worked this week?

Solution:

[1] Choose a variable for the unknown number.

Let h = the number of hours Tom worked this week.

[2] Write an equation for the problem.

Think: The hours worked this week and last week total 90.

Translate: $h \quad + \quad 44 \quad = \quad 90$

[3] Solve the equation. ⟶ $h + 44 = 90$

$$h + 44 - \mathbf{44} = 90 - \mathbf{44}$$

$$h = 46$$

[4] **Check:** Does the total of the hours Tom worked this week and last week equal 90?

Does $46 + 44 = 90$? Yes ✓

Tom worked 46 hours this week.

EXAMPLE 2

Sara bought rock concert tickets at \$12 each. The product of \$12 and the number of tickets she bought was \$192. How many tickets did she buy?

Solution:

[1] Choose a variable for the unknown number.

Let $t =$ the number of tickets she bought.

[2] Write an equation for the problem.

Think: **The product of \$12 and the number of tickets is \$192.**

Translate: $12t \qquad = 192$

[3] Solve the equation. ⟶ $12t = 192$

$$\frac{12t}{12} = \frac{192}{12}$$

$$t = 16$$

[4] **Check:** Does the product of \$12 and the number of tickets equal \$192?

Does $12 \times 16 = 192$? Yes ✓

Sara bought 16 tickets.

EXAMPLE 3

The number of cans of tomato juice in a shipment divided by 24 equals a total of 312 cartons. Find the number of cans.

Solution:

[1] Let $t =$ the number of cans of tomato juice.

[2] **Think:** **The number of cans divided by 24 equals 312 cartons.**

Translate: $\frac{t}{24} \qquad = \qquad 312$

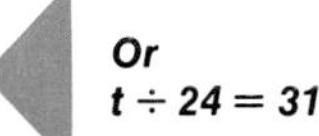

Or $t \div 24 = 312$

[3] $\frac{t}{24} = 312$

$$\frac{t}{24} \cdot \mathbf{24} = 312 \cdot \mathbf{24}$$

$$t = 7488$$

[4] The check is left for you. There are 7488 cans of tomato juice.

The steps for using algebra to solve word problems that follow at the top of the next page are a summary of the steps illustrated in Examples 1–3.

Using Algebra to Solve Word Problems

1. Choose a variable. Use the variable to represent the unknowns.
2. Write an equation for the problem.
3. Solve the equation.
4. Check your answer with the statements in the original problem. Answer the question.

CLASSROOM EXERCISES

For Exercises 1–4, match each item with one item in a–d. (Step 2, Examples 1–3)

1. The product of 8 and some number is 60.
2. The sum of a number and 8 is 60.
3. Some number minus 8 is sixty.
4. The quotient of 60 and some number is eight.

a. $s + 8 = 60$
b. $\frac{60}{p} = 8$
c. $x - 8 = 60$
d. $8t = 60$

For Exercises 5–12, write an equation that describes each word sentence. (Step 2, Examples 1–3)

5. The sum of x and 7 equals 5.
6. 42 is 3 less than n.
7. t increased by seven equals eighteen.
8. Thirty–four less y is twenty–two.
9. The difference between twenty-eight and some number is 13.
10. Four more than a number is sixty–three.
11. Ten divided by an unknown number is 42.
12. The product of $2\frac{1}{2}$ and an unknown number is 54.

WRITTEN EXERCISES

Goal: To solve and check word problems with one unknown
Sample Problem: Sam's present monthly income will be increased by $175. His new monthly income will be $2800. What is his present monthly income?
Answer: $2625

For Exercises 1–14, solve and check. (Example 1)

1. Janet added $56 to her checking account. Her new balance is $552. What was her previous balance?
2. Bill's monthly rent was increased by $50. His new rent is $525. What was his rent before the increase?

3. This year, Karla got 21 fewer hits than last year. She got 113 hits this year. How many hits did she get last year?

4. Margery wants to reduce the number of her tax clients by 12 in order to have only 50 clients. How many clients does she have?

(Example 2)

5. The product of the number of tickets sold and the price per ticket is $375. Each ticket costs $5. Find the number of tickets sold.

6. The product of the number of hours David worked last week and his hourly wage of $4 is $112. Find how many hours he worked.

(Example 3)

7. The height of a basketball player in inches divided by 12 equals the basketball player's height in feet. A basketball player is 7 feet tall. Find the player's height in inches.

8. The quotient of the number of students and 6 is the number of volleyball teams that can be formed. Fourteen teams are formed. Find the number of students.

MIXED PRACTICE

9. The number of miles hiked by Sue and Jack tripled equals the number of miles hiked by their parents. Their parents hiked 27 miles. How many miles did Sue and Jack hike?

10. On Sunday, 216 persons played tennis at a school's courts. The number of persons playing on Sunday was 34 more than the number playing on Saturday. How many persons played on Saturday?

11. The enrollment in Liberty School's ninth grade is 14 less than in the tenth grade. There are 83 students in the ninth grade. Find the number of students in the tenth grade.

12. The number of books shipped divided by 18 cartons equals the number of books in each carton. There are 60 books in each carton. Find the total number shipped.

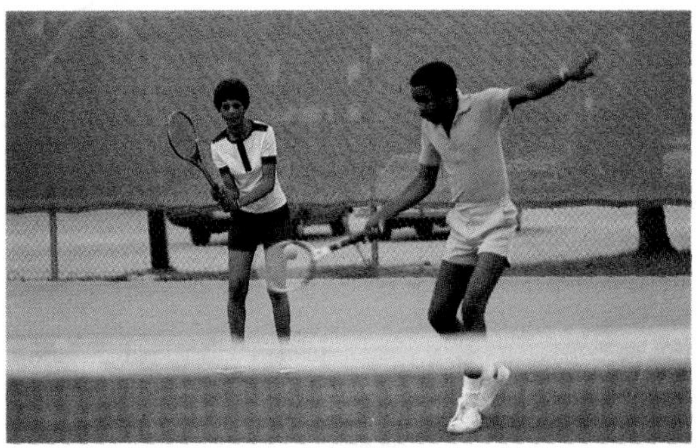

13. In April, Greg worked 19 fewer days out of town than in March. He worked out of town 9 days in April. How many days did he work out of town in March?

14. The number of women enrolled in an aerobics class is 3 times the number of men. There are 24 women in the class. How many men are there?

Decoding

Decoding is the process of translating **ciphertext** (secret code) information to **plaintext** (English) information. The reverse process, writing plaintext information in ciphertext, is called **encoding.**

One way to set up a secret code is to use the American Standard Code for Information Interchange (ASCII) shown at the right and a formula. The ASCII is a standard code used with computers.

Character	ASCII Decimal Value
A	65
B	66
C	67
D	68
E	69
F	70
G	71
H	72
I	73
J	74
K	75
L	76
M	77
N	78
O	79
P	80
Q	81
R	82
S	83
T	84
U	85
V	86
W	87
X	88
Y	89
Z	90
space	32
$	36
,	44
0	48
1	49
2	50
3	51
4	52
5	53
6	54
7	55
8	56
9	57

EXAMPLE 1: Use the key (formula) $C = A - 3$ to encode this message.

SEND MONEY 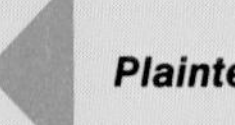 *Plaintext*

SOLUTION:

[1] Write the ASCII value for each character.

S	E	N	D	space	M	O	N	E	Y
↓	↓	↓	↓	↓	↓	↓	↓	↓	↓
83	69	78	68	32	77	79	78	69	89

[2] Use the formula to translate the ASCII characters to ciphertext characters.

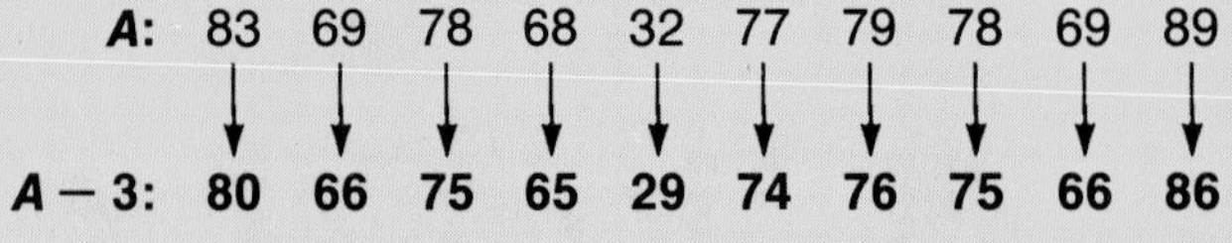

***A*:**	83	69	78	68	32	77	79	78	69	89
	↓	↓	↓	↓	↓	↓	↓	↓	↓	↓
***A* − 3:**	**80**	**66**	**75**	**65**	**29**	**74**	**76**	**75**	**66**	**86**

Ciphertext

To decode a message, you **work backward** from the C values in the formula to obtain the A values.

To encode when the key is $C = A - 3$, subtract 3 from each ASCII A-value.
To decode when the key is $C = A - 3$, add 3 to each ciphertext C-value.

EXAMPLE 2: Decode the message.
Encoding key: $C = 2A - 3$

129	167	175	61	103	85	93
93	93	61	163	141	127	161
135	163	61	127	131	151	135

SOLUTION: Work backwards. Since the key is $C = 2A - 3$, first add 3 to each ciphertext value. Then divide by 2.

[1] Add 3.

132	170	178	64	106	88	96
96	96	64	166	144	130	164
138	166	64	130	134	154	138

[2] Divide by 2.

66	85	89	32	53	44	48
48	48	32	83	72	65	82
69	83	32	65	67	77	69

[3] Refer to the table of ASCII values. Work backward from these values to obtain the plaintext.

BUY 5,000 SHARES ACME  **Decoded Message**

EXERCISES

Decode each message. Use the key given for each message.

1. $C = A + 7$

94	80	83	83	39	72	89
89	80	93	76	39	85	76
94	39	96	86	89	82	39
77	89	80	75	72	96	

2. $C = A - 12$

71	57	64	64	20	53
64	64	20	58	67	70
57	61	59	66	20	71
72	67	55	63		

3. $C = 2A + 3$

139	141	163	161	169	149	171
67	75	103	109	99	91	99
99	99	67	149	159	67	133
137	137	161	173	159	171	67
101	99	99	107	105		

4. $C = 3A - 2$

217	232	256	205	247	250	94
148	94	229	217	226	226	217
235	232	94	202	235	226	226
193	244	247	94	217	232	94
196	244	217	250	217	247	214
94	238	235	253	232	202	247

CHAPTER SUMMARY

IMPORTANT TERMS	Inverse operations *(p. 110)*	Sale price *(p. 115)*
	Equation *(p. 110)*	List price *(p. 115)*
	Solve an equation *(p. 110)*	Discount *(p. 115)*
	Solution *(p. 110)*	Distance *(p. 119)*
	Equivalent equations *(p. 110)*	Rate *(p. 119)*
	Selling price *(p. 111)*	Time *(p. 119)*
	Cost *(p. 111)*	Fuel economy *(p. 123)*
	Markup *(p. 111)*	

IMPORTANT IDEAS

1. *Subtraction Property for Equations:* Subtracting the same number from each side of an equation forms an equivalent equation.
2. *Addition Property for Equations:* Adding the same number to each side of an equation forms an equivalent equation.
3. *Division Property for Equations:* Dividing each side of an equation by the same nonzero number forms an equivalent equation.
4. *Multiplication Property for Equations:* Multiplying each side of an equation by the same nonzero number forms an equivalent equation.
5. The words sum, more than, plus, increased by, and total suggest the operation of addition.
6. The words difference, minus, decreased by, less, and less than suggest the operation of subtraction.
7. The words product, times, twice, doubled, tripled, and multiplied by suggest the operation of multiplication.
8. The words quotient and divided by suggest the operation of division.
9. Translating Word Expressions to Algebraic Expressions
   See page 126.
10. Using Algebra to Solve Word Problems
   See page 136.

CHAPTER REVIEW

SECTION 5.1

Solve and check. Show all steps.

1. $t + 28 = 73$ **2.** $a + 35 = 102$ **3.** $x + 13 = 24$

4. $42 = n + 27$ **5.** $32.1 = y + 19.3$ **6.** $t + 43.9 = 60.2$

7. A video store sells a tape of a certain movie for \$40. The store's cost for the tape is \$18. Use the formula $p = c + m$ to find the markup.

8. A bookstore sells a popular book for \$14.95. The markup on the book is \$6.25. Use the formula $p = c + m$ to find the store's cost.

SECTION 5.2

Solve and check. Show all steps.

9. $a - 59 = 23$
10. $x - 109 = 58$
11. $k - 37.8 = 58.6$
12. $47.8 = b - 23.4$
13. $361 = m - 251$
14. $t - 16 = 23$

15. An office supply store offers a discount on their calendars. The discount for one calendar is \$3.20. The sale price is \$5.75. Use the formula $s = \ell - d$ to find the list price.

16. A school bookstore is selling spiral notebooks at a discount of \$0.40. The sale price is \$2.10 each. Use the formula $s = \ell - d$ to find the list price.

SECTION 5.3

Solve and check. Show all steps.

17. $126 = 14x$
18. $22y = 132$
19. $16t = 72$
20. $105 = 9w$
21. $0.8a = 344$
22. $2.5n = 136$

23. A runner ran a 5000-meter race at a rate of 6.34 meters per second. Use the formula $d = rt$ to find the time it took to run the race. Round your answer to the nearest second.

24. A greyhound ran a 200-yard course in 10.4 seconds. Use the formula $d = rt$ to find the rate in yards per second. Round your answer to the nearest tenth.

SECTION 5.4

Solve and check. Show all steps.

25. $\frac{x}{7} = 19$
26. $\frac{a}{24} = 13$
27. $\frac{t}{18} = 12$
28. $\frac{b}{27} = 15$
29. $27 = \frac{r}{1.3}$
30. $35 = \frac{n}{9.7}$

31. The fuel economy of the Burke family's motor home is 8 miles per gallon. Use the formula $f = \frac{d}{g}$ to find how far they can drive on 48 gallons of fuel.

32. A certain limousine averages 5 kilometers per liter of fuel. Use the formula $k = \frac{d}{\ell}$ to find how far the limousine can travel on 92.7 liters of fuel.

SECTION 5.5

Write an algebraic expression. Use the variable n.

33. Twenty-five more than the number of stamps

34. The number of singers decreased by 10

35. Five centimeters less than twice the width of a rectangle

36. One-third the number of persons increased by 25,000

37. The distance in miles divided by 45

38. The product of 88 kilometers and the number of hours

SECTION 5.6

Choose a variable and write what it represents. Then write an equation for the word sentence.

39. The sum of the number of passengers and 7 is 146.

40. Three times the number of months equals 30.

41. The difference between the number of years and 15 is 18.

42. The quotient of the amount of a loan and 12 is $65.

SECTION 5.7

Solve and check.

43. The number of dogs boarded at a clinic is 12 more than the number of cats boarded. There are 27 dogs boarded. Find the number of cats.

44. The length of a rectangle equals 3 times the width of the rectangle. The length of the rectangle is 72 centimeters. Find the width.

45. The number of students who voted for Frank Sullivan in the class election is 34 less than the number who voted for Elaine Smith. Frank received 127 votes. How many students voted for Elaine?

46. Melissa swims twice a day. The quotient of the number of laps she swims in the morning and the number she swims in the afternoon is 4. She swims 38 laps in the afternoon. How many laps does she swim in the morning?

Equations and Problem Solving

CHAPTER 6

The wind tunnel shown below is used by engineers to study complex problems in the aerodynamics of an aircraft at speeds near the speed of sound. Many of these problems must be translated into algebra in order that they may be solved. You will learn some methods of solving problems in this chapter.

6.1 Equations with More Than One Operation

To solve equations such as the following

$$3n + 5 = 26, \qquad 14 = 4x - 10, \qquad \frac{x}{5} + 7 = 18$$

the first step is to get the term with the variable alone on one side of the equation.

> To solve an equation with more than one operation, addition and subtraction are performed before multiplication and division.

EXAMPLE 1 Solve and check: $3n + 5 = 26$ Show all steps.

Solution: $3n + 5 = 26$

1 Subtract 5 from each side. → $3n + 5 - \mathbf{5} = 26 - \mathbf{5}$

$3n = 21$

2 Divide each side by 3. → $\frac{3n}{\mathbf{3}} = \frac{21}{\mathbf{3}}$

$n = 7$

Check: $3n + 5 = 26$

$3(7) + 5$

$21 + 5$

26

P–1 **Solve and check.**

a. $8x + 7 = 39$ **b.** $4x + 3 = 27$ **c.** $10x + 10 = 130$

EXAMPLE 2 Solve and check: $14 = 4x - 10$ Show all steps.

Solution: $14 = 4x - 10$

1 $14 + \mathbf{10} = 4x - 10 + \mathbf{10}$

$24 = 4x$

2 $\frac{24}{\mathbf{4}} = \frac{4x}{\mathbf{4}}$

$6 = x$, or $x = 6$

Check: $14 = 4x - 10$

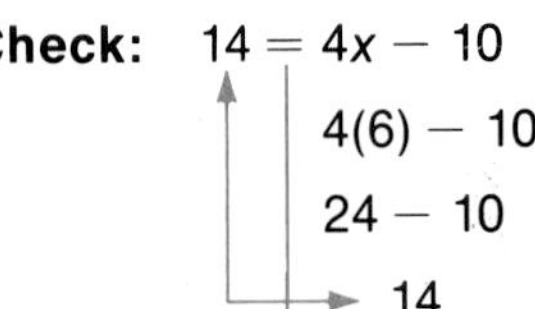

$4(6) - 10$

$24 - 10$

14

P–2 **Solve and check.**

a. $23 = 5x - 12$ **b.** $16 = 9x - 2$ **c.** $18 = 6x - 6$

EXAMPLE 3

Frank Malone drove 120 miles on the second day of a trip. This was 30 more miles than one-third the number of miles he drove on the first day. How many miles did Frank drive on the first day?

Solution:

[1] Choose a variable for the unknown number.
Let $m =$ the number of miles he drove on the first day.

[2] Write an equation for the problem.

Think: 30 more than one-third the number of miles is 120.

Translate: $\frac{m}{3} + 30 \quad = 120$

Or
$\frac{1}{3}m + 30 = 120$

[3] Solve the equation.

$$\frac{m}{3} + 30 = 120$$

$$\frac{m}{3} + 30 - \mathbf{30} = 120 - \mathbf{30}$$

$$\frac{m}{3} = 90$$

$$\frac{m}{3} \cdot \mathbf{3} = 90 \cdot \mathbf{3}$$

$$m = 270$$

[4] **Check:** Does 30 more than one-third the number of miles driven on the first day equal 120? Does $270 \div 3 + 30 = 120$? Yes ✓

He drove 270 miles on the first day.

P–3 **Solve and check.**

a. $\frac{x}{3} + 12 = 42$ **b.** $\frac{x}{6} - 9 = 10$ **c.** $90 = \frac{x}{10} + 14$

CLASSROOM EXERCISES

Write what operation to perform first in solving each equation.
(Step 1 of Examples 1–2 and Step 3 of Example 3)

1. $2y - 5 = 13$ **2.** $7 = \frac{x}{3} + 2$ **3.** $5\frac{1}{2} = 3t + 2\frac{1}{2}$

4. $\frac{r}{4} - 1.6 = 0.4$ **5.** $1.2a - 3.6 = 3.6$ **6.** $2\frac{3}{8} = \frac{1}{4}m + \frac{1}{8}$

Write the equivalent equation formed as the first step in solving each equation. (Step 1 of Examples 1–2 and Step 3 of Example 3)

7. $4n - 1 = 5$ **8.** $9y + 3 = 16$ **9.** $13 = 12s - 5$

10. $11w + 8 = 25$ **11.** $21.8 = 8a + 5.6$ **12.** $13x - 6.9 = 47.1$

13. $19 = 3b - 2\frac{1}{2}$ **14.** $5p + 6\frac{1}{4} = 12\frac{3}{4}$ **15.** $3.1 = 0.7x + 0.9$

16. $22.3y - 19.5 = 3.6$ **17.** $\frac{x}{5} - 2 = 13$ **18.** $19 = \frac{q}{12} + 7$

Decide whether the value of x shown is the solution for the given equation. Answer Yes *or* No. (Check, Examples 1–3)

19. $3x + 4 = 22$: $x = 6$ **20.** $2x + 5 = 21$: $x = 4$ **21.** $12 = 5x - 2$: $x = 2$

22. $16 = 7x - 5$: $x = 3$ **23.** $\frac{x}{3} - 5 = 8$: $x = 3$ **24.** $26 = \frac{x}{2} + 14$: $x = 24$

WRITTEN EXERCISES

Goal: To solve an equation with more than one operation
Sample Problem: $3x - 6 = 18$
Answer: 8

For Exercises 1–36, solve and check. Show all steps. (Example 1)

1. $4x + 7 = 39$ **2.** $2x + 13 = 27$ **3.** $139 = 12x + 37$

4. $143 = 8x + 27$ **5.** $0.3x + 1.5 = 3.9$ **6.** $1.4x + 2.7 = 11.1$

(Example 2)

7. $5x - 19 = 51$ **8.** $6x - 9 = 87$ **9.** $9.4 = 2.3x - 18.2$

10. $61.9 = 4.8x - 19.7$ **11.** $20x - 4.5 = 73.5$ **12.** $60x - 38.5 = 339.5$

(Example 3)

13. $\frac{x}{5} - 8 = 13$ **14.** $\frac{x}{4} - 11 = 12$ **15.** $23 = \frac{x}{12} + 5$

16. $32 = \frac{x}{9} + 13$ **17.** $24.9 = \frac{x}{8} + 12.7$ **18.** $30.3 = \frac{x}{6} + 16.9$

MIXED PRACTICE

19. $12 + 15x = 147$ **20.** $33 + 12x = 201$ **21.** $\frac{x}{5} - 7 = 5$

22. $14.2 = 8.6 + 0.7x$ **23.** $18 = 12 + 3x$ **24.** $\frac{x}{8} - 3 = 12$

25. $1.8x - 7.9 = 13.7$ **26.** $4x + 3\frac{1}{8} = 5\frac{3}{8}$ **27.** $23.1 = 17.9 + 0.4x$

28. $23 = 9 + 7x$ **29.** $3x + 2\frac{1}{4} = 7\frac{3}{4}$ **30.** $2.3x - 12.8 = 7.9$

31. $7 + 5x = 12$ **32.** $46 = 6x - 32$ **33.** $52 = 11x - 15$

34. $\frac{x}{3} + 4 = 9$ **35.** $6 = \frac{x}{5} + 2$ **36.** $34 = 6x - 5$

APPLICATIONS

For Exercises 37–40, solve and check. (Example 3)

37. Marcy scored 251 points as a senior player. This was five more than three times the number of points she scored as a sophomore player. How many points did she score as a sophomore?

38. In May, Marco Delano made 7 more sales than the quotient of the number of sales he made in April and 5. He made 12 sales in May. Find how many sales he made in April.

39. The interest earned on Melanie's savings account this year was \$12 less than half the interest earned last year. The interest earned this year was \$5. Find the interest earned last year.

40. The president of the Ski Club reported that 35 students will go on this year's trip. This is 3 fewer than twice the number of students who went on last year's trip. How many students went on last year's trip?

NON-ROUTINE PROBLEMS

41. What number gives the same result when multiplied by 6 as it does when 6 is added to it?

42. At 12:00, the two hands of a clock coincide with one another. How many times a day do the two hands coincide?

REVIEW CAPSULE FOR SECTION 6.2

Combine like terms. (Pages 40–45)

1. $4n + n$ **2.** $5t + 6t - 3t$ **3.** $1.4r - 0.9r + 0.8r$

4. $12a + 5a - a - 11$ **5.** $12 + 5y + 2y - 9$ **6.** $7m - 5m + 5m - 6$

7. $2y + 5 + y + 7$ **8.** $3x + 8 - x - 2$ **9.** $4a + 12 - 2a - 11$

Evaluate each expression when the variable is equal to 3. (Pages 5–7)

10. $4 + 3n$ **11.** $4b - 6$ **12.** $3y - y$

13. $6 + x - 2$ **14.** $5t + 3t + 4$ **15.** $6m - 2m - 1$

16. $3s + 4 + 2s$ **17.** $8p - 10 - 2p$ **18.** $9q - 4 + 3q$

6.2 More Equations: Like Terms

When you solve an equation such as

$$2x + 4x - 5 = 13,$$

you must first combine (add or subtract) the like terms.

EXAMPLE 1 Solve and check: $2x + 4x - 5 = 13$
Show all steps.

Solution: $2x + 4x - 5 = 13$

[1] Combine like terms. ⟶ $6x - 5 = 13$

[2] Add 5 to each side. ⟶ $6x - 5 + \mathbf{5} = 13 + \mathbf{5}$

$6x = 18$

[3] Divide each side by 6. ⟶ $\frac{6x}{\mathbf{6}} = \frac{18}{\mathbf{6}}$

$x = 3$

Check: $2x + 4x - 5 = 13$

$2(3) + 4(3) - 5$

$6 + 12 - 5$

$18 - 5$

13

P–1 **Solve and check.**

a. $3y + 5y + 2 = 18$ **b.** $4t + t - 4 = 16$

Sometimes you have to combine the like terms by subtracting.

EXAMPLE 2 Solve and check: $23 = 10t + 3 - 5t$
Show all steps.

Solution: $23 = 10t + 3 - 5t$

[1] Combine like terms. ⟶ $23 = 5t + 3$

[2] Subtract 3 from each side. ⟶ $23 - \mathbf{3} = 5t + 3 - \mathbf{3}$

$20 = 5t$

[3] Divide each side by 5. ⟶ $\frac{20}{\mathbf{5}} = \frac{5t}{\mathbf{5}}$

$4 = t$

Check:

$23 = 10t + 3 - 5t$

$10(4) + 3 - 5(4)$

$40 + 3 - 20$

23

P–2 **Solve and check.**

a. $24 = 12x - x - 9$ **b.** $5p - 3p + 4 = 14$

EXAMPLE 3

Solve and check: $0.8m - 3.9 + 0.4m = 5.7$
Show all steps.

Solution: $0.8m - 3.9 + 0.4m = 5.7$

[1] Combine like terms. → $1.2m - 3.9 = 5.7$

[2] Add 3.9 to each side. → $1.2m - 3.9 + \mathbf{3.9} = 5.7 + \mathbf{3.9}$

$1.2m = 9.6$

[3] Divide each side by 1.2. → $\frac{1.2m}{\mathbf{1.2}} = \frac{9.6}{\mathbf{1.2}}$

$m = 8$

Check:

$0.8m - 3.9 + 0.4m = 5.7$

$0.8(8) - 3.9 + 0.4(8)$

$6.4 - 3.9 + 3.2$

5.7

P–3 **Solve and check.**

a. $15.6 = 3.2 + 1.9r - 1.5r$

b. $0.4y + 1.2y - 3.6 = 1.2$

To Solve an Equation with Like Terms

1. Combine like terms.
2. Use the Addition or Subtraction Property for Equations.
3. Use the Multiplication or Division Property for Equations.
4. Check the solution.

CLASSROOM EXERCISES

For Exercises 1–24, write the equivalent equation formed as the first step in solving each equation. (Example 1)

1. $12r + 8r + 3 = 10$

2. $22 = 6w + 8 + w$

3. $3a + 9 + 4a + 2 = 27$

4. $9t + 6 + t = 46$

5. $56 = 7w + 5w - 4$

6. $6p + 2p + 3 = 27$

7. $n - 6 + n = 28$

8. $44 = 5 + 7s + 3 + 2s$

(Example 2)

9. $10b - 7b + 6 = 19$

10. $28 = 5s + 12 - s$

11. $14t - 8t - 3 = 15$

12. $9q - 5 - q = 35$

13. $108 = 18t + 16 - t - 13$

14. $28 = 4w - w + 7$

15. $33 = 16m - 12 - m$

16. $9y + 12 - 3y - 4 = 50$

(Example 3)

17. $2.5p + 3.5 + 1.5p = 24.5$

18. $12.2 = 1.6x - 2.8 + 3.4x$

19. $8.3t - 2.3t + 1.7 = 49.7$

20. $4m + 5 - 1.8m = 16$

21. $14 = 6.3r + 1.7r - 2$

22. $14.4 = 1.3k + 12 - 0.7k + 1.8$

23. $4.9p - 1.2p - 2.4p = 5.2$

24. $6.9y + 3.3y + 9.6 - 8.2 = 42.2$

WRITTEN EXERCISES

Goal: To solve an equation that has like terms
Sample Problem: $2x + 5 - x = 13$
Answer: 8

For Exercises 1–32, solve and check. Show all steps. (Example 1)

1. $3x + 2x + 4 = 39$

2. $4x + 3x + 5 = 47$

3. $46 = x + 5x - 8$

4. $56 = 3x + x - 12$

5. $4y + 3 + 2y = 57$

6. $8t - 4 + t = 59$

7. $62 = m + 8 + 3m - 2$

8. $132 = 5k + 13 + 3k - 9$

(Example 2)

9. $5x - 3x + 13 = 67$

10. $8x - 5x + 20 = 59$

11. $8t + 14 - 3t = 89$

12. $12n + 13 - 9n = 67$

13. $118 = 18q - 5q - q + 10$

14. $127 = 13y - y - 7y + 7$

15. $13r + 18 - 9r - 13 = 73$

16. $22w + 24 - 16w - 8 = 100$

(Example 3)

17. $1.8t + 2.6t - 3.3 = 23.1$

18. $3.1r + 2.5r + 5.3 = 24.9$

19. $7.0 = 12.3d - 9.7d + 1.8$

20. $21.2 = 15.8g - 7.3g - 4.3$

21. $28.4 = 3.2x - 4.6 - x$

22. $68.9 = 3.5x + 5.9 + x$

23. $14.6t - 5.3t + 0.7t - 6 = 44$

24. $5.9w + 2.8w - 2.7w + 12 = 84$

MIXED PRACTICE

25. $23 = 10x + 8 - 7x$

26. $11x + 15 + x = 75$

27. $0.8y - 0.4 + 0.6y = 8.0$

28. $24r - r + 2r - 8 = 192$

29. $20.5 = 0.5t + 2.5 + 3.5t$

30. $59 = 18b + 19 - 12b - 4$

31. $\frac{2}{3}n + 6 - \frac{1}{3}n = 15$

32. $17 = \frac{3}{4}c - 3 - \frac{1}{4}c$

REVIEW CAPSULE FOR SECTION 6.3

Solve and check. (Pages 43–45 and 144–147)

1. $m + m + 1 = 25$
2. $x + x + 1 + x + 2 = 36$
3. $n + n + 1 + n + 3 = 95$
4. $p + 2 + p + 4 = 50$
5. $309 = f + (f + 2) + (f + 4)$
6. $174 = b + (b + 1) + (b + 2) + (b + 3)$

PROBLEM SOLVING AND APPLICATIONS

6.3 More Than One Unknown: Consecutive Numbers

Numbers such as 5, 6, 7, 8, and 9 are *consecutive whole numbers.*

> Two whole numbers are ***consecutive*** when the sum of the smaller number and 1 equals the larger number.

Consecutive whole numbers are listed in order from smallest to largest.

EXAMPLE 1 The sum of two consecutive whole numbers is 31. Find the two numbers.

Solution:

[1] Choose a variable. Represent the two unknowns. ⟶ Let n = the smaller number. Then $n + 1$ = the larger number.

[2] Write an equation for the problem. ⟶ **Think: The sum of the numbers is 31.**

Translate: $n + (n + 1) = 31$

[3] Solve the equation. ⟶

$$2n + 1 = 31$$
$$2n = 30$$
$$n = 15$$
$$n + 1 = 16$$

Don't forget to find n + 1.

[4] **Check: a.** Are the two numbers consecutive? Does $15 + 1 = 16$? Yes ✓

b. Is the sum of the numbers 31? Does $15 + 16 = 31$? Yes ✓

The numbers are 15 and 16.

Whole numbers such as 0, 2, and 4 are consecutive even numbers.
Whole numbers such as 1, 3, 5, and 7 are consecutive odd numbers.

Two ***consecutive even numbers*** differ by 2.
Two ***consecutive odd numbers*** also differ by 2.

EXAMPLE 2 The sum of three consecutive even whole numbers is 48. Find the numbers.

Solution: [1] Choose a variable. Represent the three unknowns.

Let $q =$ the smallest number.

Then $q + 2 =$ the middle number.

And $q + 4 =$ the largest number.

[2] Write an equation.

Think: The sum of the numbers is 48.

Translate: $q + (q + 2) + (q + 4) = 48$

[3] Solve the equation. ⟶ $3q + 6 = 48$

$$3q = 42$$
$$q = 14$$
$$q + 2 = 16$$
$$q + 4 = 18$$

Don't forget to find $q + 2$ and $q + 4$.

[4] **Check: a.** Are the numbers even and consecutive?

Does $14 + 2 = 16$ and $16 + 2 = 18$? Yes ✓

b. Is the sum of the numbers 48?

Does $14 + 16 + 18 = 48$? Yes ✓

The numbers are 14, 16, and 18.

CLASSROOM EXERCISES

Write an expression for each whole number. Let t represent the smallest number. (Step 1, Examples 1–2)

1. Three consecutive whole numbers

2. Two consecutive even whole numbers

3. Four consecutive odd whole numbers

4. Five consecutive even whole numbers

Write an equation for each sentence. Let h represent the smallest whole number. (Step 2, Examples 1–2)

5. The sum of two consecutive whole numbers is 5.

6. The sum of three consecutive whole numbers is 33.

7. The sum of two consecutive odd whole numbers is 28.

8. The sum of three consecutive even whole numbers is 54.

WRITTEN EXERCISES

Goal: To solve word problems involving consecutive whole numbers
Sample Problem: The sum of three consecutive odd whole numbers is 87. Find the numbers.
Answer: 27, 29, and 31

For Exercises 1–18, solve and check. (Example 1)

1. The sum of two consecutive whole numbers is 25. Find the numbers.

2. The sum of two consecutive whole numbers is 37. Find the numbers.

3. The sum of two consecutive whole numbers is 67. Find the numbers.

4. The sum of three consecutive whole numbers is 48. Find the numbers.

5. The sum of three consecutive whole numbers is 129. Find the numbers.

6. The sum of four consecutive whole numbers is 106. Find the numbers.

(Example 2)

7. The sum of two consecutive even whole numbers is 98. Find the numbers.

8. The sum of three consecutive odd whole numbers is 159. Find the numbers.

9. Find three consecutive odd whole numbers with a sum of 105.

10. Find two consecutive even whole numbers with a sum of 206.

11. The sum of three consecutive odd whole numbers is 63. Find the numbers.

12. The sum of four consecutive even whole numbers is 372. Find the numbers.

MIXED PRACTICE

13. The sum of two consecutive whole numbers is 41. Find the numbers.

14. The sum of two consecutive whole numbers is 109. Find the numbers.

15. The sum of two consecutive even whole numbers is 242. Find the numbers.

16. The sum of two consecutive odd whole numbers is 128. Find the numbers.

17. The sum of three consecutive whole numbers is 309. Find the numbers.

18. The sum of three consecutive odd whole numbers is 255. Find the numbers.

Using Formulas and Tables

Braking Distance

City and state **police officers** learn how driving conditions and the speed of a car affect a driver's ability to stop quickly in an emergency.

The **braking distance** of a car is the distance traveled until the car comes to a stop after the brakes are applied. Braking distance depends on the rate of speed as well as the type of road surface. This formula gives an estimate for the braking distance.

$$b = \frac{r^2}{30F}$$

b = braking distance in feet
r = rate in miles per hour
F = coefficient of friction

Type of Surface	*Coefficient of Friction* Dry Road	Wet Road
Asphalt	0.85	0.65
Concrete	0.90	0.60
Gravel	0.65	0.65
Packed Snow	0.45	0.45

EXAMPLE 1: Estimate the braking distance for a car traveling at 50 miles per hour on wet asphalt. Write your answer to the nearest foot.

SOLUTION: Substitute the known values in the formula $b = \frac{r^2}{30F}$. ⟶ $b = \frac{50^2}{(30)(0.65)}$ ◀ *F = 0.65 from the table.*

Evaluate the denominator first. Store the value in the memory.

3 0 [×] . 6 5 [=] [M+] 5 0 [×] [=] [÷] [MR] [=] → 128.20512

The braking distance is about **128 feet.**

A **nomogram** can also be used to estimate braking distance.

EXAMPLE 2: Use the nomogram on the next page to estimate the braking distance for a car traveling at 60 miles per hour on dry concrete.

SOLUTION:

[1] Locate a point for 60 miles per hour on the "speed" scale. Locate a point for 0.90 (from the table above) on the "friction" scale.

2 Use a ruler to align the points found in step 1. Read the braking distance where the ruler crosses the "skid" scale.

The braking distance is about **133 feet.**

You can also use the nomogram to estimate the speed of a car.

EXAMPLE 3: Estimate the speed of a car on dry asphalt if it left 55-foot skid marks.

SOLUTION:

1 Locate the point for 0.85 (see the table on page 154) on the "friction" scale. Then locate the point for 55 feet on the "skid" scale.

2 Use a ruler to align the points found in step 1. Read the speed where the ruler crosses the "speed" scale.

The speed was about **37 miles per hour.**

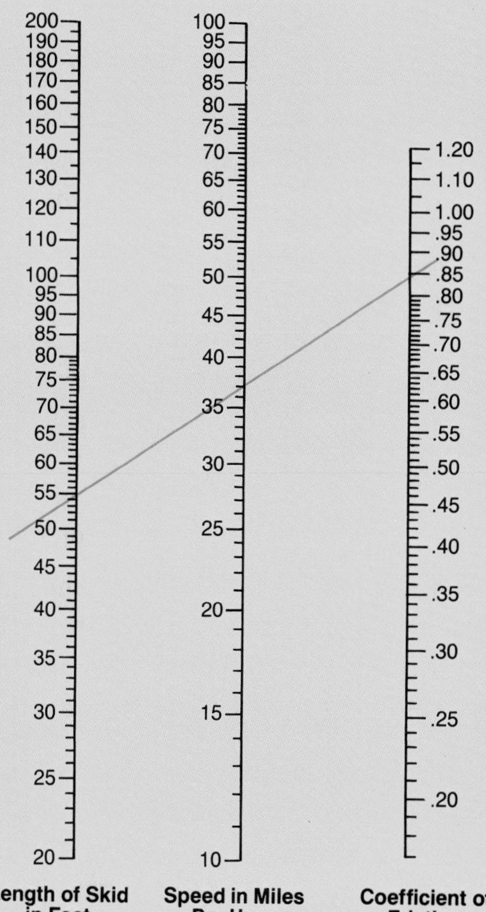

EXERCISES

For Exercises 1–4, use the formula to estimate the braking distance for a car traveling under the given conditions. Round your answer to the nearest foot.

1. 45 mi/hr on dry concrete
2. 38 mi/hr on packed snow
3. 55 mi/hr on dry asphalt
4. 42 mi/hr on gravel

For Exercises 5–8, use the nomogram.

5. A car is traveling at 50 miles per hour on dry asphalt. Estimate its braking distance.
6. A car in stopping on wet gravel left skid marks of 170 feet. Estimate its speed.
7. How is the braking distance affected if the speed of the car is increased and the driving surface remains the same?
8. How can the driver of a car reduce the braking distance required if the driving surface stays the same?

MID-CHAPTER REVIEW

Solve and check. Show all steps. (Section 6.1)

1. $24 = 4m + 4$
2. $\frac{a}{8} - 3 = 9$
3. $\frac{r}{6} + 5 = 7$
4. $1.8y + 2.7 = 18.9$
5. $10.2t - 5.6 = 55.6$
6. $29 = 8q - 13$

Solve and check. (Section 6.1)

7. Hal scored twenty-four touchdowns. This was eight more touchdowns than Roger scored. How many touchdowns did Roger score?
8. Cindy and Alexis have part-time jobs. Cindy earns $25 more than one-half the amount Alexis earns each week. Cindy earns $65 each week. How much does Alexis earn?

Solve and check. Show all steps. (Section 6.2)

9. $4n + 9 + n = 44$
10. $42 = 9w - 3w - 12$
11. $2.2r + 1.4r - 5.3 = 3.5$
12. $12k - 4k - k + 17 = 115$
13. $142 = 7d + 16 + 2d$
14. $3.6s - 4.9 + 4.4s = 148.9$

Solve and check. (Section 6.3)

15. The sum of two consecutive whole numbers is 73. Find the numbers.
16. The sum of three consecutive whole numbers is 129. Find the numbers.
17. The sum of three consecutive odd whole numbers is 261. Find the numbers.
18. The sum of two consecutive even whole numbers is 150. Find the numbers.

REVIEW CAPSULE FOR SECTION 6.4

Write each of the following as a sum or difference. (Pages 31–33)

1. $3(3 - t)$
2. $(b + 4)6$
3. $5(2c - 3)$
4. $6(12 + 3s)$
5. $9(6 + y)$
6. $12(5x - 12)$

Combine like terms. (Pages 40–45)

7. $4t - 3t$
8. $8s + 3s$
9. $12b + 6 + 4b$
10. $7g - 3g - 9$
11. $2d - d + 3d - 5$
12. $6n + 4 - 4n + 2n$

Find the value of each expression. (Pages 31–33)

13. $4(15 - 7)$
14. $3(5 \cdot 6 - 4)$
15. $2(7 \cdot 2 + 5)$

6.4 Distributive Property in Equations

In some equations you must apply the Distributive Property as the first step. Then you add or subtract.

EXAMPLE 1 Solve and check: $54 = 3(4x - 2)$ Show all steps.

Solution:

[1] Use the Distributive Property. → $54 = 3(4x - 2)$

$54 = 12x - 6$

[2] Add 6 to each side. → $54 + \mathbf{6} = 12x - 6 + \mathbf{6}$

$60 = 12x$

[3] Divide each side by 12. → $\frac{60}{\mathbf{12}} = \frac{12x}{\mathbf{12}}$

$5 = x$

Check: $54 = 3(4x - 2)$

$3(4 \cdot 5 - 2)$

$3(20 - 2)$

$3(18)$

54

P–1 **Solve and check: $42 = 6(x + 2)$**

In Example 2, you first apply the Distributive Property. Then you combine like terms.

EXAMPLE 2 Solve and check: $2(3a + 7) + a = 70$
Show all steps.

Solution:

[1] Use the Distributive Property. → $2(3a + 7) + a = 70$

$6a + 14 + a = 70$

[2] Combine like terms. → $7a + 14 = 70$

[3] Subtract 14 from each side. → $7a + 14 - \mathbf{14} = 70 - \mathbf{14}$

$7a = 56$

[4] Divide each side by 7. → $\frac{7a}{\mathbf{7}} = \frac{56}{\mathbf{7}}$

$a = 8$

Check:

$2(3a + 7) + a = 70$

$2(3 \cdot 8 + 7) + 8$

$2(24 + 7) + 8$

$2(31) + 8$

$62 + 8$

70

P–2 **Solve and check: $7(2b - 8) + b = 4$**

Using the Distributive Property to Solve an Equation

1. Use the Distributive Property.
2. Combine like terms.
3. Use the Addition or Subtraction Property for Equations.
4. Use the Multiplication or Division for Equations.
5. Check the solution.

CLASSROOM EXERCISES

For Exercises 1–16, write the equivalent equation formed as the first step in solving each equation.
(Example 1)

1. $7(n+2)=56$
2. $12(y-3)=108$
3. $(4n-9)5=144$
4. $117=(12w+5)3$
5. $0.5(8a+6)=42$
6. $24=(1.2y-0.4)5$
7. $40=12(\frac{3}{4}r-\frac{1}{2})$
8. $\frac{1}{4}(3s-8)=9$

(Example 2)

9. $2(x-3)+x=9$
10. $3(2x+7)+3x=30$
11. $19=4(5x-1)-3x$
12. $87=5+4(4x+2)$
13. $7x+3(x-5)=42$
14. $1.5(8n-4)-n=16$
15. $24=(1.4t-0.8)5+3.5t$
16. $3x+(3x-5)2+4x=126$

WRITTEN EXERCISES

Goal: To solve an equation that has parentheses and like terms
Sample Problem: $3(x+7)+5x=37$
Answer: 2

For Exercises 1–28, solve and check. Show all steps.
(Example 1)

1. $2(3x+4)=35$
2. $3(4x+5)=54$
3. $(x+5)3=57$
4. $(2x+7)4=124$
5. $73.2=6(1.3x+1.8)$
6. $133.5=5(2.5x+4.2)$
7. $15=\frac{1}{2}(6x-10)$
8. $13=\frac{1}{4}(8x-24)$

(Example 2)

9. $6(x-3)+2x=38$

10. $8(x-2)+3x=61$

11. $35=3x+2(x-5)$

12. $60=5x+3(x-4)$

13. $114.4=0.8(6x-7)-2.3x$

14. $4(0.8x-2.3)-0.6x=211.8$

15. $\frac{1}{2}(4x+6)-x=57$

16. $\frac{1}{2}(6x+2)-2x=82$

MIXED PRACTICE

17. $(9x-6)\frac{1}{3}=29$

18. $4x+2(8x-3)=95$

19. $3(2r+3)=57$

20. $(16t-28)\frac{1}{4}=24$

21. $4s+3(4s+3)=61$

22. $58=2(5w-3)-6w$

23. $2p+6(p-7)=93$

24. $39=12+(x-2)4$

25. $70=2(6x-7)$

26. $(x+2.4)3=10.8$

27. $2x+0.4(x+5)=14$

28. $6(2x+1)=36$

MORE CHALLENGING EXERCISES

29. If the expression $3(x+4)$ has a value of 10, what is the value of the expression $6(x+4)$?

30. If the expression $8(c-3)$ has a value of 16, what is the value of the expression $2(c-3)$?

31. If $a(b+c)=\frac{1}{2}a(d+e)$, how does the value of $(d+e)$ compare with the value of $(b+c)$?

32. If $p(m-n)=3p(q-r)$, how does the value of $(q-r)$ compare with the value of $(m-n)$?

REVIEW CAPSULE FOR SECTION 6.5

Substitute in the formula $p=a+b+c$. Then simplify to find a value or expression for p. (Pages 104–106)

1. $a=4$; $b=5$; $c=7$

2. $a=9$; $b=15$; $c=13$

3. $a=12.1$; $b=13.5$; $c=15.6$

4. $a=3.2$; $b=6.5$; $c=9.4$

5. $a=9$; $b=2x$; $c=x+5$

6. $a=s$; $b=3s$; $c=s-8$

Substitute in the formula $p=2(\ell+w)$. Then simplify to find a value or expression for p. (Pages 37–39)

7. $\ell=8$; $w=12$

8. $\ell=16$; $w=6$

9. $\ell=5.8$; $w=0.9$

10. $\ell=12.5$; $w=13.2$

11. $\ell=q$; $w=q+3.2$

12. $w=d$; $\ell=2d+7.8$

PROBLEM SOLVING AND APPLICATIONS

6.5 Using Geometry Formulas

The ***perimeter*** of a triangle is the sum of the lengths of its sides.

To solve a word problem involving perimeter, use the perimeter formula and the relationship between the sides of the given figure. It is helpful to draw and label a figure.

EXAMPLE 1 The longest side of a triangular steel plate is twice as long as the shortest side. The third side is one centimeter longer than the shortest side. The perimeter of the plate is 25 centimeters. Find the length of each side.

Solution: Draw and label the figure.

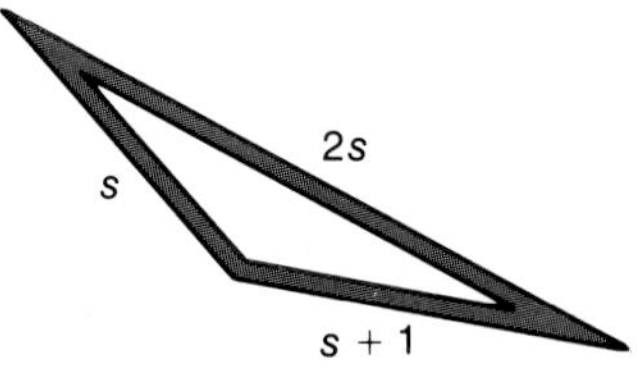

[1] Choose a variable. Represent the unknowns.

Let $s =$ the shortest side.

Then $2s =$ the longest side,

and $s + 1 =$ the third side.

[2] Use the formula for the perimeter of a triangle to write an equation. ⟶ $p = a + b + c$

Substitute for p, a, b, and c.

$$25 = s + 2s + s + 1$$

[3] Solve the equation. ⟶ $25 = 4s + 1$

$$24 = 4s$$

$$6 = s$$

$$2s = 12$$

$$s + 1 = 7$$

Don't forget to find 2s and s + 1.

[4] **Check:** **a.** Is the longest side twice as long as the shortest side?

Does $12 = 2(6)$? Yes ✔

b. Is the third side one centimeter longer than the shortest side?

Does $7 = 6 + 1$? Yes ✔

c. Is the perimeter 25 centimeters?

Does $6 + 12 + 7 = 25$? Yes ✔

The sides are 6 centimeters, 7 centimeters, and 12 centimeters long.

Recall that the formula for the perimeter of a rectangle is $p = 2(\ell + w)$.

EXAMPLE 2

The length of a rectangular sign is 1.4 meters more than the width. The perimeter of the sign is 9.6 meters. Find the length and width of the sign.

Solution: Draw and label the figure.

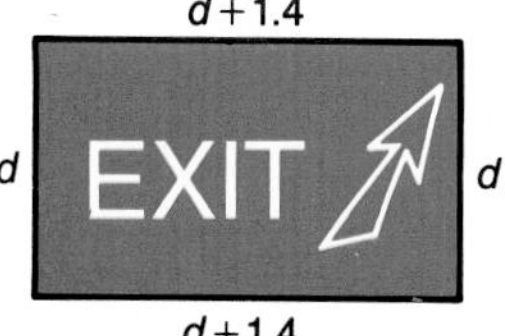

[1] Choose a variable. Represent the unknowns.

Let d = the width.

Then $d + 1.4$ = the length.

[2] Use the formula for the perimeter of a rectangle to write an equation. ⟶ $p = 2(\ell + w)$ ◀ ***Substitute for p, ℓ, and w.***

$$9.6 = 2(d + 1.4 + d)$$

[3] Solve the equation. ⟶ $9.6 = 2(2d + 1.4)$ ◀ ***$2(2d + 1.4) = 2(2d) + 2(1.4)$***

$$9.6 = 4d + 2.8$$

$$6.8 = 4d$$

$$1.7 = d$$ ◀ ***Don't forget to find $d + 1.4$.***

$$d + 1.4 = 3.1$$

[4] **Check:** **a.** Is the length 1.4 meters more than the width?

Does $3.1 = 1.7 + 1.4$? Yes ✓

b. Is the perimeter 9.6 meters?

Does $9.6 = 2(3.4 + 1.4)$? Yes ✓

The length is 3.1 meters and the width is 1.7 meters.

CLASSROOM EXERCISES

Represent the lengths of the sides of each triangle in terms of the shortest side. Let x represent the length of the shortest side. (Step 1, Example 1)

1. The longest side is 8 units longer than the shortest side. The third side is 4 units longer than the shortest side.

2. The longest side is twice the length of the shortest side. The third side is 3 units longer than the shortest side.

3. The longest side is 9 units longer than the shortest side. The third side is one half the length of the longest side.

4. The longest side is double the length of the shortest side. The third side is $\frac{3}{4}$ the length of the longest side.

Write an equation by substituting in the formula for the perimeter of a triangle. (Step 2, Example 1)

	Side a	Side b	Side c	Perimeter
5.	d	$d + 1$	$2d + 5$	60
6.	s	$2s$	$2s + 3$	88
7.	$2m - 1$	m	$3m$	72
8.	$4c$	$c - 5$	c	98

Represent the length of each rectangle in terms of the width. Let w represent the width. (Step 1, Example 2)

9. The length is 4 units longer than the width.

10. The length is double the width.

11. The length is 10 times the width, less 7 units.

12. The length is 3 units less than four times the width.

Write an equation by substituting in the formula for the perimeter of a rectangle. (Step 2, Example 2)

13. The length is 8 units longer than the width. The perimeter of the rectangle is 49 units.

14. The length is 3 units more than twice the width. The perimeter of the rectangle is 76 units.

15. The width is 5 units shorter than the length. The perimeter of the rectangle is 35 units.

16. The width is 19 units fewer than double the length. The perimeter of the rectangle is 43 units.

WRITTEN EXERCISES

Goal: To solve word problems involving perimeter

Sample Problem: The length of a rectangle is 5 more than twice the width. The perimeter is 40. Find the length and width.

Answer: The length is 15, and the width is 5.

For Exercises 1–8, solve and check each problem. (Example 1)

1. The lengths of two sides of an ***isosceles triangle*** are equal. The third side is 3 units longer than each of the other sides. The perimeter is 60 units. Find the length of each side.

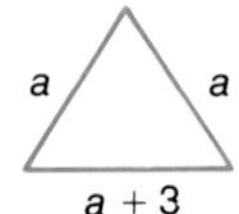

2. The length of one side of a triangle is two units more than the length of another. The third side is 5 units. The perimeter is 57 units. Find the two unknown lengths.

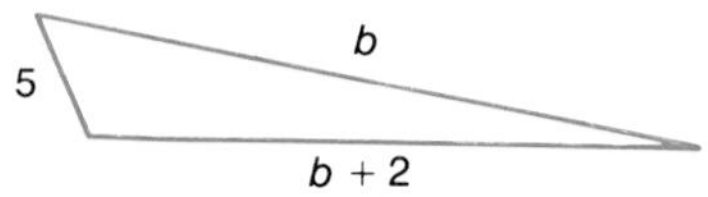

3. A private plane flies a triangular route formed by three cities. The longest leg of the route is 299 kilometers more than the shortest leg. The third leg is 674 kilometers and the total route is 1971 kilometers. Find the length of the longest leg.

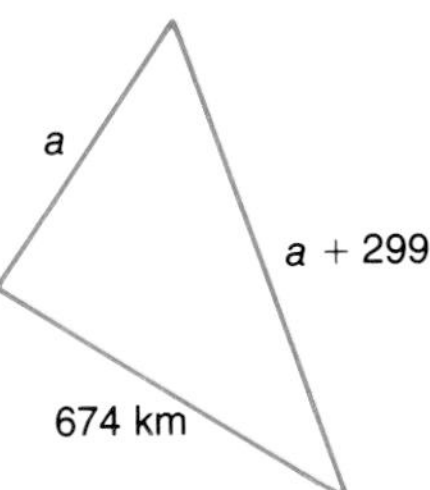

(Example 2)

4. A boat race took place on a triangular 764–kilometer course. One leg was three kilometers more than the shortest leg. The longest leg was 281 kilometers. Find the length of the shortest leg.

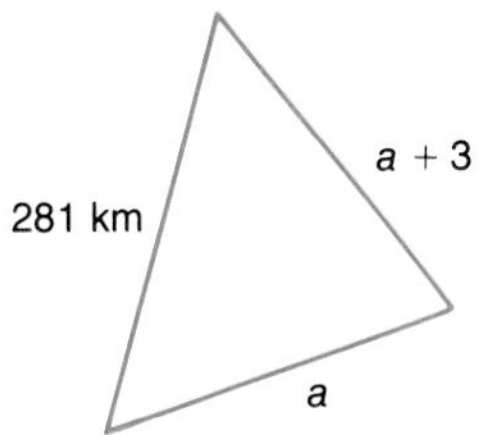

5. The length of a rectangle is six units more than the width. The perimeter is 44 units. Find the length and width.

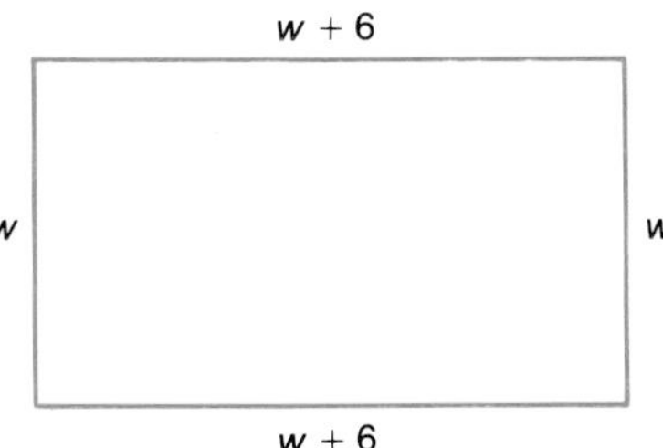

6. The length of a rectangle equals the width increased by 15 units. The perimeter is 58 units. Find the length and width.

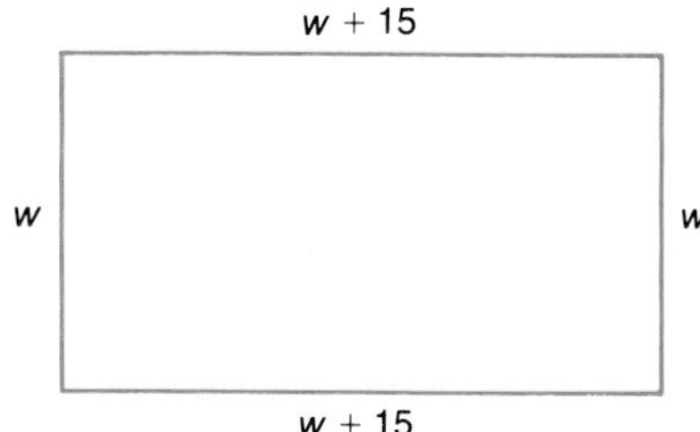

7. The width of a rectangular parking lot is 116 meters less than the length. The perimeter is 1596 meters. Find the length and width.

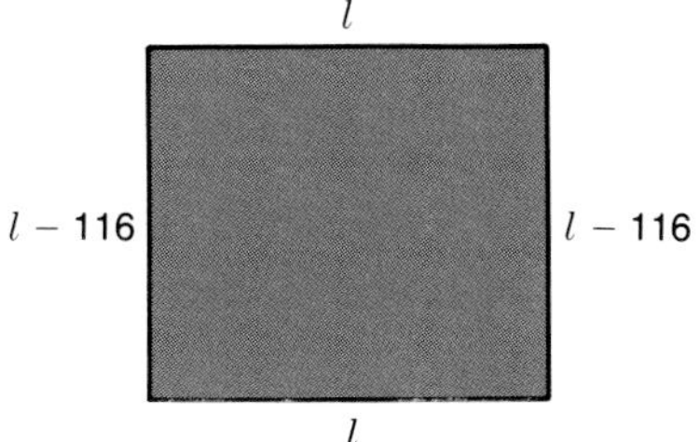

8. A soccer field is a rectangle with a perimeter of 370 meters. The width is 35 meters less than the length. Find the length and width.

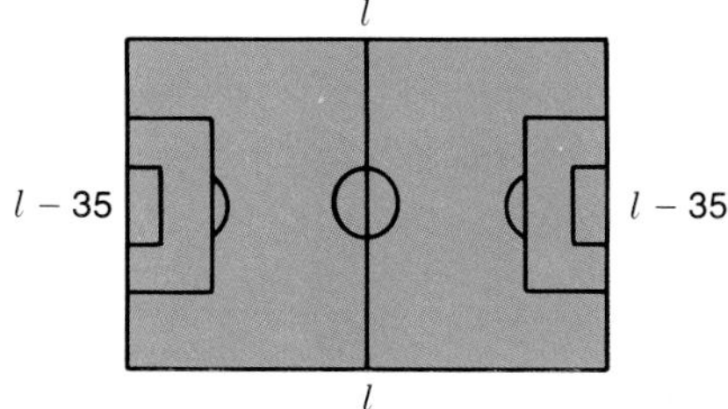

MIXED PRACTICE

For Exercises 9–12, a figure is partially labeled. Complete the figure using information in the problem. Then solve and check the problem.

9. The triangle below has a perimeter of 307.3 feet. The distance from second base to home plate is 37.3 feet more than the distance from home to first base. Find all three distances.

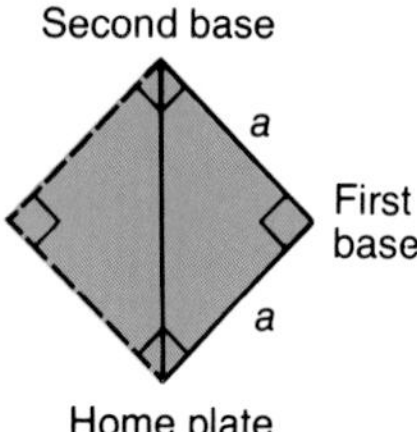

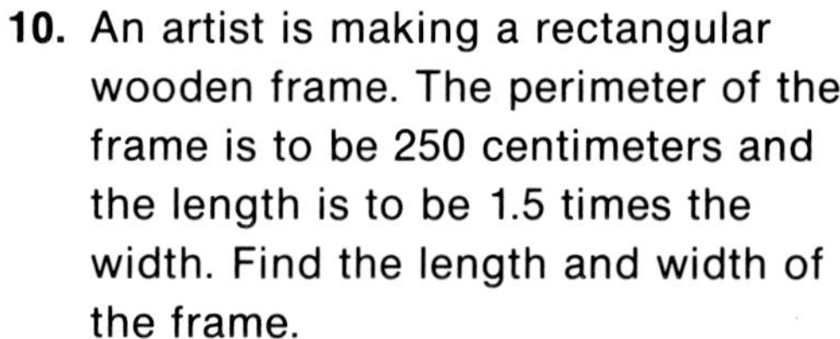

10. An artist is making a rectangular wooden frame. The perimeter of the frame is to be 250 centimeters and the length is to be 1.5 times the width. Find the length and width of the frame.

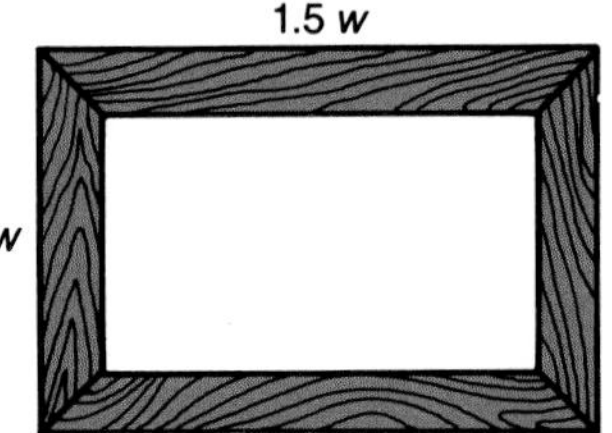

11. Surveyors working on a new freeway placed stakes at points A, B, C. The perimeter of the triangle formed is 1929.8 meters. The distance from A to C equals the distance from B to C. The distance from A to B is 415 meters less than either of the other two distances. Find all three distances.

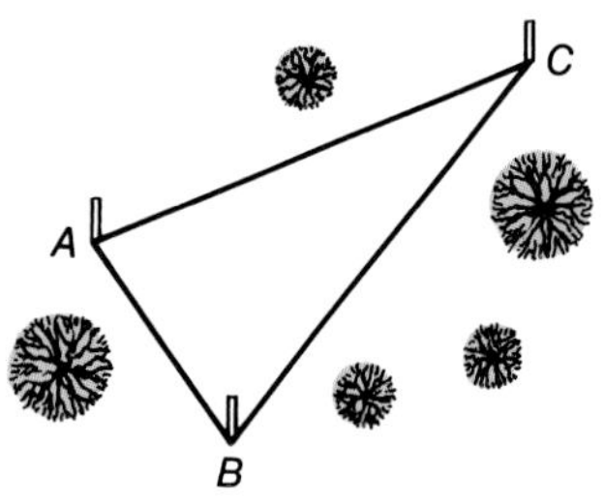

12. A metal worker bent a 19.1 inch strip of metal into a triangular support for a shelf. The shelf rests on a side of the triangular support that is 3 inches longer than the side attached to the wall. The third side is 8.1 inches long. Find the length of the shortest side.

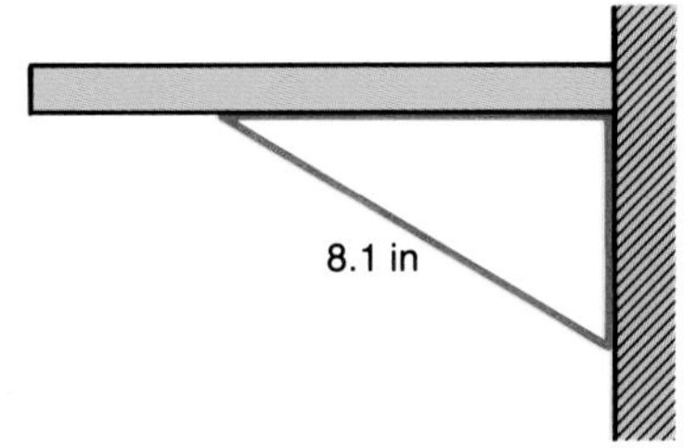

For Exercises 13–14, draw and label a figure. Then solve.

13. The triangular frame of a garage roof has two sloping sides of equal length. The bottom side of the frame is 24 feet long, and the frame's perimeter is 54 feet. Find the length of each sloping side.

14. The width of a rectangular yard is 12.7 meters less than the length. The perimeter of the yard is 48.2 meters. Find the length and width of the yard.

6.6 Too Much/Too Little Data

Sometimes word problems contain information that is not needed to solve the problem. Applying the problem–solving techniques used in the previous sections will help you to determine this.

EXAMPLE 1 The quotient of an odd number and 23 is 49.

a. Find the number.

b. What information is not needed to solve the problem?

Solutions: Decide what information to use to find the number.

[1] Choose a variable. ——→ Let $n =$ the unknown number.

[2] Write an equation for the problem. ——→ $\frac{n}{23} = 49$

[3] Solve the equation. ——→ $n = 1127$

a. The number is 1127.

b. You do not need to know that the unknown number is odd.

Sometimes you are not given enough information to solve a problem.

EXAMPLE 2 The width of a tennis court is 15.5 meters less than the length.

a. What are the length and width of the court?

b. If there is not enough information given to solve the problem, tell what information is needed.

Solutions: [1] Choose a variable.
Represent the length and width.
Let $d =$ the length.
Then $d - 15.5 =$ the width.

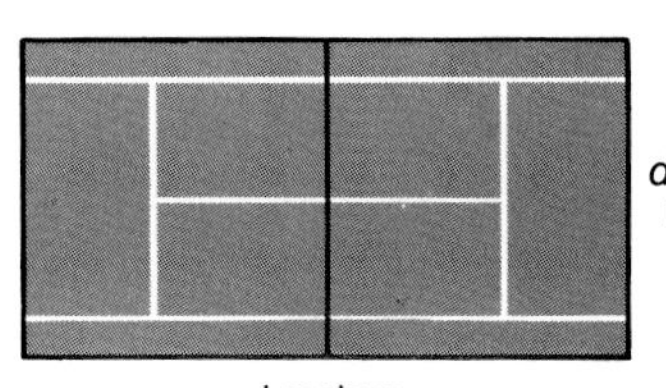

[2] Use the formula for the perimeter of a rectangle to write an equation. ——→ $p = 2(\ell + w)$ ◀ ***Substitute for p, ℓ, and w.***

$$\underline{\ ?\ } = 2(d + d - 15.5)$$

You cannot solve the problem because you need to know the perimeter of the court.

WRITTEN EXERCISES

Goal: To determine whether a problem contains too much or too little information

Sample Problem: The width of a parking lot is 17 meters less than the length. Find the length and width of the lot.

Answer: There is not enough information to solve the problem. You need to know the perimeter of the lot.

Determine whether each problem contains *too much information* *or* *too little information* *to solve the problem.*

a. *When a problem contains too much information, tell what information is not needed. Solve the problem.* (Example 1)

b. *When a problem does not contain enough information, tell what information you would need to solve the problem.* (Example 2)

1. The quotient of an unknown even number and 14 is 16. Find the number.

2. The product of some number and 68 is 612. The number is less than 68. Find the number.

3. A private plane flies a triangular route formed by three cities. The total length of the route is 8562 kilometers. Find the length of the shortest leg.

4. The length of one side of a triangle is 4 units more than the length of another. The third side is 9 units long. Find the lengths of the unknown sides.

5. Light travels at a speed of about 300,000,000 meters per second. Find the number of seconds it takes for light from the sun to reach Venus.

6. A rectangular patio is built in the backyard. The width of the patio is 12 feet less than the length. Find the length and width of the patio.

7. The sum of three consecutive odd whole numbers is 69. Each number is two more than the preceding number. Find the numbers.

8. In 1978, Janet Guthrie finished ninth in the Indianapolis 500. What was her average speed in kilometers per hour?

9. The sum of two consecutive whole numbers is 55. One of the numbers is even and the other is odd. Find the numbers.

10. At a point in Lake Superior, it takes sound waves 0.27 seconds to reach the bottom of the lake. How deep is the lake at this point?

11. A speed skater covered a course in 80 seconds. What was her average speed in meters per second?

12. The quotient of an unknown even whole number and 14 is 55. Find the number.

Focus on Reasoning: Logic Tables

Logical reasoning is used to solve many types of problems.

EXAMPLE Bill, Marie, and Jan each left school in their own car. The three cars were a red Hawk, a blue Cheetah, and a white Sunburst. Use the clues below to find the driver of each car.

CLUE A Marie waved to the driver of the Hawk.
CLUE B Bill followed the Sunburst out of the parking lot.
CLUE C Marie does not drive a white car.

Solution: Make a table to show all the possibilities.
Use an X to show that a possibility cannot be true.
Use a ✓ when you are certain that a possibility is true.

1 Read Clue A.
Is Marie the driver of the Hawk?
Place an X next to Marie's name in the Hawk column.

	Hawk	Cheetah	Sunburst
Bill			
Marie	X		
Jan			

2 Read Clue B.
Is Bill the driver of the Sunburst?
Place an X next to Bill's name in the Sunburst column.

	Hawk	Cheetah	Sunburst
Bill			X
Marie	X		
Jan			

3 Read Clue C.
Is Marie driving the Sunburst?
What car must Marie drive?
Place an X in the Sunburst column and a ✓ in the Cheetah column next to Marie's name.
Place X's next to Bill's and Jan's names in the Cheetah column.

	Hawk	Cheetah	Sunburst
Bill		X	X
Marie	X	✓	X
Jan		X	

4 Who is driving the Hawk?
Who is driving the Sunburst?

Bill: Hawk **Marie: Cheetah**
Jan: Sunburst

Check the answer with the clues.

	Hawk	Cheetah	Sunburst
Bill	✓	X	X
Marie	X	✓	X
Jan	X	X	✓

EXERCISES

For each exercise, copy the table. Use the clues to complete the table to solve each problem.

1. Ann, Beth, and Carol each took a different piece of fruit for lunch. The fruits were an apple, an orange, and a banana.

CLUE A Carol does not like bananas.

CLUE B Beth does not have to peel her fruit.

Who has each fruit?

	Apple	Orange	Banana
Ann			
Beth			
Carol			

2. Darin, Kyle, and Rita each get to school in a different way. One rides the bus, one drives a car, and one walks.

CLUE A Kyle is too young to drive.

CLUE B Darin does not walk or drive.

How does each get to school?

	Bus	Car	Walk
Darin			
Kyle			
Rita			

Solve. Make a table to show all the possibilities.

3. Jim, Sarah, and Jane are in the school play. One plays a teacher, one plays a detective, and one plays a town mayor.

CLUE A Jane and the teacher ride to rehearsals together.

CLUE B Sarah helps put on the detective's makeup.

CLUE C Sarah walks to rehearsals.

Who plays the detective?

4. Lisa, Diane, and David attend the same high school. One is a basketball player, one is a band member, and one is a cheerleader.

CLUE A Lisa and the cheerleader have the same lunch period.

CLUE B David and the basketball player's brother are friends.

CLUE C Lisa plays no sports.

Who is the cheerleader?

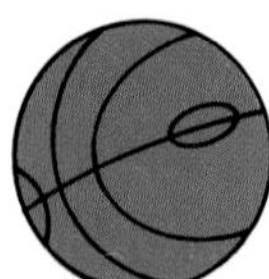
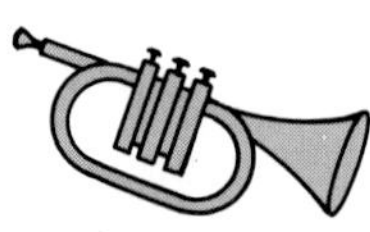

CHAPTER SUMMARY

IMPORTANT TERMS	Consecutive whole numbers *(p. 151)* Consecutive even numbers *(p. 152)* Consecutive odd numbers *(p. 152)*	Perimeter *(p. 160)* Isosceles triangle *(p. 162)*

IMPORTANT IDEAS

1. To solve an equation with more than one operation, addition and subtraction are performed before multiplication and division.

2. To Solve an Equation with Like Terms
   a. Combine like terms.
   b. Use the Addition or Subtraction Property for Equations.
   c. Use the Multiplication or Division Property for Equations.
   d. Check the solution.

3. If n represents a whole number, then $n + 1$ represents the next greater whole number.

4. If n represents an even whole number, then $n + 2$ represents the next greater even whole number.

5. If n represents an odd whole number, then $n + 2$ represents the next greater odd whole number.

6. Using the Distributive Property to Solve an Equation
See page 158.

7. When a word problem contains information that is not needed to solve the problem, solve the problem and identify the unnecessary information.

8. When a word problem does not contain enough information to solve the problem, identify the information needed.

CHAPTER REVIEW

SECTION 6.1

Solve and check. Show all steps.

1. $3x + 8 = 65$
2. $7r + 24 = 185$
3. $189 = 5.3w - 23$
4. $3.6t - 19.8 = 27$
5. $\frac{a}{14} - 28 = 9$
6. $62 = \frac{n}{9} + 43$

7. This week Brenda has worked 12 more than half the number of hours she worked last week. She worked 30 hours this week. How many hours did she work last week?

8. Scott and his father caught 15 fish on Tuesday. This was 6 less than 3 times the number of fish they caught on Monday. How many fish did they catch on Monday?

SECTION 6.2

Solve and check. Show all steps.

9. $4x + x = 38$
10. $8a + 12 - 5a = 51$
11. $41 = 3t + 6 - t$
12. $5m + 7 - 2m - 5 = 38$
13. $2.4w + 1.8 - 0.8w = 11.4$
14. $9.2n + 3.6 - 6.4n = 14.8$

SECTION 6.3

Solve and check.

15. The sum of three consecutive whole numbers is 252. Find the numbers.

16. The sum of three consecutive even whole numbers is 642. Find the numbers.

SECTION 6.4

Solve and check. Show all steps.

17. $3(2x + 5) = 45$
18. $7(y - 4) = 14$
19. $4(2r + 1) - 4r = 25$
20. $97 = t + 3(2t - 5)$

SECTION 6.5

Solve and check.

21. A triangular cross-country ski course is 12.8 kilometers long. The first leg is 2.4 kilometers longer than the second leg. The third leg is twice the length of the second leg. Find the length of each leg.

22. The width of a rectangular courtyard at the City Center is 34 feet shorter than its length. The perimeter is 268 feet. Find the length and width of the courtyard.

SECTION 6.6

Solve. Identify any information that is not needed to solve the problem. If the problem does not contain enough information, identify what information is needed.

23. In one city the average annual snowfall is 30.7 inches more than the average annual rainfall. The average windspeed is 18 miles per hour. What is the average amount of snowfall?

24. The annual cost of insurance on one family car is \$135 more than on their second car. The first car averages 17.5 miles per gallon of fuel. The total cost of insurance for the two cars is \$975. What is the cost of insurance for each car?

CUMULATIVE REVIEW: CHAPTERS 1–6

Evaluate each expression. (Section 1.2)

1. $19 + 2c$ when $c = 6$

2. $12m - 5p$ when $m = 6$ and $p = 8$

Multiply. (Section 2.2)

3. $12(4t)$

4. $(9.8x)(7w)$

5. $(11a)(12ab)$

6. $(8kn)(2kn^2)(3k)$

Write each product as a sum or difference. (Section 2.3)

7. $9(3r + 4)$

8. $(5w - 3)12$

9. $(6p - 5)2p^2$

10. $3t(11t + 4)$

Combine like terms. (Sections 2.6 and 2.7)

11. $14t^2 - t^2$

12. $6m - 7 + 5m$

13. $3v^2 + 4v - 2v^2 + 2 - v$

Write the prime factorization of each number. (Section 3.4)

14. 92

15. 88

16. 75

17. 225

Write the least common multiple. (Section 3.5)

18. 15 and 60

19. 12 and 42

20. 8, 12, and 30

21. 5, 10, and 45

Write each fraction in lowest terms. (Sections 4.1 and 4.2)

22. $\frac{12}{20}$

23. $\frac{28}{154}$

24. $\frac{10t}{26t}$

25. $\frac{14a^2b}{35ab^3}$

26. $\frac{12x^3y^5}{66x^7y^2}$

Perform the indicated operations. Write each answer in lowest terms. (Sections 4.3 through 4.6)

27. $\frac{1}{8} \cdot 12x$

28. $\frac{7x^3}{3} \cdot \frac{9}{14x}$

29. $\frac{4a^2}{3b} \div 2a^3b$

30. $\frac{6a^2}{b^3} \div \frac{21a}{b}$

31. $\frac{13a}{4c} - \frac{9h}{4c}$

32. $\frac{t}{12} + \frac{3t}{12}$

33. $\frac{5x}{8} + \frac{5x}{12}$

34. $\frac{4}{15a} - \frac{1}{6a}$

Solve each problem. (Section 4.7)

35. The design for an advertisement is triangular. The sides of the triangle are $5\frac{1}{4}$ inches, $7\frac{3}{8}$ inches, and 10 inches. Find the length of a border drawn around the triangle.

36. The Pep Club plans to make triangular school pennants to sell. Each pennant has a base of $\frac{3}{4}$ foot and a height of $2\frac{1}{3}$ feet. How many square feet of material are needed for each pennant?

Solve and check. Show all steps. (Sections 5.1 through 5.4)

37. $n + 12 = 21$

38. $78 = a + 31$

39. $21 = x - 17$

40. $c - 8.4 = 10.8$

41. $3b = 7.2$

42. $20 = 8x$

43. $7 = \frac{x}{7}$

44. $\frac{d}{12} = 70$

Write as an algebraic expression. Use the variable n. (Section 5.5)

45. The product of the height of a triangle and 9

46. Five less than the quotient of the number of students and 12

Choose a variable and write what it represents. Then write an equation for the word sentence. (Section 5.6)

47. Twice the number of miles driven is 92.

48. The sum of the number of trucks and 27 equals 100.

Solve and check. (Section 5.7)

49. Sheila earned $312 this week. The amount she earned this week is $26 less than what she earned last week. How much did she earn last week?

50. The number of sailboats tied to a pier is 13 more than the number of motorboats. There are 22 sailboats. Find the number of motorboats.

Solve and check. Show all steps. (Sections 6.1, 6.2, and 6.4)

51. $106 = 3w + 19$

52. $\frac{x}{6} - 1.7 = 9.5$

53. $0.5m + 0.7m - 13 = 83$

54. $131 = 8w + 19 - w$

55. $147 = (2n + 3)3$

56. $\frac{1}{2}(4t - 6) + 3t = 42$

Solve and check. (Sections 6.3 and 6.5)

57. The sum of three consecutive odd whole numbers is 225. Find the numbers.

58. The length of a rectangular sign is 5 centimeters longer than the width. The perimeter is 44 centimeters. Find the length and width.

Solve. Identify any information that is not needed to solve the problem. If the problem does not contain enough information, identify what information is needed. (Section 6.6)

59. A family drove 325 miles. They traveled at an average speed of 50 miles per hour. The fuel economy of their car was 20 miles per gallon. How long did the trip take?

60. A football team had 6 more wins than losses this season. Its quarterback completed 35% of his passes. He threw 220 passes. Find how many wins the team had.

Ratio, Proportion, and Per Cent

Photographers must be able to reduce and enlarge photographs. To do this, a photographer must know how to use proportions in order to determine the final size of a photograph. Photographers must also be able to use ratios to find the per cent of change in size when a photograph is reduced or enlarged.

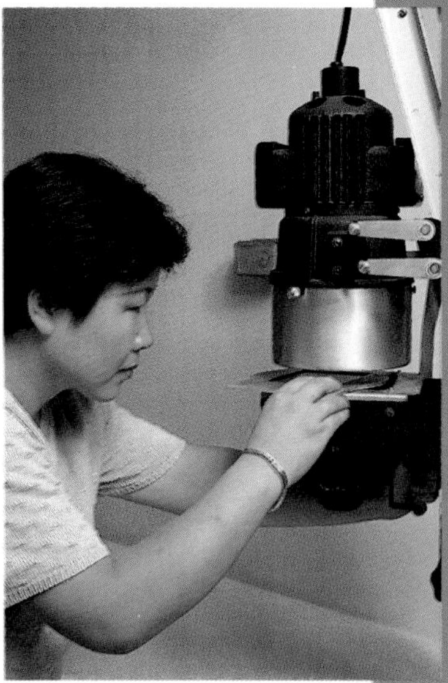

7.1 Ratio

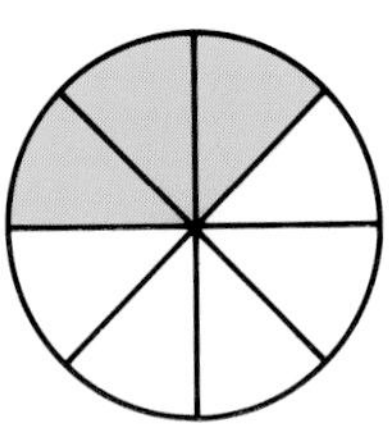

Three regions of the circle at the right are shaded. Five regions are unshaded. The comparison of "three shaded to five unshaded regions" is called a ratio. Another ratio is the comparison of "five unshaded to three shaded regions."

Definition

A ***ratio*** is a quotient that compares two numbers.

$3 : 5$ or $\frac{3}{5}$ ◀ ***Read: "three to five"***

$5 : 3$ or $\frac{5}{3}$ ◀ ***Read: "five to three"***

The ratios $3 : 5$ and $5 : 3$ do not have the same meaning. They are not equal ratios.

Definitions

In a ratio such as $a : b$ or $\frac{a}{b}$, a is the ***first term*** and b is the ***second term.*** Ratios, like fractions, can be written in lowest terms.

P–1 **Write each fraction in lowest terms.**

a. $\frac{4}{6}$ **b.** $\frac{9}{12}$ **c.** $\frac{2}{8}$ **d.** $\frac{14}{16}$ **e.** $\frac{8}{36}$ **f.** $\frac{6}{21}$

EXAMPLE 1 Write a fraction in lowest terms for the ratio "8 to 12."

Solution:

$$\frac{8}{12} = \frac{\overset{1}{\cancel{2}} \cdot \overset{1}{\cancel{2}} \cdot 2}{\underset{1}{\cancel{2}} \cdot \underset{1}{\cancel{2}} \cdot 3}, \text{ or } \frac{2}{3}$$

◀ ***As a ratio, read "$\frac{2}{3}$" as "two to three."***

P–2 **Write each ratio in lowest terms.**

a. 6 to 24 **b.** $12 : 18$ **c.** 8 to 20 **d.** $25 : 15$

Variables can be used in one or both terms of ratios. However, a variable cannot have a value that will make the second term zero.

The second term of a ratio cannot be zero.

EXAMPLE 2 Write the ratio $\frac{6n}{9n}$, $n \neq 0$, in lowest terms.

Solution:

$$\frac{6n}{9n} = \frac{2 \cdot \overset{1}{\cancel{3}} \cdot \overset{1}{\cancel{n}}}{3 \cdot \underset{1}{\cancel{3}} \cdot \underset{1}{\cancel{n}}}, \text{ or } \frac{2}{3}$$

P–3 **Write each ratio in lowest terms. None of the variables equals 0.**

a. $\frac{6y}{15y}$ **b.** $\frac{4m}{14m}$ **c.** $\frac{30r}{25r}$ **d.** $12t : 18t$

CLASSROOM EXERCISES

Write each ratio in lowest terms. (Example 1)

1. $\frac{9}{12}$ **2.** $\frac{27}{9}$ **3.** $\frac{50}{35}$ **4.** $\frac{16}{98}$ **5.** 4 to 10 **6.** 12 to 3

7. 20 to 4 **8.** 6 to 15 **9.** 10 : 100 **10.** 8 : 14 **11.** 36 : 15 **12.** 300 : 100

Write each ratio in lowest terms. (Example 2)

13. $\frac{3x}{5x}$, $x \neq 0$ **14.** $\frac{16}{8t}$, $t \neq 0$ **15.** $\frac{42p}{12}$ **16.** $\frac{24d}{16d}$, $d \neq 0$

17. $4 : 10y$, $y \neq 0$ **18.** $10a : 18a$, $a \neq 0$ **19.** $8m : 2m$, $m \neq 0$ **20.** $28c : 12$

WRITTEN EXERCISES

Goal: To write the ratio of two numbers in lowest terms

Sample Problems: **a.** 27 to 15 **b.** $\frac{8c}{14c}$, $c \neq 0$

Answers: **a.** 9 : 5, or $\frac{9}{5}$ **b.** $\frac{4}{7}$, or 4 : 7

Write each ratio in lowest terms. (Example 1)

1. 15 to 27 **2.** 14 to 44 **3.** 12 : 27 **4.** 15 : 24

The table shows the number of different instruments in an orchestra.

	String	Woodwind	Brass	Percussion
	20 Violins	3 Oboes	3 Trumpets	1 Bass drum
	8 Violas	3 Clarinets	2 Trombones	2 Kettle drums
	8 Cellos	3 Flutes	2 French horns	1 Snare drum
	6 Basses	3 Bassoons	1 Tuba	2 Cymbals
Total	42	12	8	6

For Exercises 5–31, write a fraction in lowest terms for each ratio. (Example 1)

5. Number of basses to the number of string instruments
6. Number of percussion instruments to the number of brass instruments
7. Number of cellos to the number of string instruments
8. Number of brass instruments to the number of trombones
9. Number of string instruments to the number of woodwind instruments
10. Number of brass instruments to the total number of instruments

(Example 2)

11. $\frac{10a}{12a}$, $a \neq 0$
12. $\frac{21y}{14y}$, $y \neq 0$
13. $\frac{15r}{6}$
14. $\frac{8}{20k}$, $k \neq 0$
15. $40d : 28d$, $d \neq 0$
16. $12 : 30p$, $p \neq 0$
17. $15t$ to $35t$, $t \neq 0$
18. $48c$ to 18
19. $5a$ to $12a$, $a \neq 0$

MIXED PRACTICE

20. 10 to 12
21. $17n : 34n$, $n \neq 0$
22. $3a$ to $2a$, $a \neq 0$
23. $18g : 15$
24. $\frac{1000}{100}$
25. $\frac{40}{22d}$, $d \neq 0$
26. $\frac{30}{4}$
27. $\frac{18y}{60y}$, $y \neq 0$
28. $\frac{15p}{8}$
29. $21x$ to $14x$, $x \neq 0$
30. $16 : 18$
31. 16 to $34b$, $b \neq 0$

NON-ROUTINE PROBLEM

32. The figure at the right shows a stack of 8 identical building blocks. There is no block in the center of the figure. For each of the eight building blocks, answer the following.

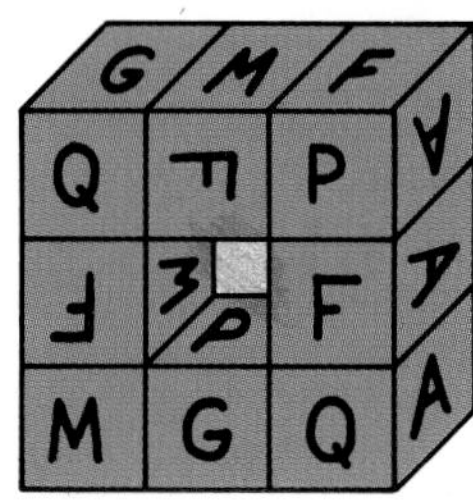

a. What letter is opposite the letter F?
b. What letter is opposite the letter P?
c. What letter is opposite the letter A?

7.2 The Probability Ratio

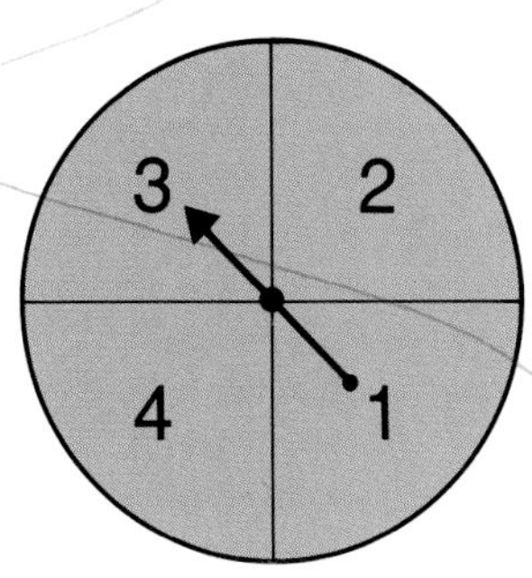

The four regions on the spinner at the right are equal in size. When the arrow is spun, it is as likely to stop on region 3 as on any of the other regions.

The arrow's stopping on one of the regions is an ***event.*** If you want to find the chance or probability of the arrow's stopping on region 3, then the arrow's stopping on region 3 is a ***favorable outcome.*** The ***possible outcomes*** of spinning the arrow are its stopping on regions 1, 2, 3 or 4.

Definition

The ***probability*** of an event E is the ratio of the number of favorable outcomes, f, of the event to the total number of possible outcomes, n, of the event.

$$P(E) = \frac{\text{number of favorable outcomes}}{\text{number of possible outcomes}} = \frac{f}{n}$$

$$P(\text{arrow stopping on region 3}) = \frac{1}{4}$$

1 ← ***Number of favorable outcomes***
4 ← ***Number of possible outcomes***

EXAMPLE 1

The spinner at the right below is equally likely to stop on any of the ten regions. Find the probability that it will stop on an even number. Write the answer as a ratio in lowest terms.

Solution:

[1] Identify the number of possible outcomes. ⟶ 10

[2] Identify the number of favorable outcomes. ⟶ 5

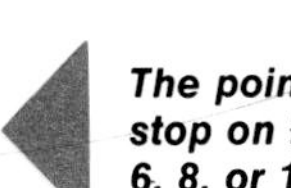

The pointer can stop on 2, 4, 6, 8, or 10.

[3] Write the probability ratio. ⟶ $P = \frac{5}{10}$, or $\frac{1}{2}$

P–1 **Find the probability that the spinner in Example 1 will stop on each of the following.**

a. A number divisible by 5 **b.** A prime number

EXAMPLE 2

The six faces of the number cube at the right are numbered from 1 to 6. Each number is equally likely to land up when the cube is tossed. Find each probability.

a. A 7 lands up. **b.** A number less than 7 lands up.

Solutions:

	a.		**b.**	
[1] Identify the number of possible outcomes. →	6		6	
[2] Identify the number of favorable outcomes. →	0	***There is no 7.***	6	***The numbers 1, 2, 3, 4, 5, or 6 can occur.***
[3] Write the probability ratio. →	$P = \frac{0}{6}$, or 0		$P = \frac{6}{6}$, or 1	

P–2 **The number cube in Example 2 is tossed. Find each probability.**

a. A two–digit number lands up.

b. A one–digit number lands up.

> The probability of an event that **cannot** happen is **0.**
>
> The probability of an event that is **certain** to happen is **1.**

CLASSROOM EXERCISES

Suppose a jar contains 3 black, 5 white, and 7 red marbles. A marble is chosen at random and then replaced before another selection is made. Find the number of favorable outcomes for each of the following events.
(Step 2, Examples 1 and 2)

1. Choosing a black marble
2. Choosing a white marble
3. Choosing a red marble
4. Choosing a blue marble
5. Choosing a marble that is <u>not</u> white
6. Choosing a marble that is <u>not</u> red

Find the probability of each of the following events. (Examples 1 and 2)

7. Choosing a black marble

8. Choosing a white marble

9. Choosing a red marble

10. Choosing a blue marble

11. Choosing a marble that is <u>not</u> white

12. Choosing a marble that is <u>not</u> red

A cube has one white face, two red faces, and three blue faces. The cube is tossed. Find the number of favorable outcomes for each of the following events. (Step 2, Examples 1 and 2)

13. A white face landing up

14. A red face landing up

15. A blue face landing up

16. A red face not landing up

For Exercises 17–20, use your answers for Exercises 13–16 to find the probability of each of the following events. (Examples 1 and 2)

17. A white face landing up

18. A red face landing up

19. A blue face landing up

20. A red face not landing up

WRITTEN EXERCISES

Goal: To write the probability ratio of the outcome of an event from the description of a probability problem

Sample Problem: Two red cards and three green cards are placed with the colored sides down and then shuffled. A card is drawn at random. Find the probability of drawing a green card.

Answer: $\frac{3}{5}$

The spinner at the right has 3 blue, 4 yellow, and 5 red regions. The pointer is equally likely to stop on any of the regions. Find the probability of the pointer's stopping on each of the following. (Example 1)

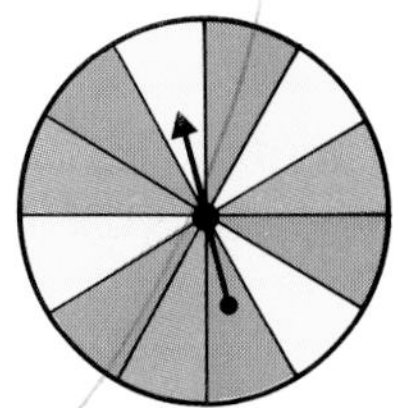

1. A blue region

2. A yellow region

3. A red region

4. A region other than blue

5. A region other than red

6. A region other than yellow

7. A green region

8. A white region

The 20 faces of the plastic marker at the right are numbered from 1 to 20. Each face is equally likely to land up when the marker is tossed. Find the probability of each event when the marker is tossed. (Example 2)

9. A number divisible by 5 lands up.
10. An odd number lands up.
11. A number less than 9 lands up.
12. A number greater than 15 lands up.
13. A prime number lands up.
14. A composite number lands up.
15. A multiple of 4 lands up.
16. A factor of 36 lands up.
17. A number greater than 25 lands up.
18. A counting number less than 21 lands up.

MORE CHALLENGING EXERCISES

A basket contains 5 white, 7 yellow, and 9 orange tennis balls. A person draws a ball at random without being able to tell the color. For Exercises 19–24, find each probability if each ball drawn is not replaced.

19. Of drawing a yellow ball on the second draw if the first ball drawn is yellow
20. Of drawing an orange ball on the second draw if the first ball drawn is orange
21. Of drawing a white ball on the second draw if the first ball drawn is orange
22. Of drawing a white ball on the second draw if the first ball drawn is yellow
23. Of drawing a yellow ball on the third draw if the first draw is white and the second draw is orange
24. Of drawing an orange ball on the third draw if the first draw is yellow and the second draw is orange
25. The probability of a certain event is n. Write an expression using n for the probability that the event will not occur.
26. The probability of randomly drawing a blue marble from a jar of marbles is $\frac{3}{5}$. Does this mean that there are 5 marbles in the jar? Explain.

REVIEW CAPSULE FOR SECTION 7.3

Write a mixed number for each fraction. Write your answer in lowest terms.

1. $\frac{16}{6}$ **2.** $\frac{22}{8}$ **3.** $\frac{132}{16}$ **4.** $\frac{154}{24}$ **5.** $\frac{92}{12}$ **6.** $\frac{308}{32}$

Solve and check. (Pages 122–125)

7. $3z = 192$ **8.** $12b = 216$ **9.** $90 = 8w$ **10.** $621 = 108q$

7.3 Proportion

An equation such as $\frac{2}{5} = \frac{4}{10}$ is called a proportion.

Definition

> A ***proportion*** is an equation that states that two ratios are equal.

A ratio has two terms, but a proportion has four terms.

$$\begin{array}{l}\textbf{First term} \longrightarrow \\ \textbf{Second term} \longrightarrow\end{array} \frac{5}{8} = \frac{10}{16} \begin{array}{l}\longleftarrow \textbf{Third term} \\ \longleftarrow \textbf{Fourth term}\end{array}$$

The product of the first and fourth terms is a ***cross product.*** The product of the second and third terms is also a cross product.

$\frac{2}{3} \times \frac{4}{6}$ $\quad 2 \cdot 6 = 12$, $\; 3 \cdot 4 = 12$

Since $2 \cdot 6 = 3 \cdot 4$, $\frac{2}{3} = \frac{4}{6}$ is a true proportion.

$\frac{3}{4} \times \frac{5}{6}$ $\quad 3 \cdot 6 = 18$, $\; 4 \cdot 5 = 20$

Since $3 \cdot 6 \neq 4 \cdot 5$, $\frac{3}{4} = \frac{5}{6}$ is a false proportion.

This suggests the following property.

> ***Cross Products Property***
>
> 1. If the cross products of a proportion are equal, it is a true proportion.
> 2. If the cross products of a proportion are not equal, it is a false proportion.

EXAMPLE 1 Determine whether $\frac{4}{5} = \frac{9}{11}$ is a true proportion.

Solution: $\frac{4}{5} \times \frac{9}{11}$

[1] Find the two cross products of the proportion. ⟶ $4 \cdot 11 = 44$

$5 \cdot 9 = 45$

[2] Determine if the cross products are equal. ⟶ $4 \cdot 11 \stackrel{?}{=} 5 \cdot 9$

$44 \neq 45$

Since the cross products are not equal, $\frac{4}{5} = \frac{9}{11}$ is a false proportion.

P–1 **Determine whether each proportion is true or false.**

a. $\frac{3}{5} = \frac{9}{15}$ b. $\frac{7}{8} = \frac{6}{7}$ c. $\frac{20}{28} = \frac{15}{21}$ d. $\frac{14}{25} = \frac{4}{7}$

A variable can appear in one or more terms of a proportion. You can solve a proportion with a variable like an equation after you find cross products.

A proportion is equivalent to the equation formed by its cross products.

$\frac{2}{3} = \frac{x}{9}$
and
$2 \cdot 9 = 3 \cdot x$
are equivalent.

EXAMPLE 2 Solve: $\frac{x}{4} = \frac{9}{12}$

Solution: $\frac{x}{4} \times \frac{9}{12}$

1 Write an equivalent equation using cross products. → $x \cdot 12 = 4 \cdot 9$

$12x = 36$

2 Divide each side by 12. → $\frac{12x}{12} = \frac{36}{12}$

$x = 3$

Check

$\frac{3}{4} \stackrel{?}{=} \frac{9}{12}$

$3 \cdot 12$	$4 \cdot 9$
36	36

EXAMPLE 3 Solve: $\frac{10}{3} = \frac{x}{2}$

Solution: 1 $10 \cdot 2 = 3 \cdot x$

$20 = 3x$

2 $\frac{20}{3} = \frac{3x}{3}$

$6\frac{2}{3} = x$

Check

$\frac{10}{3} \stackrel{?}{=} \frac{6\frac{2}{3}}{2}$

$10 \cdot 2$	$3 \cdot 6\frac{2}{3}$
20	20

P–2 **Solve.**

a. $\frac{m}{9} = \frac{14}{3}$ b. $\frac{y}{4} = \frac{12}{5}$ c. $\frac{7}{3} = \frac{x}{8}$ d. $\frac{3}{10} = \frac{w}{8}$

EXAMPLE 4

To get green paint, Amanda mixes yellow and blue paint in the ratio 4:3. She uses 12 quarts of blue paint. Find the number of quarts of yellow paint she uses.

Solution:

1 Choose a variable. ⟶ Let z = the number of quarts of yellow paint.

2 Write two ratios. ⟶ First ratio: $\frac{\text{Yellow paint}}{\text{Blue paint}} = \frac{4}{3}$

Second ratio: $\frac{\text{Number of quarts of yellow paint}}{\text{Number of quarts of blue paint}} = \frac{z}{12}$

3 Write a proportion. ⟶ $\frac{4}{3} = \frac{z}{12}$ ◀ ***4 corresponds to z; 3 corresponds to 12.***

4 Solve the proportion. ⟶ $4 \cdot 12 = 3z$

$48 = 3z$

$16 = z$

She uses 16 quarts of yellow paint.

CLASSROOM EXERCISES

Determine whether each proportion is true or false. (Example 1)

1. $\frac{3}{5} = \frac{9}{15}$ **2.** $\frac{1}{2} = \frac{4}{10}$ **3.** $\frac{1}{3} = \frac{12}{36}$ **4.** $\frac{5}{4} = \frac{20}{15}$ **5.** $\frac{5}{6} = \frac{10}{12}$ **6.** $\frac{3}{2} = \frac{9}{6}$

7. $\frac{4}{3} = \frac{2}{5}$ **8.** $\frac{3}{100} = \frac{2}{75}$ **9.** $\frac{8}{12} = \frac{2}{75}$ **10.** $\frac{4}{5} = \frac{20}{25}$ **11.** $\frac{11}{12} = \frac{7}{8}$ **12.** $\frac{3}{18} = \frac{2}{12}$

Write an equivalent equation using cross products. (Step 1, Example 2)

13. $\frac{9}{10} = \frac{2}{r}$ **14.** $\frac{2}{3} = \frac{p}{18}$ **15.** $\frac{15}{x} = \frac{7}{3}$ **16.** $\frac{y}{5} = \frac{8}{6}$ **17.** $\frac{3}{4} = \frac{m}{9}$ **18.** $\frac{x}{100} = \frac{1}{12}$

For Exercises 19–30, solve each proportion. (Example 2)

19. $\frac{x}{6} = \frac{1}{2}$ **20.** $\frac{c}{2} = \frac{3}{1}$ **21.** $\frac{a}{5} = \frac{1}{5}$ **22.** $\frac{t}{8} = \frac{3}{24}$ **23.** $\frac{w}{3} = \frac{4}{6}$ **24.** $\frac{s}{12} = \frac{2}{3}$

(Example 3)

25. $\frac{1}{3} = \frac{p}{13}$ **26.** $\frac{5}{6} = \frac{x}{4}$ **27.** $\frac{3}{4} = \frac{m}{5}$ **28.** $\frac{12}{5} = \frac{k}{10}$ **29.** $\frac{2}{7} = \frac{b}{7}$ **30.** $\frac{14}{35} = \frac{q}{10}$

WRITTEN EXERCISES

Goal: To solve proportions by using cross products

Sample Problem: $\frac{x}{6} = \frac{7}{21}$ **Answer:** 2

Determine whether each proportion is true or false. (Example 1)

1. $\frac{2}{3} = \frac{16}{24}$ **2.** $\frac{3}{4} = \frac{18}{24}$ **3.** $\frac{5}{6} = \frac{21}{25}$ **4.** $\frac{4}{7} = \frac{23}{41}$ **5.** $\frac{3}{18} = \frac{4}{24}$ **6.** $\frac{7}{8} = \frac{20}{24}$

7. $\frac{6}{32} = \frac{8}{48}$ **8.** $\frac{9}{12} = \frac{12}{16}$ **9.** $\frac{22}{32} = \frac{32}{48}$ **10.** $\frac{14}{32} = \frac{21}{48}$ **11.** $\frac{23}{57} = \frac{19}{48}$ **12.** $\frac{17}{29} = \frac{33}{42}$

For Exercises 13–36, solve each proportion. (Example 2)

13. $\frac{x}{30} = \frac{3}{5}$ **14.** $\frac{y}{24} = \frac{5}{6}$ **15.** $\frac{n}{10} = \frac{27}{15}$ **16.** $\frac{g}{8} = \frac{18}{72}$ **17.** $\frac{x}{12} = \frac{45}{36}$ **18.** $\frac{s}{24} = \frac{12}{16}$

(Example 3)

19. $\frac{1}{3} = \frac{p}{15}$ **20.** $\frac{2}{5} = \frac{w}{45}$ **21.** $\frac{1}{4} = \frac{m}{28}$ **22.** $\frac{9}{16} = \frac{x}{8}$ **23.** $\frac{3}{4} = \frac{k}{15}$ **24.** $\frac{7}{25} = \frac{z}{12}$

MIXED PRACTICE

25. $\frac{2}{3} = \frac{q}{15}$ **26.** $\frac{7}{9} = \frac{28}{z}$ **27.** $\frac{16}{5} = \frac{a}{2}$ **28.** $\frac{s}{6} = \frac{4}{10}$ **29.** $\frac{m}{11} = \frac{15}{33}$ **30.** $\frac{5}{p} = \frac{15}{24}$

31. $\frac{7}{5} = \frac{6}{n}$ **32.** $\frac{c}{10} = \frac{8}{25}$ **33.** $\frac{5}{9} = \frac{d}{18}$ **34.** $\frac{f}{12} = \frac{5}{4}$ **35.** $\frac{7}{y} = \frac{8}{15}$ **36.** $\frac{9}{16} = \frac{6}{h}$

APPLICATIONS (Example 4)

37. The ratio of a softball pitcher's wins to losses one year was 8:3. The pitcher lost 9 games that year. Find the number of games won.

38. The ratio of pets to families in one town is 7:9. There are 639 families in the town. Find the number of pets.

39. The ratio of the number of boys to the number of girls in a school is 3:2. If 1200 students are boys, how many are girls?

40. The ratio of seniors to juniors on a football team is 2:3. There are 24 juniors. Find the number of seniors on the team.

REVIEW CAPSULE FOR SECTION 7.4

Divide. Write a decimal for the answer.

1. 8 ÷ 100 **2.** 35 ÷ 100 **3.** 2 ÷ 100 **4.** 0.2 ÷ 100 **5.** 0.5 ÷ 100

6. 250 ÷ 100 **7.** 0 ÷ 100 **8.** 61 ÷ 100 **9.** 0.003 ÷ 100 **10.** 98 ÷ 100

7.4 Meaning of Per Cent

The ratio $\frac{17}{100}$ can be read as "17 per cent." It can also be shown as 17%.

Definition

> Per cent is a special ratio. A ***per cent*** is a ratio in which the second term is 100.
>
> $9\% = \frac{9}{100}$
>
> $a\% = \frac{a}{100}$

P–1 **Write a per cent for each ratio.**

a. $\frac{12}{100}$ **b.** $\frac{2.8}{100}$ **c.** $\frac{12\frac{1}{2}}{100}$ **d.** $\frac{33\frac{1}{3}}{100}$

If the second term of a ratio is not 100, you can use a proportion to write the ratio as a per cent.

EXAMPLE 1 Write a per cent for $\frac{1}{6}$.

Solution: [1] Choose a variable. ⟶ Let t = the first term of the per cent. 100 is the second term.

[2] Write a proportion. ⟶ $\frac{1}{6} = \frac{t}{100}$

[3] Solve the proportion. ⟶ $100 = 6t$

$$\frac{100}{6} = t$$

$$16\frac{2}{3} = t$$

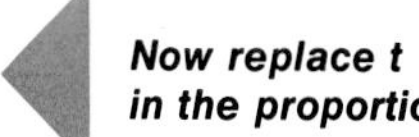

Now replace t in the proportion.

Therefore, $\frac{1}{6} = \frac{16\frac{2}{3}}{100}$, or $16\frac{2}{3}\%$.

If the first and second terms of a ratio are the same, the ratio equals 100%.

$$\frac{100}{100} = 100\% \qquad \frac{1.8}{1.8} = 100\% \qquad \frac{n}{n} = 100\%,\ n \neq 0$$

If the first term of a ratio is greater than the second term, the ratio is greater than 100%.

EXAMPLE 2 Write a per cent for $\frac{9}{8}$.

Solution: [1] Let m = the first term of the per cent.

[2] $$\frac{9}{8} = \frac{m}{100}$$

[3] $$9(100) = 8m$$

$$900 = 8m$$

$$\frac{900}{8} = m$$

$$112\frac{1}{2} = m$$

Now replace m in the proportion.

Therefore, $\frac{9}{8} = \frac{112\frac{1}{2}}{100}$, or $112\frac{1}{2}\%$.

P–2 **Write a per cent for each ratio.**

a. $\frac{1}{4}$ **b.** $\frac{3}{8}$ **c.** $\frac{5}{12}$ **d.** $\frac{7}{3}$ **e.** $\frac{24}{8}$ **f.** $\frac{0}{5}$

To write a decimal as a per cent, first write the decimal as a fraction with 100 as the denominator.

In Table 1, note that 0.3 can also be written as 0.30.

Table 1

Decimal	Fraction	Per Cent
0.18	$\frac{18}{100}$	18%
0.3	$\frac{30}{100}$	30%
0.002	$\frac{0.2}{100}$	0.2%
2.4	$\frac{240}{100}$	240%

P–3 **Write a per cent for each decimal.**

a. 0.04 **b.** 0.52 **c.** 0.301 **d.** 5.59 **e.** 6.3

To write a per cent as a decimal, first write the per cent as a ratio with 100 as the second term.

Table 2

Per Cent	Ratio	Decimal
12%	$\frac{12}{100}$	0.12
5%	$\frac{5}{100}$	0.05
0.3%	$\frac{0.3}{100}$	0.003
350%	$\frac{350}{100}$	3.50

P–4 **Write a decimal for each per cent.**

a. 2% **b.** 83% **c.** 16.9% **d.** 101% **e.** 0.7%

Tables 1 and 2 suggest the following rules.

1. To write a decimal as a per cent, multiply by 100 (move the decimal point two places to the right). Then write the per cent symbol. — 7.20 = 720%; 0.01 = 1%
2. To write a per cent as a decimal, divide by 100 (move the decimal point two places to the left). Then write the decimal without the per cent symbol. — 01.42% = 0.0142; 56% = 0.56

CLASSROOM EXERCISES

For Exercises 1–12, write a proportion as the first step in writing a per cent for each ratio. (Step 2, Examples 1 and 2)

1. $\frac{3}{8}$ **2.** $\frac{5}{12}$ **3.** $\frac{2}{3}$ **4.** $\frac{5}{6}$ **5.** $\frac{13}{15}$ **6.** $\frac{3}{16}$

7. $\frac{5}{4}$ **8.** $\frac{8}{5}$ **9.** $\frac{20}{19}$ **10.** $\frac{11}{8}$ **11.** $\frac{20}{7}$ **12.** $\frac{24}{11}$

Write a fraction with 100 as the denominator for each decimal. (Table 1)

13. 0.43 **14.** 0.6 **15.** 0.87 **16.** 0.05 **17.** 2.7 **18.** 0.113

Write a ratio with 100 as the second term for each per cent. (Table 2)

19. 26% **20.** 2% **21.** 4.7% **22.** 112% **23.** 0.13% **24.** 0.8%

WRITTEN EXERCISES

Goal: a. To write ratios and decimals in per cent form, and
b. to write per cents in decimal form

Sample Problems: a. 0.09 **b.** 89.5%

Answers: a. 9% **b.** 0.895

For Exercises 1–12, write a per cent for each ratio. (Example 1)

1. $\frac{1}{4}$ **2.** $\frac{3}{10}$ **3.** $\frac{9}{16}$ **4.** $\frac{5}{8}$ **5.** $\frac{5}{6}$ **6.** $\frac{4}{9}$

(Example 2)

7. $\frac{12}{10}$ **8.** $\frac{14}{10}$ **9.** $\frac{30}{20}$ **10.** $\frac{15}{8}$ **11.** $\frac{21}{16}$ **12.** $\frac{17}{6}$

Write a per cent for each decimal. (Table 1)

13. 0.4 **14.** 3.6 **15.** 9.4 **16.** 0.15 **17.** 0.203 **18.** 4.67

Write a decimal for each per cent. (Table 2)

19. 15% **20.** 18% **21.** 66.1% **22.** 3% **23.** 101% **24.** 7.7%

MIXED PRACTICE

Complete the table below. Write all fractions in lowest terms.

	Per Cent	Decimal	Fraction		Per Cent	Decimal	Fraction
25.	?	0.17	?	**26.**	?	0.041	?
27.	22%	?	?	**28.**	176%	?	?
29.	?	?	$\frac{19}{20}$	**30.**	?	?	$\frac{17}{25}$
31.	40%	?	?	**32.**	?	3.022	?
33.	?	?	$\frac{7}{10}$	**34.**	52.4%	?	?

MORE CHALLENGING EXERCISES

35. When the per cent for the fraction $\frac{1}{8}$ is known, what is a shortcut for finding the per cent for $\frac{3}{8}$? Show that your answer is correct.

36. When the fraction $\frac{a}{b}$ is expressed as a per cent, the per cent is greater than 100. How does the value of a compare with the value of b?

37. The per cent for the fraction $\frac{g}{100}$ is g%. Write a per cent for the fraction $\frac{g}{25}$.

38. The per cent for the fraction $\frac{x}{100}$ is x%. Write a per cent for the fraction $\frac{x}{300}$.

7.5 Proportions and Per Cent

There are three basic types of per cent problems. Each type can be solved by using a proportion. To write a proportion, it is helpful to write the problem first in general form as illustrated in the table below. Note that the per cent comes first in the statement of general form.

Per Cent Problem	General Form
1. A number is 15% of 25.	15% of 25 is some number.
2. What per cent of 28 is 42?	Some % of 28 is 42.
3. 38 is 91% of some number.	91% of some number is 38.

The following statement is also helpful in writing the proportion for a per cent problem.

100% of a number is the number.

EXAMPLE 1 A number is 25% of 64. Find the number.

Solution:

1. Write the problem in general form. → 25% of 64 is some number.
2. Use a variable to represent the unknown. → 25% of 64 is n
3. Write a statement using 100% of 64. → 100% of 64 is 64.
4. Use the statements in steps 2 and 3 to write a proportion. → $\frac{25}{100} = \frac{n}{64}$ ◀ ***n* corresponds to 25; 64 corresponds to 100.**
5. Solve the proportion. → $(25)(64) = 100n$

$$1600 = 100n$$

$$16 = n$$

16 is 25% of 64.

Steps for Solving Per Cent Problems by Proportion

1. Write the problem in the general form.
2. Use a variable to represent the unknown.
3. Write a statement using 100%.
4. Use the statements in steps 2 and 3 to write a proportion.
5. Solve the proportion.

In the second type of per cent problem, the per cent is unknown.

EXAMPLE 2 Find what per cent 36 is of 80.

Solution:

[1] Write the problem in general form. → Some % of 80 is 36.

[2] Use a variable to represent the unknown. → p % of 80 is 36.

[3] Write a statement using 100% of 80. → 100% of 80 is 80.

[4] Use the statement in steps 2 and 3 to write a proportion. → $\frac{p}{100} = \frac{36}{80}$ ◀ ***p corresponds to 36; 100 corresponds to 80.***

[5] Solve the proportion. → $80p = 3600$

$p = 45$

36 is 45% of 80.

Example 3 shows the third type of per cent problem.

EXAMPLE 3 13.2 is 24% of what number?

Solution: [1] 24% of some number is 13.2.

[2] 24% of k is 13.2.

[3] 100% of k is k.

[4] $\frac{24}{100} = \frac{13.2}{k}$

[5] $24k = 1320$

$k = 55$

13.2 is 24% of 55.

CLASSROOM EXERCISES

For Exercises 1–9, write each sentence in general form. (Step 2, Examples 1, 2, and 3)

1. x is 25% of 36. **2.** r is 12% of 512. **3.** 8% of 90 is g.

4. 17 is $y\%$ of 82.

5. 23 is $t\%$ of 156.

6. $w\%$ of 15 is 92.

7. 10 is 15% of a.

8. 19 is 80% of b.

9. 26 is 15% of n.

For Exercises 10–18, first write a statement using 100%. Then write a proportion. (Steps 3 and 4, Examples 1, 2, and 3)

10. 0.4% of 19 is n.

11. k is 300% of 86.

12. $3\frac{1}{2}\%$ of 20 is p.

13. 70.8 is $v\%$ of 25.2.

14. 10 is $q\%$ of $2\frac{1}{2}$.

15. $w\%$ of 1 is $\frac{1}{4}$.

16. 45 is 128% of t.

17. 5.2% of z is 14.

18. 0.3% of s is 98.7.

WRITTEN EXERCISES

Goal: To solve per cent problems by using proportions
Sample Problem: Write a proportion and solve: 1.96 is $x\%$ of 56.
Answer: 1.96 is 3.5% of 56.

For Exercises 1–32, write a proportion and solve. (Example 1)

1. 25% of 28 is x.

2. A number is 40% of 35.

3. 12% of 13 is a certain number.

4. p is 8% of 72.

5. $66\frac{2}{3}\%$ of $126 = b$.

6. Find $16\frac{2}{3}\%$ of 114.

7. What number is 0.8% of 92?

8. 4.5% of 108 is some number.

(Example 2)

9. $r\%$ of 92 is 69.

10. What per cent of 65 is 39?

11. 150 is $x\%$ of 100.

12. 250 is a certain per cent of 100.

13. 15 is what per cent of 40?

14. 30 is $y\%$ of 48.

15. Find what per cent 58.24 is of 52.

16. $f\%$ of $1.599 = 78$.

(Example 3)

17. 15 is 20% of some number.

18. 30% of b is 21.

19. $14 = 12\frac{1}{2}\%$ of t.

20. $37\frac{1}{2}\%$ of what number is 36?

21. 20 is 0.8% of a certain number.

22. 7 is 0.5% of r.

23. 84 is $83\frac{1}{3}\%$ of what number?

24. $87\frac{1}{2}\%$ of $h = 112$.

MIXED PRACTICE

25. 42 is 96% of a number.

26. a is 21.5% of 108.

27. What per cent of $1\frac{1}{2}$ is 3?

28. A number is 0.4% of 40.

29. $28 = 115\%$ of k.

30. $12\frac{1}{2}$ is $w\%$ of 5.

31. 12% of c is 2.5.

32. Find what per cent 206.4 is of 480.

MID-CHAPTER REVIEW

Write a fraction in lowest terms for each ratio. (Section 7.1)

1. 12 to 15 **2.** 42 to 28 **3.** 16 : 10 **4.** 5 : 25

5. $\frac{12x}{45}$ **6.** $\frac{18t}{75t}$, $t \neq 0$ **7.** $7n$ to $42n$, $n \neq 0$ **8.** 16 to $24k$, $k \neq 0$

Solve each proportion. (Section 7.3)

9. $\frac{4}{12} = \frac{5}{d}$ **10.** $\frac{15}{k} = \frac{40}{16}$ **11.** $\frac{9}{12} = \frac{x}{8}$ **12.** $\frac{a}{21} = \frac{2}{15}$

Solve each problem. (Sections 7.2 and 7.3)

13. There are 2 red pencils, 3 blue pencils, and 5 black pencils in a desk drawer. A pencil is chosen at random. Find the probability of choosing a black pencil.

14. The ratio of people to cars in one town is 9:5. There are 873 people in the town. Find the number of cars.

Write a per cent for each of the following. (Section 7.4)

15. $\frac{95}{100}$ **16.** $\frac{13}{16}$ **17.** $\frac{7}{8}$ **18.** 2.8 **19.** 1.92 **20.** 0.005

Write a decimal for each per cent. (Section 7.4)

21. 25% **22.** 39% **23.** 4% **24.** 7.7% **25.** 304% **26.** 99.9%

Write a proportion and solve. (Section 7.5)

27. 42% of 14 is a number.

28. What per cent of 55 is 3.3?

29. 102 is 120% of what number?

30. $42 = y\%$ of 336.

REVIEW CAPSULE FOR SECTION 7.6

Multiply.

1. $39 \cdot 100$ **2.** $68 \cdot 100$ **3.** $10.8 \cdot 100$ **4.** $0.09 \cdot 100$ **5.** $45.50 \cdot 100$

Round each decimal to the nearest whole number.

6. 13.2 **7.** 12.9 **8.** 27.53 **9.** 109.085 **10.** 0.81

Round each amount to the nearest dollar.

11. $18.20 **12.** $23.87 **13.** $59.50 **14.** $119.09 **15.** $246.75

7.6 Per Cents/Proportions

Refer to the steps on page 189 to solve word problems with per cent.

EXAMPLE 1 In a recent election at Crescent High School, 65% of the students voted. There are 1240 students in the school. How many voted?

Solution:

1 Write the problem in general form. → 65% of 1240 is some number.

2 Use a variable to represent the unknown. → 65% of 1240 is z.

3 Write a statement using 100% of 1240. → 100% of 1240 is 1240.

4 Use the statements in Steps 2 and 3 to write a proportion. → $\frac{65}{100} = \frac{z}{1240}$

5 Solve the proportion. → $65(1240) = 100z$

$80{,}600 = 100z$

$806 = z$ 806 students voted in the election.

Cruising range is the distance a car can travel on a full tank of fuel.

EXAMPLE 2 The cruising range of a car is 450 miles for city driving and 680 miles for highway driving. Find what per cent the city cruising range is of the highway cruising range. Round your answer to the nearest whole per cent.

Solution: 1 Write the problem in general form. → Some % of 680 is 450.

2 Use a variable to represent the unknown. → p% of 680 is 450.

3 Write a statement using 100% of 680. → 100% of 680 is 680.

4 Use the statements in Steps 2 and 3 to write a proportion. → $\frac{p}{100} = \frac{450}{680}$

5 Solve the proportion. → $680p = 45{,}000$

$p = 66.176$

The city cruising range is about 66% of the highway cruising range.

Wholesale price is the price that a store pays for items it will sell to consumers. Consumers pay the ***retail price*** for these items.

EXAMPLE 3 The wholesale price a store pays for a certain watch is 58% of the retail price. The wholesale price of the watch is $84.00. Find the retail price the store charges for the watch. Round your answer to the nearest dollar.

Solution: [1] 58% of some number is 84.

[2] 58% of t is 84.

[3] 100% of t is t.

[4] $\frac{58}{100} = \frac{84}{t}$

[5] $58t = 8400$

$t = 144.827$ ◀ ***Round to the nearest dollar.***

The retail price of the watch is $145.00.

CLASSROOM EXERCISES

For Exercises 1–6, write two statements and a proportion for each problem. Use y as the variable. (Steps 2, 3, and 4, Examples 1, 2, and 3)

1. A student answers 75% of 120 questions on an exam correctly. How many answers are correct?

2. A softball player got a hit 30% of 40 times at bat. How many hits did the player get?

3. The total cost of repairs on a car was $65.00. The cost of labor alone was $45.50. Find what per cent the labor costs are of the total cost.

4. Mario paid $28.60 in taxes on earnings of $130. Find what per cent the taxes were of his earnings.

5. Lin paid a sales tax of $2.03 on a sweater. The tax in her state is 5%. Find the price of the sweater to the nearest cent.

6. There are 259 eighth grade students in a school. The number of eighth graders is 35% of the total number of students. Find the total number of students.

WRITTEN EXERCISES

Goal: To solve word problems involving per cents by using proportions

Sample Problem: Anna answered 17 out of 20 questions on a test correctly. Find the per cent of questions that she answered correctly.

Answer: 85%

For Exercises 1–10, solve each problem. (Example 1)

1. There are 12 girls on the tennis team of a school. Twenty-five per cent of the girls are seniors. How many seniors are on the team?

2. A quarterback completes 60% of his passes in a football game. How many passes does he complete if he throws 25 passes?

(Example 2)

3. A worker who earns $4.00 per hour receives a raise of $0.20 per hour. What is the per cent of increase in the wages?

4. Of the 120 students in a school band, 24 play woodwind instruments. What per cent play woodwind instruments?

(Example 3)

5. A reporter spends 20% of her monthly salary for rent. How much is her monthly salary if she spends $192 for rent?

6. A baseball player got a hit 32% of the times he batted. If he batted 75 times, how many hits did he get?

MIXED PRACTICE

7. A concert by the Supersonics had an audience of 20,000 people. The next night a concert by the group drew 24,000 people. Find what per cent the smaller crowd was of the larger crowd.

8. The longest song on a record plays for $4\frac{1}{5}$ minutes. This is 20% of the total time it takes all the songs on the record to play. Find the total time the record takes to play.

9. A certain car has a city cruising range of 336 miles. The car's cruising range for highway driving is 150% of the city cruising range. Find the highway cruising range.

10. Jan paid a sales tax of $1.20 when buying gloves. The sales tax in her state is 6%. Find the price of the gloves to the nearest cent.

REVIEW CAPSULE FOR SECTION 7.7

Multiply. (Pages 27–30)

1. $(0.01d)(15)$ **2.** $(0.01z)(35.4)$ **3.** $(0.01m)(412)$ **4.** $(0.01s)(3245)$

PROBLEM SOLVING AND APPLICATIONS

7.7 Per Cents and Equations

In Section 7.4, you wrote decimals for per cents.

$15\% = \frac{15}{100} = 0.15$ $300\% = \frac{300}{100} = 3.0$ $0.5\% = \frac{0.5}{100} = 0.005$

You can also write a decimal for $p\%$.

$$p\% = \frac{p}{100} \text{ or } \frac{1}{100} \cdot p \text{ or } 0.01p$$

P–1 **Write a decimal for each per cent.**

a. 9% **b.** 15.2% **c.** $33\frac{1}{3}\%$ **d.** $q\%$ **e.** $r\%$

The three basic types of per cent problems can also be solved by using equations.

Per Cent Problem	General Form	Equation
1. A number is 25% of 60.	25% of 60 is some number.	$0.25 \cdot 60 = n$
2. What per cent of 48 is 36?	Some % of 48 is 36.	$0.01p \cdot 48 = 36$
3. 18 is 30% of some number.	30% of some number is 18.	$0.30 \cdot s = 18$

P–2 **Write an equation for each of the following. Use c as the variable.**

a. 18% of 250 is some number. **b.** Some % of 96 is 32.

EXAMPLE 1 A family budgets 24% of its take–home pay for food. The family's monthly take–home pay is $1840. How much is the monthly food budget? Round your answer to the nearest dollar.

Solution:

1. Write a per cent statement in general form for the problem. → 24% of 1840 is some number.
2. Write an equation for the statement. Use a variable to represent the unknown. → $0.24 \cdot 1840 = y$
3. Solve the equation. → $441.60 = y$

The family budgets $442 each month for food.

Rounded to the nearest dollar.

Example 2 shows the second type of per cent problem. You are asked to find what per cent one number is of another.

EXAMPLE 2

A waiter in a restaurant received a tip of \$2.00 from a customer. The customer's bill for the meal was \$13.25. Find what per cent the tip was of the bill. Round your answer to the nearest whole per cent.

Solution:

[1] Write a per cent statement in general form for the problem. → Some % of \$13.25 is \$2.00.

[2] Write an equation for the statement. Use a variable to represent the unknown. → $0.01q \cdot 13.25 = 2.00$

[3] Solve the equation. → $0.1325q = 2.00$

$$q = \frac{2.00}{0.1325}$$

$$q = 15.09$$

The tip was about 15% of the bill.

Rounded to the nearest whole per cent.

The third type of per cent problem is illustrated in Example 3. You are asked to find a number when a per cent of it is known.

EXAMPLE 3

In the 1984 presidential election, about 53% of the registered voters voted. About 92,000,000 people voted that year. Find the total number of registered voters in 1984. Round your answer to the nearest million.

Solution: [1] 53% of some number is 92,000,000.

[2] $0.53 \cdot s = 92,000,000$

[3] $0.53s = 92,000,000$

$$s = \frac{92,000,000}{0.53}$$

$$s = 173,584,905.7$$

Round to the nearest million.

There were about 174,000,000 registered voters in 1984.

You can follow these steps to solve a word problem involving per cents by using an equation.

Steps for Solving Per Cent Problems by Writing an Equation

1. Write a per cent statement in general form for the problem.
2. Use a variable to represent the unknown. Then write an equation for the statement.
3. Solve the equation.

CLASSROOM EXERCISES

For Exercises 1–20, write each problem in general form. Then write an equation for the problem.
(Table, Steps 1 and 2 of Examples 1, 2, and 3)

1. A number is 14% of 21.

2. 36% of 56 is t.

3. What number is 120% of 83?

4. 225% of 41 is what number?

5. b is 5% of 42.

6. 9% of 120 is a number.

7. 70% of 6400 is g.

8. 125% of 56,500 = n.

9. 14 is x% of 10.

10. What per cent of 125 is 36?

11. r% of 15.8 = 10.

12. 4.2 is w% of 0.8.

13. 500 is what per cent of 2600?

14. 375 is some per cent of 480.

15. 10,000 = z% of 6500.

16. a% of 132 is 14.

17. 15% of a number is 9.

18. 16 = 10% of f.

19. 74% of w is 18.

20. 9 is 60% of what number?

WRITTEN EXERCISES

Goal: To solve word problems involving per cents by using equations
Sample Problem: A car that cost $6500 when new lost 22% of its original value after one year. Find how much the car lost in value.
Answer: $1430

For Exercises 1–12, solve each problem. (Example 1)

1. A real estate agent gets a 6.5% commission on sales. Find the commission on a sale of a house for $60,000.

2. There are 248 sophomores at a school. 42% of the sophomores are on at least one athletic team. How many sophomores are on a team?

(Example 2)

3. A salesperson received commissions of $324 during one year. The sales for the year were $7200. Find what per cent commission the salesperson received.

4. 350 people attended a dance. What per cent of the 840 people invited to the dance actually attended?

(Example 3)

5. A survey showed that 15% of the students in a class are left–handed. There are three left–handed students in the class. Find the number of students in the class.

6. In a certain city, 62,000 people voted in an election. This is 12.5% of the number of registered voters. How many registered voters are there in the city?

MIXED PRACTICE

7. In a basketball season, Linda made 120 free throws out of 150 attempts. Find what per cent of attempted free throws she made.

8. The sales tax in Joe's state is 4%. He paid a sales tax of $5.50 on an overcoat. Find the price of the coat to the nearest dollar.

9. Luis budgets 30% of his monthly take-home pay for rent. His monthly take–home pay is $1020. How much does he budget each month for rent?

10. A census shows that the population of a city of 2,500,000 people dropped by 400,000 in ten years. Find what per cent of its original population the city lost.

11. A shopper bought a pair of gloves at a "15%–off" sale. The price of the gloves was reduced by $2.40. Find the original price of the gloves.

12. A certain automobile loses 14% of its original value in a year. Its cost when new is $8700. Find how much value the car loses in a year.

REVIEW CAPSULE FOR SECTION 7.8

Write each fraction in lowest terms. (Pages 78–82)

1. $\frac{2}{12}$　2. $\frac{9}{12}$　3. $\frac{12}{12}$　4. $\frac{15}{12}$　5. $\frac{26}{12}$　6. $\frac{42}{12}$

Write a fraction in lowest terms for each ratio. (Pages 174–176)

7. 2 months to 12 months　8. 7 months to 12 months　9. 6 months to 12 months
10. 18 months to 12 months　11. 28 months to 12 months　12. 39 months to 12 months

Write a decimal for each per cent. (Pages 185–188)

13. 5%　14. 12%　15. 10%　16. $9\frac{1}{2}$%　17. $21\frac{1}{2}$%　18. $12\frac{1}{4}$%

7.8 Using Formulas: Simple Interest

One important use of per cent is in computing interest.

Definitions

Interest is the amount of money that is charged for the use of borrowed money. The amount of money borrowed is called the ***principal.*** Interest is also paid on money deposited in certain bank accounts.

Interest is usually expressed as a per cent of the principal. This per cent is called the ***interest rate.***

The following word rule and formula show how to compute simple interest.

Word Rule: The amount of simple interest equals the principal times the rate times the time (in years).

Formula: $i = prt$ — p = principal, r = rate, t = time

EXAMPLE 1

Compute the interest on $200 for one year at a rate of 9%.

Solution:

1. Write the formula. ⟶ $i = prt$
2. Identify known values. ⟶ $p = 200$; $r = 0.09$; $t = 1$
3. Substitute and multiply. ⟶ $i = (200)(0.09)(1)$

$i = 18$

The interest on $200 for one year at a rate of 9% is $18.00.

P–1 **How much is the interest if $1000 is borrowed for one year at 12%.**

When using the interest formulas, you must express time in months as a fractional part of a year.

Months	Fractional Part of a Year
5	$\frac{5}{12}$
8	$\frac{8}{12}$ or $\frac{2}{3}$
30	$\frac{30}{12}$ or $\frac{5}{2}$

EXAMPLE 2

Mr. and Mrs. Diaz opened a savings account for \$1200. The account pays interest at a yearly rate of 7%. How much interest does the account earn in three months?

Solution: [1] Write the formula. ⟶ $i = prt$

[2] Identify known values. ⟶ $p = \$1200; r = 0.07, t = \frac{1}{4}$

[3] Substitute and multiply. ⟶ $i = (1200)(0.07)(\frac{1}{4})$

$i = 84 \cdot \frac{1}{4}$

$i = 21$

3 months is one quarter of a year.

The interest earned by the savings account is \$21.

EXAMPLE 3

Chris borrows \$44,000 from his credit union to open a cooking school. The yearly interest rate for the loan is $9\frac{1}{2}\%$. Compute the interest if he pays the loan back in five years.

Solution: [1] $i = prt$

[2] $p = 44{,}000;\ r = 0.095;\ t = 5$

[3] $i = (44{,}000)(0.095)(5)$

$i = 4180 \cdot 5$

$i = 20{,}900$

$9\frac{1}{2}\% = 0.095$

He pays \$20,900 in interest.

CLASSROOM EXERCISES

Replace the variables in $i = prt$ with the known values. (Step 3, Examples 1, 2, and 3)

1. Interest on \$850 at a rate of 8% for one year
2. Interest on \$1250 at a rate of 9% for one year
3. Interest on \$1956 at a yearly rate of $8\frac{1}{2}\%$ for 6 months
4. Interest on \$5280 at a yearly rate of 7.3% for 3 months
5. Interest on \$2575 borrowed for 2 years at a yearly rate of $9\frac{1}{2}\%$
6. Interest on \$835 borrowed for 18 months at a yearly rate of 12%

For Exercises 7–20, find the simple interest. (Example 1)

7. \$100 borrowed for one year at 5%
8. \$100 borrowed for one year at 6%
9. \$200 borrowed for one year at 6%
10. \$300 borrowed for one year at 7%
11. \$100 borrowed for one year at $5\frac{1}{2}$%
12. \$100 borrowed for one year at $6\frac{1}{2}$%

(Example 2)

13. \$100 borrowed for 3 months at 12%
14. \$200 borrowed for 6 months at 8%
15. \$1000 borrowed for 2 months at 12%
16. \$1000 borrowed for 4 months at 9%

(Example 3)

17. \$500 borrowed for 2 years at 5%
18. \$300 borrowed for 5 years at 6%
19. \$100 borrowed for 18 months at 10%
20. \$1000 borrowed for 30 months at 8%

WRITTEN EXERCISES

Goal: To use the formula $i = prt$ for computing simple interest

Sample Problem: Find the simple interest on \$900 borrowed for 3 years at a yearly rate of 8%.

Answer: \$216

For Exercises 1–18, find the simple interest. (Example 1)

1. \$500 borrowed for one year at 8%
2. \$800 borrowed for one year at 12%
3. \$100 borrowed for one year at $8\frac{1}{2}$%
4. \$200 borrowed for one year at $6\frac{1}{2}$%
5. \$200 borrowed for one year at 5.6%
6. \$300 borrowed for one year at 6.2%

(Example 2)

7. \$600 borrowed for 6 months at 5%
8. \$800 borrowed for 3 months at 6%
9. \$1200 borrowed for 10 months at 8%
10. \$1600 borrowed for 4 months at 9%
11. \$800 borrowed for 9 months at $6\frac{1}{2}$%
12. \$900 borrowed for 8 months at $5\frac{1}{2}$%

(Example 3)

13. \$500 borrowed for 2 years at 8%
14. \$1800 borrowed for 18 months at 12%
15. \$7500 borrowed for 3 years at 8.5%
16. \$10,000 borrowed for 2 years at 9.2%
17. \$12,400 borrowed for $2\frac{1}{2}$ years at $9\frac{1}{2}$%
18. \$3750 borrowed for $3\frac{1}{2}$ years at 10.4%

MIXED PRACTICE

19. Mr. and Mrs. Lightfoot borrow \$15,000 to start a business. The yearly interest rate is 11%. How much interest is due every 3 months?

20. Maria and Leslie borrow \$750 to operate a stand at the state fair. The yearly interest rate is 18%. How much is the interest for one month?

21. The Ohira family gets a loan to buy a car priced at \$9500. The yearly interest rate on the loan is 12% Find the interest due when they repay the loan after three years.

22. A club has \$2500 in a savings account. The account pays a yearly interest rate of 6.5%. How much interest will the account earn in a year?

MORE CHALLENGING EXERCISES

Solve each problem.

23. Jack paid \$18 interest on \$300 that he borrowed for one year. What is the yearly interest rate?

24. A bank pays \$65 in interest for each \$1000 invested for a year. What is the interest rate that the bank pays?

NON-ROUTINE PROBLEMS

25. Trace the figure below. Then divide the figure into four pieces that have the same size and shape.

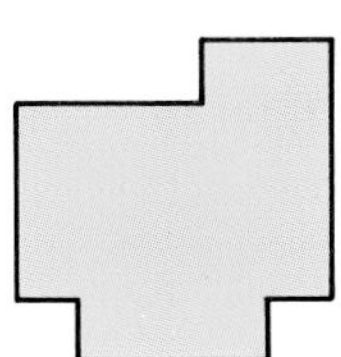

26. A jet leaves St. Louis for New York at the same time a twin-engine plane leaves New York for St. Louis. The jet travels at a speed of 600 miles per hour. The twin-engine plane travels at a speed of 300 miles per hour. Which is closer to St. Louis when they meet?

27. Four people are running in a marathon. Rob is 30 blocks behind Mary. Mary is 75 blocks ahead of Kathy. Kathy is 20 blocks behind Tim. How far behind Rob is Tim?

28. There are five oranges in a bag, and there are five people in a room. Is there any way to give each person an orange and also have one orange remain in the bag?

REVIEW CAPSULE FOR SECTION 7.9

Write each per cent in decimal form. (Pages 185–188)

1. 60%	2. 6.3%	3. 101%	4. 9%	5. 41.6%	6. 250%
7. 4%	8. 25%	9. 8.5%	10. 0.6%	11. x%	12. k%

7.9 Using Formulas: Discount

A ***discount*** is the amount that an article of merchandise is reduced in price.

The regular price of an article is the ***list price.*** The price after the discount is deducted is the ***net price*** or ***sale price.***

The list price of the tennis racket at the right is \$35.00. The sale price is \$27.50. The discount (difference between the list and sale prices) is \$7.50.

P–1 **What is the discount on a suit with a list price of \$130.00 and a sale price of \$95.00?**

Discounts are often expressed as a certain per cent of the list price. This per cent is called the ***rate of discount.***

The following word rule and formula show how to compute the amount of discount when the rate is known.

Word Rule: The amount of discount equals the rate of discount times the list price.

Formula: $d = rp$

d = amount of discount
r = rate of discount
p = list price

Below are some examples that show the relationship between list price, rate of discount, amount of discount, and sale price.

List Price	Rate of Discount	Discount	Net Price or Sale Price
\$50	15%	(0.15)(50) = \$7.50	\$50 − \$7.50 = \$42.50
\$400	25%	(0.25)(400) = \$100	\$400 − \$100 = \$300
\$500	20%	(0.20)(500) = \$100	\$500 − \$100 = \$400
\$900	$33\frac{1}{3}$%	$(\frac{1}{3})$(900) = \$300	\$900 − \$300 = \$600

EXAMPLE 1

Compute the discount on a radio if the rate of discount is 22% and the list price is $65.50.

Solution: First, estimate the discount.

22% is about 20%. $65.50 is about $70.

Estimated discount: $0.2 \times 70 = 14$

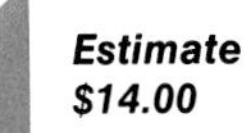

Estimate: $14.00

Find the actual amount of discount.

[1] Write the formula. ⟶ $d = rp$

[2] Identify known values. ⟶ $r = 0.22;\ p = 65.50$

[3] Substitute and multiply. ⟶ $d = (0.22)(65.50)$

$d = 14.41$

Since the estimate is $14.00, the answer is reasonable.

The amount of discount is $14.41.

Example 2 below involves finding the sale price of an item when the list price and rate of discount are known. The amount of discount must be computed before the sale price of the item can be found.

EXAMPLE 2

A calculator with a list price of $26.50 is on sale at a discount of 40%. Find the net price.

Solution: First, find the discount.

[1] Write the formula. ⟶ $d = rp$

[2] Identify known values. ⟶ $r = 0.40;\ p = 26.50$

[3] Substitute and multiply. ⟶ $d = (0.40)(26.50)$

$d = 10.60$

The amount of discount is $10.60.

[4] Find the net price. ⟶

List price	−	Discount	=	Net price
↓		↓		↓
$26.50	−	$10.60	=	$15.90

The net price of the calculator is $15.90.

Example 3 involves finding the rate of discount of an item whose list price and sale price are known. The amount of discount must be computed before the rate can be found.

EXAMPLE 3 A jacket with a list price of \$90 is on sale for \$63. Find the rate of discount.

Solution:

1. Find the amount of discount. ⟶ **List price − Sale price = Discount**

   \$90 − \$63 = \$27

2. Choose a variable. ⟶ Let $x\%$ = the rate of discount.
3. Write the formula. ⟶ $d = rp$
4. Identify known values. ⟶ $d = 27$; $r = 0.01x$; $p = 90$
5. Substitute and solve. ⟶ $27 = (0.01x)(90)$

$27 = (0.01 \cdot 90)x$

$27 = 0.9x$

$30 = x$ The rate of discount is 30%.

x% can be written as 0.01x.

CLASSROOM EXERCISES

Estimate the amount of discount on each item. Then find the actual amount. (Example 1)

1. A baseball glove listed at \$18.15 with a discount rate of 9%

2. A hat listed at \$12.39 with a discount rate of 24%

3. A \$59.65 dress on sale at a 21% discount

4. A refrigerator listed at \$589 with a 33% discount

Find the net price of each item to the nearest dollar. (Example 2)

5. A \$49.98 table with a 10% discount

6. A \$99.99 radio reduced by 30%

7. A coat listed at \$39.95 with a discount of 40%

8. An automobile listed at \$5995 with a 5% rebate (discount)

Find the rate of discount on each item. (Example 3)

9. Sneakers listed at \$25 and reduced by \$10

10. A \$360 color television set reduced by \$120

11. A chair listed at \$240 and discounted by \$60

12. A \$3200 sailboat with a \$400 discount

WRITTEN EXERCISES

Goal: To use $d = rp$ for solving word problems that involve discount

Sample Problem: Find the sale price of a camera listed at $60 if the rate of discount is 15%.

Answer: $51

Estimate the amount of discount on each item. Then find the actual amount. (Example 1)

1. A typewriter listed at $310 with a 38% discount
2. A $79.50 fishing rod reduced by 18%
3. A stereo set listed at $879 reduced by 34%
4. A $39.90 sweater with a discount of 9%

Find the net price of each item. (Example 2)

5. A house listed at $70,000 reduced by 20%
6. A $96 dress with a discount of $33\frac{1}{3}$%
7. A $576 motorbike with a $12\frac{1}{2}$% discount
8. A lamp listed at $42 reduced by 10%

Find the rate of discount on each item. (Example 3)

9. A basketball listed at $25 reduced by $7
10. A $95 table with a $19 discount
11. A record album listed at $39.50 discounted by $5.53
12. $49.50 warm–up suit reduced by $14.85

MIXED PRACTICE *Complete the table below.*

	List Price	Discount	Net Price	Rate of Discount
13.	$100	$15	?	?
14.	$125	?	$80	?
15.	?	$30	$370	?
16.	$729	?	?	$33\frac{1}{3}$%
17.	$195	?	?	15%
18.	?	$15	?	10%
19.	$259	$51.80	?	?
20.	?	$35	$175	?

Circle Graphs

Graphic artists create the art used in advertising, publications, packaging, television, statistical reports, and so on. A knowledge of proportion, per cent, and geometry is essential for artists who prepare graphs for advertising and for statistical reports.

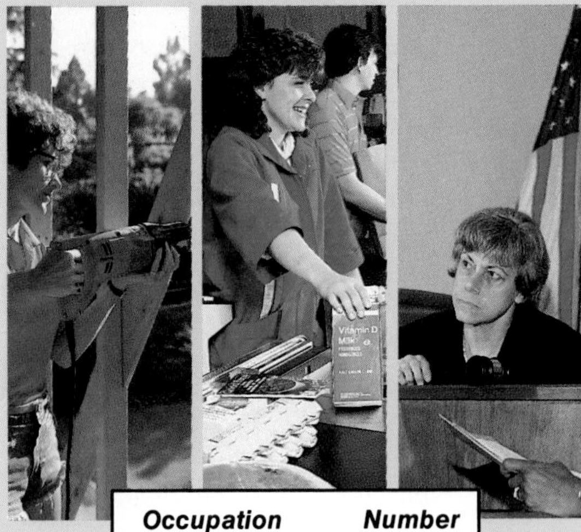

EXAMPLE: The table at the right shows how many non-farm workers were employed in four major occupations in Capital County. Make a circle graph to show the data.

Occupation	*Number*
Industry	4,575
Services	19,825
Government	9,550
Other	825

SOLUTION:

[1] Find what per cent of the total each occupation is. Round per cents to the nearest half.

Total: 4 5 7 5 [+] 1 9 8 2 5 [+] 9 5 5 0 [+] 8 2 5 [=] → 34775.

Industry: 4 5 7 5 [÷] 3 4 7 7 5 [%] → 13.156002 ⟶ 13.0%

Services: 1 9 8 2 5 [÷] 3 4 7 7 5 [%] → 57.009345 ⟶ 57.0%

Government: 9 5 5 0 [÷] 3 4 7 7 5 [%] → 27.462257 ⟶ 27.5%

Other: 8 2 5 [÷] 3 4 7 7 5 [%] → 2.3723939 ⟶ 2.5%

[2] For each occupation, find the number of angle degrees in the circle graph. There are 360° in a circle. Round to the nearest whole degree.

Industry: 3 6 0 [×] 1 3 [%] → 46.8 ⟶ 47°

Services: 3 6 0 [×] 5 7 [%] → 205.2 ⟶ 205°

Government: 3 6 0 [×] 2 7 . 5 [%] → [99.] → 99°

Other: 3 6 0 [×] 2 . 5 [%] → [9.] → 9°

[3] Use a protractor to draw the circle graph. Figures 1 and 2 show how to measure an angle of 47° followed by an angle of 99°. Figure 3 shows the completed graph.

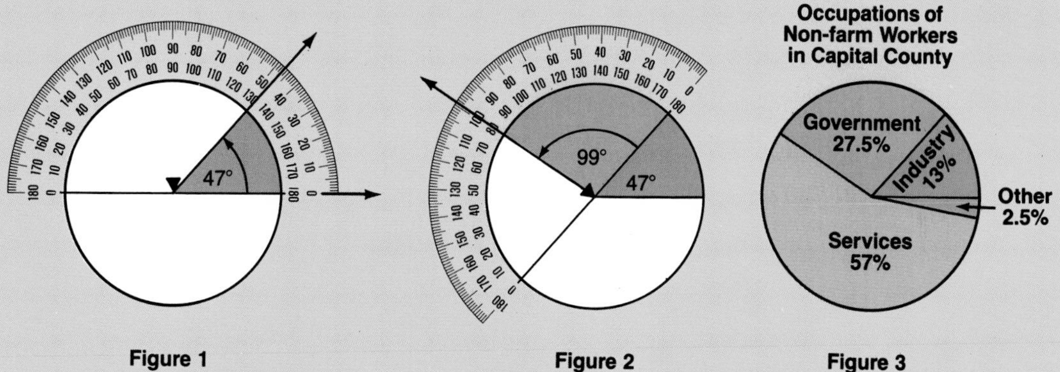

Figure 1 Figure 2 Figure 3

EXERCISES

For Exercises 1–2, draw circle graphs to show the given data.

1. ***How Time is Spent in Football Practice***

Warm-up	12 minutes
Team Drills	15 minutes
Special Drills	23 minutes
Scrimmage	40 minutes
Team Meeting	10 minutes

2. 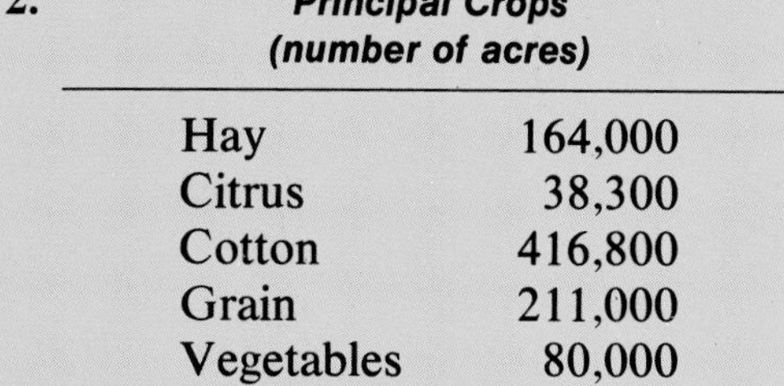

Principal Crops (number of acres)

Hay	164,000
Citrus	38,300
Cotton	416,800
Grain	211,000
Vegetables	80,000

For Exercises 3–6, refer to the circle graph at the right.

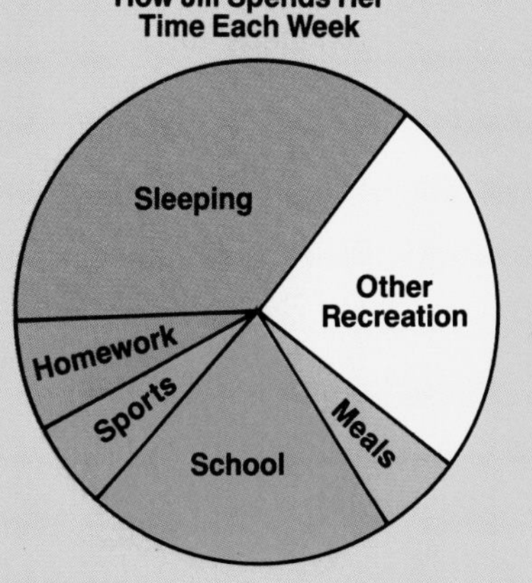

3. Estimate what per cent of her time Jill spends sleeping.
4. Estimate what per cent of her time Jill spends in school.
5. Estimate what per cent of her time Jill spends in sports and recreation.
6. Estimate the number of hours per week Jill spends on homework.

CHAPTER SUMMARY

IMPORTANT TERMS

Ratio *(p. 174)*
Terms of a ratio *(p. 174)*
Event *(p. 177)*
Favorable outcome *(p. 177)*
Possible outcome *(p. 177)*
Probability *(p. 177)*
Proportion *(p. 181)*
Cross product *(p. 181)*
Per cent *(p. 185)*
Cruising range *(p. 193)*
Wholesale price *(p. 194)*
Retail price *(p. 194)*
Interest *(p. 200)*
Principal *(p. 200)*
Interest rate *(p. 200)*
Discount *(p. 204)*
List price *(p. 204)*
Net or sale price *(p. 204)*
Rate of discount *(p. 204)*

IMPORTANT IDEAS

1. The second term of a ratio cannot be zero.
2. The probability of an event is the ratio of the number of favorable outcomes to the total number of possible outcomes of the event.
3. *Cross Products Property*
   a. If the cross products of a proportion are equal, it is a true proportion.
   b. If the cross products of a proportion are not equal, it is a false proportion.
4. A proportion is equivalent to the equation formed by its cross products.
5. If the first and second terms of a ratio are the same, the ratio equals 100%. If the first term of a ratio is greater than the second term, the ratio is greater than 100%.
6. To write a decimal as a per cent, move the decimal point two places to the right. Then write the per cent symbol.
7. To write a per cent as a decimal, move the decimal point two places to the left. Then write the decimal without the % symbol.
8. General form of per cent problems

   $$x\% \text{ of } y = z$$

9. 100% of a number is the number.
10. Steps for Solving Per Cent Problems by Proportion
    a. Write the problem in the general form.
    b. Use a variable to represent the unknown.
    c. Write a statement using 100%.
    d. Use the two statements to write a proportion.
    e. Solve the proportion.
11. Steps for Solving Per Cent Problems by Writing an Equation
    a. Write a per cent statement in general form for the problem.
    b. Use a variable to represent the unknown. Then write an equation for the statement.
    c. Solve the equation.

CHAPTER REVIEW

SECTION 7.1

Write a fraction in lowest terms for each ratio.

1. 10 to 80

2. 44 : 4

3. 18 to $42c$, $c \neq 0$

4. $24x : 30x$, $x \neq 0$

5. $\frac{14y}{24}$

6. $\frac{16}{24}$

SECTION 7.2

Solve each problem.

7. The six faces of a number cube are numbered from 1 to 6. Each face has a number on it. The cube is tossed once. Find the probability that the cube lands with an even number on the top face.

8. The eight regions of a spinner are numbered from 1 to 8. The pointer is equally likely to stop on any of the regions. Find the probability that the pointer will stop on a number greater than 2.

SECTION 7.3

Solve each proportion.

9. $\frac{5}{6} = \frac{p}{12}$

10. $\frac{x}{5} = \frac{28}{35}$

11. $\frac{12}{m} = \frac{3}{5}$

12. $\frac{1}{z} = \frac{13}{16}$

13. $\frac{5}{8} = \frac{c}{12}$

14. $\frac{16}{12} = \frac{12}{m}$

Solve each problem.

15. The ratio of the number of people moving into a town to the number of people moving out was 5 : 7 one year. 182 people moved away from the town that year. Find the number of people who moved into the town.

16. The ancient Greeks often built temples with a ratio of length to width of 5 : 3. Find the length of such a temple that is 33 meters wide.

SECTION 7.4

Write a per cent for each number.

17. $\frac{83}{100}$

18. $\frac{5}{12}$

19. $\frac{4}{7}$

20. 0.18

21. 3.09

22. 0.058

Write a decimal for each per cent.

23. 17%

24. 57%

25. 1%

26. 80.1%

27. 150%

28. 241.2%

SECTION 7.5

Write a proportion and solve.

29. 33% of 28 is what number?

30. Some number is 12% of 48.

31. $12 = r\%$ of 36.

32. 23 is a certain per cent of 276.

33. 3 is 15% of what number?

34. 1.2 is 0.8% of x.

SECTION 7.6

Solve each problem.

35. About 15,000 people attended a soccer match in a stadium that can hold 25,000 people. What per cent of the stadium was filled?

36. The members of the Elastic Band pay 15% of their earnings to their manager. Their earnings for one concert were $72,000. How much was the manager's percentage?

SECTION 7.7

Solve each problem.

37. The Hernandez family's heating bill one month was $41.50. The oil company added a surcharge of 1.8%. How much was the surcharge? Round your answer to the nearest cent.

38. The world's total oil reserves were recently estimated to be 642 billion barrels. The United States has an estimated reserve of 28 billion barrels. What per cent of the world's reserves is this? Round your answer to the nearest per cent.

SECTIONS 7.8 AND 7.9

Solve each problem.

39. The yearly interest rate on a loan of $742 is 18%. Find the interest due after one month.

40. Customers who pay their bills promptly receive a $2\frac{1}{2}$% discount for a certain company. Find the net amount due if a bill of $290 is paid promptly.

CHAPTER 8

Real Numbers

At temperatures far below zero, most materials become brittle and can be shattered easily. A carnation that has been frozen in liquefied nitrogen at a temperature of approximately 315°F below zero shatters when it is dropped.

8.1 Integers

The set of whole numbers is represented on this number line.

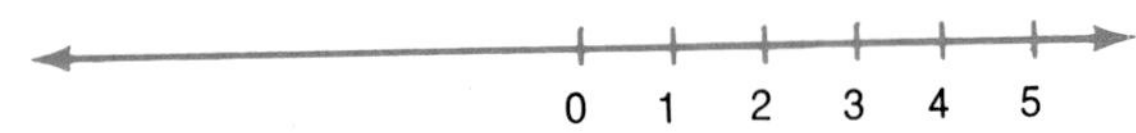

Whole Numbers: {0, 1, 2, 3, 4, 5, · · ·}

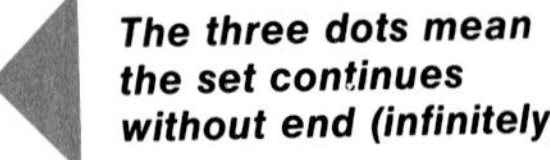

The three dots mean the set continues without end (infinitely).

The whole numbers to the right of zero are ***positive integers.***

Positive Integers: {1, 2, 3, 4, 5, · · ·}

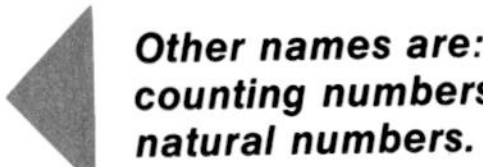

The points labeled 0 and 1 determine the length of a unit. On the following number line, this unit is used to mark points to the left of 0.

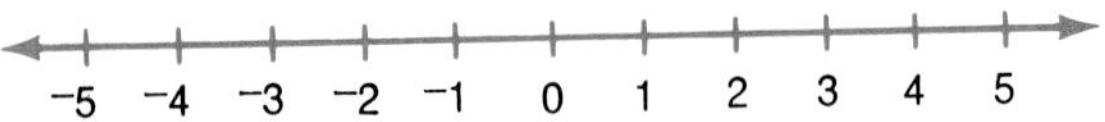

The first point to the left of zero is labeled $^-1$ and is read "negative one." The next point is $^-2$ ("negative two"), the next is $^-3$ ("negative three"), and so on. These are ***negative integers.***

Negative Integers: {· · ·, $^-5$, $^-4$, $^-3$, $^-2$, $^-1$}

Definition

> The set of ***integers*** is made up of the positive integers, zero, and the negative integers.
>
> **Integers:** {· · ·, $^-3$, $^-2$, $^-1$, 0, 1, 2, 3, · · ·}

To compare integers, it may help to graph them.

EXAMPLE 1 **a.** Graph: 5, $^-2$, 0, $^-10$, $^-1$

b. Write the numbers in order from smallest to largest.

Solutions: **a.** Each dot is the ***graph*** of a number. The number paired with each point is its ***coordinate.***

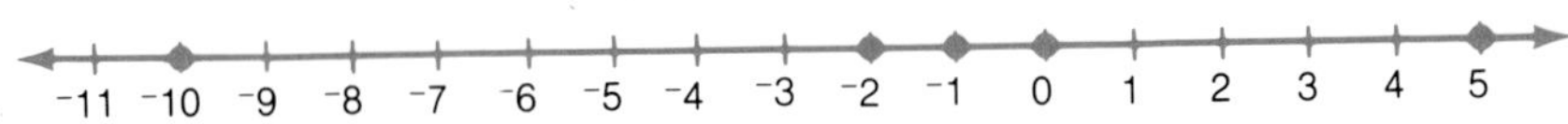

b. Write the integers from left to right: $^-10$, $^-2$, $^-1$, 0, 5

P–1 **Write in order from smallest to largest.**

a. $^{-}6$, $^{-}7$, $^{-}9$, 2, $^{-}1$ **b.** 11, 9, $^{-}12$, $^{-}6$, 0

Integers are used in several ways in everyday situations, as the table below shows. The key words that mean positive or negative are underlined.

Table

In Words	Integers	In Words	Integers
33° below 0	$^{-}33$	112 meters above sea level	112
15° above 0	15	62 meters below sea level	$^{-}62$
A loss of 3 pounds	$^{-}3$	A deposit of $85	85
A gain of 5 yards	5	A withdrawal of $44	$^{-}44$

P–2 **Write an integer for each word description.**

a. An increase of $175 **b.** A drop of 500 meters

c. 16 fathoms below the surface **d.** A growth of 19 inches

Definition

The symbols > and < are used to compare numbers.

">" means "is greater than." "<" means "is less than."

Statements that use > or < are ***inequalities.***

EXAMPLE 2

Write two inequalities to compare $^{-}10$ and $^{-}2$.

Solution:

Think: $^{-}10$ is to the left of $^{-}2$ on a number line. **Answer:** $^{-}10 < {}^{-}2$

Think: $^{-}2$ is to the right of $^{-}10$ on a number line. **Answer:** $^{-}2 > {}^{-}10$

Note that the inequality symbol always points toward the smaller number.

P–3 **Write two inequalities for each of the following.**

a. $^{-}7$ and $^{-}17$ **b.** 8 and $^{-}1$ **c.** $^{-}5$ and 0 **d.** $^{-}6$ and $^{-}5$

CLASSROOM EXERCISES

For Exercises 1–9, graph the numbers. (Example 1)

1. 0, 3, -3, 6
2. -6, 2, 4, -4
3. 0, 3, -2, 1
4. 3, 1, -2, 2
5. -4, 8, 1, 3, -2
6. 5, -2, 1, -1, 0
7. 2, 10, 6, -4, -7
8. 10, -1, 2, 0, -5
9. 11, 8, -7, 6, 5

For Exercises 10–18, write the numbers in order from smallest to largest. (Example 1)

10. -3, 2, 7, 1, 3
11. 4, 8, -7, 3, 0
12. 2, 0, 11, -8, -1
13. 12, -15, 3, 0
14. 6, 12, -12, 0, 13
15. 11, 7, -9, 8, -14
16. 13, 6, -6, -8, 4
17. 12, -9, 10, -6, 0
18. 14, -24, -2, 3, 15

Write an integer for each word description. (Table)

19. A gain of 10 yards
20. A deposit of $93
21. 25 feet below sea level
22. A loss of 75 pounds
23. 37° above 0
24. A drop of 14°

Write two inequalities for each pair of numbers. (Example 2)

25. -6 and -1
26. 3 and -5
27. -4 and 8
28. 0 and -6
29. -10 and 10
30. 4 and 0
31. -21 and -12
32. 11 and -17

WRITTEN EXERCISES

Goal: To graph and compare integers

Sample Problems: **a.** Graph: 2, 5, -1, 4, -4. Write the numbers in order from smallest to largest.

b. Write two inequalities to compare -9 and 8.

Answers: **a.**

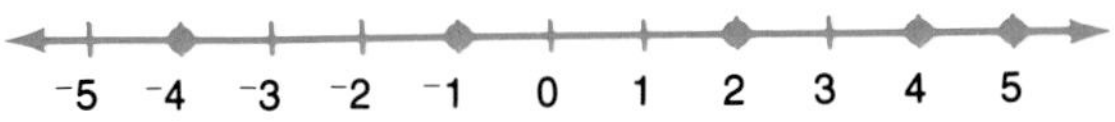

; -4, -1, 2, 4, 5

b. $-9 < 8$; $8 > -9$

For Exercises 1–18, graph the numbers. Then write the numbers in order from smallest to largest. (Example 1)

1. 2, 4, -3, 6, -1
2. 8, -10, 2, 0, -7
3. -9, 3, 7, -5, 5
4. 1, -4, -1, 0, 2
5. 7, -4, -3, 2, 0
6. 11, -12, 9, 8, -3
7. -4, 6, -6, 8, 2
8. 0, -1, 3, -5, 9
9. 7, 2, -3, -4, -7

10. $^-5, 8, ^-9, 12, ^-3$
11. $5, ^-6, 7, ^-8, 9$
12. $^-2, ^-1, ^-3, ^-10, 8$
13. $1, ^-3, 7, ^-6, ^-5$
14. $2, 4, 8, ^-9, ^-3$
15. $^-3, ^-7, ^-2, 0, ^-10$
16. $13, ^-2, ^-9, 0, 9, 12$
17. $4, 11, ^-1, 7, ^-7, 8$
18. $0, ^-8, 3, 2, ^-5, 10$

Write an integer for each word description. (Table)

19. A growth of 2 inches
20. A loss of $16
21. A raise of $1000
22. A drop of 200 meters
23. A fall of 105 feet
24. 5 fathoms below sea level
25. 17 floors above ground
26. A deposit of $253
27. 7° below 0
28. A $15 decrease
29. A gain of 17 kilograms
30. 19° above 0

Write two inequalities for each pair of numbers. (Example 2)

31. $^-11$ and 4
32. 9 and $^-11$
33. 2 and $^-12$
34. 16 and $^-3$
35. $^-3$ and $^-7$
36. $^-14$ and $^-1$
37. $^-9$ and $^-6$
38. $^-15$ and $^-24$
39. 0 and 6
40. $^-3$ and 0
41. 0 and $^-14$
42. 11 and 0
43. $^-7$ and $^-2$
44. 18 and $^-9$
45. $^-13$ and 23
46. $^-19$ and $^-21$
47. $^-21$ and 21
48. $^-37$ and $^-45$
49. $^-72$ and 27
50. $^-92$ and $^-95$

NON-ROUTINE PROBLEMS

51. Alvin had $1.19 in change. None of the coins was a dollar. Louise asked him for change for a dollar but he did not have the correct change. What coins did he have?

52. Alicia Robbins drove 70 kilometers per hour on a trip. She arrived at her destination one hour earlier than she would have if she had driven 60 kilometers per hour. How far did she drive?

REVIEW CAPSULE FOR SECTION 8.2

Write a mixed number for each fraction.

1. $\frac{11}{3}$
2. $\frac{33}{5}$
3. $\frac{21}{2}$
4. $\frac{19}{9}$
5. $\frac{31}{8}$
6. $\frac{16}{7}$
7. $\frac{39}{10}$
8. $\frac{55}{6}$
9. $\frac{250}{100}$
10. $\frac{834}{400}$

Write a mixed number for each decimal.

11. 1.7
12. 2.3
13. 3.9
14. 2.5
15. 8.1
16. 4.1
17. 6.4
18. 10.8
19. 13.02
20. 4.06

8.2 Rational Numbers

Note that 3 and -3 are the same distance from 0 on a number line.

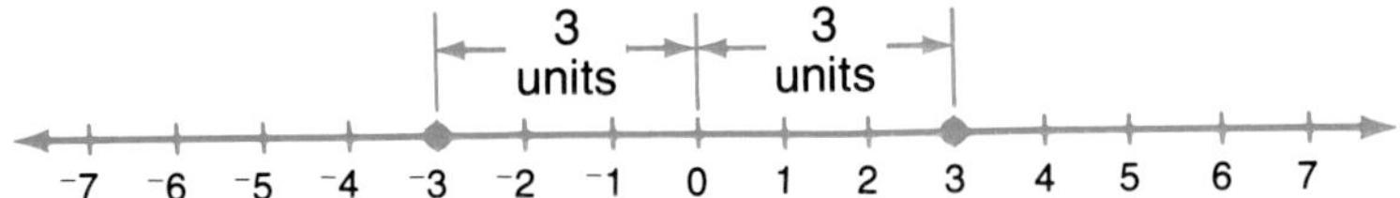

Therefore, 3 and -3 are <u>opposites</u>. The number 0 is its own opposite.

Definition

> Two numbers that are the same distance from 0 <u>and</u> are in opposite directions from 0 are ***opposites.***
>
> 3 is the opposite of -3. $\quad$ -3 is the opposite of 3.

A dash, $-$, is used to represent "the opposite of."

$3 = -(^-3)$  ***Read: "3 equals the opposite of negative 3."***

$^-3 = -3$ ◀ ***Read: "Negative 3 equals the opposite of 3."***

Since $^-3 = -3$, the symbol for "the opposite of" can also be used to indicate a negative number. Thus, -3 can be read "negative 3." To simplify numerals that involve opposites, it may be helpful to say "the opposite of" in certain cases. Also, think of a given problem as a question, as shown in the Example.

EXAMPLE

Simplify: **a.** $-(7)$ $\quad$ **b.** $-(-5)$

Solutions: **a. Think:** What is the <u>opposite of 7?</u> $\quad$ **Answer:** -7

b. Think: What is the <u>opposite of negative 5?</u> $\quad$ **Answer:** 5

P–1 **Simplify.**

a. $-(11)$ $\quad$ **b.** $-(0)$ $\quad$ **c.** $-(-24)$ $\quad$ **d.** $-(-35)$

Clearly, there are points on a number line that are between the points having integers as coordinates. For example, the point halfway between the points labeled 0 and 1 has the coordinate $\frac{1}{2}$. All the points to the right of 0 that can be labeled with fractions represent the ***positive rational numbers.***

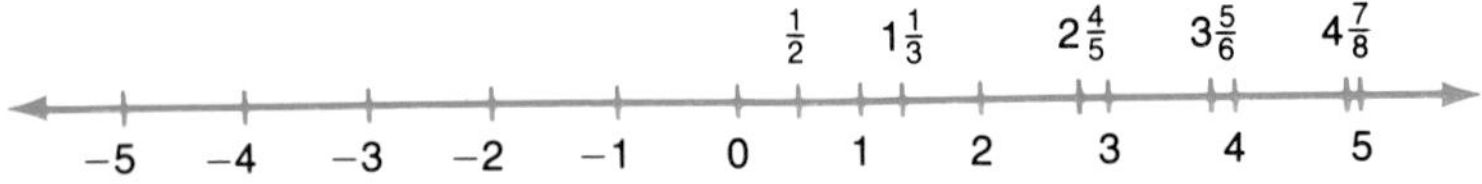

Mixed numbers and whole numbers can be written as fractions. Thus, mixed numbers and whole numbers are rational numbers.

P–2 **Write a fraction for each of the following.**

a. $1\frac{3}{4}$ **b.** 2.5 **c.** 4 **d.** 0

Each positive rational number can be paired with its opposite, a ***negative rational number.***

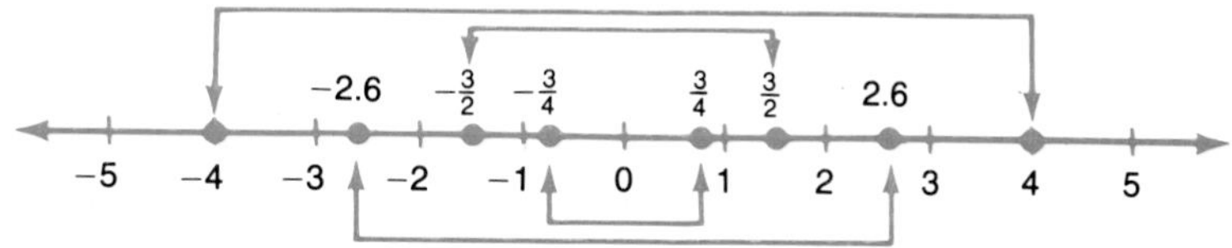

Definition

> The set of ***rational numbers*** consists of the negative rational numbers, zero, and the positive rational numbers.

You can use rational numbers and opposites to represent everyday situations.

Table

In Words	In Symbols
5° below 0 is the opposite of 5° above 0.	$-5 = -(5)$
A gain of $6\frac{1}{2}$ pounds is the opposite of a loss of $6\frac{1}{2}$ pounds.	$6\frac{1}{2} = -(-6\frac{1}{2})$
The opposite of a withdrawal of \$10.75 is a deposit of \$10.75.	$-(-10.75) = 10.75$
The opposite of an enlargement of 3.2 centimeters is a reduction of 3.2 centimeters.	$-(3.2) = -3.2$

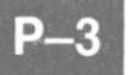

P–3 **Use symbols to represent each sentence.**

a. A descent of 4 meters is the opposite of a rise of 4 meters.

b. The opposite of a gain of 2.6 seconds is a loss of 2.6 seconds.

CLASSROOM EXERCISES

Graph each number and its opposite.

1. 5 **2.** −6 **3.** $3\frac{1}{3}$ **4.** −1 **5.** 2.3

6. −8 **7.** −2.6 **8.** $-2\frac{2}{3}$ **9.** $\frac{3}{4}$ **10.** $1\frac{1}{2}$

Simplify. (Example)

11. $-(12)$ **12.** $-(2.7)$ **13.** $-(0)$ **14.** $-(\frac{7}{10})$ **15.** $-(\frac{3}{4})$

16. $-(-1.5)$ **17.** $-(-34)$ **18.** $-(-\frac{1}{2})$ **19.** $-(-9)$ **20.** $-(2\frac{1}{5})$

Use symbols to represent each sentence. (Table)

21. Depositing $36.55 in a checking account is the opposite of withdrawing $36.55.

22. The opposite of taking an elevator down 7 floors is taking an elevator up 7 floors.

23. A fall of 8° in temperature is the opposite of a rise of 8°.

24. The opposite of a profit of $3500 is a loss of $3500.

WRITTEN EXERCISES

Goal: a. To simplify numerals involving opposites
b. To use symbols to represent a word sentence

Sample Problems: a. Simplify $-(-30)$. **Answer:** 30
b. 45° below 0 is the opposite of 45° above 0.
Answer: $-45 = -(45)$

Simplify. (Example)

1. $-(9)$ **2.** $-(6.2)$ **3.** $-(-\frac{1}{4})$ **4.** $-(-41)$ **5.** $-(-5.3)$

6. $-(2\frac{1}{3})$ **7.** $-(-10)$ **8.** $-(17)$ **9.** $-(-3\frac{1}{5})$ **10.** $-(10.4)$

11. $-(0)$ **12.** $-(1.4)$ **13.** $-(2\frac{3}{8})$ **14.** $-(17.9)$ **15.** $-(-\frac{7}{4})$

Use symbols to represent each sentence. (Table)

16. An altitude of 20.5 meters above sea level is the opposite of a depth of 20.5 meters below sea level.

17. An investment loss of $$1\frac{1}{4}$ per share is the opposite of an investment gain of $$1\frac{1}{4}$ per share.

18. The opposite of owing $100 is saving $100.

19. The opposite of taking 5 steps forward is taking 5 steps backward.

20. The opposite of losing 10 minutes is gaining 10 minutes.

21. The opposite of climbing 500 feet is descending 500 feet.

REVIEW CAPSULE FOR SECTION 8.3

Write a decimal for each rational number.

1. $\frac{5}{8}$ **2.** $\frac{3}{5}$ **3.** 2 **4.** $1\frac{1}{4}$ **5.** $\frac{7}{16}$ **6.** $\frac{11}{12}$

7. $3\frac{11}{16}$ **8.** $6\frac{3}{4}$ **9.** $\frac{9}{20}$ **10.** $3\frac{1}{2}$ **11.** $\frac{3}{10}$ **12.** $\frac{2}{25}$

8.3 Decimal Forms of Rational Numbers

You can use this table to **find** the decimal equivalents for certain fractions and their opposites.

Table

Fraction	Decimal	Fraction	Decimal
$\frac{1}{2}$	0.5	$\frac{1}{8}$	0.125
$\frac{1}{4}$	0.25	$\frac{3}{8}$	0.375
$\frac{3}{4}$	0.75	$\frac{5}{8}$	0.625
$\frac{1}{5}$	0.2	$\frac{7}{8}$	0.875
$\frac{2}{5}$	0.4	$\frac{1}{16}$	0.0625
$\frac{3}{5}$	0.6	$\frac{3}{16}$	0.1875
$\frac{4}{5}$	0.8	$\frac{5}{16}$	0.3125

Since $\frac{1}{2}$ and $-\frac{1}{2}$ are opposites, and $\frac{1}{2} = 0.5$, it follows that

$$-\frac{1}{2} = -0.5.$$

P–1 **Write a decimal for each of the following.**

a. $-\frac{1}{5}$ **b.** $-\frac{1}{8}$ **c.** $-\frac{3}{5}$ **d.** $-\frac{5}{16}$

To write a decimal for a negative fraction, first find the decimal for its opposite.

EXAMPLE 1 Write a decimal for $-\frac{9}{16}$.

Solution: Write a decimal for $\frac{9}{16}$. ⟶ $16\overline{)9.0000}$ with quotient 0.5625

Since $\frac{9}{16} = 0.5625$, $-\frac{9}{16} = -0.5625$.

P–2 **Write a decimal for each of the following.**

a. $-\frac{1}{25}$ **b.** $-\frac{3}{20}$ **c.** $-\frac{23}{80}$ **d.** $-\frac{7}{20}$

EXAMPLE 2 Write a decimal for $-3\frac{7}{8}$.

Solution: 1 Write a decimal for $\frac{7}{8}$. ⟶ $\frac{7}{8} = 8\overline{)7.000}$ with quotient 0.875

2 Write a decimal for $3\frac{7}{8}$. ⟶ $3\frac{7}{8} = 3.875$

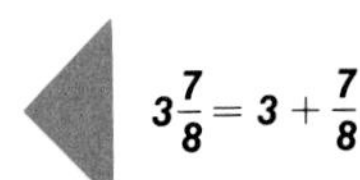
$3\frac{7}{8} = 3 + \frac{7}{8}$

Since $-3\frac{7}{8}$ is the opposite of $3\frac{7}{8}$, $-3\frac{7}{8} = -3.875$.

P–3 **Write a decimal for each of the following.**

a. $-2\frac{2}{5}$ **b.** $5\frac{3}{16}$ **c.** $8\frac{1}{12}$ **d.** $-6\frac{5}{8}$

For many fractions, the decimal equivalent repeats without end. The three dots mean the number pattern continues without end.

$\frac{1}{3} = 0.333 \cdots$ $\frac{2}{3} = 0.666 \cdots$

$\frac{1}{9} = 0.111 \cdots$, or $0.\overline{1}$ $\frac{1}{11} = 0.0909 \cdots$, or $0.\overline{09}$

The bar indicates the repeating pattern.

P–4 **Write the decimal equivalents with a bar.**

a. $\frac{1}{3} = 0.333 \cdots$ **b.** $\frac{13}{33} = 0.3939 \cdots$ **c.** $\frac{2}{7} = 0.285714285714 \cdots$

The decimal equivalents with repeating patterns are ***infinite repeating decimals.*** "Infinite" means without end.

EXAMPLE 3 Write a decimal for $\frac{9}{11}$.

Solution: $\frac{9}{11}$ ⟶

```
     0.8181
11 ) 9.0000 · · ·
     8 8
       20
       11
        90
        88
         20
         11
          9
```

$\frac{9}{11} = 0.8181 \cdots$, or $\frac{9}{11} = 0.\overline{81}$

P–5 **Write a decimal for each of the following.**

a. $\frac{5}{33}$ **b.** $-\frac{5}{6}$ **c.** $\frac{4}{9}$ **d.** $\frac{6}{7}$ **e.** $-\frac{1}{11}$

The decimals in the table on page 221 do not appear to have repeating patterns. However, they can be written as infinite repeating decimals.

$$\frac{3}{4} = 0.75 = 0.75000 \cdots \qquad \frac{3}{16} = 0.1875 = 0.1875000 \cdots$$

Every rational number can be written as an infinite repeating decimal.	$-14 = -14.000 \cdots$ $\frac{1}{3} = 0.333 \cdots$

CLASSROOM EXERCISES

For Exercises 1–48, write a decimal for each rational number. (Table)

1. $\frac{1}{16}$ **2.** $\frac{3}{4}$ **3.** $\frac{7}{8}$ **4.** $\frac{1}{2}$ **5.** $\frac{3}{8}$ **6.** $\frac{4}{5}$
7. $\frac{1}{5}$ **8.** $\frac{3}{16}$ **9.** $\frac{2}{5}$ **10.** $\frac{5}{8}$ **11.** $\frac{1}{4}$ **12.** $\frac{5}{16}$

(Example 1)

13. $-\frac{3}{20}$ **14.** $-\frac{1}{50}$ **15.** $-\frac{9}{10}$ **16.** $-\frac{4}{25}$ **17.** $-\frac{7}{16}$ **18.** $-\frac{7}{10}$
19. $-\frac{17}{25}$ **20.** $-\frac{3}{40}$ **21.** $-\frac{19}{50}$ **22.** $-\frac{1}{10}$ **23.** $-\frac{7}{40}$ **24.** $-\frac{3}{80}$

(Example 2)

25. $-4\frac{1}{8}$ **26.** $5\frac{5}{16}$ **27.** $-8\frac{1}{4}$ **28.** $-1\frac{1}{2}$ **29.** $2\frac{4}{5}$ **30.** $-3\frac{1}{50}$
31. $7\frac{1}{40}$ **32.** $6\frac{3}{25}$ **33.** $-10\frac{7}{10}$ **34.** $5\frac{9}{16}$ **35.** $11\frac{4}{25}$ **36.** $-9\frac{11}{16}$

(Example 3)

37. $\frac{3}{7}$ **38.** $\frac{4}{15}$ **39.** $\frac{5}{14}$ **40.** $-\frac{3}{11}$ **41.** $\frac{4}{9}$ **42.** $\frac{5}{6}$
43. $-\frac{1}{15}$ **44.** $\frac{5}{12}$ **45.** $-\frac{6}{7}$ **46.** $-\frac{5}{11}$ **47.** $\frac{13}{33}$ **48.** $\frac{2}{27}$

WRITTEN EXERCISES

Goal: To write a decimal for a rational number
Sample Problem: $-4\frac{3}{8}$
Answer: -4.375

For Exercises 1–48, write a decimal for each rational number.
(Table and Example 1)

1. $-\frac{8}{25}$ **2.** $\frac{5}{8}$ **3.** $\frac{9}{40}$ **4.** $-\frac{3}{20}$ **5.** $\frac{21}{50}$ **6.** $\frac{11}{20}$
7. $-\frac{47}{80}$ **8.** $\frac{13}{25}$ **9.** $-\frac{3}{10}$ **10.** $-\frac{17}{20}$ **11.** $\frac{49}{50}$ **12.** $-\frac{15}{16}$

(Example 2)

13. $-3\frac{1}{4}$ 14. $7\frac{1}{2}$ 15. $-9\frac{1}{8}$ 16. $-12\frac{3}{5}$ 17. $1\frac{15}{16}$ 18. $-5\frac{1}{10}$

19. $6\frac{9}{25}$ 20. $4\frac{5}{16}$ 21. $3\frac{7}{40}$ 22. $8\frac{1}{25}$ 23. $-9\frac{13}{20}$ 24. $-3\frac{7}{8}$

(Example 3)

25. $\frac{1}{6}$ 26. $\frac{5}{12}$ 27. $-\frac{11}{15}$ 28. $\frac{23}{33}$ 29. $-\frac{3}{7}$ 30. $3\frac{2}{11}$

31. $\frac{2}{9}$ 32. $-\frac{6}{11}$ 33. $1\frac{5}{9}$ 34. $-\frac{7}{12}$ 35. $\frac{8}{15}$ 36. $\frac{41}{99}$

MIXED PRACTICE

37. $-\frac{11}{16}$ 38. $7\frac{3}{4}$ 39. $-5\frac{9}{16}$ 40. $1\frac{7}{9}$ 41. $-\frac{10}{11}$ 42. $\frac{13}{20}$

43. $\frac{3}{25}$ 44. $-\frac{7}{10}$ 45. $7\frac{13}{20}$ 46. $-5\frac{3}{8}$ 47. $2\frac{4}{7}$ 48. $\frac{2}{33}$

MID-CHAPTER REVIEW

For Exercises 1–6, graph the numbers. Then write the numbers in order from smallest to largest. (Section 8.1)

1. −8, 0, 8, 4, −4 2. 1, −1, 6, −3, −2 3. 0, 2, −4, −8, −6

4. 7, −3, 4, 8, 1 5. −10, −6, −12, −14, −2 6. 5, −4, −5, −1, 2

Write two inequalities for each pair of numbers. (Section 8.1)

7. −3 and 3 8. 1 and −6 9. 0 and −4 10. −5 and −6

11. −86 and −88 12. 9 and −10 13. −12 and 21 14. −15 and 0

Write an integer for each word description. (Section 8.1)

15. A drop of 1700 meters 16. A deposit of $713.80

17. 5° above 0 18. A loss of $2500

Simplify. (Section 8.2)

19. $-(-10)$ 20. $-(28)$ 21. $-(-\frac{3}{5})$ 22. $-(6.9)$ 23. $-(-\frac{9}{2})$

24. $-(0)$ 25. $-(-1)$ 26. $-(8\frac{1}{3})$ 27. $-(-12\frac{9}{10})$ 28. $-(-0.01)$

Write a decimal for each rational number. (Section 8.3)

29. $-\frac{3}{5}$ 30. $\frac{7}{16}$ 31. $-\frac{15}{22}$ 32. $-1\frac{3}{8}$ 33. $17\frac{1}{4}$ 34. $\frac{11}{12}$

REVIEW CAPSULE FOR SECTION 8.4

Write a fraction for each decimal. (Pages 185–188)

1. 0.03 2. 0.07 3. 0.65 4. 0.72 5. 1.14 6. 3.08

7. 0.01 8. 0.04 9. 0.09 10. 0.16 11. 0.25 12. 0.36

8.4 Square Root

Recall that to square a number, you multiply it by itself.

EXAMPLE 1 Square each of the following.

a. 24 **b.** 0.7 **c.** $1\frac{1}{8}$

Solutions: **a.** $24^2 = 24 \cdot 24$

$= 576$

b. $(0.7)^2 = 0.7 \cdot 0.7$

$= 0.49$

c. $(1\frac{1}{8})^2 = \left(\frac{9}{8}\right)^2$

$= \frac{9}{8} \cdot \frac{9}{8}$

$= \frac{81}{64} = 1\frac{17}{64}$

P–1 **Square each of the following.**

a. 1 **b.** 12 **c.** 0.3 **d.** $2\frac{3}{4}$

The inverse operation of squaring a number is taking a ***square root.*** Since $8 \cdot 8 = 64$, it is correct to say that 8 is a square root of 64. In symbols,

$$\sqrt{64} = 8$$

Read: "The positive square root of 64 equals 8."

The expression $\sqrt{64}$ is a ***radical.*** The $\sqrt{\ }$ is a ***radical symbol,*** and it represents a positive square root. The number under the radical symbol, 64, is the ***radicand.***

Every positive square root can be paired with its opposite.

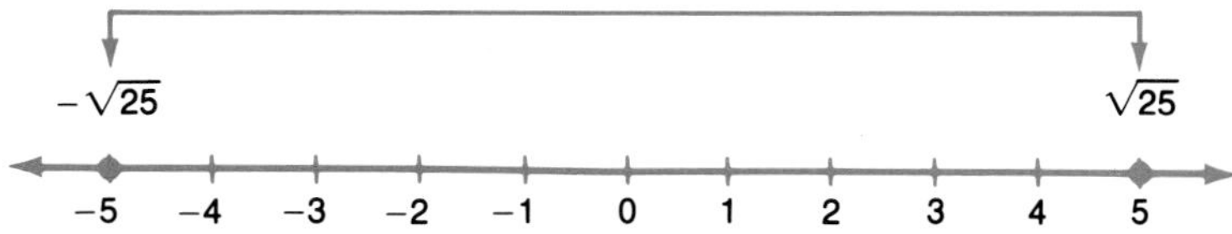

A negative square root is represented by $-\sqrt{\ }$.

$$-\sqrt{25} = -5$$

In the **Table of Squares** at the top of the next page, you will find the squares of the integers 0 through 25. Also, each number is the positive square root of its square. Reading from right to left, the positive square root of 441 is 21.

Table of Squares	Number	Square	Number	Square
	0	0	13	169
	1	1	14	196
	2	4	15	225
	3	9	16	256
	4	16	17	289
	5	25	18	324
	6	36	19	361
	7	49	20	400
	8	64	21 ← 441	441
	9	81	22	484
	10	100	23	529
	11 → 121	121	24	576
	12	144	25	625

$11^2 = 121$

$\sqrt{441} = 21$

P–2 **Use the Table of Squares to find each of the following.**

a. 14^2 **b.** 23^2 **c.** $\sqrt{484}$ **d.** $\sqrt{169}$

EXAMPLE 2

Use the Table of Squares to find $\sqrt{\frac{25}{49}}$.

Solution: Use the Table to write two equal factors for the radicand. ⟶ $\frac{25}{49} = \frac{5}{7} \cdot \frac{5}{7}$

Therefore, $\sqrt{\frac{25}{49}} = \frac{5}{7}$.

P–3 **Use the Table of Squares to find each square root.**

a. $\sqrt{\frac{64}{81}}$ **b.** $\sqrt{\frac{144}{36}}$ **c.** $\sqrt{\frac{4}{196}}$ **d.** $\sqrt{\frac{0}{529}}$

You can also use the Table to find a negative square root.

EXAMPLE 3

Use the Table of Squares to find:

a. $-\sqrt{100}$ **b.** $-\sqrt{\frac{16}{36}}$

Solutions: **a.** Since $\sqrt{100} = 10$, $-\sqrt{100} = -10$.

b. Since $\sqrt{\frac{16}{36}} = \frac{4}{6}$ or $\frac{2}{3}$, $-\sqrt{\frac{16}{36}} = -\frac{2}{3}$.

P–4 **Use the Table of Squares to find each square root.**

a. $-\sqrt{9}$ **b.** $-\sqrt{256}$ **c.** $-\sqrt{\frac{4}{25}}$ **d.** $-\sqrt{\frac{225}{169}}$

EXAMPLE 4 Use the Table of Squares to find $-\sqrt{0.81}$.

Solution: [1] Write a fraction for the radicand. ⟶ $0.81 = \frac{81}{100}$

[2] Use the Table to write two equal factors for the radicand. ⟶ $\frac{81}{100} = \frac{9}{10} \cdot \frac{9}{10}$

Since $\sqrt{0.81} = \frac{9}{10}$ or 0.9, $-\sqrt{0.81} = -0.9$

P–5 **Use the Table of Squares to find each square root.**

a. $-\sqrt{0.16}$ **b.** $-\sqrt{0.49}$ **c.** $-\sqrt{1.44}$ **d.** $-\sqrt{5.76}$

CLASSROOM EXERCISES

Square each of the following. (Example 1)

1. 5.2 **2.** 1.4 **3.** 16 **4.** 25 **5.** 0.1 **6.** 0.3
7. 0.15 **8.** 0.12 **9.** $1\frac{1}{2}$ **10.** $2\frac{2}{3}$ **11.** $\frac{1}{2}$ **12.** $\frac{3}{4}$

Find each of the following. (Table)

13. 16^2 **14.** 24^2 **15.** 0^2 **16.** 12^2 **17.** 21^2 **18.** 13^2
19. 18^2 **20.** 17^2 **21.** 19^2 **22.** 22^2 **23.** 7^2 **24.** 15^2

Write two equal factors for each of the following. (Example 2)

25. $\frac{4}{9}$ **26.** $\frac{1}{36}$ **27.** $\frac{1}{25}$ **28.** $\frac{9}{16}$ **29.** $\frac{36}{49}$ **30.** $\frac{25}{144}$
31. $\frac{16}{121}$ **32.** $\frac{9}{169}$ **33.** $\frac{4}{225}$ **34.** $\frac{100}{289}$ **35.** $\frac{25}{81}$ **36.** $\frac{49}{81}$

Find each square root. (Example 3)

37. $\sqrt{144}$ **38.** $\sqrt{324}$ **39.** $-\sqrt{81}$ **40.** $-\sqrt{49}$ **41.** $\sqrt{576}$ **42.** $\sqrt{289}$
43. $-\sqrt{169}$ **44.** $-\sqrt{9}$ **45.** $\sqrt{400}$ **46.** $-\sqrt{400}$ **47.** $\sqrt{\frac{16}{25}}$ **48.** $-\sqrt{\frac{16}{25}}$

Write two equal fractions as factors of each number. (Step 2, Example 4)

49. 0.36 **50.** 1.21 **51.** 0.09 **52.** 0.04 **53.** 0.16 **54.** 0.25
55. 1.44 **56.** 2.25 **57.** 5.76 **58.** 0.01 **59.** 0.64 **60.** 5.29

WRITTEN EXERCISES

Goal: To find the squares and square roots of rational numbers
Sample Problems: Evaluate: **a.** $(1\frac{2}{3})^2$ **b.** $-\sqrt{3.24}$
Answers: a. $2\frac{7}{9}$ **b.** -1.8

Square each of the following. (Example 1)

1. $\frac{3}{8}$ **2.** $\frac{1}{5}$ **3.** 12 **4.** 14 **5.** $\frac{1}{7}$ **6.** $\frac{1}{6}$
7. 1.8 **8.** 1.7 **9.** 0.12 **10.** 0.14 **11.** 0.08 **12.** 0.07

For Exercises 13–60, find each square root. (Example 2)

13. $\sqrt{\frac{16}{9}}$ **14.** $\sqrt{\frac{49}{81}}$ **15.** $\sqrt{\frac{25}{36}}$ **16.** $\sqrt{\frac{16}{49}}$ **17.** $\sqrt{\frac{49}{100}}$ **18.** $\sqrt{\frac{81}{169}}$
19. $\sqrt{\frac{4}{121}}$ **20.** $\sqrt{\frac{121}{144}}$ **21.** $\sqrt{\frac{169}{400}}$ **22.** $\sqrt{\frac{1}{9}}$ **23.** $\sqrt{\frac{1}{4}}$ **24.** $\sqrt{\frac{196}{625}}$

(Example 3)

25. $-\sqrt{169}$ **26.** $-\sqrt{400}$ **27.** $-\sqrt{\frac{16}{25}}$ **28.** $-\sqrt{\frac{9}{49}}$ **29.** $-\sqrt{576}$ **30.** $-\sqrt{441}$
31. $-\sqrt{\frac{25}{81}}$ **32.** $-\sqrt{\frac{1}{16}}$ **33.** $-\sqrt{\frac{144}{169}}$ **34.** $-\sqrt{\frac{100}{441}}$ **35.** $-\sqrt{\frac{36}{361}}$ **36.** $-\sqrt{\frac{1}{400}}$

(Example 4)

37. $-\sqrt{0.49}$ **38.** $-\sqrt{0.25}$ **39.** $-\sqrt{0.64}$ **40.** $-\sqrt{0.81}$ **41.** $-\sqrt{0.04}$ **42.** $-\sqrt{0.01}$
43. $-\sqrt{0.09}$ **44.** $-\sqrt{0.16}$ **45.** $-\sqrt{6.25}$ **46.** $-\sqrt{1.69}$ **47.** $-\sqrt{1.21}$ **48.** $-\sqrt{4.84}$

MIXED PRACTICE

49. $-\sqrt{\frac{9}{16}}$ **50.** $-\sqrt{1.44}$ **51.** $\sqrt{\frac{16}{625}}$ **52.** $-\sqrt{361}$ **53.** $\sqrt{\frac{36}{225}}$ **54.** $-\sqrt{\frac{1}{25}}$
55. $-\sqrt{1.96}$ **56.** $-\sqrt{\frac{225}{529}}$ **57.** $\sqrt{\frac{4}{529}}$ **58.** $\sqrt{\frac{324}{49}}$ **59.** $-\sqrt{4.41}$ **60.** $-\sqrt{121}$

REVIEW CAPSULE FOR SECTION 8.5

Evaluate each expression. (Pages 27–30)

1. $3^2 + 4^2$ **2.** $5^2 + 6^2$ **3.** $11^2 + 12^2$ **4.** $9^2 + 13^2$
5. $15^2 + 18^2$ **6.** $(1.2)^2 + (3.4)^2$ **7.** $(0.8)^2 + (1.9)^2$ **8.** $(\frac{1}{4})^2 + (\frac{5}{4})^2$

PROBLEM SOLVING AND APPLICATIONS

8.5 Rule of Pythagoras

A right triangle is a triangle with one right angle (90°). The symbol ⌝ is used to indicate that an angle is a right angle. The ***hypotenuse*** of a right triangle is the side opposite the right angle. It is the longest side of the right triangle. The other two sides are the **legs.**

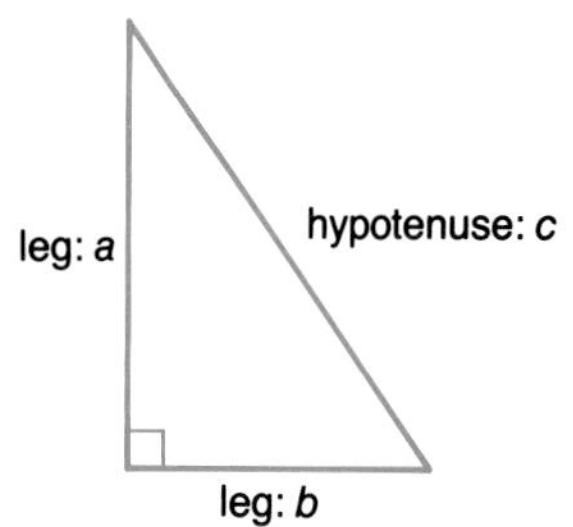

The ***Right Triangle Rule*** or the ***Rule of Pythagoras*** is stated below.

Rule of Pythagoras

In any right triangle, the square of the length of the hypotenuse, c, equals the sum of the squares of the lengths of the legs, a and b.

$$\mathbf{c^2 = a^2 + b^2}$$

EXAMPLE 1

In triangle ABC, $a = 1.5$ centimeters, $b = 2.0$ centimeters, and $c = 2.5$ centimeters. Use the Rule of Pythagoras to determine whether the triangle is a right triangle.

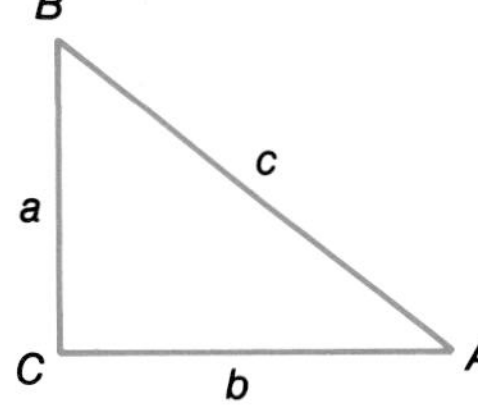

Solution:

1. Write the formula. ⟶ $c^2 = a^2 + b^2$
2. Identify values of the variables. ⟶ $c = 2.5$; $a = 1.5$; $b = 2.0$
3. Replace the variables with their values. ⟶ $(2.5)^2 \stackrel{?}{=} (1.5)^2 + (2.0)^2$
4. Perform the operations. ⟶ $6.25 \stackrel{?}{=} 2.25 + 4.00$

$6.25 = 6.25$ ◀ ***The sentence is true.***

Since the sentence is true, triangle ABC is a right triangle.

P–1 **The lengths of the three sides of a triangle are given. Determine whether the triangle is a right triangle.**

a. $a = 4.5$ m; $b = 6$ m; $c = 7.5$ m

b. $a = 5$ cm; $b = 9$ cm; $c = 10.3$ cm

When the lengths of two legs of a right triangle are known, the Rule of Pythagoras can be used to find the length of the hypotenuse.

EXAMPLE 2

A carpenter cuts a brace for two boards that form a right angle. The lengths of the boards are 5 feet and 12 feet. Find the length of the brace.

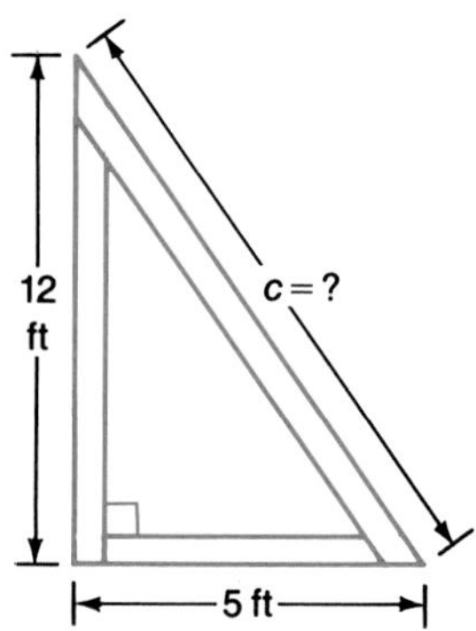

Solution:

Think: Since the triangle is a right triangle, use the Rule of Pythagoras.

1 Write the formula. ⟶ $c^2 = a^2 + b^2$

2 Identify values of the variables. ⟶ Let $a = 5$ and $b = 12$

3 Replace the variables with their values. ⟶ $c^2 = 5^2 + 12^2$

4 Perform the operations. ⟶ $c^2 = 25 + 144$

$c^2 = 169$

5 Find the square root. ⟶ $c = \sqrt{169}$

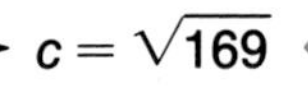

Use the table on page 226.

$c = 13$

The length of the brace is 13 feet.

P–2 **For each of the following, the lengths of the two legs of a right triangle are given. Find the length of the hypotenuse, c. Use the table on page 226.**

a. $a = 2.4$; $b = 1.8$; $c =$ __?__

b. $a = 1.4$; $b = 4.8$; $c =$ __?__

When the lengths of the hypotenuse and one leg of a right triangle are known, the Rule of Pythagoras can be used to find the length of the other leg.

EXAMPLE 3

A 25-foot wire is staked to the ground 15 feet from the base of the pole. Find the height of the pole.

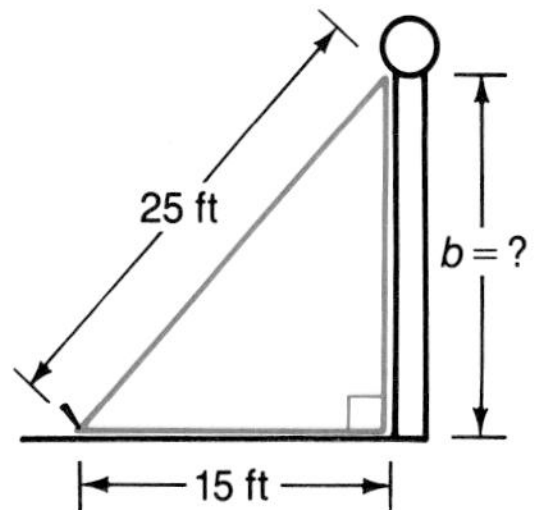

Solution:

Think: Since the triangle is a right triangle, use the Rule of Pythagoras to find the leg, b.

1. Write the formula. ⟶ $c^2 = a^2 + b^2$
2. Identify values of the variables. ⟶ Let $c = 25$ and $a = 15$.
3. Replace the variables with their values. ⟶ $25^2 = 15^2 + b^2$
4. Perform the operations. ⟶ $625 = 225 + b^2$

$$625 - \mathbf{225} = 225 - \mathbf{225} + b^2$$

$$400 = b^2$$

5. Find the square root. ⟶ $\sqrt{400} = b$ ◀ ***Use the table on page 226.***

$$20 = b$$

The height of the pole is 20 feet.

P–3 **For each of the following, the length of the hypotenuse, c, and the length of one leg of a right triangle are given. Find the unknown length. Use the table on page 226.**

a. $a = 3$; $c = 5$; $b = \underline{\ ?\ }$

b. $a = 8$; $c = 17$; $b = \underline{\ ?\ }$

CLASSROOM EXERCISES

For Exercises 1–4, the lengths of three sides of a triangle are given. Determine whether the triangle is a right triangle. (Example 1)

1. $a = 30$ cm; $b = 40$ cm; $c = 50$ cm

2. $a = 10$ ft; $b = 13$ ft; $c = 16$ ft

3. $a = 27$ m; $b = 36$ m; $c = 45$ m

4. $a = 2.5$ in; $b = 6$ in; $c = 6.5$ in

For Exercises 5–6, find the length of the hypotenuse, c, of each right triangle. Use the table on page 226. (Example 2)

5. $a = 9$; $b = 12$; $c = \underline{\ ?\ }$

6. $a = 12$; $b = 16$; $c = \underline{\ ?\ }$

For Exercises 7–9, the length of the hypotenuse, c, and the length of one leg of the right triangle are given. Find the unknown length. Use the table on page 226.
(Example 3)

7.

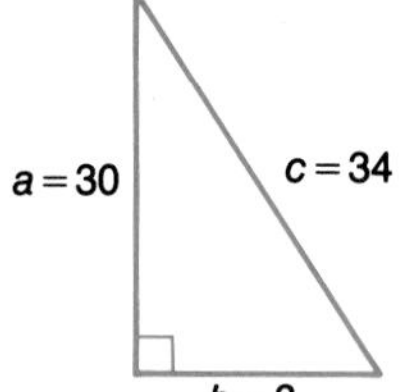

8.

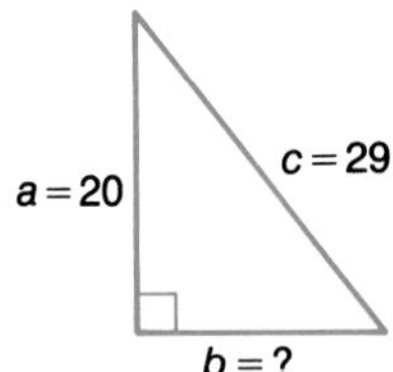

9.

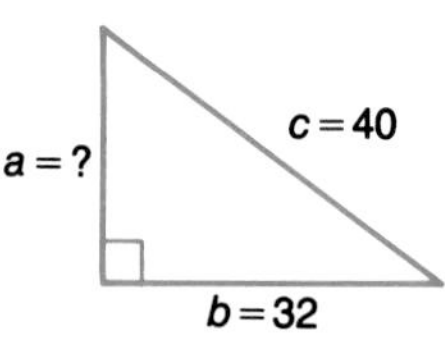

WRITTEN EXERCISES

Goal: To use the Rule of Pythagoras to find the length of a side of a right triangle

Sample Problem: The length of one leg of a right triangle is 12 inches. The length of the other leg is 16 inches. Find the length of the hypotenuse.

Answer: 20 inches

For Exercises 1–4, the lengths of three sides of a triangle are given. Determine whether the triangle is a right triangle. (Example 1)

1. $a = 6$ ft; $b = 9$ ft; $c = 11$ ft

2. $a = 20$ cm; $b = 25$ cm; $c = 32$ cm

3. $a = 1.8$ m; $b = 2.4$ m; $c = 3.0$ m

4. $a = 1$ in; $b = 1.5$ in; $c = 3.25$ in

For Exercises 5–10, refer to the right triangle DEF with hypotenuse f and legs d and e. Find the unknown length. Use the table on page 226.
(Example 2)

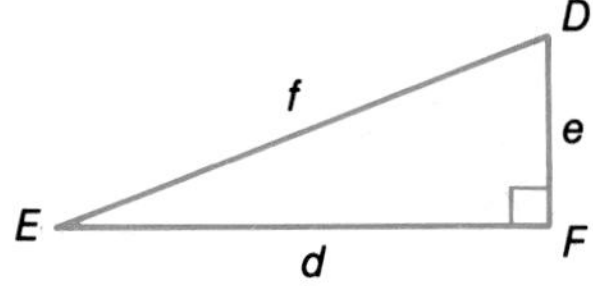

5. $d = 8$; $e = 15$; $f =$ __?__

6. $d = 3$; $e = 4$, $f =$ __?__

(Example 3)

7. $d = 10$; $f = 26$; $e =$ __?__

8. $e = 35$; $f = 37$; $d =$ __?__

9. $e = 1.6$; $f = 3.4$; $d =$ __?__

10. $d = 2.1$; $f = 2.9$; $e =$ __?__

Solve. (Examples 2 and 3)

11. A ladder is placed against the side of a house as shown below. Find the length of the ladder.

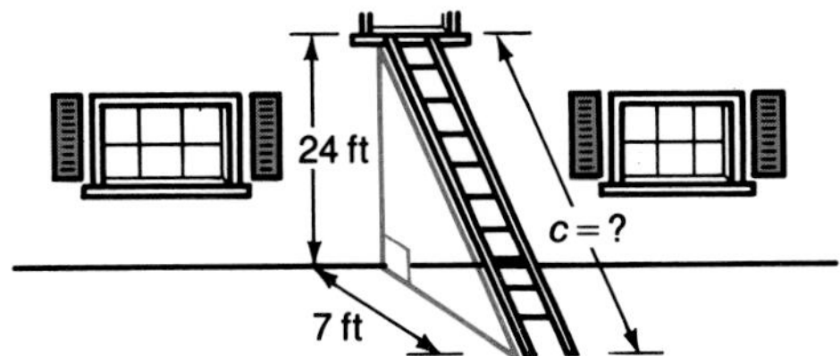

12. The flag pole below has a 30-meter wire fastened at the top of the pole and staked into the ground 18 meters from the base of the pole. Find the height of the flag pole.

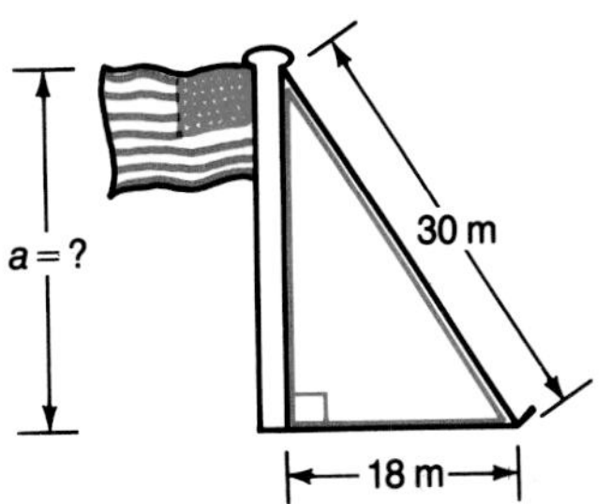

13. A sailboat sails from the marina to an island. It sails north a distance of 8 kilometers. Then it sails west a distance of 6 kilometers. Find the straight-line distance from the marina to the island.

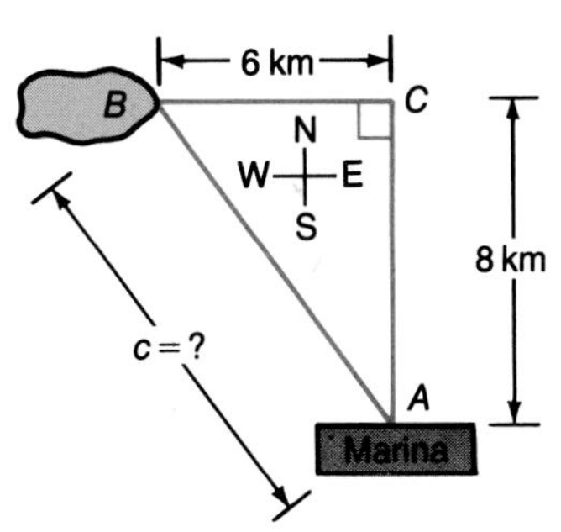

14. A ramp into a building is 39 feet long. The ramp begins at a distance of 36 feet from the building. Find the height in feet the ramp rises.

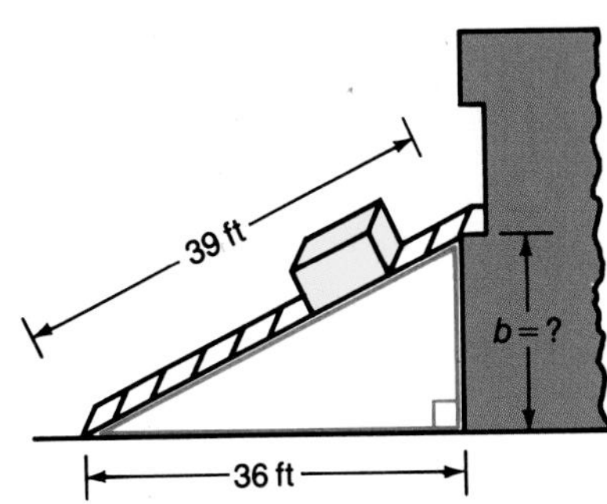

REVIEW CAPSULE FOR SECTION 8.6

Write an infinite repeating decimal for each of the following. (Pages 221–224)

1. $\frac{5}{8}$ **2.** $\frac{9}{5}$ **3.** 2 **4.** $\frac{13}{33}$ **5.** $1\frac{2}{5}$ **6.** $4\frac{1}{6}$

7. 0 **8.** $7\frac{3}{4}$ **9.** $\frac{3}{16}$ **10.** $\frac{1}{3}$ **11.** $7\frac{1}{2}$ **12.** $\frac{1}{12}$

Square each of the following. (Pages 225–228)

13. 15 **14.** 1.5 **15.** 4 **16.** 0.4 **17.** $\frac{2}{3}$ **18.** $\frac{3}{4}$

8.6 Real Numbers

Rational numbers such as the following are perfect squares.

$16 \qquad \frac{4}{9} \qquad 0.16 \qquad 144 \qquad 1\frac{7}{9}$

Definition

A rational number is a ***perfect square*** if it is the square of a rational number.

$\mathbf{\frac{4}{9}} = \frac{2}{3} \cdot \frac{2}{3}$

$\mathbf{0.16} = (0.4)(0.4)$

$\mathbf{144} = 12 \cdot 12$

Rational numbers such as 5, $\frac{3}{8}$, and 0.7 are not perfect squares. This means there is no rational number whose square is 5, no rational number whose square is $\frac{3}{8}$, and no rational number whose square is 0.7. Thus, $\sqrt{5}$, $\sqrt{\frac{3}{8}}$, and $\sqrt{0.7}$ are irrational numbers.

A rational number that is not a perfect square has an ***irrational*** square root.

EXAMPLE 1 Identify each square root as rational or irrational. Give a reason.

a. $\sqrt{15}$ **b.** $-\sqrt{49}$ **c.** $\sqrt{3}$

Solutions: Check whether the radicand is a perfect square.

a. Since 15 is not a perfect square, $\sqrt{15}$ is **irrational.**

b. Since 49 is a perfect square, $-\sqrt{49}$ is **rational.**

c. Since 3 is not a perfect square, $\sqrt{3}$ is **irrational.**

P–1 **Identify each square root as rational or irrational.**

a. $\sqrt{36}$ **b.** $-\sqrt{122}$ **c.** $-\sqrt{169}$ **d.** $\sqrt{50}$

EXAMPLE 2 Identify each square root as rational or irrational. Give a reason.

a. $-\sqrt{0.81}$ **b.** $\sqrt{\frac{9}{16}}$ **c.** $\sqrt{1\frac{2}{3}}$

Solutions: **a.** Since 0.81 is a perfect square, $-\sqrt{0.81}$ is **rational.**

b. Since $\frac{9}{16}$ is a perfect square, $\sqrt{\frac{9}{16}}$ is **rational.**

c. $\sqrt{1\frac{2}{3}} = \sqrt{\frac{5}{3}}$. Since $\frac{5}{3}$ is not a perfect square, $\sqrt{1\frac{2}{3}}$ is **irrational.**

P–2 **Identify each square root as rational or irrational.**

a. $-\sqrt{\frac{1}{9}}$ **b.** $-\sqrt{\frac{4}{7}}$ **c.** $\sqrt{0.49}$ **d.** $\sqrt{0}$

Recall that every rational number can be written as an infinite repeating decimal.

$$2\tfrac{1}{2} = 2.5000\cdots \qquad \frac{2}{3} = 0.666\cdots \qquad \frac{3}{11} = 0.\overline{27}$$

The decimal equivalents of irrational numbers are non–repeating.

Some Irrational Numbers

$\sqrt{2} = 1.41421\cdots$ $\sqrt{3} = 1.73205\cdots$ $\sqrt{5} = 2.23606\cdots$

$\sqrt{6} = 2.44948\cdots$ $\sqrt{7} = 2.64575\cdots$ $-\sqrt{8} = -2.82842\cdots$

There is no repeating pattern.

Numbers such as 2, $-\sqrt{2}$, 14, -3.4, 0, and $0.\overline{6}$ are real numbers.

Definition

The set of ***real numbers*** contains all the rational numbers and all the irrational numbers.

CLASSROOM EXERCISES

Identify each square root as rational or irrational. (Example 1)

1. $\sqrt{12}$
2. $-\sqrt{4}$
3. $-\sqrt{23}$
4. $\sqrt{64}$
5. $\sqrt{5}$
6. $-\sqrt{49}$
7. $\sqrt{81}$
8. $\sqrt{9}$
9. $-\sqrt{7}$
10. $\sqrt{21}$
11. $-\sqrt{15}$
12. $\sqrt{144}$

Identify each square root as rational or irrational. (Example 2)

13. $\sqrt{1\frac{9}{16}}$
14. $-\sqrt{6\frac{1}{4}}$
15. $-\sqrt{0.25}$
16. $\sqrt{0.16}$
17. $\sqrt{\frac{25}{169}}$
18. $-\sqrt{\frac{16}{49}}$
19. $\sqrt{3\frac{1}{5}}$
20. $-\sqrt{2\frac{1}{8}}$
21. $\sqrt{0.01}$
22. $-\sqrt{1.72}$
23. $-\sqrt{\frac{9}{21}}$
24. $\sqrt{\frac{36}{225}}$

WRITTEN EXERCISES

Goal: To identify a real number as rational or irrational

Sample Problem: $-\sqrt{\frac{9}{144}}$

Answer: Rational. $\frac{9}{144}$ is a perfect square.

For Exercises 1–36, identify each square root as rational or irrational. Give a reason. (Example 1)

1. $\sqrt{25}$ **2.** $-\sqrt{3}$ **3.** $\sqrt{100}$ **4.** $\sqrt{22}$ **5.** $-\sqrt{18}$ **6.** $-\sqrt{529}$

7. $-\sqrt{96}$ **8.** $\sqrt{121}$ **9.** $\sqrt{7}$ **10.** $-\sqrt{400}$ **11.** $\sqrt{529}$ **12.** $-\sqrt{1}$

(Example 2)

13. $-\sqrt{\frac{1}{10}}$ **14.** $\sqrt{\frac{1}{100}}$ **15.** $-\sqrt{2.56}$ **16.** $\sqrt{1.44}$ **17.** $\sqrt{4\frac{1}{6}}$ **18.** $-\sqrt{5\frac{1}{4}}$

19. $-\sqrt{\frac{7}{12}}$ **20.** $-\sqrt{\frac{8}{9}}$ **21.** $\sqrt{\frac{9}{36}}$ **22.** $\sqrt{0.81}$ **23.** $\sqrt{1.69}$ **24.** $-\sqrt{1\frac{1}{4}}$

MIXED PRACTICE

25. $-\sqrt{5}$ **26.** $\sqrt{5\frac{1}{8}}$ **27.** $\sqrt{\frac{81}{25}}$ **28.** $-\sqrt{130}$ **29.** $\sqrt{3.61}$ **30.** $-\sqrt{0.0324}$

31. $\sqrt{112}$ **32.** $\sqrt{0.05}$ **33.** $-\sqrt{\frac{1}{36}}$ **34.** $\sqrt{\frac{9}{47}}$ **35.** $\sqrt{3\frac{1}{5}}$ **36.** $-\sqrt{64}$

MORE CHALLENGING EXERCISES

a. *Select one or more letters below to describe each of the following numbers.*
b. *Simplify each rational square root.*

A. *whole number* **B.** *integer* **C.** *rational number*
D. *irrational number* **E.** *real number*

37. -15 **38.** $\sqrt{92}$ **39.** -108 **40.** $-2\frac{3}{4}$ **41.** 246 **42.** $\sqrt{50}$

43. $-\sqrt{289}$ **44.** -13.8 **45.** $-\sqrt{1\frac{19}{81}}$ **46.** -69 **47.** 5.09 **48.** $\sqrt{\frac{10}{25}}$

REVIEW CAPSULE FOR SECTION 8.7

Simplify. (Pages 218–220)

1. $-(9)$ **2.** $-(1.5)$ **3.** $-(-\frac{1}{3})$ **4.** $-(-80)$ **5.** $-(-4.9)$

6. $-(6\frac{1}{4})$ **7.** $-(-12\frac{1}{8})$ **8.** $-(0)$ **9.** $-(-\frac{10}{3})$ **10.** $-(0.05)$

11. $-(-14.1)$ **12.** $-(-2.7)$ **13.** $-(-\frac{1}{7})$ **14.** $-(0.1)$ **15.** $-(-\frac{2}{3})$

8.7 Absolute Value

As you can see on the number line below, numbers that are opposites, such as 4 and -4, are the same distance from 0.

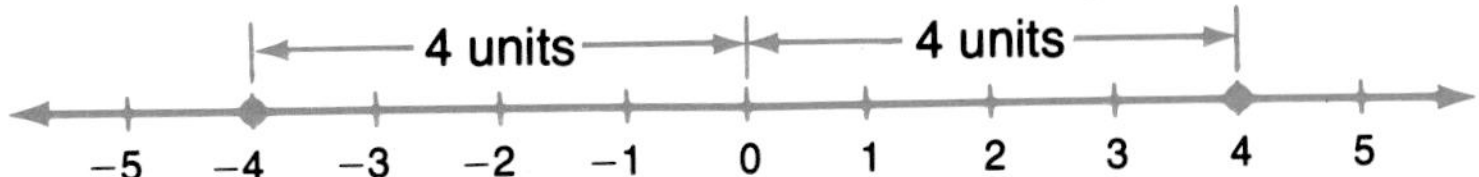

To indicate distance, but not direction, from zero, you use absolute value.

The absolute value of -4 is 4.
The absolute value of 4 is 4.

Distances are positive numbers.

The symbol | | is used to represent absolute value.

Read: "The absolute value of negative four equals four.

$|-4| = 4$ $\qquad$ $|4| = 4$

Read: "The absolute value of four equals four."

The ***absolute value*** of a positive number equals the number. The absolute value of a negative number equals the opposite of the number. The absolute value of 0 equals 0.

$|6.7| = 6.7$

$\left|-\frac{2}{13}\right| = \frac{2}{13}$

$|0| = 0$

The table below illustrates this definition.

Table

Absolute Value	Meaning	Value
$|13|$	Distance of 13 from 0	13
$|-6.7|$	Distance of -6.7 from 0	6.7
$|0|$	Distance of 0 from 0	0

P–1 **Evaluate each of the following.**

a. $|-14|$ **b.** $|2\frac{3}{4}|$ **c.** $|0.8|$ **d.** $|-12.04|$

Numerical expressions often involve absolute value.

To evaluate a numerical expression containing absolute values

1. Find each absolute value.
2. Then perform the indicated operation(s).

EXAMPLE Evaluate: **a.** $|-14| + |-2|$ **b.** $|-12| - |-7|$

Solutions: **a.** [1] $|-14| + |-2| = 14 + 2$
[2] $= 16$

b. [1] $|-12| - |7| = 12 - 7$
[2] $= 5$

P–2 **Evaluate each expression.**

a. $|15.8| - |-6.3|$ **b.** $|-3\frac{1}{2}| + |2\frac{1}{4}|$ **c.** $|-10| + |-12| - |5|$

CLASSROOM EXERCISES

For Exercises 1–20, evaluate. (Table)

1. $|8|$ **2.** $|-15|$ **3.** $|-1.5|$ **4.** $|0.9|$ **5.** $|\frac{3}{8}|$ **6.** $|-1\frac{2}{3}|$

7. $|14.7|$ **8.** $|-8.3|$ **9.** $|-2\frac{1}{2}|$ **10.** $|9\frac{1}{4}|$ **11.** $|-\sqrt{21}|$ **12.** $|\sqrt{19}|$

(Example)

13. $|-6| - |-3|$ **14.** $|5| - |4|$ **15.** $|-2| + |3|$ **16.** $|-2| + |-3|$

17. $|0| + |12|$ **18.** $|0| + |-12|$ **19.** $|-21.1| - |2.9|$ **20.** $|-8.7| - |-4.7|$

WRITTEN EXERCISES

Goal: To evaluate expressions containing absolute values
Sample Problem: Evaluate: $|-7| - |-4|$ **Answer:** 3

For Exercises 1–20, evaluate. (Table)

1. $|-0.8|$ **2.** $|3.5|$ **3.** $|6|$ **4.** $|-6|$ **5.** $|-3\frac{1}{2}|$ **6.** $|-13.5|$

7. $|16|$ **8.** $|-100|$ **9.** $|-18|$ **10.** $|8\frac{3}{4}|$ **11.** $|-\sqrt{11}|$ **12.** $|\sqrt{47}|$

(Example)

13. $|14| - |-5|$ **14.** $|-3| + |18|$ **15.** $|-3| + |10|$ **16.** $|7| + |-13|$

17. $|-2| + |-11|$ **18.** $|-41| - |-41|$ **19.** $|-7| - |0|$ **20.** $|0| + |-7|$

MORE CHALLENGING EXERCISES

Write True or False for each statement. When a statement is false, tell why it is false.

21. If x is any real number, then $|x| = x$.

22. If y is any real number, then $|y| = |-y|$.

23. If x and y are real numbers and $x < y$, then $|x| < |y|$.

Displacement

Automobile mechanics must be able to understand both customary and metric measures, because automotive specifications are given in both systems. For example, displacement is given in both cubic centimeters and cubic inches.

Displacement is the total volume of all an engine's cylinders. The diameter of an engine cylinder is called the **bore.** The distance the piston moves is called the **stroke.** The following formula relates bore, stroke, and displacement.

$$D = \pi\left(\frac{b}{2}\right)^2(s)(n)$$

D = displacement
b = bore in cm
s = stroke in cm
n = number of cylinders

EXAMPLE: A certain 4-cylinder engine has a bore of 84 mm and a stroke of 84.4 mm. Compute its displacement to the nearest cubic centimeter.

SOLUTION: Convert mm to cm.
84 mm = 8.4 cm; 84.4 mm = 8.44 cm

Use $\pi = 3.14$; $b = 8.4$ cm; $s = 8.44$ cm; $n = 4$.

$$D = 3.14\left(\frac{8.4}{2}\right)^2(8.44)(4)$$

$$D = 3.14(17.64)(8.44)(4)$$

Use a calculator.

$D = 1869.9528$, or **1870 cubic centimeters**

EXERCISES

For Exercises 1–4, find the displacement to the nearest cubic centimeter.

1. Bore: 70 mm; Stroke: 73 mm; Cylinders: 4
2. Bore: 90 mm; Stroke: 58.86 mm; Cylinders: 4
3. Bore: 85.5 mm; Stroke: 69 mm; Cylinders: 4
4. Bore: 92 mm; Stroke: 79 mm; Cylinders: 8
5. If the stroke of an engine is decreased and the displacement remains the same, how will the bore have to change?
6. If the stroke of an engine is increased and the bore is decreased, can you tell how the displacement will be affected? Explain.

CHAPTER SUMMARY

IMPORTANT TERMS

Positive integers *(p. 214)*
Negative integers *(p. 214)*
Integers *(p. 214)*
Graph *(p. 214)*
Coordinate *(p. 214)*
Inequalities *(p. 215)*
Opposites *(p. 218)*
Positive rational numbers *(p. 218)*
Negative rational numbers *(p. 219)*
Rational numbers *(p. 219)*
Infinite repeating decimals *(p. 222)*
Square root *(p. 225)*
Radical *(p. 225)*
Radical symbol *(p. 225)*
Radicand *(p. 225)*
Right triangle *(p. 229)*
Hypotenuse *(p. 229)*
Leg *(p. 229)*
Perfect square *(p. 234)*
Irrational numbers *(p. 234)*
Real numbers *(p. 235)*
Absolute value *(p. 237)*

IMPORTANT IDEAS

1. The inequality symbols $>$ and $<$ are used to compare numbers.
2. Every rational number can be written as an infinite repeating decimal.
3. If a rational number is not a perfect square, its square root is irrational.
4. *Rule of Pythagoras:* The sum of the squares of the lengths of the legs of a right triangle equals the square of the length of the hypotenuse.
5. The decimal equivalents of irrational numbers are non-repeating.

CHAPTER REVIEW

SECTION 8.1

For Exercises 1–6, graph the numbers. Then write the numbers in order from smallest to largest.

1. 0, −1, 7, 3, −5
2. 4, 8, −9, 2, −2
3. 6, 5, −4, 0, −3
4. −4, −3, 5, −6, 10
5. 12, 2, −11, 5, −4
6. 3, 7, 11, −8, −2, 0

Write two inequalities for each pair of numbers.

7. 3 and −5
8. 17 and −15
9. 0 and −8
10. 4 and 0
11. −7 and 9
12. −13 and 16
13. −12 and −15
14. −6 and −16

SECTION 8.2

Simplify.

15. $-(-12)$
16. $-(-4)$
17. $-(7.1)$
18. $-(-\frac{1}{3})$
19. $-(2\frac{1}{5})$
20. $-(-14)$
21. $-(-2.3)$
22. $-(\frac{6}{5})$
23. $-(-9\frac{1}{2})$
24. $-(6.7)$

Use symbols to represent each sentence.

25. A decrease in speed of 9.1 kilometers per hour is the opposite of an increase in speed of 9.1 kilometers per hour.

26. A weight gain of $3\frac{1}{2}$ pounds is the opposite of a weight loss of $3\frac{1}{2}$ pounds.

27. The opposite of a loss of 82.3 meters in altitude is a gain in altitude of 82.3 meters.

28. The opposite of climbing four flights of stairs is descending four flights of stairs.

SECTION 8.3

Write a decimal for each of the following.

29. $\frac{9}{20}$ **30.** $-\frac{4}{11}$ **31.** $1\frac{7}{16}$ **32.** $-11\frac{2}{5}$ **33.** $3\frac{2}{9}$ **34.** $9\frac{8}{9}$

35. $-\frac{11}{16}$ **36.** $\frac{21}{50}$ **37.** $-3\frac{3}{4}$ **38.** $-\frac{9}{10}$ **39.** $\frac{4}{33}$ **40.** $4\frac{1}{12}$

SECTION 8.4

Square each of the following.

41. 14 **42.** $\frac{2}{5}$ **43.** 1.9 **44.** 0.13 **45.** 17 **46.** $3\frac{2}{3}$

47. 0.3 **48.** $2\frac{1}{2}$ **49.** 11 **50.** $\frac{4}{9}$ **51.** 0.15 **52.** 1.2

Find each square root.

53. $\sqrt{\frac{16}{81}}$ **54.** $-\sqrt{144}$ **55.** $-\sqrt{\frac{36}{121}}$ **56.** $\sqrt{\frac{225}{49}}$ **57.** $\sqrt{1.96}$ **58.** $-\sqrt{0.64}$

59. $-\sqrt{400}$ **60.** $\sqrt{\frac{1}{16}}$ **61.** $-\sqrt{0.25}$ **62.** $-\sqrt{4.41}$ **63.** $\sqrt{\frac{16}{529}}$ **64.** $-\sqrt{625}$

SECTION 8.5

The lengths of three sides of a triangle are given. Determine whether the triangle is a right triangle.

65. $a = 27$m; $b = 36$ m; $c = 45$ m

66. $a = 18$ cm; $b = 20$ cm; $c = 26$ cm

67. $a = 10$ ft; $b = 23$ ft; $c = 25$ ft

68. $a = 24$ in; $b = 32$ in; $c = 40$ in

Find the length of the hypotenuse, c, of each right triangle. Use the table on page 226.

69. $a = 12$; $b = 16$; $c =$ ___?___

70. $a = 8$; $b = 15$; $c =$ ___?___

71. $a = 9$; $b = 12$; $c =$ ___?___

72. $a = 15$; $b = 20$; $c =$ ___?___

Solve each problem.

73. A utility pole 48 feet tall has a guy wire that is 52 feet long. How far is the guy wire from the base of the pole?

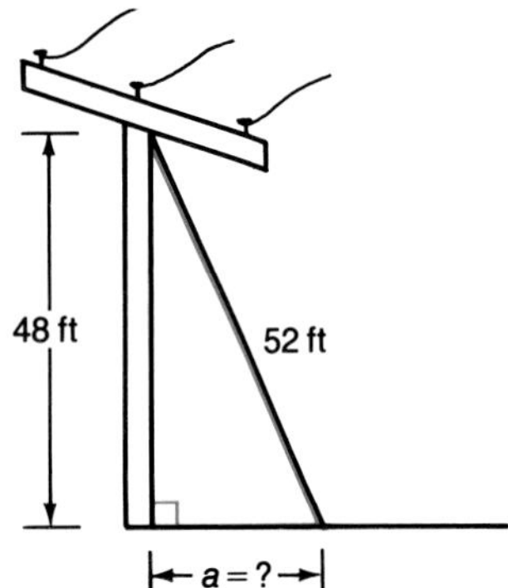

74. A carpenter is making a rectangular frame for a concrete patio. The distance from A to B is 65 feet. The distance from C to B is 60 feet. What length must side AC equal in order for triangle ABC to be a right triangle?

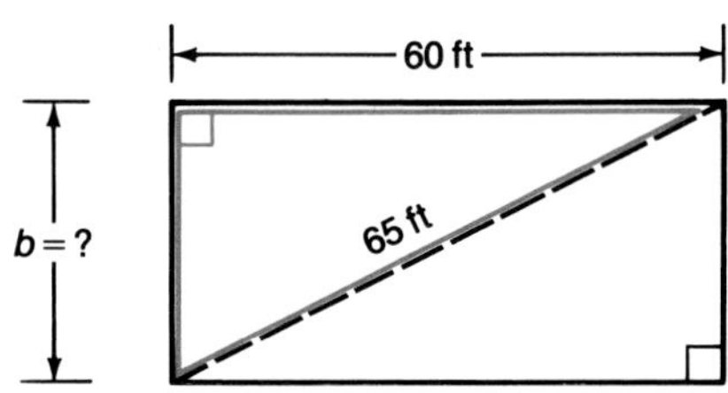

SECTION 8.6

Identify each square root as rational or irrational. Give a reason.

75. $\sqrt{95}$ **76.** $-\sqrt{0.49}$ **77.** $\sqrt{2\frac{1}{3}}$ **78.** $-\sqrt{9}$ **79.** $\sqrt{\frac{100}{169}}$ **80.** $\sqrt{1.42}$

81. $-\sqrt{5}$ **82.** $\sqrt{0.07}$ **83.** $-\sqrt{1}$ **84.** $-\sqrt{5\frac{2}{5}}$ **85.** $\sqrt{6.5}$ **86.** $\sqrt{0.68}$

SECTION 8.7

Evaluate.

87. $|8|$ **88.** $|-\sqrt{4}|$ **89.** $|3\frac{1}{5}|$ **90.** $|-6.3|$ **91.** $|-11\frac{1}{9}|$ **92.** $|2.9|$

93. $|-2| + |-7|$ **94.** $|5| - |0|$ **95.** $|-41| + |11|$ **96.** $|-16| - |-13|$

CHAPTER 9

Real Numbers: Addition and Subtraction

Renaissance explorers used instruments such as the *astrolabe* shown at the right to determine the position of a ship relative to lines of longitude and latitude. In sailing from one position to another, an explorer added or subtracted the number of degrees of longitude and latitude.

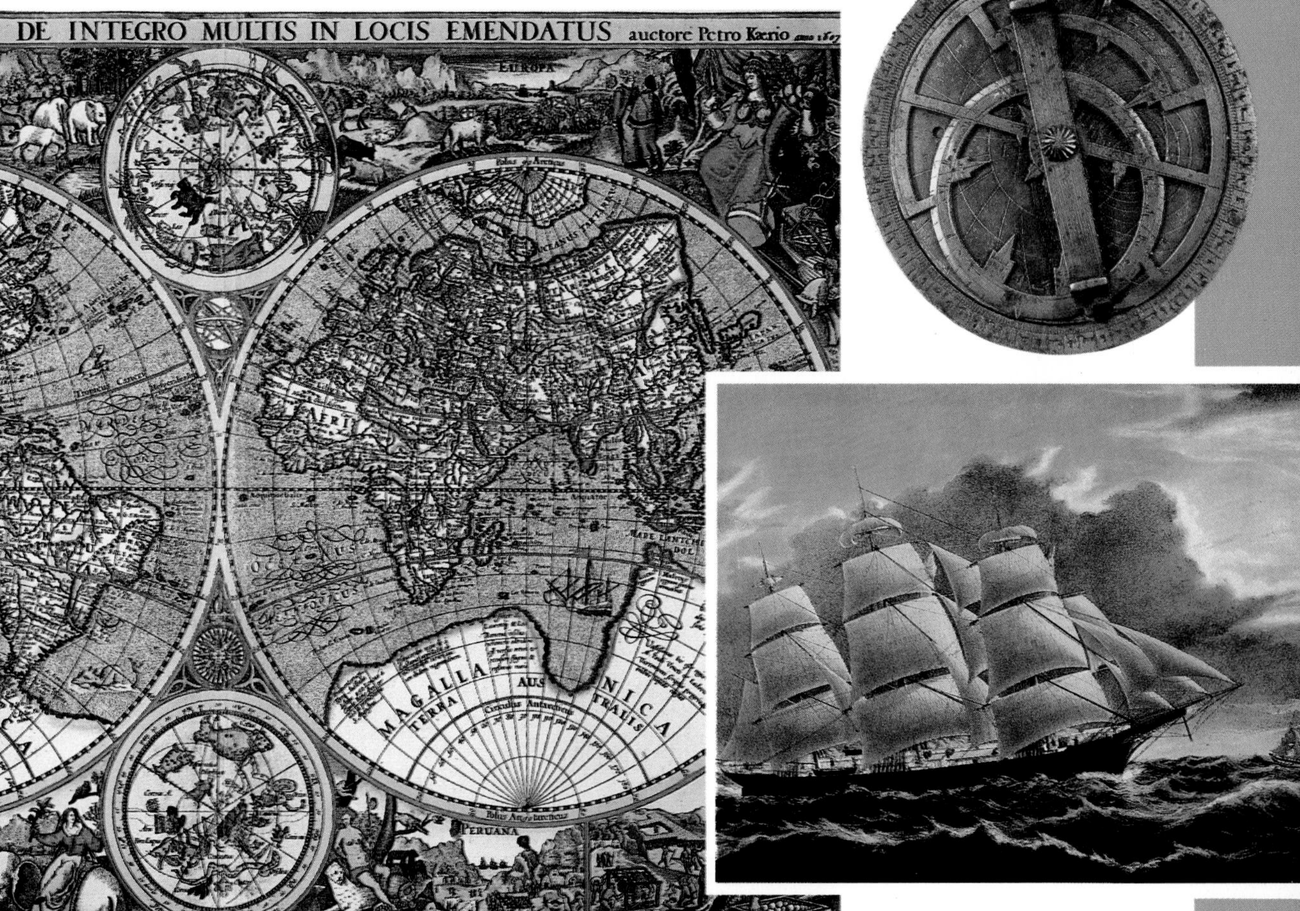

9.1 Adding on a Number Line

Recall from Section 8.2 that a number line can be used to represent rational numbers.

P–1 **Graph the numbers below on one number line.**

a. 4 **b.** $-\frac{1}{4}$ **c.** -4.5 **d.** $3\frac{1}{2}$ **e.** $-\frac{8}{3}$

By using direction arrows, you can use a number line to add two real numbers. Examples 1, 2, and 3 show how this is done.

EXAMPLE 1 Use a number line to add: $4 + (-8)$

Solution:

1. Graph the first addend.
2. From this point, draw an arrow to the left for the second addend.
3. Read the coordinate of the point where the arrow ends, **−4.**

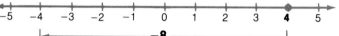

$$4 + (-8) = \mathbf{-4}$$

P–2 **Use a number line to add.**

a. $3 + (-6)$ **b.** $2 + (-9)$ **c.** $5 + (-5)$ **d.** $12 + (-6)$

EXAMPLE 2 Use a number line to add: $(-4) + 9$

Solution:

1. Graph the first addend.
2. From this point, draw an arrow to the right for the second addend.

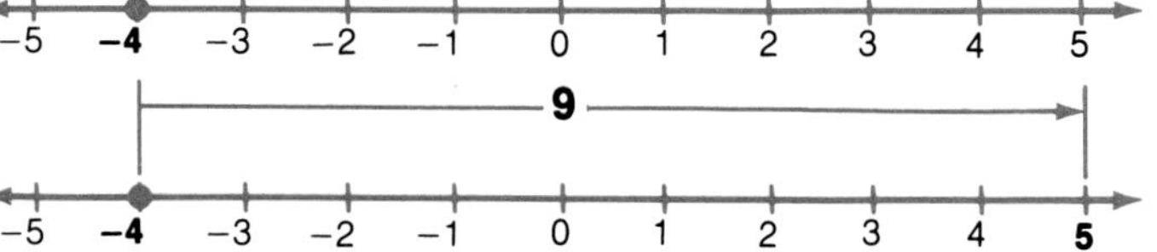

3. Read the coordinate of the point where the arrow ends, **5.**

$$(-4) + 9 = \mathbf{5}$$

P–3 **Use a number line to add.**

a. $(-4) + 8$ **b.** $(-1) + 15$ **c.** $(-5) + 3$ **d.** $(-10) + 9$

EXAMPLE 3

Use a number line to add: $(-6) + (-7)$

Solution:

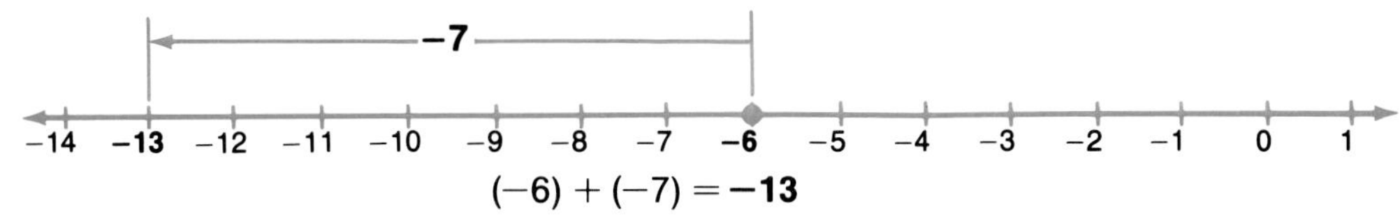

$(-6) + (-7) = \mathbf{-13}$

P–4 **Use a number line to add.**

a. $(-3) + (-4)$ **b.** $(-5) + (-5)$ **c.** $(-5) + (-9)$ **d.** $(-9) + (-4)$

EXAMPLE 4

A football team gained 6 yards on one play, but lost 9 yards on the next play. Show this on a number line. Find the team's position after the two plays.

Solution:

[1] Graph the position of the team after the first play.

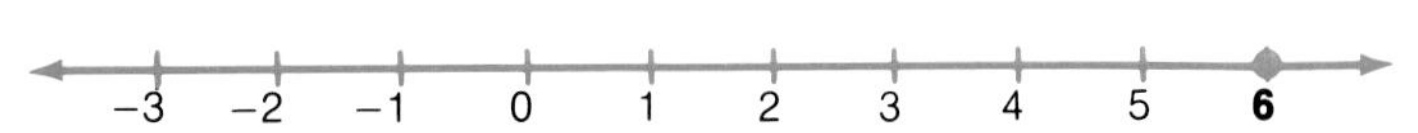

0 marks the team's original position.

[2] From this point, draw an arrow to the left to show a loss of 9 yards.

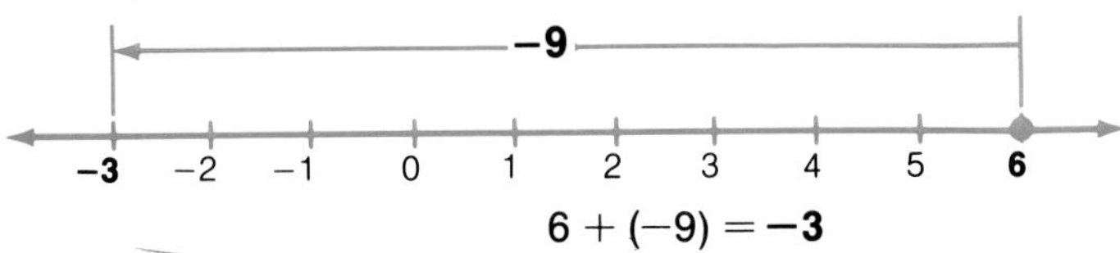

$6 + (-9) = \mathbf{-3}$

The team is now 3 yards behind its position before the first play.

CLASSROOM EXERCISES

For Exercises 1–24, use a number line to add. (Example 1)

1. $2 + (-7)$ **2.** $5 + (-8)$ **3.** $1 + (-4)$ **4.** $9 + (-12)$

5. $3 + (-2)$ **6.** $6 + (-5)$ **7.** $10 + (-7)$ **8.** $8 + (-4)$

(Example 2)

9. $(-7) + 10$ **10.** $(-3) + 8$ **11.** $(-1) + 3$ **12.** $(-6) + 12$

13. $(-9) + 6$ **14.** $(-5) + 1$ **15.** $(-12) + 9$ **16.** $(-9) + 8$

(Example 3)

17. $(-2) + (-4)$ 18. $(-7) + (-4)$ 19. $(-3) + (-3)$ 20. $(-1) + (-7)$
21. $(-4) + (-6)$ 22. $(-8) + (-5)$ 23. $(-5) + (-2)$ 24. $(-6) + (-2)$

WRITTEN EXERCISES

Goal: To add integers using a number line
Sample Problem: Use a number line to add: $(-6) + 8$
Answer:

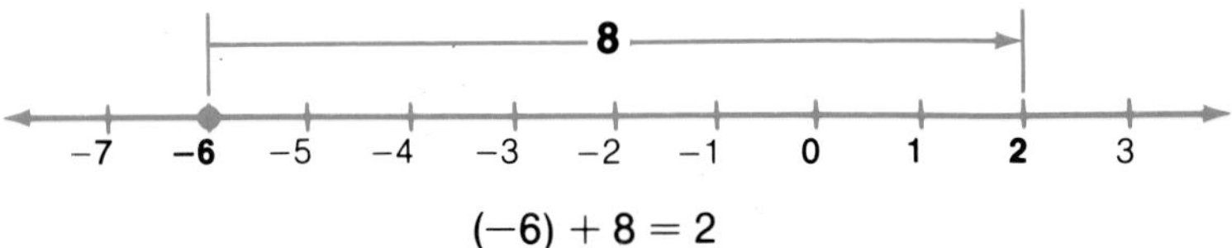

$(-6) + 8 = 2$

For Exercises 1–36, use a number line to add. (Example 1)

1. $3 + (-5)$ 2. $4 + (-7)$ 3. $5 + (-3)$ 4. $6 + (-2)$
5. $1 + (-6)$ 6. $2 + (-6)$ 7. $7 + (-2)$ 8. $5 + (-1)$

(Example 2)

9. $(-4) + 7$ 10. $(-3) + 6$ 11. $(-5) + 2$ 12. $(-6) + 4$
13. $(-2) + 8$ 14. $(-1) + 5$ 15. $(-7) + 3$ 16. $(-8) + 5$

(Example 3)

17. $(-5) + (-4)$ 18. $(-6) + (-5)$ 19. $(-1) + (-6)$ 20. $(-4) + (-1)$
21. $(-3) + (-7)$ 22. $(-8) + (-4)$ 23. $(-4) + (-4)$ 24. $(-6) + (-6)$

MIXED PRACTICE

25. $(-8) + 5$ 26. $(-7) + (-2)$ 27. $6 + (-6)$ 28. $(-3) + 3$
29. $(-2) + (-7)$ 30. $(-4) + 7$ 31. $(-10) + 4$ 32. $(-6) + (-4)$
33. $(-7) + 7$ 34. $2 + (-2)$ 35. $(-6) + (-7)$ 36. $(-8) + 3$

APPLICATIONS

Use a number line to solve. (Example 4)

37. A football team gained 12 yards on one play, and lost 7 yards on the next play. Find the team's position after the second play.

38. A football team lost 10 yards on one play, and gained 2 yards on the next play. Find the team's position after the second play.

REVIEW CAPSULE FOR SECTION 9.2

Write an integer for each word description. (Pages 214–217)

1. A temperature of 20° below 0

2. A drop of 6° in temperature

3. A stock increase of 2 points

4. A loss of \$75

Write the opposite. (Pages 218–220)

5. 12 **6.** 9.6 **7.** $-3\frac{1}{8}$ **8.** -0.6 **9.** $\frac{3}{10}$

Find the absolute value. (Pages 237–238)

10. -15 **11.** 28 **12.** $-14\frac{1}{4}$ **13.** -8.5 **14.** $-\frac{7}{2}$

9.2 Addition

In the previous lesson, you learned how to find the sum of two negative numbers by using the number line.

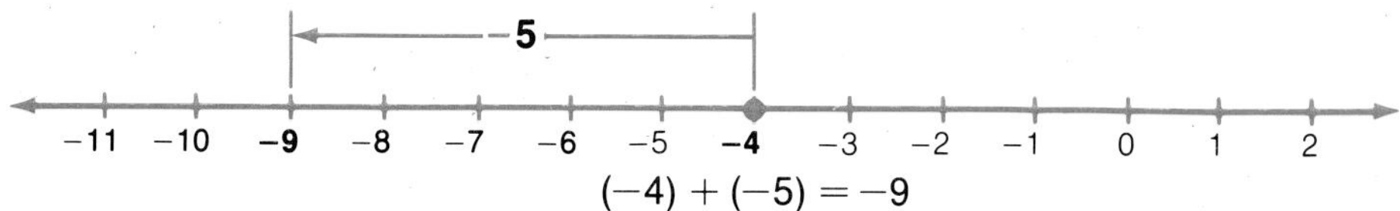

$(-4) + (-5) = -9$

The following steps show how to find the sum of two negative numbers without using the number line.

> ***Steps for Adding Two Negative Numbers***
>
> 1. Add the absolute values.
> 2. Write the opposite of the result.

EXAMPLE 1 Add: $(-4) + (-5)$

Solution: $(-4) + (-5)$

[1] Add the absolute values. ⟶ $|-4| + |-5| = 4 + 5$

$= 9$

[2] Write the opposite of the result. ⟶ -9

$(-4) + (-5) = -9$

P–1 **Add.**

a. $(-6) + (-9)$ **b.** $(-7) + (-11)$ **c.** $(-8) + (-3)$ **d.** $(-12) + (-2)$

Steps for Adding a Positive and a Negative Number

1. Subtract the absolute values.
2. If the positive addend has the greater absolute value, write the difference.
   If the negative addend has the greater absolute value, write the opposite of the difference.

EXAMPLE 2 Add: $8 + (-3)$

Solution: $8 + (-3)$

[1] Subtract the absolute values. ⟶ $|8| - |-3| = 8 - 3$

[2] Since $|8| > |-3|$, the difference is the result. ⟶ $= 5$

$8 + (-3) = 5$  ***The positive number has the greater absolute value.***

P–2 **Add.**

a. $9 + (-4)$ **b.** $5 + (-1)$ **c.** $12 + (-10)$ **d.** $14 + (-3)$

In Example 3, the negative addend has the greater absolute value.

EXAMPLE 3 Add: $6 + (-14)$

Solution: $6 + (-14)$

[1] Subtract the absolute values. ⟶ $|-14| - |6| = 14 - 6$
$= 8$

[2] Since $|-14| > |6|$, take the opposite of the result. ⟶ -8

$6 + (-14) = -8$ ***The negative number has the greater absolute value.***

P–3 **Add.**

a. $(-13) + 6$ **b.** $12 + (-14)$ **c.** $3 + (-6)$ **d.** $(-12) + 4$

EXAMPLE 4 During one day, the temperature rose 8° from midnight to noon. It dropped 6° from noon to midnight. Find the net change in temperature that day.

Solution:

[1] Write an expression for the problem. ⟶ $8 + (-6)$

[2] Subtract the absolute values. ⟶ $|8| - |-6| = 8 - 6$

[3] Since $|8| > |-6|$, the difference is the result. ⟶ $= 2$

The net change in temperature was 2°.

"rose" indicates a positive number; "dropped" indicates a negative number.

A positive number indicates an increase.

CLASSROOM EXERCISES

For Exercises 1–24, add the given numbers. (Example 1)

1. $(-6) + (-12)$ **2.** $(-9) + (-4)$ **3.** $(-8) + (-13)$ **4.** $(-5) + (-15)$

5. $(-40) + (-20)$ **6.** $(-10) + (-62)$ **7.** $(-8.4) + (-1.6)$ **8.** $(-2.1) + (-7)$

(Example 2)

9. $6 + (-4)$ **10.** $7 + (-1)$ **11.** $(-4) + 15$ **12.** $(-5) + 21$

13. $20 + (-16)$ **14.** $31 + (-11)$ **15.** $(-2.4) + 6$ **16.** $8 + (-4.4)$

(Example 3)

17. $12 + (-20)$ **18.** $6 + (-9)$ **19.** $(-16) + 6$ **20.** $(-22) + 4$

21. $24 + (-84)$ **22.** $13 + (-67)$ **23.** $(-9.2) + 5.6$ **24.** $(-7.1) + 2$

WRITTEN EXERCISES

Goal: To add two real numbers
Sample Problem: $(-7) + (-8)$ **Answer:** -15

For Exercises 1–48, add the given numbers.
(Example 1)

1. $(-11) + (-5)$ **2.** $(-9) + (-3)$ **3.** $(-6) + (-22)$ **4.** $(-15) + (-3)$

5. $(-32) + (-24)$ **6.** $(-48) + (-10)$ **7.** $(-7.6) + (-9.1)$ **8.** $(-5) + (-3.9)$

(Example 2)

9. $17 + (-12)$ **10.** $8 + (-7)$ **11.** $(-4) + 12$ **12.** $(-1) + 41$

13. $(-81) + 82$ **14.** $(-16) + 98$ **15.** $8.7 + (-4.8)$ **16.** $9 + (-1.4)$

(Example 3)

17. $(-12) + 7$ **18.** $(-21) + 14$ **19.** $8 + (-20)$ **20.** $4 + (-19)$

21. $(-47) + 16$ **22.** $(-77) + 71$ **23.** $1.1 + (-2.2)$ **24.** $8 + (-9.8)$

MIXED PRACTICE

25. $(-26) + 31$ **26.** $(-17) + 70$ **27.** $(-84) + (-41)$ **28.** $(-2) + (-52)$

29. $(-13) + 8$ **30.** $(-86) + 7$ **31.** $27 + (-12)$ **32.** $54 + (-50)$

33. $(-99) + (-9)$ **34.** $(-28) + (-28)$ **35.** $(-60) + 30$ **36.** $(-78) + 52$

37. $(-2) + (-9)$ **38.** $(-4) + (-7)$ **39.** $88 + (-40)$ **40.** $39 + (-32)$

41. $(-8.4) + (2.7)$ **42.** $(-8) + (2.7)$ **43.** $6\frac{3}{4} + (-2\frac{1}{4})$ **44.** $9 + (-3\frac{2}{3})$

45. $(-1.4) + (-9.6)$ **46.** $(-4.5) + (-6.9)$ **47.** $8\frac{7}{8} + (-6\frac{3}{8})$ **48.** $7\frac{5}{6} + (-4\frac{2}{3})$

APPLICATIONS

Solve. (Example 4)

49. During one day, the temperature rose 11° from midnight to 3:00 p.m. It dropped 15° from 3:00 p.m. to midnight. What was the net change in temperature for the 24 hours?

50. During one day, the temperature fell 20° from midnight to 6:00 a.m., and rose 5° between 6:00 a.m. and midnight. Find the net change in temperature for the day.

51. During one business day, the price of a share of stock fell \$1.00 in the morning, and fell \$2.00 more in the afternoon. Find the net change in price for that day.

52. During one business day, the trading price of an ounce of gold rose \$4.00, and then dropped \$7.00 in the afternoon. Find the net change in price for the day.

REVIEW CAPSULE FOR SECTION 9.3

Evaluate expressions **a–c.** *Find the expression that does not equal the given expression.* (Pages 43–45)

1. $(16 + 24) + (-19)$

a. $(24 + 16) + (-19)$

b. $(16 + (-19)) + (24 + 16)$

c. $16 + (24 + (-19))$

2. $(3\frac{1}{4} + (-1\frac{1}{8})) + 2\frac{7}{8}$

a. $3\frac{1}{4} + (-1\frac{1}{8} + 2\frac{7}{8})$

b. $(-1\frac{1}{8} + 3\frac{1}{4}) + 2\frac{7}{8}$

c. $(3\frac{1}{4} + 2\frac{7}{8}) + (-1\frac{1}{8} + 2\frac{7}{8})$

3. $4.2 + (-5.8 + (-4.9))$

a. $(-4.2 + 5.8) + (-4.9)$

b. $4.2 + (-4.9 + (-5.8))$

c. $-4.9 + (4.2 + (-5.8)$

9.3 Addition: More Than Two Numbers

The properties of addition discussed in Chapter 2 also apply to addition with real numbers.

Addition Property of Zero

For any real number a,

$$\mathbf{a+0=a} \text{ and } \mathbf{0+a=a}$$

$6+0=6$

$0+(-3.8)=-3.8$

Addition Property of Opposites

For any real number a,

$$\mathbf{a+(-a)=0} \text{ and } \mathbf{(-a)+a=0}$$

$79+(-79)=0$

$(-3\frac{1}{2})+3\frac{1}{2}=0$

Commutative Property of Addition

For any real numbers a and b,

$$\mathbf{a+b=b+a}$$

$6\frac{1}{3}+(-7\frac{2}{3})=-7\frac{2}{3}+6\frac{1}{3}$

Associative Property of Addition

For any real numbers a, b, and c,

$$\mathbf{(a+b)+c=a+(b+c)}$$

$(1.5+6.2)+4=1.5+(6.2+4)$

The Commutative Property means that you can add real numbers in any order. The Associative Property means that you can group real numbers in any way to add them.

Addition of three or more real numbers is based on the Commutative and Associative Properties of Addition.

EXAMPLE 1 Add: $(-3)+8+(-6)$

Solution: **Method 1**

$(-3)+8+(-6)$

$5+(-6)$

-1

Add from left to right.

Method 2

$(-3)+8+(-6)$

$(-3)+(-6)+8$

$(-9)+8$

-1

First add the negative numbers. Then add 8 to the sum.

EXAMPLE 2 Add: $7 + (-1) + 5 + (-3)$

Solution:

Method 1

$7 + (-1) + 5 + (-3)$

$6 + 2$

8

Method 2

$7 + (-1) + 5 + (-3)$

$7 + 5 + (-1) + (-3)$

$12 + (-4)$

8

Group the positive and negative numbers. Then add.

P–1 **Add.**

a. $2 + (-5) + (-7)$ **b.** $9 + (-3) + 7$
c. $(-4) + 9 + 6 + (-5)$ **d.** $(-3) + 4 + (-5) + 8$

CLASSROOM EXERCISES

For Exercises 1–18, add the given numbers. (Example 1)

1. $12 + (-3) + 2$ **2.** $9 + (-6) + 3$ **3.** $(-6) + 4 + (-8)$
4. $(-8) + 3 + (-1)$ **5.** $(-23) + 58 + (-14)$ **6.** $(-45) + (-38) + (-53)$
7. $4 + (-1) + (-2)$ **8.** $3 + (-13) + 5$ **9.** $(-5) + 20 + (-10)$
10. $18 + (-15) + 7$ **11.** $8.4 + (-6.5) + (-0.4)$ **12.** $(-5.5) + 6.2 + 1.4$

(Example 2)

13. $1 + (-3) + (-2) + 4$ **14.** $(-12) + 7 + 13 + (-8)$ **15.** $(-10) + 9 + 14 + (-16)$
16. $(-2) + 1 + (-1) + 9$ **17.** $(-6) + (-8) + (-11) + (-9)$ **18.** $8 + (-9) + 13 + (-3)$

WRITTEN EXERCISES

Goal: To add more than two real numbers
Sample Problem: $57 + (-63) + (-89)$ **Answer:** -95

For Exercises 1–34, add the given numbers. (Example 1)

1. $(-8) + 6 + (-3)$ **2.** $(-12) + 5 + (-2)$ **3.** $7 + (-13) + 8$
4. $5 + (-12) + 13$ **5.** $(-6) + (-9) + (-14)$ **6.** $(-3) + (-9) + (-15)$
7. $(-51) + 27 + (-40)$ **8.** $(-43) + 16 + (-90)$ **9.** $(-48) + 72 + (-35)$
10. $(-72) + 27 + (-39)$ **11.** $(-\frac{5}{16}) + (-\frac{9}{16}) + \frac{3}{16}$ **12.** $3\frac{5}{8} + (-2\frac{7}{8}) + (-4\frac{3}{8})$

(Example 2)

13. $(-5) + (-2) + (-7) + 4$

14. $(-6) + 2 + (-11) + 8$

15. $33 + (-29) + 16 + 13$

16. $42 + (-56) + 16 + 19$

17. $38 + (-43) + 19 + (-27)$

18. $29 + (-37) + 23 + (-46)$

19. $(-39) + 35 + (-23) + 14$

20. $(-78) + 65 + (-86) + 29$

21. $(-3\frac{1}{2}) + 5\frac{1}{2} + (-9\frac{1}{2}) + 3\frac{1}{2}$

22. $(-3\frac{3}{4}) + 6\frac{1}{4} + (-8\frac{1}{4}) + 2\frac{3}{4}$

MIXED PRACTICE

23. $(-14) + 23 + (-45) + 30 + (-19)$

24. $(-42) + 35 + (-52) + 12 + 76$

25. $(-35) + 25 + (-15)$

26. $(-12) + 14 + (-6)$

27. $(-12.1) + (-8.8) + 14.7$

28. $5 + 9 + (-5) + 17 + (-9)$

29. $(-8) + 8 + (-12) + (-4)$

30. $(-40) + 15 + (-25)$

31. $0.3 + (-5.8) + 4.2 + (-9.4)$

32. $8.6 + (-6.3) + 2.4 + (-7.7)$

33. $(-5\frac{2}{3}) + (-3\frac{1}{3}) + 1\frac{2}{3}$

34. $4\frac{5}{12} + (-2\frac{7}{12}) + (-1\frac{1}{12})$

APPLICATIONS

You can use a calculator with a "+/–" or "sign change" key to operate with negative numbers. The advantage of this key is that it enables you to add real numbers from left to right, without grouping.

EXAMPLE Add: $-6.9 + (-2.3) + 4.01 + (-7.2)$

SOLUTION Press the sign change key for any negative addend after you enter the value of the addend.

6.9 [+/–] [+] 2.3 [+/–] [+] 4.01 [+] 7.2 [+/–] [=] → -12.39

Add.

35. $-17 + 52 + (-94) + (-38)$

36. $-52 + (-16) + 31 + (-64)$

37. $-6.0 + 3.23 + (-5.2) + 7.3$

38. $-0.2 + 0.41 + 0.37 + (-0.25)$

39. $-96 + 102 + 112 + (-204)$

40. $-157 + (-203) + 301 + 98$

NON-ROUTINE PROBLEM

41. Trace the figure at the right. Number the dots on the triangle 1–9 so that the sums of the four numbers shown on each side of the triangle are equal. No number may be used more than once.

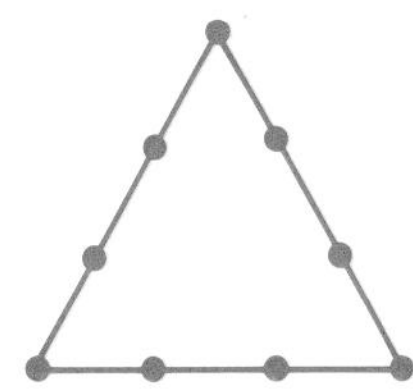

Scale Factor

Commercial artists create artwork for advertising, publications, packaging, and television. Artists must often reduce or enlarge drawings or photographs from their original size.

A reduced or enlarged picture is a **reproduction** of the original. You can use a proportion to determine the dimensions of a reproduction.

$$\frac{\textbf{Length of Original}}{\textbf{Width of Original}} = \frac{\textbf{Length of Reproduction}}{\textbf{Width of Reproduction}}$$

You can use a calculator or pencil and paper to find an unknown dimension of a reproduction.

EXAMPLE 1: A photograph 10 centimeters long and 6 centimeters wide is to be reduced to fit a space 4 centimeters long. Find the width of the reduced photograph.

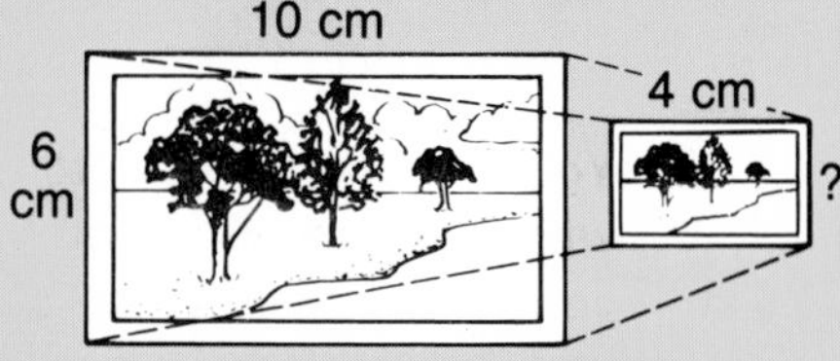

SOLUTION: Choose a variable for the unknown number.

Let w = the unknown width.

$$\frac{10}{6} = \frac{4}{w}$$

$$10w = 4 \cdot 6$$

$$w = \frac{4 \cdot 6}{10}$$

[4] [×] [6] [÷] [1] [0] [=] 2.4

The width of the reduced photograph is **2.4 centimeters.**

When the size of a drawing or photograph is changed, the artist will often give the photographer the per cent of reduction or enlargement. This per cent is the **scale factor.**

$$\textbf{Scale Factor} = \frac{\textbf{Dimension of Reproduction}}{\textbf{Corresponding Dimension of Original}}$$

The photographer will set the camera at this per cent, and the resulting reproduction will be the desired size.

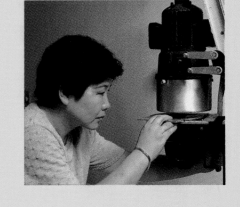

EXAMPLE 2: An artist prepares a drawing 24.8 centimeters high for an advertisement. The drawing must be reduced to fit a space 8.9 centimeters high. Find the scale factor to the nearest whole per cent.

SOLUTION: Scale Factor $= \frac{8.9}{24.8}$

8 . 9 ÷ 2 4 . 8 = 0.3588709

The scale factor is about **36%**.

EXERCISES

1. Will the width of the reproduction be greater than or less than 10 cm?
   Length of drawing: 16 cm
   Width of drawing: 10 cm
   Length of reproduction: 3 cm

2. Will the length of the copy be greater than or less than 25.5 cm?
   Photo dimensions: 22.5 cm × 16.5 cm
   Width of copy: 25.5 cm

3. The scale factor for producing a copy of a photo is 125%. Will the copy be larger or smaller than the original?

4. The length of a drawing is 18.6 centimeters. The scale factor for making a copy is 75%. Will the length of the copy be greater than or less than 18.6 cm?

5. A drawing 24 centimeters long and 16 centimeters wide is to be reduced to fit a space 9 centimeters long. Find the width of the reduced drawing.

6. A design 15 centimeters long and 10 centimeters wide is to be enlarged to become a poster 125 centimeters long. Find the width of the poster to the nearest tenth.

7. A snapshot 12.7 centimeters long will be enlarged for framing. The enlarged photo will be 30 centimeters long. Find the scale factor to the nearest whole per cent.

8. Floor plans for a house are 105 centimeters long and 48 centimeters wide. They are reduced to be read by the architect's clients. The reduced plans are 21.5 centimeters wide. Find the scale factor to the nearest whole per cent.

9.4 Using Real Numbers

You can use the special properties of addition from Section 9.3 to help you solve word problems involving addition of real numbers.

EXAMPLE 1

The **excess** of rainfall in a city is the amount of rain that is above average. It is shown as a positive number. The **deficiency** (amount of rain below average) is represented as a negative number. Find the net excess or deficiency for a six-month period.

Sept.	Oct.	Nov.	Dec.	Jan.	Feb.
2.7 cm	−0.8 cm	−2.9 cm	−3.4 cm	1.8 cm	4.1 cm

Solution: [1] Add the negative amounts. ⟶ $(-0.8) + (-2.9) + (-3.4) = -7.1$

[2] Add the positive amounts. ⟶ $2.7 + 1.8 + 4.1 = 8.6$

[3] Add the two results. ⟶ $(-7.1) + 8.6 = 1.5$

There was an excess of 1.5 centimeters of rainfall.

EXAMPLE 2

The price of a certain stock was $16\frac{1}{2}$ (\$16.50) at the beginning of a week. The net change in price for each day is shown below. Find the price at the end of the week.

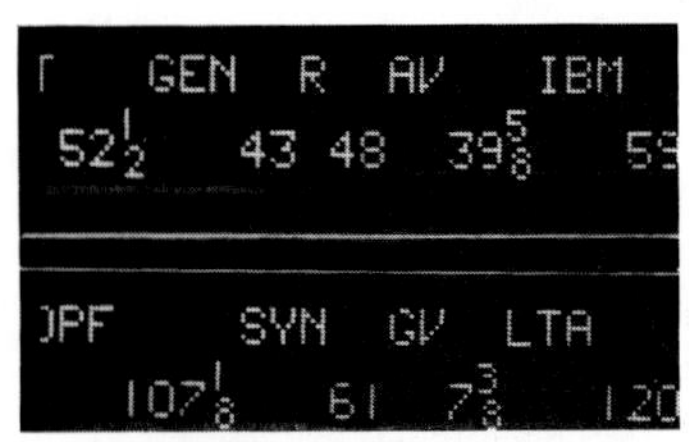

Stock Market Prices

M	T	W	Th	F
$-\frac{1}{4}$	$\frac{3}{8}$	$\frac{1}{4}$	$-\frac{5}{8}$	$-\frac{1}{2}$

Solution: First find the net change for the week.

[1] Add the negative numbers. ⟶ (M, Th, F)

$$-\frac{1}{4} + \left(-\frac{5}{8}\right) + \left(-\frac{1}{2}\right) = \left(-\frac{2}{8}\right) + \left(-\frac{5}{8}\right) + \left(-\frac{4}{8}\right)$$

$$= -\frac{11}{8}$$

[2] Add the positive numbers.	⟶	T W $\frac{3}{8}+\frac{1}{4}=\frac{3}{8}+\frac{2}{8}$ $=\frac{5}{8}$
[3] Add the two results.	⟶	$-\frac{11}{8}+\frac{5}{8}=-\frac{6}{8}$ or $-\frac{3}{4}$
[4] Add the beginning price and the net change.	⟶	$16\frac{1}{2}+\left(-\frac{3}{4}\right)=15\frac{3}{4}$

The price was $15.75 at the end of the week.

CLASSROOM EXERCISES

Find the net excess or deficiency of rainfall. (Example 1)

1.

Jan.	Feb.	March	April	May	June
1.2 cm	2.3 cm	−0.6 cm	1.2 cm	−1.5 cm	−0.9 cm

2.

July	Aug.	Sept.	Oct.	Nov.	Dec.
−1.5 cm	−0.2 cm	1.1 cm	1.4 cm	−2.7 cm	3.3 cm

3.

April	May	June	July	Aug.	Sept.
0.8 cm	−1.5 cm	2.3 cm	1.2 cm	−3.5 cm	0.4 cm

4.

Oct.	Nov.	Dec.	Jan.	Feb.	March
−2.5 cm	1.8 cm	−0.7 cm	−1.3 cm	1.6 cm	−2.6 cm

Find the stock's price at the end of the week. Give your answers to the nearest cent. (Example 2)

5. Beginning price: $35\frac{1}{4}$

M	T	W	Th	F
$-1\frac{1}{4}$	$\frac{3}{4}$	$-\frac{7}{8}$	$\frac{1}{4}$	$-\frac{5}{8}$

6. Beginning price: $67\frac{3}{8}$

M	T	W	Th	F
$\frac{7}{8}$	$-\frac{3}{4}$	$-2\frac{1}{2}$	$\frac{5}{8}$	$1\frac{1}{8}$

7. Beginning price: $12\frac{1}{8}$

M	T	W	Th	F
$-\frac{3}{8}$	$1\frac{1}{4}$	$-\frac{1}{2}$	-1	$\frac{3}{4}$

8. Beginning price: 108

M	T	W	Th	F
$-2\frac{1}{8}$	$\frac{3}{4}$	$-\frac{1}{4}$	$-\frac{7}{8}$	$-\frac{1}{4}$

WRITTEN EXERCISES

Goal: To solve word problems involving addition of several real numbers

Sample Problem: A scuba diver descended 53 meters below the surface of the ocean, rose 37 meters, and then descended 6 meters. How far was she from the surface then?

Answer: She was 22 meters below the surface.

Solve. (Example 1)

1. Find the net excess or deficiency of rainfall.

March	April	May	June	July	Aug.
2.2 cm	−1.6 cm	−2.1 cm	−0.5 cm	1.1 cm	1.4 cm

2. Find the net excess or deficiency of rainfall.

Sept.	Oct.	Nov.	Dec.	Jan.	Feb.
−0.6 cm	1.3 cm	−0.8 cm	−1.2 cm	0.9 cm	−1.7 cm

3. A football coach asks each player to report for training camp at a certain weight. The first seven players he checks are over *(O)* or under *(U)* their weights as shown. Find the net amount in kilograms by which the seven players are over or under the desired weights.

Player	1	2	3	4	5	6	7
Weight	6.3*(O)*	1.8*(U)*	2.5*(U)*	2.9*(O)*	2.9*(O)*	3.6*(U)*	0.7*(U)*

4. A zoo keeper keeps a record of the mass of a chimpanzee each week. She records the increase or decrease each week for five weeks. Find the net increase or decrease in mass.

Week	1	2	3	4	5
Mass	−2.1 kg	1.3 kg	−0.9 kg	−1.3 kg	0.7 kg

Solve. (Example 2)

For Exercises 5–6, find the stock's price at the end of the week.

5. Beginning price: $45\frac{5}{8}$

M	T	W	Th	F
$-\frac{3}{8}$	$-1\frac{1}{2}$	$-\frac{1}{4}$	$1\frac{1}{8}$	$\frac{1}{4}$

6. Beginning price: $9\frac{1}{4}$

M	T	W	Th	F
$\frac{1}{8}$	$-\frac{3}{8}$	-1	$\frac{5}{8}$	$-\frac{7}{8}$

7. An athlete had a long jump of 19 feet $9\frac{3}{4}$ inches in the first meet of the season. The table below shows how her best jump changed in each successive meet. Find her best jump in the sixth meet.

Meet 2	Meet 3	Meet 4	Meet 5	Meet 6
$-1\frac{1}{2}$ in.	$4\frac{3}{8}$ in.	$2\frac{1}{8}$ in.	$-3\frac{5}{8}$ in.	$\frac{9}{16}$ in.

8. A scientist records the change in temperature of a solution every four hours as compared with the previous reading. The beginning temperature is $55\frac{1}{2}$ degrees. Find the temperature at the end of 24 hours.

After 4 hours	After 8 hours	After 12 hours	After 16 hours	After 20 hours	After 24 hours
$-\frac{3}{4}^\circ$	$\frac{3}{8}^\circ$	$-\frac{1}{2}^\circ$	$-\frac{1}{8}^\circ$	$\frac{1}{4}^\circ$	$\frac{5}{8}^\circ$

MIXED PRACTICE

9. The enrollment in a school at the beginning of 1984 was 2140, and showed the following gains *(G)* or losses *(L)* over a five-year period. Find the enrollment at the end of 1988.

1984	1985	1986	1987	1988
158*(G)*	192*(G)*	16*(L)*	33*(L)*	96*(L)*

10. Mrs. Ortega has $840 in her checking account on November 1. She makes the following deposits *(D)* and withdrawals *(W)* during November. Find the balance of her account on December 1.

$19.49*(W)*, $208.72*(W)*, $87.12*(D)*, $56.70*(W)*

$150*(D)*, $306.87*(W)*, $96.22*(W)*, $210.50*(D)*

11. A car dealer records the increase *(I)* or decrease *(D)* in the number of cars sold each month as compared with the previous year's sales. Find the net increase or decrease for the six-month period shown below.

Jan.	Feb.	March	April	May	June
28*(D)*	5*(I)*	12*(D)*	19*(I)*	27*(I)*	35*(D)*

12. A business shows the following increase *(I)* or decrease *(D)* in sales compared with the same six-month period of the previous year. Find the net increase or decrease for the sixth-month period.

Jan. $36,000 *(I)* **Feb.** $18,000 *(D)* **March** $25,000 *(D)*

April $8500 *(I)* **May** $1300 *(I)* **June** $12,500 *(D)*

NON-ROUTINE PROBLEMS

13. Draw a circle. Then divide the interior of the circle into seven regions, drawing just 3 straight lines.

14. Find the greatest four-digit number that is divisible by 1, 2, 3, and 4.

15. Three bananas were given to two mothers who were with their daughters. Each person had a banana to eat. How is this possible?

16. Frank had 3 piles of leaves in the front yard and 2 piles in the back yard. If he put them all together, how many piles would he have?

MID-CHAPTER REVIEW

Use a number line to add. (Section 9.1)

1. $2 + (-9)$ **2.** $8 + (-7)$ **3.** $(-2) + 4$ **4.** $(-6) + 10$

5. $(-5) + (-5)$ **6.** $(-2) + (-1)$ **7.** $(-3) + (-7)$ **8.** $(-4) + (-5)$

Add. (Section 9.2)

9. $-6 + 12$ **10.** $-14 + (-7)$ **11.** $(-18) + (-92)$ **12.** $(-51) + (-10)$

13. $(312) + (-256)$ **14.** $(-297) + (-308)$ **15.** $29 + (-35)$ **16.** $(-48) + 31$

Solve. (Section 9.2)

17. One day, the temperature rose 8° from midnight to noon, and then fell 17° by midnight. Find the net change in temperature for the day.

18. During one business day, the price of a share of stock fell $3.00, and then rose $5.00. Find the net change in price for the day.

Add. (Section 9.3)

19. $(-10) + 4 + (-12)$

20. $(-5) + (-12) + (-9)$

21. $12 + (-10) + (-32) + 60$

22. $(-4) + 15 + 16 + (-24)$

23. $(-2) + 14 + 6 + (-7)$

24. $5 + (-4) + 16 + (-19)$

Solve. (Section 9.4)

25. Find the net excess or deficiency of rainfall.

May	June	July	Aug.	Sept.	Oct.
−1.1 cm	2.1 cm	0.9 cm	1.6 cm	−0.8 cm	−1.3 cm

26. Find Friday's stock price to the nearest cent.

Beginning price: $41\frac{1}{4}$

M	T	W	Th	F
$-\frac{3}{8}$	$1\frac{1}{8}$	$\frac{1}{4}$	$-\frac{5}{8}$	$-\frac{1}{8}$

9.5 Subtraction

Notice that the answers to the two problems below are the same.

Subtraction: $12 - 9 = 3$ **Addition:** $12 + (-9) = 3$

In words, the ***difference,*** $12 - 9$, is the same as the sum, $12 + (-9)$. This suggests the following.

> To subtract b from a, add the opposite of b to a.
>
> $$\boldsymbol{a - b = a + (-b)}$$
>
> $0 - 23 = 0 + (-23)$
>
> $1.2 - 0.5 = 1.2 + (-0.5)$
>
> $3\frac{3}{4} - (-\frac{1}{4}) = 3\frac{3}{4} + \frac{1}{4}$

P–1 **Write as a sum.**

a. $6 - 14$ **b.** $24 - 37$ **c.** $0.5 - 2.1$

EXAMPLE 1 Subtract: $7 - 12$

Solution: 1 Write as a sum. ⟶ $7 - 12 = 7 + (-12)$

2 Add. ⟶ $= -5$

P–2 **Subtract.**

a. $6 - 7$ **b.** $106 - 212$ **c.** $4.6 - 5.3$

EXAMPLE 2 Subtract: $13 - (-18)$

Solution: 1 Write as a sum. ⟶ $13 - (-18) = 13 + 18$

2 Add. ⟶ $= 31$

P–3 **Subtract.**

a. $12 - (-14)$ **b.** $43 - (-51)$ **c.** $8.8 - (-12.1)$

EXAMPLE 3 Subtract: $(-18) - (-9)$

Solution:

$$\begin{aligned}(-18) - (-9) &= (-18) + 9 \\ &= -9\end{aligned}$$

P–4 **Subtract.**

a. $-16 - (-12)$ **b.** $-231 - (-48)$ **c.** $(-6.6) - (-3.6)$

Subtraction is indicated when you are asked to find the temperature difference in the following word problem.

EXAMPLE 4 The low temperature on a winter day was −20.3°C. The day's high temperature was −4.9°C. How much greater was the high temperature than the low temperature?

Solution:

[1] Subtract the low temperature from the high temperature. ⟶ $(-4.9) - (-20.3)$

[2] Write as a sum. ⟶ $(-4.9) - (-20.3) = (-4.9) + 20.3$

[3] Add. ⟶ $= 15.4$

The day's high temperature was 15.4°C greater than the low temperature.

The dash symbol, −, is used to show subtraction. You know that it is also used to show a negative number and the opposite of a number.

Summary of the "Dash" Symbol

1. In $a - b$, the dash means subtraction.
   "$a - b$" means "a minus b" or "a less b."
2. In $a + (-b)$, the dash means "the opposite of."
   "$a + (-b)$" means "a plus the opposite of b."
3. The dash is also used to indicate a negative number.
   "−5" is the number "negative five."

CLASSROOM EXERCISES

Write as a sum. (Step 1, Examples 1–3)

1. $14 - 5$ **2.** $2 - 9$ **3.** $5 - (-14)$ **4.** $(-6) - 8$

5. $16 - (-7)$ **6.** $(-1) - (-10)$ **7.** $0 - 17$ **8.** $0 - (-8)$

9. $13 - 19$ **10.** $(-4) - (-23)$ **11.** $2.6 - (-14.9)$ **12.** $8.1 - (-3.3)$

For Exercises 13–36, subtract the given numbers. (Example 1)

13. $5 - 8$	**14.** $3 - 13$	**15.** $0 - 15$	**16.** $27 - 30$
17. $5 - 17$	**18.** $6 - 10$	**19.** $2.3 - 4.7$	**20.** $3.5 - 6.9$

(Example 2)

21. $4 - (-5)$	**22.** $12 - (-4)$	**23.** $(-14) - 3$	**24.** $(-9) - 11$
25. $6 - (-6)$	**26.** $3 - (-2)$	**27.** $8.4 - (-3.2)$	**28.** $0.9 - (-1.8)$

(Example 3)

29. $(-6) - (-3)$	**30.** $(-4) - (-10)$	**31.** $(-13) - (-13)$	**32.** $(-12) - (-8)$
33. $(-7) - (-2)$	**34.** $(-1) - (-3)$	**35.** $(-0.8) - (-0.9)$	**36.** $(-5.8) - (-2.5)$

Solve. (Example 4)

37. Subtract −15 from 17.

38. Subtract −24 from 11.

39. Find a number that is 16 less than 7.

40. Find a number that is 25 less than 9.

41. Find how much greater −6 is than −21.

42. Find how much greater −12 is than −35.

WRITTEN EXERCISES

Goal: To subtract real numbers
Sample Problem: $(-8) - 2$ **Answer:** -10

For Exercises 1–60, subtract the given numbers. (Example 1)

1. $11 - 15$	**2.** $17 - 22$	**3.** $19 - 26$	**4.** $16 - 24$
5. $0 - 24$	**6.** $0 - 16$	**7.** $18 - 43$	**8.** $37 - 84$
9. $0.24 - 0.93$	**10.** $0.53 - 0.87$	**11.** $3\frac{1}{2} - 5\frac{1}{4}$	**12.** $4\frac{3}{4} - 6\frac{1}{4}$

(Example 2)

13. $18 - (-7)$	**14.** $20 - (-12)$	**15.** $24 - (-29)$	**16.** $31 - (-39)$
17. $0 - (-36)$	**18.** $0 - (-50)$	**19.** $9 - (-48)$	**20.** $5 - (-76)$
21. $9.5 - (-13.9)$	**22.** $8.9 - (-17.3)$	**23.** $3\frac{3}{4} - (-1\frac{3}{4})$	**24.** $5\frac{1}{2} - (-7\frac{3}{4})$

(Example 3)

25. $(-13) - (-5)$	**26.** $(-21) - (-7)$	**27.** $(-18) - (-43)$	**28.** $(-26) - (-33)$
29. $(-3) - (-2)$	**30.** $(-5) - (-2)$	**31.** $(-9) - (-23)$	**32.** $(-11) - (-25)$
33. $(-0.9) - (-1.7)$	**34.** $(-0.8) - (-2.7)$	**35.** $(-\frac{5}{8}) - (-\frac{3}{4})$	**36.** $(-\frac{3}{8}) - (-\frac{13}{16})$

MIXED PRACTICE

37. $(-127) - (-127)$
38. $(-4) - (-18)$
39. $0 - (-7)$
40. $(-3) - (-2)$
41. $(-128) - 43$
42. $34 - (-41)$
43. $29 - (-106)$
44. $(-6) - (-59)$
45. $76 - (-58)$
46. $0 - (-5)$
47. $(-9) - (-12)$
48. $48 - 236$
49. $(-7) - (-2)$
50. $48 - (-134)$
51. $62 - (-50)$
52. $(-254) - 97$
53. $(-0.9) - (-2.8)$
54. $(-64) - (-64)$
55. $39 - 124$
56. $4.8 - (-6.7)$
57. $2\frac{3}{8} - 5\frac{1}{4}$
58. $1\frac{7}{16} - 3\frac{7}{8}$
59. $5.4 - (-2.9)$
60. $3.1 - (-0.8)$

APPLICATIONS

Write an expression using subtraction. Then solve. (Example 4)

61. The record low temperature for a certain city is $-30.1°C$. Last year the low temperature was $-25.8°C$. How much less is the record temperature than last year's low temperature?

62. The low temperature for a city on a winter day was $-5.7°C$. The high temperature for the same day was $-1.4°C$. How much less was the low than the high temperature?

63. The net number of yards gained by a football player in a game was -16. The net number of yards gained by a second player was 27. How many more yards did the second player gain than the first?

64. In a game, a football player gained 24 yards. In the next game he gained -12 yards. How many more yards did the player gain in the first game than in the second game?

REVIEW CAPSULE FOR SECTION 9.6

Write the missing numerals in each addition exercise below. (Pages 251–253)

1. $8 + (-3) + (-2)$
   $8 + \underline{\ ?\ }$
   $\underline{\ ?\ }$

2. $12 + (-6) + (-10)$
   $6 + \underline{\ ?\ }$
   $\underline{\ ?\ }$

3. $5 + (-8) + (-1)$
   $\underline{\ ?\ } + (-1)$
   $\underline{\ ?\ }$

4. $24 + (-18) + (-14) + (-5)$
   $6 + \underline{\ ?\ }$
   $\underline{\ ?\ }$

5. $60 + (-30) + (-10) + (-1)$
   $60 + \underline{\ ?\ } + (-1)$
   $\underline{\ ?\ }$

6. $1\frac{1}{2} + (-2\frac{3}{4}) + (-5\frac{1}{4}) + 7$
   $1\frac{1}{2} + \underline{\ ?\ } + 7$
   $\underline{\ ?\ }$

7. $(-\frac{3}{5}) + 2\frac{1}{3} + (-1\frac{2}{3}) + 5\frac{2}{5}$
   $1\frac{11}{15} + \underline{\ ?\ }$
   $\underline{\ ?\ }$

9.6 Subtraction Practice

Since you can write a difference as a sum, you can use the Commutative and Associative Properties of Addition as a basis for subtracting more than two real numbers. It is important to write all expressions as sums before applying the properties.

EXAMPLE 1 Subtract: $3-5-7$

Solution: Two methods are shown. In both methods the first step is to write $3-5-7$ as a sum.

Method 1

$$3-5-7=3+(-5)+(-7)$$
$$=(-2)+(-7)$$
$$=-9$$

Add from left to right.

Method 2

$$3-5-7=3+(-5)+(-7)$$
$$=3+(-12)$$
$$=-9$$

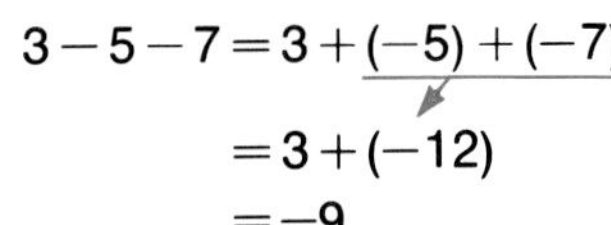

First add the negative numbers.

EXAMPLE 2 Subtract: $14-7-6-22$

Solution: Method 1

$$14-7-6-22=14+(-7)+(-6)+(-22)$$
$$=\quad 7 \quad + \quad (-28)$$
$$=-21$$

Method 2

$$14-7-6-22=14+(-7)+(-6)+(-22)$$
$$=14+(-13)+(-22)$$
$$=14+(-35)$$
$$=-21$$

P–1 **Subtract.**

a. $12-15-4$ **b.** $81-45-51-22$ **c.** $1.7-3.3-4.4-1.5$

CLASSROOM EXERCISES

For Exercises 1–15, subtract the given numbers. (Example 1)

1. $11-16-5$ **2.** $14-20-6$ **3.** $30-50-5$

4. $12-7-19$ **5.** $(-8)-4-(-7)$ **6.** $3-(-10)-(-7)$

7. $22-(-11)-3$ **8.** $14-5-(-1)$ **9.** $4.3-(-2.5)-3.2$

(Example 2)

10. $12 - 15 - 5 - 4$ **11.** $29 - 9 - 3 - 8$ **12.** $1 - 2 - 3 - 5$

13. $2 - 3 - 9 - 1$ **14.** $40 - (-50) - 20 - (-5)$ **15.** $1.1 - 2.1 - (-3.1) - 4.1$

WRITTEN EXERCISES

Goal: To subtract more than two real numbers
Sample Problem: $6 - 9 - (-4)$ **Answer:** 1

For Exercises 1–32, subtract the given numbers. (Example 1)

1. $12 - 29 - 6$ **2.** $20 - 43 - 13$ **3.** $8 - (-13) - 29$

4. $4 - (-29) - 37$ **5.** $(-14) - (-12) - 18$ **6.** $(-21) - (-15) - 24$

7. $(-9) - (-43) - 57$ **8.** $(-7) - (-35) - 88$ **9.** $37 - 53 - (-128)$

10. $19 - 42 - (-75)$ **11.** $2.7 - 5.4 - (-9.5)$ **12.** $1.7 - 4.6 - (-10.3)$

(Example 2)

13. $3 - 4 - 1 - 7$ **14.** $8 - 4 - 7 - 2$

15. $8 - 13 - 24 - 2$ **16.** $5 - 10 - 3 - 18$

17. $85 - 18 - 41 - 96$ **18.** $123 - 29 - 48 - 57$

19. $12 - (-28) - 17 - (-39)$ **20.** $(-14) - 29 - (-11) - 23$

21. $-0.6 - (-4.9) - 1.6 - 4.3$ **22.** $-1.7 - (-0.4) - (-8.3) - 5.6$

23. $-1\frac{1}{4} - (-2\frac{1}{2}) - 1\frac{3}{4} - 9\frac{3}{4}$ **24.** $3\frac{1}{8} - 1\frac{5}{8} - (-2\frac{1}{2}) - (-3\frac{7}{8})$

MIXED PRACTICE

25. $-18 - (-23) - 42 - 59$ **26.** $-67 - 139 - 21 - 10$

27. $24.8 - 50.3 - 10.5 - 6.2$ **28.** $23 - 51 - 11$

29. $12 - 14 - 7$ **30.** $0 - 28.3 - (-28.3)$

31. $(-5\frac{1}{2}) - 4\frac{1}{4} - (-2\frac{1}{4})$ **32.** $(-2\frac{3}{8}) - 3\frac{1}{4} - (-4\frac{3}{8})$

REVIEW CAPSULE FOR SECTION 9.7

Find the expression, **a, b,** *or* **c,** *that is not equivalent to the given expression.* (Pages 35–36)

1. $2x + (-5x)$
a. $(2 + (-5))x$
b. $-3x$
c. $3x$

2. $(-1.3t) - (-4.8t)$
a. $t(-1.3 - (-4.8))$
b. $(-1.3 - (-4.8))t$
c. $6.1t$

3. $-5y^2 + 3y$
a. $-2y^3$
b. $(-5y + 3)y$
c. $3y + (-5y^2)$

4. $(-5.7r^2) + (-r^2)$
a. $-5.7r^2$
b. $(-5.7 + (-1))r^2$
c. $-6.7r^2$

9.7 Simplifying Algebraic Expressions

Algebraic expressions can sometimes be simplified by writing differences as sums or by writing sums as differences.

EXAMPLE 1 Simplify: $-8.6 - (-n)$

Solution: Write the expression as a sum: $-8.6 - (-n) = -8.6 + n$

The expression $-8.6 + n$ is in simplest form because it has fewer symbols than $-8.6 - (-n)$.

P–1 **Simplify: $-7 - (-y)$**

Recall from Section 2.6 that like terms can be combined in algebraic expressions. ***Like terms*** have the same variables, and the same powers of these variables.

EXAMPLE 2 Simplify: $9n - 12 - 4n$

Solution:

1. Write a sum for each difference. ⟶ $9n - 12 - 4n = 9n + (-12) + (-4n)$
2. Group like terms. Combine like terms. ⟶ $= 9n + (-4n) + (-12)$
3. Simplify. ⟶ $= 5n + (-12)$

$= 5n - 12$ ◀ *Simplest form*

P–2 **Simplify: $12p - 21 - 2p$**

EXAMPLE 3 Simplify: $7 + x - 10 - 5x$

Solution:

1. Write a sum for each difference. ⟶ $7 + x - 10 - 5x = 7 + x + (-10) + (-5x)$
2. Group like terms. Combine like terms. ⟶ $= (7 + (-10)) + (x + (-5x))$
3. Simplify. ⟶ $= -3 + (-4x)$

$= -3 - 4x$

P–3 **Simplify.**

a. $4n-10-12n-16$ **b.** $-2z-(-1)-1-z$ **c.** $-1.5-0.8c-9.9c-5$

EXAMPLE 4

Simplify: $-8n^2-n-5n+3n^2-28$

Solution:
$$\begin{aligned}-8n^2-n-5n+3n^2-28 &= -8n^2+(-n)+(-5n)+3n^2+(-28)\\ &= (-8n^2+3n^2)+(-n+-5n)+(-28)\\ &= (-5n^2)+(-6n)+(-28)\\ &= -5n^2-6n-28\end{aligned}$$

P–4 **Simplify.**

a. $t^2+6t-12-4t^2-9t$ **b.** $4.6+3.8f-f^2-5.1f-f^2$

CLASSROOM EXERCISES

For Exercises 1–32, simplify.
(Example 1)

1. $14-(-y)$ **2.** $(-19)-(-t)$ **3.** $u-(-27)$ **4.** $(-r)-(-26)$
5. $4-(-a)$ **6.** $k-(-4)$ **7.** $(-4.2)-(-p)$ **8.** $-(-w)-(-1.9)$

(Example 2)

9. $3y-2y-4$ **10.** $p-3-5p$ **11.** $x+(-3)-2x$ **12.** $2n+6-n$
13. $3a-1-5a$ **14.** $-2b-1+3b$ **15.** $1.7-1.3t-2.5t$ **16.** $-0.7q-1.4-0.3q$

(Example 3)

17. $2m+6-m-5$
18. $z-5-3z-7$
19. $-s-1-2s-3$
20. $-12-5y-y+25$
21. $6x-4x+1-3$
22. $-4-2y-y-2$
23. $6.1g-1.9g-2.3-3.5$
24. $-0.9c+1.2-3.9-2.3c$

(Example 4)

25. y^2-7-2y^2+3
26. p^2-8-p^2+5
27. $9s-8s^2+2s^2-4s-11$
28. $2x^2-x-5x^2-2x+5$
29. $g^2-g-3+g-g^2$
30. $-k^2-k+1+k-k^2$
31. $2.5w+3.6-7w^2-4.3w-1.4w^2$
32. $3.2b^2-0.9+7.3b-4.5-12.2b$

WRITTEN EXERCISES

Goal: To simplify algebraic expressions using subtraction
Sample Problem: $7 - 2x - 10 - 4x$
Answer: $-6x - 3$

For Exercises 1–48, simplify.
(Example 1)

1. $52 - (-k)$
2. $(-65) - (-t)$
3. $m - (-4)$
4. $(-v) - (-21)$
5. $r - (-86)$
6. $(-193) - (-y)$
7. $-4\frac{1}{2} - (-t)$
8. $n - (-1\frac{7}{8})$

(Example 2)

9. $5k - 7 - 4k$
10. $-8p - 3 - 2p$
11. $-3s + 7 - 12s$
12. $13y - 5y + 6$
13. $-15x - 6 + 8x$
14. $24a + 7 - 15a$
15. $9.6n - 3.4 - 11.2n$
16. $-4.3m - 1.8 + 3.9m$

Example 3)

17. $2a - 3 - 5a - 8$
18. $5t - 1 - 9t - 5$
19. $-12y - 15 - y - 28$
20. $-15k - 23 - 49 - k$
21. $9 - 13p - 24 - 38p$
22. $28q - 8 - 19q - 34$
23. $-3\frac{1}{4}w - \frac{1}{2} - 6\frac{1}{2}w - \frac{3}{4}$
24. $-7\frac{5}{8} - \frac{5}{8}h - \frac{7}{8} - 3\frac{3}{8}h$

(Example 4)

25. $-x^2 - 3x - x - 12x^2$
26. $-3j^2 - 14j - j - j^2$
27. $-12s - s^2 - 11s^2 - s - 13$
28. $-23m - 9m^2 - 9m - m^2 - 25$
29. $-7r^2 - 3r - r - 9r^2 - 2$
30. $-5w - 14w^2 - 10w - 3 - 11w^2$
31. $19.3n^2 - 8.7n - 9.2 - 8.9n^2 - 4.5n$
32. $-0.8r^2 - 12.6r - 1.9r - 13.6 - 6.3r$

MIXED PRACTICE

33. $5x - 8 - x - 12$
34. $-q - (-1.2)$
35. $1.3t - 0.5 - 9.6t$
36. $11r - 5r - 18r$
37. $0 - (-5w)$
38. $87s - 59 - 149s - (-108)$
39. $0.5y - 1.2 - 9.3y - 2.5$
40. $112a - a^2 - 94 - 203a - 3a^2$
41. $3a - 5a - 9 - 14$
42. $r - 10r - 13 - 14$
43. $16 - (-2t)$
44. $0.1w - 1.5w - 3.9$
45. $\frac{2}{3}w - (-\frac{1}{6}w) - 4$
46. $-\frac{7}{6}n - \frac{5}{3} - \frac{17}{12}n - \frac{1}{3}$
47. $\frac{1}{2}c - (-19) - \frac{7}{8}c$
48. $\frac{3}{4}x^2 - \frac{7}{2}x - \frac{1}{4} - \frac{5}{4}x^2 - \frac{9}{2}$

Focus on Reasoning: Comparisons

Many problems involving the **comparison of two quantities** can be solved by logical reasoning. Little or no computation with paper and pencil may be necessary.

Refer to these instructions for the Examples and Exercises.

Each problem consists of two quantities, one in Column I and one in Column II. Compare the quantities.

- Write **A** if the quantity in Column I is greater.
- Write **B** if the quantity in Column II is greater.
- Write **C** if the two quantities are equal.
- Write **D** if there is not enough information to determine how the two quantities are related.

EXAMPLE 1

Column I	Column II
$9 \times 682 \times 7$	$10 \times 682 \times 6$

Solution: Since 682 appears in both products, compare (9×7) and (10×6). Since (9×7) is greater than (10×6), then $9 \times 682 \times 7$ is greater than $10 \times 682 \times 6$.

Answer: **A**

EXERCISES

	Column I	Column II
1.	$7 \times 5 \times 8 \times 9$	$4 \times 10 \times 63$
2.	$(5)(121)(6)$	$(11^2)(5^2)$
3.	$(2)(4)(6)(8)(10)(12)(14)$	$(16)(14)(12)(10)(8)(6)$
4.	$\dfrac{8 \cdot 8 \cdot 8}{8 + 8 + 8}$	1
5.	0.3	$\dfrac{0.7}{2}$
6.	$\dfrac{0.1}{2}$	$\dfrac{2}{0.1}$

Sometimes additional information is given about the quantities being compared.

EXAMPLE 2

Column I		Column II
	$8x = 128$	
$\frac{x}{8}$		2

Solution: Since $8x = 128$, $x = 16$.

Since $x = 16$, $\frac{x}{8} = 2$.

Then the quantities in Columns 1 and 2 are equal.

Answer: **C**

EXERCISES

	Column I		Column II
7.		$5x = 120$	
	5		$\frac{x}{12}$
8.		$x^2 = xy$ and $x \neq 0$	
	x		y
9.		$\frac{x}{y} = 1$ and $y \neq 0$	
	x^2		y^2
10.		z is greater than y, $y \neq 0$, $z \neq 0$	
	$\frac{y}{z}$		$\frac{y}{z} \cdot \frac{z}{y}$
11.		Six gold coins and one silver coin have the same value as eight silver coins and five gold coins.	
	8 silver coins		1 gold coin
12.		n is the reciprocal of the smallest number divisible by 2, 4, 5, and 10.	
	$10n$		$n + 0.1$

CHAPTER SUMMARY

IMPORTANT TERMS

Difference *(p. 261)*

Like terms *(p. 267)*

IMPORTANT IDEAS

1. Steps for Adding Two Negative Numbers
   - **a.** Add the absolute values.
   - **b.** Write the opposite of the result.
2. Steps for Adding a Positive and Negative Number
   - **a.** Subtract the absolute values.
   - **b.** If the positive addend has the greater absolute value, write the difference.
     If the negative addend has the greater absolute value, write the opposite of the difference.
3. *Addition Property of 0:* For any real number a, $a + 0 = a$ and $0 + a = a$.
4. *Addition Property of Opposites:* For any real number a, $a + (-a) = 0$ and $(-a) + a = 0$.
5. *Commutative Property of Addition:* For any real numbers a and b, $a + b = b + a$.
6. *Associative Property of Addition:* For any real numbers a, b, and c, $(a + b) + c = a + (b + c)$.
7. If a and b are any real numbers, then $a - b = a + (-b)$. To subtract b from a, add the opposite of b to a.
8. Summary of the "Dash" symbol
   See page 262.

CHAPTER REVIEW

SECTION 9.1

Use a number line to add.

1. $5 + (-7)$ **2.** $4 + (-9)$ **3.** $6 + (-8)$ **4.** $3 + (-6)$

5. $(-5) + 5$ **6.** $(-3) + 0$ **7.** $-9 + 7$ **8.** $(-7) + 6$

9. $(-3) + 8$ **10.** $(-5) + 9$ **11.** $(-5) + (-7)$ **12.** $(-6) + (-8)$

SECTION 9.2

Add.

13. $(-6) + (-11)$ **14.** $(-21) + (-19)$ **15.** $(-11) + (-13)$ **16.** $(-43) + (-75)$

17. $61 + (-42)$ **18.** $133 + (-87)$ **19.** $(-39) + 78$ **20.** $(-50) + 72$

21. $14.2 + (-21.1)$ **22.** $17.4 + (-38.2)$ **23.** $(-7\frac{1}{2}) + 5\frac{2}{3}$ **24.** $(-12\frac{3}{4}) + 10\frac{2}{3}$

Solve.

25. During one day, the temperature fell 8° between midnight and noon, and rose 14° between noon and midnight. Find the net change in temperature that day.

26. During one business day, the trading price of an ounce of silver fell $3.00 in the morning, and fell $4.00 more in the afternoon. Find the net change in price for the day.

SECTION 9.3

Add.

27. $(-12) + 17 + (-8)$

28. $(-16) + 23 + (-11)$

29. $(-12) + (-19) + (-27)$

30. $(-18) + (-25) + (-37)$

31. $(-13) + 27 + (-16) + (-9)$

32. $(-20) + (-33) + 24 + (-17)$

33. $(-42.7) + (-18.3) + 19.6 + (-5.9)$

34. $34.6 + (-22.3) + (-19.6) + 12.5$

35. $6\frac{1}{4} + (-5\frac{1}{2}) + (-2\frac{3}{4}) + (\frac{3}{4})$

36. $3\frac{3}{4} + (-2\frac{1}{4}) + (-1\frac{3}{4}) + \frac{1}{2}$

SECTION 9.4

Solve.

37. Find the net excess or deficiency of rainfall.

Feb.	March	April	May	June	July
−0.9 cm	1.1 cm	−2.1 cm	−1.4 cm	0.6 cm	0.8 cm

38. Find the stock's price at the end of the week. Give your answers to the nearest cent.

Beginning price: 28

M	T	W	Th	F
$-1\frac{1}{2}$	$\frac{3}{4}$	$-\frac{5}{8}$	$\frac{7}{8}$	$\frac{1}{4}$

39. An accountant enters the following amounts in an account record. Consider a <u>debit</u> *(DR)* as positive and a <u>credit</u> *(CR)* as negative. Find the net sum of the amounts and whether it is a debit or credit.

$21.56 *(DR)*; $8.73 *(CR)*; $102.19 *(CR)*; $145.24 *(DR)*; $89.51 *(DR)*

40. A company showed the following <u>Profit</u> *(P)* and <u>Loss</u> *(L)* results for each of the first six months of a year. Find the net profit or loss for the six months.

Jan.	Feb.	March	April	May	June
$12,337 *(L)*	$8419 *(L)*	$1234 *(P)*	$3506 *(P)*	$4836 *(L)*	$2946 *(P)*

SECTION 9.5

Subtract.

41. $25 - 18$

42. $23 - 37$

43. $123 - 58$

44. $79 - 103$

45. $13 - 9$

46. $10 - 15$

47. $15 - (-8)$

48. $24 - (-5)$

49. $5.8 - (-12.3)$

50. $11.9 - (-13.2)$

51. $-1\frac{3}{4} - (-3\frac{1}{2})$

52. $-3\frac{2}{3} - (-4\frac{1}{6})$

Solve.

53. The highest point in California is 14,494 feet above sea level (+). The lowest point is 282 feet below sea level (−). Find the difference in altitude between these two points.

54. The lowest temperature ever recorded in a certain city was −16°F. The highest temperature ever recorded was 114°F. How much less was the low than the high temperature?

SECTION 9.6

Subtract.

55. $26 - 39 - 5$

56. $19 - (-17) - 47$

57. $-12 - 15 - 31 - 16$

58. $-16 - 17 - 18 - (-19)$

59. $2.3 - 4.5 - 6.8$

60. $7.5 - 10.2 - (-9.4)$

61. $-3\frac{1}{2} - 2\frac{1}{3} - 5\frac{5}{6}$

62. $4\frac{3}{4} - (-2\frac{1}{2}) - 8\frac{1}{4}$

SECTION 9.7

Simplify.

63. $16 - (-2s)$

64. $(-p) - (-45)$

65. $5a^2 - 12a + 6 - a^2 - 4a$

66. $2x^2 - 7 - x^2 - 8x + 3$

67. $0.3w - 12 - 1.2w$

68. $-0.8k - 6.1 - 2.3k$

69. $\frac{1}{2}x - \frac{4}{5} + 1\frac{2}{3} - \frac{4}{5}x - \frac{1}{3}$

70. $3\frac{1}{2}y + 1\frac{3}{8} - \frac{5}{8} - 1\frac{4}{5}y - 2\frac{1}{9}y^2$

CUMULATIVE REVIEW: CHAPTERS 1–9

Evaluate each expression. (Section 1.2)

1. $15a - 3c$ when $a = 2$ and $c = 5$

2. $t - 16v \div 4$ when $t = 48$ and $v = 2$

Multiply. (Section 2.2)

3. $7w(8)$

4. $(0.4t)(15m)$

5. $(18a)(2ab)$

6. $(9h^2k)(7hk)$

Write each product as a sum or difference. (Section 2.3)

7. $8(9 - 3n)$

8. $(7t + 6)5$

9. $(3x + 4)5x$

10. $2p(7p - 9)$

Perform the indicated operations. Write each answer in lowest terms. (Sections 4.3 through 4.6)

11. $\frac{3x}{4} \cdot \frac{8}{9x}$

12. $\frac{4m}{15h} \cdot \frac{5h}{2m^2}$

13. $\frac{8m^2}{3k} \div 24mk$

14. $\frac{5b^4}{2a^2} \div \frac{b^3}{10a}$

15. $\frac{6}{7e} - \frac{4}{7e}$

16. $\frac{3x}{10} + \frac{2x}{10}$

17. $\frac{7y}{10} + \frac{y}{6}$

18. $\frac{4}{5d} - \frac{1}{2d}$

Solve and check. (Section 5.7)

19. Dave drove 457 miles yesterday. The number of miles that Dave drove is 3 times the number of miles that Alan drove. How many miles did Alan drive?

20. Sue earned \$45 less than Maria earned last week. Sue earned \$286 last week. How much did Maria earn?

Solve and check. Show all steps. (Sections 6.1, 6.2, and 6.4)

21. $23 = \frac{w}{8} + 17$

22. $109 = 12k + 7 - 9k$

23. $27 = \frac{1}{3}(6x - 9) + 2x$

A jar contains 7 red pens, 3 blue pens, and 12 green pens. A pen is chosen at random. Find the probability of choosing each of the following. (Section 7.2)

24. A blue pen

25. A green pen

26. A pen that is not green

Solve each proportion. (Section 7.3)

27. $\frac{3}{x} = \frac{9}{15}$

28. $\frac{14}{35} = \frac{y}{10}$

29. $\frac{c}{10} = \frac{8}{25}$

Write a per cent for each number. (Section 7.4)

30. $\frac{3}{25}$

31. $\frac{8}{5}$

32. $\frac{5}{12}$

33. 0.17

34. 0.075

Write a proportion and solve. (Section 7.5)

35. 72 is 60% of what number?

36. What number is $33\frac{1}{3}$% of 96?

Solve each problem. (Sections 7.6 through 7.9)

37. Of the 80 questions on a Spanish test, Fred answered 52 correctly. What per cent of the items did he answer correctly?

38. A basketball player made 80% of her attempted free throws during a game. She made 12 of her attempted free throws. How many did she attempt?

39. The yearly interest rate on a loan of $1500 is 8%. Find the interest due after 2 years.

40. A snowmobile listed at $1250 is reduced by $500. Find the rate of discount.

Find each square root. (Section 8.4)

41. $\sqrt{64}$

42. $\sqrt{\frac{9}{16}}$

43. $-\sqrt{576}$

44. $\sqrt{4.84}$

Solve each problem. (Section 8.5)

45. A 15-foot ladder leans against a house and touches the house 12 feet above the ground. How far is the base of the ladder from the house?

46. A boat sails 10 miles east from Pearl Bay, then 24 miles south to Hudson Harbor. Find the straight-line distance from Pearl Bay to Hudson Harbor.

Add. (Sections 9.1 through 9.3)

47. $(-5) + (-8)$

48. $5 + (-7)$

49. $(-6) + 16$

50. $(-2.7) + (-19.3)$

51. $(-7) + 9 + (-120)$

52. $42 + (-56) + 16$

Solve. (Section 9.4)

53. An activities fund had a balance of $230 on Monday. The following deposits (*D*) and withdrawals (*W*) were made during the week. Find the balance of the account at the end of the week.

$121.00(*D*); $73.00(*W*); $83.50(*D*); $186.25(*W*); $36.15(*D*)

Subtract. (Sections 9.5 and 9.6)

54. $9 - (-9)$

55. $23 - 70$

56. $1.6 - (-9.6)$

57. $13 - 20 - 19$

58. $-7 - (-21) - 30$

59. $(-15) - 9 - (-12)$

Simplify. (Section 9.7)

60. $5x - 9 - (-13) - 7x$

61. $31a - 6a^2 - (-10a) - 19a^2$

CHAPTER 10

Real Numbers: Multiplication and Division

Stocks and bonds are bought and sold in the **stock exchange.** Prices of stock change according to business conditions. The net change in the price is shown with positive and negative real numbers. A net change of $-1\frac{1}{4}$ means that the price of a stock went down \$1.25.

Stocks

52-Week High	52-Week Low	Stock	Div	Yld %	PE Ratio	Sales 100s	High	Low	Last	Chg.
		M N O P								
16⅜	11⅝	MACOM	.24	1.6	9	507	15⅝	15¼	15¼	− ¼
17⅞	10⅛	MAIBF		...	12	147	17¾	17⅜	17½	+ ⅛
22⅞	18½	MBIA n	.16	.7	12	162	21½	21⅛	21½	...
64½	37⅞	MCA	.68	1.1	29	3769	61¾	60¼	60½	−1
21¼	7½	MCorp		...	...	875	7¾d	7⅜	7¾	+ ⅛
43	32⅝	MCor pf	3.50	10.8	...	3	32½d	32½	32½	− ⅛
19¼	13¾	MDCA n	1.15e	6.2	...	504	18⅝	18⅛	18⅝	...
17⅛	8¼	MDC	.40	4.0	7	330	10⅛	9⅞	10	...
27½	20	MDU	1.42	6.5	15	213	21⅞	21½	21⅞	+ ¼
10	6⅞	MEI		...	119	122	7¼	7	7⅛	+ ⅛
10¾	9½	MMT n	1.23	12.3	...	1329	10	9⅞	10	...
10¾	8⅞	MFM n	.73	7.7	...	259	9½	9⅜	9½	+ ⅛
10¼	9⅞	MGI n	1.18	11.9	...	1024	10⅛	9⅞	9⅞	− ¼
14⅛	8	MGMUA		...	...	430	10⅜	10	10⅜	− ⅛
4	⅞	MHI Gp		...	...	277	3¼	3	3⅛	− ⅛
74¼	39½	Macmil	.72	1.0	29	1314	73¾	71¼	73¼	+2¼
17½	11⅝	Malaysa n		...	...	318	14⅞	14⅝	14⅝	+ ¼
16¾	11½	Manhln	.20b	1.3	...	124	15⅞	15⅝	15⅝	− ⅛
12½	7¼	ManhNt		...	...	13	7⅞	7⅞	7⅞	...
23⅞	15¾	ManrCr	.13	.7	17	912	17⅝	17⅛	17½	+ ¼
82⅛	33½	Manpwr		...	38	3102	82⅛	82	82	...
49⅛	37¼	MfrHan	3.28	7.5	...	1660	44¼	43¼	43½	+ ⅛
53¾	45½	MfrH pf	4.16e	8.6	...	210	49	48⅝	48⅝	...
52½	41¾	MfrH pf	3.82e	8.9	...	15	43¾	43⅛	43⅛	− ⅛
5⅜	1⅝	vjManvl		...	1	308	4	3¾	3⅞	...
29	17⅜	vjMnvl pf		...	...	95	27⅛	27	27⅛	...
66⅝	46¾	MAPCO	1.00	1.7	22	83	59⅝	59	59	− ¼
7⅝	15/16	Marcde		...	...	167	5⅜	5⅛	5⅜	+ ⅛
78	44⅜	MarMid	2.04	2.6	...	532	78	77½	78	+ ½
41½	16¾	Marion s	.20	.5	64	7203	40¼	38⅞	39¼	+ ¼
9⅞	8⅞	Martr n	.25e	2.8	...	25	9	9	9	...
26	10⅛	MarkC	.32	1.2	39	633u	26¾	25⅝	26⅜	+ ½
34¾	16⅜	Mark pf	1.20	3.4	...	29u	35½	34½	35½	+1
43¾	27⅜	Marriot	.16	.4	29	3220	39½	38½	39¼	+ ⅝
76¾	56	MrshMc	2.40	3.4	18	2042	70¾	69¾	70	− ⅜
20⅜	8⅞	Msrhln s		...	30	117u	20¾	20⅜	20½	+ ¼
56⅛	37⅜	MartM	1.10	2.1	14	893	53⅞	53	53⅜	− ⅜
40⅞	23½	Masco	.40	1.0	22	5242	39¼	38⅜	38¾	+ ½
50⅜	39¼	MasCp	3.60	9.1	...	26	40	39¾	39¾	...
14¼	10⅞	MasInc	1.14e	10.4	...	x95	11 d	10¾	11	+ ⅛
178¼	93¼	MatsuE	.48e	.3	...	431	171½	169¾	171½	+3½
13½	7¾	Mattel		...	...	1342	12¾	12½	12¾	+ ⅜
11½	9⅞	MauLoa	1.05e	9.9	11	55	10½	10¼	10⅜	+ ¼
16	11	Maxus n		...	...	3379	14	13⅝	13⅞	+ ¼
46¾	38	Maxus pf	4.00	9.1	...	5	43¾	43½	43¾	− ¼
14½	8⅜	Maxam		...	...	53	13½	13⅜	13⅜	...
50⅞	32½	MayDS	1.14	2.3	18	6784	49½	48¼	49⅛	+1⅛

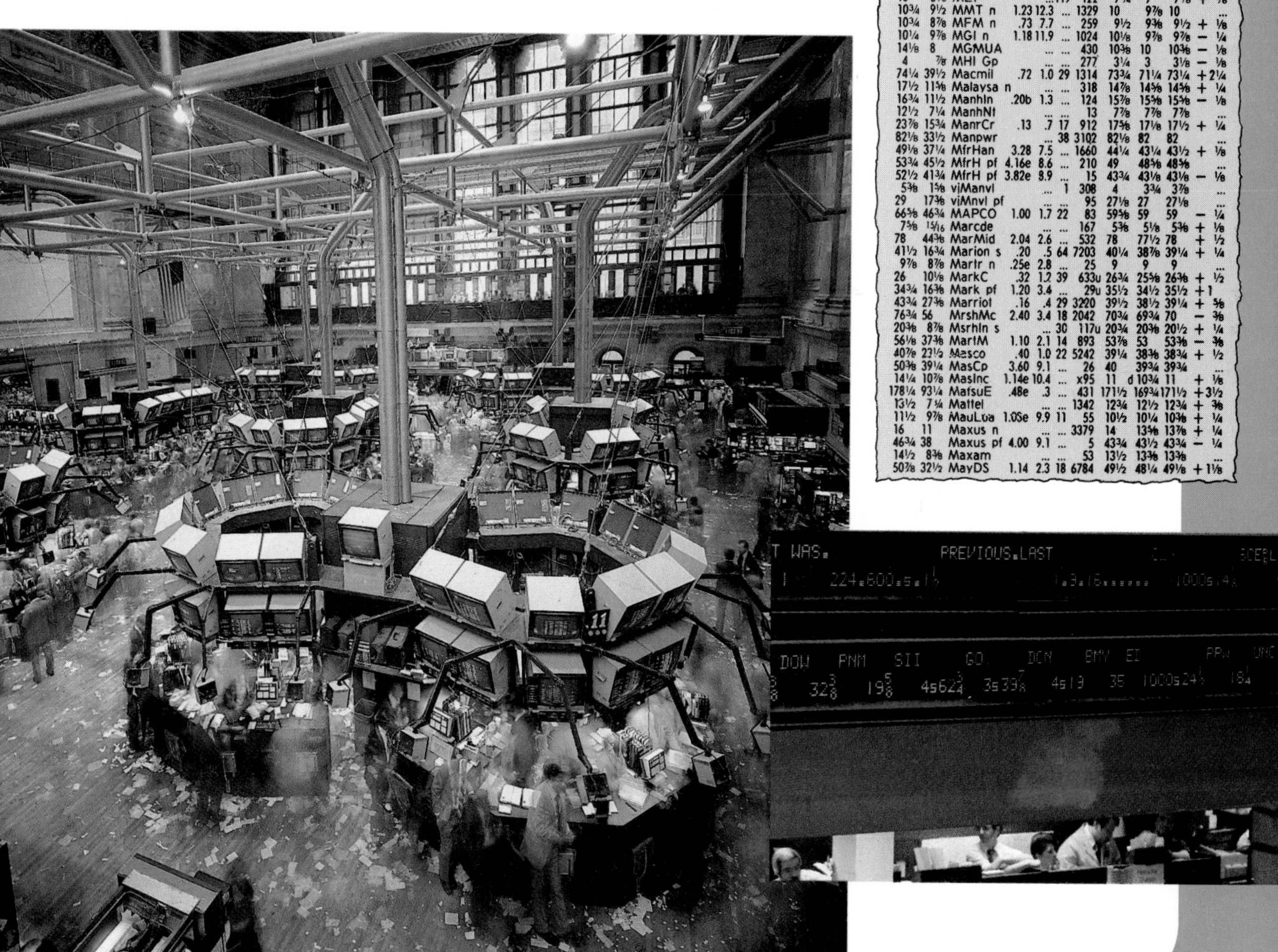

10.1 Products of Positive and Negative Numbers

Look at the pattern of the following multiplication problems.

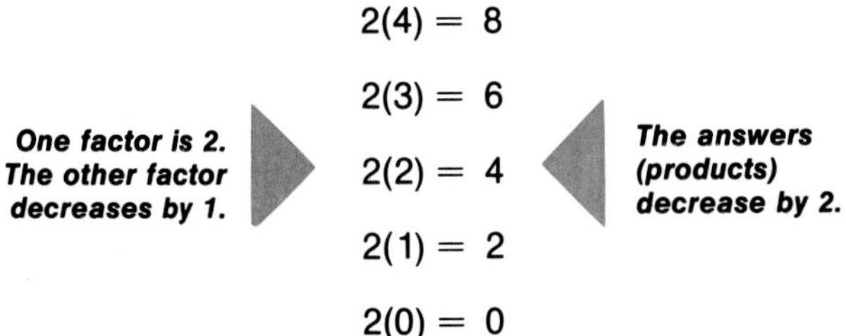

2(4) = 8
2(3) = 6
One factor is 2. The other factor decreases by 1.
2(2) = 4
The answers (products) decrease by 2.
2(1) = 2
2(0) = 0

Since the answers decrease by 2 at each step, you would expect the answer after 0 to be -2, and the one after -2 to be -4, and so forth.

$$2(-1) = -2$$

$$2(-2) = -4$$

$$2(-3) = -6$$

P–1 **Find the missing products.**

a	b	c
$3(3) = 9$	$6(3) = 18$	$10(1\frac{1}{2}) = 15$
$3(2) = 6$	$6(2) = 12$	$10(1) = 10$
$3(1) = 3$	$6(1) = 6$	$10(\frac{1}{2}) = 5$
$3(0) = 0$	$6(0) = 0$	$10(0) = 0$
$3(-1) =$ ___?___	$6(-1) =$ ___?___	$10(-\frac{1}{2}) =$ ___?___
$3(-2) =$ ___?___	$6(-2) =$ ___?___	$10(-1) =$ ___?___
$3(-3) =$ ___?___	$6(-3) =$ ___?___	$10(-1\frac{1}{2}) =$ ___?___

These patterns suggest the following rule.

Steps for Multiplying a Positive Number and a Negative Number

1. Multiply the absolute values.
2. Write the opposite of the result.

EXAMPLE Multiply: $(-12)(3)$

Solution: [1] Multiply the absolute values. ⟶ $|-12| \cdot |3| = 12 \cdot 3$

[2] Write the opposite of this. ⟶ $= 36$

$(-12)(3) = -36$

As the Example shows, the product of a positive number and a negative number is a negative number.

P–2 **Multiply.**

a. $4(-9)$ **b.** $(-7)(12)$ **c.** $(-1.4)(5)$ **d.** $(15.2)(-8)$

The following properties studied in Section 2.1 are true for real numbers.

Multiplication Property of Zero

For any real number *a*, $0 \cdot (-4) = 0$

$\mathbf{a \cdot 0 = 0}$ and $\mathbf{0 \cdot a = 0}$ $(-2.6)(0) = 0$

Multiplication Property of One

For any real number *a*, $(1)(-2.6) = -2.6$

$\mathbf{a \cdot 1 = a}$ and $\mathbf{1 \cdot a = a}$ $\left(-\frac{2}{3}\right)(1) = -\frac{2}{3}$

P–3 **Multiply.**

a. $(-8)(0)$ **b.** $(-4)(1)$ **c.** $(-0)(-18.7)$ **d.** $(2.1)(-8)$

CLASSROOM EXERCISES

Multiply. (Example)

1. $-1(6)$ **2.** $3(-2)$ **3.** $4(-5)$ **4.** $-7(8)$

5. $-7(4)$ **6.** $-8(9)$ **7.** $7(-6)$ **8.** $5(-5)$

9. $-4(8)$ **10.** $6(-4)$ **11.** $-9(3)$ **12.** $3(-8)$

13. $-12(4)$ **14.** $2(-7)$ **15.** $4(-11)$ **16.** $-6(7)$

17. $-2(9)$ 18. $6(-9)$ 19. $9(-4)$ 20. $-11(3)$
21. $-4(0)$ 22. $0(-19)$ 23. $(-400)3$ 24. $-4(250)$
25. $2(-8)$ 26. $5(-12)$ 27. $(-2)16$ 28. $(-4)12$
29. $-2(20)$ 30. $-1(59)$ 31. $22(-3)$ 32. $12(-5)$
33. $(-\frac{1}{2})(\frac{2}{3})$ 34. $-1(5.9)$ 35. $0(-19.6)$ 36. $(\frac{3}{8})(-\frac{5}{2})$
37. $(-1.5)(6)$ 38. $(-2\frac{1}{2})(20)$ 39. $1\frac{1}{4}(-24)$ 40. $(-0.6)(12)$

WRITTEN EXERCISES

Goal: To multiply a positive real number and a negative real number
Sample Problem: $(0.6)(-4)$ **Answer:** -2.4

Multiply. (Example)

1. $-3(4)$ 2. $-5(6)$ 3. $7(-5)$ 4. $4(-7)$
5. $-6(5)$ 6. $-4(9)$ 7. $-9(7)$ 8. $-7(6)$
9. $9(-6)$ 10. $8(-4)$ 11. $12(-8)$ 12. $12(-7)$
13. $8(-8)$ 14. $6(-6)$ 15. $-9(6)$ 16. $-8(12)$
17. $600(-5)$ 18. $-356(6)$ 19. $(-2)(28)$ 20. $(-3)(36)$
21. $-6(16)$ 22. $(-8)(14)$ 23. $(4)(-24)$ 24. $(11)(-12)$
25. $-3(52)$ 26. $-4(28)$ 27. $(-86)(7)$ 28. $(-10)(8)$
29. $\frac{1}{3}(-15)$ 30. $\frac{1}{3}(-21)$ 31. $(-1.3)(0.3)$ 32. $(-3.7)(0.4)$
33. $(-\frac{1}{4})(\frac{2}{5})$ 34. $(-\frac{2}{3})(\frac{6}{7})$ 35. $0(-23.6)$ 36. $0(-9.8)$
37. $12(-35)$ 38. $15(-42)$ 39. $(-22)(29)$ 40. $(-34)(27)$
41. $\frac{1}{2}(-14)$ 42. $\frac{1}{2}(-12)$ 43. $(-8.1)0$ 44. $(-6.3)0$

REVIEW CAPSULE FOR SECTION 10.2

Multiply.

1. $(12)(8)(2)$ 2. $2(35)(10)$ 3. $\frac{2}{3}(12)(\frac{1}{8})$ 4. $\frac{3}{4}(228)(5)$
5. $(\frac{5}{8})(\frac{4}{15})(\frac{3}{2})$ 6. $(3\frac{1}{8})(4\frac{4}{5})(\frac{1}{3})$ 7. $(\frac{9}{16})(2\frac{2}{3})(\frac{4}{3})$ 8. $\frac{13}{16}(592)(\frac{19}{37})$

Evaluate. (Pages 234–236)

9. $|8|$ 10. $|-12|$ 11. $|-1.7|$ 12. $|0.3|$
13. $\left|-2\frac{1}{4}\right|$ 14. $\left|7\frac{1}{8}\right|$ 15. $\left|-\frac{5}{4}\right|$ 16. $\left|\frac{17}{9}\right|$

10.2 Products of Negative Numbers

Look at the pattern of the following multiplication problems.

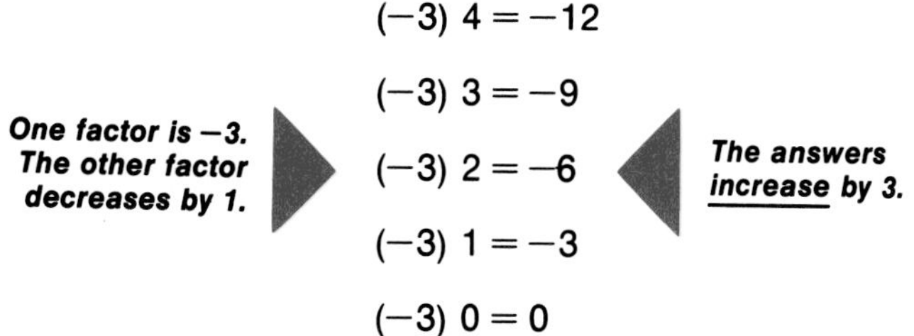

(−3) 4 = −12
(−3) 3 = −9
One factor is −3. The other factor decreases by 1.
(−3) 2 = −6
The answers increase by 3.
(−3) 1 = −3
(−3) 0 = 0

Since the answers <u>increase</u> by 3 at each step, you would expect the answer after 0 to be 3, the answer after 3 to be 6, and so forth.

$$(-3)(-1) = 3$$

$$(-3)(-2) = 6$$

$$(-3)(-3) = 9$$

$$(-3)(-4) = 12$$

P–1 **Find the missing products.**

a	b	c
$-4(3) = -12$	$-5(3) = -15$	$-10(1.5) = -15$
$-4(2) = -8$	$-5(2) = -10$	$-10(1.0) = -10$
$-4(1) = -4$	$-5(1) = -5$	$-10(0.5) = -5$
$-4(0) = 0$	$-5(0) = 0$	$-10(\ 0\ ) = 0$
$-4(-1) = \underline{\ ?\ }$	$-5(-1) = \underline{\ ?\ }$	$-10(-0.5) = \underline{\ ?\ }$
$-4(-2) = \underline{\ ?\ }$	$-5(-2) = \underline{\ ?\ }$	$-10(-1.0) = \underline{\ ?\ }$
$-4(-3) = \underline{\ ?\ }$	$-5(-3) = \underline{\ ?\ }$	$-10(-1.5) = \underline{\ ?\ }$

These patterns suggest the following rule.

Steps for Multiplying Two Negative Numbers

1. Multiply the absolute values.
2. Write the result.

EXAMPLE 1 Multiply: $(-24)(-9)$

Solution: [1] Multiply the absolute values. ⟶ $|-24| \cdot |-9| = (24)(9)$
$= 216$

Thus, $(-24)(-9) = 216$.

As the Example shows, the product of two negative numbers is a positive number.

P-2 **Multiply.**

a. $(-7)(-8)$ **b.** $(-10)(-1)$ **c.** $(-2.5)(-4)$ **d.** $(-\frac{1}{5})(-20)$

The Sign of a Product

1. The product of two positive real numbers is a positive real number.
2. The product of a positive and a negative real number is a negative real number.
3. The product of two negative real numbers is a positive real number.

To multiply more than two numbers, simply work from left to right.

EXAMPLE 2 Multiply: $(-4)(5)(-3)$

Solution: First find $(-4)(5)$.

$$(-4)(5)(-3) = (-20)(-3)$$
$$= 60$$

CLASSROOM EXERCISES

For Exercises 1–40, multiply. (Example 1)

1. $(-7)(-5)$ **2.** $(-4)(-8)$ **3.** $(-1)(-9)$ **4.** $(-3)(-6)$

5. $(-5)(-9)$
6. $(-8)(-3)$
7. $(-10)(-8)$
8. $(-11)(-4)$
9. $(-6)(-6)$
10. $(-9)(-8)$
11. $(-6)(-3)$
12. $(-7)(-2)$
13. $(-3)(-22)$
14. $(-14)(-2)$
15. $(-9)(-3)$
16. $(-5)(-12)$
17. $(-\frac{1}{2})(-14)$
18. $(-\frac{1}{3})(-18)$
19. $(-1.2)(-12)$
20. $(-3.2)(-2)$
21. $(-\frac{3}{4})(-\frac{1}{2})$
22. $(-\frac{1}{4})(-\frac{5}{4})$
23. $(-0.6)(-0.9)$
24. $(-2.3)(-0.2)$

(Example 2)

25. $(-2)(3)(-2)$
26. $(-1)(5)(2)$
27. $(-1)(-2)(-3)$
28. $(-2)(-2)(-2)$
29. $(6)(-1)(-5)$
30. $(-1)(8)(-3)$
31. $(-5)(2)(-4)$
32. $6(-2)(3)$
33. $(-5)(-8)(3)$
34. $(0)(-5)(-6)$
35. $(-3)(-4)(-2)$
36. $(9)(-5)(1)$
37. $(4)(-0.2)(-5)$
38. $(-6)(-0.7)(8)$
39. $(-3)(-2)(-\frac{1}{3})$
40. $(-\frac{1}{2})(-\frac{4}{5})(-1)$

WRITTEN EXERCISES

Goal: To multiply two negative real numbers
Sample Problem: $(-\frac{1}{5})(-15)$
Answer: 3

For Exercises 1–64, multiply.
(Example 1)

1. $(-6)(-4)$
2. $(-5)(-7)$
3. $(-3)(-8)$
4. $(-4)(-9)$
5. $(-12)(-7)$
6. $(-8)(-12)$
7. $(-9)(-8)$
8. $(-6)(-9)$
9. $(-16)(-8)$
10. $(-14)(-9)$
11. $(-18)(-7)$
12. $(-24)(-8)$
13. $(-6)(-18)$
14. $(-8)(-20)$
15. $(-20)(-5)$
16. $(-4)(-12)$
17. $(-22)(-5)$
18. $(-4)(-16)$
19. $(-9)(-15)$
20. $(-7)(-11)$
21. $(-0.6)(-18)$
22. $(-20.8)(-5.6)$
23. $(-0.2)(-0.7)$
24. $(-0.4)(-0.9)$
25. $(-\frac{2}{5})(-40)$
26. $(-\frac{3}{5})(-30)$
27. $(-1\frac{2}{5})(-2\frac{3}{4})$
28. $(-2\frac{4}{5})(-3\frac{1}{2})$

(Example 2)

29. $(7)(-3)(2)$
30. $(8)(-5)(3)$
31. $(-4)(-6)(-2)$
32. $(-3)(-5)(-4)$
33. $(-6)(3)(5)$
34. $(-8)(1)(-6)$
35. $(12)(-3)(-5)$
36. $(15)(-2)(-4)$
37. $(-19)(-1)(2)$
38. $(-24)(3)(-1)$
39. $(0.5)(12)(-5)$
40. $(-0.2)(20)(-8)$
41. $(-\frac{1}{4})(-8)(-5)$
42. $(-\frac{1}{3})(-9)(-7)$
43. $(-15)(-\frac{2}{3})(-11)$
44. $(-\frac{3}{4})(-20)(-12)$
45. $(-1.2)(3)(4)$
46. $(-2.3)(-4)(3)$

MIXED PRACTICE

47. $(-280)(-45)$
48. $(-0.9)(-1.04)$
49. $(-2)(5)(-6.7)$
50. $(-290)(80)$
51. $(-9)(-8)(-3)$
52. $(-12)(-1)(4)$
53. $(-64)(-3)$
54. $(18)(-4)(9.5)$
55. $(30)(-1.5)(2)$
56. $(18)(-40)$
57. $(-\frac{3}{4})(\frac{1}{2})$
58. $(-360)(-55)$
59. $(-25)(93)$
60. $(-3\frac{1}{8})(-\frac{4}{5})$
61. $(\frac{3}{8})(-240)$
62. $(18)(-4)(12)$
63. $(-37)(-509)$
64. $(-7)(-2)(5)$

MORE CHALLENGING EXERCISES

Evaluate each expression.

65. $(-8+5)(-15)$
66. $(-3+(-12))(-15)$
67. $(-10-6)(-3+8)$
68. $(-20-(-3))(-5-14)$
69. $3(-7)+(-6)(-12)+4(-11)$
70. $(-3)(-8)+(-4)(-10)+(-9)(-14)$
71. $8(-9)-(-3)(15)+(-12)(-5)$
72. $(-2)(-5)-(-4)(-7)-9(2)$
73. $-9(-4+5)+5(-1+11)$
74. $-8(-7+(-8))+(12+(-16))$

Let p be any positive number, let n be any negative number, and let z be zero.

Write positive, negative, or zero for the value of each expression. If the value cannot be determined, write cannot be determined.

75. $p+z$
76. $n+z$
77. $n+p$
78. $p+(-p)$
79. $-n+n$
80. $p-n$
81. $z-p$
82. $z-n$
83. np
84. nz
85. $n(-p)$
86. $-np$

REVIEW CAPSULE FOR SECTION 10.3

Multiply. (Pages 31–33)

1. $6(8+x)$
2. $10(z+\frac{1}{2})$
3. $\frac{1}{2}(92+32y)$
4. $t(8+s)$
5. $(16+s)\frac{1}{4}$
6. $(\frac{1}{3}+\frac{2}{3}r)6r$
7. $2x(\frac{1}{2}+4x)$
8. $(x^2+1)30$

Multiply. (Pages 278–280)

9. $-8(5)$
10. $12(-6)$
11. $10(-15)$
12. $-7(12)$
13. $-1.4(0.4)$
14. $2.5(-8)$
15. $\frac{1}{3}(-21)$
16. $-\frac{1}{2}(\frac{3}{4})$

10.3 Multiplication and Addition

The properties below were discussed in Chapter 2. They also apply to real numbers.

Commutative Property of Multiplication

For any real numbers a and b,

$ab = ba.$ $(-8)(4) = (4)(-8)$

Associative Property of Multiplication

For any real numbers a and b,

$(ab)c = a(bc).$ $(-5 \cdot 6)7 = -5(6 \cdot 7)$

Distributive Property

For any real numbers a, b, and c,

$a(b + c) = ab + ac$ $2(-4 + 3) = 2(-4) + 2(3)$

and

$(b + c)a = ba + ca.$ $(x + 0.7)4 = x(4) + 0.7(4)$

P–1 **Name the property illustrated in each sentence.**

a. $(-7 \cdot 8)5 = -7(8 \cdot 5)$

b. $6(-3 + -5) = 6(-3) + 6(-5)$

c. $(-4 \cdot 6)(-7 + 9) = (-7 + 9)(-4 \cdot 6)$

EXAMPLE 1 Multiply: **a.** $-3p(-12r)$ **b.** $7t(-8tw)$

Solutions:

	a.	**b.**
1 Use the Commutative and Associative Properties. ⟶	$(-3 \cdot -12)(p \cdot r)$	$(7 \cdot -8)(t \cdot t \cdot w)$
2 Multiply. Write in simplest form. ⟶	$36pr$	$-56t^2w$

P–2 **Multiply.**

a. $-12(-4k)$ **b.** $7c(-d)$ **c.** $-9c(3d)$ **d.** $-4m(-11mn)$

You can use the Distributive Property to express a product as a sum.

$$2(x + y) = 2x + 2y$$
$$-3(a + b) = (-3a) + (-3b)$$

EXAMPLE 2 Multiply: $-3(2x + 5)$

Solution:

[1] Use the Distributive Property. ⟶ $-3(2x + 5) = -3(2x) + (-3(5))$

[2] Multiply. Write in simplest form. ⟶ $= -6x + (-15)$

$$= -6x - 15$$

P–3 **Multiply.**

a. $3(6n + 9)$ **b.** $-5(-8t + 12)$ **c.** $(p + 4)(-2)$ **d.** $(4k + 3)(-4)$

EXAMPLE 3 Multiply: $-3x(-8x + 5)$

Solution: $-3x(-8x + 5) = (-3x)(-8x) + (-3x)(5)$

$$= 24x^2 + (-15x)$$
$$= 24x^2 - 15x$$

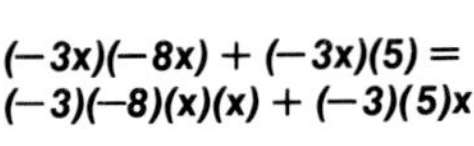

P–4 **Multiply.**

a. $(4p + 7)(-3p)$ **b.** $(r + 12)\frac{3}{4}r$ **c.** $2x(4x + (-3))$ **d.** $-s(-4s + 6)$

EXAMPLE 4 Multiply: $(4y - 8)9y$

Solution: $(4y - 8)9y = (4y + (-8))9y$

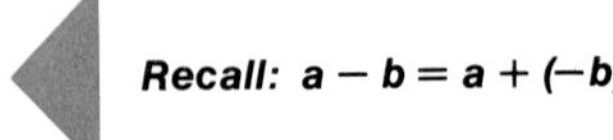

$$= 36y^2 + (-72y)$$
$$= 36y^2 - 72y$$

P–5 **Multiply.**

a. $6(4x - 3)$ **b.** $-2x(\frac{1}{2}x - 6)$ **c.** $(3 - x^2)2x$ **d.** $(3x - 1)x^2$

CLASSROOM EXERCISES

For Exercises 1–24, multiply. (Example 1)

1. $8(-9y)$
2. $-6(-7q)$
3. $15r(-s)$
4. $-3pq(-9q)$
5. $(12s)(-6s^2)$
6. $(-\frac{1}{2}rt)(-16rw)$

(Example 2)

7. $-2(x+5)$
8. $(a+3)(-6)$
9. $-1(n+3)$
10. $-3(x+8)$
11. $(2x+3)(-5)$
12. $-5(-3x+1)$

(Example 3)

13. $-3r(r+2)$
14. $2t(5t+7)$
15. $\frac{1}{2}k(-4k+10t)$
16. $(2m+(-3n))(-2m)$
17. $(-10w+1)(-w)$
18. $(-9p+(-6))(\frac{1}{3}p)$

(Example 4)

19. $5(y-6)$
20. $12(x^2-4)$
21. $-3(a-2)$
22. $-4(6a^2-a)$
23. $(8-x)x$
24. $(9-2x)x^3$

WRITTEN EXERCISES

Goal: To multiply using the Distributive Property
Sample Problem: Multiply: $-12n(-3n+5)$
Answer: $36n^2-60n$

For Exercises 1–36, multiply. (Example 1)

1. $-4(-6g)$
2. $7(-9s)$
3. $-11n(8t)$
4. $7r(-w)$
5. $(-\frac{1}{4}tw)(-8tr)$
6. $(15pq)(-6pq^2)$

(Example 2)

7. $-2(8a+5)$
8. $-3(10x+6)$
9. $-5(-3r+(-7))$
10. $-4(-7n+(-9))$
11. $(5+3x)(-7)$
12. $(5a+(-3))(-5)$

(Example 2)

13. $x(5+2x)$
14. $y(3+2y)$
15. $(-3a+(-2))2a$
16. $(5r+(-3))3r$
17. $-3n(2n+(-5))$
18. $-5t(-3t+1)$

(Example 4)

19. $12(x - 7)$
20. $5(y^2 - 3)$
21. $-8(3 - x)$
22. $-13(y^2 - 2)$
23. $(27x^2 - x)\frac{1}{3}$
24. $(16y^3 - 4y)(\frac{1}{2}y)$

MIXED PRACTICE

25. $\frac{1}{2}a(4a + 2)$
26. $8x(x - 8)$
27. $-4(r^2 - 3r)$
28. $(15y + 10)\frac{1}{5}$
29. $3(15y + 6)$
30. $(x^3 - 4)(-x)$
31. $-x(x^2y^2 - 4)$
32. $-\frac{1}{5}(25x - 10)$
33. $2x(-8x + (-15))$
34. $(18xy - 4y)(-5x)$
35. $-5t(9t + 7r)$
36. $(-7m + 6n)(-3n)$

MORE CHALLENGING EXERCISES

An expression with two terms, such as $2x + 3y$, is a **binomial.** You can multiply two binomials by finding the product of the first, outside, inside, and last terms. Combine any like terms.

EXAMPLE: $(t + 2)(t + 3)$ **SOLUTION:** $(t + 2)(t + 3) = t \cdot t + t \cdot 3 + 2 \cdot t + 2 \cdot 3$

$= t^2 + 3t + 2t + 6$

$= t^2 + 5t + 6$

(F, O, I, L brackets mark the first, outside, inside, and last terms.)

Multiply. Combine like terms.

37. $(x + 2)(x + 1)$
38. $(y + 5)(y + 3)$
39. $(r + 3)(r + 10)$
40. $(w + 5)(w + 2)$
41. $(x + 2)(x - 1)$
42. $(z - 4)(z - 4)$

NON-ROUTINE PROBLEMS

43. It takes Eric 6 minutes to cut a log into 3 pieces. At that rate, how long would it take him to cut a log into 4 pieces?

44. A special ball is dropped from the top of a building 64 meters high. Each time the ball bounces, it bounces $\frac{1}{2}$ the distance that it fell. The ball is caught when it bounces 1 meter high. Find the total distance the ball travels before being caught.

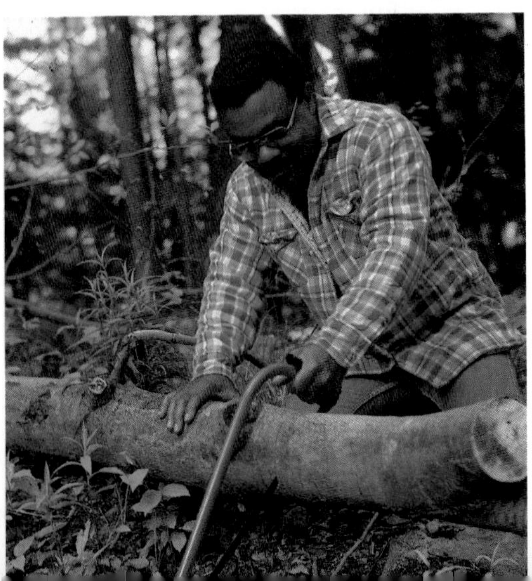

MID-CHAPTER REVIEW

Multiply. (Section 10.1)

1. $(-250)(8)$
2. $(14)(-56)$
3. $(0)(-129)$
4. $(-342)(0)$
5. $(24)(-14.5)$
6. $(-36)(22.5)$
7. $(-3)(26)$
8. $(-5)(19)$
9. $(0.6)(-9)$
10. $(-0.8)(7)$
11. $(-18)(2)$
12. $(24)(-3)$

Multiply. (Section 10.2)

13. $(-5)(-28)$
14. $(-32)(-48)(2)$
15. $(-\frac{1}{2})(-168)$
16. $(-6)(-3)(2)$
17. $(-14.5)(-2)$
18. $(-17.5)(-4)$
19. $(-12)(-7)$
20. $(-8)(-15)$
21. $(-2)(-0.8)$
22. $(-0.9)(7)$
23. $(-104)(-8)$
24. $(-7)(-209)$

Multiply. (Section 10.3)

25. $-14(-3y)$
26. $20m(-m^2n)$
27. $-\frac{1}{4}a(-12b)$
28. $(-7cd)(-8bc)$
29. $-8(9x+7)$
30. $-10(12n+9)$
31. $4m(-8m+7)$
32. $7k(-9k+12)$
33. $(18c+12d)\frac{1}{3}c$
34. $(24r+40s)\frac{1}{4}r$
35. $-2.5(-14t+2)$
36. $-4.2(-15w+35)$
37. $-12p(9-3p)$
38. $-14a(2-8a)$
39. $\frac{1}{2}g(18g-5)$
40. $\frac{1}{4}s(20s-27)$

REVIEW CAPSULE FOR SECTION 10.4

Write Product or Sum to identify each expression. Note: The last operation you perform determines whether an expression is a product or a sum. (Pages 2–4)

1. $5(5)+8$
2. $12(15)+14(18)$
3. $29(13+17)$
4. $(12+41)8$
5. $\frac{1}{2}(5)+\frac{1}{2}(31)$
6. $\frac{1}{3}(7)+\frac{1}{4}(7)$
7. $8(m+4)$ when $m=6$
8. $18t+7$ when $t=12$
9. $6d+2d$ when $d=3$
10. $6(w+2w)$ when $w=3$

For Exercises 11–20, match each expression in Column 1 with an equivalent expression from Column 2. (Pages 35–36)

Column 1	Column 2	Column 1	Column 2
11. $3x+18$	$2(2x+1)$	**16.** $24y+36x$	$(3y+2x)6$
12. $6x+12$	$4(3+x)$	**17.** $18y+12x$	$2x(y+6)$
13. $12+4x$	$6(x+2)$	**18.** $2xy+12x$	$12(2y+3x)$
14. $6+2x$	$3(x+6)$	**19.** $2xy+6$	$6(2x+y)$
15. $4x+2$	$2(3+x)$	**20.** $12x+6y$	$2(xy+3)$

10.4 Factoring

In Section 10.3 you used the Distributive Property to express a product as a sum. The Distributive Property is also used to express a sum as a product. Writing a sum as a product is called ***factoring.***

$$2x + 2y = 2(x + y)$$

2 and (x + y) are factors.

To factor a sum, find all the factors (numbers and variables) shared by the terms of the sum.

Definition

The **greatest common factor** (GCF) of two or more terms is the product of all the factors shared by the terms.

$2x + 2y = \mathbf{2}(x + y)$

$3x^2 + 6x = \mathbf{3x}(x + 2)$

Sum	Greatest Common Factor	Product
$cf + cd$	c	$c(f + d)$
$2g + gk$	g	$g(2 + k)$
$2d - 2dk$	$2d$	$2d(1 - k)$
$24 + 4r$	4	$4(6 + r)$
$14x - 7x^2$	$7x$	$7x(2 - x)$
$-9y + (-15y^2)$	$-3y$	$-3y(3 + 5y)$

EXAMPLE 1 Factor: $(-2x) + (-2y)$

Solution:

1. Identify the greatest common factor. ⟶ -2

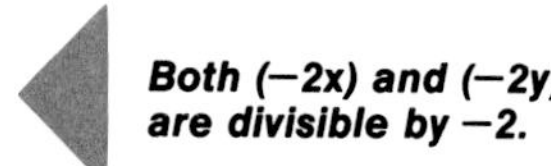
Both (−2x) and (−2y)
are divisible by −2.

2. Write a product using the GCF as a factor. ⟶ $-2(x + y)$

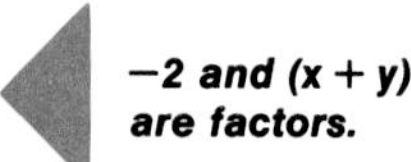
−2 and (x + y)
are factors.

P–1 **Factor.**

a. $-4p + (-4q)$ **b.** $5h + 15g$ **c.** $-q(1) + (-q)(t)$ **d.** $-7n + (-7)$

EXAMPLE 2 Factor: $-5rx + (-5x)$

Solution: $-5rx + (-5x)$

1. Identify the greatest common factor. ⟶ $-5x$
2. Write a product using the GCF as a factor. ⟶ $(-5x)(r + 1)$

P–2 **Factor.**

a. $-3p + (-3pq)$ **b.** $12p + 12p^3$ **c.** $-6h^2 + 6h$ **d.** $ab + (-2a)$

EXAMPLE 3 Factor: $3x^2 - 12x$

Solution: $3x^2 - 12x$

1. Identify the greatest common factor. ⟶ $3x$
2. Write a product using the GCF as a factor. ⟶ $3x(x - 4)$

P–3 **Factor.**

a. $8xy - 4x$ **b.** $9f^2 - 3g^2$ **c.** $-2 - 4a$ **d.** $-12y^2 - 6yz$

CLASSROOM EXERCISES

Identify the greatest common factor of each pair. (Table)

1. $2a$, $2d$ **2.** $-3b$, $-3f$ **3.** $-3d$, dg **4.** 16, $8q$

5. $-14p$, -7 **6.** $-t^2$, $-3t$ **7.** m^2, $6m$ **8.** $-2r^2$, $-4r$

For Exercises 9–26, factor. (Example 1)

9. $3y + 3x$ **10.** $a(5) + b(5)$ **11.** $-10(r) + (-10(7))$

12. $k(8) + (-5(8))$ **13.** $-4(t) + (-4(1))$ **14.** $m(-100) + n(-100)$

(Example 2)

15. $5ax + 3x$ **16.** $12ny + 12nx$ **17.** $xm + (-2xn)$

18. $-3st + (-3t)$ **19.** $13tr + (-13rs)$ **20.** $(-7pq) + apq$

(Example 3)

21. $-12m - 12$
22. $k - 2kx$
23. $15k - 15$
24. $-3xt - 3t$
25. $-4r - r$
26. $ab - 5ab$

WRITTEN EXERCISES

Goal: To factor using the Distributive Property
Sample Problem: Factor: $-7ax + 9bx$
Answer: $x(-7a + 9b)$

For Exercises 1–36, factor. (Example 1)

1. $7x + 7y$
2. $15a + 15b$
3. $-8t + (-8s)$
4. $-11p + (-11q)$
5. $12r + 12$
6. $-7m + (-7)$

(Example 2)

7. $5rw + 5rx$
8. $ab + 5a$
9. $-9kn + (-9kt)$
10. $5ax + (-6ay)$
11. $-13rt + 5st$
12. $-3xm + (-3my)$

(Example 3)

13. $18x^2 - 9x$
14. $15y - 5xy$
15. $-3a^2b - 9b$
16. $9t - 9$
17. $5st - s$
18. $a^2b - ab^2$

MIXED PRACTICE

19. $-8r + (-8)$
20. $144s^2 - 12s$
21. $-5y + (-5)$
22. $24x^2y + 8xy^2$
23. $3pq - 9p^2$
24. $-13rt + 5st$
25. $13x - 13y$
26. $-t^2 + (-4t)$
27. $-34s + s$
28. $-10r + (-10s)$
29. $25mn + 10n^2$
30. $16w - 16w^2$
31. $13(8) - 4g(8)$
32. $-12ay + 7by$
33. $m + (-12am)$
34. $-6t + (-6)$
35. $-x - x^2$
36. $-4x + (-4)$

REVIEW CAPSULE FOR SECTION 10.5

Divide.

1. $186 \div 3$
2. $375 \div 15$
3. $376 \div 47$
4. $3230 \div 95$
5. $90 \div \frac{1}{3}$
6. $87 \div \frac{3}{5}$
7. $108 \div 2\frac{1}{4}$
8. $132 \div 5\frac{1}{2}$

10.5 Division

For each multiplication problem you can write two division problems.

$$3 \cdot 7 = \mathbf{21} \qquad \mathbf{21} \div 7 = 3 \qquad \mathbf{21} \div 3 = 7$$

Look at the following multiplication problems and their related division problems.

$-7(-3) = 21$	$\mathbf{21} \div (-7) = -3$	$\mathbf{21} \div (-3) = -7$
$-9(-5) = 45$	$\mathbf{45} \div (-9) = -5$	$\mathbf{45} \div (-5) = -9$

Now look at the quotients in Columns 1 and 2 below.

	Column 1	**Column 2**
$-2(8) = -16$	$-16 \div 8 = -2$	$-16 \div (-2) = 8$
$-7(4) = -28$	$-28 \div 4 = -7$	$-28 \div (-7) = 4$

Column 1 suggests that a negative number divided by a positive number is a negative number. Column 2 suggests that the quotient of two negative numbers is a positive number.

> The quotient of a positive number and a negative number is a negative number.
>
> The quotient of two negative numbers is a positive number.
>
> $-18 \div 6 = -3$
>
> $42 \div (-3) = -14$
>
> $-18 \div (-6) = 3$
>
> $\frac{-24}{-4} = 6$

EXAMPLE 1 Divide.

a. $\frac{-36}{4}$ **b.** $42 \div (-7)$ **c.** $-21 \div (-3)$

Solutions: **a.** $\frac{-36}{4}$ means $-36 \div 4$. Thus, $\frac{-36}{4} = -9$.

b. $42 \div (-7) = -6$ **c.** $-21 \div (-3) = 7$

P–1 **Divide.**

a. $\frac{-27}{3}$ **b.** $4 \div (-4)$ **c.** $-13 \div (-1)$ **d.** $0 \div (-6)$

Recall that two numbers are reciprocals if their product is 1.

$$\frac{4}{1} \cdot \frac{1}{4} = 1 \qquad \left(-\frac{5}{6}\right)\left(-\frac{6}{5}\right) = 1 \qquad \left(-\frac{9}{5}\right)\left(-\frac{5}{9}\right) = 1$$

Reciprocals: 4 and $\frac{1}{4}$; $-\frac{5}{6}$ and $-\frac{6}{5}$; $-\frac{9}{5}$ and $-\frac{5}{9}$

P–2 **What is the reciprocal of each of the following?**

a. 3 **b.** $\frac{1}{5}$ **c.** $-\frac{4}{3}$ **d.** $-3\frac{1}{3}$ **e.** 0

Recall from Section 4.4 that dividing by a number is the same as multiplying by its reciprocal. This is true for all real numbers except zero. The number, 0, cannot have a reciprocal.

> For any real numbers a and b, $b \neq 0$,
>
> $$\boldsymbol{a \div b = a \cdot \frac{1}{b}}. \qquad 7 \div (-3) = 7 \cdot \left(-\frac{1}{3}\right)$$

EXAMPLE 2 Divide.

a. $-30 \div \frac{1}{5}$ **b.** $-8 \div \left(-\frac{4}{3}\right)$

Solutions: **a.** $-30 \div \frac{1}{5} = -30 \cdot 5$

$= -150$

b. $-8 \div \left(-\frac{4}{3}\right) = -\frac{\overset{2}{\cancel{8}}}{1} \cdot \left(-\frac{3}{\underset{1}{\cancel{4}}}\right)$

$= -\frac{2}{1} \cdot \left(-\frac{3}{1}\right)$

$= \frac{6}{1}$, or 6

P–3 **Divide.**

a. $-21 \div \frac{1}{3}$ **b.** $36 \div (-\frac{12}{5})$ **c.** $-48 \div (-1\frac{1}{5})$ **d.** $-1 \div (-1\frac{1}{4})$

CLASSROOM EXERCISES

Divide. (Example 1)

1. $\frac{-54}{9}$ **2.** $\frac{-24}{8}$ **3.** $\frac{63}{-7}$ **4.** $\frac{81}{-9}$

5. $\frac{-100}{-10}$ **6.** $\frac{-27}{-3}$ **7.** $52 \div (-4)$ **8.** $(-64) \div 8$

9. $(-18) \div (-2)$ **10.** $42 \div (-6)$ **11.** $-121 \div (11)$ **12.** $-99 \div (-9)$

Write the reciprocal. (For further review, see Section 4.4)

13. -5 **14.** -23 **15.** $-\frac{7}{8}$ **16.** $-\frac{5}{6}$ **17.** $-\frac{3}{2}$ **18.** $-\frac{5}{2}$
19. $-4\frac{1}{2}$ **20.** $-3\frac{2}{3}$ **21.** $-\frac{13}{4}$ **22.** $-\frac{11}{6}$ **23.** -0.9 **24.** -1.7

Write a product for each quotient. (Step 1, Example 2)

25. $-16 \div \frac{1}{3}$ **26.** $-9 \div \frac{4}{7}$ **27.** $25 \div (-\frac{3}{5})$ **28.** $17 \div (-\frac{2}{3})$
29. $10 \div (-\frac{5}{2})$ **30.** $18 \div (-\frac{12}{7})$ **31.** $-6 \div \frac{4}{3}$ **32.** $-14 \div \frac{10}{3}$

Divide. (Example 2)

33. $-3 \div \frac{1}{4}$ **34.** $-5 \div \frac{1}{3}$ **35.** $6 \div (-\frac{3}{5})$ **36.** $8 \div (-\frac{4}{5})$
37. $8 \div (-\frac{5}{2})$ **38.** $16 \div (-\frac{11}{4})$ **39.** $-7 \div (-\frac{14}{9})$ **40.** $-10 \div (-\frac{15}{7})$
41. $-16 \div (-\frac{4}{7})$ **42.** $15 \div (-\frac{3}{8})$ **43.** $51 \div (-\frac{17}{20})$ **44.** $52 \div (-\frac{13}{16})$

WRITTEN EXERCISES

Goal: To divide real numbers
Sample Problem: $-16 \div \frac{3}{5}$
Answer: $-26\frac{2}{3}$

For Exercises 1–44, divide. (Example 1)

1. $\frac{-8}{2}$ **2.** $\frac{-12}{3}$ **3.** $\frac{20}{-2}$ **4.** $\frac{30}{-5}$
5. $\frac{-98}{-2}$ **6.** $\frac{-84}{-4}$ **7.** $51 \div (-3)$ **8.** $32 \div (-16)$
9. $-44 \div 11$ **10.** $-24 \div 6$ **11.** $-68 \div (-4)$ **12.** $-72 \div (-6)$

(Example 2)

13. $16 \div (-\frac{8}{5})$ **14.** $12 \div (-\frac{6}{5})$ **15.** $-4 \div \frac{10}{7}$ **16.** $-9 \div \frac{12}{7}$
17. $10 \div (-\frac{12}{5})$ **18.** $14 \div (-\frac{7}{5})$ **19.** $-8 \div \frac{4}{3}$ **20.** $-12 \div \frac{16}{9}$
21. $-\frac{1}{4} \div (\frac{5}{4})$ **22.** $\frac{7}{8} \div (-\frac{7}{8})$ **23.** $-\frac{8}{9} \div \frac{4}{3}$ **24.** $-\frac{12}{5} \div \frac{8}{5}$
25. $-12 \div (-1\frac{1}{5})$ **26.** $-20 \div (-3\frac{1}{3})$ **27.** $-\frac{7}{12} \div (-2\frac{1}{3})$ **28.** $-\frac{9}{4} \div (-1\frac{1}{3})$

MIXED PRACTICE

29. $-10 \div (\frac{2}{5})$ **30.** $-46 \div (-23)$ **31.** $-19 \div (-1)$ **32.** $-12 \div \frac{6}{7}$
33. $-\frac{4}{3} \div (-1\frac{7}{9})$ **34.** $102 \div (-17)$ **35.** $21 \div (-\frac{7}{10})$ **36.** $-169 \div 13$
37. $-200 \div (40)$ **38.** $\frac{4}{3} \div (-\frac{2}{9})$ **39.** $39 \div (-13)$ **40.** $8 \div (-\frac{4}{5})$
41. $111 \div (-3)$ **42.** $42 \div (-\frac{6}{11})$ **43.** $-196 \div 14$ **44.** $6 \div (-\frac{15}{16})$

REVIEW CAPSULE FOR SECTION 10.6

Add or divide as indicated.

1. $151 + 232 + 204 + 161$ **2.** $112 + 114 + 208 + 173$ **3.** $135 + 139 + 145 + 162$

4. $1.2 + 0.8 + 3.1 + 2.4$ **5.** $4.1 + 0.9 + 0.6 + 2.3$ **6.** $1.4 + 1.6 + 2.4 + 0.8$

7. $930 \div 6$ **8.** $855 \div 5$ **9.** $4.34 \div 7$ **10.** $6.02 \div 7$

PROBLEM SOLVING AND APPLICATIONS

10.6 Mean, Median, and Mode

The arithmetic ***mean*** (or ***average***) of a collection of numbers, scores, measures, and so on, is the sum of the items in the collection divided by the number of items.

EXAMPLE 1 The low temperatures for one week in a city are shown below in degrees Celsius. Find the mean low temperature for the week. Give your answer to the nearest tenth of a degree Celsius.

S	M	T	W	Th	F	S
$-2.4°$	$-3.5°$	$4.3°$	$2.6°$	$-2.6°$	$1.4°$	$-5.6°$

Solution:

1. Add the negative numbers. ⟶ $(-2.4) + (-3.5) + (-2.6) + (-5.6) = -14.1$
2. Add the positive numbers. ⟶ $4.3 + 2.6 + 1.4 = 8.3$
3. Find their sum. ⟶ $(-14.1) + (8.3) = -5.8$
4. Divide by the number of temperatures. ⟶ $(-5.8) \div 7 \approx -0.8$

The mean low temperature is approximately $-0.8°$ Celsius.

P–1 **For each of the following, find the mean low temperature for the week in degrees Celsius.**

a.

S	M	T	W	Th	F	S
$1.3°$	$-2.2°$	$0.4°$	$-1.1°$	$2.5°$	$2.8°$	$-0.9°$

b.

S	M	T	W	Th	F	S
$-1.5°$	$2.0°$	$3.1°$	$-0.1°$	$-1.9°$	$-1.4°$	$-0.9°$

Some other common averages are the median and mode.

The ***median*** of a collection of data is the middle item when the items are arranged in order of size.

When there is an even number of items, the median is the mean of the two middle items.

The ***mode*** is the item in the collection of data that occurs most often.

If no item occurs more than once, there is no mode.

There may be more than one mode.

EXAMPLE 2 The final scores of 22 card players in a tournament are listed in order. Find the median and the mode of the scores.

−5, −4, −3, −3, −2, −1, −1, 0, 0, 0, 1,

2, 2, 2, 2, 2, 3, 4, 4, 4, 5, 6

Solution: Since there is an even number of scores, identify the two middle scores.

[1] Find the mean of the two middle scores. ⟶ $\frac{1+2}{2}=\frac{3}{2}=1.5$  ***Median***

[2] Identify the score that occurs most often. ⟶ The score, 2, occurs five times.  ***Mode***

The median score is 1.5 and the mode is 2.

In Example 2, the mean score is about 0.8. You can see the mean, median, and mode are not necessarily the same number.

The mean is the most commonly used "average." The median is useful when the mean might be influenced by extreme values in the data. The mode is useful when analyzing data to determine the most common choices.

EXAMPLE 3

The owner of a small business has four employees. Their yearly earnings are $25,000, $15,000, $14,000, and $13,500. The owner pays himself a yearly wage of $50,000.

a. Find the mean and the median of the yearly wages.

b. Which of these, the mean or the median, better represents the average yearly earnings of a wage earner in the business?

Solutions:

a. [1] Find the mean. ⟶ $50{,}000 + 25{,}000 + 15{,}000 + 14{,}000 + 13{,}500 = 117{,}500$

$$\frac{117{,}500}{5} = 23{,}500$$

[2] Find the median. ⟶ 15,000

b. Determine which average gives a better description of the data.

Three of the wages are much less than the mean wage of $23,500. The median wage of $15,000 is probably a better description of the data.

CLASSROOM EXERCISES

The given temperatures are in degrees Celsius. Find the mean low temperature for the week. (Example 1)

1.

S	M	T	W	Th	F	S
−1.5°	2.0°	−2.1°	−1.6°	−0.2°	1.4°	0.6°

2.

S	M	T	W	Th	F	S
2.2°	4.1°	3.5°	1.2°	−0.8°	−1.7°	−0.8°

3.

S	M	T	W	Th	F	S
1.5°	−2.2°	0.8°	−1.4°	3.2°	−0.9°	−1.7°

4.

S	M	T	W	Th	F	S
−3.5°	−4.0°	−1.5°	0.2°	−0.4°	−1.3°	−1.4°

Find the median and the mode or modes for each collection of data. (Example 2)

5. Card game scores: −4, −3, −3, −2, −2, −2, −1, 0, 1, 2, 2

6. Bowling scores: 202, 190, 178, 160, 158, 148, 130, 130, 121

7. Runs scored by a baseball team in 15 games: 10, 9, 6, 5, 5, 5, 4, 3, 3, 2, 2, 2, 1, 0, 0

8. Low temperatures in degrees Fahrenheit for a 2–week period: 28, 15, 12, 12, 10, 4, 4, −1, −3, −5, −8, −8, −9, −11

For each exercise, find the mean and the median. Decide whether the mean or the median would best represent the data. Give a reason for each answer. (Example 3)

9. Deficiency in rainfall (in centimeters) for six months:

April: −6.2; May: −1.5; June: −1.9; July: −1.7; August: −0.9; September: −1.5

10. Points scored on free throws by a basketball player in 15 games:

2 4 5 11 2
1 2 12 2 1
12 2 1 3 4

WRITTEN EXERCISES

Goal: To solve word problems involving the mean, median, and mode

Sample Problem: The low temperatures for one week in a city are shown below in degrees Celsius. Find the mean, median, and mode of the low temperatures for the week.

S	M	T	W	Th	F	S
3°	2°	−1°	−2°	−2°	0°	4°

Answers: Mean 0.6°C, Median 0°C, Mode −2°C

The low temperatures for the day given below are in degrees Celsius. Find the mean low temperature for the week. Give your answer to the nearest tenth of a degree. (Example 1)

1.

S	M	T	W	Th	F	S
−4.0°	−2.1°	1.3°	8.2°	−3.0°	−0.5°	−0.6°

2.

S	M	T	W	Th	F	S
−1.1°	−0.6°	0.3°	−1.2°	−2.3°	−0.7°	−0.4°

For Exercises 3–4, solve each problem. (Example 1)

3. The table below shows the change (in feet per second per second) in the average speed of a moving car over a five-minute interval. Positive numbers represent an increase in speed, and negative numbers represent a reduction in speed. Find the mean change in the average speed of the car between 8:01 and 8:06. Give your answer to the nearest tenth of a unit.

Time Interval	**8:01–8:02**	**8:02–8:03**	**8:03–8:04**	**8:04–8:05**	**8:05–8:06**
Change in Speed	1.3	−0.3	0.2	−0.8	0.9

4. The table below shows the early arrivals, recorded as +, and the late arrivals, recorded as −, of a long-distance bus for one week. Find the mean number of minutes the bus was early or late for the seven days. Give your answer to the nearest tenth of a minute.

S	**M**	**T**	**W**	**Th**	**F**	**S**
+51 min	−28 min	−72 min	+35 min	+12 min	−56 min	+42 min

(Example 2)

5. The heights of 12 players on a professional basketball team are given to the nearest inch. Find the median and the mode of the players' heights.

Player	**1**	**2**	**3**	**4**	**5**	**6**
Height	82 in	79 in	84 in	81 in	77 in	85 in
Player	**7**	**8**	**9**	**10**	**11**	**12**
Height	83 in	79 in	80 in	78 in	82 in	79 in

6. The weights of 10 linemen on a football team are given in pounds. Find the median and mode of the weights of the players.

Player	**1**	**2**	**3**	**4**	**5**
Weight	251 lb	265 lb	240 lb	278 lb	251 lb
Player	**6**	**7**	**8**	**9**	**10**
Weight	283 lb	255 lb	243 lb	247 lb	256 lb

For each exercise, find the mean and the median. Decide whether the mean or the median would best represent the data. Give a reason for each answer. (Example 3)

7. Ages of 30 players on a football team

Ages					
27	22	32	29	20	28
32	35	22	19	27	32
34	30	27	29	27	22
34	35	29	24	22	24
27	22	24	20	30	34

8. Low wind chill temperatures for a five–day period

Day	Temperatures
Monday	−10°C
Tuesday	−34°C
Wednesday	2°C
Thursday	3°C
Friday	−1°C

MIXED PRACTICE

For Exercises 9–12, find the mean, median, and mode for each collection of data.

9. The deficiency or excess of rainfall for each month given in centimeters

Jan.	Feb.	Mar.	Apr.	May	June
+0.8	+1.2	−0.3	−1.4	+1.2	−0.3

July	Aug.	Sept.	Oct.	Nov.	Dec.
−0.4	−0.8	−3.8	−4.2	−2.8	−3.1

10. Change in the national inflation rate for seven years

Year	1	2	3	4	5	6	7
Rate	−0.3%	−0.2%	+0.8%	+1.2%	−0.6%	+0.5%	+1.4%

11. A company reported the following profit (shown as a positive amount) and loss (shown as a negative amount) for each of the first six months of a year.

Jan.	Feb.	March	April	May	June
$53,080	$12,425	$−6,240	$−3,465	$−12,476	$7,946

12. The table below shows the sales above quota (recorded as +) and below quota (recorded as −) of a salesperson over a six-month period.

July	Aug.	Sept.	Oct.	Nov.	Dec.
+$5,250	+$3,400	−$1,050	−$4,280	+$2,420	−$1,750

Using Comparisons

Medicine

One major responsibility of a **registered nurse** is to give medicine to patients under a doctor's direction. To do this a nurse must have a knowledge of measurement and be able to use proportions in calculations.

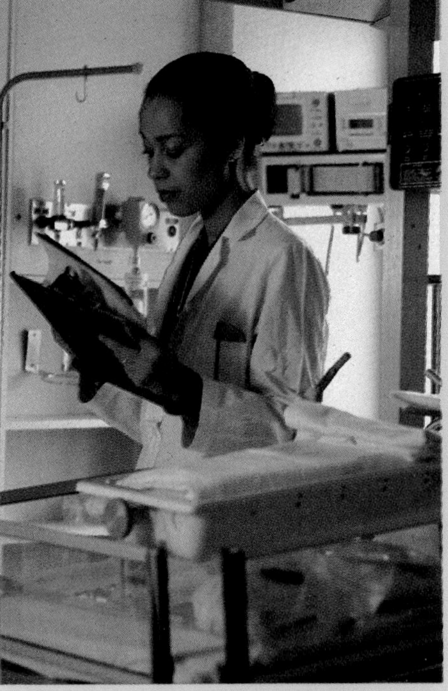

EXAMPLE: A doctor wants a patient to have 25 milligrams of a medicine per dose. A solution is available which has 500 milligrams of the medicine in 10 milliliters of solution. How many milliliters of solution should a nurse measure for each dose?

SOLUTION: Let m = the number of milliliters of solution.

$$\frac{25}{m} = \frac{500}{10}$$

$$25 \cdot 10 = 500m$$

$$\frac{25 \cdot 10}{500} = m$$

[2] [5] [×] [1] [0] [÷] [5] [0] [0] [=]   0.5

The nurse measures **0.5 milliliters** for each dose.

EXERCISES

1. A patient is to receive 1.5 milligrams of a medicine per dose. A solution of 9 milligrams of medicine in 30 milliliters of solution is available. How many milliliters of solution should the patient receive?

2. A doctor prescribes 15 milligrams of a medicine per dose. A solution that has 225 milligrams of the medicine in 9 milliliters of solution is available. How many milliliters of solution should the nurse measure for each dose?

3. A doctor prescribes 225 milligrams of a medicine to be given every four hours. The medicine comes in 75-milligram tablets. How many tablets should be given in each dose?

CHAPTER SUMMARY

IMPORTANT TERMS

Factoring *(p. 290)*
Greatest common factor *(p. 290)*
Mean *(p. 296)*
Median *(p. 297)*
Mode *(p. 297)*

IMPORTANT IDEAS

1. The product of a positive real number and a negative real number is a negative real number.
2. The product of two negative real numbers is a positive real number.
3. *Distributive Property:* For any real numbers a, b, and c,
   $$a(b + c) = ab + ac \text{ and } (b + c)a = ba + ca.$$
4. The reciprocal of any real number a, $a \neq 0$, can be written as $\frac{1}{a}$.
5. If a and b are any real numbers, $b \neq 0$, then $a \div b = a \cdot \frac{1}{b}$.
   To divide a by b, multiply a by the reciprocal of b.
6. The quotient of a positive real number and a negative real number is a negative real number.
7. The quotient of two negative real numbers is a positive real number.
8. The mean of a collection of numbers, scores, measures, and so on, is the sum of the items in the collection divided by the number of items.
9. The median of a collection of data is the middle item when the items are arranged in order of size.
10. The mode is the item in the collection of data that occurs most often.

CHAPTER REVIEW

SECTION 10.1

Multiply.

1. $(-12)(9)$
2. $(14)(-10)$
3. $(-8)(26)$
4. $(43)(-5)$
5. $\frac{1}{4}(-24)$
6. $(-3)(51)$
7. $(340)(-11)$
8. $(-4)(40)$

SECTION 10.2

Multiply.

9. $-8 \cdot (-15)$
10. $(-24)(-4)(2)$
11. $(-0.6)(-45)$
12. $(-\frac{1}{4})(-28)$
13. $(-18)(-6)$
14. $(-36)(-3)$
15. $(-4)(-9)(-1)$
16. $(-8)(-14)$

SECTION 10.3

Multiply.

17. $-12(3r+2)$ **18.** $(-4t+9)(-15)$ **19.** $4(-8s+(-15))$

20. $y(-9y+12)$ **21.** $-8x(3x-9)$ **22.** $-6k(-5p-8q)$

SECTION 10.4

Factor.

23. $19r+19s$ **24.** $-25m+(-25n)$ **25.** c^2+3c

26. $-3rs+9rt$ **27.** $-12an-12nk$ **28.** $2k-3km$

SECTION 10.5

Divide.

29. $-30 \div 6$ **30.** $\frac{-24}{4}$ **31.** $-27 \div (-3)$ **32.** $\frac{-34}{-2}$

33. $-16 \div \frac{1}{4}$ **34.** $-20 \div \frac{1}{3}$ **35.** $-15 \div (-\frac{5}{8})$ **36.** $-9 \div (-\frac{3}{4})$

SECTION 10.6

For Exercises 37–42, find the mean, the median, and the mode for each collection of data.

37. Team A: 162, 173, 185, 156 **38.** Team X: 145, 163, 159, 145

39. Team B: 144, 144, 112, 160 **40.** Team Y: 138, 158, 140, 160

41. Team C: 204, 180, 180, 192 **42.** Team Z: 108, 210, 133, 173

43. The table below shows the monthly rainfall over a five–month period. Find the mean monthly rainfall for the five months. Give your answer to the nearest tenth of a centimeter.

March	April	May	June	July
1.5 cm	2.4 cm	1.6 cm	0.7 cm	2.1 cm

44. Find the mean low temperature for the week. The given temperatures are in degrees Celsius. Give your answer to the nearest tenth of a degree.

S	M	T	W	Th	F	S
2.0°	−1.1°	−1.4°	−2.1°	−1.9°	−0.8°	1.5°

Solving Equations: Real Numbers

CHAPTER 11

One of the problems an engineer must solve when designing a steel frame building is determining the strength of the steel beams. The engineer has to use the length, depth, and cross-sectional area of a beam to calculate the greatest safe load the beam can bear.

11.1 Addition and Subtraction Properties

The Addition and Subtraction Properties for Equations studied in Chapter 5 also apply to real numbers.

> ***Addition Property for Equations***
>
> Adding the same real number to each side of an equation forms an equivalent equation.
>
> $$p - 4.6 = 1.8$$
> $$p - 4.6 + \mathbf{4.6} = 1.8 + \mathbf{4.6}$$
> $$p = 6.4$$

P–1 **What number should be added to each side of the following equations in order to get the variable alone?**

a. $x - 3.2 = 7.1$ **b.** $y - 8 = 2$ **c.** $s - 2\frac{1}{4} = 8$

EXAMPLE 1 Solve and check: $x - \frac{1}{4} = -5\frac{1}{2}$

Solution: To solve an equation of the form $x - a = b$ or $b = x - a$, first add "a" to each side.

$$x - \frac{1}{4} = -5\frac{1}{2}$$
$$x - \frac{1}{4} + \frac{1}{4} = -5\frac{1}{2} + \frac{1}{4}$$
$$x = -5\frac{1}{4}$$

$a = \frac{1}{4}$

Check: $x - \frac{1}{4} = -5\frac{1}{2}$

$-5\frac{1}{4} - \frac{1}{4}$

$-5\frac{1}{2}$

P–2 **Solve and check.**

a. $x - \frac{1}{3} = -6\frac{2}{3}$ **b.** $x - 12 = -15$ **c.** $x - 4.8 = 7.0$

> ***Subtraction Property for Equations***
>
> Subtracting the same real number from each side of an equation forms an equivalent equation.
>
> $$s + 7\frac{2}{3} = 4\frac{1}{2}$$
> $$s + 7\frac{2}{3} - \mathbf{7\frac{2}{3}} = 4\frac{1}{2} - \mathbf{7\frac{2}{3}}$$
> $$s = -3\frac{1}{6}$$

P–3 **What number should be subtracted from each side of the following equations in order to get the variable alone?**

a. $x + 14 = 12$ **b.** $x + 3.2 = 7.6$ **c.** $4\frac{1}{2} = x + 5\frac{3}{4}$

EXAMPLE 2

Solve and check: $1.8 = k + 4.7$

Solution: To solve an equation of the form $x + a = b$ or $b = x + a$, first subtract "a" from each side.

$$1.8 = k + 4.7$$
$$1.8 - \mathbf{4.7} = k + 4.7 - \mathbf{4.7} \quad a = 4.7$$
$$-2.9 = k$$

Check: $1.8 = k + 4.7$

$-2.9 + 4.7$

1.8

P–4 **Solve and check.**

a. $y + 7.3 = 3.3$ **b.** $-2\frac{1}{2} = 2\frac{1}{2} + x$ **c.** $z + 12 = 12$

> Any real number a can be eliminated from $x + a = b$ by adding the opposite of a to each side. This is the same as subtracting a from each side.
>
> $$x + 5 = -3$$
> $$x + 5 + \mathbf{(-5)} = -3 + \mathbf{(-5)}$$
> $$x = -8$$

EXAMPLE 3

The temperature in a city decreased by 7°F from 6:00 P.M. to midnight. The temperature at midnight was −13°F. Find the temperature at 6:00 P.M.

Solution:

1 Choose a variable. ⟶ Let t = the temperature at 6:00 P.M.

2 Write an equation for the problem.

Think: **The temperature at 6:00 P.M. decreased by 7° was −13°F.**

Translate: $t - 7 = -13$

1 Solve the equation. ⟶ $t - 7 + \mathbf{7} = -13 + \mathbf{7}$

$$t = -6$$

4 **Check:** If −6°F is decreased by 7°, is the difference −13°F?

Does $-6 - 7 = -13$? Yes ✓

The temperature at 6:00 P.M. was −6°F.

CLASSROOM EXERCISES

Write an equivalent equation as the first step in solving each equation. (Examples 1 and 2)

1. $y + 18 = 7$
2. $x - 24 = 20$
3. $a + 5 = -22$
4. $q - 18 = -30$
5. $k - \frac{3}{4} = -4\frac{1}{2}$
6. $t + 3\frac{1}{5} = \frac{9}{10}$
7. $28 = -17 + m$
8. $-15 = 23 + b$
9. $-12.8 = y - 3.9$
10. $-7.5 = y + 8.8$
11. $15.6 = -0.9 + w$
12. $-8.7 = n - 4.8$
13. $43 + f = 24$
14. $-19 + g = 49$
15. $-56 = 27 + z$

For Exercises 16–27, solve and check. (Example 1)

16. $x - 10 = 16$
17. $y - 42 = 42$
18. $-41 = a - 61$
19. $-25 = c - 20$
20. $c - \frac{1}{4} = -2\frac{1}{2}$
21. $d - \frac{1}{2} = -1\frac{1}{2}$

(Example 2)

22. $h + 2 = -8$
23. $k + 8 = -2$
24. $-4.6 = n + 4.7$
25. $-91.9 = n + 19.9$
26. $t + 4\frac{1}{3} = -9\frac{2}{3}$
27. $w + 5\frac{1}{5} = 2$

WRITTEN EXERCISES

Goal: To solve equations by using the Addition and Subtraction Properties
Sample Problem: Solve: $16.2 + x = 7.8$
Answer: $x = -8.4$

For Exercises 1–42, solve and check. (Example 1)

1. $x - 14 = 25$
2. $a - 9 = 17$
3. $-15 = d - 24$
4. $-13 = k - 18$
5. $p - 12 = -15$
6. $p - 9 = -14$
7. $-6.2 = q - 4.4$
8. $-9.8 = q - 5.5$
9. $y - 5\frac{3}{4} = -3\frac{1}{4}$
10. $x - 8\frac{5}{6} = -5\frac{1}{6}$
11. $m - 2\frac{1}{5} = -5\frac{3}{5}$
12. $-21\frac{2}{3} = n - 9\frac{1}{3}$

(Example 2)

13. $16 = p + 29$
14. $27 = q + 35$
15. $t + 8 = -19$
16. $m + 12 = -23$
17. $-6 = x + 24$
18. $-7 = y + 10$
19. $c + 9.4 = 5.1$
20. $b + 13.7 = 8.6$
21. $0 = t + 11.1$
22. $-8.6 = u + 8.6$
23. $y + 7\frac{1}{2} = -11\frac{1}{2}$
24. $s + 5\frac{2}{3} = 6\frac{1}{3}$

MIXED PRACTICE

25. $148 + j = 123$

26. $-38 + z = -5$

27. $-29 = x - 47$

28. $-10 = g + 4$

29. $-0.7 = f - 1.6$

30. $7.1 = s - 1.5$

31. $8.7 + c = -19.5$

32. $n + 22.9 = 15.4$

33. $5\frac{3}{4} = y + 3\frac{1}{4}$

34. $b - 34 = -18$

35. $-32 + y = 32$

36. $a - 11.9 = -27.4$

37. $2.7 = m - 4.6$

38. $h - (-6) = 8$

39. $n + 6.2 = -3.9$

40. $33.4 = 22.9 + k$

41. $\frac{5}{12} = x - (-\frac{13}{12})$

42. $\frac{7}{8} = -\frac{15}{8} + x$

APPLICATIONS

Solve. (Example 3)

43. The wind chill temperature at a ski resort was −21°F. The wind chill temperature was 18°F less than the thermometer reading. Find the thermometer reading.

44. The temperature aboard a ship in the North Atlantic is 20°F. This is 38°F warmer than the wind chill temperature aboard the ship. Find the wind chill temperature.

45. The thermometer reading at halftime of a football game was −4°C. This was 11°C less than the reading at the start of the game. Find the thermometer reading at the start of the game.

46. At the end of the summer, the average surface temperature of the Arctic Ocean is 7°C. This is 4°C higher than the average surface temperature during the spring. Find the temperature during the spring.

REVIEW CAPSULE FOR SECTION 11.2

Combine like terms. (Pages 43–45)

1. $18 + 6x - 5x$

2. $6p - 12 - 4p$

3. $-n + 1.9 + 0.2n$

4. $6.4x - 2 + 2.1x$

5. $7p + 2 - 10p$

6. $6q - 8 - q$

7. $\frac{3}{4}x + \frac{5}{4}x + \frac{1}{4}$

8. $3\frac{1}{2} - 2\frac{1}{2}x - 7\frac{1}{2}x$

9. $3\frac{1}{2}x - 2\frac{1}{2}x - 7\frac{1}{2}$

Multiply. (Pages 285–288)

10. $-8(3d - 4)$

11. $-12(6 - g)$

12. $-\frac{3}{4}(-12h + 8)$

13. $-\frac{5}{6}(12a + 3)$

14. $-5(-5w - \frac{5}{3})$

15. $-9(-p - \frac{1}{3})$

16. $(-\frac{3}{4}h - \frac{9}{8})(-\frac{2}{3})$

17. $(\frac{4}{5} - 4c)(-1\frac{3}{4})$

18. $-0.2(1.4q - 2.5)$

19. $4(6x - 3)$

20. $5(-3m - 2)$

21. $0.5(12g + 8)$

11.2 Like Terms in Equations

When solving an equation, combine like terms on each side of the equation before applying the Addition and Subtraction Properties.

EXAMPLE 1 Solve and check: $-13 + 27 = -n + 19 + 2n$

Solution: [1] Combine like terms. ⟶ $-13 + 27 = -n + 19 + 2n$

$$14 = 19 + n$$

[2] Subtract 19 from each side. ⟶ $14 \mathbf{- 19} = 19 + n \mathbf{- 19}$

$$-5 = n \text{ or } n = -5$$

Check: $-13 + 27 = -n + 19 + 2n$

14	$-(-5) + 19 + 2(-5)$
	$5 + 19 + (-10)$
	$24 - 10$
↑	→ 14

When solving equations, apply the Distributive Property <u>before</u> combining like terms.

EXAMPLE 2 Solve and check: $-3(2m + 5) + 7m = 23 - 37$

Solution:

[1] Use the Distributive Property. ⟶ $-3(2m + 5) + 7m = 23 - 37$

[2] Combine like terms. ⟶ $-6m - 15 + 7m = 23 - 37$

$$-15 + m = -14$$

[3] Add 15 to each side. ⟶ $-15 + m \mathbf{+ 15} = -14 \mathbf{+ 15}$

$$m = 1$$

The check is left for you.

P–1 **Solve and check.**

a. $-8 + 12 = -2x + 42 + 3x$

b. $12.4 - 2.6 = -2(4x - 1) + 9x$

CLASSROOM EXERCISES

Combine like terms on both sides of each equation. (Step 1, Example 1; Step 2, Example 2)

1. $4p - 10 - 3p = 4 - 9$
2. $-19 + h + 13h = -1 - 12$
3. $-2z + 6 + 3z = -6 - 5$
4. $2 - 3 = 4c - 3 - 3c$
5. $4.2m - 3.2m + 5.1 = -4.2 - 1.1$
6. $1.8x - 2.5 - (-0.2x) = 1.9 - 5.9$
7. $2(y + 3) - y = 19$
8. $-3(d + 4) + 4d = -8 - 7$

For Exercises 9–16, solve. (Example 1)

9. $1 - 4 = 3x - 5 - 2x$
10. $14g - 13g + 2 = 19 - (-7)$
11. $-x - 5 + 2x = 9 - 15$
12. $-12 + 16 = -7n + 4 + 8n$

(Example 2)

13. $3(t - 4) - 2t = 9 - 34$
14. $3 - 7 = -4(q - 12) + 5q$
15. $3(x + 6) - 2x = 18$
16. $12 - 15 = -2(y - 4) + 3y$

WRITTEN EXERCISES

Goal: To solve an equation that contains like terms
Sample Problem: Solve: $0.4 = 5(j - 0.1) - 4j$ **Answer:** $j = 0.9$

For Exercises 1–20, solve and check. (Example 1)

1. $6q - 19 - 5q = -16 + 13$
2. $-12z - 23 + 13z = 18 - 25$
3. $5.6 - 8.3 = -4.2d + 2.6 + 5.2d$
4. $7.9 - 10.4 = -8.1r + 3.3 + 9.1r$

(Example 2)

5. $3(y - 4) - 2y = -27 + 19$
6. $4(b - 5) - 3b = -33 + 12$
7. $-14.7 + 6.9 = 5(h - 2) - 4h$
8. $12.8 - 17.3 = 10(m - 1) - 9m$

MIXED PRACTICE

9. $8s - 3 - 7s = 5 - 11$
10. $-33 + 26 = 17x - 21 - 16x$
11. $4y - 3(y + 9) = 35 - 43$
12. $3x - 2(3 + x) = 47 - 57$
13. $-n - 2(3 - n) = 49 + 3$
14. $28x + 3(6 - 9x) = 13 - 27$
15. $16 - 18 = -3r + 4(r - 6)$
16. $-12 + 7 = 7(g - 3) - 6g$
17. $18.4 - 23.7 = 6.8k - 9.2 - 5.8k$
18. $7.6 - (-11.2) = -2.1t - 16.4 + 3.1t$
19. $-\frac{1}{4}r - 3\frac{1}{2}r + 4\frac{3}{4}r = 2\frac{3}{8} - \frac{7}{8}$
20. $4\frac{3}{4} - 9\frac{1}{4} = 3\frac{5}{8}c - 11 - 2\frac{5}{8}c$

11.3 Multiplication and Division Properties

The Multiplication and Division Properties for Equations studied in Chapter 5 also apply to real numbers.

Multiplication Property for Equations

Multiplying each side of an equation by the same nonzero number forms an equivalent equation.

$$\frac{m}{2} = -3$$

$$\frac{m}{2}(\mathbf{2}) = -3(\mathbf{2})$$

$$m = -6$$

EXAMPLE 1 Solve and check: $\frac{x}{12} = -9$

Solution: To solve an equation of the form $\frac{x}{a} = b$, $a \neq 0$, first multiply each side by "a."

$$\frac{x}{12} = -9$$

$$\mathbf{12}\left(\frac{x}{12}\right) = \mathbf{12}(-9)$$

$$x = -108$$

Check: $\frac{x}{12} = -9$

$$\frac{-108}{12}$$

$$-9$$

P–1 **Solve and check.**

a. $\frac{x}{7} = -6$ **b.** $14 = \frac{y}{-8}$ **c.** $\frac{d}{-0.6} = 3.2$ **d.** $\frac{f}{-5} = 2\frac{1}{5}$

Remember that $-x$ means $-1x$. To get x alone in an equation of the form $-x = a$, multiply each side by -1.

$$-x = 8$$

$$(\mathbf{-1})(-1x) = (\mathbf{-1})8$$

$$x = -8$$

Another way to solve $-x = 8$ is to "take the opposite of each side."

$-x = 8$ is equivalent to $x = -8$.

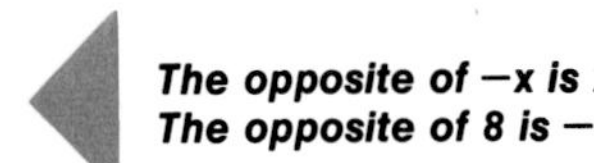

The opposite of $-x$ is x.
The opposite of 8 is -8.

P–2 **Solve and check.**

a. $16 = -q$ **b.** $-n = -\frac{3}{4}$ **c.** $2.2 = -b$ **d.** $-y = 193$

Division Property for Equations

Dividing each side of an equation by the same nonzero number forms an equivalent equation.

$$27 = -3w$$
$$27 \div \mathbf{(-3)} = -3w \div \mathbf{(-3)}$$
$$-9 = w$$

EXAMPLE 2

Solve and check: $-4x = 15$

Solution: To solve an equation of the form $ax = b$, $a \neq 0$, first divide each side by "a."

$$-4x = 15$$
$$\frac{-4x}{\mathbf{-4}} = \frac{15}{\mathbf{-4}}$$
$$x = -3\tfrac{3}{4}$$

Check: $-4x = 15$

$$-4(-3\tfrac{3}{4})$$
$$-4\left(-\frac{15}{4}\right)$$
$$15$$

P–3 **Solve and check.**

a. $-2y = 17$ **b.** $18.6 = -0.3d$ **c.** $137p = -411$ **d.** $-7.52 = 1.6t$

To solve an equation of the form $\frac{a}{b} \cdot x = c$ where $a \neq 0$ and $b \neq 0$, multiply each side by the reciprocal of $\frac{a}{b}$.

EXAMPLE 3

Solve and check: $\frac{4}{3}x = -12$

Solution:

$$\frac{4}{3}x = -12$$

Multiply each side by $\frac{3}{4}$. ⟶

$$\mathbf{\frac{3}{4}}\left(\frac{4}{3}x\right) = \mathbf{\frac{3}{4}}(-12)$$
$$\left(\frac{3}{4} \cdot \frac{4}{3}\right)x = \frac{3}{4}(-12)$$
$$1x = -9$$
$$x = -9$$

Check: $\frac{4}{3}x = -12$

$$\frac{4}{3}(-9)$$
$$-\frac{\overset{}{4}}{\underset{1}{\cancel{3}}} \cdot \frac{\overset{3}{\cancel{9}}}{1}$$
$$-12$$

P–4 **Solve and check.**

a. $\frac{2}{3}n = -18$ **b.** $-84 = \frac{4}{5}p$ **c.** $36 = -\frac{2}{7}t$ **d.** $\frac{3}{4}x = -12$

Any nonzero real number a can be eliminated from $ax = b$ by multiplying each side by the reciprocal of a. This is the same as dividing by a.

EXAMPLE 4 The deficiency in rainfall for a year divided by 12 equals the average deficiency in rainfall per month. One year, the average monthly deficiency in rainfall was −4.2 centimeters. Find the deficiency in rainfall for that year.

Solution:

1 Choose a variable. ⟶ Let $d =$ the deficiency in rainfall for the year in centimeters.

2 Write an equation for the problem.

Think: The deficiency in rainfall for a year divided by 12 is −4.2.

Translate: $d \div 12 = -4.2$

3 Solve the equation. ⟶ $\mathbf{12}\left(\frac{d}{12}\right) = \mathbf{12}(-4.2)$

$d = -50.4$

4 **Check:** If −50.4 is divided by 12, is the quotient −4.2?

Does $\frac{-50.4}{12} = -4.2$? Yes ✓

The deficiency in rainfall for the year was −50.4 centimeters.

CLASSROOM EXERCISES

Write an equivalent equation as the first step in solving each equation. (Examples 1–2)

1. $\frac{x}{4} = -7$ **2.** $\frac{n}{-3} = -19$ **3.** $\frac{r}{-7} = 10$ **4.** $13 = \frac{a}{5}$

5. $12y = -48$ **6.** $-72 = -8k$ **7.** $-6p = 42$ **8.** $-r = 16$

For Exercises 9–32, solve and check. (Example 1)

9. $\frac{p}{12} = -8$ **10.** $\frac{q}{-3} = -7$ **11.** $-12 = \frac{s}{5}$ **12.** $16 = \frac{k}{-4}$

13. $8 = \frac{x}{-2.1}$ **14.** $-4 = \frac{y}{5}$ **15.** $\frac{a}{-4} = -8.1$ **16.** $\frac{b}{-2.5} = -5$

(Example 2)

17. $4c = -32$
18. $18 = -6y$
19. $-132 = 12x$
20. $15g = -105$
21. $8.1 = -3z$
22. $3x = -4.2$
23. $9w = -\frac{3}{7}$
24. $-\frac{3}{8} = 6b$

(Example 3)

25. $17 = -\frac{1}{2}r$
26. $21 = -\frac{3}{4}t$
27. $\frac{2}{5}x = -18$
28. $\frac{1}{3}s = -14$
29. $-\frac{7}{8}n = -28$
30. $-\frac{7}{11}u = 42$
31. $40 = \frac{5}{12}v$
32. $-56 = -\frac{2}{9}c$

WRITTEN EXERCISES

Goal: To solve equations by using the Multiplication and Division Properties

Sample Problem: Solve: $-\frac{5}{4}r = -\frac{5}{12}$

Answer: $r = \frac{1}{3}$

For Exercises 1–44, solve and check. (Example 1)

1. $\frac{a}{6} = -19$
2. $\frac{j}{-9} = -4$
3. $-12 = \frac{d}{4.3}$
4. $\frac{c}{-0.4} = 5.2$
5. $\frac{2}{3} = \frac{x}{-6}$
6. $\frac{t}{-4} = \frac{7}{8}$
7. $1\frac{2}{5} = \frac{b}{-7}$
8. $\frac{h}{9} = -2\frac{1}{3}$

(Example 2)

9. $-12 = -3z$
10. $28 = -7x$
11. $1.4d = -9.8$
12. $1.5k = -10.2$
13. $4h = -\frac{2}{5}$
14. $-6p = \frac{3}{8}$
15. $4\frac{1}{2} = -3x$
16. $-3\frac{2}{3} = 11t$

(Example 3)

17. $\frac{1}{2}y = -9$
18. $-\frac{1}{5}t = -14$
19. $-\frac{11}{4}d = 33$
20. $-\frac{9}{8}n = 27$
21. $-35 = -\frac{5}{4}m$
22. $-24 = -\frac{8}{7}d$
23. $-\frac{6}{5}d = 96$
24. $\frac{4}{3}x = -84$

MIXED PRACTICE

25. $-72 = 9a$
26. $-26 = -13p$
27. $-39y = 13$
28. $-42x = 21$
29. $-28 = \frac{7}{8}m$
30. $-30 = \frac{5}{4}g$
31. $-77 = 11s$
32. $-b = 8.76$
33. $1\frac{3}{4}c = -\frac{7}{8}$
34. $-1\frac{7}{8}t = \frac{5}{16}$
35. $-8r = -48$
36. $-d = -34$
37. $\frac{1}{2}k = -14$
38. $\frac{1}{4}k = -11$
39. $-8r = -\frac{4}{9}$
40. $-\frac{1}{4}c = \frac{3}{8}$
41. $\frac{c}{5} = -20$
42. $8 = \frac{d}{1.5}$
43. $2.4 = \frac{m}{-0.3}$
44. $\frac{n}{-6} = -12$

APPLICATIONS

Solve. (Example 4)

45. The average daily temperature for Fairbanks, Alaska, in January multiplied by -5 equals the average daily temperature for July which is 60°F. Find the average daily temperature for January.

46. The number of points lost by a certain stock on Wednesday divided by 3 equals the loss of the stock on Thursday which was $-\frac{5}{8}$. Find how many points the stock lost on Wednesday.

MID-CHAPTER REVIEW

Solve and check. (Section 11.1)

1. $96 = c + 128$
2. $21 = z + 89$
3. $n - 19.8 = -24.3$
4. $q - 15.2 = -0.8$
5. $140 = p - 87$
6. $-118 = x - 434$
7. $-1\frac{2}{3} = t + 4\frac{5}{6}$
8. $7\frac{1}{3} = s + 2\frac{2}{9}$
9. $h - 0.9 = -12.5$

10. On flying from Calcutta, India, to Sydney, Australia, the temperature decreased by 14°C. The temperature upon arrival in Sydney was 18°C. Find the temperature in Calcutta at the beginning of the flight.

11. The record low temperature for St. Paul, Minnesota, is 22°F higher than for Barrow, Alaska. The record low temperature for St. Paul is -34°F. Find the record low temperature for Barrow.

Solve and check. (Section 11.2)

12. $-4d - 10 + 5d = -92 - 45$
13. $-11r + 12r + 15 = -86 - 127$
14. $247 - 159 = 73q - 123 - 72q$
15. $148 - 233 = -43m + 84 + 44m$
16. $12(3b - 4) - 35b = 128 - 143$
17. $109t - 12(9t - 7) = 138 - 273$

Solve and check. (Section 11.3)

18. $-42 = 14x$
19. $-108 = 18z$
20. $-\frac{6}{5}y = 36$
21. $\frac{7}{11}r = 49$
22. $-3t = \frac{2}{3}$
23. $4m = -\frac{9}{10}$
24. $\frac{15}{16} = \frac{d}{-8}$
25. $-\frac{17}{54} = \frac{b}{-9}$
26. $-2\frac{3}{4}k = \frac{11}{16}$

27. The number of points gained by a stock on Tuesday multiplied by -4 equals the stock's loss on Wednesday which was $-1\frac{1}{2}$. Find how many points the stock gained on Tuesday.

28. The temperature reading from an outdoor thermometer divided by -0.2 equals the wind chill temperature of -30°F. Find the thermometer reading.

REVIEW CAPSULE FOR SECTION 11.4

Solve and check. (Pages 144–147)

1. $4g + 2 = 18$ **2.** $8u - 5 = 35$ **3.** $21 = 2h - 3$

4. $18 = 3x + 6$ **5.** $19 = \frac{k}{2} + 1$ **6.** $\frac{c}{4} - 5 = 3$

11.4 Equations With More Than One Operation

For equations with more than one operation, use the Addition or Subtraction Properties before the Multiplication or Division Properties.

EXAMPLE 1

Solve and check: $3n + 11 = 5$

Solution:

$$3n + 11 = 5$$

1 Subtract 11 from each side. ⟶ $3n + 11 - \mathbf{11} = 5 - \mathbf{11}$

$$3n = -6$$

2 Divide each side by 3. ⟶ $\frac{3n}{\mathbf{3}} = \frac{-6}{\mathbf{3}}$

$$n = -2$$

Check

$3n + 11 = 5$

$3(-2) + 11$

$-6 + 11$

5

P–1 **Solve and check.**

a. $7n + 20 = 13$ **b.** $12x - 11 = -35$ **c.** $81 = -5y + 6$

EXAMPLE 2

Solve and check: $23 = \frac{r}{-4} - 13$

Solution:

$$23 = \frac{r}{-4} - 13$$

1 Add 13 to each side. ⟶ $23 + \mathbf{13} = \frac{r}{-4} - 13 + \mathbf{13}$

$$36 = \frac{r}{-4}$$

2 Multiply each side by -4. ⟶ $36\ \mathbf{(-4)} = \frac{r}{-4}\mathbf{(-4)}$

$$-144 = r$$

$$r = -144$$

Check

$23 = \frac{r}{-4} - 13$

$\frac{-144}{-4} - 13$

$36 - 13$

23

P–2 **Solve and check.**

a. $\frac{s}{3} - 6 = -14$ b. $1 = \frac{t}{5} + 7$ c. $\frac{n}{-9} + 14 = 10$

EXAMPLE 3

Twice a company's net loss for one month increased by 2500 dollars equals 350 dollars. Find the net loss.

Solution:

[1] Choose a variable. ⟶ Let n = the net loss.

[2] Write an equation for the problem.

Think: **Twice a company's net loss increased by $2500 equals $350.**

Translate: $2n \quad + \quad 2500 \quad = \quad 350$

[3] Solve the equation. ⟶ $2n + 2500 - \mathbf{2500} = 350 - \mathbf{2500}$

$$2n = -2150$$

$$\frac{2n}{\mathbf{2}} = \frac{-2150}{\mathbf{2}}$$

$$n = -1075$$

The – indicates a loss.

[4] **Check:** If twice –$1075 is increased by $2500, is the result $350?

Does $2(-1075) + 2500 = 350$? Does $-2150 + 2500 = 350$? Yes ✓

The net loss was $1075 (–$1075).

CLASSROOM EXERCISES

Write an equivalent equation as the first step in solving each equation. (Step 1, Examples 1 and 2)

1. $3k + 2 = 5$
2. $12 = -2y + 8$
3. $4n - 6 = -19$
4. $7b - 1 = 15$
5. $12 = 8 - 3w$
6. $-13 = 5d + 7$
7. $5.8 = 12.4 - 2z$
8. $3t - 1.2 = -3.9$
9. $\frac{m}{3} - 5 = -22$
10. $\frac{a}{4} + 7 = -5$
11. $-3.5 = 6x + 1.9$
12. $-9f + 0.3 = 12.6$
13. $-\frac{3}{5}v - 1\frac{2}{3} = 4\frac{1}{3}$
14. $\frac{3}{4}j - \frac{1}{8} = -\frac{5}{2}$
15. $-3\frac{1}{2} = \frac{g}{-2} - 8\frac{1}{4}$
16. $12 = \frac{c}{-1.3} + 15$
17. $0.6r + 1.3 = -5.3$
18. $-0.3s + 1.8 = 3.4$
19. $\frac{1}{2}n - 3 = -6$
20. $5 = -\frac{1}{3}y + 9$
21. $\frac{d}{0.2} + 9 = 3$

WRITTEN EXERCISES

Goal: To solve an equation with more than one operation

Sample Problem: $-14.2 = \frac{p}{-6} - 9.7$

Answer: $p = 27$

For Exercises 1–24, solve and check. (Example 1)

1. $4m + 13 = -7$
2. $3p + 12 = -15$
3. $-16 = 5t - 6$
4. $-43 = 6y - 7$
5. $-6a - 7.8 = 19.2$
6. $-4s - 3.6 = 12.0$

(Example 2)

7. $\frac{b}{4} + 19 = 16$
8. $21 = \frac{f}{8} + 24$
9. $-21 = \frac{c}{-6} - 14$
10. $-25 = \frac{z}{-5} - 19$
11. $-\frac{3}{4}d + 12 = -15$
12. $-\frac{5}{8}m + 14 = -26$

MIXED PRACTICE

13. $7.2 = 13.8 - 2p$
14. $\frac{v}{1.2} + 28 = -45$
15. $22 - \frac{q}{7} = 17$
16. $18.26 = -6.23 - 3.1n$
17. $24 - 8x = 36$
18. $-84 = -18 + \frac{h}{12}$
19. $15.4 = -4.4 - 2.4r$
20. $14.9 - 3y = 5.3$
21. $\frac{b}{2.6} + 19 = -37$
22. $33 - \frac{w}{9} = 25$
23. $-158 = -26 + \frac{a}{15}$
24. $17 - 6n = 32$

APPLICATIONS

Solve. (Example 3)

25. Four times the length of the first commercial telephone line decreased by 5 miles equals 7 miles. Find the length of the first commercial telephone line.

26. Twice the record low temperature for Omaha, Nebraska, increased by 96°F equals 32°F, the freezing point of water. Find the record low temperature for Omaha.

27. A halfback's net number of yards gained or lost this year is divided by 10 games. This quotient decreased by 23.9 equals 14.7. Find the halfback's net number of yards this year.

28. When the number of dialing areas in the United States and Canada is divided by 3 and 240 is added to the quotient, the result is 280. Find the total number of dialing areas.

11.5 From Words to Equations

Translating from words to symbols is an important step in solving word problems. Table 1 reviews some words or expressions that suggest the various operations on numbers.

Table 1

Operation	Words or Expressions
Addition	sum, more than, plus, increased by, added to, total
Subtraction	difference, minus, decreased by, diminished by, subtracted from, less than, less
Multiplication	product, times, twice, doubled, tripled, multiplied by
Division	quotient, divided by

EXAMPLE 1 Write an algebraic expression for this word expression.

A record low temperature less $-32°$

Solution:

[1] Choose a variable. → Let $t =$ a record low temperature.

[2] Identify the operation. → subtraction ("less")

[3] Write an algebraic expression. → $t - (-32)$, or $t + 32$ ◀ $a - b = a + (-b)$

Sometimes more than one operation is needed to represent a word expression.

P–1 **Write an algebraic expression for this word expression.**

Four pounds more than $\frac{1}{2}$ times a loss of weight, w

Translating from word sentences to equations is another important step in solving word problems. Table 2 shows some word sentences and their corresponding equations.

Table 2

Word Sentence	Equation
The sum of some number, n, and -7 **totals** -26.	$n + (-7) = -26$
Twice the number of stamps, s, decreased by -4 **equals** 93	$2s - (-4) = 93$
The product of some number, q, and -0.5 **is** 48.	$-0.5q = 48$
Three times a number, p, divided by -7 **equals** -9.	$\frac{3p}{-7} = -9$
Six dollars less than one-half the cost of a new television set, c, **is** \$139.	$\frac{1}{2}c - 6 = 139$

EXAMPLE 2 Write an equation for this word sentence.

An increase in temperature plus $-12°F$ equals $-3°F$.

Solution:

1 Choose a variable. ⟶ Let t = an increase in temperature.

2 Identify the operation. ⟶ addition ("plus")

3 Write an equation.

Think: An increase in temperature plus $-12°F$ equals $-3°F$.

Translate: $t \quad + (-12) \quad = \quad -3$

The equation is $t + (-12) = -3$, or $t - 12 = -3$.

CLASSROOM EXERCISES

For Exercises 1–6, write an algebraic expression for each word expression. Use the variable n. (Table 1 and Example 1)

1. Five strokes less than Anna's golf score
2. A depth below sea level in meters divided by -1000
3. The amount of a bank deposit increased by sixty dollars
4. Triple the deficiency of rainfall for a year
5. Thirty-two degrees more than the product of 1.8 and a number of degrees Celsius
6. The product of $\frac{1}{2}$ and the number of kilometers on a car's odometer decreased by -4

Match each word sentence in Exercises 7–14 with a corresponding equation from a–f below. (Table 2 and Example 2)

a. $3q = -11$ **b.** $-9w + 4 = -77$ **c.** $-77 + 3(11) = h$

d. $9k = -6$ **e.** $\frac{-6}{t} = 9$ **f.** $\frac{77}{x} = -11$

7. Seventy-seven divided by some number **equals** -11.

8. The sum of -77 and three times 11 **is** some number.

9. Three times some number **is** -11.

10. -11 **is** triple some number.

11. Negative six **is** nine times some number.

12. The quotient of -6 and some number **is** 9.

13. Four more than -9 times some number **is** -77.

14. The product of some number and 9 **is** -6.

WRITTEN EXERCISES

Goal: To write an algebraic equation for a word sentence

Sample Problems: **a.** Three times the cost, c, of a concert ticket is \$24.
b. One-half the cost, c, of a concert ticket increased by one equals \$5.

Answers: **a.** $3c = 24$
b. $\frac{1}{2}c + 1 = 5$

For Exercises 1–10:
a. *Choose a variable and write what it represents.*
b. *Write an algebraic expression for the word expression.*
(Example 1)

1. Last Tuesday's low temperature less seven degrees

2. Three strokes more than Saturday's golf score

3. \$4.50 times the number of players on a team

4. A gain of 16 yards on a play minus a loss

5. The quotient of 100 points and the number of questions

6. The number of passengers increased by a flight crew of 7

7. 20.4 more than twice the number of millimeters of rain

8. Fifteen less the quotient of the number of students and -6

9. 2.5 times the profit of a business decreased by a loss of \$2685

10. -6 points less than the total value divided by 12

For Exercises 11–18:
a. *Choose a variable and write what it represents.*
b. *Write an equation for the word sentence.*
(Example 2)

11. The quotient of \$570 and a number of payments is \$47.50.

12. The number of animals in a pet store minus fourteen is 57.

13. The product of the number of yards gained per game and 11 games is 619.3.

14. A pitcher's earned run average of 3.8 less last year's average equals −1.7.

15. −10 less than four times a number of tugboats equals 26.

16. The sum of three times a number of votes and 95 totals 308.

17. Thirty-five degrees less than the average daily temperature of a city in June equals 30°F, the average daily temperature in January.

18. The deficiency in rainfall of −8.6 inches increased by several inches of rain equals a deficiency of −1.8 inches.

MIXED PRACTICE

For Exercises 19–26, write an algebraic expression or equation for each word expression or word sentence.

19. A thermometer reading less 29°F equals a wind chill temperature of −29°F.

20. A stock loss followed by a gain of $2\frac{1}{2}$ points equals a change of $-1\frac{1}{2}$ points.

21. The total of bank deposits in a week less \$1200 in withdrawals

22. Sixteen centimeters more than five times the width of a board

23. The difference between the high temperature of 35°F for a day and the low temperature

24. The deficiency in rainfall for the first six months added to this month's deficiency of −5.3 inches

25. Twice the number of hours diminished by fifteen equals −3.

26. Four times the number of artists divided by −15 is −128.

NON-ROUTINE PROBLEM

27. The figure at the right can be formed by using 16 toothpicks. Move exactly two of the toothpicks to form exactly four squares of the same size. Each of the four squares must touch at least one other square.

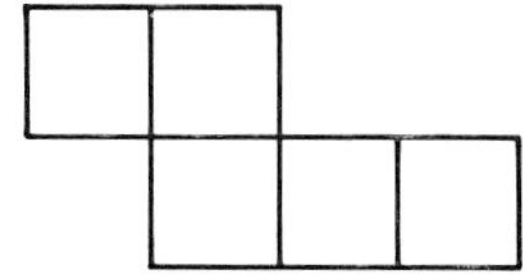

PROBLEM SOLVING AND APPLICATIONS

11.6 Using Conditions

To solve a word problem, translate the conditions of the problem into an algebraic expression and an equation. That is, use one condition to represent the unknowns by using a variable. Call this Condition 1. Use the other condition to write an equation for the problem. Call this Condition 2.

EXAMPLE 1 On the Harris family's telephone bill for May, the cost of long-distance calls was \$41.96 more than the cost for local service (Condition 1). The total bill was \$63.80 (Condition 2). Find the cost for long-distance calls.

Solution: [1] Use Condition 1 to represent the unknowns.

Let $c =$ the cost for local service.

Then $c + 41.96 =$ the cost for long-distance calls.

[2] Use Condition 2 to write an equation.

Think: **Cost of local service** + **Cost of long-distance calls** = **Total bill**

Translate: $c + c + 41.96 = 63.80$

[3] Solve the equation. ⟶

$$2c + 41.96 = 63.80$$
$$2c = 21.84$$
$$c = 10.92$$
$$c + 41.96 = 52.88$$

Don't forget to find $c + 41.96$.

Check: [4] Condition 1 Is the cost of long-distance calls \$41.96 more than the cost of local service?

Does \$52.88 = \$10.92 + \$41.96? Yes ✓

Condition 2 Does the total bill amount to \$63.80?

Does \$10.92 + \$52.88 = \$63.80? Yes ✓

The cost for long-distance calls was \$52.88.

Always check your answer(s) in the original conditions.

EXAMPLE 2

The airline distance from Chicago to Cleveland is 175 kilometers less than the distance from Cleveland to New York (Condition 1). The total distance from Chicago to New York is 1185 kilometers (Condition 2). Find the distance from Chicago to Cleveland.

Solution:

[1] Use Condition 1 to represent the unknowns.

Let $d =$ the number of kilometers from Cleveland to New York.

Then $d - 175 =$ the number of kilometers from Chicago to Cleveland.

[2] Use Condition 2 to write an equation.

Think: **Chicago to Cleveland** + **Cleveland to New York** = **Chicago to New York**

Translate: $d - 175 \quad + \quad d \quad = \quad 1185$

[3] Solve the equation. ⟶ $2d - 175 = 1185$

$$2d = 1360$$

$$d = 680$$

$$d - 175 = 505$$

Don't forget to find d − 175.

[4] **Check:** Condition 1 Is the distance from Chicago to Cleveland 175 kilometers less than the distance from Cleveland to New York?

Does $505 = 680 - 175$? Yes ✓

Condition 2 Is the distance from Chicago to New York equal to 1185 kilometers?

Does $505 + 680 = 1185$? Yes ✓

The distance from Chicago to Cleveland is 505 kilometers.

Steps for Solving Word Problems

1. Identify Condition 1. Use Condition 1 to represent the unknowns.
2. Identify Condition 2. Use Condition 2 to write an equation.
3. Solve the equation.
4. Check the results with the two conditions. Answer the questions.

CLASSROOM EXERCISES

With each sentence in Exercises 1–3, match the corresponding item in a–c. (Step 1, Examples 1 and 2)

1. One number is 6 less than 4 times another.

2. One number is 4 less than 6 times another.

3. One number is 6 more than 4 times another.

a. Let m = first number. Then $6m - 4$ = second number.

b. Let p = first number. Then $4p + 6$ = second number.

c. Let h = first number. Then $4h - 6$ = second number.

For Exercises 4–9, Condition 1 is underscored once. Condition 2 is underscored twice.

a. *Use Condition 1 to represent the unknowns. Use x as the variable.*

b. *Use Condition 2 to write an equation for the problem.* (Example 1)

4. Nora earned $5 more each week than her brother Ned. Together, Nora and Ned's salaries totaled $365 per week.

5. At a sale, a jacket sold for $46 more than a matching pair of slacks. The total for both items was $155.

6. The cost for labor on a car repair bill is $3.75 more than the cost for parts. The total bill is $92.85.

7. A doctor drives 120 kilometers farther on Tuesday than on Monday. She drives 960 kilometers in the two days.

(Example 2)

8. A family's electric bill for one month was $13.76 less than the gas bill. The total bill for gas and electric service was $98.92.

9. The interest on Jody's savings account is $10.11 less this year than last year. The total interest for the past two years is $58.32.

WRITTEN EXERCISES

Goal: To represent the conditions of a word problem and to solve the problem

Sample Problem: A mid-size car has 79 cubic feet more interior space than trunk space (Condition 1). The combined interior and trunk space is 113 cubic feet (Condition 2). Find the interior space in cubic feet.

Answer: The interior space is 96 cubic feet.

For Exercises 1–20, Condition 1 is underscored once. Condition 2 is underscored twice.

a. *Use Condition 1 to represent the unknowns.*

b. *Use Condition 2 to write an equation for the problem.*

c. *Solve.* (Example 1)

1. In a summer marathon, there are 31 more runners this year than last year. The total number of runners for the last two years is 279. Find the number of runners last year.

2. The Milltown library has 93 more science books than the Haverton library. The total number in both libraries is 657. Find the number of science books in each library.

3. The amount of snow that fell in Detroit in a recent year was 1.8 centimeters more than the amount of rain. The total amount of rain and snow was 159.2 centimeters. How much snow and how much rain fell that year?

4. A company spends $4,000,000 more on T.V. commercials than it does on magazine advertisements. The total cost for advertising is $26,000,000. How much is spent on each type of advertising?

5. Max has $238.57 more in his savings account than in his checking account. Altogether, he has $1526.95. How much does he have in each account?

6. Minh can type 19 words per minute faster than Juan. Their combined typing speed is 97 words per minute. Find Minh's typing speed.

(Example 2)

7. The airline distance from Fort Worth to New Orleans is 366 kilometers less than the distance from New Orleans to Miami. The total distance from Fort Worth to Miami is 1788 kilometers. Find the distance from Fort Worth to New Orleans.

8. The airline distance from Boston to New York is 12 miles less than the distance from New York to Washington, D.C. The total distance from Boston to Washington is 430 miles. Find the distance from Washington to New York.

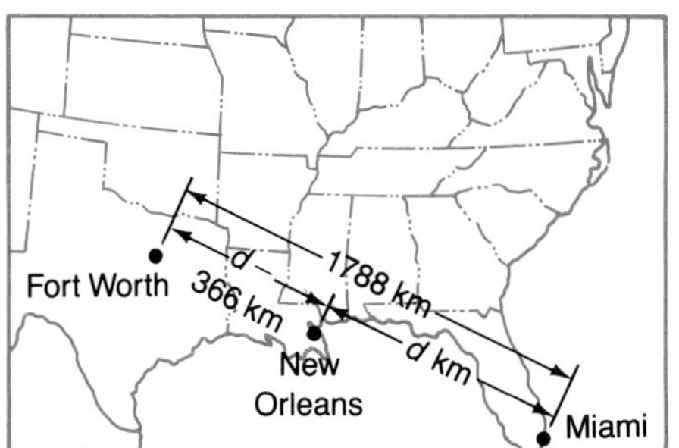

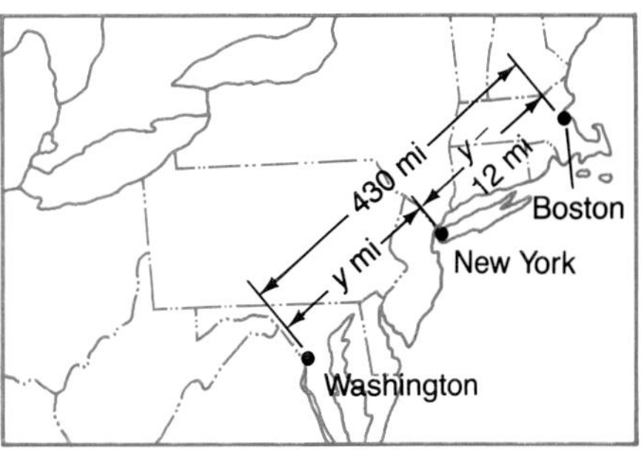

9. There are 9 fewer boys than girls in a karate class. The class has 37 students in all. Find the number of boys and the number of girls in the class.

10. A theater has 250 fewer balcony seats than orchestra seats. The theater has a total of 1200 seats. Find the number of balcony seats and the number of orchestra seats.

11. Mr. Harris pays $37.50 less to fly to Dallas during the week than to fly back on Saturday. The total cost of both flights is $246.90. Find the cost of the return flight.

12. Maria's coin collection has 28 fewer foreign coins than U.S. coins. There are a total of 218 coins in the collection. Find the number of each kind of coin in the collection.

MIXED PRACTICE

13. The Browns' sedan holds 13.6 fewer gallons of fuel than their station wagon. It takes 36.8 gallons of gas to completely fill the tanks of both cars. Find the capacity of the fuel tank of each car.

14. While Felipe was on vacation, he spent $69 less on travel expenses than on food. Travel and food costs amounted to $837. Find how much Felipe spent on travel and how much he spent on food.

15. Myra's movie ticket cost $2.00 less than her father's. Her father paid $7.00 for both tickets. Find the cost of each ticket.

16. Barbara works $2\frac{1}{2}$ fewer hours per week than Kerrick. Together they work $77\frac{1}{2}$ hours per week. Find the number of hours each person works.

17. An auto mechanic earned $14.75 more for labor on a repair job than her assistant. The total charge for labor was $89.25. Find the amount that each person received.

18. On Saturday the Lightfeather Bakery sold 55 fewer pies than cakes. The total number of pies and cakes sold was 187. Find the number of cakes sold on Saturday.

19. In a survey, the group using Toothpaste A had 176 fewer perfect checkups than the group using Toothpaste B. In all there were 1418 perfect checkups. Find the number of perfect checkups for each toothpaste.

20. On Friday night there were 85 more customers at the Rounders Roller Rink than on Thursday night. The total number of customers for both nights was 555. Find the number of customers on Friday night.

Using Statistics

Sampling

To find how many defective batteries can be expected in a large shipment, a **quality control technician** inspects and tests a certain number of batteries chosen **at random.** The batteries that are actually tested are called a **sample.**

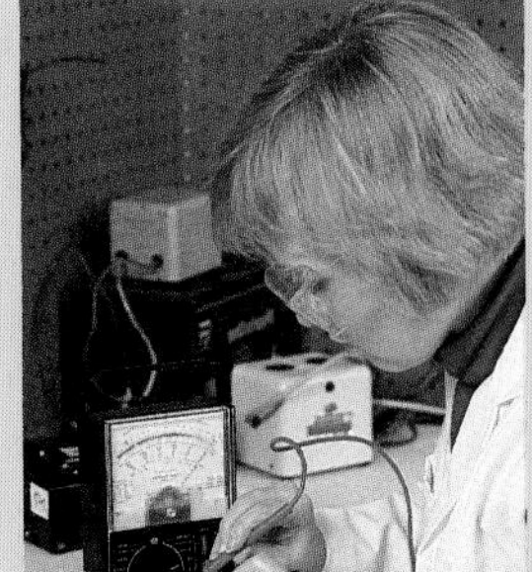

EXAMPLE: A technician found that 6 out of every 120 batteries tested were defective. About how many defective batteries can be expected in a shipment of 80,000?

SOLUTION: **a.** Find the percent of defective batteries in the sample.

6 out of 120 $= \frac{6}{120} = \frac{1}{20} = 0.05 = \mathbf{5\%}$ *5% are defective.*

b. Since 5% of the batteries in the sample are defective, it is reasonable to expect that about 5% of the batteries in the shipment will be defective.

Find 5% of 80,000.

5% of 80,000 $= 0.05 \times 80{,}000 = \mathbf{4000}$  *Expected number of defective batteries*

EXERCISES

For Exercises 1–4, complete the table.

	Product	Defective Items in Sample	Per Cent of Defective Items in Sample	Total Number of Items in Shipment	Expected Number of Defective Items in Shipment
1.	Stereos	2 out of 50	?	4000	?
2.	Watches	3 out of 60	?	6800	?
3.	Calculators	9 out of 300	?	800	?
4.	Light bulbs	7 out of 140	?	13,400	?

5. A sample of 400 cameras contains 20 that are defective. About how many defective cameras can be expected in a shipment of 26,000?

6. In a sample of 250 radios, five were found to be defective. About how many defective radios can be expected in a shipment of 1450?

CHAPTER SUMMARY

IMPORTANT IDEAS

1. The Addition, Subtraction, Multiplication, and Division Properties for Equations stated in Chapter 5 also apply to real numbers.
2. To solve an equation of the form $x - a = b$ or $b = x - a$, first add a to each side.
3. To solve an equation of the form $x + a = b$ or $b = x + a$, first subtract a from each side.
4. Any real number a can be eliminated from $x + a = b$ by adding the opposite of a to each side. This is the same as subtracting a from each side.
5. To solve an equation of the form $\frac{x}{a} = b$ or $b = \frac{x}{a}$, $a \neq 0$, first multiply each side by a.
6. To solve an equation of the form $ax = b$ or $b = ax$, $a \neq 0$, first divide each side by a.
7. Any nonzero real number a can be eliminated from $ax = b$ by multiplying each side by the reciprocal of a. This is the same as dividing by a.
8. The words sum, total, plus, increased by, more than, and added to suggest the operation of addition.
9. The words difference, less, less than, minus, decreased by, diminished by, and subtracted from suggest the operation of subtraction.
10. The words product, multiplied by, times, twice, doubled, and tripled suggest the operation of multiplication.
11. The words quotient, divided by, and half suggest the operation of division.
12. Steps for Solving Word Problems
    - **a.** Identify Condition 1. Use Condition 1 to represent the unknowns.
    - **b.** Identify Condition 2. Use Condition 2 to write an equation.
    - **c.** Solve the equation.
    - **d.** Check the result with the two conditions. Answer the questions.

CHAPTER REVIEW

SECTION 11.1

Solve and check.

1. $p - 33 = -19$ **2.** $n - 47 = -28$ **3.** $a + 29 = -15$

4. $y + 36 = -18$ **5.** $-1.8 = 3.2 + g$ **6.** $4.7 = 8.1 + b$

7. The temperature in a greenhouse decreased by 18°F from 5:00 P.M. to 10:00 P.M. The temperature at 10:00 P.M. was 59°F. Find the temperature at 5:00 P.M.

8. The thermometer reading at the ranger station was 22°F more than the wind chill temperature. The thermometer reading was 7°F. Find the wind chill temperature.

SECTION 11.2

Solve and check.

9. $-12m + 14 + 13m = 21 - 13$

10. $-8c - 9 + 9c = 5 + 18$

11. $-3.7d + 8 + 4.7d = -12 + 18$

12. $3.8f + 11.9 - 2.8f = -8.5 + 13.9$

13. $-5 + 19 = 2(x - 4) - x + 3$

14. $-19 + 33 = -4(q + 11) + 5q + 9$

15. $-42 + 3(9 - w) + 4w - 8 = 28 + 4$

16. $3(2z - 3) - 5z + 4 = -4 - 12$

SECTION 11.3

Solve and check.

17. $-46 = 8n$

18. $69 = -12s$

19. $-\frac{c}{4.6} = -7$

20. $-\frac{w}{1.9} = 12$

21. $49 = -\frac{7}{2}y$

22. $-72 = \frac{8}{3}z$

23. $1\frac{9}{16}d = -1\frac{3}{5}$

24. $-1\frac{7}{8}b = -3\frac{1}{5}$

25. The record low temperature in Louisville, Kentucky, multiplied by −5.5 equals the record high temperature of 110°F for Des Moines, Iowa. Find the record low temperature for Louisville.

26. The total withdrawals from an account for a year divided by 12 equals −\$512.50, the average total of withdrawals per month. Find the total withdrawals for the year.

SECTION 11.4

Solve and check.

27. $6m + 15 = -9$

28. $4c - 27 = -19$

29. $-26 = \frac{r}{8} - 23$

30. $9 = \frac{d}{12} + 15$

31. $1.4z - 2.7 = 4.3$

32. $-2.6y - 5.4 = -10.6$

33. Twice the net gain of a restaurant's revenue for one month decreased by 4500 dollars equals 2000 dollars. Find the net gain in revenue for this month.

34. Triple the record low temperature for a city increased by 188°F equals the record high temperature of 107°F. Find the record low temperature.

SECTION 11.5

Choose a variable and write what it represents. Write an equation for the word sentence.

35. The deficiency in rainfall in a city for six months divided by six equals −1.2 inches, the average deficiency per month.

36. Deposits of $1200 to an account added to the total withdrawals equals a net change of −$75 in the account balance.

37. Twenty times a quarterback's number of net yards running in a game decreased by four equals 156, his number of net yards passing.

38. One degree more than eight times the thermometer reading equals the wind chill temperature of −25°F.

SECTION 11.6

For Exercises 39–42, Condition 1 is underscored once. Condition 2 is underscored twice.

a. *Use Condition 1 to represent the unknowns.*

b. *Use Condition 2 to write an equation for the problem.*

c. *Solve.*

39. A hiker took 95 minutes more to climb to the top of a mountain than to come down. The total time for the hike was 245 minutes. How long did it take to climb to the top?

40. An airline flight had 147 fewer passengers in first class than in tourist class. The total number of passengers on the flight was 171. Find the number of passengers in first class and the number of passengers in tourist class.

41. The number of persons entering the A–Z Department Store at 10:00 A.M. one day was 235 fewer than at 10:00 A.M. the next day. The total for both days was 885. Find the number of persons who entered the store at 10:00 A.M. each day.

42. The Harris family's telephone bill for October was $75.60. The cost for long-distance calls was $43.46 more than the cost for the local calls. Find the cost for the long-distance calls.

Equations and Problem Solving: Real Numbers

Solving mixture problems is important to industry. The strength and the durability of steel are determined by the mixture of iron with other elements such as carbon, manganese, and aluminum. Mixtures are also important in everyday activities such as mixing colors of paint to produce a different color.

PROBLEM SOLVING AND APPLICATIONS

12.1 Mixture Problems

Recall that a ratio is a way to compare numbers. When you know that the ratio of the ingredients (parts) in a mixture is 4:3, you can represent the amounts by $4x$ and $3x$.

EXAMPLE 1 The ratio of vinegar to oil in a salad dressing is 4:3 (Condition 1). How much of each ingredient is there in 882 milliliters (Condition 2) of this mixture?

Solution: [1] Let $4x$ = number of milliliters of vinegar.
Then $3x$ = number of milliliters of oil.

Condition 1

[2] Write an equation.

Think: **Milliliters of vinegar** + **Milliliters of oil** = **Total in milliliters**

Translate: $4x + 3x = 882$

Condition 2

[3] Solve the equation. ⟶ $7x = 882$

$$\frac{7x}{7} = \frac{882}{7}$$

$$x = 126$$

Don't forget to find 4x and 3x.

$$4x = 504$$

$$3x = 378$$

[4] **Check:** Condition 1 Is the ratio of vinegar to oil 4:3?
Does $\frac{504}{378} = \frac{4}{3}$? Yes ✓

Condition 2 Are there 882 milliliters of the mixture?
Does $504 + 378 = 882$? Yes ✓

The mixture contains 504 milliliters of vinegar and 378 milliliters of oil.

Recall that the numbered steps in the Examples refer to the steps for solving word problems stated on page 325. You can use the methods of Example 1 to solve related problems.

EXAMPLE 2 A town voted to spend an amount of money on streets, parks, and the library in the ratio 4:3:2 (Condition 1). The total amount spent was $810,000 (Condition 2). How much was spent for each purpose?

Solution:

1 Let $4x$ = amount spent on streets.

Then $3x$ = amount spent on parks.

And $2x$ = amount spent on the library.

Condition 1

2 Write an equation.

Think: **Amount for streets** + **Amount for parks** + **Amount for library** = **Total spent**

Translate: $4x + 3x + 2x = 810{,}000$

Condition 2

3 Solve the equation. ⟶

$$9x = 810{,}000$$
$$x = 90{,}000$$
$$4x = 360{,}000$$
$$3x = 270{,}000$$
$$2x = 180{,}000$$

Don't forget to find 4x, 3x, and 2x.

4 **Check:** Condition 1 Are the amounts spent in the ratio 4:3:2?

Does $\frac{360,000}{270,000} = \frac{4}{3}$? Yes ✓ Does $\frac{270,000}{180,000} = \frac{3}{2}$? Yes ✓

Condition 2 Does the total amount spent equal $810,000?

Does $360{,}000 + 270{,}000 + 180{,}000 = 810{,}000$? Yes ✓

Thus, $360,000 was spent on streets, $270,000 was spent on parks, and $180,000 was spent on the library.

CLASSROOM EXERCISES

Represent the unknowns in each problem. Use the variable n. (Examples 1–2, Step 1)

1. The ratio of sand to cement in a mixture of concrete is 2:1.
2. The ratio of votes for Candidate A to votes for Candidate B is 5:8.
3. The ratio of games won to games lost is 3:2.
4. The ratio of hydrogen to oxygen in water is 2:1.

For Exercises 5–8, Condition 1 is underscored once. Condition 2 is underscored twice. (Examples 1–2, Steps 1–2)

a. *Use Condition 1 to represent the unknowns.*

b. *Use Condition 2 to write an equation for the problem.*

5. The ratio of alcohol to iodine in a bottle of tincture of iodine is 7:1. The bottle contains 720 milliliters.
6. Gwen made a meat loaf that was 2 parts beef, 2 parts pork, and 1 part veal. She used 2 kilograms of meat.

7. A town council voted to spend money on the town's schools, library, and hospital in the ratio 4:2:2. The total amount spent was $555,000.

8. A mixture of sunflower seeds, pumpkin seeds, and carob was in ratio 2:3:1. Dina bought 18 ounces of the mixture.

WRITTEN EXERCISES

Goal: To solve word problems involving mixtures

Sample Problem: A clerk mixes raisins, almonds, and cashews in the ratio 8:5:3 (Condition 1). How much of each ingredient is needed for a mixture of 12.8 kilograms (Condition 2)?

Answers: 6.4 kilograms of raisins, 4 kilograms of almonds, and 2.4 kilograms of cashews.

For Exercises 1–8, Condition 1 is underscored once. Condition 2 is underscored twice.

a. *Use Condition 1 to represent the unknowns.*

b. *Use Condition 2 to write an equation for the problem.*

c. *Solve.* (Example 1)

1. The ratio of copper to zinc in an alloy is 11:9. There are 400 kilograms of the alloy. How much of each metal is there?

2. The ratio of orange concentrate to water in a mixture is 1:3. How much concentrate and how much water are there in 12 liters of the mixture?

3. The ratio of brown sugar to flour in a crumb topping for apple pie is 5:3. How much flour is there in one cup of topping?

4. One part of oxygen and 4 parts of nitrogen are combined to form 500 cubic meters of a mixture. How many cubic meters of nitrogen are used?

(Example 2)

5. A businessman invested three sums of money in the ratio 6:5:4. The total amount invested was $390,000 What were the amounts invested?

6. In an election, three candidates received votes in the ratio 4:3:2. There were 396,000 votes cast. How many votes did each receive?

MIXED PRACTICE

7. A grocer made a blend of coffee using Brands A, B, and C in the ratio 2:5:11. How much of each brand did he use in 54 kilograms of the mixture?

8. A soap solution mixes palm-oil, glycerin, and water in the ratio 1:4:8. How much of each ingredient is there in 520 grams of the solution?

PROBLEM SOLVING AND APPLICATIONS

12.2 Geometry: Using a Figure

This theorem from geometry gives the sum of the measures of the angles of any triangle.

The sum of the measures of the angles of a triangle is 180°.

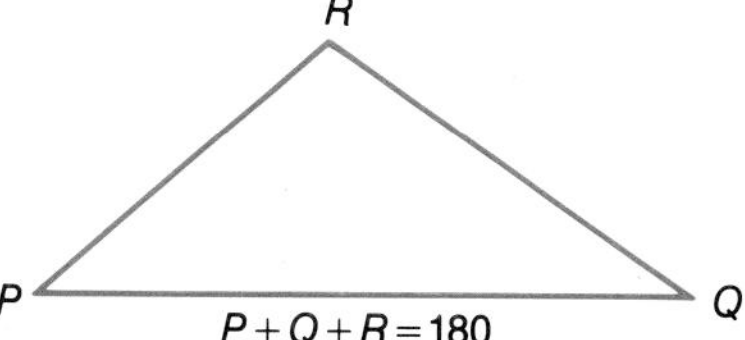

The theorem is part of Condition 2 in the problems of this section. In problems, words such as "in triangle *ABC*" refer to this theorem.

EXAMPLE 1 In triangle *ABC* (Condition 2), the measure of angle *B* is three times the measure of angle *C* (Condition 1). The measure of angle *A* is 120° (Condition 2). Find the measure of angle *B*.

Solution: Draw a figure.

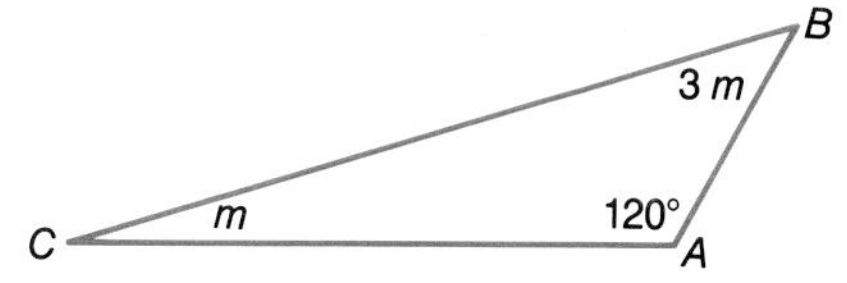

[1] Let $m =$ the measure of angle *C*.

Then $3m =$ the measure of angle *B*.

[2] Write an equation.

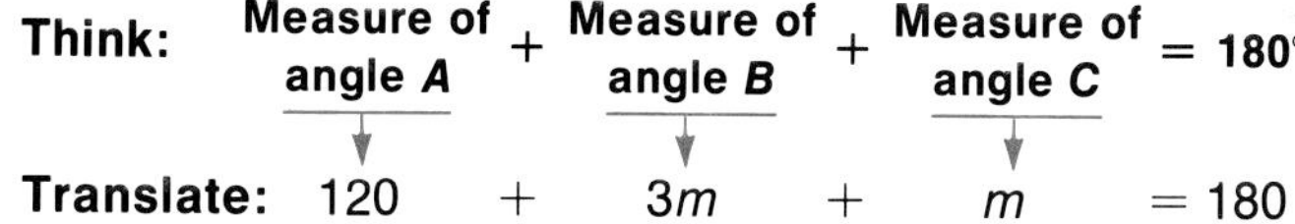

Think: **Measure of angle *A*** + **Measure of angle *B*** + **Measure of angle *C*** = **180°**

Translate: $120 \quad + \quad 3m \quad + \quad m \quad = 180$

[3] Solve the equation. ⟶ $120 + 4m = 180$

$$4m = 60$$

$$m = 15$$

$$3m = 45$$

Don't forget to find 3m.

[4] **Check:** Condition 1 Is the measure of angle *B* three times the measure of angle *C*? Does $45 = 3(15)$? Yes ✓

Condition 2 Does the sum of the measures of the angles equal 180°? Does $15 + 45 + 120 = 180$? Yes ✓

The measure of angle *B* is 45°.

Recall that the numbered steps in the Examples refer to the steps used to solve word problems stated at the end of Section 11.6.

Sometimes Condition 2 is shown in the given figure for a problem.

EXAMPLE 2

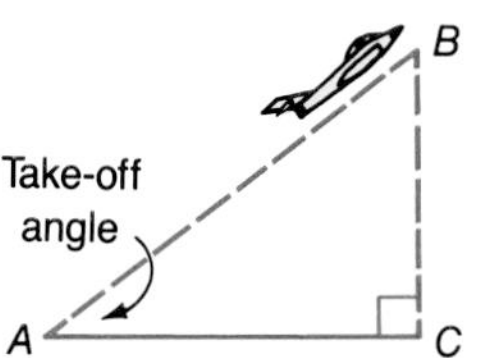

The **angle of take-off** for a plane is the angle between the ground and the flight path of the plane. In the figure at the left (Condition 2), the measure of the angle of take–off is 75° less than twice the measure of angle B (Condition 1). The measure of angle C is 90° Condition 2). Find the angle of take–off.

Solution: [1] Let y = the measure of angle B.

Then $2y - 75$ = the measure of angle A (take–off angle).

Think: **Measure of angle A** + **Measure of angle B** + **Measure of angle C** = **180°**

[2] $2y - 75 + y + 90 = 180$

[3] $3y + 15 = 180$

$3y = 165$

$y = 55$

$2y - 75 = 2(55) - 75 = 35$

Don't forget to find 2y − 75.

[4] **Check:** Condition 1 Is the measure of angle A 75° less than twice the measure of angle B?

Does $35 = 2(55) - 75$? Yes ✓

Condition 2 Does the sum of the measures of the angles equal 180°?

Does $55 + 35 + 90 = 180$? Yes ✓

The measure of the angle of take–off is 35°.

CLASSROOM EXERCISES

The measure of angle R in triangle PQR is 25°. Write an equation that can be used to find the measures of angles P and Q. (Step 2, Examples 1–2)

1. $P = x$; $Q = x + 17$ **2.** $P = y$; $Q = 4y$ **3.** $Q = t$; $P = 3t - 1$

WRITTEN EXERCISES

Goal: To solve word problems involving the angles of a triangle

Sample Problem: In triangle ABC (Condition 2), the measure of angle C is four times the measure of angle A (Condition 1). The measure of angle B is 60° (Condition 2). Find the measure of angle C.

Answer: 96°

For Exercises 1–14, Condition 1 is underscored once. Condition 2 is underscored twice. (When no figure is provided for a problem, you may find it helpful to draw one.)

a. *Use Condition 1 to represent the unknown angle measures.*

b. *Use Condition 2 to write an equation for the problem.*

c. *Solve.* (Example 1)

1. In triangle *ABC*, the measure of angle *A* is 5 times the measure of angle *B*. The measure of angle *C* is 150°. Find the measure of angle *A*.

2. In triangle *ABC*, the measure of angle *C* is 1.5 times the measure of angle *B*. The measure of angle *A* is 105°. Find the measure of angle *C*.

3. In triangle *ABC*, the measure of angle *B* is 3° more than two times the measure of angle *C*. The measure of angle *A* is 27°. Find the measure of angle *B*.

4. In triangle *ABC*, the measure of angle *C* is 5° less than 6 times the measure of angle *B*. The measure of angle *A* is 98°. Find the measure of angle *C*.

5. In triangle *ABC*, the measure of angle *A* is 14° more than three times the measure of angle *B*. The measure of angle *C* is 106°. Find the measure of angle *A*.

6. In triangle *ABC*, the measure of angle *B* is 9° more than half the measure of angle *C*. The measure of angle *A* is 156°. Find the measure of angle *B*.

7. One angle of a triangle is 70°. The measures of the other two angles are equal. Find the measures of the equal angles.

8. One angle of a triangle is 90°. The measures of the other two angles are equal. Find the measures of the equal angles.

(Example 2)

9. The ramp for a delivery truck forms triangle *XYZ* as shown below. The measure of angle *Z* is 8° less than 25 times the measure of angle *X*. The measure of angle *Y* is 90°. Find the measure of angle *X*.

10. The figure below shows how a pilot changed his flight path to avoid a thunderstorm. The measure of angle *Q* is 16° less than four times the measure of angle *R*. Find the measure of angle *Q*.

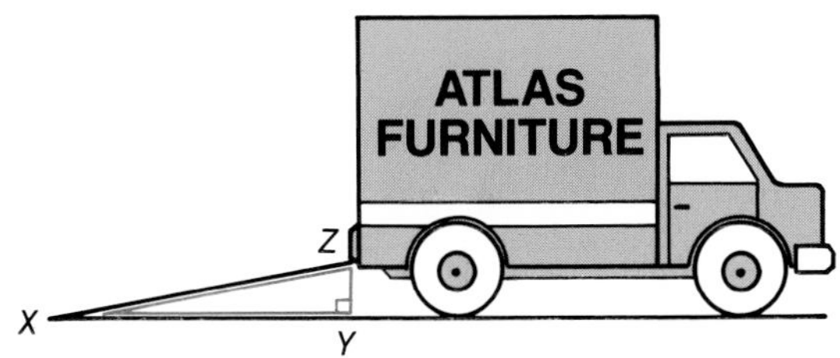

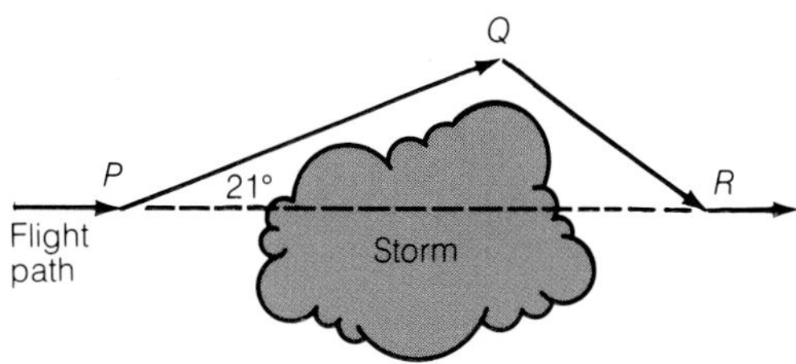

11. In the figure below, a kite string forms angle A with the ground. The measure of angle C is three times the measure of angle A. The measure of angle B is 90°. Find the measure of angle A.

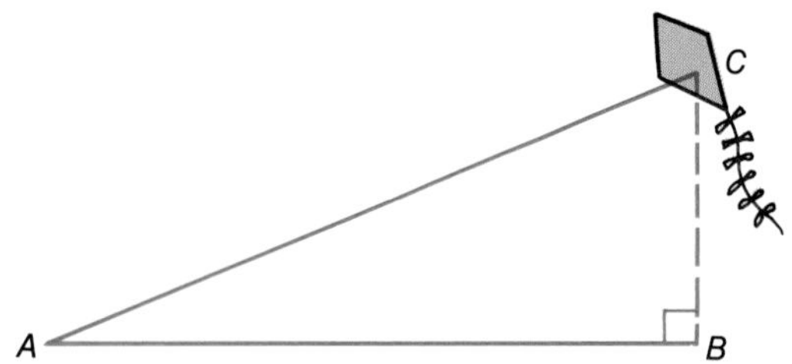

12. In the figure below, a submersible is descending to the ocean floor. The measure of angle A is one half the measure of angle C. The measure of angle B is 90°. Find the measure of angle A.

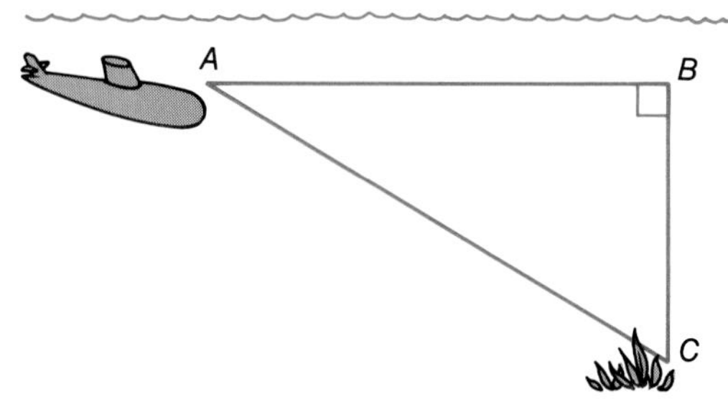

13. The figure below shows the path of a ferry as it travels from point A to point C. The measure of angle A is 10° more than 4 times the measure of angle C. The measure of angle B is 90°. Find the measure of angle A.

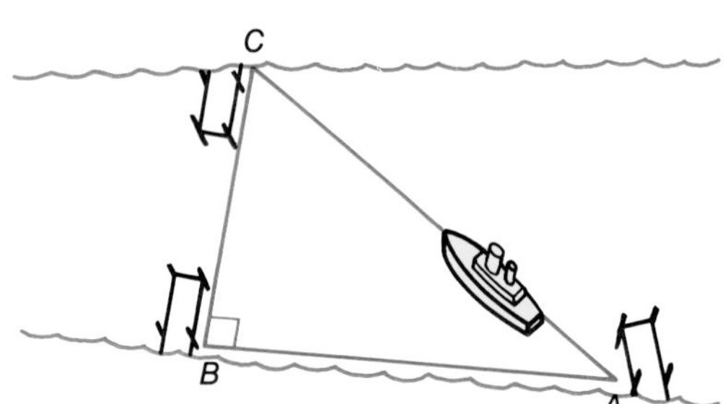

14. Triangle ABC below shows that angle A is formed by the floor of the house and the hill. The measure of angle A is 3° less than the measure of angle C. The measure of angle B is 90°. Find the measure of angle A.

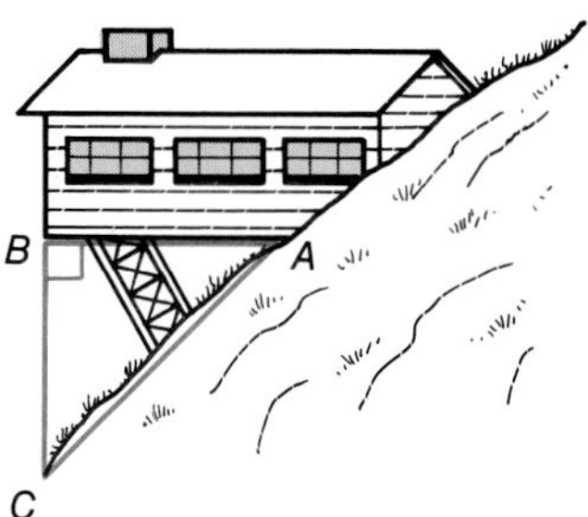

REVIEW CAPSULE FOR SECTION 12.3

Solve and check. (Pages 148–150)

1. $3(2t + 1) - 7 = 50$
2. $4(r - 3) + 3r = 16$
3. $10q + 5(15 - q) = 135$
4. $5(19 - p) + 10p = 135$
5. $15b - 2(b + 6) = 14$
6. $28z - 6(3z - 5) = 40$
7. $14(10 - m) + 4m = 75$
8. $12f - 3(f + 2) = 114$
9. $2a + 8(4a - 7) = 318$
10. $19g + 7(2 + g) = 40$
11. $21(p - 11) - 7p = 7$
12. $19d - 4(4d - 9) = 81$

12.3 Organizing Data: Money Problems

In problems involving coins, Condition 1 expresses the number of each kind of coin. Condition 2 expresses the total value of the coins. Making a table will help you to organize this information and to write an equation for the problem.

EXAMPLE 1

Rosa saves nickels, dimes and quarters. The number of nickels is 30 fewer than the number of dimes. There are 54 quarters (Condition 1). The total value of the coins is \$22.95 (Condition 2). How many nickels does Rosa have?

Solution: Make a table.

[1]

Coins	Number of Each Coin (Condition 1)	Value of Each Coin	Total Value in Cents (Condition 2)
Dimes	n	10¢	$10n$
Nickels	$n - 30$	5¢	$5(n - 30)$
Quarters	54	25¢	1350

[2] Write an equation.

Think: Value of dimes + Value of nickels + Value of quarters = Total value

Translate: $10n + 5(n - 30) + 1350 = 2295$ ◀ **\$22.95 = 2295¢**

[3] Solve the equation. ⟶

$$10n + 5n - 150 + 1350 = 2295$$
$$15n + 1200 = 2295$$
$$15n = 1095$$
$$n = 73$$
$$n - 30 = 43$$

◀ ***73 dimes 43 nickels***

Check:

[4] Condition 1 Is the number of nickels 30 fewer than the number of dimes?

Does $43 = 73 - 30$? Yes ✓

Condition 2 Does the total value of the coins equal \$22.95?

Does \$0.10(73) + \$0.05(43) + \$0.25(54) = \$22.95? Yes ✓

Rosa has 43 nickels.

Example 2 is another kind of money problem. When solving problems, always be sure to answer the question that was asked. Thus, in Example 1, you were asked to find the number of nickels. In Example 2, you are asked to find the number of 25–cent stamps and the number of 6–cent stamps.

EXAMPLE 2 The number of 25–cent stamps on a package is one fewer than the number of 6–cent stamps (Condition 1). The total value of the stamps is 99¢ (Condition 2). Find the number of each kind of stamp.

Solution: Make a table.

	Stamps	Number of Each Stamp (Condition 1)	Value of Each Stamp in Cents	Total Value in Cents (Condition 2)
1	6–cent	n	6¢	$6n$
	25–cent	$n-1$	25¢	$25(n-1)$

Think: **Value of 6¢ stamps** + **Value of 25¢ stamps** = **Total value**

2 **Translate:** $6n + 25(n-1) = 99$

3
$$6n + 25n - 25 = 99$$
$$31n - 25 = 99$$
$$31n = 124$$
$$n = 4$$
$$n - 1 = 3$$

Don't forget to find (n − 1).

Check:

4 Condition 1 Is the number of 25–cent stamps one fewer than the number of 6–cent stamps? Does $3 = 4 - 1$? Yes ✓

Condition 2 Is the total postage 99¢?

Does 6¢(4) + 25¢(3) = 99¢? Yes ✓

There are four 6–cent stamps and three 25–cent stamps.

CLASSROOM EXERCISES

Choose the algebraic expression that gives the value in cents for each number of coins or stamps. (Example 1, Step 1)

1. $(n-5)$ nickels

a. $n-5$ **b.** $0.5(n-5)$
c. $0.05(n-5)$ **d.** $5n-25$

2. d dimes and $(d+7)$ pennies

a. $11d+7$ **b.** $d+(d+7)$
c. $2d+7$ **d.** $10d+7$

3. d nickels and $(d + 2)$ quarters
   a. $2d + 2$
   b. $5d + 25d + 2$
   c. $30d + 50$
   d. $5d + d + 2$

4. d dimes, $(d + 7)$ nickels, and 3 quarters
   a. $d + (d + 7) + 3$
   b. $10d + 75$
   c. $15(d + 7) + 75$
   d. $15d + 110$

(Example 2, Step 1)

5. t 25–cent stamps
   a. $0.25t$
   b. $25(t - 1)$
   c. $t \div 25$
   d. $25t$

6. 12 five–cent stamps and $(n - 2)$ 15–cent stamps
   a. $12 + 15(n - 2)$
   b. $75n - 30$
   c. $60 + 15(n - 2)$
   d. $75(n - 2)$

7. x 15–cent stamps and $(x + 3)$ 13–cent stamps
   a. $28x + 39$
   b. $28x + 13$
   c. $28(x + 3)$
   d. $28x + 3$

8. s 15–cent stamps, $(s + 6)$ 5–cent stamps, and six 25–cent stamps
   a. $15s + 5(s + 6)$
   b. $170s$
   b. $20s + 180$
   d. $20s + 12$

WRITTEN EXERCISES

Goal: To solve word problems involving money

Sample Problem: The number of dimes Peter saved is 5 more than 6 times the number of quarters (Condition 1). The total value of the coins is $14.10 (Condition 2). How many dimes and how many quarters did he save?

Answer: 16 quarters and 101 dimes

For Exercises 1–4, Condition 1 is underscored once. Condition 2 is underscored twice.

a. *Use the given information (Condition 1 and Condition 2) to complete the table.*

b. *Use Condition 2 to write an equation for the problem.*

c. *Solve.* (Example 1)

1. *Juan paid a taxi fare with half–dollars and nickels. He gave the driver two more nickels than half–dollars. The fare was $3.40. How many nickels and how many half–dollars did Juan give the driver?*

Coins	Number of Each Coin (Condition 1)	Value of Each Coin in Cents	Total Value in Cents (Condition 2)
Half–Dollars	h	?	?
Nickels	$h + 2$	?	?

2. Melanie has the exact change for her bus fare. She has two more nickels than quarters, and she has 5 dimes. The bus fare is $1.20. How many nickels does she have?

Coins	Number of Each Coin (Condition 1)	Value of Each Coin in Cents	Total Value in Cents (Condition 2)
Quarters	q	?	?
Nickels	?	?	?
Dimes	5	?	?

(Example 2)

3. The number of 6–cent stamps on a package is three less than the number of 25–cent stamps. The total value of the stamps is $1.06. How many 6–cent stamps and how many 25–cent stamps are there?

Stamps	Number of Each Stamp (Condition 1)	Value of Each Stamp in Cents	Total Value in Cents (Condition 2)
25–cent	x	?	?
6–cent	?	?	?

4. The number of 15–cent stamps a postal clerk sold on Saturday was 150 more than 3 times the number of 8–cent stamps. The total value of the stamps was $75.50. How many of each kind of stamp did she sell?

Stamps	Number of Each Stamp (Condition 1)	Value of Each Stamp in Cents	Total Value in Cents (Condition 2)
8–cent	y	?	?
15–cent	?	?	?

MIXED PRACTICE *Make a table and solve each problem.*

5. A cash register contains 13 fewer quarters than nickels. The value of the quarters and nickels is $4.25. How many quarters are there?

6. Nona bought twice as many 13–cent stamps as 1–cent stamps, and four 15–cent stamps. She paid $2.22 for the stamps. How many of each kind of stamp did she buy?

7. On a letter there were five more 1–cent stamps than 3–cent stamps, and 2 more 10–cent stamps than 3–cent stamps. The total postage amounted to 53¢. Find the number of 10 cent stamps.

8. Jim counted the money from his paper route. He had 3 times as many quarters as nickels, and twice as many dimes as nickels. The total amount was $40. How many coins of each kind did he have?

MID-CHAPTER REVIEW

For Exercises 1–12, Condition 1 is underscored once, and Condition 2 is underscored twice.

Use Condition 1 to represent the unknowns. Use Condition 2 to write an equation for the problem. Solve. (Section 12.1)

1. Bronze is an alloy containing copper and tin in the ratio 4:1. How much of each metal is there in 832 grams of bronze?

2. An alloy contains copper and nickel in the ratio 3:7. How much of each metal is in 48 ounces of the alloy?

3. Two partners share in the profits of a business venture in the ratio 2:5 How much does each receive if the profit is $21,315?

4. A worker does correspondence, order processing, and accounting in the ratio 2:3:5. How much time is spent on each task in a 40-hour work week?

(Section 12.2)

5. In triangle *ABC*, the measure of angle *A* is 8° more than twice the measure of angle *B*. The measure of angle *C* is 34°. Find the measure of angle *A*.

6. In triangle *ABC*, the measure of angle *B* is 5° less than three times the measure of angle *C*. The measure of angle *A* is 61°. Find the measure of angle *B*.

7. One angle of a triangle is 48°. The measures of the other two angles are equal. Find the measures of the equal angles.

8. One angle of a triangle is 136°. The measures of the other two angles are equal. Find the measures of the equal angles.

(Section 12.3)

9. Sonia has the exact change to pay for her lunch. She has three more nickels than quarters, and she has one dime. The total cost of her lunch is $1.15. How many nickels does she have?

10. Jose counted the number of coins in his pocket. He had two more dimes than nickels, and he had two quarters. The total value of the coins was $1.00. How many dimes did he have?

11. The number of 10-cent stamps on a package was 2 more than three times the number of 2-cent stamps. The total value of the stamps was $1.16. How many of each kind of stamp were there on the package?

12. The number of 22-cent stamps sold at a post office one morning was 34 more than two times the number of 13-cent stamps sold. The total value of these stamps was $19.45. How many of each kind of stamp were sold that morning?

Using Formulas and Tables **Heat Transfer**

In warm weather, outside temperatures are higher than inside temperatures. The following formula is used to estimate the rate at which heat is transferred from the exterior to the interior of a building through the walls and windows. **Heat transfer** is measured in Btu's (British thermal units) per hour.

Heat transfer $= A \cdot U(i - o)$

A: surface area (ft^2)
U: heat transfer factor
i: inside temperature (°F)
o: outside temperature (°F)

The variable U in the formula is a number that varies according to the type of surface through which the heat passes.

Insulation is installed in houses to reduce heat transfer. Proper insulation can help keep heat inside in winter and outside in summer.

Surface	*Value of U*
Concrete, 6 inches thick	0.58
Glass, single pane	1.13
Brick, 8 inches thick	0.41
Wood, 2 inches thick	0.43

EXAMPLE: Estimate the rate of heat transfer through a 200-ft^2 wall of 6-inch concrete. The indoor temperature is 65°F and the outdoor temperature is 80°F.

SOLUTION: Heat transfer $= A \cdot U(i - o)$
$= (200)(0.58)(65 - 80)$
$= 116(-15)$
$=$ **−1740 Btu's per hour**

The negative number means that the interior gains heat.

EXERCISES

1. Surface: glass
   $A = 150$ ft^2; $i = 78$°F; $o = 90$°F
   Rate of heat transfer: ___?___

2. Surface: wood
   $A = 400$ ft^2; $i = 72$°F; $o = 85$°F
   Rate of heat transfer: ___?___

3. Which is greater, heat transfer through a concrete wall 6 inches thick, or through a brick wall 8 inches thick? Explain.

4. If the outside temperature is 30°F, and the inside temperature is 71°F, will there be a heat loss or a heat gain in the interior? Explain.

REVIEW CAPSULE FOR SECTION 12.4

Add or subtract. (Pages 43–45)

1. $4n + (13 - n)$
2. $(-m) + (-5m - 11)$
3. $(3.8 - 2.9w) + (-0.8w)$
4. $(5y - 13) - 8y$
5. $(18x + 17) - 18x$
6. $(0.4c - 9) - 0.4c$

What expression should be added to each side of each of the following equations to eliminate the variable from the left side? (Pages 114–117)

7. $11 - x = 7 + 3x$
8. $4 - 9a = 2 + a$
9. $14 - 5b = 9 - 4b$
10. $7 - q = q - 5$
11. $1.4 - 2.1p = 6 - 0.5p$
12. $12 - 5.5j = 2.4j - 7$
13. $1\frac{1}{3} - \frac{2}{3}z = 3z - 4\frac{5}{9}$
14. $\frac{3}{4} - \frac{3}{4}t = 1\frac{1}{2} - \frac{2}{3}t$
15. $4 - r = 1\frac{1}{3}r - 12\frac{1}{6}$

12.4 Equations: Variable on Both Sides

When you solve an equation with a variable on both sides, you can use the Addition and Subtraction Properties to eliminate the variable from one side.

EXAMPLE 1 Solve and check: $12 - 4r = 27 - 3r$

Solution:

$$12 - 4r = 27 - 3r$$

[1] Add $4r$ to each side. ⟶ $12 - 4r + \mathbf{4r} = 27 - 3r + \mathbf{4r}$

$$12 = 27 + r$$

[2] Subtract 27 from each side. ⟶ $12 - \mathbf{27} = 27 + r - \mathbf{27}$

$$-15 = r$$

Check: $12 - 4r = 27 - 3r$

$12 - 4(-15)$	$27 - 3(-15)$
$12 - (-60)$	$27 - (-45)$
$12 + 60$	$27 + 45$
72	72

P–1 **Solve and check.**

a. $8 - 2x = 32 - x$ b. $45 - x = 79 - 2x$ c. $12 + 3x = 16 + 4x$

Combine like terms on each side of an equation before applying the Addition and Subtraction Properties.

EXAMPLE 2 Solve and check: $5d - (3d - 7) = 6 + 3d - 14$

Solution: To use the Distributive Property in step [1], think of $-(3d - 7)$ as $-1(3d - 7)$. Then, $-1(3d - 7) = -3d + 7$.

[1] Use the Distributive Property. ⟶ $5d - (3d - 7) = 6 + 3d - 14$

[2] Combine like terms. ⟶ $5d - 3d + 7 = 6 + 3d - 14$

$$2d + 7 = 3d - 8$$

[3] Subtract $2d$ from each side. ⟶ $2d + 7 - \mathbf{2d} = 3d - 8 - \mathbf{2d}$

$$7 = d - 8$$

[4] Add 8 to each side. ⟶ $7 + \mathbf{8} = d - 8 + \mathbf{8}$

$$15 = d$$

The check is left for you.

P-2 **Solve and check: $6n - (4n + 2) = 5 + n - 13$**

CLASSROOM EXERCISES

For Exercises 1–10, solve and check. (Example 1)

1. $10 - 3n = -4n + 6$
2. $7 - x = 12 - 2x$
3. $7a + 2 = 6a - 7$
4. $12b - 8 = 13b + 6$
5. $0.1x - 1 = -5 - 0.9x$
6. $1.3y + 2 = -5.6 + 0.3y$

(Example 2)

7. $12y + 3 = 11y - (6 - 2y)$
8. $8m - (6m + 1) = m - 2$
9. $8x - 4 - 2x = -3 + 5x$
10. $2y + 11 = -7y + 12 + 10y$

Evaluate both sides of each of the following equations. Write Yes or No to indicate whether the given value of the variable is a solution. (Check, Examples 1 and 2)

11. $4c - 3 = 3c + 5$ when $c = 8$
12. $-y + 10 = 14 - 2y$ when $y = 4$
13. $5 - 3d = -4d - 2$ when $d = -7$
14. $2m + 6 = m - 4$ when $m = 10$
15. $\frac{1}{2}f + 8 = 6 - \frac{1}{2}f$ when $f = 2$
16. $-\frac{3}{4}p - 2 = \frac{1}{4}p + 6$ when $p = -8$

WRITTEN EXERCISES

Goal: To solve an equation with the variable on both sides
Sample Problem: Solve: $6.2 + 14.7z = 13.7z - 3.8$
Answer: $z = -10$

For Exercises 1–34, solve and check. (Example 1)

1. $a - 5 = 2a + 7$
2. $3u + 2 = 2u - 11$
3. $8 - 3g = -4g - 12$
4. $-9y - 3 = 12 - 10y$
5. $-24m - 5 = 12 - 23m$
6. $7 - 14x = -15x - 16$
7. $0.2b - 1.7 = 4.9 - 0.8b$
8. $3.4 - 0.6c = 0.4c - 5.9$
9. $\frac{3}{8}q - 6 = -\frac{5}{8}q + 10$
10. $-\frac{3}{4}h + 3 = \frac{1}{4}h - 5$

(Example 2)

11. $3k - (k - 3) = k - 7$
12. $-4v - (7 - v) = 4 - 2v$
13. $10p - 5 - 12p = 10 - p$
14. $-2z + 5 - 5z = 8 - 6z$
15. $-3(2d - 1) + d = 5 - 4d$
16. $5(-x + 3) + 2x = -2x + 7$
17. $3n - 7 - 8n = -17 - 4n$
18. $12q - 4 - 9q = 6 + 4q$

MIXED PRACTICE

19. $18f + 26 = -58 + 17f$
20. $y + 2 = 14 + 2y$
21. $8m - (5m - 7) = 6 + 4m - 15$
22. $6(14 - 9r) = -123r + 39 + 68r$
23. $5 - 4w + 12 = -2(3w - 8) + w$
24. $15 - 4n = -3n - 1$
25. $5a + 1 = 4a - 7$
26. $\frac{1}{3}(5y - 6) = -\frac{1}{6}(24 - 4y)$
27. $3(4.5 - 1.2v) = -8.3v + 6.8 + 3.7v$
28. $k - 3(4k - 2) = -7k + 5 - 3k$
29. $-\frac{3}{4}b - 19 = 12 - \frac{7}{4}b$
30. $4.9 - 6.5g = -7.5g - 2.2$
31. $7c + 8 = -12 + 8c$
32. $-6(6x - 9) = 7(8 - 5x)$
33. $-2(7m - 12) = 5(19 - 3m)$
34. $5h + 8 - h = 10h - (7h + 12)$

MORE CHALLENGING EXERCISES

For Exercises 35–42, let a and b be positive numbers where $a > b$.
a. *Solve each equation for x.*
b. *Then write positive or negative for the value of x.*

35. $x - a = b$
36. $x + b = a$
37. $-x - a = -b$
38. $3x + a = 2x + b$
39. $5x - b = 4x - a$
40. $-2x - b = a - 3x$
41. $3x - (x - a) = x - b$
42. $-4x + b + x = a - 2x$

REVIEW CAPSULE FOR SECTION 12.5

For Exercises 1–9, write the expression that should be added to each side of the given equation to eliminate the variable from the right side. (Pages 114–117)

1. $3x - 20 = 5 - 2x$ **2.** $4y + 7 = -3y + 21$ **3.** $9 - 10 = 4 - 11z$

4. $3\frac{1}{2} - 4\frac{1}{4}p = 7 - 8\frac{3}{4}p$ **5.** $9r + 5\frac{1}{3} = 7\frac{1}{5} - 4\frac{4}{5}r$ **6.** $-8\frac{1}{6} + 3\frac{1}{2}t = -4\frac{5}{6}t + 9\frac{1}{4}$

7. $5.2f - 3 = -1.2f + 6$ **8.** $4 + 2.8m = 12 - 4.8m$ **9.** $3 - 0.9s = 0{:}5 - 0.6s$

Use the Distributive Property. Then combine like terms. (Pages 310–311)

10. $-5(3x - 6) + 7x$ **11.** $-3(x + 8) - x$ **12.** $-12n + 6(4 - 3n)$

13. $-t + (-5 - 2t)2$ **14.** $-\frac{1}{2}(4r - 10) - r + 3$ **15.** $-\frac{1}{3}(6 - 9w) - 4w + 7$

16. $1.3(-2m - 6) - 3.5m$ **17.** $4.2y - (3y - 5)(0.8)$ **18.** $\frac{3}{8}p + \frac{3}{4}(8p - 12)$

12.5 More Equations: Variable on Both Sides

You can solve equations with a variable on both sides by following the rules studied in Section 11.4 on page 317.

EXAMPLE 1 Solve and check: $5m + 7 = 2m - 8$

Solution: First, try to get the variable on one side of the equation.

$$5m + 7 = 2m - 8$$

$$5m + 7 \mathbf{- 2m} = 2m - 8 \mathbf{- 2m}$$

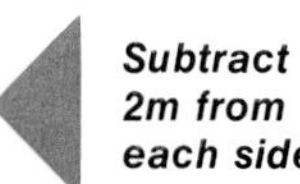

Subtract 2m from each side.

$$3m + 7 = -8$$

$$3m + 7 \mathbf{- 7} = -8 \mathbf{- 7}$$

Subtract 7 from each side.

$$3m = -15$$

$$\frac{3m}{\mathbf{3}} = \frac{-15}{\mathbf{3}}$$

$$m = -5$$

Check

$5m + 7 = 2m - 8$

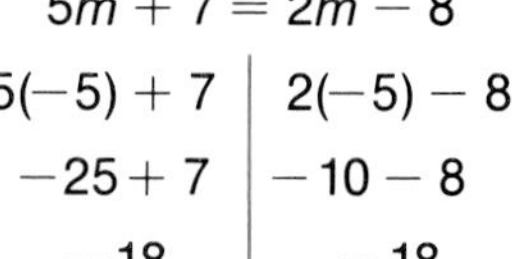

$5(-5) + 7$	$2(-5) - 8$
$-25 + 7$	$-10 - 8$
-18	-18

P–1 **Solve and check.**

a. $22 + 16m = 4 + 7m$ **b.** $12x + 8 = -18 - x$ **c.** $21y - 9 = 6y - 39$

EXAMPLE 2 Solve and check: $-3(4d+2)+5d=d-(9+4d)$

Solution: First, use the Distributive Property.

$$-3(4d+2)+5d=d-(9+4d)$$

$$-12d-6+5d=d-9-4d$$

◀ *Combine like terms.*

$$-7d-6=-3d-9$$

$$-7d-6+\mathbf{3d}=-3d-9+\mathbf{3d}$$

$$-4d-6=-9$$

$$-4d-6+\mathbf{6}=-9+\mathbf{6}$$

$$-4d=-3$$

$$\frac{-4d}{\mathbf{-4}}=\frac{-3}{\mathbf{-4}}$$

$$d=\frac{3}{4}$$

Check

$$-3(4d+2)+5d=d-(9+4d)$$

Left side	Right side
$-3\left(4\cdot\frac{3}{4}+2\right)+5\left(\frac{3}{4}\right)$	$\frac{3}{4}-\left(9+4\cdot\frac{3}{4}\right)$
$-3(3+2)+\frac{15}{4}$	$\frac{3}{4}-(9+3)$
$-3(5)+3\frac{3}{4}$	$\frac{3}{4}-12$
$-15+3\frac{3}{4}$	$-11\frac{1}{4}$
$-11\frac{1}{4}$	

P–2 **Solve and check:** $-4(2x+3)=-8x+4(2+x)$

Steps for Solving an Equation

1. Use the Distributive Property.
2. Combine like terms.
3. Use the Addition or Subtraction Property for Equations.
4. Use the Multiplication or Division Property for Equations.
5. Check the solution in the original equation.

CLASSROOM EXERCISES

Write an equivalent equation as the first step in solving each equation. (Step 1, Example 1)

1. $4-m=6-5m$ **2.** $9+3x=6-2x$ **3.** $12+6y=9y+15$

4. $14+b=13-3b$ **5.** $20n-6=18n+4$ **6.** $-13-3z=27-5z$

7. $2.8+0.5f=3.7-4.5f$ **8.** $12-\frac{3}{4}h=\frac{5}{4}h-15$ **9.** $\frac{1}{2}a+3=\frac{1}{4}a-2$

(Step 1, Example 2)

10. $3(5-2b)-b=3-2b-10$ **11.** $3q-q-5q=4q-12-8q$

12. $3g-(5-g)=5-g-19$ **13.** $-5(2r-3)+3r=4r-r$

WRITTEN EXERCISES

Goal: To solve an equation with the variable on both sides

Sample Problem: $2p - 3 = 5p + 7$ **Answer:** $p = -3\frac{1}{3}$

For Exercises 1–22, solve and check. (Example 1)

1. $6m + 8 = 2m - 12$
2. $8t + 3 = 5t - 18$
3. $6.7 - 1.8n = 3.7n - 9.8$
4. $8.4y - 2.9 = -13.1y + 14.3$
5. $\frac{1}{2}w - 16 = 14 - \frac{3}{2}w$
6. $\frac{1}{4}c - 12 = 28 - \frac{7}{4}c$

(Example 2)

7. $-3(2x - 3) = 4(5 - x)$
8. $2(5 - 3a) = -4(a + 4)$
9. $3c - (5 + c) = -4c - 6 + 9c$
10. $2w + 8 - 7w = 12 - (3w - 1)$
11. $-\frac{3}{4}(16k - 12) = \frac{2}{3}(18 - 9k)$
12. $\frac{5}{4}(12 - 8r) = -\frac{4}{3}(9r - 21)$

MIXED PRACTICE

13. $4q - 3 = 15 - 2q$
14. $3.8z - 5.3 = 7.9z - 25.8$
15. $4n - 5(n + 2) = 6 - 3(3 - n)$
16. $s - 6 - 6s = (2s - 1) - (5s - 3)$
17. $2.3 + 1.4b = 14.9 - 2.8b$
18. $8 - 5f = 3f - 24$
19. $(p - 3) - (2p + 2) = 5 + 2p - 19$
20. $\frac{1}{6}(12 - 8e) = \frac{1}{2}(10e - 14)$
21. $\frac{1}{3}(6g - 9) = \frac{1}{4}(24 - 8g)$
22. $13 - 2(3n - 3) = -5n + 3(n - 1)$

APPLICATIONS

You can use a calculator to check solutions of equations. First calculate the value of the left side of the equation and store it in memory. Then calculate the right side of the equation and compare it with the value in memory.

EXAMPLE Check the answer to Example 1 on page 350.

SOLUTION Equation: $5m + 7 = 2m - 8$ Answer to be checked: -5

Left side:  [5] [×] [5] [+/−] [+] [7] [=] -18. ◀ ***Store this in memory.***

Right side: [2] [×] [5] [+/−] [−] [8] [=] -18.

Since the two sides check, the answer, −5, is correct.

Check the given "answer" to each equation below.

23. $8y - 4.8 = 4y + 19.6$; 6.1
24. $2t - 3 = 5t + 7$; −3
25. $2.7x - 18.5 = -3.2x + 5.1$; 4
26. $3z - 13 = 10z + 15$; −2
27. $4x + 14 = 2x - 7$; −10.5
28. $1.3p + 2.5 = -2.9p + 15.1$; 3

REVIEW CAPSULE FOR SECTION 12.6

Solve and check. (Pages 347–349)

1. $4t - 7 = 3t + 8$

2. $7m - 6 = 8m - 14$

3. $3(r - 6) = 2r + 11$

4. $4x + 17 = 5(x - 5)$

5. $5(y + 3) = 6y - 1$

6. $5w - 10 = 4(w + 10)$

PROBLEM SOLVING AND APPLICATIONS

12.6 Age Problems

To solve age problems, you often have to represent a person's present age, that person's age a number of years ago, or that person's age a number of years from now. The table shows how to do this.

Table

	Age Now	Age 6 Years Ago	Age 12 Years From Now
Manuel	11	$11 - 6$, or 5	$11 + 12$, or 23
Beth	n	$n - 6$	$n + 12$
Erin	$2t$	$2t - 6$	$2t + 12$

Thus, to represent a person's past age, you subtract from the present age. To represent age in the future, you add to the present age.

In the following examples, Condition 1 is underscored once. Condition 2 is underscored twice.

EXAMPLE 1 Theo is now three times as old as Vanessa. Five years from now, Theo will be twice as old as Vanessa. How old are Theo and Vanessa now?

Solution: Make a table.

[1] Use Condition 1 to make a table.

	Age Now	Age Five Years From Now
Vanessa	y	$y + 5$
Theo	$3y$	$3y + 5$

[2] Use the table and Condition 2 to write an equation.

Think: **Theo's age five years from now** | **will be** | **twice Vanessa's age then.**

Translate: $3y + 5 \quad = \quad 2(y + 5)$

[3] Solve the equation. ⟶ $3y + 5 = 2y + 10$

$$3y + 5 - \mathbf{2y} = 2y + 10 - \mathbf{2y}$$
$$y + 5 = 10$$
$$y + 5 - \mathbf{5} = 10 - \mathbf{5}$$
$$y = 5$$
$$3y = 15$$

$2(y + 5) = 2 \cdot y + 2 \cdot 5$

Don't forget to find Theo's age.

[4] **Check:** Condition 1 Is Theo now three times as old as Vanessa? Does $3(5) = 15$? Yes ✓

Condition 2 In five years will Theo be twice as old as Vanessa? Does $15 + 5 = 2(5 + 5)$? Does $20 = 2(10)$? Yes ✓

Theo is now 15 years of age and Vanessa is 5.

EXAMPLE 2

Mr. Davidson is now four times as old as Julie. Four years ago, Mr. Davidson's age was seven times Julie's age then. How old are Mr. Davidson and Julie now?

Solution: [1]

	Age Now	Age 4 Years Ago
Julie	x	$x - 4$
Mr. Davidson	$4x$	$4x - 4$

Think: **Mr. Davidsons's age 4 years ago** | **was** | **7 times Julie's age then.**

[2] $4x - 4 \quad = \quad 7(x - 4)$

[3] $4x - 4 = 7x - 28$

$$4x - 4 - \mathbf{4x} = 7x - 28 - \mathbf{4x}$$
$$-4 = 3x - 28$$
$$-4 + \mathbf{28} = 3x - 28 + \mathbf{28}$$
$$24 = 3x$$
$$\frac{24}{\mathbf{3}} = \frac{3x}{\mathbf{3}}$$
$$8 = x$$
$$4x = 32$$

Julie is 8 years old.
Mr. Davidson is 32 years old.

[4] **Check:** The check is left for you.

CLASSROOM EXERCISES

For Exercises 1–4, write an algebraic expression for eac
(Examples 1–2, Step 1)

1. Miyako is r years old now. Represent her age 6 years from now.
2. Gina is t years old now. Represent her age 10 years now.
3. David is $3m$ years old now. Represent his age 5 years ago.
4. Carlos is $4d$ years old now. Represent his age 11 years ago.

For Exercises 5–8, Mae is n years old. Steve is twice as old as Mae.
(Examples 1–2, Step 1)

5. Represent Mae's age 4 years from now.
6. Represent Mae's age 5 years ago.
7. Represent Steve's age 2 years ago.
8. Represent Steve's age 3 years from now.

WRITTEN EXERCISES

Goal: To solve word problems involving age

Sample Problem: Skip is now five times as old as Paul (Condition 1). In 4 years, Skip will be three times as old as Paul will be then (Condition 2). How old is each now?

Answer: Paul is 4 years of age and Skip is 20.

For Exercises 1–4, Condition 1 is underscored once. Condition 2 is underscored twice. (Example 1)

a. *Use Condition 1 to complete the table.*

b. *Use the table and Condition 2 to write an equation for the problem.*

c. *Solve.*

1. Joe is now six times as old as his niece, Lisa. One year from now, his age will be five times Lisa's age then. How old are Joe and Lisa now?

	Age Now	Age 1 Year From Now
Lisa	x	?
Joe	$6x$	?

2. Anna is now four times as old as her brother, Ramon. In 4 years, her age will be twice Ramon's age then. How old are Anna and Ramon now?

	Age Now	Age 4 Years From Now
Ramon	n	?
Anna	?	?

(Example 2)

Carl is now three times as old as his nephew, Pete. Three years ago, Carl was four times as old as Pete was then. How old are Carl and Pete now?

	Age Now	Age 3 Years Ago
Pete	y	?
Carl	?	?

4. Judy is now three times as old as her niece, Jill. Seven years ago, she was five times as old as Jill was then. How old are Judy and Jill now?

	Age Now	Age 7 Years Ago
Jill	d	?
Judy	?	?

MIXED PRACTICE

For Exercises 5–8, first make a table. Then solve.

5. Mrs. Montez is now twice as old as her neighbor, Mrs. Cooke. Twelve years ago, she was three times as old as Mrs. Cooke was then. How old is each now?

6. Mr. Cole is now four times as old as his grandson, Bruce. In 8 years, his age will be three times Bruce's age then. How old is each now?

7. Mr. Adams is now four times as old as his daughter, Luann. In 20 years, he will be twice as old as Luann will be then. How old is each now?

8. Mrs. Jones is now six times as old as her son, Allan. Three years ago, her age was eleven times Allan's age then. How old is each now?

MORE CHALLENGING EXERCISES

Complete the table. Then solve.

9. Mr. Conners is 40 years old and Robert is 8 years old. In how many years will Mr. Conners' age be exactly three times Robert's age?

	Age Now	Age n Years From Now
Mr. Conners	?	?
Robert	?	?

10. Maria is 9 years old and Louellen is 15 years old. How many years ago was Louellen exactly twice as old as Maria?

	Age Now	Age n Years Ago
Maria	?	?
Louellen	?	?

Focus on Reasoning: Mental Computation

Logical reasoning can help you to solve problems more efficiently. Sometimes it will enable you to solve problems using **mental computation** only. The ability to identify efficient strategies to solve problems is an important test-taking skill.

EXAMPLE 1 If $14 - r \cdot t = 52$, evaluate $14 - r \cdot t - 9$.

Solution: Since $14 - r \cdot t = 52$, $\overline{14 - r \cdot t} - 9 = 52 - 9$
$= \mathbf{43}$

EXAMPLE 2 If $9 - d + f = 111$, evaluate $8(9 - d + f)$.

Solution: Since $9 - d + f = 111$, $8\overline{(9 - d + f)} = 8 \cdot 111$
$= \mathbf{888}$

EXERCISES

1. If $15y + 8x = 53$, evaluate $15y + 8x + 21$.
2. If $p + 8q = 84$, evaluate $p + 8q + 91$.
3. If $14 - r \cdot t = 21$, evaluate $(14 - r \cdot t) \div 3$.
4. If $(p + q) \div 3 = 12$, evaluate $[(p + q) \div 3] \div 4$.
5. If $a + 6b = 2.4$, evaluate $2(a + 6b)$.
6. If $b - 4c = 0.5$, evaluate $4(b - 4c)$.
7. If $21 + x \cdot y = 288$, evaluate $\frac{1}{4}(21 + x \cdot y)$.
8. If $y^2 - k = 624$, evaluate $\frac{1}{2}(y^2 - k)$.
9. If $p + 8q = 84$, evaluate $(p + 8q) \div 12 + 13$.
10. If $x \div y - 8 = 30$, evaluate $(x \div y - 8) \div 5 + 50$.
11. If $8t - u = 22$, evaluate $\frac{1}{2}(8t - u) - 9$.
12. If $x \div y - 8 = 12$, evaluate $\frac{1}{4}(x \div y - 8) - 2$.
13. If $x \div y - 8 = 1$, evaluate $x \div y - 9$.
14. If $14 - r \cdot t = 1$, evaluate $13 - r \cdot t$.

Examples 3 and 4 apply logical reasoning and the **Distributive Property** to problems involving **factoring.**

EXAMPLE 3 If $2b + 6c = 72$, find the value of $b + 3c$.

Solution: Since $2b + 6c = 72$, $2(b + 3c) = 72$.

Since $2(b + 3c) = 72$, $(b + 3c) = \frac{1}{2} \cdot 72$

$= \mathbf{36}$

EXAMPLE 4 If $\frac{1}{4}z - \frac{1}{4}y = 52$, find the value of $z - y$.

Solution: Since $\frac{1}{4}z - \frac{1}{4}y = 52$, $\frac{1}{4}(z - y) = 52$.

Since $\frac{1}{4}(z - y) = 52$, $(z - y) = 4 \cdot 52$

$= \mathbf{208}$

EXERCISES

15. If $6b + 12a = 108$, find the value of $b + 2a$.

16. If $5a - 15b = 40$, find the value of $a - 3b$.

17. If $4m - 8t = 60$, find the value of $m - 2t$.

18. If $9x + 27y = 36$, find the value of $x + 3y$.

19. If $\frac{1}{2}r + \frac{1}{2}t = 8$, find the value of $r + t$.

20. If $\frac{1}{2}a - \frac{1}{2}b = 13$, find the value of $a - b$.

21. If $3a - 3b = 9$, find the value of $a - b$.

22. If $10r + 10t = 120$, find the value of $r + t$.

23. If $\frac{1}{2}t + \frac{1}{2}w = 300$, find the value of $(t + w) \div 50$.

24. If $\frac{1}{2}c - \frac{1}{2}d = 100$, find the value of $(c - d) \div 40$.

25. If $8t - 40y = 0$, find the value of $(t - 5y) + 6$.

26. If $7b + 35c = 63$, find the value of $(b + 5c) + 1$.

27. If $2q + 6r = 132$, find the value of $(q + 3r) - 5$.

28. If $9t - 18x = 243$, find the value of $(t - 2x) + 1$.

29. If $5t + 15r = 27.5$, find the value of $(t + 3r) + 2$.

30. If $7n - 154p = 31.5$, find the value of $(n - 22p) + 3$.

CHAPTER SUMMARY

IMPORTANT IDEAS

1. If the ratio of the amounts of two parts of a mixture is 2:3, the parts can be represented by $2x$ and $3x$.
2. The sum of the measures of the angles of a triangle is 180°.
3. Combine like terms on each side of an equation before applying the Addition and Subtraction Properties.
4. *Steps for Solving an Equation*
   a. Use the Distributive Property.
   b. Combine like terms.
   c. Use the Addition or Subtraction Property for Equations.
   d. Use the Multiplication or Division Property for Equations.
   e. Check the solution in the original equation.

CHAPTER REVIEW

SECTION 12.1

Use Condition 1 to represent the unknowns. Use Condition 2 to write an equation for the problem. Solve.

1. A tank of bottled gas contains propane and butane in the ratio 3:2 by weight. The total gas mixture weighs 25 pounds. Find the number of pounds of each gas.

2. A company invested in stocks, bonds, and real estate in the ratio 2:3:5. The total investment amounted to $75,000. How much did the company invest in bonds?

SECTION 12.2

Use Condition 1 to represent the unknown angle measures. Use Condition 2 to write an equation for the problem. Solve.

3. Some shoals near a ship channel are marked by triangle ABC on the navigation chart below. The measure of angle A is 18° less than the measure of angle C. The measure of angle B is 72°. Find the measure of an angle C.

4. The measure of angle A in triangle ABC below is the **glide slope** of a plane in its approach for a landing. The measure of angle B in the figure is 85° more than the glide slope. The measure of angle C is 90°. Find the glide slope.

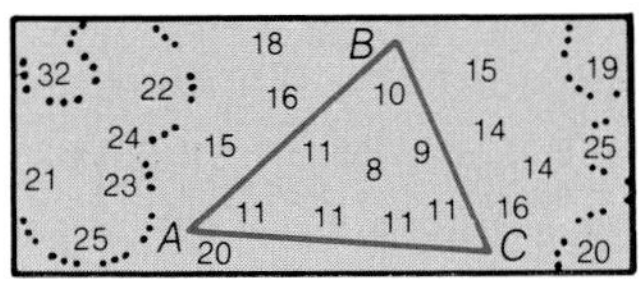

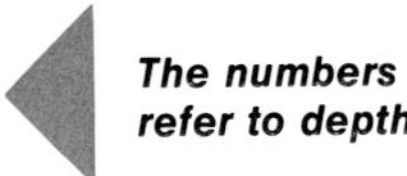

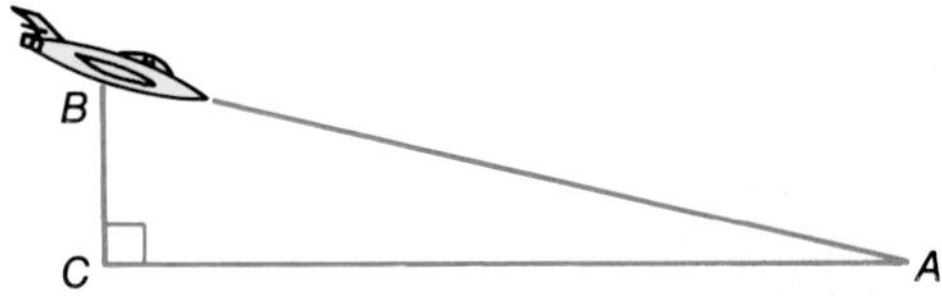

SECTION 12.3

Use Condition 1 to represent the unknowns. Use Condition 2 to write an equation for the problem. Solve.

5. The number of 25-cent stamps on a package is two more than the number of 15-cent stamps on it. The total value of the stamps is \$2.90. How many 25-cent stamps are on the package?

6. Maura counted the coins that she had saved. She had 4 times as many quarters as nickels and 7 times as many dimes as nickels. She saved a total of \$3.50. How many coins of each kind did she have?

SECTION 12.4

Solve and check.

7. $3p - 8 = 4p - 17$
8. $12 - 6h = -5h - 8$
9. $7.4 - 0.6w = -1.9 + 0.4w$
10. $0.9m - 13.3 = -5.9 - 0.1m$
11. $-4(r - 8) + 3r = 17 - 2r$
12. $27 - 4v = 8(3 - v) + 5v$

SECTION 12.5

Solve and check.

13. $4f - 6 = 7f + 9$
14. $10s + 5 = 6s - 19$
15. $-7(3 - p) + 3p = 2(4p + 3)$
16. $6(3 - 2k) - 12 = -4(5k - 4)$
17. $3(8x - 5) - 6x = x - (5x - 15)$
18. $12 - p - 3p = \frac{1}{2}(2 - 4p) + 3p$

SECTION 12.6

Use Condition 1 to make a table. Use the table and Condition 2 to write an equation for the problem. Solve.

19. Maria is now four times as old as her sister Ann. Six years from now, Maria will be twice as old as Ann will be then. How old are Marie and Ann now?

20. Jim's present age is three times Tony's present age. Three years ago, Jim's age was 5 times Tony's age then. What is each boy's present age?

CUMULATIVE REVIEW: CHAPTERS 1–12

Evaluate each expression. (Section 1.2)

1. $6 + 9m \div 3$ when $m = 4$

2. $9p + 3pq$ when $p = 4$ and $q = 7$

Multiply. (Section 2.2)

3. $(5c)3$

4. $(7d)(0.5e)$

5. $(3p)(6hp)$

6. $(3fg)(2f^2g)(f)$

Perform the indicated operations. Write each answer in lowest terms. (Sections 4.3 through 4.6)

7. $\frac{3t}{4} \cdot \frac{8}{9t}$

8. $\frac{7ab}{12c} \cdot \frac{9c^2}{14a}$

9. $6ac \div \frac{2a}{3c}$

10. $\frac{6a^2}{b^3} \div \frac{21a}{b}$

11. $\frac{5x}{3} - \frac{2y}{3}$

12. $\frac{3}{12n} + \frac{7}{12n}$

13. $\frac{2}{15f} + \frac{1}{6f}$

14. $\frac{5p}{6} - \frac{p}{4}$

Solve and check. (Sections 5.7 and 6.3)

15. Chris sold 63 fewer tickets than Nancy sold. Chris sold 312 tickets. How many tickets did Nancy sell?

16. The sum of three consecutive even whole numbers is 654. Find the numbers.

A bag contains 4 blue whistles, 3 red whistles, and 5 white whistles. A whistle is drawn at random. Find the probability of choosing each of the following. (Section 7.2)

17. A white whistle

18. A red whistle

19. A whistle that is not red

Solve each problem. (Sections 7.6 and 7.7)

20. Sales tax in a city is 6%. Find the amount of sales tax paid on a purchase of $65.

21. The commission for selling a stereo that sells for $450 is $58.50. Find the rate of commission.

For Exercises 22–23, the lengths of two legs of a right triangle are given. Find the length of the hypotenuse, c. (Section 8.5)

22. $a = 5$; $b = 12$; $c = \underline{\ ?\ }$

23. $a = 12$; $b = 16$; $c = \underline{\ ?\ }$

Simplify. (Section 9.7)

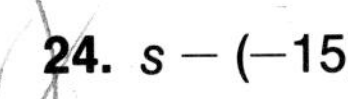

24. $s - (-15)$

25. $3.7k + 6 - 2.9k$

26. $12c - 27c - 2$

27. $-3y + 16 - (-9y)$

28. $-a - 10 + 3 - (-7a)$

29. $3a^2 - 6a - 4a - 5a^2$

Multiply. (Sections 10.1 through 10.3)

30. $5(-18)$ **31.** $(-0.4)(-35)$ **32.** $-7(8-5x)$ **33.** $-4a(-3a+5a^2)$

Factor. (Section 10.4)

34. $17a+17b$ **35.** $-6x+(-6x^2)$ **36.** $5c^2-30c$ **37.** $8ab-6a^2$

Divide. (Section 10.5)

38. $\frac{-21}{7}$ **39.** $-15 \div (-3)$ **40.** $-168 \div (-6)$ **41.** $8 \div (-\frac{4}{3})$

Find the mean, median, and mode for each collection of data. (Section 10.6)

42. Football scores: 7, 7, 0, 24, 35, 6, 7, 24

43. Celsius temperatures: 11°, 7°, −10°, −3°, 9°, −7°, 0°, 10°, 9°

Solve and check. (Sections 11.1 through 11.4)

44. $-95=c+123$ **45.** $5x-(4x+8)=-12$ **46.** $203=-7x$

47. $-\frac{3}{8}b=-48$ **48.** $11-2x=35$ **49.** $-12=\frac{r}{3}-7$

Use Condition 1 to represent the unknowns. Use Condition 2 to write an equation for the problem. Solve. (Sections 11.6 through 12.3)

50. The number of boys in class is 2 fewer than the number of girls. The class has 28 students in all. Find the number of boys and the number of girls in the class.

51. Red paint and white paint are mixed in a ratio of 8:3 to make the color pink. How many gallons of each color are needed to make 88 gallons of pink paint?

52. In triangle *ABC*, the measure of angle *A* is twice the measure of angle *B*. Angle *C* has a measure of 27°. Find the measure of angle *A*.

53. Kirk has 7 more dimes than quarters. The value of the dimes and quarters is $9.10. Find the number of dimes Kirk has.

Solve and check. (Section 12.5)

54. $-1.6-3n=5n+1.6$ **55.** $-4(x+2)=2(3x+1)$

Use Condition 1 to represent the unknowns. Use Condition 2 to write an equation for the problem. Solve. (Section 12.6)

56. Frank is now 4 times as old as Karl. In 10 years, he will be twice as old as Karl will be then. How old is each now?

57. Beth is now 3 times as old as Carol. Six years ago, Beth was 9 times as old as Carol was then. How old are Beth and Carol now?

CHAPTER 13

Inequalities

Archeologists can compute the age of the remains of ancient plants and animals by measuring the amount of carbon present and determining the ratio of these two types. Ratios are one way to compare numbers. In this chapter, you will use $>$, $<$, and $=$ to show comparisons.

13.1 Inequalities and Graphs

You learned the meaning of the inequality symbols $<$ and $>$ in Chapter 8.

$-2 < 5$ **_Read: "−2 is less than 5."_**

$-5 > -6$ **_Read: "−5 is greater than −6."_**

When an inequality involving $<$ or $>$ is true, then $<$ or $>$ always points toward the smaller number.

An inequality with a variable may have one or more solutions. The solution depends on the replacement set. The ***replacement set*** is the set of numbers from which replacements for the variable are chosen. Graphing an inequality using different replacement sets helps to show this.

EXAMPLE 1 Graph $x > -6$ using these replacement sets.
a. {negative integers} **b.** {integers}

Solutions: **a.** The only negative integers that are greater than −6 are −5, −4, −3, −2, and −1. Graph these numbers.

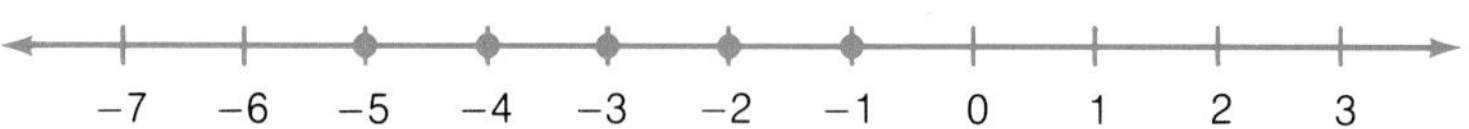

b. The solution set includes 0, 1, 2, 3, and so on, as well as −5, −4, −3, −2, and −1. Clearly, you cannot graph the complete solution set.

The "etc." at the right portion of the graph means that the graph continues without end.

P–1 **Which of the following are solutions for $x < 5$ when the replacement set is the set of integers?**

a. 4 **b.** −4 **c.** $2\frac{1}{2}$ **d.** 0 **e.** 5

EXAMPLE 2

Graph these inequalities when the replacement set is the set of real numbers.

a. $n > -6$ **b.** $n \geq -6$

Solutions: **a.** All real numbers greater than -6 are to the right of -6 on a number line. Since -6 is not in the solution set, draw an open circle at -6. Indicate that the graph continues to the right by drawing a heavy line in that direction.

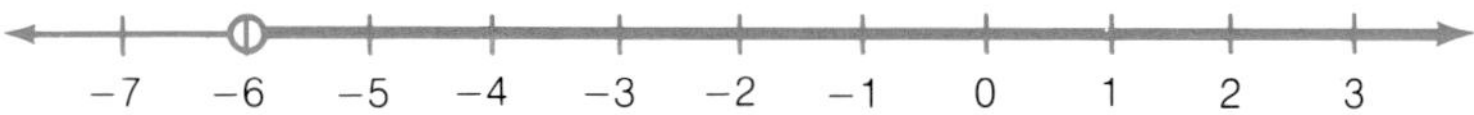

b. Since the solution set is any number n that is greater than or equal to -6, the graph will indicate that -6 is included in the solution set. Therefore, draw a large closed dot at -6.

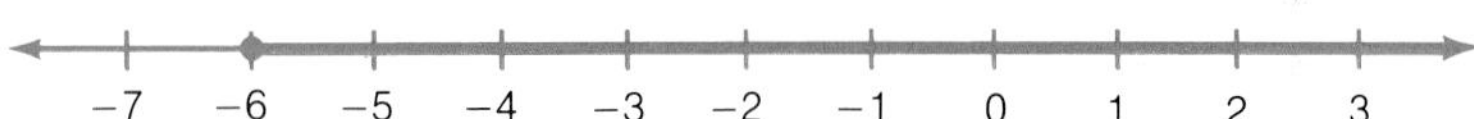

Given the graph of an inequality, you can write the inequality.

EXAMPLE 3

Write an inequality using < for each graph below. Identify the replacement set for each.

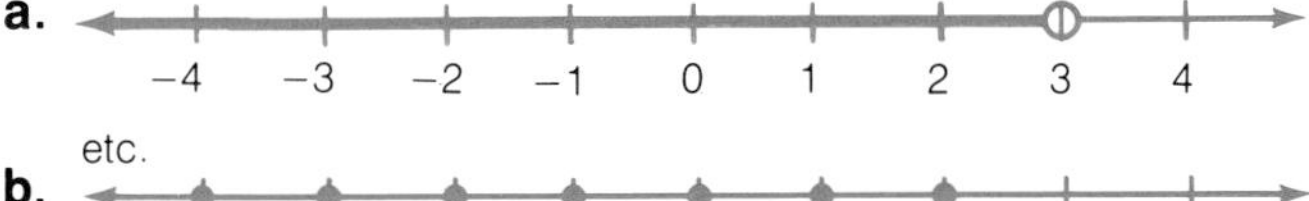

Solutions: **a.** $q < 3$; Replacement set: {real numbers}

b. $q < 3$; Replacement set: {integers}

P–2 **Which inequality is graphed below?**

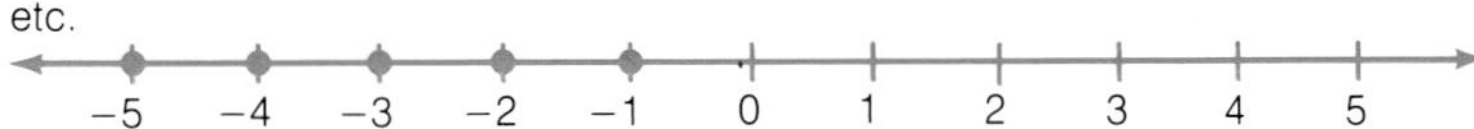

a. $r < 0$; Replacement set: {integers}

b. $r < 0$; Replacement set: {real numbers}

c. $r < -1$; Replacement set: {negative integers}

CLASSROOM EXERCISES

Which of the following are solutions for $d > -2$ when the replacement set is the set of integers? (Example 1)

1. -4 **2.** 3 **3.** $\frac{1}{2}$ **4.** 0 **5.** 2.1 **6.** $-6\frac{1}{3}$

Which of the following are solutions for $n < 6.2$ when the replacement set is the set of positive integers? (Example 1)

7. 3 **8.** -3 **9.** $2\frac{1}{2}$ **10.** -4.5 **11.** 0 **12.** -1

Which of the following are solutions for $j \leq 2$ when the replacement set is the set of real numbers? (Example 2)

13. -1 **14.** -3.5 **15.** $3\frac{1}{4}$ **16.** -9 **17.** 0 **18.** 2

Write an inequality for each graph. Identify the replacement set. Let f represent the variable. (Example 3)

19. 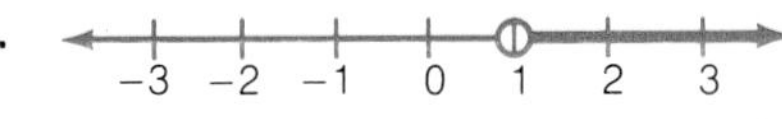

20. 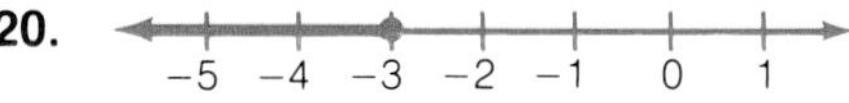

21. etc.
-4 -3 -2 -1 0 1 2

22. etc.
-4 -3 -2 -1 0 1 2

23. -1.5
-4 -3 -2 -1 0 1 2

24. $2\frac{2}{3}$
-2 -1 0 1 2 3 4

25. etc.
-3 -2 -1 0 1 2 3

26. π
-1 0 1 2 3 4 5

WRITTEN EXERCISES

Goal: a. To graph an inequality, and **b.** to write an inequality from a graph and to identify its replacement set

Sample Problem: Graph $p < 4\frac{1}{3}$. The replacement set is the set of real numbers.

Answer: 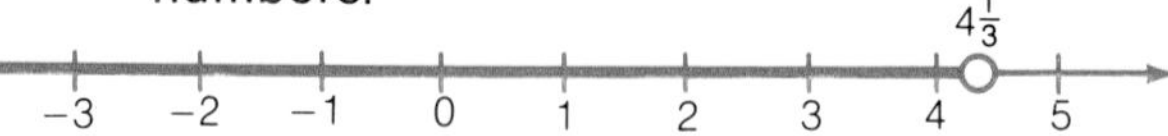

Graph each inequality using the set of integers as the replacement set. (Example 1)

1. $a > 2$ **2.** $n < 3$ **3.** $s < -1$ **4.** $y > -4$

5. $m < 4.2$ **6.** $e > -1.8$ **7.** $b > \frac{3}{4}$ **8.** $h > -\frac{7}{8}$

Graph each inequality using the set of real numbers as the replacement set. (Example 2)

9. $c < 2$ **10.** $f > -3$ **11.** $z > -5$ **12.** $g < -1$

13. $w \leq -2$ **14.** $a \geq 0$ **15.** $x > -1\frac{1}{3}$ **16.** $t < 1.5$

Write an inequality for each graph. Identify the replacement set. Let q represent the variable. (Example 3)

17. 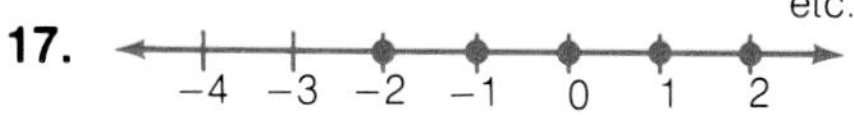

18. etc. −2 −1 0 1 2 3 4

19. 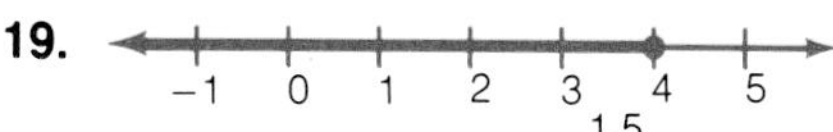

20. 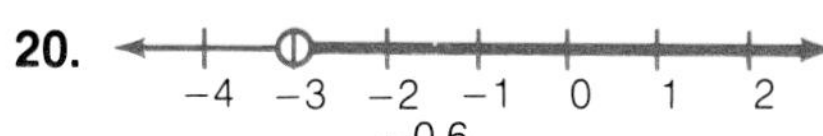

21. 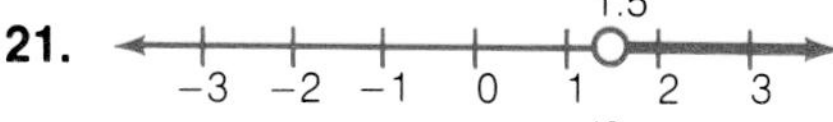

22. −0.6 −3 −2 −1 0 1 2 3

23. 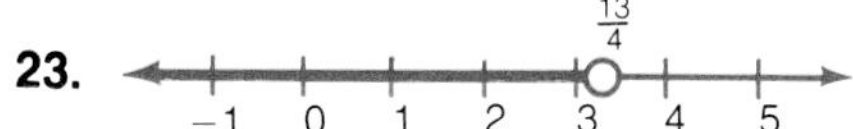

24. $-\frac{19}{4}$ −6 −5 −4 −3 −2 −1 0

MIXED PRACTICE

Select the inequality and the replacement set that matches the given graph in each exercise.

25. 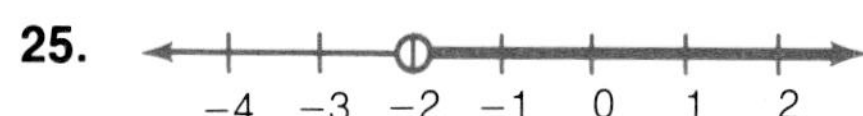

a. $x < -2$; {real numbers}
b. $x > -2$; {integers}
c. $x > -2$; {real numbers}

26. 

a. $q < 0$; {integers}
b. $q > -4$; {negative integers}
c. $q < 0$; {real numbers}

27. 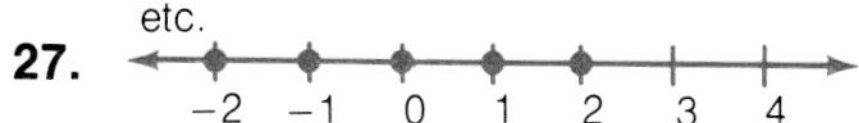

a. $w < 2$; {integers}
b. $w < 3$; {integers}
c. $w < 2$; {whole numbers}

28. −4 −3 −2 −1 0 1 2

a. $c \leq -1$; {real numbers}
b. $c \geq -1$; {real numbers}
c. $c \leq -1$; {negative integers}

29. −3.5 −5 −4 −3 −2 −1 0 1

a. $s > -3.5$ {negative integers}
b. $s < 0$; {negative integers}
c. $s > -3$; {negative integers}

30. −2.3 −3 −2 −1 0 1 2 3

a. $d < -2.3$; {real numbers}
b. $d > -3$; {integers}
c. $d > -2.3$; {real numbers}

MORE CHALLENGING EXERCISES

Write an inequality relating y and w. If not enough information is given, write Not Possible. Give an example to illustrate your answer.

31. $x < y$ and $x > w$ **32.** $x > y$ and $x > w$

13.2 Addition and Subtraction Properties

Note that the solution set for each pair of inequalities below is the same. The replacement set is $\{0, 1, 2, 3, 4, 5, 6\}$.

Table

Inequalities	Solution Set	Graph
a. $c > 3$	$\{4, 5, 6\}$	−2 −1 0 1 2 3 4 5 6 7
b. $c + 2 > 3 + 2$	$\{4, 5, 6\}$	−2 −1 0 1 2 3 4 5 6 7
c. $r < 4$	$\{0, 1, 2, 3\}$	−2 −1 0 1 2 3 4 5 6 7
d. $r + 3 < 4 + 3$	$\{0, 1, 2, 3\}$	−2 −1 0 1 2 3 4 5 6 7

Inequalities that have the same solution set are called ***equivalent inequalities.*** Thus, inequalities **a** and **b** are equivalent. Similarly, **c** and **d** are equivalent. This suggests the following property.

Addition Property for Inequalities

Adding the same real number to each side of an inequality forms an equivalent inequality.

$$k - 5 < 7$$
$$k - 5 + \mathbf{5} < 7 + \mathbf{5}$$
$$k < 12$$

P–1 **What number should be added to each side of the following inequalities to get the variable alone?**

a. $d - 5 < 9$ **b.** $8 > w - 6$ **c.** $8.2 > p - 6.5$

From now on, the replacement set will be the set of real numbers, unless otherwise indicated.

EXAMPLE 1 Solve and graph: $f - 3 > -5$

Solution:

$$f - 3 > -5$$

[1] Add 3 to each side. ⟶ $f - 3 + \mathbf{3} > -5 + \mathbf{3}$

$$f > -2$$

[2] Graph the inequality. ⟶

−4 −3 −2 −1 0 1

The solution set consists of all real numbers greater than −2.

P–2 **Solve.**

a. $h - 6 > 7$ **b.** $g - 9 < -11$ **c.** $1.8 < m - 0.7$

Just as there is a Subtraction Property for Equations, there is a similar property for inequalities of the form $x + a > b$ or $x + a < b$.

> ***Subtraction Property for Inequalities***
>
> Subtracting the same real number from each side of an inequality forms an equivalent inequality.
>
> $q + 6 > 8$
>
> $q + 6 - \mathbf{6} > 8 - \mathbf{6}$
>
> $q > 2$

EXAMPLE 2 Solve and graph: $5 < x + 4$

Solution: [1] Subtract 4 from each side. ⟶ $5 - \mathbf{4} < x + 4 - \mathbf{4}$

$1 < x$, or $x > 1$

[2] Graph the inequality. ⟶

The solution set consists of all real numbers greater than 1.

P–3 **Solve: a.** $\mathbf{11 < x + 7}$ **b.** $\mathbf{c + 47 > 81}$ **c.** $\mathbf{r + 1\frac{1}{2} < 5}$

EXAMPLE 3

After 12:00 noon, the temperature in a city dropped 15 degrees to reach the low temperature for the day (Condition 1). This low temperature was higher than the record low temperature of −33°F (Condition 2). What was the lowest possible reading (integer) for the temperature at noon?

Solution:

[1] Let $t =$ the temperature at 12:00 noon.
Then $t - 15 =$ the low temperature for that day. ◀ ***Condition 1***

[2] Use Condition 2 to write the inequality.

Think: **The day's low temperature was higher than −33.**

Translate: $t - 15$ $>$ -33

[3] Solve the inequality. ⟶ $t - 15 + \mathbf{15} > -33 + \mathbf{15}$

$t > -18$

Since $t > -18$, the lowest possible temperature at noon was −17°F.

[4] **Check:** Is $(-17) - 15 > -33$? Is $-32 > -33$? Yes ✓

CLASSROOM EXERCISES

Write Yes or No to show whether the two inequalities are equivalent. (Table)

1. $g-5>7;\ g>2$
2. $p+3<60;\ p<57$
3. $t-8<11;\ t<19$
4. $-4>w-1;\ w<-3$
5. $9.6>x+5.1;\ 14.7>x$
6. $8.7<s+9.9;\ 1.2<s$

Write an equivalent inequality as the first step in solving each inequality. (Step 1, Examples 1 and 2)

7. $r+5<3$
8. $t-7<-5$
9. $w-1>2$
10. $y+2>-2$
11. $-13>a+7$
12. $-9<p-11$
13. $k-234>-189$
14. $-248<m+135$
15. $-3\frac{1}{4}>t-5\frac{1}{8}$
16. $1\frac{1}{6}+h>-4\frac{2}{3}$
17. $-5.6<1.8+c$
18. $-0.8+x<5.9$

WRITTEN EXERCISES

Goal: To solve and graph inequalities by using the Addition and Subtraction Properties

Sample Problem: Solve and graph: $t-3>-6$

Answer: $t>-3$

For Exercises 1–33, solve and graph. (Example 1)

1. $t-3<-10$
2. $r-7>-13$
3. $p-3>-3$
4. $x-\frac{1}{2}<-2$
5. $-4.5<b-1.2$
6. $-3.4>q-0.3$
7. $w-\frac{3}{4}>1\frac{5}{8}$
8. $y-\frac{2}{3}>-3\frac{1}{6}$
9. $-16<s-7$

(Example 2)

10. $18+b>3$
11. $r+9>-13$
12. $n+12<-5$
13. $-89>63+d$
14. $-65<78+p$
15. $m+5.9>-8.7$
16. $w+4.3<-8.9$
17. $3\frac{1}{8}<t+6\frac{3}{4}$
18. $2\frac{1}{4}>k+5\frac{1}{2}$

MIXED PRACTICE

19. $19<t-8$
20. $44<d-81$
21. $n+5>-3$
22. $r+5>-4$
23. $y-254>-431$
24. $m-319>-507$
25. $-198<-309+f$
26. $-104<-210+b$
27. $19.3<t-7.9$
28. $24.8<w-9.7$
29. $-25.4+x<-14.6$
30. $-8.56+q>-13.05$
31. $n+5\frac{3}{8}>-2\frac{7}{8}$
32. $r+4\frac{3}{4}>-3\frac{3}{4}$
33. $37<m-88$

APPLICATIONS

For Exercises 34–37, Condition 1 is underscored once. Condition 2 is underscored twice. Use these conditions to solve the problem. (Example 3)

34. In the season's final game, a high school pitcher struck out three more batters than the greatest number in any previous game. However, this number of strikeouts was still fewer than the school record of 13. What is the greatest number of strikeouts recorded this season prior to this final game?

35. On a winter day, the wind chill temperature was 35°F below the thermometer reading. At 6 P.M., the TV weather forecaster reported that the wind chill temperature had not yet dropped to −28°F. What was the lowest possible thermometer reading before 6 P.M. on that day?

36. After 6:00 P.M., the temperature dropped 23 degrees to reach the low temperature for the day. This temperature was higher than the record low temperature of −46°F. What was the lowest possible reading (integer) for the temperature at 6:00 P.M.?

37. Jeff's score on a math test was 10 points higher than the scores on any of his previous math tests. His score was lower than the highest possible score of 150 points. What was his highest possible score on any previous math test?

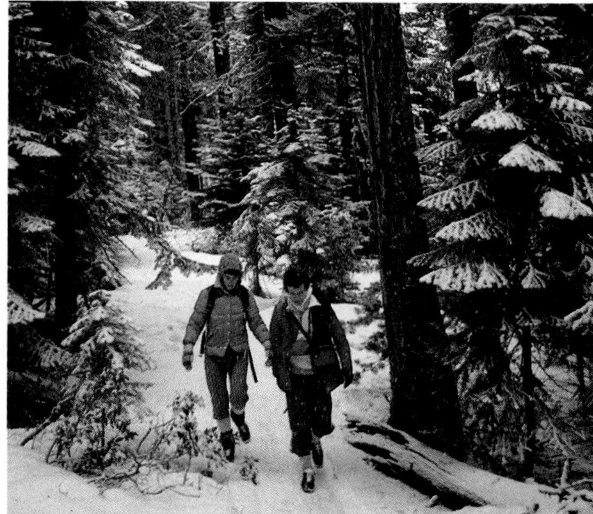

REVIEW CAPSULE FOR SECTION 13.3

Multiply both sides of each inequality by 5. Then write an inequality to compare the products. (Pages 214–217)

1. $-4 < 8$ **2.** $-6 < -2$ **3.** $3 > 2$

4. $5 > -5$ **5.** $-1.2 < 2.6$ **6.** $-5.8 > -7.4$

Multiply both sides of each inequality by −10. Then write an inequality to compare the products. (Pages 214–217)

7. $2 > -8$ **8.** $5 > 1$ **9.** $-4 > -10$

10. $-4 < 3$ **11.** $-1.8 > -3.2$ **12.** $-6.9 < -2.7$

13.3 Multiplication and Division Properties

Each row of Table 1 shows an inequality and the inequality that results when each side is multiplied by the same positive number.

Table 1

Inequality	Multiply Each Side by	Resulting Inequality
$-3 < 7$	2	$-6 < 14$
$1\frac{1}{2} < 3$	2	$3 < 6$
$-3 > -5$	4	$-12 > -20$
$3 > -8$	$\frac{1}{2}$	$1\frac{1}{2} > -4$
$8 > 1$	5	$40 > 5$
$-12 < 0$	$\frac{3}{4}$	$-9 < 0$

P–1 **What inequality results when each side is multiplied by 3?**

a. $-5 > -6$ **b.** $-\frac{1}{3} < \frac{2}{3}$ **c.** $0 > -1.2$ **d.** $12 > -12$

Table 2 shows an inequality and the inequality that results when each side is multiplied by the same negative number.

Table 2

Inequality	Multiply Each Side by	Resulting Inequality
$-3 < 7$	-2	$6 > -14$
$1\frac{1}{2} < 3$	-2	$-3 > -6$
$-3 > -5$	-4	$12 < 20$
$3 > -8$	$-\frac{1}{2}$	$-1\frac{1}{2} < 4$
$8 > 1$	-5	$-40 < -5$
$-12 < 0$	$-\frac{3}{4}$	$9 > 0$

Note that the direction of the inequality has changed. The inequality symbol is reversed.

P–2 **What inequality results when each side is multiplied by −6?**

a. $5 < 9$ **b.** $-7 < -2$ **c.** $0 > -5$ **d.** $-\frac{1}{3} > -6$

Tables 1 and 2 suggest the following property for inequalities.

Multiplication Property for Inequalities

1. The direction of an inequality is unchanged when each side is multiplied by the same positive number.

$$\frac{1}{2}p > 12$$
$$\mathbf{2}(\tfrac{1}{2}p) > \mathbf{2}(12)$$
$$p > 24$$

2. The direction of an inequality is reversed when each side is multiplied by the same negative number.

$$6 < -2g$$
$$(-\tfrac{1}{2})6 > (-\tfrac{1}{2})(-2g)$$
$$-3 > g \text{ or } g < -3$$

EXAMPLE 1 Solve and graph: $\frac{1}{3}n < 15$

Solution:

$$\frac{1}{3}n < 15$$

[1] Multiply each side by 3. ⟶ $\mathbf{3}\left(\frac{1}{3}n\right) < \mathbf{3}(15)$

$$n < 45$$

[2] Graph the inequality. ⟶

45

−20 −10 0 10 20 30 40 50 60

The solution set consists of all real numbers less than 45.

P–3 **Solve.**

a. $\frac{1}{4}n < 16$ **b.** $18 > \frac{1}{2}t$ **c.** $35 < -\frac{1}{5}r$ **d.** $-16 > \frac{1}{9}d$

The Division Property for Inequalities is similar to the Multiplication Property for Inequalities.

Division Property for Inequalities

1. The direction of an inequality is unchanged when each side is divided by the same positive number.

$$5y < 15$$
$$\frac{5y}{\mathbf{5}} < \frac{15}{\mathbf{5}}$$
$$y < 3$$

2. The direction of an inequality is reversed when each side is divided by the same negative number.

$$-3b > 9$$
$$\frac{-3b}{\mathbf{-3}} < \frac{9}{\mathbf{-3}}$$
$$b < -3$$

EXAMPLE 2 Solve and graph: $-24 < 12x$

Solution:

$$-24 < 12x$$

[1] Divide each side by 12. ⟶ $\frac{-24}{\mathbf{12}} < \frac{12x}{\mathbf{12}}$

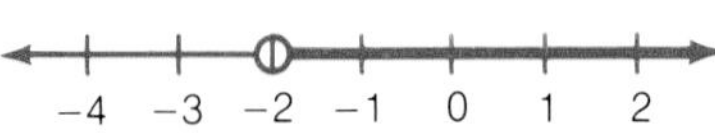

The direction of the inequality is unchanged.

$-2 < x$ or

$x > -2$

[2] Graph the inequality. ⟶

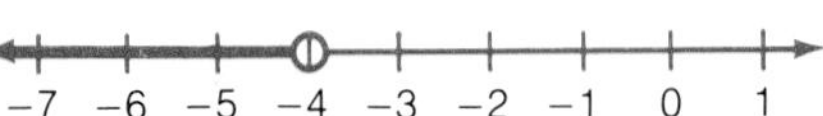

The solution set consists of all real numbers greater than -2.

EXAMPLE 3 Solve and graph: $20 < -5x$

Solution:

$$20 < -5x$$

[1] Divide each side by -5. ⟶ $\frac{20}{\mathbf{-5}} > \frac{-5x}{\mathbf{-5}}$

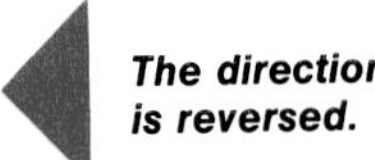

The direction is reversed.

$-4 > x$ or

$x < -4$

[2] Graph the inequality. ⟶

The solution set consists of all real numbers less than -4.

P–4 **Solve.**

a. $3p < 48$ **b.** $-54 > 9b$ **c.** $6a < -24$ **d.** $18 > -3f$

EXAMPLE 4

In a school election, the number of seniors who voted was two-thirds of the number of juniors who voted (Condition 1). The election committee reported that fewer than 286 seniors voted (Condition 2). Is it possible that 420 juniors voted in the election?

Solution: [1] Use Condition 1 to represent the unknowns.

Let $n =$ the number of juniors who voted.

Then $\frac{2}{3}n =$ the number of seniors who voted.

[2] Use Condition 2 to write an inequality.

Think: **The number of seniors who voted was fewer than 286.**

Translate: $\frac{2}{3}n \quad < \quad 286$

[3] Solve the inequality. ⟶ $\frac{3}{2}\left(\frac{2}{3}n\right) < \frac{3}{2}(286)$

$$1 \cdot n < \frac{3}{2} \cdot \frac{286}{1}$$

$$n < 429$$

Since $n < 429$, it is possible that 420 juniors voted.

[4] **Check:** Is $\frac{2}{3}(420) < 286$? Is $280 < 286$? Yes ✓

CLASSROOM EXERCISES

Multiply each side of the following inequalities by 5. Write the resulting inequality. (Table 1)

1. $-2 < 7$ **2.** $4 > -8$ **3.** $3 < 5$ **4.** $-6 > -9$ **5.** $-\frac{1}{5} < 2$ **6.** $-\frac{3}{10} > -\frac{3}{6}$

Multiply each side of the following inequalities by −4. Write the resulting inequality. (Table 2)

7. $5 > -1$ **8.** $-2 < 3$ **9.** $-6 > -8$ **10.** $-\frac{1}{4} < 3$ **11.** $-\frac{5}{8} > -3$ **12.** $-\frac{3}{4} < -\frac{3}{8}$

Write an equivalent inequality as the first step in solving each inequality. (Step 1, Examples 1–3)

13. $\frac{1}{3}x < -2$ **14.** $-4n > 24$ **15.** $3t < -12$ **16.** $-\frac{1}{2}k > -6$

17. $4m < -21$ **18.** $-28 > -6a$ **19.** $-\frac{1}{3}b < 2\frac{1}{4}$ **20.** $-\frac{3}{4} > \frac{1}{4}c$

WRITTEN EXERCISES

Goal: To solve and graph inequalities by using the Multiplication and Division Properties

Sample Problems: Solve and graph: **a.** $-\frac{5}{6}r > 10$ **b.** $27 > 3m$

Answers: **a.** $r < -12$

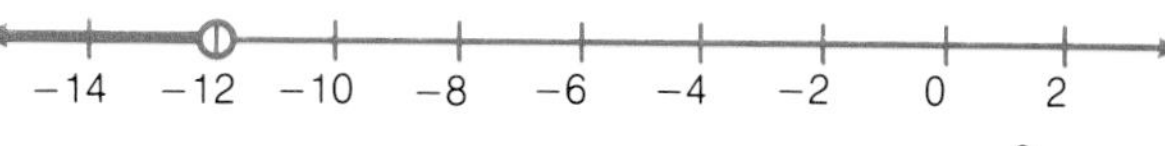

b. $9 > m$ or $m < 9$

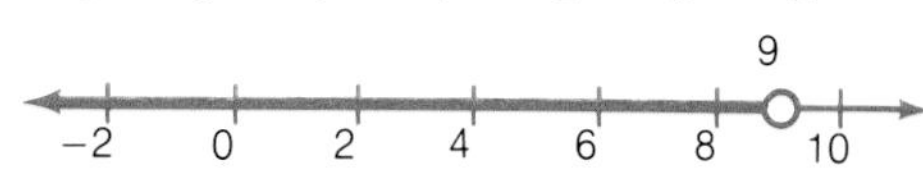

For Exercises 1–44, solve and graph.
(Example 1)

1. $\frac{1}{3}y < 15$
2. $\frac{1}{5}x > 30$
3. $\frac{1}{2}w > -36$
4. $\frac{1}{9}n < -6$
5. $\frac{1}{2} > -\frac{1}{4}d$
6. $\frac{1}{3} < -\frac{1}{5}t$
7. $-\frac{3}{4}r < -3$
8. $-5 > \frac{5}{6}t$

(Example 2)

9. $5y < 30$
10. $8x < 72$
11. $16t > 48$
12. $6s > 90$
13. $7p < -49$
14. $9m > -81$
15. $-40 > 3p$
16. $-31 < 11q$

(Example 3)

17. $-7s < 21$
18. $-11p > 44$
19. $-20 < -10t$
20. $-8z < -56$
21. $-12 < -5a$
22. $21 < -3x$
23. $-4w > -21$
24. $-29 < -6p$

MIXED PRACTICE

25. $\frac{2}{3}w < -4$
26. $-\frac{1}{8}x < 3$
27. $96 < -12y$
28. $-35 < 7k$
29. $\frac{3}{4}r < -\frac{3}{8}$
30. $\frac{2}{3}w > -\frac{2}{9}$
31. $-\frac{2}{3}w < -6$
32. $\frac{3}{5}p > -9$
33. $-43 > -8g$
34. $90 > 15d$
35. $-\frac{1}{4}r > 7$
36. $-\frac{1}{5}n > 10$
37. $-16 > -\frac{1}{2}b$
38. $81 < -\frac{1}{3}y$
39. $\frac{4}{5}g < 16$
40. $\frac{3}{5}r > 15$
41. $-72 > 18t$
42. $-56 > 14w$
43. $-\frac{2}{3}q < 1$
44. $\frac{7}{8}j < 14$

APPLICATIONS

For Exercises 45–48, Condition 1 is underscored once. Condition 2 is underscored twice. Use these conditions to solve the problem. (Example 4)

45. The average monthly auto expenses of a company equals the yearly expenses divided by 12. The company accountant wants to keep the average monthly expenses under $2500. Is it possible that the yearly auto expenses total more than $31,500?

46. The average monthly grocery bill for a household equals the total of all the monthly grocery bills divided by 12. The Lazears want to keep their average monthly grocery bill under $450. Is it possible that the total of all their monthly grocery bills is $5500?

47. In a school survey, the number of juniors polled was three-fourths of the number of sophomores polled. The committee conducting the survey reported that fewer than 120 juniors were polled. Is it possible that 148 sophomores were polled?

48. The Hanes family bought a radio and a television set. The television cost four times as much as the radio. The television cost less than $300. Is it possible that the radio cost $65?

MID-CHAPTER REVIEW

Graph each inequality using **a.** *the set of integers and* **b.** *the set of real numbers as the replacement sets.* (Section 13.1)

1. $n > -2$ **2.** $r < 5\frac{1}{2}$ **3.** $t < 2.6$ **4.** $y > -4.3$

Write an inequality for each graph. Identify the replacement set.

5. etc.
−3 −2 −1 0 1 2 3

6.
−3 −2 −1 0 1 2 3

7.
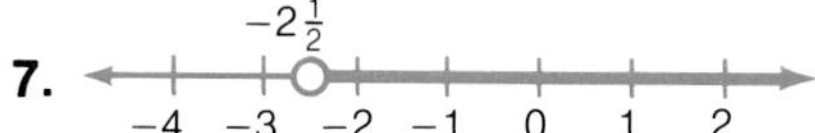

8.
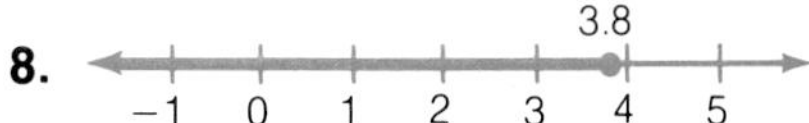

Select the inequality and replacement set that matches the graph.

9. etc.
−4 −3 −2 −1 0 1 2

a. $x < -2$; {integers}
b. $-2 < x$; {real numbers}
c. $x > -2\frac{1}{2}$; {integers}

10.
−2 −1 0 1 2 3 4

a. $p \leq 2$; {integers}
b. $2 \geq p$; {real numbers}
c. $p < 2$; {real numbers}

Solve and graph each inequality. (Section 13.2)

11. $-185 > w - 75$ **12.** $-215 > t - 219$ **13.** $p + 3 > 1$ **14.** $-3 + k > -1$
15. $-15 + r < 9$ **16.** $-37 + m > -41$ **17.** $-2 > x - 3$ **18.** $17 < y + 4$

Solve and graph each inequality. (Section 13.3)

19. $\frac{2}{3}r < -4$ **20.** $\frac{1}{5}s < -1$ **21.** $-4w < -15$ **22.** $-6t < 18$
23. $7n < -56$ **24.** $37m > -148$ **25.** $30 > -6g$ **26.** $-36 > -12k$

For Exercises 27–28, Condition 1 is underscored once. Condition 2 is underscored twice. Use these conditions to solve the problem. (Sections 13.2 and 13.3)

27. In their third game, a bowling team rolls eight more strikes than their greatest number in either of the first two games. This number is still fewer than their high of 28 strikes in one game. What is the greatest number of strikes rolled in either of the first two games?

28. The average monthly electric bill for a household equals the total of all the monthly bills divided by 12. The Smiths want to keep their average monthly bill under $115. Is it possible that the total of all their monthly bills is $1405?

Using Statistics

Frequency Tables/ Histograms

Statisticians use tables, charts, and graphs to organize and display data. One way to organize data is to use a **frequency table.**

EXAMPLE: The highway mileage ratings in miles per gallon for forty compact cars are listed at the left below.

a. Make a frequency table for the data.

SOLUTION: Organize the data in intervals of 5 units.

Mileage Ratings (miles per gallon)

24	35	27	23	26	23	28	29
35	33	34	24	24	35	34	42
18	26	26	26	24	33	28	26
35	26	21	28	33	33	32	36
24	34	42	32	28	22	22	40

FREQUENCY TABLE

Interval	*Midpoint*	*Tally*	*Frequency*
15–19	17	|	1
20–24	22	𝍸 𝍸	10
25–29	27	𝍸 𝍸 ||	12
30–34	32	𝍸 ||||	9
35–39	37	𝍸	5
40–44	42	|||	3

b. Use the midpoints of the intervals to compute the approximate mean of the data. In statistics, $\bar{x}$ (read: x bar) is used to represent the mean.

SOLUTION: $$\bar{x} \approx \frac{1(17) + 10(22) + 12(27) + 9(32) + 5(37) + 3(42)}{40}$$

17 [+] 220 [+] 324 [+] 288 [+]

185 [+] 126 [÷] 40 [=] 29.

The mean, $\bar{x}$, is about **29.**

c. Use the frequency table to draw a **histogram** for the data.

SOLUTION: The horizontal axis shows the gas mileage data.
The vertical axis shows the frequencies.
The interval midpoint is the center of each bar.

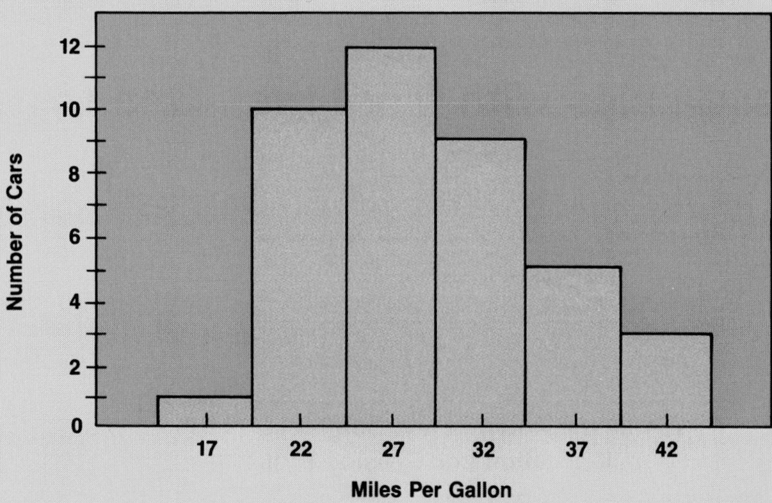

EXERCISES

For Exercises 1–2, make a frequency table using the suggested interval width. Then draw a histogram for the data.

1. ***Quiz Scores of Thirty Students***

5	7	3	3	9	6
5	4	6	8	5	1
8	6	5	1	9	8
3	4	10	7	2	5
3	4	5	6	8	6

Interval width: 1

2. ***Number of Runs Scored by 50 Hitters***

19	16	26	18	22	28	17	20	22	23
16	18	19	25	29	12	8	23	22	20
18	17	29	23	18	12	11	20	17	12
23	18	10	15	18	19	22	6	11	14
19	17	14	20	18	16	17	19	20	22

Interval width: 3

3. Compute the approximate mean of the data in Exercise 2 by using the midpoints of the intervals as values.

4. On Monday's test, the mean of 18 girls' scores was 70. The mean of 18 boys' scores was 75. What was the mean for the entire class?

5. On Friday's test, the mean of 16 girls' scores was 80 and the mean of 16 boys' scores was 72. What was the mean for the entire class?

REVIEW CAPSULE FOR SECTION 13.4

Solve. (Pages 312–316)

1. $-g = 8$ **2.** $-k = -4$ **3.** $13 = -c$ **4.** $-21 = -b$ **5.** $\frac{3}{8} = -z$ **6.** $-m = -1\frac{2}{3}$

13.4 Inequalities with More than One Operation

The rules for solving inequalities are similar to the rules for solving equations.

Rules for Solving Inequalities

1. Simplify each side of the inequality by combining like terms.
2. Use the Addition and Subtraction Properties before using the Multiplication and Division Properties.

P–1 **What number should be subtracted from both sides of each inequality to get 2*a* by itself on one side of the inequality?**

a. $2a + 6 < 14$ **b.** $12 > 5 + 2a$ **c.** $2a + 2 < -12$

EXAMPLE 1 Solve and graph: $5s + 2 - 3s > -12$

Solution:

[1] Combine like terms. ⟶ $5s + 2 - 3s > -12$

$$2s + 2 > -12$$

[2] Subtract 2 from each side. ⟶ $2s + 2 \mathbf{- 2} > -12 \mathbf{- 2}$

$$2s > -14$$

[3] Multiply each side by $\frac{1}{2}$. ⟶ $\left(\frac{1}{2}\right)2s > \left(\frac{1}{2}\right)(-14)$

$$s > -7$$

[4] Graph the inequality. ⟶

−9 −8 −7 −6 −5 −4 −3 −2 −1 0 1 2

The solution set consists of all real numbers greater than −7.

P–2 **Solve.**

a. $6v + 7 - 2v < 35$ **b.** $20 < 5t - 4 - 8t$

Example 2 shows two methods for solving an inequality when there is a variable on each side.

EXAMPLE 2

Solve and graph: $2f - 5 > 3f + 2$

Solution: Method 1

$$2f - 5 > 3f + 2$$
$$2f - 5 - \mathbf{2f} > 3f + 2 - \mathbf{2f}$$
$$-5 > f + 2$$
$$-5 - \mathbf{2} > f + 2 - \mathbf{2}$$
$$-7 > f \quad \text{or}$$
$$f < -7$$

Method 2

$$2f - 5 > 3f + 2$$
$$2f - 5 - \mathbf{3f} > 3f + 2 - \mathbf{3f}$$
$$-f - 5 > 2$$
$$-f - 5 + \mathbf{5} > 2 + \mathbf{5}$$
$$-f > 7$$
$$(\mathbf{-1})(-f) < (\mathbf{-1})7$$
$$f < -7$$

Multiply each side by −1. This reverses the inequality symbol.

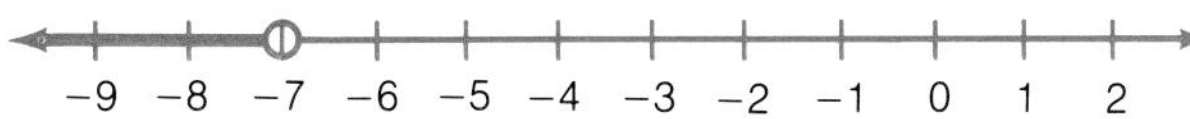

The solution set consists of all real numbers less than −7.

In Example 2, Method 1 is shorter because subtracting $2f$ in the first step resulted in f. In Method 2, subtracting $3f$ resulted in $-f$. Since $-f$ means $-1f$, it is necessary to multiply each side by −1 to get f. This also reverses the direction of the inequality.

P–3 **Solve.**

a. $5h - 4 > 6h + 1$ **b.** $8g + 2 > 7g - 4$

The properties for solving inequalities also apply to inequalities that use $\leq$ or $\geq$.

CLASSROOM EXERCISES

For Exercises 1–18, write an equivalent inequality as the first step in solving each inequality. (Step 1, Example 1)

1. $5n - 13 - 2n < -17$
2. $8t + 14 - 11t > 25$
3. $3r - 6 - r < -4$
4. $12 < 6x - 8 - 2x$
5. $-5y - 23 + y > -17$
6. $12w - 15 - w < 28$
7. $-28 > -8m + 17 - 15m$
8. $43 < 29 - 14p - 19p$
9. $18b - 42 - 19b < -56$

(Step 1, Example 2)

10. $4p - 3 < 5p + 6$
11. $8 - 3y > -4y - 2$
12. $3x + 5 > 2x - 2$
13. $4 - 6d < 5d - 3$
14. $-3c - 1 < 8 - 2c$
15. $6p + 4 > 7p - 12$
16. $7y - 3 > -4y + 8$
17. $12 - t < -8 - 5t$
18. $-3q + 23 > -q + 26$

Solve. (Step 3 in Method 2, Example 2)

19. $-x > 5$
20. $10 > -f$
21. $-3 < -g$
22. $-h < -6.8$
23. $0 > -z$
24. $\frac{5}{8} > -p$

WRITTEN EXERCISES

Goal: To solve and graph inequalities with combined operations

Sample Problem: Solve and graph: $5c + 6 - c < c - 12$

Answer: $c < -6$

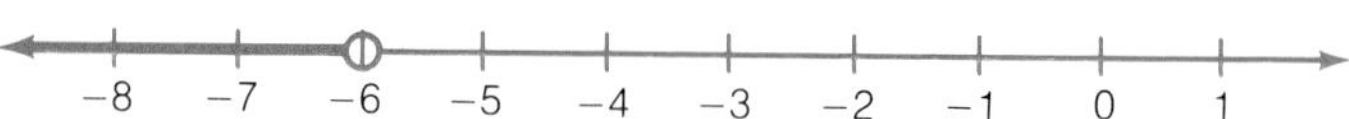

For Exercises 1–24, solve and graph.
(Example 1)

1. $-3 < -7n + 2 + 8n$
2. $8 > 4q - 3 - 3q$
3. $3a - 2 - 2a < -7$
4. $-5t + 7 + 6t < -9$
5. $12y - \frac{3}{4} - 9y > 11\frac{1}{4}$
6. $-7r + \frac{7}{8} + 11r < -3\frac{1}{8}$

(Example 2)

7. $4w - 6 < 5w + 12$
8. $5s + 2 < 4s + 8$
9. $3x - 1 > 5x - 3$
10. $-2y + 3 > y - 6$
11. $0.6m - 0.7 < 0.4m + 0.3$
12. $1.8n - 4.8 < 0.8n + 5.2$

MIXED PRACTICE

13. $3t + 5 > -7t - 10$
14. $26r - 41 > r - 116$
15. $18k + 2 - 21k < k - 14$
16. $g + 18 - 3g < 3g + 3$
17. $-3w - 5 < -2w + 10$
18. $6m - 9 < -2m + 21$
19. $p - 15 - 2p > 2p + 12$
20. $17 - 12a - 24 > -8a - 15$
21. $1.8d - 3.2d > 0.6d - 14.2$
22. $\frac{6}{5} - 4n > 4n - \frac{2}{5}$
23. $-2n + \frac{8}{3} > 2n - \frac{4}{3}$
24. $1.4x - 3.6x > 0.8x - 1.5$

NON-ROUTINE PROBLEMS

25. Mike has 12 blue socks and 8 black socks in a drawer. Without looking, how many socks would he need to take out to be sure he gets two socks of the same color?

26. After spending $\frac{1}{4}$ of her money on food, $\frac{1}{2}$ of her money on clothing, and $\frac{1}{6}$ of her money on records, Julie had six dollars left. How much money did she have at first?

13.5 Using Inequalities

The steps for solving word problems involving inequalities are similar to those for solving word problems involving equations.

Steps for Solving Word Problems Involving Inequalities

1. Identify Condition 1. Use Condition 1 to represent the unknowns.
2. Identify Condition 2. Use Condition 2 to write an inequality.
3. Solve the inequality. Answer the question.
4. Check your answer with the statements in the problem.

EXAMPLE 1

In a store the cost of a dress is \$40 more than the cost of a pair of shoes (Condition 1). Joann plans to buy both items for less than \$110 (Condition 2). What is the greatest amount Joann can spend for the shoes?

Solution:

[1] Let $s =$ the cost of the shoes.
Then $s + 40 =$ the cost of the dress. ◀ ***Condition 1***

[2] Write an inequality for the problem.

Think:	**Cost of the shoes**	+	**Cost of the dress**	**Is less than**	**\$110**	◀ ***Condition 2***
	↓		↓	↓		
Translate:	s	+	$s + 40$	$<$	110	

[3] Solve the inequality. ⟶

$$s + s + 40 < 110$$
$$2s + 40 < 110$$
$$2s + 40 - \mathbf{40} < 110 - \mathbf{40}$$
$$2s < 70$$
$$\frac{2s}{\mathbf{2}} < \frac{70}{\mathbf{2}}$$
$$s < 35$$

Since $s < 35$, Joann can spend no more than \$34.99 for the shoes.

[4] **Check:** Is $34.99 + 34.99 + 40 < 110$? Is $109.98 < 110$? Yes ✓

P–1 **What is the greatest amount of money in dollars and cents that satisfies each inequality?**

a. $x < \$5.00$ **b.** $y < \$15.50$ **c.** $\$19.95 > d$

Recall from Section 10.6 that the ***mean*** of two or more scores is the sum of the scores divided by the number of scores.

In Example 2, Condition 1 is underscored once. Condition 2 is underscored twice. Note that Condition 1 is stated after Condition 2 in this problem.

EXAMPLE 2

Raoul's scores on four tests were 80, 77, 81, and 84. He wants his mean score on five tests to be greater than 80. Find the lowest score Raoul can get on the fifth test.

Solution:

1. Let $x =$ Raoul's score on the fifth test.

2. Write an inequality for the problem.

Think: **The mean of the five tests** **must be greater than** **80.**

Translate: $\frac{80 + 77 + 81 + 84 + x}{5} > 80$

3. Solve the inequality. ⟶ $\mathbf{5}\left(\frac{80 + 77 + 81 + 84 + x}{5}\right) > \mathbf{5}\,(80)$

$$80 + 77 + 81 + 84 + x > 400$$

$$322 + x > 400$$

$$322 + x - \mathbf{322} > 400 - \mathbf{322}$$

$$x > 78$$

Since $x > 78$, Raoul's score must be at least 79.

4. **Check:** Is $\frac{80 + 77 + 81 + 84 + 79}{5} > 80$? Is $\frac{401}{5} > 80$? Yes ✓

P–2 **What is the mean of the scores?**

a. 82, 85, 80, 81 **b.** 91, 88, 76, 84, 81 **c.** 79, 81, x

CLASSROOM EXERCISES

Write an inequality for each sentence. Use b as the variable. (Step 2, Example 1)

1. The cost of a lamp is greater than $45.
2. The number of persons decreased by 50 is less than 2900.
3. The sum of the cost of a sofa and $275 is less than $1050.
4. The quotient of an amount of money and 14 is greater than $125.50.

Write an inequality for each sentence. (Step 2, Example 2)

5. The mean of 87, 92, and z is less than 91.
6. The mean of 85, 87, 84, and y is greater than 86.
7. The mean of 65, 70, 68, and x is greater than 66.
8. The mean of 68, 70, 72, 73, and y is less than 71.

WRITTEN EXERCISES

Goal: To solve word problems involving inequalities

Sample Problem: A family plans to buy a television set and a video tape recorder. They estimate that the tape recorder will cost $300 more than the TV set (Condition 1). They want to spend no more than $1400 in all (Condition 2). What is the greatest amount they can spend on the TV set?

Answer: $550

For Exercises 1–8, Condition 1 is underscored once. Condition 2 is underscored twice. Use these conditions to solve the problem.

Solve. (Example 1)

1. Two people share an apartment. They estimate that the amount spent each month for rent should be twice the amount spent for food. The amount they can pay each month for food and rent must be less than $540. What is the greatest amount they can pay each month for food?
2. The Bligh family plans to buy a new sofa and chair. They estimate that the cost of the sofa will be four times the cost of the chair. They want to spend less than $1100 for the sofa and the chair. What is the greatest amount they can spend for the chair?

3. The Chiu family's yearly rent budget is three times their clothing budget. They estimate that they should spend less than $6400 for rent and clothing per year. Find the greatest amount they can spend for clothing.

4. Bill estimates that a sweater costs $30 less than a certain jacket. He plans to buy both items for less than $70. Find the greatest amount he can spend for the sweater.

Solve. (Example 2)

5. Ava's scores on 3 quizzes were 78, 81, and 93. She wants her mean score on four quizzes to be greater than 85. Find the lowest score she can get on the fourth quiz.

6. A student's grades on four tests are 68, 82, 78, and 80. He would like to have a mean score of 80 or more on five tests. What is the lowest grade he can get on the fifth test?

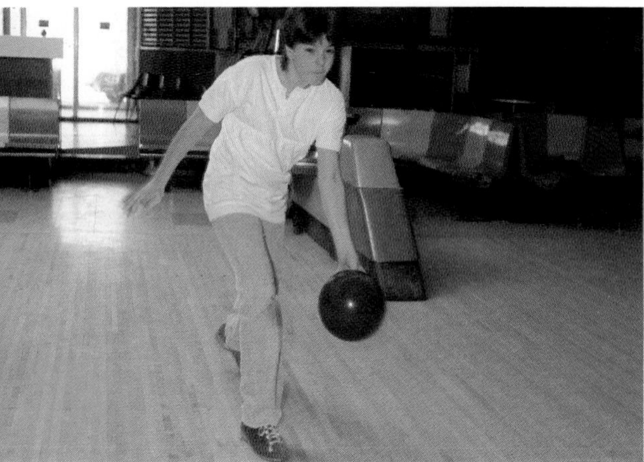

7. Rocco's scores for three games in bowling were 76, 103, and 121. In order to have a mean score greater than 100 for four games, what is the lowest score he can bowl on the next game?

8. During the month of July, Lillian wants her mean golf score to be less than 76. She has already shot rounds of 78, 81, 73, and 76. Find the highest score she can shoot on her fifth round of golf.

NON-ROUTINE PROBLEMS

9. A photographer wishes to take a picture of four people standing side by side. In how many ways can these four people be arranged?

10. Rob has two jars of equal size. One is full of nickels. The other is half full of dimes. Which jar has more money in it? Explain.

11. In a marching formation, there is always one majorette marching in front of two majorettes and one majorette marching behind two majorettes. Another majorette marches between two majorettes. What is the least possible number of majorettes marching?

12. Grace has two minutes to get to the bus station before her bus departs. The bus station is two miles from her home. She drives at a speed of 30 miles per hour for the first mile traveled. Will she get to the bus station before her bus departs? Explain.

CHAPTER SUMMARY

IMPORTANT TERMS	Replacement set *(p. 364)* Equivalent inequalities *(p. 368)*
IMPORTANT IDEAS	1. *Addition Property for Inequalities:* Adding the same real number to each side of an inequality forms an equivalent inequality. 2. *Subtraction Property for Inequalities:* Subtracting the same real number from each side of an inequality forms an equivalent inequality. 3. *Multiplication Property for Inequalities* **a.** The direction of an inequality is unchanged when each side is multiplied by the same positive number. **b.** The direction of an inequality is reversed when each side is multiplied by the same negative number. 4. *Division Property for Inequalities* **a.** The direction of an inequality is unchanged when each side is divided by the same positive number. **b.** The direction of an inequality is reversed when each side is divided by the same negative number. 5. Rules for Solving Inequalities See page 380. 6. Steps for Solving Word Problems Involving Inequalities See page 383.

CHAPTER REVIEW

SECTION 13.1

Graph each inequality using **a.** *the set of integers and* **b.** *the set of real numbers as the replacement sets.*

1. $x < 5$ **2.** $y > -5$ **3.** $t \geq -3.9$ **4.** $n \leq 1\frac{1}{2}$

Write an inequality for each graph. Identify the replacement set. Let p represent the variable.

5. 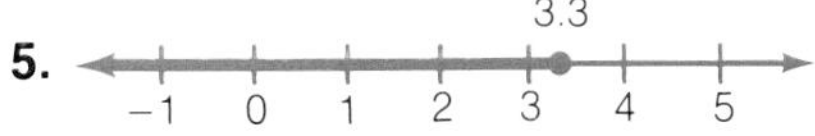

6. 

SECTION 13.2

Solve and graph.

7. $n - 18 < -23$ **8.** $t - 21 < -24$ **9.** $3 + p > 5$

10. $-9 + r > -5$ **11.** $-24.3 < t - 19.4$ **12.** $-7.3 < k - 3.6$

For Exercises 13–14, Condition 1 is underscored once. Condition 2 is underscored twice. Use these conditions to solve the problem.

13. The present temperature in San Diego is three degrees higher than the present temperature in Atlanta. The present temperature in San Diego is less than 72°F. What is the highest possible temperature in Atlanta?

14. A baseball player stole 16 more bases his second year in the league than during his first year. The number of bases stolen the second year was fewer than the team record of 56. What is the greatest number of bases stolen the first year?

SECTION 13.3

Solve and graph.

15. $4r < 23$

16. $5t < 34$

17. $-6n > -21$

18. $-8n > -42$

19. $-4 < -\frac{4}{7}x$

20. $-3 < -\frac{3}{4}w$

For Exercises 21–22, Condition 1 is underscored once. Condition 2 is underscored twice. Use these conditions to solve the problem.

21. The length of a building is 85 meters and the area equals the product of the length and width. The area must be less than 5695 square meters. Could the building's width be 70 meters?

22. The high thermometer reading on a certain day equals the quotient of the wind chill temperature and −5. The reading is less than the record high of 6°F. Could the wind chill temperature be −37°F?

SECTION 13.4

Solve and graph.

23. $-10 < 4d - 8 - 3d$

24. $6 - 6k + 7k < 2$

25. $12m - 5 > 10m + 8$

26. $8q - 9 < 5q + 3$

27. $-12 < -6p + 12 - 3p$

28. $4x - 3 - 10x > 15$

29. $6 - 4c > c - 19$

30. $5\frac{1}{2}w + 3\frac{1}{4} > -2\frac{1}{2}w - 1\frac{3}{4}$

31. $-5\frac{3}{8} - 2\frac{1}{4}g < 2\frac{5}{8} + 3\frac{3}{4}g$

SECTION 13.5

For Exercises 32–33, Condition 1 is underscored once. Condition 2 is underscored twice. Use these conditions to solve the problem.

32. A basketball team scored these points in their last five games: 102, 121, 118, 98, 113. What is the least number the team would need to score in its sixth game in order to have a mean of more than 109 points for the six games?

33. Gail wants to buy a new car. She will make a down payment of $750 on the car and pay the remainder of the cost in 36 equal monthly payments. What is the most she can pay each month if she wants the total cost to be under $8000?

Relations and Functions

CHAPTER 14

An aerial view of Manhattan shows the rectangular grid of streets that locates any particular intersection. For example, the ordered pair A(2nd Avenue, 1st Street) describes the location of building A (see bottom right).

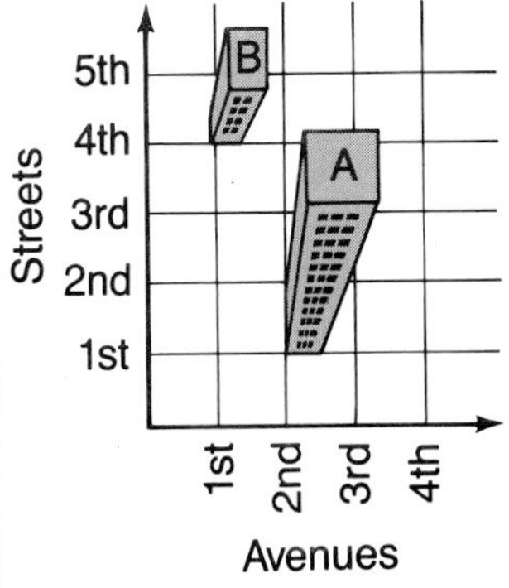

14.1 Graphs in a Plane

There are two intersecting number lines called ***axes*** in the figure at the right. The horizontal number line is the ***x axis.*** The vertical number line is the ***y axis.***

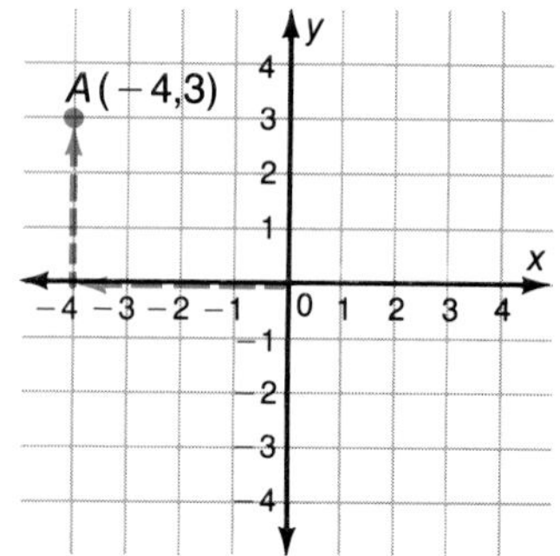

Note that point *A* is 4 units to the left of the *y* axis. It is also 3 units above the *x* axis.

The numbers −4 and 3 are the ***coordinates*** of point *A*. They are also called the *x* value and *y* value of the point. The dot at *A* is the ***graph*** of the point with coordinates (−4, 3).

P–1 **What is the *x* value in each pair of numbers below? What is the *y* value?**

a. (0, 3) **b.** (6, −2) **c.** (−5, 8) **d.** (7, 0)

The point with coordinates (0, 0) is called the ***origin.*** It is the point where the *x* axis and the *y* axis cross.

EXAMPLE 1 Graph *B*(3, −4).

"B(3, −4)" means "the point B with coordinates (3, −4)."

Solution:

1. Start at the origin. Move 3 units to the right. This is the *x* value.

Move right, because x is positive.

2. Move down 4 units. This is the *y* value.

Move down, because y is negative.

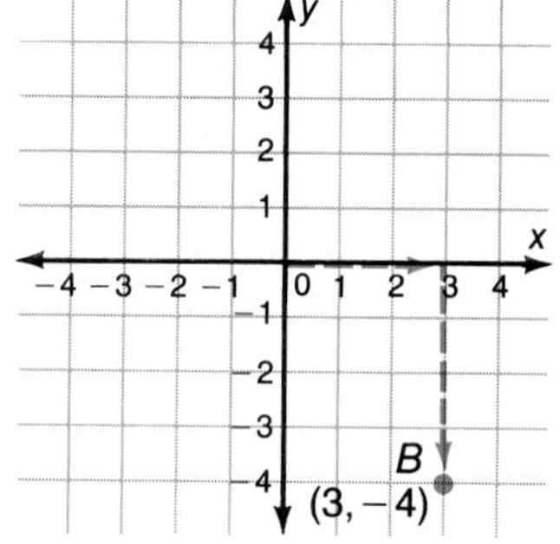

The graph of (3, −4) is the dot at *B*.

The points *A*(−4, 3) and *B*(3, −4) were graphed using the same two numbers, −4 and 3. Thus, the order of the numbers is important. A number pair such as (−4, 3) is called an ***ordered pair.***

EXAMPLE 2 Write the coordinates of these points.

a. P **b.** Q

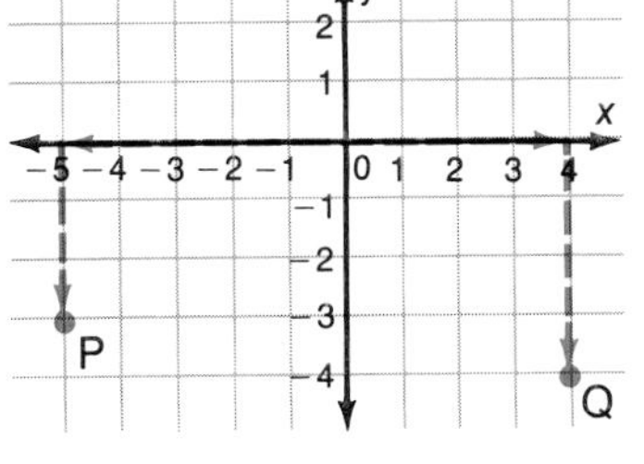

Solutions: **a.** [1] Start at the origin. Move left to the vertical line containing P. The x value is -5.

[2] Move down to point P. The y value is -3.

The coordinates of P are $(-5, -3)$.

b. [1] Start at the origin. Move right to the vertical line containing Q. The x value is 4.

[2] Move down to Q. The y value is -4.

The coordinates of Q are $(4, -4)$.

P–2 **Write the coordinates of each point.**

a. R **b.** S **c.** T

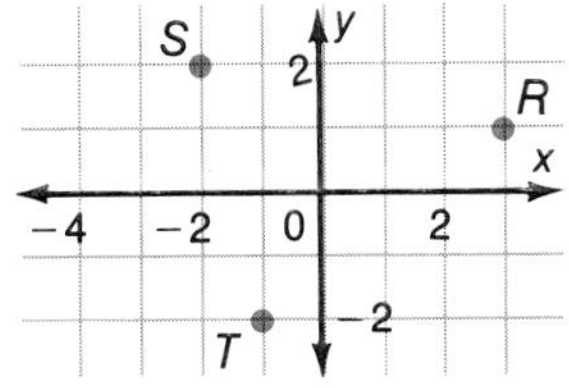

CLASSROOM EXERCISES

Graph each point. (Example 1)

1. $(3, -6)$ **2.** $(4, 8)$ **3.** $(-1, -7)$ **4.** $(-12, 6)$ **5.** $(8, -9)$

6. $(15, 10)$ **7.** $(-12, -2)$ **8.** $(10, -5)$ **9.** $(-5, 5)$ **10.** $(0, -8)$

11. $(-9, 0)$ **12.** $(-8, 0)$ **13.** $(0, 3)$ **14.** $(-8, -8)$ **15.** $(6, -6)$

Write the coordinates of each point. (Example 2)

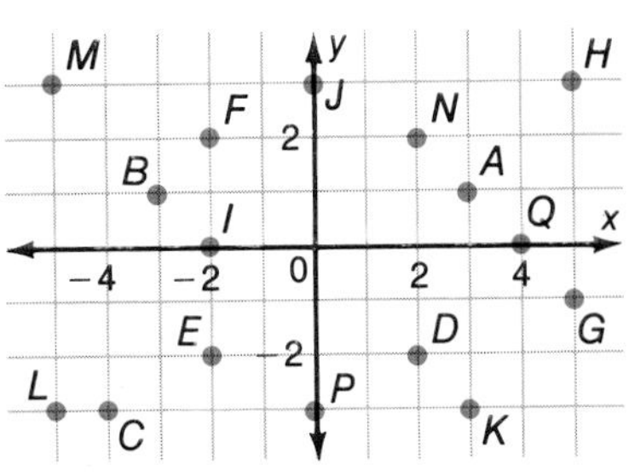

16. A **17.** B **18.** C **19.** D

20. E **21.** F **22.** G **23.** H

24. I **25.** J **26.** K **27.** L

28. M **29.** N **30.** P **31.** Q

WRITTEN EXERCISES

Goal: To write the coordinates of a point
Sample Problem: Write the coordinates of R.
Answer: $(-4, 2)$

Goal: To graph a point
Sample Problem: Graph $J(3, -4)$.
Answer: See the figure at the right.

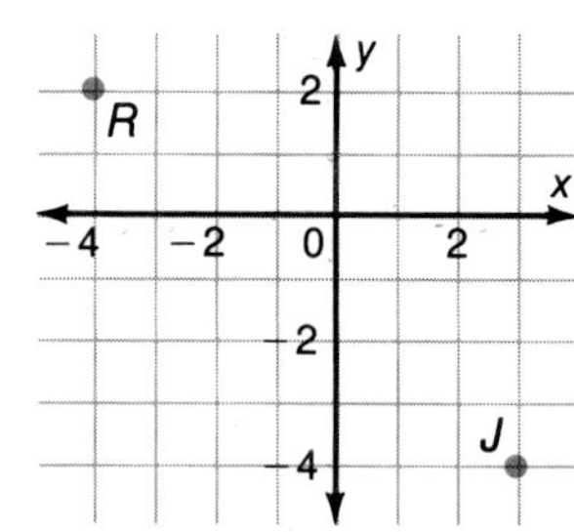

Graph each point on the same set of axes. (Example 1)

1. $A(4, 2)$ **2.** $B(3, 5)$ **3.** $C(-3, -5)$ **4.** $D(-2, -4)$
5. $E(4, -5)$ **6.** $F(6, -3)$ **7.** $G(-5, 4)$ **8.** $H(-2, 6)$
9. $I(6, 0)$ **10.** $J(-4, 0)$ **11.** $K(0, -5)$ **12.** $L(0, 6)$

Write the coordinates of each point shown below. (Example 2)

13. R **14.** S **15.** F
16. P **17.** K **18.** W
19. T **20.** Q **21.** B
22. V **23.** N **24.** A

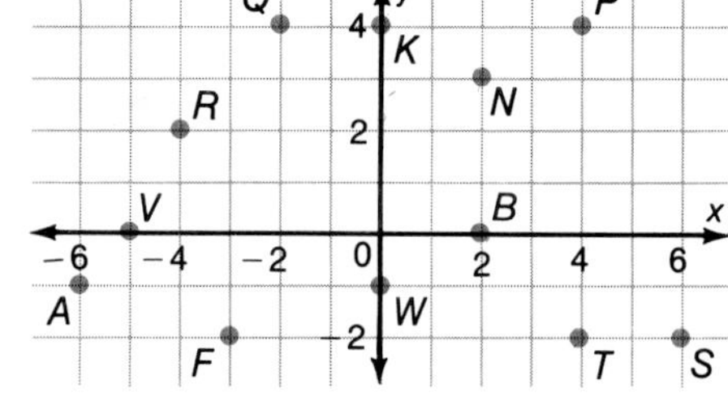

MIXED PRACTICE *For Exercises 25–28, graph each point. For Exercises 29–32, write the coordinates of the points shown on the graph.*

25. $A(-4, -1)$ **26.** $B(-2, -3)$
27. $C(-2, 3)$ **28.** $D(3, 3)$
29. E __?__ **30.** F __?__
31. G __?__ **32.** H __?__

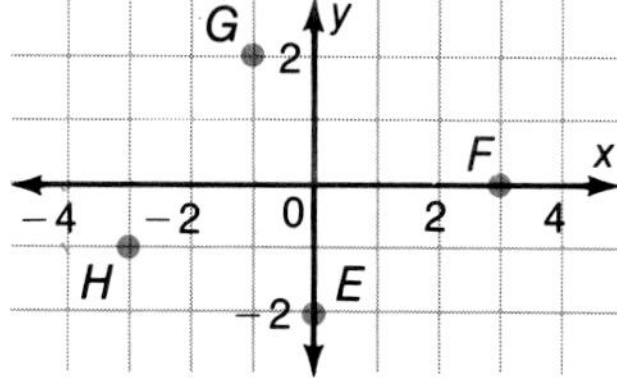

REVIEW CAPSULE FOR SECTION 14.2

Evaluate each expression. (Pages 5–7)

1. $y + 7$ when $y = 4$ **2.** $n - 3$ when $n = 15$ **3.** $z - \frac{1}{2}$ when $z = -7$
4. $4x - 1$ when $x = 3$ **5.** $\frac{1}{2}k + 2$ when $k = 3$ **6.** $12m - 3$ when $m = -1$
7. $-r + 9$ when $r = 2$ **8.** $-\frac{3}{4}g$ when $g = \frac{2}{3}$ **9.** $-5f - \frac{1}{7}$ when $f = \frac{4}{5}$

14.2 Relations and Graphs

The x axis and y axis are part of an infinite flat surface, or plane, that associates coordinates with points. The ***coordinate plane*** has four regions, or ***quadrants,*** which are separated by the x and y axes. Point C below is in Quadrant III.

In which quadrant does each point lie?

a. B **b.** D **c.** A

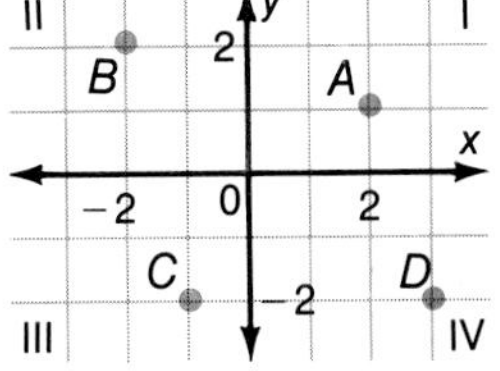

The coordinates of A, B, C, and D at the right form a set of ordered pairs of numbers:

$$\{(2, 1), (-2, 2), (-1, -2), (3, -2)\}$$

Such a set is called a relation.

Definition

A ***relation*** is a set of ordered pairs.

An ordered pair is an ***element*** of the relation.

EXAMPLE 1

Graph the following relation.

$$\{(-2, 2), (-3, -1), (2, 1\tfrac{1}{2}), (1\tfrac{1}{2}, -1), (2, 0)\}$$

Solution:

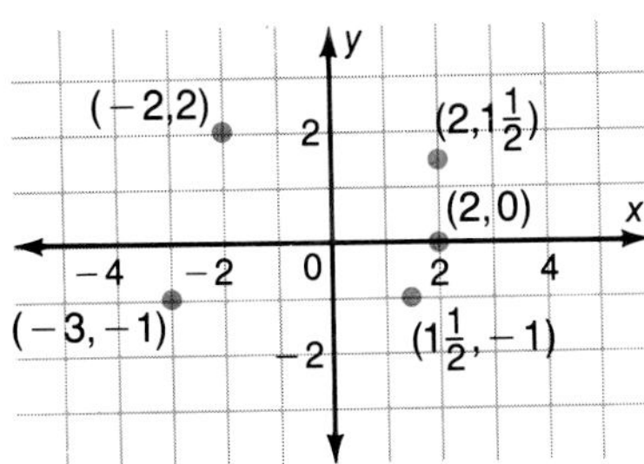

Example 1 shows two ways to describe a relation—by a list of its elements and by a graph. A relation can also be described by a word rule or by a table.

EXAMPLE 2

Use the word rule to complete the table. Then graph the relation.

Word Rule: "Multiply each x value by $\frac{1}{3}$ and add -2."

x	−9	−6	−3	0	3	6	9
y	?	?	?	?	?	?	?

Solution:

x	$\frac{1}{3}x + (-2)$	y
−9	$\frac{1}{3}(-9) + (-2) = -3 - 2$	or −5
−6	$\frac{1}{3}(-6) + (-2) = -2 - 2$	or −4
−3	$\frac{1}{3}(-3) + (-2) = -1 - 2$	or −3
0	$\frac{1}{3}(0) + (-2) = 0 - 2$	or −2
3	$\frac{1}{3}(3) + (-2) = 1 - 2$	or −1
6	$\frac{1}{3}(6) + (-2) = 2 - 2$	or 0
9	$\frac{1}{3}(9) + (-2) = 3 - 2$	or 1

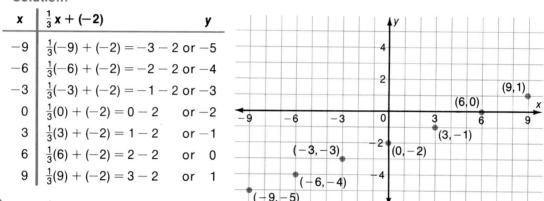

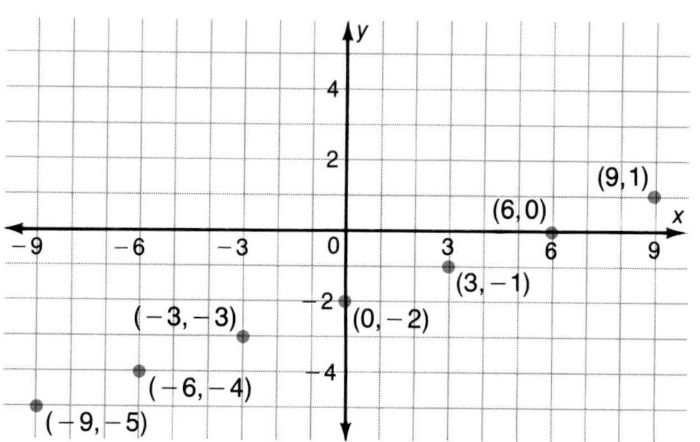

Sometimes you can discover a word rule for a relation. To do this, look for a relationship between the x value and the y value.

EXAMPLE 3

Write a word rule for the relation below.

x	−2	−1	0	1	2	3
y	0	1	2	3	4	5

To get each y value, add two to each x value.

Solution: Word Rule: "Each y value is two more than each x value."

P–2 **Which word rule below describes the table at the right?**

x	−2	−1	0	1	2
y	−4	−2	0	2	4

a. "Multiply each x value by one–half."

b. "Add two to each x value."

c. "Multiply each x value by two."

CLASSROOM EXERCISES

Graph each relation. (Example 1)

1. $\{(-2, 3), (1, 1), (0, 0), (3, -4)\}$

2. $\{(4, 4), (-2, 1), (2, -3), (0, 5)\}$

3. $\{(1, -3), (-2, 4), (0, 3), (-5, -3\frac{1}{2})\}$

4. $\{(5, -1), (-4, 3), (0, -2\frac{1}{2}), (3, -3\frac{1}{2})\}$

Use the word rule to complete the table. Then graph each relation. (Example 2)

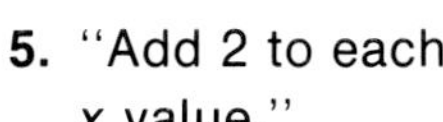

5. "Add 2 to each x value."

x	y
−5	?
−1	?
0	?
4	?

6. "Add −1 to each x value."

x	y
−3	?
−1	?
0	?
3	?

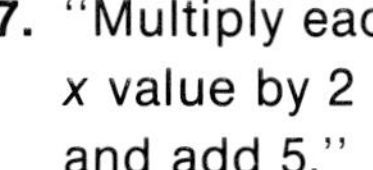

7. "Multiply each x value by 2 and add 5."

x	y
−3	?
−1	?
0	?
$1\frac{1}{2}$	?

8. "Take the opposite of each x value and subtract 3."

x	y
−5	?
−2	?
0	?
3	?

Write a word rule for each relation. (Example 3)

9.

x	−3	0	2	3
y	−6	0	4	6

10.

x	−2	0	1	4
y	4	0	1	16

11.

x	−1	1	$5\frac{3}{4}$	10
y	2	4	$8\frac{3}{4}$	13

12.

x	−10	0	$1\frac{1}{2}$	1.6
y	10	0	$-1\frac{1}{2}$	−1.6

WRITTEN EXERCISES

Goal: To describe a relation by a graph, a word rule or a table

Sample Problem: Describe the following relation by

a. a word rule,

b. a table, and

c. a graph: {(0, 1), (3, 7), (−2, −3)}

Answer: a. "Multiply each x value by 2 and add 1."

b.

x	−2	0	3
y	−3	1	7

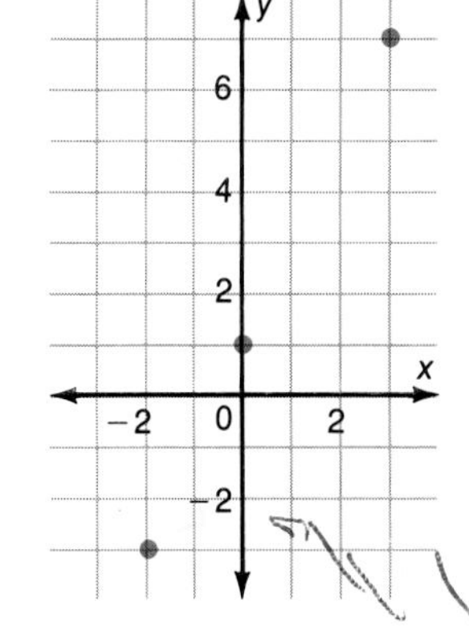

Graph each relation. (Example 1)

1. {(−3, 2), (4, −2), (0, 3), (2, 5)}
2. {(5, 2), (0, −4), (−2, 4), (3, 0)}
3. {(2, −2), (−4, −5), (0, −3), (4, −1)}
4. {(−2, 5), (0, 1), (3, −5), (1, −1)}
5. {(2, 3), (−4, −1), $(0, -3\frac{1}{2})$, $(4\frac{1}{2}, 2\frac{1}{2})$}
6. {(−4, 0), (1, 3), $(3\frac{1}{2}, -5)$, $(-1\frac{1}{2}, -2\frac{1}{2})$}

For each given word rule, complete the table at the right. Then graph the relation. (Example 2)

x	−4	0	$1\frac{1}{2}$	3
y	?	?	?	?

7. Multiply each x value by −2.
8. Multiply each x value by −3.
9. Multiply each x value by 2; add −2.
10. Multiply each x value by −2; add 1.
11. Multiply each x value by $\frac{1}{2}$; add $-1\frac{1}{2}$.
12. Multiply each x value by $-\frac{1}{2}$; add $2\frac{1}{2}$.

Write a word rule for each relation. (Example 3)

13.

x	−4	−3	0	1	3
y	−6	−5	−2	−1	1

14.

x	−3	−2	−1	0	3	4
y	0	1	2	3	6	7

15.

x	−5	−1.7	−1.5	0	0.75	3
y	5	1.7	1.5	0	−0.75	−3

16.

x	−4	−3.6	0	3.5	4.5	6
y	8	7.2	0	−7	−9	−12

MIXED PRACTICE *Solve each exercise.*

17. Graph this relation.

x	−2	$-\frac{1}{2}$	0	2	3
y	−5	−2	−1	3	5

18. Write a word rule for this relation.

x	−2	−1	0	1	2
y	−3	−1	1	3	5

19. Write a word rule for this relation.

x	−3	−1	0	2	4
y	−9	−3	0	6	12

20. Graph this relation.

x	−3	$-1\frac{1}{2}$	0	2	4
y	5	$-3\frac{1}{2}$	2	0	−2

21. Use the word rule to complete the table. Then graph the relation. Word Rule: "Each *y* value is 1 less than half each *x* value."

x	−4	−1	0	2	5
y	?	?	?	?	?

22. Use the word rule to complete the table. Then graph the relation. Word Rule: "Each *y* value is 2 less than the opposite of each *x* value."

x	4	2	−3	0	$-1\frac{1}{2}$
y	?	?	?	?	?

NON-ROUTINE PROBLEMS

23. Three friends ate lunch at a restaurant and split the total cost of the lunch equally. If they had paid separately, Mary would have paid \$0.10 less, Sue would have paid \$0.80 more, and Audrey would have paid \$2.20. What was the total cost of the lunch?

24. Trace the figure below. Connect all the points in the figure by drawing exactly four line segments without lifting your pencil off the paper.

• • •
• • •
• • •

REVIEW CAPSULE FOR SECTION 14.3

For each exercise, write Yes *or* No *to indicate whether two or more ordered pairs have the same x value.* (Pages 390–392)

1. {(1, 2), (3, 4)}

2. {(7, 0), (7, −1)}

3. {(3, 4), (4, 3)}

4. {(−2, 7), (5, 7), (9, 0)}

5. {(2, 1), (1, 2), (−2, 1)}

6. {(−3, 0), (−3, 1), (8, 7)}

14.3 Functions

In the relation at the left below, the pairs (2, 3) and (2, 4) have the same x value, 2.

$$\{(\mathbf{2},\ 3),\ (1,\ 5),\ (\mathbf{2},\ 4)\} \qquad \{(\mathbf{4},\ 1),\ (\mathbf{5},\ 6),\ (\mathbf{7},\ 2)\}$$

In the relation at the right above, the three x values, 4, 5, and 7, are different. This relation is an example of a function.

Definition

> A ***function*** is a relation in which no two ordered pairs have the same x value.

EXAMPLE 1 Write Yes or No to indicate whether each relation is a function.
If the answer is No, give a reason.

a. $\{(0, 1), (1, 0), (-1, 0), (0, -1)\}$
b. $\{(\frac{1}{2}, \frac{3}{4}), (0, -\frac{1}{2}), (-\frac{1}{2}, \frac{1}{2})\}$
c. $\{(5.1, 1.1), (3.4, 2.1), (-1.7, 4.3)\}$

Solutions: **a.** No, because the pairs (0, 1) and (0, −1) have the same x value.
b. Yes
c. Yes

P–1 **Write Yes or No to indicate whether each relation is a function. If the answer is No, give a reason.**

a. $\{(-4, 2), (-4, 3), (5, 6)\}$ **b.** $\{(-1, 2), (0, 3), (1, 4)\}$

Every function has a domain (set of x values) and a range (set of y values) as illustrated below.

Function: $\{(2, 1), (1, 0), (0, 1), (-1, 0)\}$
Domain: $\{2, 1, 0, -1\}$ **Range:** $\{1, 0\}$

Definitions

> The ***domain*** of a function is the set of x values of its ordered pairs. The ***range*** of a function is the set of y values of its ordered pairs.

EXAMPLE 2

Write the domain and range of the following function.

$$\{(1, 3), (5, \tfrac{2}{3}), (-7, 0)\}$$

Solutions: Domain: $\{1, 5, -7\}$ Range: $\{3, \frac{2}{3}, 0\}$

P–2 **Write the domain and range of each function.**

a. $\{(-1, 1), (3, 6), (2, 0)\}$ **b.** $\{(0, 0), (1, 0), (3.2, -\frac{1}{2})\}$

Sometimes a function can be described by an equation as well as a word rule.

Word Rule	Equation
"Subtract 3 from each x value."	$y = x - 3$
"Add 5 to twice each x value."	$y = 2x + 5$
"Take the absolute value of each x value and subtract 1."	$y = \|x\| - 1$

Given the domain (the x values) for a function and the rule or equation of the function, you can find the range (the y values).

EXAMPLE 3

Compute the range for the function described by $y = 3x + 1$. The domain is $\{-1, 0, 1, 2\}$.

Solution: In the equation, replace x by elements in the domain and compute.

x	$3x + 1$		y
-1	$3(-1) + 1 = -3 + 1$	or	-2
0	$3(0) + 1 = 0 + 1$	or	1
1	$3(1) + 1 = 3 + 1$	or	4
2	$3(2) + 1 = 6 + 1$	or	7

The range is $\{-2, 1, 4, 7\}$.

CLASSROOM EXERCISES

Write Yes or No to indicate whether each relation is a function. If the answer is No, give a reason. (Example 1)

1. $\{(3, 1), (5, -2)\}$

2. $\{(4, 0), (0, 2), (0, -1)\}$

3. $\{(\frac{1}{2}, 2), (\frac{1}{2}, -2), (\frac{1}{2}, 1)\}$

4. $\{(-3, 1), (-2, -1), (-1, 1)\}$

5. $\{(0.1, 2), (0.2, 3), (0.3, 4)\}$

6. $\{(1, 1), (-1, 1), (-1, -1), (1, -1)\}$

Write the domain and range of each function. (Example 2)

7. $\{(1, 4), (-3, 5), (-1, 4)\}$

8. $\{(3.4, 5), (1.3, 2), (-2, 2.5)\}$

9. $\{(\frac{1}{2}, 3), (-\frac{1}{4}, 2), (7, \frac{3}{4})\}$

10. $\{(2, -\frac{1}{2}), (3, \frac{1}{2}), (0, 5)\}$

11.

x	0	−1	3	5
y	5	6	−2	−1

12.

x	5.2	0.5	−1.3	0.8
y	−1.2	1.7	4.2	−1.2

Compute the range of each function. The domain is $\{-1, 0, 1, 2\}$. (Example 3)

13. $y = 5x$

14. $y = x + 3$

15. $y = -2x$

16. $y = -x + 1$

17. $y = 3x - 1$

18. $y = -x - 2$

19. $y = x$

20. $y = \frac{1}{2}x + \frac{1}{2}$

WRITTEN EXERCISES

Goal: To compute the range of a function by using its equation and the given domain

Sample Problem: Compute the range for the function described by $y = -x + 1$. The domain is $\{-3, -1, 0, 2, 5\frac{1}{2}\}$

Answer: See the row of y values in the table at the right.

x	−3	−1	0	2	$5\frac{1}{2}$
y	4	2	1	−1	$-4\frac{1}{2}$

Write Yes or No to indicate whether each relation is a function. If the answer is No, give a reason. (Example 1)

1. $\{(5, 9), (23, 3), (8, 2)\}$

2. $\{(3, 16), (0, 10), (5, 25)\}$

3. $\{(1, 0), (0, -1), (-1, 0), (0, 1)\}$

4. $\{(2, -2), (-2, 2), (1, -1), (-1, 1)\}$

5. $\{(\frac{1}{2}, 1), (-\frac{1}{2}, \frac{1}{2}), (0.5, 2)\}$

6. $\{(\frac{1}{4}, 0), (\frac{2}{8}, -1), (-1, 0)\}$

Write the domain and range of each function. (Example 2)

7. $\{(5, 0), (3, -2), (8, 3)\}$

8. $\{(-2, 1), (4, -2), (5, -6)\}$

9. $\{(-1, 8), (5, 2), (0, 2), (2, -3)\}$

10. $\{(7, 0), (-5, 2), (-1, 3), (0, 2)\}$

11.

x	1	−2	−3	−4
y	2	3	4	0

12.

x	−1	1	2	0
y	3	−2	−6	5

Compute the range for each described function and given domain. (Example 3)

13. $y = 4x + 1$ Domain: $\{-3, -1, 0, 2\}$

14. $y = 2x + 5$ Domain: $\{-2, -1, 0, 7\}$

15. $y = -2x + 3$ Domain: $\{-2, 0, 2, 3\}$

16. $y = -x + 5$ Domain: $\{-3, -1, 0, 3\}$

17. $y = -x - 2$ Domain: $\{-5, 0, \frac{1}{2}, 2\}$

18. $y = -3x - 2$ Domain: $\{-1, -\frac{1}{2}, 0, \frac{1}{3}\}$

MID-CHAPTER REVIEW

Graph the points in Exercises 1–8 in the same coordinate plane. (Section 14.1)

1. $R(-3, 4)$ **2.** $S(4, -2)$ **3.** $T(0, -4)$ **4.** $W(-2, -3)$

5. $X(-5, 0)$ **6.** $Y(5, 2)$ **7.** $O(0, 0)$ **8.** $N(-5, 5)$

Write the coordinates of each point shown below. (Section 14.1)

9. B **10.** F **11.** R

12. K **13.** W **14.** T

15. M **16.** V **17.** P

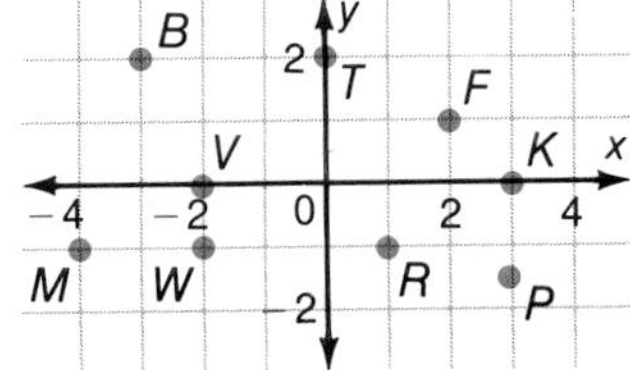

Match the correct table, list of elements, or graph with each word rule. (Section 14.2)

18. "Twice each x value increased by 1 equals its corresponding y value."

19. "The opposite of each x value decreased by 2 equals its corresponding y value."

20. "Three more than one-fourth of each x value equals its corresponding y value."

A.

x	0	−2	2	−4
y	−2	0	−4	2

B.

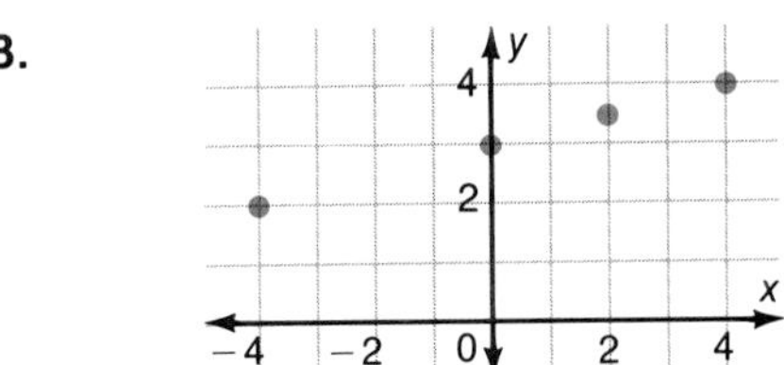

C. $\{(2, 5), (-2, -3), (0, 1), (-3, -5)\}$

Compute the y values for each described function and given domain. (Section 14.3)

21. $y = -x + 5$ Domain: $\{-5, 0, 3, -2\}$

22. $y = 3x - 1$ Domain: $\{1, 2, -10, -1\}$

23. $y = x - 3$ Domain: $\{-2, 5, 0, -\frac{1}{2}\}$

24. $y = -x + 2$ Domain: $\{-4, 4, 3, -5\}$

25. $y = \frac{1}{3}x - 1$ Domain: $\{-6, 3, -3, 0\}$

26. $y = \frac{1}{2}x + 3$ Domain: $\{-4, -2, 2, 0\}$

REVIEW CAPSULE FOR SECTION 14.4

Graph each relation. (Pages 393–396)

1. $\{(-3, 5), (-1, 5), (0, 5), (4, 5)\}$

2. $\{(6, 2), (1, -1), (3, 0), (4, 1)\}$

3. $\{(0, 1), (4, 6), (2, 3), (-1, -1)\}$

4. $\{(2, -1), (2, 0), (2, 2), (2, 5)\}$

5. $\{(0, 1), (0, 6), (0, -4), (0, -3)\}$

6. $\{(4, -4), (-\frac{1}{2}, \frac{1}{2}), (1, -1), (2, -2)\}$

14.4 Graphs of Functions

A certain function is described in three ways below—by an equation, by a table, and by a graph.

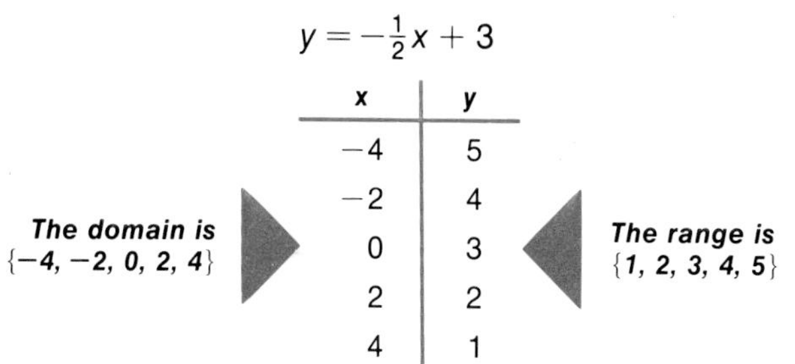

$$y = -\tfrac{1}{2}x + 3$$

x	y
−4	5
−2	4
0	3
2	2
4	1

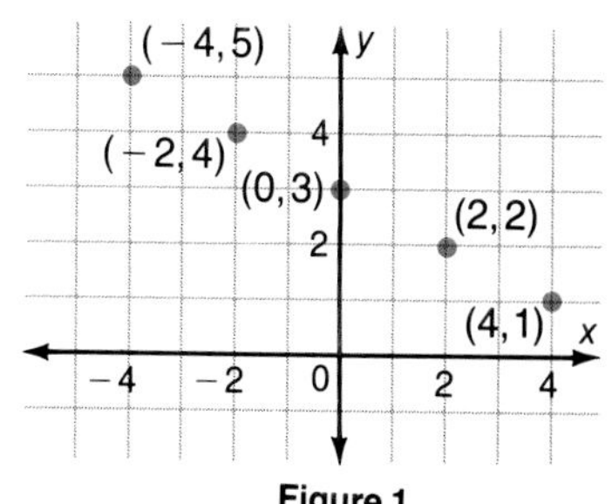

Figure 1

The points graphed in Figure 1 are represented in Figure 2, and a line is drawn containing the points.

The five dots of Figure 1 are the graph of a function. The entire line of Figure 2 is also the graph of a function. The domain for the graph in Figure 2 is the set of real numbers.

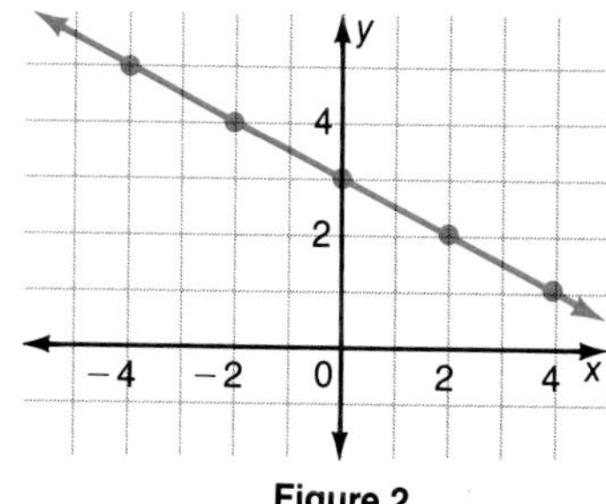

Figure 2

EXAMPLE 1 Complete a table. Then graph the function described by $y = -x + 2$. The domain is the set of real numbers.

Solution:

1 Select some numbers from the domain. ⟶ $\{-3, -1, 0, 3\}$

2 Compute the y values.

x	−3	−1	0	3
$-x + 2$	$3 + 2$	$1 + 2$	$0 + 2$	$-3 + 2$
or	or	or	or	or
y	5	3	2	−1

3 Draw the graph.

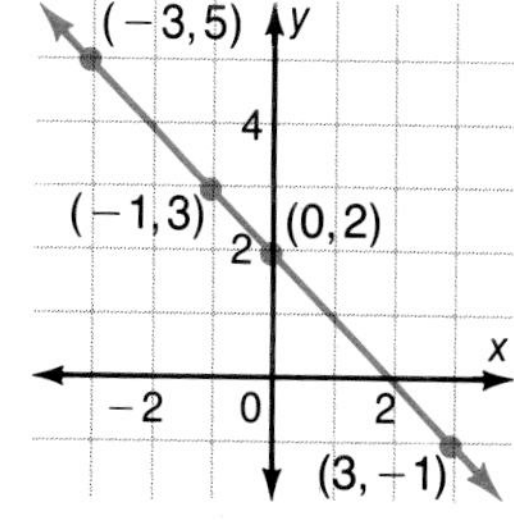

P–1 **Complete each table.**

a.

x	x + 1
−2	−2 + 1 = −1
2	?
4	?

b.

x	2x
−3	?
4	2(4) = 8
7	?

c.

x	2x + 1
−1	?
3	?
5	2(5) + 1 = 11

Since there are an infinite number of points on a line, there are an infinite number of ordered pairs in the function of Example 1.

The functions graphed on page 401 are called linear functions.

Definition

A ***linear function*** is a function in which the points of its graph lie in a straight line. The rule of a linear function is of the form $y = mx + b$ in which m and b are real numbers.

When you graph a linear function, the domain is always the set of real numbers unless indicated otherwise.

EXAMPLE 2 Graph the function defined by $y = \frac{1}{2}x - 4$.

Solution: Select a few values for x.

x	$y = \frac{1}{2}x - 4$
−4	$\frac{1}{2}(-4) - 4 = -6$
−2	$\frac{1}{2}(-2) - 4 = -5$
0	$\frac{1}{2}(0) - 4 = -4$
2	$\frac{1}{2}(2) - 4 = -3$
4	$\frac{1}{2}(4) - 4 = -2$

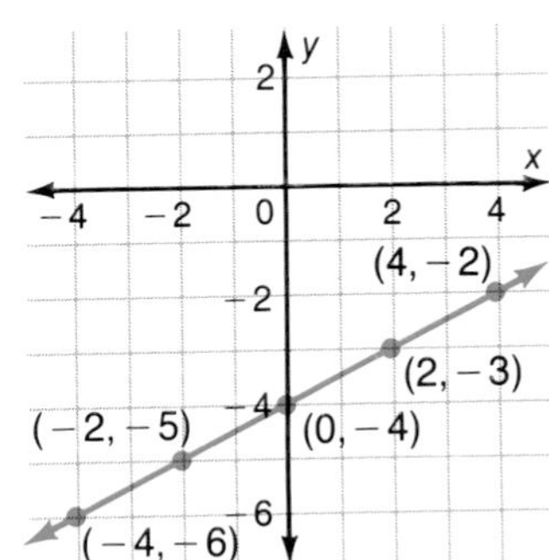

CLASSROOM EXERCISES

Find the y value for each given x value. (Step 2 of Examples 1 and 2)

1. $y = x + 4$
   $x = -6$
2. $y = 2x - 5$
   $x = 2$
3. $y = -4x + 1$
   $x = \frac{1}{2}$
4. $y = 2 - x$
   $x = -3$

5. $y = x + 3$
$x = -4$

6. $y = -2x + 25$
$x = 0$

7. $y = x + \frac{1}{2}$
$x = \frac{1}{2}$

8. $y = \frac{3}{4}x + \frac{1}{4}$
$x = 1$

9. $y = x + 2$

x	y
−2	?
−1	?
0	?
1	?
2	?

10. $y = x - 3$

x	y
−2	?
−1	?
0	?
1	?
2	?

11. $y = -x + 1$

x	y
−2	?
−1	?
0	?
1	?
2	?

12. $y = -2x + 3$

x	y
−2	?
−1	?
0	?
1	?
2	?

13. $y = 3x + 1$

x	−2	−1	0	1	2
y	?	?	?	?	?

14. $y = -x - 2$

x	−2	−1	0	1	2
y	?	?	?	?	?

WRITTEN EXERCISES

Goal: To graph a linear function

Sample Problem: Graph $y = -\frac{1}{2}x + 1$

Answer:

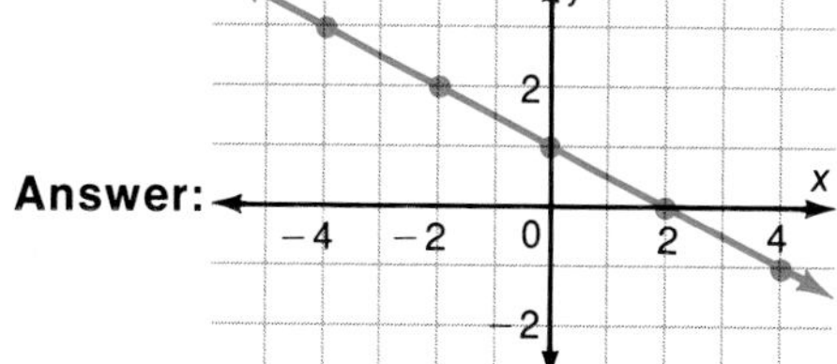

Complete each table. Then graph the given function. (Example 1)

1. $y = -x - 3$

x	y
−4	?
−2	?
0	?
2	?

2. $y = -x + 1$

x	y
−3	?
−1	?
2	?
4	?

3. $y = -2x + 1$

x	y
−2	?
0	?
2	?
3	?

4. $y = -\frac{1}{2}x + 3$

x	y
−4	?
−2	?
0	?
2	?

5. $y = x + 2$

x	y
−4	?
−2	?
0	?
2	?

6. $y = x - 2$

x	y
−3	?
−1	?
2	?
4	?

7. $y = 2x - 3$

x	y
−1	?
0	?
2	?
4	?

8. $y = \frac{1}{2}x - 1$

x	y
−4	?
−2	?
0	?
2	?

Graph each function. (Example 2)

9. $y = x - 5$

10. $y = -x + 3$

11. $y = 2x + 4$

12. $y = 4x - 2$

Graph each function. (Example 2)

13. $y = -x + 4$ **14.** $y = -x - 1$ **15.** $y = 2x - 1$ **16.** $y = 3x + 1$

17. $y = -\frac{1}{2}x + 2$ **18.** $y = -\frac{3}{2}x + 2$ **19.** $y = \frac{3}{2}x - 1$ **20.** $y = \frac{1}{2}x - 3$

MORE CHALLENGING EXERCISES

Complete each table. Then graph the given function.

21. $y = 0x + 1$

x	y
−4	?
−2	?
0	?
2	?

22. $y = 0x - 3$

x	y
−3	?
−1	?
0	?
1	?

23. $y = 2$

x	y
−2	?
−1	?
0	?
1	?

24. $y = 3$

x	y
−1	?
0	?
1	?
2	?

APPLICATIONS

You can use a calculator with a sign change key to compute y values for a function, given the equation and the domain of the function.

EXAMPLE Compute the range for the function described by $y = \frac{3}{4}x - 2$. The domain is $\{-7, \frac{1}{2}, 4, 5\}$.

SOLUTION Substitute each number in the domain for x in the equation.

3 ÷  4 × 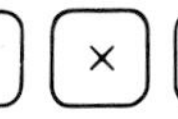 7  +/− −  2 =  -7.25

Follow the procedure above for $x = \frac{1}{2}$, $x = 4$, and $x = 5$.
The range is $\{-7\frac{1}{4}, -1\frac{5}{8}, 1, 1\frac{3}{4}\}$.

Compute the y values for each described function and given domain.

25. $y = -4x + \frac{1}{2}$ Domain: $\{\frac{1}{4}, 1, 3, 7\frac{1}{2}\}$ **26.** $y = -x + 8$ Domain: $\{-9, -\frac{4}{5}, 0, 3\}$

27. $y = 5x - 10$ Domain: $\{-6, -4, -\frac{2}{5}, 3\}$ **28.** $y = -\frac{5}{2}x + 3$ Domain: $\{-7, -\frac{1}{2}, \frac{1}{5}, 2\}$

REVIEW CAPSULE FOR SECTION 14.5

Find the value of y in each equation if x equals zero. (Pages 24–26)

1. $y = 3x - 4$ **2.** $y = -x + 5$ **3.** $y = -\frac{1}{2}x + 1$ **4.** $y = 4x$

Find the value of x in each equation if y equals zero. (Pages 24–26)

5. $y = 2x + 2$ **6.** $y = 3x - 3$ **7.** $y = -x + 1$ **8.** $y = \frac{1}{3}x - 2$

14.5 The Intercepts

Point A in Figure 1 is on the y axis. Therefore, its x value is 0. Its y value, -3, is called the y intercept of the line.

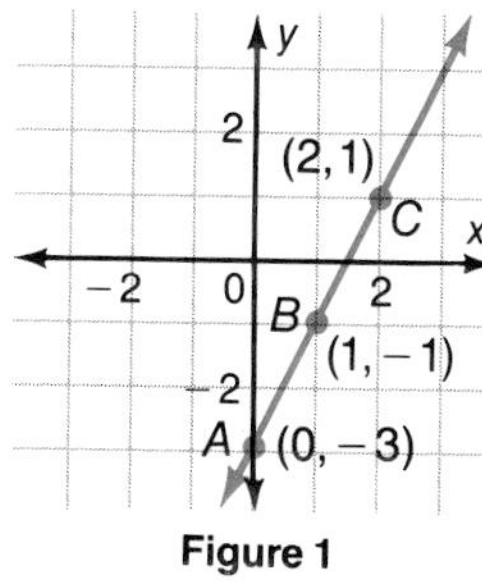

Figure 1

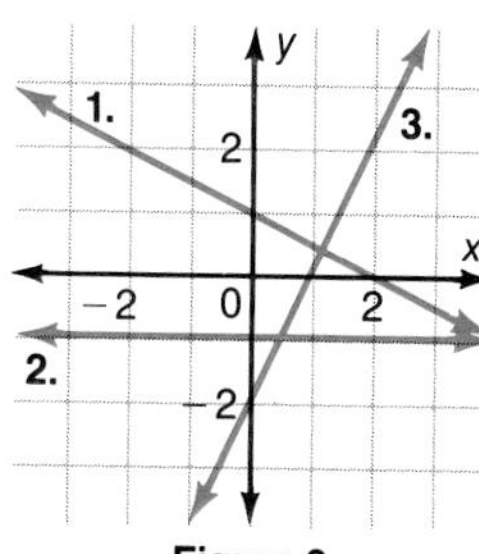

Figure 2

Definition The ***y intercept*** of a line is the y value of its point on the y axis.

P–1 **What is the *y* intercept for each line?**

a. Line **1** of Figure 2 **b.** Line **2** of Figure 2

Any linear function can be described by $y = mx + b$. This equation is called the ***intercept form*** because b is the y intercept.

You can find the y intercept of an equation written in the intercept form by replacing x with 0.

$$y = 2x - 3$$
$$y = 2 \cdot 0 - 3$$
$$y = 0 - 3 \quad \text{or} \quad \mathbf{y = -3}$$

Note that the y intercept is -3 and also that b is -3 in the equation $y = 2x - 3$. Therefore, you can read the y intercept directly from an equation that is written in the intercept form.

EXAMPLE 1 Find the y intercept.

a. $y = \frac{1}{2}x + 4$ **b.** $y = -x + \frac{13}{4}$ **c.** $y = 5x - 10$

Solutions: Since each equation is in the intercept form, read the value of b. This is the y intercept.

a. 4 **b.** $\frac{13}{4}$ **c.** -10

Point A at the right is <u>on</u> the x axis. Therefore, its y value is 0. Its x value, -3, is the <u>x intercept</u> of the line.

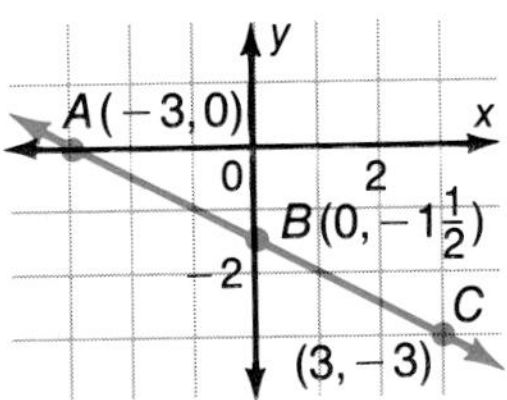

Definition The ***x intercept*** of a line is the x value of its point on the x axis.

Example 2 shows how to find the x intercept.

EXAMPLE 2 Find the x intercept of $y = -2x + 4$.

Solution:

1. Replace y with 0. ⟶ $0 = -2x + 4$
2. Subtract 4 from each side. ⟶ $0 - \mathbf{4} = -2x + 4 - \mathbf{4}$

   $-4 = -2x$
3. Divide each side by -2. ⟶ $\frac{-4}{\mathbf{-2}} = \frac{-2x}{\mathbf{-2}}$

   $2 = x$

The x intercept is 2.

P–2 **Find the x intercept of each line.**

a. $y = x - 2$ **b.** $y = 2x - 6$ **c.** $y = \frac{1}{2}x - 3$

EXAMPLE 3 Graph $y = x + 3$ by using the x and y intercepts.

Solution:

1. Find the y intercept. ⟶ The y intercept is 3.

Since the equation is in the intercept form, find the value of b.

2. Find the x intercept. ⟶ $y = x + 3$

   $0 = x + 3$

   $-3 = x$

To find the x intercept, replace y with 0.

Therefore, the y intercept is 3, and the x intercept is -3. Graph and connect the two points with coordinates (0, 3) and (-3, 0).

CLASSROOM EXERCISES

Find the y intercept. (Example 1)

1. $y = x - 2$
2. $y = 2x + 3$
3. $y = -x + 10$
4. $y = -\frac{1}{2}x - 6$
5. $y = 4x - \frac{1}{2}$
6. $y = 2 + x$
7. $y = 0.5x + 0.7$
8. $y = -5x - 1.8$

Find the x intercept. (Example 2)

9. $y = x - 1$
10. $y = x + 1$
11. $y = x - 5$
12. $y = x + 4$
13. $y = 2x - 4$
14. $y = 3x + 6$
15. $y = 2x - 1$
16. $y = 2x + 1$

WRITTEN EXERCISES

Goal: To graph a linear function using the x and y intercepts
Sample Problem: Graph $y = x + 3$. **Answer:** See Example 3.

Find the y intercept. (Example 1)

1. $y = x + 5$
2. $y = x - 3$
3. $y = -3x - 4$
4. $y = 2x + 3$
5. $y = \frac{5}{2}x - 1$
6. $y = \frac{7}{4}x - 2$
7. $y = \frac{3}{2}x - \frac{1}{2}$
8. $y = \frac{1}{4}x - \frac{3}{4}$
9. $y = 1.3x - 2.1$
10. $y = -1.9x + 1.2$
11. $y = 0.5x + 6.9$
12. $y = -4.7x - 2.5$

Find the x intercept. (Example 2)

13. $y = x - 3$
14. $y = x - 2$
15. $y = x + 10$
16. $y = x + 15$
17. $y = -x - 2$
18. $y = -x - 4$
19. $y = 3x - 1$
20. $y = 4x - 3$
21. $y = 2x + 5$
22. $y = 3x - 7$
23. $y = 6x + 14$
24. $y = 8x - 33$

Graph each function by using the x and y intercepts. (Example 3)

25. $y = x - 3$
26. $y = -x + 2$
27. $y = x + 5$
28. $y = x - 4$
29. $y = 2x - 4$
30. $y = 2x + 2$
31. $y = 3x - 6$
32. $y = -3x + 9$
33. $y = -\frac{3}{2}x + 3$
34. $y = \frac{3}{4}x - 3$
35. $y = \frac{1}{2}x - 2$
36. $y = \frac{3}{5}x + 3$

REVIEW CAPSULE FOR SECTION 14.6

Write the value of m and b in each linear function. (Pages 401–404)

1. $y = 3x + 7$
2. $y = -5x - 2$
3. $y = x - 3$
4. $y = -x + 6$
5. $y = \frac{1}{2}x + 3$
6. $y = -\frac{1}{3}x - \frac{4}{3}$
7. $y = 13x + 13$
8. $y = -2x + \frac{7}{5}$

14.6 Intersection of Two Lines

When you graph two different linear functions on the same set of axes, the resulting lines may or may not intersect.

Definition The ***intersection*** of two lines is the set of points common to both lines.

EXAMPLE 1 Graph the following lines. Write their intersection.

a. $y = -x + 1$ **b.** $y = x + 3$

Solutions: **a.**

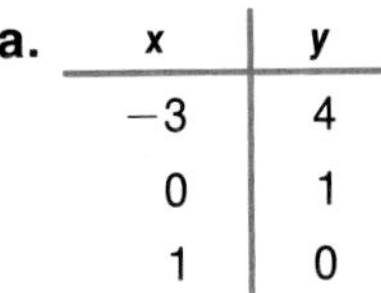

x	y
−3	4
0	1
1	0

b.

x	y
1	4
0	3
−3	0

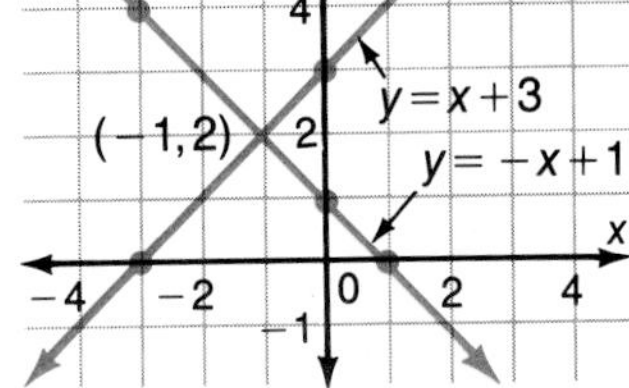

The point of intersection is (−1, 2).

If the graphs of two different linear functions do not intersect, the lines are parallel to each other. Note in Example 2 that the set of points common to both lines is the ***empty set.***

EXAMPLE 2 Graph the following lines. Write their intersection.

a. $y = x + 2$ **b.** $y = x - 1$

Solutions: **a.**

x	y
0	2
−2	0
−4	−2

b.

x	y
3	2
1	0
−1	−2

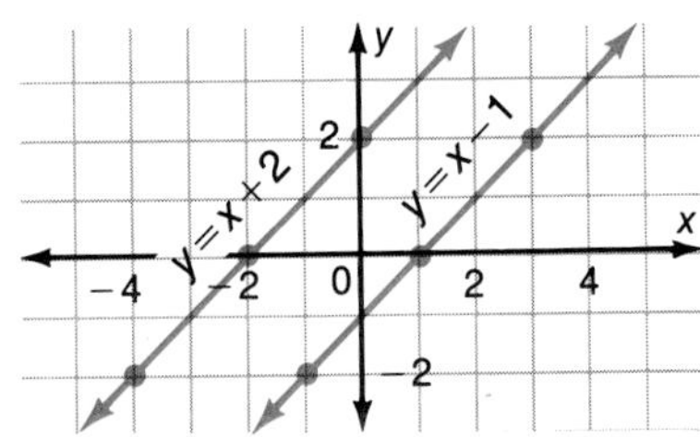

The intersection of parallel lines is the empty set, ϕ.

You can tell when the graphs of two linear functions are parallel. Examine the functions in their $y = mx + b$ form. If the values of m are equal but the values of b are unequal, the lines are parallel.

CLASSROOM EXERCISES

Write the intersection of the two lines. (Examples 1 and 2)

1. 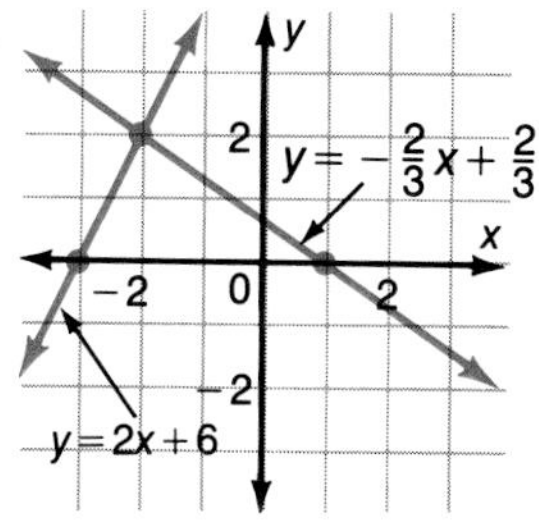

2. 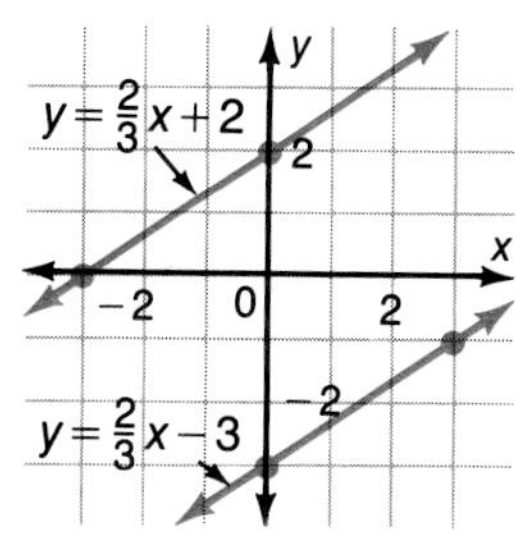

3. 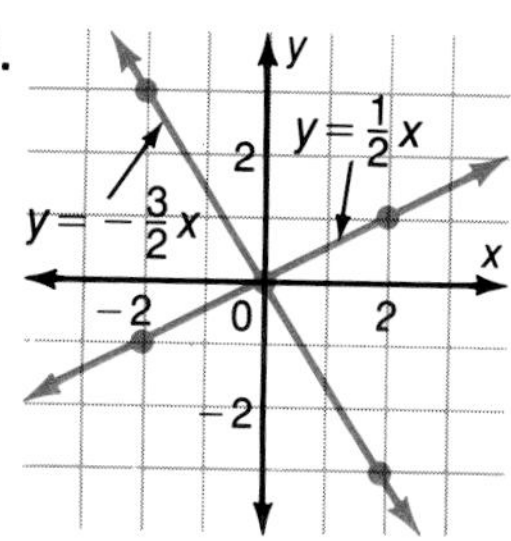

Write Yes or No to indicate whether each given set is the intersection of the two given lines. (Examples 1 and 2)

4. $\{(-1, 3)\}$: $y = x + 4$; $y = -x + 2$

5. $\{(2, 3)\}$: $y = 2x - 1$; $y = x + 4$

6. $\{(-1, 3)\}$: $y = x + 4$; $y = -2x + 1$

7. ϕ: $y = -x + 3$; $y = -x - 2$

8. $\{(0, -1)\}$: $y = \frac{1}{2}x - 1$; $y = -\frac{1}{2}x + 1$

9. $\{(\frac{1}{2}, -2)\}$: $y = 2x - 3$; $y = -4x + 4$

WRITTEN EXERCISES

Goal: To write the intersection of two lines
Sample Problem: Graph $y = -x + 1$ and $y = x + 3$. Write their intersection.
Answer: See Example 1.

Complete each table. Then graph the two lines and write the intersection. (Example 1)

1. $y = x - 2$

x	y
−2	?
0	?
4	?

$y = -x + 1$

x	y
−3	?
0	?
4	?

2. $y = -x + 3$

x	y
−2	?
0	?
4	?

$y = x + 1$

x	y
−4	?
−2	?
3	?

3. $y = 2x - 1$

x	y
−2	?
0	?
2	?

$y = x + 1$

x	y
−4	?
−2	?
3	?

4. $y = x - 1$

x	y
−2	?
0	?
2	?

$y = -2x + 2$

x	y
−3	?
−1	?
2	?

Complete each table. Then graph the two lines and write their intersection. (Example 2)

5. $y = 2x + 2$

x	y
−1	?
1	?
3	?

$y = 2x - 3$

x	y
−3	?
−1	?
3	?

6. $y = -x + 3$

x	y
−2	?
0	?
4	?

$y = -x - 2$

x	y
−4	?
−2	?
2	?

7. $y = -3x - 1$

x	y
−2	?
0	?
2	?

$y = -3x + 1$

x	y
−1	?
1	?
3	?

8. $y = \frac{1}{2}x - 2$

x	y
−4	?
−2	?
0	?

$y = \frac{1}{2}x + 1$

x	y
−2	?
0	?
4	?

MIXED PRACTICE *Graph the two lines and write their intersection.*

9. $y = 2x + 1$
$y = -2x + 5$

10. $y = 3x - 3$
$y = -x + 1$

11. $y = \frac{1}{2}x + 3$
$y = \frac{1}{2}x - 1$

12. $y = -\frac{1}{2}x - 1$
$y = x - 4$

MORE CHALLENGING EXERCISES

Examine the two linear functions and indicate whether the lines intersect or are parallel.

EXAMPLE: $x + y = 2$; $x + y = 6$

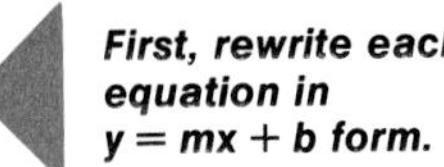

First, rewrite each
equation in
y = mx + b form.

SOLUTION: $x + y = 2$ $\quad$ $x + y = 6$

$y = -x + 2$ $\quad$ $y = -x + 6$

Since the values of m are equal and the values of b are unequal, the lines are parallel.

13. $\frac{1}{3}x - y = 2$
$2x - y = 2$

14. $x + y = 2$
$x - y = -\frac{7}{2}$

15. $3x + y = 1$
$\frac{1}{2}x + y = 0$

16. $\frac{1}{2}x + y = 2$
$\frac{1}{2}x + y = -3$

NON-ROUTINE PROBLEM

17. Eight square sheets of paper, all the same size, are placed on top of each other. They overlap as shown in the drawing at the right. The top sheet, labeled 1, is shown completely. Number the other sheets in order from top to bottom.

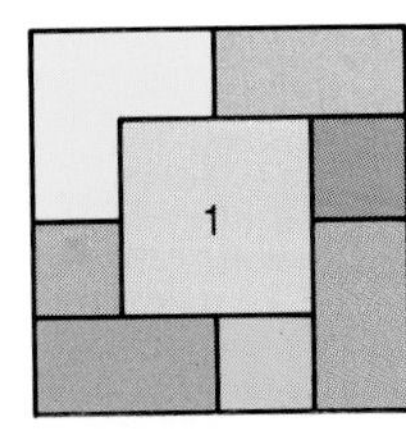

PROBLEM SOLVING AND APPLICATIONS

14.7 Using Graphs for Predicting

Examples of functions can be found in everyday activities such as driving an automobile. One such function is the reaction-distance function. ***Reaction distance*** is the distance through which an automobile travels while the driver's foot is moving from the accelerator to the brake pedal.

The reaction distance depends upon the speed of the automobile just before it starts to slow down.

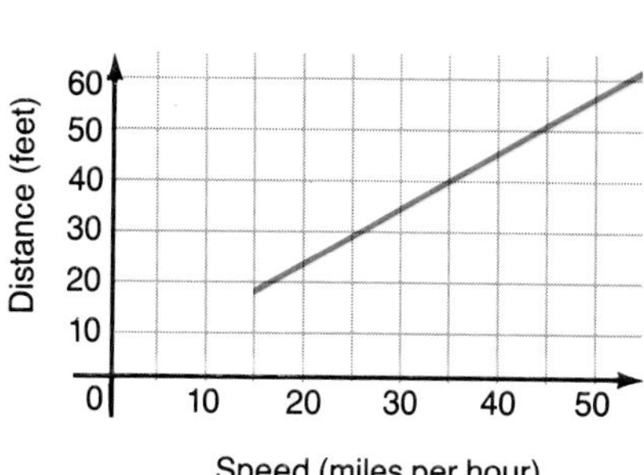

The graph at the right describes the reaction–distance function. The domain is the set of speeds between 15 miles per hour (mi/hr) and 55 mi/hr, inclusive. The range is the set of distances from 17 feet to 61 feet.

EXAMPLE 1

Use the graph above to find the approximate reaction distance, in feet, for each given speed.

a. 30 mi/hr **b.** 55 mi/hr

Solutions: **a.** 33 ft **b.** 61 feet

Example 2 shows a function whose graph is a curve, not a straight line.

EXAMPLE 2

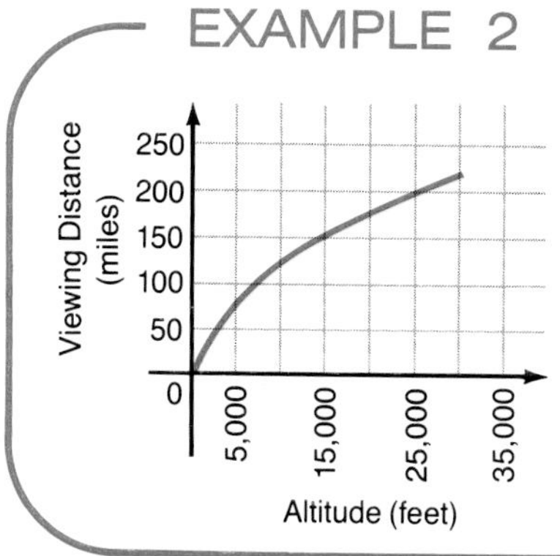

This graph describes the maximum distance in miles that the pilot of a plane can see to the horizon for a given altitude in feet. The domain of the function is the set of altitudes from 0 to 30,000 feet. The range is the set of distances from 0 to 210 miles. Find the approximate viewing distance, in miles, for each of the following altitudes.

a. 10,000 ft **b.** 20,000 feet

Solutions: **a.** 125 miles **b.** 175 miles

By examining the change in values at regular intervals on a graph, predictions can be made for points not shown on the graph.

EXAMPLE 3

This graph describes the linear relation between the width, w, of an image on a screen and the distance, d, the projector is from the screen.

a. Use the graph to find w for $d = 5$, 10, 15, and 20.

b. By how much does w increase when d increases by 5 feet?

c. Use your answer to **b** to predict the width of the image if the projector is 25 feet from the screen.

Solutions:

a.

d	5	10	15	20
w	1	2	3	4

b. When d increases 5 feet, w increases 1 foot.

c. Thus, when the projector is 25 feet away, the image will be about 5 feet wide.

CLASSROOM EXERCISES

Refer to the graph in Example 1 to find the approximate reaction distance for an automobile traveling at the given speed. (Example 1)

1. 50 mi/hr **2.** 35 mi/hr **3.** 20 mi/hr **4.** 40 mi/hr

Refer to the graph in Example 2 to find the approximate viewing distance at the given altitude. (Example 2)

5. 15,000 feet **6.** 25,000 feet **7.** 5,000 feet **8.** 30,000 feet

Use the graph in Example 3 to predict the width of the image for each of the following distances. (Example 3)

9. 30 feet **10.** 45 feet **11.** 40 feet **12.** 35 feet

WRITTEN EXERCISES

Goal: To use the graph of a function to solve word problems

Sample Problem: From the graph of reaction distances in Example 1, find the approximate reaction distance for an automobile that is traveling at 42 miles per hour.

Answer: 46 feet

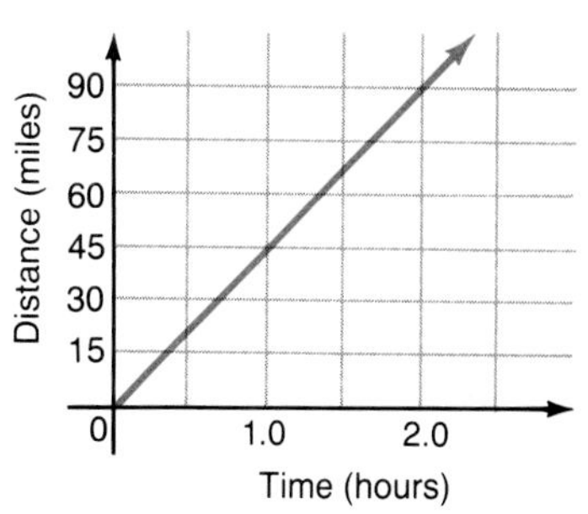

The graph at the right describes the time–distance function for an automobile traveling at 45 miles per hour. The domain is the set of time measurements in hours.

Use the graph to find the distance traveled by an automobile traveling at 45 miles per hour in the time indicated. (Example 1)

1. 2 hours **2.** 1 hour **3.** 1.25 hours

4. 0.75 hours **5.** $\frac{1}{2}$ hour **6.** $1\frac{1}{2}$ hours

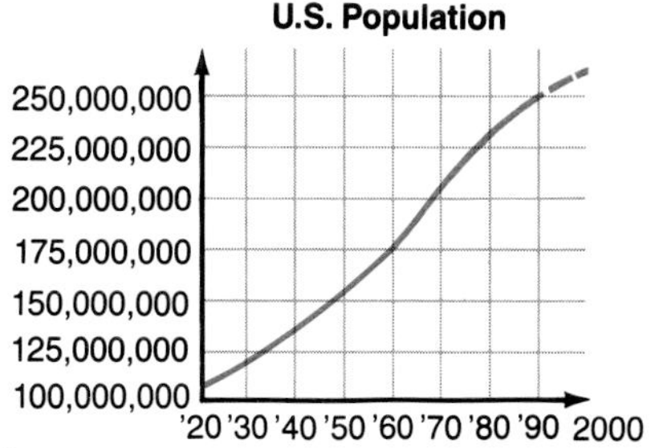

The graph at the right describes the United States population function, based on Census Bureau figures. The domain is the time in years from 1920 to 2000. Note that the function from 1990 to 2000 is a prediction.

Find each of the following. (Examples 2 and 3)

7. The approximate 1930 U.S. population

8. The approximate 1970 U.S. population

9. The approximate 1940 U.S. population

10. The approximate 1980 U.S. population

11. The predicted U.S. population in 2000

12. The predicted U.S. population in 1990

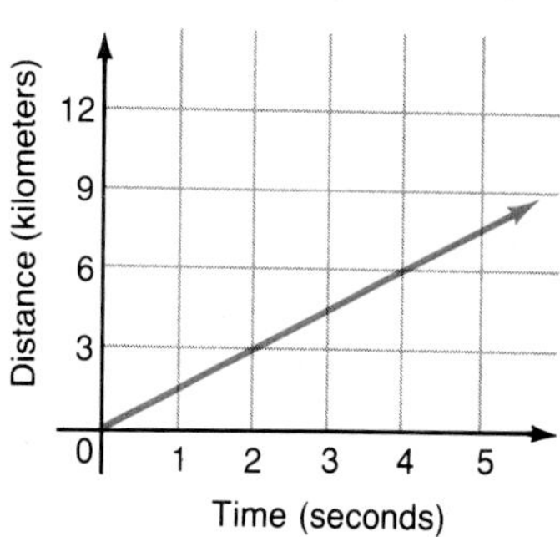

The graph at the right describes the rate at which sound travels in water. The domain is the set of time measurements in seconds.

Use the graph to predict the approximate distance that sound will travel in water for each of the following times. (Example 3)

13. 6 seconds **14.** 7 seconds

15. 10 seconds **16.** 12 seconds

Using Tables and Graphs

Wind Chill

In cold weather, a combination of low temperature and strong wind can make a person feel colder than the actual temperature. **Wind chill** is an estimate of how cold the wind makes a person feel under these conditions.

Wind chill does not measure cold exactly, but gives a better estimate of how cold it feels than a thermometer reading alone.

The National Weather Service publishes a table for estimating wind chill. Part of that table is shown below.

EQUIVALENT WIND CHILL TEMPERATURES

Wind Speed in mi/hr	*Thermometer Reading in °F*						
	30	*20*	*10*	*0*	*−10*	*−20*	*−30*
5	27	16	7	−6	−15	−26	−35
10	16	2	−9	−22	−31	−45	−58
20	3	−9	−24	−40	−52	−68	−81
30	−2	−18	−33	−49	−64	−79	−93
40	−5	−21	−37	−53	−69	−84	−100

EXAMPLE: Find the wind chill temperature when the thermometer reading is 10°F and the wind speed is 20 mi/hr.

SOLUTION: Find 20 in the "wind speed" column.
Look along that row to the number in the "10" column.
The wind chill temperature is **−24°F.**

The graph at the top of the next page shows the relation of wind chill temperatures to thermometer readings for various wind speeds.

EXAMPLE: Use the graph to estimate the wind chill temperature when the wind speed is 10 miles per hour and the thermometer reading is 15°F.

SOLUTION:

1. Locate the point for 15° on the horizontal (x) axis.

2. Draw a vertical line to locate a point on the graph for 10 miles per hour.

3. Write the y value of the point. This is the wind chill temperature.

   The wind chill temperature is about **−3°F.**

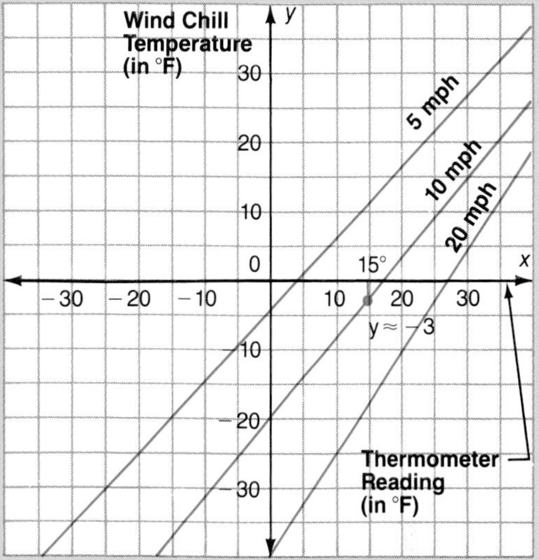

EXERCISES

For Exercises 1–4, refer to the table on page 414.

1. Wind speed: 20 mi/hr
   Thermometer reading: 0°F
   Wind chill temperature: __?__

2. Wind speed: 10 mi/hr
   Thermometer reading: −20°F
   Wind chill temperature: __?__

3. Which conditions will produce the colder wind chill temperature?
   a. Wind speed: 10 mi/hr
      Thermometer reading: 0°F
   b. Wind speed: 20 mi/hr
      Thermometer reading: 10°F

4. A TV weather forecaster reports a wind chill temperature of −31°F. The thermometer reading is −10°F. Find the approximate wind speed.

For Exercises 5–7, refer to the graph above.

5. Estimate the wind chill temperature when the wind speed is 5 miles per hour and the thermometer reading is −15°F.

6. Estimate the wind chill temperature when the wind speed is 20 miles per hour and the thermometer reading is 5°F.

7. Draw a graph similar to the one above. Have the graph show the relation of wind chill temperature to thermometer readings for wind speeds of 30 miles per hour and 40 miles per hour. Refer to the table on page 414 for the data you need to draw the graph.

CHAPTER SUMMARY

IMPORTANT TERMS

Axes *(p. 390)*
x axis *(p. 390)*
y axis *(p. 390)*
Coordinates *(p. 390)*
Graph *(p. 390)*
Origin *(p. 390)*
Ordered pair *(p. 390)*
Coordinate plane *(p. 393)*
Quadrants *(p. 393)*
Relation *(p. 393)*
Element *(p. 393)*
Function *(p. 397)*
Domain *(p. 397)*
Range *(p. 397)*
Linear function *(p. 402)*
y intercept *(p. 405)*
Intercept form *(p. 405)*
x intercept *(p. 406)*
Intersection *(p. 408)*
Empty set *(p. 408)*
Reaction distance *(p. 411)*

IMPORTANT IDEAS

1. A point in the coordinate plane is represented by an ordered pair of numbers.
2. Elements of a set can be ordered pairs of numbers.
3. Relations and functions can be described
   a. by a graph in a coordinate plane,
   b. by a list of elements,
   c. by a word rule,
   d. by a table of ordered pairs, or
   e. by an equation.
4. The intercept form of the rule for any linear function is $y = mx + b$.
5. The y intercept has a corresponding x value of 0. The x intercept has a corresponding y value of 0.
6. The y intercept of a linear function is the value of b in the rule $y = mx + b$.
7. The intersection of two functions whose graphs are parallel lines is the empty set.
8. If two rules of the form $y = mx + b$ have m values that are equal but b values that are unequal, the graphs of the two functions are parallel.

CHAPTER REVIEW

SECTION 14.1

Write the coordinates of each point shown at the right.

1. *A* **2.** *B* **3.** *C* **4.** *D* **5.** *E* **6.** *F*

Graph each point.

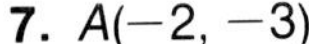

7. $A(-2, -3)$ **8.** $B(2, -2)$ **9.** $C(2, 4)$ **10.** $D(-3, 1)$
11. $E(0, 2)$ **12.** $F(3, 0)$ **13.** $G(0, -4)$ **14.** $H(-1, 3)$

SECTION 14.2

Graph each relation.

15. $\{(2, 3), (-4, 2), (0, -3), (-2, -2)\}$

16. $\{(-1, -3), (2, 4), (3, -3), (-4, 0)\}$

17. $\{(2, 1), (-3, 2\frac{1}{2}), (-3\frac{1}{2}, -4), (-\frac{1}{2}, 0)\}$

18. $\{(-3, 2), (0, -1\frac{1}{2}), (3\frac{1}{2}, -2), (\frac{1}{2}, 3)\}$

For each given word rule, complete the table at the right. Then graph the relation.

x	-4	0	$1\frac{1}{2}$	3
y	?	?	?	?

19. Subtract 5 from each x value.

20. Multiply each x value by -3; add 1.

21. Add -3 to each x value.

22. Divide each x value by 2; subtract 3.

Write a word rule for each relation.

23.

x	-3	-1	0	2	4	7
y	-8	-6	-5	-3	-1	2

24.

x	-4	0	2	3	6	7
y	-2	0	1	$\frac{3}{2}$	3	$\frac{7}{2}$

SECTION 14.3

Write Yes or No to indicate whether each relation is a function. If the answer is No, give a reason.

25. $\{(5, -1), (0, 2), (3, 2)\}$

26. $\{(-1, 2), (0, 5), (\frac{1}{2}, 2)\}$

27. $\{(3, 1), (4, 2), (3, -1), (2, 4)\}$

28. $\{(-1, 1), (2, 1), (3, 1), (7, 1)\}$

Write the domain and range of each function.

29. $\{(7, -2), (4, 0), (-6, 1)\}$

30. $\{(3, -1), (0, 0), (5, 2)\}$

31. $\{(-\frac{1}{2}, 1), (3, -4), (0, 1), (2, -1)\}$

32. $\{(4, 2), (0, -3), (5, -3), (7, 3)\}$

Compute the range for each described function and given domain.

33. $y = x + 3$ Domain: $\{-4, -1, 0, 1\}$

34. $y = x - 1$ Domain: $\{-2, 1, 3, 5\}$

35. $y = 3x - 1$ Domain: $\{-3, -1, \frac{2}{3}, \frac{1}{4}\}$

36. $y = -4x + 2$ Domain: $\{-1, 0, \frac{3}{8}, \frac{1}{2}\}$

SECTION 14.4

Graph each function.

37. $y = -x - 2$

38. $y = -2x + 1$

39. $y = -\frac{1}{2}x - 2$

40. $y = \frac{1}{2}x + 3$

SECTION 14.5

Find the y intercept.

41. $y = 5x - 6$ **42.** $y = -\frac{3}{2}x + \frac{1}{4}$ **43.** $y = 15 - 2x$ **44.** $y = 3.8 - 1.5x$

Find the x intercept.

45. $y = x - 12$ **46.** $y = -x + 7$ **47.** $y = -x - 6$ **48.** $y = 3x + 2$

SECTION 14.6

Graph the two lines and write their intersection.

49. $y = 2x - 1 \quad y = x - 3$ **50.** $y = -x + 1 \quad y = -3x - 3$

SECTION 14.7

The graph at the right describes the speed of a train. The domain is the set of time measurements in hours. For Exercises 54–56, assume the speed of the train is the same as for the first two hours.

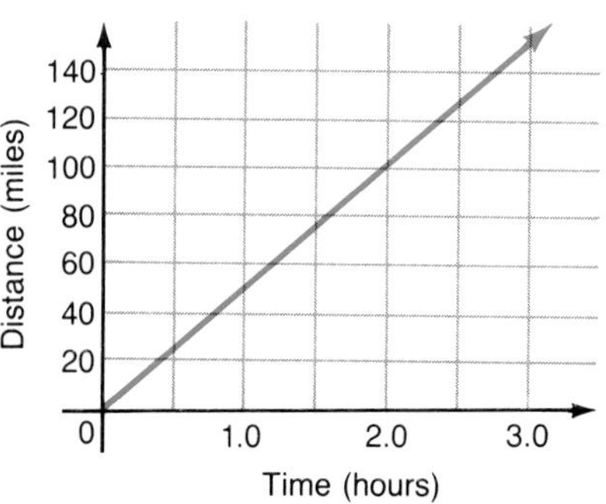

Use the graph to determine each of the following.

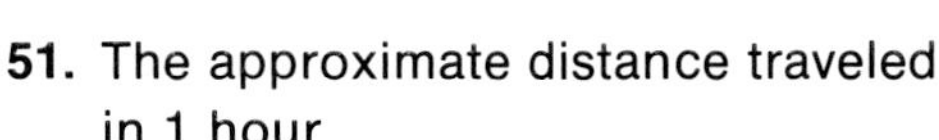

51. The approximate distance traveled in 1 hour

52. The approximate distance traveled in 2 hours

53. The approximate distance traveled in 3 hours

54. A prediction for the distance traveled in 4 hours

55. A prediction for the distance traveled in 6 hours

56. A prediction for the distance traveled in 10 hours

Equations with Two Variables

CHAPTER 15

As part of the Gemini space program, the manned space capsule docked with an unmanned vehicle while both orbited the earth. To find the meeting point, you must solve a system of equations.

15.1 Graphical Method

The equation $x + y = 2$ is an example of a ***linear equation*** in two variables.

Definition The ***standard form*** of a linear equation is $Ax + By = C$ in which A, B, and C represent integers. (A and B are not both 0.)

P–1 **What are the values of A, B, and C in $x + y = 2$?**

A pair of linear equations is a system of equations.

Definition Two equations in the same two variables form a ***system of equations.*** $\begin{cases} x - 2y = -3 \\ 3x + 4y = 1 \end{cases}$

To solve a system of equations means to find the coordinates of the point of intersection of the graphs of the two equations.

EXAMPLE Graph this system of equations. Write the intersection of the lines. $\begin{cases} -x + y = 2 \\ x + 3y = -6 \end{cases}$

Solution: Write each equation in intercept form. Then graph each equation.

$$-x + y = 2$$
$$-x + y + \mathbf{x} = 2 + \mathbf{x}$$
$$y = x + 2$$

x	−4	0	1
y	−2	2	3

$$x + 3y = -6$$
$$x + 3y - \mathbf{x} = -6 - \mathbf{x}$$
$$3y = -x - 6$$
$$y = -\tfrac{1}{3}x - 2$$

x	−3	0	3
y	−1	−2	−3

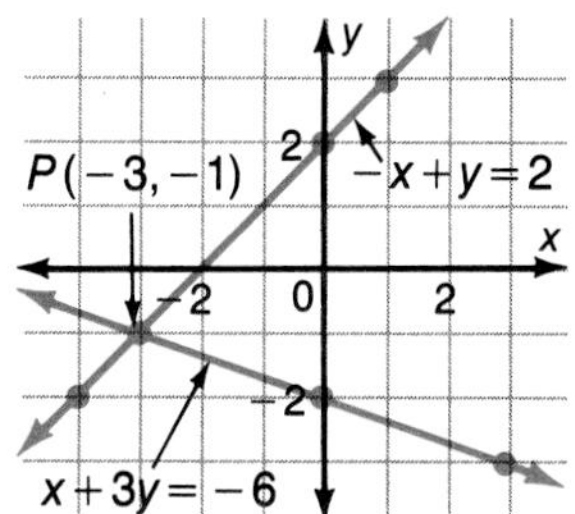

Check: Substitute −3 for x and −1 for y in both equations.

$$-\mathbf{x} + \mathbf{y} = 2 \qquad \mathbf{x} + 3\mathbf{y} = -6$$
$$-(\mathbf{-3}) + (\mathbf{-1}) \qquad \mathbf{-3} + 3(\mathbf{-1})$$
$$3 - 1 \qquad -3 - 3$$
$$2 \qquad -6$$

The intersection of the two lines is the point $P(-3, -1)$.

The coordinates of the intersection in the Example form the solution set of the system. The solution set is written as $\{(-3, -1)\}$.

Definition The ***solution set of a system of equations*** is the set of ordered pairs that makes both equations true.

CLASSROOM EXERCISES

Write the intercept form of each equation.

1. $-x + y = -1$
2. $x - y = -3$
3. $-2x + y = 5$
4. $3x + 4y = 1$
5. $x + y = 6$
6. $4x + 2y = -6$
7. $4x + y = 3$
8. $5y + 7 = 0$

Substitute to determine whether the given set is the solution set of the system of equations. Answer Yes or No. (Example)

9. $\{(1, 1)\}$; $\begin{cases} x + y = 2 \\ 2x - y = 1 \end{cases}$
10. $\{(5, -7)\}$; $\begin{cases} x + 2y = 3 \\ x + y = -2 \end{cases}$
11. $\{(-4, 3)\}$; $\begin{cases} 3x - y = -15 \\ x + 2y = 2 \end{cases}$
12. $\{(-2, 4)\}$; $\begin{cases} x - 3y = -14 \\ -x + 2y = 10 \end{cases}$
13. $\{(2, 5)\}$; $\begin{cases} -2x + y = 1 \\ 4x - y = 13 \end{cases}$
14. $\{(-1, -3)\}$; $\begin{cases} 2x - y = -5 \\ x + 2y = -7 \end{cases}$

WRITTEN EXERCISES

Goal: To solve a system of equations by the graphical method

Sample Problem: Graph $\begin{cases} \textbf{1.}\ x - y = -2 \\ \textbf{2.}\ x + 3y = -6 \end{cases}$ and write the solution set.

Answer: See the example.

Graph the system of equations in the same coordinate plane. Write the solution set of the system. (Example)

1. $\begin{cases} x + y = 2 \\ 2x - y = 1 \end{cases}$
2. $\begin{cases} 2x - y = -1 \\ x - 2y = 1 \end{cases}$
3. $\begin{cases} x - 2y = -2 \\ x - y = -3 \end{cases}$
4. $\begin{cases} x + y = 2 \\ x - 3y = 6 \end{cases}$
5. $\begin{cases} x - 2y = -2 \\ x + y = -2 \end{cases}$
6. $\begin{cases} x + 4y = -4 \\ 2x - y = 1 \end{cases}$
7. $\begin{cases} 2x - y = 0 \\ x + 3y = 0 \end{cases}$
8. $\begin{cases} x + y = 0 \\ 2x - 3y = 0 \end{cases}$
9. $\begin{cases} 2x + 3y = 3 \\ x - 2y = -2 \end{cases}$
10. $\begin{cases} x - 3y = -1 \\ x - y = 1 \end{cases}$
11. $\begin{cases} 2x - y = -2 \\ x - 2y = 2 \end{cases}$
12. $\begin{cases} x + 3y = 4 \\ x - 2y = -6 \end{cases}$

15.2 Addition Method

The solution set of the system below can be read from the graph at the right.

$$\begin{cases} x = 2 \\ y = 3 \end{cases} \quad \text{Solution set: } \{(2, 3)\}$$

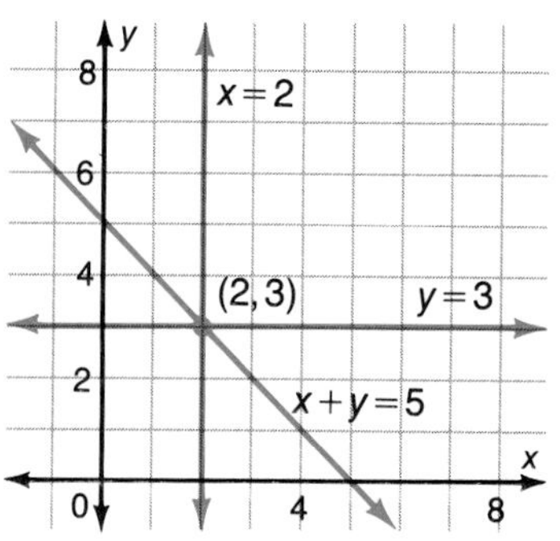

When the corresponding sides of the two equations of a system are added, a sum equation is formed. For the system $\begin{cases} x = 2 \\ y = 3 \end{cases}$, the sum equation is $x + y = 5$. The graph shows that the solution set of the equation $x + y = 5$ also includes $\{(2, 3)\}$.

To solve a system of equations algebraically, you can add the corresponding sides of the equations. This method is called the addition method. As shown in Example 1, the addition method is useful when the coefficients of either the x terms or of the y terms are opposites.

EXAMPLE 1 Solve: $\begin{cases} x + y = 5 \\ x - y = -1 \end{cases}$

Solution:

[1] Add the corresponding sides of both equations to eliminate y. →

$$\begin{array}{r} x + y = 5 \\ x - y = -1 \\ \hline 2x \quad = 4 \\ x = 2 \end{array}$$

Solve for x.

[2] Substitute 2 for x in one of the equations. →

$$\begin{aligned} \mathbf{x} + y &= 5 \\ \mathbf{2} + y &= 5 \\ y &= 3 \end{aligned}$$

Equation 1 is used.

[3] **Check:** Substitute $x = 2$ and $y = 3$ in the original equations.

$$\begin{array}{ll} \mathbf{x + y} = 5 & \quad \mathbf{x - y} = -1 \\ \mathbf{2 + 3} & \quad \mathbf{2 - 3} \\ 5 & \quad -1 \end{array}$$

The solution set is $\{(2, 3)\}$.

P–1 **Add the corresponding sides of both equations.**

a. $x + y = -3$
$x - y = 9$

b. $x + y = -4$
$-x + y = -8$

c. $x + y = 7$
$-x + y = 5$

EXAMPLE 2

Solve: $\begin{cases} 2x + y = -5 \\ -2x + 3y = 1 \end{cases}$

Solution:

$$\begin{aligned} 2x + y &= -5 \\ -2x + 3y &= 1 \\ \hline 0 + 4y &= -4 \\ y &= -1 \end{aligned}$$

Add the corresponding sides of the equations. This eliminates the x terms.

$$\begin{aligned} 2x + \mathbf{y} &= -5 \\ 2x + (\mathbf{-1}) &= -5 \\ 2x &= -4 \\ x &= -2 \end{aligned}$$

Substitute −1 for y in one of the equations.

Check: Substitute $y = -1$ and $x = -2$ in $2x + y = -5$ and in $-2x + 3y = 1$.

The solution set is $\{(-2, -1)\}$.

P–2 **Add the corresponding sides of both equations.**

a. $\begin{aligned} 2x + y &= 0 \\ -2x + y &= 4 \end{aligned}$ **b.** $\begin{aligned} 3x + y &= -1 \\ -3x + 2y &= 7 \end{aligned}$ **c.** $\begin{aligned} x + 2y &= 0 \\ x - 2y &= 0 \end{aligned}$

Steps for the Addition Method

1. Add to eliminate one of the variables. Solve the resulting equation.
2. Substitute the known value of one variable in one of the original equations of the system. Solve for the other variable.
3. Check the solution in both equations of the system.

CLASSROOM EXERCISES

Write the equation that results if you add corresponding sides of the two equations in each system. (P-1 and P-2)

1. $\begin{cases} x + 2y = 3 \\ -x + y = -2 \end{cases}$ **2.** $\begin{cases} 2x - 3y = -2 \\ x + 3y = 4 \end{cases}$ **3.** $\begin{cases} 4x - 5y = -2 \\ x + 5y = 8 \end{cases}$ **4.** $\begin{cases} x + 4y = 7 \\ 2x - 4y = -3 \end{cases}$

5. $\begin{cases} -2x + 3y = 7 \\ 2x - 5y = -3 \end{cases}$ **6.** $\begin{cases} 10x - 3y = 0 \\ -7x + 3y = -5 \end{cases}$ **7.** $\begin{cases} -3x - y = -8 \\ 3x - 2y = 7 \end{cases}$ **8.** $\begin{cases} 9x - 2y = 4 \\ -9x + 8y = -1 \end{cases}$

Add the corresponding sides of the equations. Then solve for x or y. (Step 1 of Examples 1 and 2)

9. $\begin{aligned} 5x - y &= 2 \\ 3x + y &= 14 \end{aligned}$ **10.** $\begin{aligned} x + y &= 2 \\ x - y &= -6 \end{aligned}$ **11.** $\begin{aligned} 2x + y &= 5 \\ -2x + y &= -7 \end{aligned}$ **12.** $\begin{aligned} -2x + y &= 1 \\ 2x + y &= 3 \end{aligned}$

WRITTEN EXERCISES

Goal: To solve a system of equations by the addition method

Sample Problem: Solve: $\begin{cases} 5x - 2y = 8 \\ 4x + 2y = 10 \end{cases}$ **Answer:** $\{(2, 1)\}$

For Exercises 1–21, solve each system by the addition method. Check. (Examples 1 and 2)

1. $\begin{cases} x + y = 1 \\ x - y = -3 \end{cases}$
2. $\begin{cases} -2x + y = 12 \\ 2x + y = -4 \end{cases}$
3. $\begin{cases} 3x - 2y = -3 \\ -3x + y = -3 \end{cases}$
4. $\begin{cases} x - 5y = 2 \\ 2x + 5y = 4 \end{cases}$
5. $\begin{cases} 2x - y = -1 \\ -2x + 2y = 2 \end{cases}$
6. $\begin{cases} 3x + 2y = 5 \\ -3x - y = -4 \end{cases}$
7. $\begin{cases} x - y = 4 \\ x + y = 6 \end{cases}$
8. $\begin{cases} x - y = 8 \\ x + y = 0 \end{cases}$
9. $\begin{cases} x + 5y = 10 \\ -x + 4y = 8 \end{cases}$
10. $\begin{cases} 7x + 3y = 7 \\ -7x + 2y = -7 \end{cases}$
11. $\begin{cases} 2x - 3y = 8 \\ 2x + 3y = 4 \end{cases}$
12. $\begin{cases} 7x + 3y = 19 \\ 4x - 3y = 3 \end{cases}$
13. $\begin{cases} -3x + 7y = -1 \\ 2x - 7y = 3 \end{cases}$
14. $\begin{cases} -4x - y = 3 \\ 2x + y = 3 \end{cases}$
15. $\begin{cases} x - 3y = 1 \\ -x + y = -5 \end{cases}$
16. $\begin{cases} 3x - 2y = 2 \\ x + 2y = 2 \end{cases}$
17. $\begin{cases} x - 4y = 5 \\ -2x + 4y = -2 \end{cases}$
18. $\begin{cases} 2x - 3y = 0 \\ -x + 3y = 0 \end{cases}$
19. $\begin{cases} x - 3y = -19 \\ -x + 5y = 31 \end{cases}$
20. $\begin{cases} 5x - 4y = -12 \\ x + 4y = -12 \end{cases}$
21. $\begin{cases} 6x - 4y = 18 \\ -6x + 2y = -21 \end{cases}$

NON-ROUTINE PROBLEMS

22. A train is one mile long and is traveling at a rate of one mile per minute. How long will it take the train to pass through a tunnel one mile long?

23. Alberto has a 3–liter jar and a 5–liter jar. Using these two jars only, how can he measure exactly 4 liters of water?

24. A square handkerchief is folded in half horizontally as shown below. The folded rectangle has a perimeter of 42 inches. What is the area of the square handkerchief?

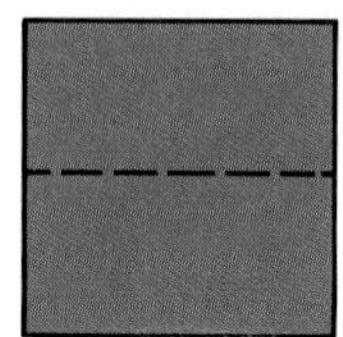

25. A man is shipwrecked on an island and a crate of apples is washed ashore with him. He calculates that if he eats one apple per day, the apples will last 14 days longer than if he eats two apples per day for two days and no apples on the third day. How many apples are in the crate?

REVIEW CAPSULE FOR SECTION 15.3

Write the numerical coefficient for each expression. (Pages 40–42)

1. $7x^2$ **2.** $\frac{1}{4}y^3$ **3.** $14x^2y$ **4.** xyz

Multiply. (Pages 285–288)

5. $4(x - 2y)$ **6.** $2(3x - 4y)$ **7.** $-3(-2x + y)$

8. $-6(-x + 3y)$ **9.** $-3(2x - y - 4)$ **10.** $-2(-x + 3y - 1)$

15.3 Multiplication/Addition Method

When you add the corresponding sides of the equations of a system, you may get an equation that has two variables rather than only one. This occurs because the ***coefficients*** of the variables are not opposites.

Note in the following examples that the ***multiplication/addition method*** is used to solve such a system of equations.

EXAMPLE 1 Solve: $\begin{cases} \textbf{1.}\ x - 5y = 1 \\ \textbf{2.}\ 2x - 4y = 14 \end{cases}$

Solution:

[1] Multiply each side of equation **1** by −2. ⟶ $-2x + 10y = -2$

[2] Add equation **2.** Solve. ⟶ $\underline{2x - 4y = 14}$

$$6y = 12$$

$$y = 2$$

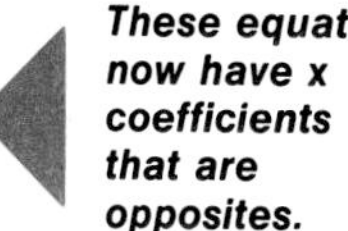
These equations now have x coefficients that are opposites.

[3] Substitute 2 for y in either equation **1** or **2.** Solve for x. ⟶ $x - 5\mathbf{y} = 1$

$$x - 5(\mathbf{2}) = 1$$

$$x - 10 = 1$$

$$x = 11$$

[4] **Check:** Substitute $x = 11$ and $y = 2$ in $x - 5y = 1$ and in $2x - 4y = 14$.

The solution set is $\{(11, 2)\}$.

P–1 **Multiply each side of equation 1 by a number that will form coefficients of x which are opposites.**

a. $\begin{cases} \textbf{1.}\ x - 2y = 4 \\ \textbf{2.}\ x + 3y = 14 \end{cases}$ **b.** $\begin{cases} \textbf{1.}\ -x + 2y = 5 \\ \textbf{2.}\ 3x + 5y = 29 \end{cases}$ **c.** $\begin{cases} \textbf{1.}\ 5x - 4y = 19 \\ \textbf{2.}\ x + 5y = 31 \end{cases}$

EXAMPLE 2

Solve: $\begin{cases} \textbf{1.}\ 3x - 2y = -1 \\ \textbf{2.}\ -2x + 3y = 4 \end{cases}$

Solution:

[1] Multiply each side of **1** by 3. ⟶ $9x - 6y = -3$

Multiply each side of **2** by 2. ⟶ $\underline{-4x + 6y = 8}$

[2] Add and solve for x. ⟶ $5x \qquad = 5$

$x = 1$

These equations now have y coefficients that are opposites.

[3] Substitute 1 for x in either equation **1** or **2**. Solve for y. ⟶

$$3\mathbf{x} - 2y = -1$$
$$3(\mathbf{1}) - 2y = -1$$
$$3 - 2y = -1$$
$$-2y = -4$$
$$y = 2$$

[4] **Check:** Substitute $x = 1$ and $y = 2$ in $3x - 2y = -1$ and in $-2x + 3y = 4$.

The solution set is $\{(1, 2)\}$.

Steps for the Multiplication/Addition Method

1. Check for opposite coefficients of one variable. If necessary, multiply each side of either or both equations by numbers that will make opposite coefficients for one variable.
2. Add to eliminate one of the variables. Solve the resulting equation.
3. Substitute the known value of one variable in one of the original equations of the system. Solve for the other variable.
4. Check the solution in both equations of the system.

CLASSROOM EXERCISES

By what number should you multiply each side of each equation in order to eliminate x by addition? (Step 1 of Examples 1 and 2)

1. $\begin{cases} x + y = -2 \\ -2x + 3y = -1 \end{cases}$

2. $\begin{cases} 3x - y = -5 \\ x + 2y = 1 \end{cases}$

3. $\begin{cases} -3x + 2y = 5 \\ x - y = -3 \end{cases}$

4. $\begin{cases} 2x - y = -2 \\ -3x + y = 1 \end{cases}$

5. $\begin{cases} 2x - y = -3 \\ 5x + 2y = 1 \end{cases}$

6. $\begin{cases} 4x - y = 2 \\ -3x + 2y = 1 \end{cases}$

For each of Exercises 7–10, write A, B, C, or D to show which of the systems below will give the described result. (Steps 1 and 2 of Examples 1 and 2)

A. $\begin{cases} \textbf{1. } 3x - 2y = -14 \\ \textbf{2. } x + y = 2 \end{cases}$

B. $\begin{cases} \textbf{1. } x - y = -3 \\ \textbf{2. } -3x - 2y = -6 \end{cases}$

C. $\begin{cases} \textbf{1. } 2x + 4y = 16 \\ \textbf{2. } -3x + 2y = -16 \end{cases}$

D. $\begin{cases} \textbf{1. } 2x + 3y = 2 \\ \textbf{2. } 6x - y = 26 \end{cases}$

7. Multiply each side of equation **1** by 3. Add the corresponding sides. Solve for y. The result is $y = 3$.

8. Multiply each side of equation **1** by -3. Add the corresponding sides. Solve for y. The result is $y = -2$.

9. Multiply each side of equation **2** by -2. Add the corresponding sides. Solve for x. The result is $x = 6$.

10. Multiply each side of equation **2** by 2. Add the corresponding sides. Solve for x. The result is $x = -2$.

WRITTEN EXERCISES

Goal: To solve a system of equations by the multiplication/addition method

Sample Problem: Solve: $\begin{cases} \textbf{1. } 3x + 4y = 10 \\ \textbf{2. } 2x - y = -8 \end{cases}$ **Answer:** $\{(-2, 4)\}$

Solve each system by the multiplication/addition method. Check. (Example 1)

1. $\begin{cases} -x + 2y = 3 \\ 3x - 4y = -1 \end{cases}$

2. $\begin{cases} x - 3y = 1 \\ 2x - 3y = -4 \end{cases}$

3. $\begin{cases} x + y = 3 \\ 2x + y = -1 \end{cases}$

4. $\begin{cases} 4x - 3y = -1 \\ -3x + y = -3 \end{cases}$

5. $\begin{cases} 2x + 3y = -3 \\ -x - 2y = 2 \end{cases}$

6. $\begin{cases} 4x + 3y = 17 \\ -2x - y = -7 \end{cases}$

(Example 2)

7. $\begin{cases} 5x + 9y = 6 \\ 6x + 5y = 13 \end{cases}$

8. $\begin{cases} 2x - 5y = 7 \\ 3x - 2y = -17 \end{cases}$

9. $\begin{cases} 2x + 7y = 5 \\ 3x - 5y = 23 \end{cases}$

10. $\begin{cases} 2x + 2y = 8 \\ 5x - 3y = 4 \end{cases}$

11. $\begin{cases} 2x + 3y = -1 \\ 5x - 2y = -12 \end{cases}$

12. $\begin{cases} 5x - 2y = 11 \\ 3x + 5y = 19 \end{cases}$

MIXED PRACTICE

13. $\begin{cases} -4x + y = 2 \\ 3x - 2y = 1 \end{cases}$

14. $\begin{cases} 2x + 3y = 9 \\ 4x + 3y = 9 \end{cases}$

15. $\begin{cases} 2x + 2y = 10 \\ 7x - 3y = 15 \end{cases}$

16. $\begin{cases} 2x - 3y = 8 \\ 3x - 7y = 7 \end{cases}$

17. $\begin{cases} 5x + y = 15 \\ 3x + 2y = 9 \end{cases}$

18. $\begin{cases} 4x + 3y = -2 \\ 3x + 2y = 1 \end{cases}$

REVIEW CAPSULE FOR SECTION 15.4

Suppose x is the smaller and y is the greater of two unknown numbers. Write an equation for each sentence. (Pages 134–137)

1. The sum of the two unknown numbers is -10.

2. The difference of the two unknown numbers is 5.

3. The smaller of the two unknown numbers is 4 less than the greater number.

4. The greater of the two unknown numbers is 12 more than the smaller number.

PROBLEM SOLVING AND APPLICATIONS

15.4 Number Problems

The addition method and multiplication/addition method of solving a system of equations can be used to solve word problems. First, you use Condition 1 and two variables to represent the unknowns. You also use Condition 2 to write one equation of the system. You use Condition 2 to write the second equation for the system.

P–1 **If x and y are two numbers, how can you represent their sum? their difference?**

EXAMPLE 1

The sum of two numbers is 12 (Condition 1). The difference of the two numbers is 1 (Condition 2). Find the two numbers.

Solution:

1. Use Condition 1 to represent the unknowns.

   Let $x =$ the greater number.

   Let $y =$ the smaller number.

2. Write the equations.

   Think: **The sum of two numbers is 12.** ◀ *Condition 1*

   Translate: $x + y = 12$

   Think: **The difference of the two numbers is 1.** ◀ *Condition 2*

   Translate: $x - y = 1$

[3] Solve. Use the addition method. ⟶

$$\begin{aligned} x + y &= 12 \\ x - y &= 1 \\ \hline 2x \quad &= 13 \\ x &= 6\tfrac{1}{2} \\ \mathbf{x} + y &= 12 \\ \mathbf{6\tfrac{1}{2}} + y &= 12 \\ y &= 5\tfrac{1}{2} \end{aligned}$$

Don't forget to find y.

[4] **Check:** Condition 1 Does $6\frac{1}{2} + 5\frac{1}{2} = 12$? Yes ✓

Condition 2 Does $6\frac{1}{2} - 5\frac{1}{2} = 1$? Yes ✓

The two numbers are $5\frac{1}{2}$ and $6\frac{1}{2}$.

Note that Example 2 is another type of number problem.

EXAMPLE 2

Twice the record high temperature for a state is 275°F more than the record low temperature (Condition 1). Twice the record low temperature is 10 more than the opposite of the record high temperature (Condition 2). Find each record temperature.

Solution:

[1] Represent the unknowns. ⟶ Let $x =$ the record low temperature.
Let $y =$ the record high temperature.

[2] Write the equations in standard form.

$$\begin{cases} \mathbf{1.}\ 2y = x + 275, \text{ or } -x + 2y = 275 \\ \mathbf{2.}\ 2x = (-y) + 10, \text{ or } 2x + y = 10 \end{cases}$$

Equation 1 represents Condition 1. Equation 2 represents Condition 2.

[3] To solve the system, multiply equation **1** by 2. ⟶

$$\begin{aligned} -2x + 4y &= 550 \\ 2x + \ \ y &= 10 \\ \hline 5y &= 560 \\ y &= 112 \\ 2x + \mathbf{y} &= 10 \\ 2x + \mathbf{112} &= 10 \\ 2x &= -102 \\ x &= -51 \end{aligned}$$

Equation 2

Don't forget to find x.

The record low temperature is -51°F.

The record high temperature is 112°F.

The check is left for you.

CLASSROOM EXERCISES

Select one variable for the smaller number and another variable for the greater number. Write an equation that describes the sentence. (Examples 1 and 2)

1. One number is 5 less than the other.
2. The sum of two numbers is 15.
3. The difference between two numbers is -3.
4. One number is one third as great as the other number.
5. One number is 3 more than twice the other number.
6. The opposite of the smaller number is 12 more than the greater number.
7. One temperature is 16°F more than the other temperature.
8. The greater temperature is 10°F less than 5 times the lesser.
9. Twice the smaller number is 2 more than the greater number.
10. Three times the sum of two numbers is 52.

WRITTEN EXERCISES

Goal: To solve a word problem about numbers using a system of equations in two variables

Sample Problem: The sum of two numbers is -5 (Condition 1). Twice the greater number equals ten more than three times the smaller number (Condition 2). Find the unknown numbers.

Answer: The two numbers are -1 and -4.

For Exercises 1–18, Condition 1 is underscored once. Condition 2 is underscored twice.

a. *Use Condition 1 and two variables to represent the unknowns.*

b. *Use Conditions 1 and 2 to write a system of equations for the problem.*

c. *Solve.* (Example 1)

1. The sum of two numbers is 31. The difference of the two numbers is 3. Find the unknown numbers.
2. The sum of two numbers is 25. The difference of the two numbers is 4. Find the unknown numbers.
3. The sum of two numbers is 5. The difference of the two numbers is -2. Find the unknown numbers.
4. The sum of two numbers is -6. The difference of the two numbers is 11. Find the unknown numbers.
5. The sum of two numbers is 40. The difference of the two numbers is 2. Find the unknown numbers.
6. The sum of two numbers is -8. The difference of the two numbers is 14. Find the unknown numbers.

(Example 2)

7. The difference of the high and low record temperatures for Los Angeles is 82°F. The sum of the high and low record temperatures is 138°F. Find the two record temperatures.

8. The record high temperature for New York City less the record low temperature is 109°F. The sum of the high and low temperatures is 105°F. Find the temperatures.

9. On a certain day the low temperature reading in Dallas is 2°C less than twice the low temperature reading in Seattle. The difference in the two temperature readings is 13°C. Find the low temperature reading in Dallas and the low temperature reading in Seattle.

10. On a certain day the high temperature reading in Miami is 5°C more than three times the high temperature reading in Kansas City. The sum of the two temperature readings is 45°C. Find the high temperature readings in Miami and in Kansas City.

11. At an altitude of 12 kilometers in summer, the temperature of the atmosphere is 63°C less than at an altitude of 1 kilometer. Four times the temperature at the lower altitude is 3°C less than the opposite of the temperature at the higher altitude. Find the temperatures at altitudes of 1 kilometer and 12 kilometers.

12. At an altitude of 4 kilometers, the temperature of the atmosphere in winter is 28°C more than the temperature at twice the altitude. Three times the temperature at an altitude of 4 kilometers is 2°C less than the temperature at 8 kilometers. Find the temperatures at altitudes of 4 kilometers and 8 kilometers.

MORE CHALLENGING EXERCISES

13. One number is 13 more than another number. The sum of the numbers is 38. Find the unknown numbers.

14. One number is 15 less than another number. The sum of the numbers is 24. Find the unknown numbers.

15. One number exceeds another number by 8. Twice the smaller number is 4 more than the greater number. Find the unknown numbers.

16. One number less another number equals 18. Twice the smaller number is one less than the greater number. Find the unknown numbers.

17. The sum of two numbers is −20. Three times the greater number equals 10 more than twice the smaller number. Find the unknown numbers.

18. The sum of two numbers is −44. Four times the greater number is 8 less than three times the smaller number. Find the unknown numbers.

MID-CHAPTER REVIEW

Graph the system of equations in the same coordinate plane. Write the solution set of the system. (Section 15.1)

1. $\begin{cases} x - y = 1 \\ 3x - y = 3 \end{cases}$

2. $\begin{cases} -x - 2y = -4 \\ x - y = 1 \end{cases}$

3. $\begin{cases} x + y = -4 \\ 2x - y = -2 \end{cases}$

4. $\begin{cases} 3x - y = 2 \\ 2x - y = 3 \end{cases}$

5. $\begin{cases} 2x + 3y = 3 \\ 3x - 2y = -2 \end{cases}$

6. $\begin{cases} 2x + y = -1 \\ 5x + 3y = -2 \end{cases}$

Solve each system by the addition method. Check. (Section 15.2)

7. $\begin{cases} 2x - y = -3 \\ -2x + 3y = 1 \end{cases}$

8. $\begin{cases} -3x - 3y = -5 \\ 4x + 3y = 2 \end{cases}$

9. $\begin{cases} 5x + y = -19 \\ -3x - y = 11 \end{cases}$

10. $\begin{cases} 4x - y = 30 \\ 3x + y = 26 \end{cases}$

11. $\begin{cases} 3x + y = 1 \\ -3x + 2y = 29 \end{cases}$

12. $\begin{cases} 5x + y = 8 \\ -5x - 6y = 2 \end{cases}$

Solve each system by the multiplication/addition method. Check. (Section 15.3)

13. $\begin{cases} 2x - 6y = -12 \\ x + y = 6 \end{cases}$

14. $\begin{cases} 2x - y = 8 \\ x - 8y = 4 \end{cases}$

15. $\begin{cases} 2x - 5y = 1 \\ -x + 2y = 4 \end{cases}$

16. $\begin{cases} -3x + 4y = 2 \\ 2x - y = -1 \end{cases}$

17. $\begin{cases} -4x + 5y = -3 \\ 3x - 2y = 1 \end{cases}$

18. $\begin{cases} 6x - 3y = 1 \\ -5x + 2y = -3 \end{cases}$

For Exercises 19–22, Condition 1 is underscored once, and Condition 2 is underscored twice.

Use Condition 1 to represent the unknowns. Use Condition 1 and Condition 2 to write a system of equations for the problem. Solve. (Section 15.4)

19. The sum of two numbers is 13. The difference of the two numbers is 7. Find the unknown numbers.

20. The sum of two numbers is −10. The difference of the two numbers is 2. Find the unknown numbers.

21. On a certain day, the low temperature reading in Milwaukee is 54°F less than the low temperature reading in Miami. The sum of the two temperature readings is 42°F. Find the low temperature reading in Milwaukee and the low temperature reading in Miami.

22. The record low temperature for Corinth, Mississippi less twice the record low temperature for Lebanon, Kansas is 61°F. The sum of the two temperatures is −59°F. Find the record low temperature for Corinth and the record low temperature for Lebanon.

Using Formulas and Tables

Automobile Speed

A car that brakes quickly in an emergency will leave skid marks on the road. You can use the following formula to estimate the speed of the car before braking.

$$s = 5.5\sqrt{dF}$$

s = speed in mi/hr
d = length of skid marks in ft
F = coefficient of friction

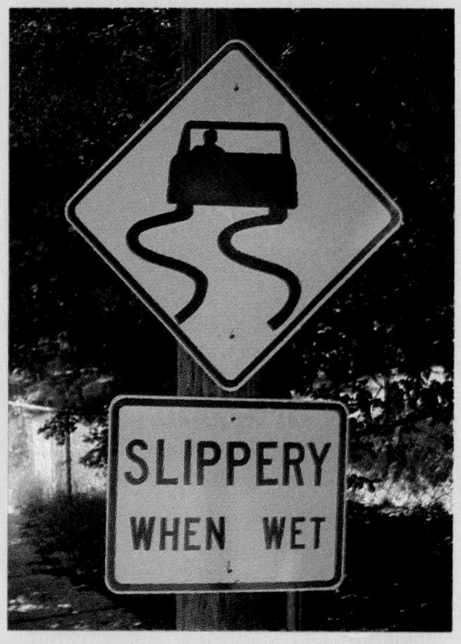

The **coefficient of friction** is a number that varies according to driving conditions.

Type of Surface	*Coefficient of Friction* *Dry Road*	*Wet Road*
Asphalt	0.85	0.65
Concrete	0.90	0.60
Gravel	0.65	0.65
Packed Snow	0.45	0.45

EXAMPLE: Estimate the speed of an automobile that leaves skid marks 140 feet long on a wet asphalt street. Write your answer to the nearest mile per hour.

SOLUTION: Substitute the known values in the formula.

$$s = 5.5\sqrt{dF}$$
$$s = 5.5\sqrt{(140)(0.65)}$$
$$s = 5.5\sqrt{91}$$

F = 0.65 from the table at the right.

Thus, the speed is about **52 miles per hour.**

EXERCISES

For Exercises 1–3, estimate the speed for an automobile using the given length of the skid marks and the driving conditions. Write your answers to the nearest mile per hour.

1. 75 feet; wet concrete
2. 40 feet; packed snow
3. 140 feet; gravel
4. A car leaves skid marks 100 feet long on a gravel road. If the car had been traveling on wet concrete when it skidded, would the skid marks be longer or shorter? Explain.

REVIEW CAPSULE FOR SECTION 15.5

Solve each equation for x. (Pages 157–159)

1. $3x - 2(x - 5) = 6$

2. $53 = 3(x + 1) + 2x$

3. $3(4x + 5) - 7 = 4x$

4. $-5(2x - 3) + 2x = -7$

5. $16 = 5(-x - 6) + 7x$

6. $8(2x + 1) - 3 = x$

15.5 Substitution Method

Example 1 shows the ***substitution method*** of solving linear equations.

P–1 **What is an equation that is equivalent to $2x + y = 1$ and that has only y on one side? only x on one side?**

EXAMPLE 1 Solve: $\begin{cases} \textbf{1.}\ 2x + y = 1 \\ \textbf{2.}\ 3x + y = -2 \end{cases}$

Solution: [1] Solve one equation for one of the variables.

1. $2x + y = 1$

$y = -2x + 1$

y is chosen in 1 because its coefficient is 1.

[2] Substitute $(-2x + 1)$ for y in equation **2.** Solve for x.

2.
$$\begin{aligned} 3x + \mathbf{y} &= -2 \\ 3x + (\mathbf{-2x + 1}) &= -2 \\ 3x - 2x + 1 &= -2 \\ x + 1 &= -2 \\ x &= -3 \end{aligned}$$

[3] Substitute -3 for x in one of the equations. Solve for y.

1.
$$\begin{aligned} 2\mathbf{x} + y &= 1 \\ 2(\mathbf{-3}) + y &= 1 \\ -6 + y &= 1 \\ y &= 7 \end{aligned}$$

Check: [4] Substitute -3 for x and 7 for y in both equations.

1. $2\mathbf{x} + \mathbf{y} = 1$

$2(\mathbf{-3}) + \mathbf{7}$

$-6 + 7$

1

2. $3\mathbf{x} + \mathbf{y} = -2$

$3(\mathbf{-3}) + \mathbf{7}$

$-9 + 7$

-2

The solution set is $\{(-3, 7)\}$.

In Example 1, the number 2 in $2x + y = 1$ is the coefficient of x. The coefficient of y is 1 since $2x + y = 1$ can be written as $2x + 1y = 1$. In the substitution method, try to solve first for a variable that has 1 or -1 as its coefficient.

P–2 **In each of the following, use equation 1 to substitute for x or y in equation 2.**

a. $\begin{cases} \textbf{1.}\ y = x + 2 \\ \textbf{2.}\ x + y = 6 \end{cases}$ b. $\begin{cases} \textbf{1.}\ x = y + 1 \\ \textbf{2.}\ x + 2y = -2 \end{cases}$ c. $\begin{cases} \textbf{1.}\ y - x = 4 \\ \textbf{2.}\ 2x + y = 7 \end{cases}$

EXAMPLE 2

Solve: $\begin{cases} \textbf{1.}\ 2x - 3y = -4 \\ \textbf{2.}\ -x + 2y = 3 \end{cases}$

Solution: [1] Solve one equation for one of the variables.

2.
$$-x + 2y = 3$$
$$-x + 2y + \mathbf{x} = 3 + \mathbf{x}$$
$$2y = x + 3$$
$$2y - \mathbf{3} = x + 3 - \mathbf{3}$$
$$2y - 3 = x$$

x is chosen in 2 because its coefficient is −1.

[2] Substitute $(2y - 3)$ for x in equation **1.** Solve for y.

1.
$$2\mathbf{x} - 3y = -4$$
$$2\mathbf{(2y - 3)} - 3y = -4$$
$$4y - 6 - 3y = -4$$
$$y - 6 = -4$$
$$y = 2$$

[3] Substitute 2 for y in either equation. Solve for x.

2.
$$2x - 3\mathbf{y} = -4$$
$$2x - 3\mathbf{(2)} = -4$$
$$2x - 6 = -4$$
$$2x = 2$$
$$x = 1$$

Check: [4] Substitute 1 for x and 2 for y in both equations.

The solution set is $\{(1, 2)\}$.

Steps for the Substitution Method

1. Solve one equation for one of the variables.
2. Substitute the resulting expression in the other equation. Solve the equation.
3. Substitute the value of the variable from Step 2 in either equation. Solve the resulting equation.
4. Check by substituting both values in both equations.

CLASSROOM EXERCISES

Solve the equation for one of the variables.
(Step 1 of Examples 1 and 2)

1. $x + y = 10$
2. $x - 3y = -4$
3. $-3x - y = -1$
4. $4x + y = 5$
5. $2y - x = 7$
6. $x + y = -9$
7. $3x - y = 9$
8. $4x + 2y = 6$

Write the next step in solving each equation. (Step 2 of Examples 1 and 2)

9. $3x - 5(x + 2) = 3$
10. $-2(y - 3) + 3y = 1$
11. $3(2y - 2) - 4y = -5$
12. $-x + 2(-2x - 3) = -4$
13. $-3x - (4x - 2) = 6$
14. $2(\frac{1}{2}y - 1) + y = 5$

WRITTEN EXERCISES

Goal: To solve a system of equations by the substitution method

Sample Problem: Solve: $\begin{cases} \textbf{1.}\ 2x + 3y = 1 \\ \textbf{2.}\ x - 2y = -3 \end{cases}$ **Answer:** $\{(-1, 1)\}$

Solve each system of equations by the substitution method. Check.
(Examples 1 and 2)

1. $\begin{cases} 2x + y = 1 \\ 5x + 2y = 4 \end{cases}$
2. $\begin{cases} -x + y = -3 \\ 3x - 2y = 5 \end{cases}$
3. $\begin{cases} x - 3y = 1 \\ -2x + 4y = 2 \end{cases}$
4. $\begin{cases} x + 2y = 3 \\ -2x - y = 3 \end{cases}$
5. $\begin{cases} x + 3y = -2 \\ -x - y = -4 \end{cases}$
6. $\begin{cases} 3x + y = 5 \\ -4x - 2y = -2 \end{cases}$
7. $\begin{cases} 2x - y = 4 \\ 2x - \frac{1}{2}y = -5 \end{cases}$
8. $\begin{cases} -x + 4y = 2 \\ -\frac{1}{2}x - 2y = 5 \end{cases}$
9. $\begin{cases} 4x - y = 3 \\ 6x - 2y = 3 \end{cases}$
10. $\begin{cases} -2x + y = 5 \\ -4x + 3y = -10 \end{cases}$
11. $\begin{cases} x - y = 4 \\ 2x - 3y = -2 \end{cases}$
12. $\begin{cases} x - y = 3 \\ 5x + y = -15 \end{cases}$
13. $\begin{cases} x + 2y = -4 \\ 2x - 2y = -5 \end{cases}$
14. $\begin{cases} -3x + 4y = 6 \\ x - 4y = -8 \end{cases}$
15. $\begin{cases} x - 3y = -6 \\ 4x + 3y = -4 \end{cases}$

REVIEW CAPSULE FOR SECTION 15.6

Write an algebraic expression for the total cost of the items in each exercise. (Pages 341–344)

1. x records cost \$6 each and y records cost \$5 each.
2. x tickets cost \$7 each and y tickets cost \$9 each.
3. x pens cost \$1.25 each and y pens cost \$0.85 each.
4. x books cost \$1.95 each and y books cost \$2.25 each.

15.6 Money Problems

Problems that involve the cost of two items can often be solved by using a system of equations in two variables. In solving such problems, it is helpful to make a table showing the number of items (Condition 1) and the value of the items (Condition 2).

EXAMPLE 1

A store had a special one-week sale on two models of digital watches. Fifty-four watches were sold (Condition 1). One model sold for \$24 and the other for \$30. The total amount of sales, not including tax, was \$1494 (Condition 2). How many of each model were sold?

Solution: Set up a table to help you write the equations.

	Number of Watches (Condition 1)	**Price per Watch**	**Total Sales (Condition 2)**
Less Expensive	x	\$24	$24x$
More Expensive	y	\$30	$30y$
Total	54		\$1494

Equations: **1.** $x + y = 54$ **2.** $24x + 30y = 1494$

[1] Solve equation **1** for y. ⟶ $x + y = 54$

$$y = 54 - x$$

[2] Substitute for y in equation **2.** Solve for x. ⟶ $24x + 30\mathbf{y} = 1494$

$$24x + 30\mathbf{(54 - x)} = 1494$$
$$24x + 1620 - 30x = 1494$$
$$-6x = -126$$
$$x = 21$$

[3] Substitute for x in equation **1.** Solve for y. ⟶ $\mathbf{x} + y = 54$

$$\mathbf{21} + y = 54$$
$$y = 33$$

[4] **Check:** Condition 1 Were 54 watches sold? Does $21 + 33 = 54$? Yes ✓

Condition 2 Was the total amount of sales \$1494?
Does $(24)21 + (30)33 = 1494$?
Does $504 + 990 = 1494$? Yes ✓

Twenty-one \$24 watches and thirty-three \$30 watches were sold.

EXAMPLE 2

A certain portion of \$7500 was invested for one year at a yearly interest rate of 8%. The remaining portion was invested at a yearly rate of 9% (Condition 1). The interest earned on the total investment was \$660 (Condition 2). Find the amount invested at each rate.

Solution: Make a table.

	Amount Invested (Condition 1)	Yearly Interest Rate	Amount of Interest (Condition 2)
Lower Rate	x	8%	$0.08x$
Higher Rate	y	9%	$0.09y$
Total	\$7500		\$660

$$\begin{cases} \textbf{1.}\ x + y = 7500 \\ \textbf{2.}\ 0.08x + 0.09y = 660 \end{cases} \quad \text{or} \quad \begin{cases} \textbf{1.}\ y = 7500 - x \\ \textbf{2.}\ 0.08x + 0.09y = 660 \end{cases}$$

$0.08x + 0.09\mathbf{(7500 - x)} = 660$ ◀ ***Replace y in 2 with 7500 − x.***

$0.08x + 675 - 0.09x = 660$

$-0.01x = -15$

$x = 1500$ ◀ ***Amount invested at 8%.***

$y = 7500 - \mathbf{x}$ ◀ ***Replace x in 1 with 1500.***

$y = 7500 - \mathbf{1500}$

$y = 6000$ ◀ ***Amount invested at 9%.***

Check: Condition 1 Is the total amount invested \$7500?
Does $1500 + 6000 = 7500$? Yes ✓

Condition 2 Is the total amount of interest \$660?
Does $(0.08)1500 + (0.09)6000 = 660$?
Does $120 + 540 = 660$? Yes ✓

\$1500 is invested at 8% and \$6000 is invested at 9%.

CLASSROOM EXERCISES

Complete the table for each problem. Then write two equations that could be used to solve for x and y. (Examples 1 and 2)

1. Eight stamps on a package consisted of x 25-cent stamps and y 18-cent stamps with a value of \$1.65.

	Number of Stamps	Value of Each Stamp	Total Value
25-cent stamps	x	\$0.25	?
18-cent stamps	y	?	?

2. An investor bought 350 shares of stock consisting of x shares of Aztec stock at \$18 per share and y shares of Acme stock at \$25 per share at a total cost of \$7000.

	Number of Shares	Price per Share	Total Cost
Aztec	x	?	$18x$
Acme	y	?	?

3. A club of 48 members has x senior members and y junior members. The receipts from dues are \$216 with \$5 from each senior member and \$2 from each junior member.

	Number of Members	Dues	Total Receipts
Senior	x	?	?
Junior	y	?	?

4. Dan and Jan walked a total of 28 miles for the March of Dimes and earned \$5.00. Dan walked x miles for 15 cents a mile, and Jan walked y miles for 20 cents a mile.

	Number of Miles	Rate per Mile	Total Earnings
Dan	x	?	?
Jan	y	?	?

5. Partners in a business took out two loans of \$$x$ and \$$y$ in the total amount of \$12,000. The annual interest rates were 12% and 15% and the total annual interest was \$1560.

	Amount of Loan	Interest Rate	Total Interest
First Loan	x	12%	?
Second Loan	y	15%	?

6. A couple invested x dollars at an annual interest rate of 7% and y dollars at an annual rate of 10%. The total investment of \$14,500 earned \$1270 interest in one year.

	Amount Invested	Interest Rate	Total Interest
Investment 1	x	?	?
Investment 2	y	10%	?

WRITTEN EXERCISES

Goal: To solve a word problem about money using a system of equations in two variables.

Sample Problem: A student sold 54 tickets for the school play (Condition 1). Student tickets cost $1.50 and adult tickets cost $3.00. The amount paid for the 54 tickets was $111.00 (Condition 2). How many tickets of each kind were sold?

Answer: 34 student tickets and 20 adult tickets

For Exercises 1–6, Condition 1 is underscored once. Condition 2 is underscored twice.

a. *Use the given information (Condition 1 and Condition 2) to make a table.*

b. *Use Conditions 1 and 2 to write a system of equations for the problem.*

c. *Solve.* (Examples 1 and 2)

1. A shopper purchased 8 shirts. One style cost $22 each and the other style cost $28 each. The total cost of the shirts was $206. How many of each style were purchased?

2. A total of 6500 tickets were sold for a concert. Tickets sold for $8 and $6 each. The total receipts for the tickets were $45,250. How many of each kind were sold?

3. Mayumi invested $5000 for one year. Part was invested at an annual rate of 8% and part at an annual rate of 10%. The total earned for one year was $456. How much was invested at each rate?

4. A salesperson sold 27 magazine subscriptions. The magazine *Now* sold for $36.00 and the magazine *World* sold for $48.50. The total sales for the subscriptions were $1122. How many subscriptions were sold for each magazine?

5. An attorney got two loans in order to open her law office. The interest rate on one loan was 12% for one year, and the rate on the other loan was 15%. The total amount borrowed was $25,000. Her total interest expense for one year was $3510. Find the amounts of the two loans.

6. A ski shop had a sale on two models of skis. In the three-day sale, 48 pairs of these skis were sold. One model was on sale for $162.50 and the other was on sale for $178.50. The total receipts for the 48 pairs of skis were $8152. How many pairs of each kind of skis were sold?

15.7 Geometry/Age/Distance

The techniques of solving systems of equations can also be used to solve word problems that deal with geometry, age, and distance.

EXAMPLE 1

The length of a rectangle is 6 centimeters more than the width (Condition 1). The perimeter is 40 centimeters (Condition 2). Find the width, x, and the length, y.

Solution: $\begin{cases} \mathbf{1.}\ y = x + 6 \\ \mathbf{2.}\ 2x + 2y = 40 \end{cases}$

1. Condition 1
2. Condition 2

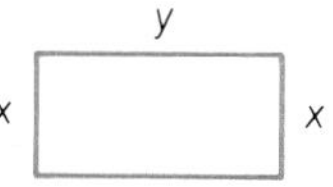

[1] Substitute $x + 6$ for y in equation **2.** ⟶ $2x + 2(\mathbf{x + 6}) = 40$

[2] Solve for x. ⟶ $2x + 2x + 12 = 40$

$$4x + 12 = 40$$
$$4x = 28$$
$$x = 7$$

[3] Substitute 7 for x in equation **1.** ⟶ $y = \mathbf{x} + 6$

$$y = \mathbf{7} + 6 \text{ or } 13$$

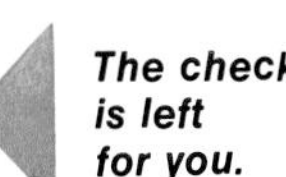

The check is left for you.

The width is 7 cm. The length is 13 cm.

EXAMPLE 2

A father's age is three times the age of his son (Condition 1). The sum of their ages is 56 (Condition 2). Find y, the father's age, and x, the son's age.

Solution: $\begin{cases} \mathbf{1.}\ y = 3x \\ \mathbf{2.}\ x + y = 56 \end{cases}$

1. Condition 1
2. Condition 2

[1] Substitute $3x$ for y in equation **2.** ⟶ $x + \mathbf{y} = 56$

$$x + \mathbf{3x} = 56$$

[2] Solve for x. ⟶ $4x = 56$

$$x = 14$$

[3] Substitute 14 for x in equation **1.** ⟶ $y = 3\mathbf{x}$

$$y = 3(\mathbf{14}) \text{ or } 42$$

The check is left for you.

The son is 14. The father is 42.

EXAMPLE 3

The airline distance from New York to London is 545 kilometers more than twice the distance from London to Moscow (Condition 1). The sum of the two distances is 8105 kilometers (Condition 2). Find the distance from New York to London and the distance from London to Moscow.

Solution: Let $x =$ the distance in kilometers from New York to London and $y =$ the distance in kilometers from London to Moscow.

$$\begin{cases} \textbf{1.}\ x = 2y + 545 \\ \textbf{2.}\ x + y = 8105 \end{cases}$$

1. Condition 1
2. Condition 2

$$\begin{aligned} \mathbf{(2y + 545)} + y &= 8105 \\ 3y + 545 &= 8105 \\ 3y &= 7560 \\ y &= 2520 \text{ km} \end{aligned}$$

Replace x in equation 2 with 2y + 545.

$$\begin{aligned} x &= 2\mathbf{y} + 545 \\ x &= 2\mathbf{(2520)} + 545 \\ x &= 5040 + 545 \\ x &= 5585 \text{ km} \end{aligned}$$

Replace y in equation 1 with 2520.

The distance from New York to London is 5585 kilometers.
The distance from London to Moscow is 2520 kilometers.

The check is left for you.

CLASSROOM EXERCISES

Write an equation that describes each problem. (Examples 1–3)

1. The perimeter of a rectangle is 36 feet. The length is 12 feet more than the width.

2. A man spent x hours at his office and y hours playing golf. The total length of time spent was $11\frac{1}{2}$ hours.

3. A scout troop hiked x kilometers on Saturday and y kilometers on Sunday. The total distance hiked was 37 kilometers.

4. Mrs. Scott's age is x years and her daughter's age is y years. The daughter is 26 years younger than her mother.

5. A tennis coach buys x cans of Brand *A* tennis balls and y cans of Brand *B*. She buys a total of 81 tennis balls (3 balls per can).

6. The ones digit of a number is x. The tens digit is 7, and the hundreds digit is y. The sum of the digits is 19.

7. On an exam, x students scored 92 points, and y students scored 85 points. The total number of points scored was 1225 points.

8. The distance from New York to Chicago is 1145 miles less than the distance from Chicago to San Francisco. The sum of the distances is 2573 miles.

WRITTEN EXERCISES

Goal: To solve a word problem using a system of equations in two variables
Sample Problem: The width of a rectangle is 6 inches less than the length (Condition 1). The perimeter is 28 inches (Condition 2). Find the length and the width.
Answer: The length is 10 inches and the width is 4 inches.

Write a system of equations in two variables for each problem. Then solve. For some problems, Condition 1 is underscored once and Condition 2 is underscored twice.

(Example 1)

1. The width, x, of a rectangle is 5 millimeters less than the length, y. The perimeter is 54 millimeters. Find the length and width.

2. The sum of the length, x, and width, y, of a rectangle is 21 centimeters. The length is 7 centimeters more than the width. Find the length and width.

3. Two sides of a triangle are equal. Each of the two equal sides is 5 inches longer than the third side. The perimeter is 35 inches. Find the length of each side.

4. Each of two equal sides of a triangle is 12 centimeters longer than the third side. The perimeter is 174 centimeters. Find the length of each side.

(Example 2)

5. Mrs. Johner's age is twice that of her son. The sum of their ages is 60. Find the ages of Mrs. Johner and her son.

6. An oak tree's age is 10 more than twice the age of a pine tree. The sum of their ages is 19. Find the age of each tree.

7. John's age is two years more than twice his brother's age. The difference of the ages of John and his younger brother is 6 years. Find the age of both boys.

8. Enrico's mother is two years older than twice his age. The difference of their ages is 17. Find Enrico's age and the age of his mother.

(Example 3)

9. The distance from St. Louis to Chicago is 111 miles more than the distance from Chicago to Minneapolis. The sum of the two distances is 697 miles. Find each unknown distance.

10. The airline distance from Dallas to Houston is 850 kilometers less than the distance from Houston to Mexico City. The sum of the two distances is 1550 kilometers. Find each unknown distance.

Focus on Reasoning: Number Patterns

A **number sequence** is a succession of numbers that follow a fixed pattern. Each number in the sequence is related to the preceding number according to a definite plan. **Identifying that plan** is the key step in determining other numbers in each sequence.

EXAMPLE Write the next two numbers in each sequence.

a. $-5, 10, -20, 40, -80, \cdots$

b. $5, 2, 5, 4, 5, 6, 5, \cdots$

Solution:

a. Think: $-5 \cdot (-2) = 10 \qquad 10 \cdot (-2) = -20 \qquad -20 \cdot (-2) = 40$

Rule: Multiply the preceding number by (−2).

Next two numbers: **160, −320**

b. Think: 5, 2, 5, 4, 5, 6, 5 (2 to 4: +2; 4 to 6: +2)

Rule: Add 2 to every other term. Otherwise write 5.

Next two numbers: **8, 5**

EXERCISES

Write the next four numbers in each sequence.

1. $1, 5, 9, 13, 17, \cdots$

2. $400, 373, 346, 319, 292, \cdots$

3. $7\frac{3}{4}, 7\frac{5}{8}, 7\frac{1}{2}, 7\frac{3}{8}, 7\frac{1}{4}, \cdots$

4. $0.1, 0.4, 1.6, 6.4, 25.6, \cdots$

5. $-2, -4, -8, -16, -32, \cdots$

6. $-26, -20, -14, -8, -2, \cdots$

7. $2.52, 3.02, 3.52, 4.02, \cdots$

8. $-3, 3, -3, 3, -3, \cdots$

9. $5, 6, 8, 11, 15, \cdots$

10. $0, 3, 8, 15, 24, \cdots$

11. $9, 9, 18, 18, 27, 27, \cdots$

12. $2, 5, 15, 18, 54, \cdots$

13. $100, 81, 64, 49, 36, \cdots$

14. $4, 16, 5, 25, 6, \cdots$

15. $1, \frac{1}{4}, \frac{1}{9}, \frac{1}{16}, \frac{1}{25}, \cdots$

16. $3, 7, 23, 87, 343, \cdots$

Focus on Reasoning: Geometric Patterns

Sometimes a sequence is a succession of geometric figures. Look for what changes and what remains the same. This will help you to identify the pattern.

EXAMPLE Find the next figure in the sequence at the right.

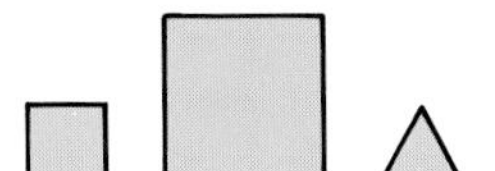
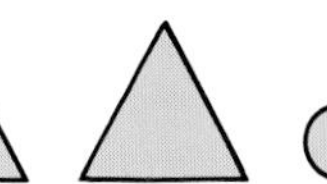

Think: The figures change.
The size changes.

Next figure:

EXERCISES

Identify the pattern. Then find the next figure in each sequence.

1.

2. 

3.

4. 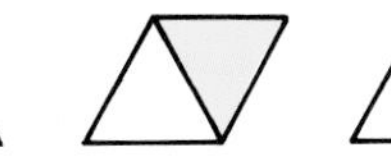

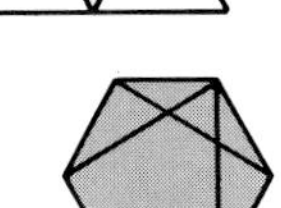

5. 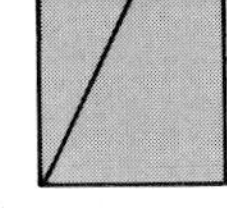

6. 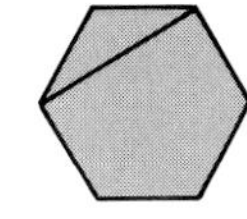 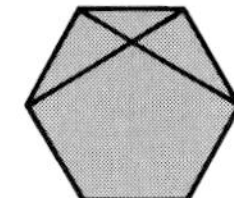 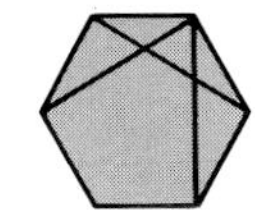

7. 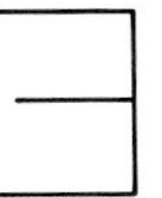 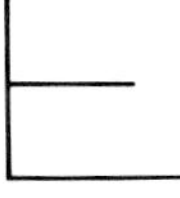

8.   

9. 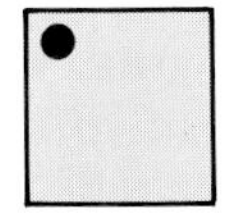 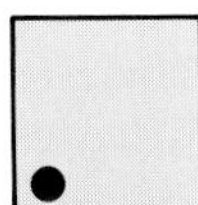 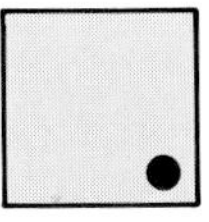

10. 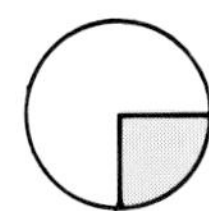 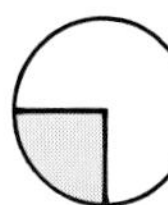 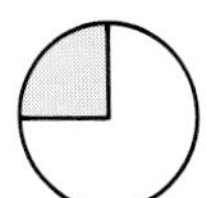

11. 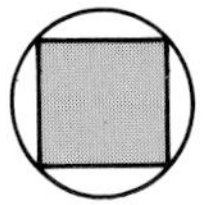 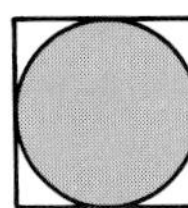 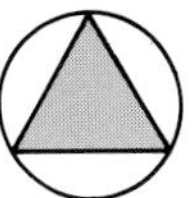

12. 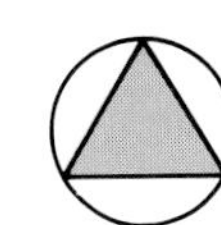 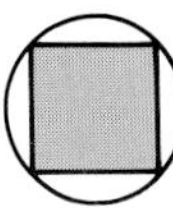 

CHAPTER SUMMARY

IMPORTANT TERMS

Linear equation *(p. 420)*
Standard form *(p. 420)*
System of equations *(p. 420)*
Solution set of a system of equations *(p. 421)*
Addition method *(p. 422)*
Multiplication/addition method *(p. 425)*
Substitution method *(p. 434)*

IMPORTANT IDEAS

1. Steps for the Multiplication/Addition Method
   - **a.** Check for opposite coefficients of one variable. If necessary, multiply each side of either or both equations by numbers that will make opposite coefficients for one variable.
   - **b.** Add to eliminate one of the variables. Solve the resulting equation.
   - **c.** Substitute the known value of one variable in one of the original equations of the system. Solve for the other variable.
   - **d.** Check the solution in both equations of the system.
2. Steps for the Substitution Method
   - **a.** Solve one equation for one of the variables.
   - **b.** Substitute the resulting expression in the other equation. Solve the equation.
   - **c.** Substitute the value of the variable in step ***b*** in either equation. Solve the resulting equation.
   - **d.** Check by substituting both values in both equations.

CHAPTER REVIEW

SECTION 15.1

Graph the two equations of each system in the same coordinate plane. Write the solution set of the system.

1. $\begin{cases} x - y = -1 \\ x + 2y = -4 \end{cases}$ **2.** $\begin{cases} x + y = 2 \\ 2x - y = 1 \end{cases}$ **3.** $\begin{cases} x - 2y = -3 \\ x - y = -2 \end{cases}$

4. $\begin{cases} x + y = 1 \\ 2x - y = -1 \end{cases}$ **5.** $\begin{cases} 2x + 3y = 3 \\ x - 2y = -2 \end{cases}$ **6.** $\begin{cases} 3x + y = 2 \\ 6x + y = -4 \end{cases}$

SECTION 15.2

Solve each system of equations by the addition method. Check.

7. $\begin{cases} x - 3y = -2 \\ 2x + 3y = -4 \end{cases}$ **8.** $\begin{cases} 3x + 2y = 5 \\ -3x - y = -4 \end{cases}$ **9.** $\begin{cases} 4x - y = 3 \\ -4x + 3y = -5 \end{cases}$

10. $\begin{cases} 4x + y = -8 \\ -6x - y = 18 \end{cases}$

11. $\begin{cases} 7x + 3y = 19 \\ 4x - 3y = 3 \end{cases}$

12. $\begin{cases} -x + 2y = 1 \\ x + y = 5 \end{cases}$

SECTION 15.3

Solve each system of equations by the multiplication/addition method. Check.

13. $\begin{cases} 2x + y = -2 \\ 2x + 3y = 6 \end{cases}$

14. $\begin{cases} 4x - 3y = -1 \\ -3x + y = -3 \end{cases}$

15. $\begin{cases} 2x + y = 1 \\ 4x - 2y = 6 \end{cases}$

16. $\begin{cases} 5x + 3y = 7 \\ 3x + 5y = 1 \end{cases}$

17. $\begin{cases} 2x + 3y = -15 \\ x - 4y = -2 \end{cases}$

18. $\begin{cases} 7x - 2y = -2 \\ -9x + 5y = 5 \end{cases}$

SECTION 15.4

For each problem, Condition 1 is underscored once. Condition 2 is underscored twice.

a. *Use Condition 1 and two variables to represent the unknowns.*

b. *Use Conditions 1 and 2 to write a system of equations for the problem.*

c. *Solve.*

19. The sum of two numbers is 32. The difference of the two numbers is 7. Find the unknown numbers.

20. The sum of two numbers is 11. The difference of the two numbers is 35. Find the unknown numbers.

21. Twice the higher temperature is 1°F less than 9 times the lower temperature. The sum of both temperatures is 104°F. Find the two temperatures.

22. In one city the record high temperature less the record low temperature is 98°F. The sum of the high and low temperatures is 106°F. Find the two record temperatures.

SECTION 15.5

Solve each system of equations by the substitution method. Check.

23. $\begin{cases} 2x + y = -3 \\ 3x - 2y = -1 \end{cases}$

24. $\begin{cases} -x + y = -3 \\ 3x - 2y = 5 \end{cases}$

25. $\begin{cases} 2x - 4y = 3 \\ -x + 3y = -1 \end{cases}$

26. $\begin{cases} x - 2y = -2 \\ 2x - 3y = 2 \end{cases}$

27. $\begin{cases} x - y = 4 \\ 2x - 3y = -2 \end{cases}$

28. $\begin{cases} 2x + y = 5 \\ 8x - y = 45 \end{cases}$

SECTION 15.6

For each problem, Condition 1 is underscored once. Condition 2 is underscored twice.

a. *Use Condition 1 and two variables to represent the unknowns.*
b. *Use Conditions 1 and 2 to write a system of equations for the problem.*
c. *Solve.*

29. The senior class sold 205 tickets to their play. Tickets cost \$2 for students and \$3 for adults. They took in \$485. How many tickets of each type did they sell?

30. Miyoko purchased a total of 6 shirts on a recent shopping spree. The short-sleeved shirts cost \$16. The long-sleeved shirts cost \$18. She spent \$100 on shirts. How many of each kind did she buy?

31. Les and Fran sold 30 glasses of cold drinks at their stand. Iced tea sold for 15 cents a glass and lemonade for 20 cents a glass. They took in \$5.10. How many glasses of each drink were sold?

32. Abel bought 32 stamps at the post office for his sister. The stamps were eight-cent stamps and ten-cent stamps. The total cost was \$2.98. How many of each kind of stamp did he buy?

SECTION 15.7

For each problem, write a system of equations in two variables. Then solve. For some problems, Condition 1 is underscored once and Condition 2 is underscored twice.

33. The width of a rectangle is 10 centimeters less than twice the length. The perimeter is 28 centimeters. Find the length and width.

34. The difference of the length and width of a rectangle is 3 meters. Four times the width equals three times the length. Find the length and width.

35. Pat's teacher is 29 years older than Pat. The sum of their ages is 61. Find the ages of Pat and her teacher.

36. Karl's age is 18 years less than twice his cousin's age. The sum of their ages is 60. Find the ages of Karl and his cousin.

37. The distance from San Francisco to Portland is 105 miles more than twice the distance from Portland to Seattle. The sum of the two distances is 685 miles. Find each unknown distance.

38. The distance from San Diego to Phoenix is 300 kilometers more than the distance from San Diego to Los Angeles. The sum of the two distances is 670 kilometers. Find each unknown distance.

CUMULATIVE REVIEW: CHAPTERS 1–15

Perform the indicated operations. Write each answer in lowest terms. (Sections 4.3 through 4.6)

1. $\frac{6a^2}{x} \cdot \frac{x^2}{18a}$ **2.** $\frac{8a^3}{3c} \div \frac{10a}{9c}$ **3.** $\frac{17t}{11} + \frac{5t}{11}$ **4.** $\frac{3}{5x} - \frac{1}{10x}$

Solve each problem. (Sections 7.7 and 7.9)

5. Last year Juan earned \$12,500. He paid \$2000 in taxes. What per cent of his earnings did he pay in taxes?

6. Find the sale price of a VCR listed at \$350 if the rate of discount is 35%.

Simplify. (Section 9.7)

7. $17k - 29k - k$ **8.** $-2x - 5 + 7x - (-4)$ **9.** $4a^2 - 6a - 9a^2 - 2a$

Factor. (Section 10.4)

10. $3a - 3b$ **11.** $-4y^2 + (-4y)$ **12.** $9pq - 18p$

Use Condition 1 to represent the unknowns. Use Condition 2 to write an equation for the problem. Solve. (Sections 12.1 and 12.6)

13. The ratio of milk to cream in a mixture is 15:2. How much milk and how much cream are there in 102 gallons of the mixture?

14. Tom is now twice as old as Sue. Two years ago, Tom's age was three times Sue's age then. How old are Tom and Sue now?

Graph each inequality using the set of real numbers as the replacement set. (Section 13.1)

15. $n > -2$ **16.** $a < 5$ **17.** $x \leq -3$ **18.** $y \geq 2.5$

Solve and graph. (Sections 13.2 through 13.4)

19. $3.2 > a + 1.2$ **20.** $-8t < -16$ **21.** $-6x + 2 < -4(x + 2)$

Use Condition 1 to represent the unknowns. Use Condition 2 to write an inequality. Solve the problem. (Section 13.5)

22. Larry estimates that a shirt costs \$28 less than a certain sweater. He plans to buy both for less than \$100. Find the greatest amount he can spend for the shirt.

23. Hilda's scores on three tests are 78, 93, and 81. She would like to have a mean score greater than 85. What is the lowest score she can get on the fourth test?

Graph the following points. (Section 14.1)

24. $A(-2,1)$ **25.** $B(3,-2)$ **26.** $C(0,-3)$ **27.** $D(1,0)$

Write the domain and range of each function. (Section 14.3)

28. $\{(5,2), (7,2), (-3,4)\}$ **29.** $\{(-3,0), (5,2), (7,11)\}$

Graph each function. (Section 14.4)

30. $y = x + 2$ **31.** $y = -3x + 1$ **32.** $y = \frac{1}{2}x + 1$

Find the X intercept and the Y intercept. (Section 14.5)

33. $y = x - 3$ **34.** $y = 3x - 6$ **35.** $y = \frac{1}{4}x + 2$

Graph the system of equations in the same coordinate plane. Write the solution set of the system. (Section 15.1)

36. $\begin{cases} -x + y = 1 \\ -2x + y = -2 \end{cases}$ **37.** $\begin{cases} x - y = -2 \\ x + 3y = -6 \end{cases}$ **38.** $\begin{cases} 4x + 2y = 0 \\ 3x + 2y = 2 \end{cases}$

Solve each system by the multiplication/addition method. Check. (Section 15.3)

39. $\begin{cases} 2x + y = 3 \\ 7x - 4y = 18 \end{cases}$ **40.** $\begin{cases} 3x + 2y = 11 \\ 2x + y = 7 \end{cases}$ **41.** $\begin{cases} 2x + 5y = 18 \\ 3x - 4y = 4 \end{cases}$

Solve each system by the substitution method. Check. (Section 15.5)

42. $\begin{cases} 2x + y = -2 \\ 5x + 3y = -1 \end{cases}$ **43.** $\begin{cases} x - y = 4 \\ 2x - 5y = 2 \end{cases}$ **44.** $\begin{cases} 3x - 4y = 7 \\ x + 2y = 4 \end{cases}$

Use Condition 1 and two variables to represent the unknowns. Use Condition 1 and Condition 2 to write a system of equations for the problem. Solve. (Sections 15.4, 15.6, and 15.7)

45. The sum of two numbers is 52. The difference of the two numbers is 20. Find the unknown numbers.

46. Sam and Greta worked a total of 19 hours. Sam was paid \$4 per hour, and Greta was paid \$5 per hour. Together, they earned \$83. How many hours did each work?

47. Emily and Janet walked a total of 42 miles in a recent walkathon. Emily earned 30 cents a mile, and Janet earned 20 cents a mile. They earned a total of \$10.40. How many miles did each walk?

48. The length of a rectangle is three feet longer than twice the width. The perimeter is 114 feet. Find the length and width.

CHAPTER 16

Radicals

Gravity causes the Foucault pendulum to swing from side to side in uniform time. The time T that it takes for a pendulum to swing from one side to the other and back again is expressed by the formula $T = 2\pi q$, where q is the square root of the quotient of the length of the cord holding the pendulum and the force of gravity.

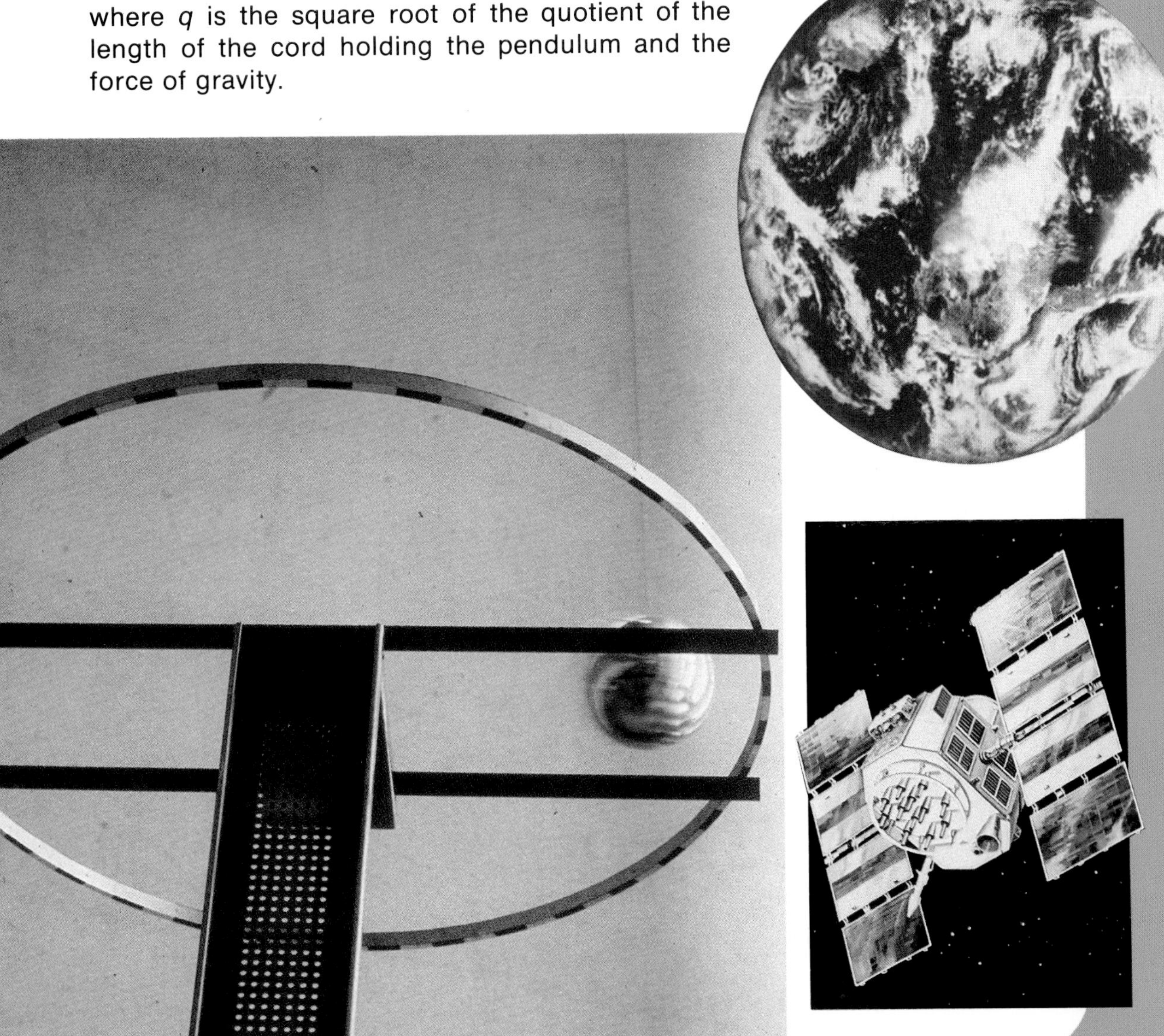

16.1 Simplifying Radicals

Recall from Chapter 8 that numerals such as the following are ***radicals.*** The number under the radical sign, such as "5" in "$\sqrt{5}$" is the ***radicand.***

$$\sqrt{5} \qquad \sqrt{8} \qquad 4\sqrt{2}$$

Compare the product of two radicals, $\sqrt{9} \cdot \sqrt{16}$, with the single radical, $\sqrt{9 \cdot 16}$.

$\sqrt{9} \cdot \sqrt{16}$	$\sqrt{9 \cdot 16}$
$3 \cdot 4$	$\sqrt{144}$
12 =	12

This suggests the following property.

> ***Product Property of Radicals***
>
> For any nonnegative numbers a and b,
>
> $\sqrt{\boldsymbol{a}} \cdot \sqrt{\boldsymbol{b}} = \sqrt{\boldsymbol{ab}}$
>
> $\sqrt{5} \cdot \sqrt{3} = \sqrt{15}$
>
> $\sqrt{4} \cdot \sqrt{16} = \sqrt{64}$

A product such as $\sqrt{5} \cdot \sqrt{3}$ is read

"radical 5 times radical 3" or
"square root of 5 times square root of 3."

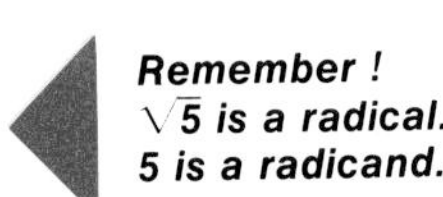
Remember !
$\sqrt{5}$ is a radical.
5 is a radicand.

EXAMPLE 1 Write one radical for each product.

a. $\sqrt{4} \cdot \sqrt{25}$ **b.** $\sqrt{9} \cdot \sqrt{17}$ **c.** $\sqrt{7} \cdot \sqrt{11}$

Solutions: **a.** $\sqrt{4} \cdot \sqrt{25} = \sqrt{100}$ **b.** $\sqrt{9} \cdot \sqrt{17} = \sqrt{153}$ **c.** $\sqrt{7} \cdot \sqrt{11} = \sqrt{77}$

P–1 **Write one radical for each product.**

a. $\sqrt{16} \cdot \sqrt{9}$ **b.** $\sqrt{6} \cdot \sqrt{25}$ **c.** $\sqrt{5} \cdot \sqrt{7}$

The table at the top of the next page shows that the Product Property of Radicals can be used to write a single radical as a product.

Single Radical	Product of Two Radicals
$\sqrt{17 \cdot 3}$	$\sqrt{17} \cdot \sqrt{3}$
$\sqrt{38}$	$\sqrt{2} \cdot \sqrt{19}$
$\sqrt{42}$	$\sqrt{6} \cdot \sqrt{7}$; $\sqrt{3} \cdot \sqrt{14}$; $\sqrt{2} \cdot \sqrt{21}$

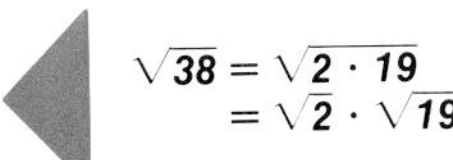

$\sqrt{38} = \sqrt{2 \cdot 19}$
$= \sqrt{2} \cdot \sqrt{19}$

There is more than one way to write the product.

EXAMPLE 2 Write as a product of two radicals.

a. $\sqrt{65}$ **b.** $\sqrt{70}$

Solutions: **a.** $\sqrt{65} = \sqrt{5 \cdot 13}$
$= \sqrt{5} \cdot \sqrt{13}$

b. $\sqrt{70} = \sqrt{7 \cdot 10}$
$= \sqrt{7} \cdot \sqrt{10}$

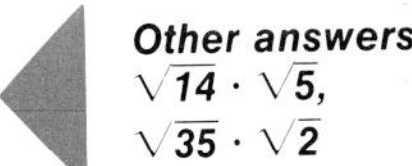

Other answers:
$\sqrt{14} \cdot \sqrt{5}$,
$\sqrt{35} \cdot \sqrt{2}$

P–2 **Write as a product of two radicals.**

a. $\sqrt{33}$ **b.** $\sqrt{10}$ **c.** $\sqrt{42}$ **d.** $\sqrt{6 \cdot 9}$

A radical can be simplified whenever the radicand has a factor that is a perfect square (a perfect square factor). Recall some of the numbers that are perfect squares.

$1^2 = \mathbf{1}$	$2^2 = \mathbf{4}$	$3^2 = \mathbf{9}$	$4^2 = \mathbf{16}$	$5^2 = \mathbf{25}$
$6^2 = \mathbf{36}$	$7^2 = \mathbf{49}$	$8^2 = \mathbf{64}$	$9^2 = \mathbf{81}$	$10^2 = \mathbf{100}$

EXAMPLE 3 Simplify $\sqrt{18}$.

Solution:
1. Write 18 as a product of 9 (a perfect square) and 2. ⟶ $\sqrt{18} = \sqrt{9 \cdot 2}$
2. Use the Product Property of Radicals. ⟶ $= \sqrt{9} \cdot \sqrt{2}$
3. Simplify the perfect square radical. ⟶ $= 3\sqrt{2}$

A radical is ***simplified*** if the radicand has no perfect square factor other than 1. $\sqrt{14}$, $2\sqrt{5}$, $\sqrt{33}$

P–3 **Simplify, if possible.**

a. $\sqrt{44}$ **b.** $\sqrt{21}$ **c.** $\sqrt{71}$ **d.** $\sqrt{63}$

Steps for Simplifying a Radical

1. Write the radicand as a product of a perfect square factor and another number.
2. Apply the Product Property of Radicals.
3. Write a square root of the perfect square radicand.
4. If necessary, repeat the above steps until no radicand has any perfect square factors other than 1.

In Example 4, Method 1 uses the greatest perfect square factor. Method 2 shows that more steps are needed when the greatest perfect square factor is not used.

EXAMPLE 4 Simplify $\sqrt{48}$.

Solution:

Method 1

16 is the greatest perfect square factor of 48.

$$\sqrt{48} = \sqrt{16 \cdot 3} = \sqrt{16} \cdot \sqrt{3} = 4\sqrt{3}$$

Method 2

$$\sqrt{48} = \sqrt{4 \cdot 12} = \sqrt{4} \cdot \sqrt{12} = 2\sqrt{4 \cdot 3} = 2\sqrt{4} \cdot \sqrt{3} = 2 \cdot 2 \cdot \sqrt{3} = 4\sqrt{3}$$

4 is a perfect square. $\sqrt{12}$ can be simplified.

$\sqrt{3}$ cannot be simplified.

P–4 **Simplify.**

a. $\sqrt{72}$ **b.** $\sqrt{162}$ **c.** $\sqrt{176}$ **d.** $\sqrt{75}$

CLASSROOM EXERCISES

Write one radical for each product. (Example 1)

1. $\sqrt{2} \cdot \sqrt{6}$ **2.** $\sqrt{3} \cdot \sqrt{5}$ **3.** $\sqrt{7} \cdot \sqrt{5}$ **4.** $\sqrt{6} \cdot \sqrt{11}$

5. $\sqrt{4} \cdot \sqrt{7}$ **6.** $\sqrt{6} \cdot \sqrt{6}$ **7.** $\sqrt{9} \cdot \sqrt{10}$ **8.** $\sqrt{12} \cdot \sqrt{5}$

Write as a product of two radicals. (Example 2)

9. $\sqrt{15}$ **10.** $\sqrt{77}$ **11.** $\sqrt{66}$ **12.** $\sqrt{54}$

13. $\sqrt{2 \cdot 36}$ **14.** $\sqrt{5 \cdot 25}$ **15.** $\sqrt{4 \cdot 13}$ **16.** $\sqrt{17 \cdot 16}$

Simplify. (Examples 3 and 4)

17. $\sqrt{12}$ 18. $\sqrt{8}$ 19. $\sqrt{20}$ 20. $\sqrt{18}$
21. $\sqrt{27}$ 22. $\sqrt{40}$ 23. $\sqrt{54}$ 24. $\sqrt{50}$

WRITTEN EXERCISES

Goal: To simplify a radical or a product of radicals
Sample Problem: Simplify: $\sqrt{72}$
Answer: $6\sqrt{2}$

Write one radical for each product. (Example 1)

1. $\sqrt{2} \cdot \sqrt{7}$ 2. $\sqrt{3} \cdot \sqrt{11}$ 3. $\sqrt{5} \cdot \sqrt{9}$ 4. $\sqrt{6} \cdot \sqrt{9}$
5. $\sqrt{8} \cdot \sqrt{7}$ 6. $\sqrt{10} \cdot \sqrt{7}$ 7. $\sqrt{12} \cdot \sqrt{3}$ 8. $\sqrt{5} \cdot \sqrt{11}$

Write as a product of two radicals. (Example 2)

9. $\sqrt{6}$ 10. $\sqrt{22}$ 11. $\sqrt{64}$ 12. $\sqrt{52}$
13. $\sqrt{9 \cdot 4}$ 14. $\sqrt{7 \cdot 8}$ 15. $\sqrt{11 \cdot 15}$ 16. $\sqrt{9 \cdot 12}$

For Exercises 17–36, simplify. (Examples 3 and 4)

17. $\sqrt{24}$ 18. $\sqrt{28}$ 19. $\sqrt{45}$ 20. $\sqrt{54}$
21. $\sqrt{80}$ 22. $\sqrt{32}$ 23. $\sqrt{112}$ 24. $\sqrt{252}$

MIXED PRACTICE

25. $\sqrt{199}$ 26. $\sqrt{3} \cdot \sqrt{19}$ 27. $\sqrt{180}$ 28. $\sqrt{150}$
29. $\sqrt{6} \cdot \sqrt{5}$ 30. $\sqrt{68}$ 31. $\sqrt{13} \cdot \sqrt{11}$ 32. $\sqrt{117}$
33. $\sqrt{200}$ 34. $\sqrt{108}$ 35. $\sqrt{92}$ 36. $\sqrt{7} \cdot \sqrt{30}$

MORE CHALLENGING EXERCISES

Simplify.

37. $\sqrt{3} \cdot \sqrt{45}$ 38. $\sqrt{13} \cdot \sqrt{32}$ 39. $\sqrt{19} \cdot \sqrt{50}$ 40. $\sqrt{63} \cdot \sqrt{35}$

NON-ROUTINE PROBLEM

41. Which triangle at the right has the greater area? (Hint: Use the Pythagorean Theorem.)

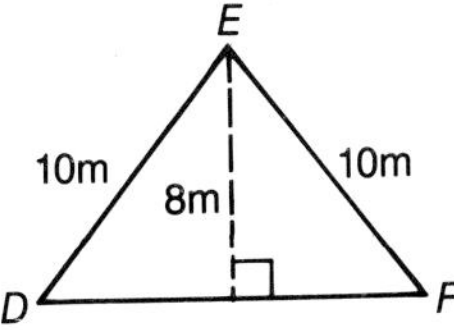

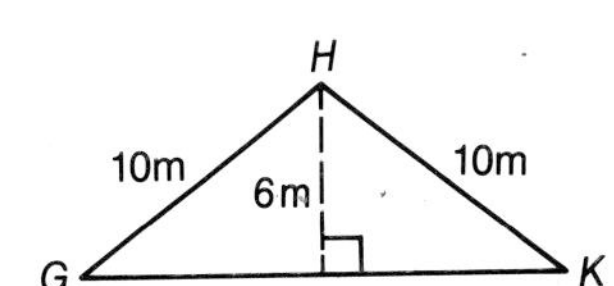

16.2 Approximations of Square Roots

A portion of a Table of Squares and Square Roots is shown below. The square roots are accurate to three decimal places.

Table of Squares and Square Roots

Number	Square	Square Root
1	1	1.000
2	4	1.414
3	9	1.732
4	16	2.000
5	25	2.236
32	1024	5.657
33	1089	5.745
34	1156	5.831
35	1225	5.916
36	1296	6.000

$3^2 = 9$ $\sqrt{3} \approx 1.732$

$32^2 = 1024$ $\sqrt{32} \approx 5.657$

P–1 **What is the square of each of the following? Use the table above.**

a. 3 **b.** 4 **c.** 35 **d.** 36

EXAMPLE 1 Approximate each square root to three decimal places. Use the table above.

a. $\sqrt{2}$ **b.** $\sqrt{5}$ **c.** $\sqrt{34}$

Solutions: **a.** $\sqrt{2} \approx 1.414$ **b.** $\sqrt{5} \approx 2.236$ **c.** $\sqrt{34} \approx 5.831$

P–2 **Use the Table of Squares and Square Roots on page 458 to write each square root.**

a. $\sqrt{33}$ **b.** $\sqrt{76}$ **c.** $\sqrt{118}$ **d.** $\sqrt{129}$

In the table, the numbers in the "Number" column are the positive square roots of the numbers in the "Square" column.

$\sqrt{1764} = 42$
$\sqrt{1849} = 43$

Number	Square	Square Root
42	1764	6.481
43	1849	6.557

Thus, many square roots of perfect squares, such as 1764 and 1849, can be read directly from the "Number" column, not the "Square Root" column.

EXAMPLE 2 Use the Table of Square Roots on page 458 to write each square root.

a. $\sqrt{4624}$ b. $\sqrt{2209}$ c. $\sqrt{17{,}689}$

Solutions: **a.** $\sqrt{4624} = 68$ **b.** $\sqrt{2209} = 47$ **c.** $\sqrt{17{,}689} = 133$

P–3 **Use the table on page 458 to write each square root.**

a. $\sqrt{441}$ **b.** $\sqrt{19{,}321}$ **c.** $\sqrt{9025}$ **d.** $\sqrt{2704}$

The table on page 458 shows squares and square roots of the positive integers from 1 to 150. The table can also be used to find square roots of some numbers that are greater than 150.

EXAMPLE 3 Approximate $\sqrt{250}$ to three decimal places.

Solution:
1. Write 250 as a product of 25 (a perfect square) and 10. ⟶ $\sqrt{250} = \sqrt{25 \cdot 10}$
2. Use the Product Property of Radicals. ⟶ $= \sqrt{25} \cdot \sqrt{10}$
3. Simplify. ⟶ $= 5\sqrt{10}$
4. Approximate $\sqrt{10}$ from the table. ⟶ $\approx 5(3.162)$

≈ 15.810

P–4 **Approximate to three decimal places.**

a. $\sqrt{200}$ **b.** $\sqrt{1314}$ **c.** $\sqrt{245}$ **d.** $\sqrt{207}$

Table of Squares and Square Roots

No.	Square	Square Root	No.	Square	Square Root	No.	Square	Square Root
1	1	1.000	**51**	2601	7.141	**101**	10,201	10.050
2	4	1.414	**52**	2704	7.211	**102**	10,404	10.100
3	9	1.732	**53**	2809	7.280	**103**	10,609	10.149
4	16	2.000	**54**	2916	7.348	**104**	10,816	10.198
5	25	2.236	**55**	3025	7.416	**105**	11,025	10.247
6	36	2.449	**56**	3136	7.483	**106**	11,236	10.296
7	49	2.646	**57**	3249	7.550	**107**	11,449	10.344
8	64	2.828	**58**	3364	7.616	**108**	11,664	10.392
9	81	3.000	**59**	3481	7.681	**109**	11,881	10.440
10	100	3.162	**60**	3600	7.746	**110**	12,100	10.488
11	121	3.317	**61**	3721	7.810	**111**	12,321	10.536
12	144	3.464	**62**	3844	7.874	**112**	12,544	10.583
13	169	3.606	**63**	3969	7.937	**113**	12,769	10.630
14	196	3.742	**64**	4096	8.000	**114**	12,996	10.677
15	225	3.873	**65**	4225	8.062	**115**	13,225	10.724
16	256	4.000	**66**	4356	8.124	**116**	13,456	10.770
17	289	4.123	**67**	4489	8.185	**117**	13,689	10.817
18	324	4.243	**68**	4624	8.246	**118**	13,924	10.863
19	361	4.359	**69**	4761	8.307	**119**	14,161	10.909
20	400	4.472	**70**	4900	8.367	**120**	14,400	10.954
21	441	4.583	**71**	5041	8.426	**121**	14,641	11.000
22	484	4.690	**72**	5184	8.485	**122**	14,884	11.045
23	529	4.796	**73**	5329	8.544	**123**	15,129	11.091
24	576	4.899	**74**	5476	8.602	**124**	15,376	11.136
25	625	5.000	**75**	5625	8.660	**125**	15,625	11.180
26	676	5.099	**76**	5776	8.718	**126**	15,876	11.225
27	729	5.196	**77**	5929	8.775	**127**	16,129	11.269
28	784	5.292	**78**	6084	8.832	**128**	16,384	11.314
29	841	5.385	**79**	6241	8.888	**129**	16,641	11.358
30	900	5.477	**80**	6400	8.944	**130**	16,900	11.402
31	961	5.568	**81**	6561	9.000	**131**	17,161	11.446
32	1024	5.657	**82**	6724	9.055	**132**	17,424	11.489
33	1089	5.745	**83**	6889	9.110	**133**	17,689	11.533
34	1156	5.831	**84**	7056	9.165	**134**	17,956	11.576
35	1225	5.916	**85**	7225	9.220	**135**	18,225	11.619
36	1296	6.000	**86**	7396	9.274	**136**	18,496	11.662
37	1369	6.083	**87**	7569	9.327	**137**	18,769	11.705
38	1444	6.164	**88**	7744	9.381	**138**	19,044	11.747
39	1521	6.245	**89**	7921	9.434	**139**	19,321	11.790
40	1600	6.325	**90**	8100	9.487	**140**	19,600	11.832
41	1681	6.403	**91**	8281	9.539	**141**	19,881	11.874
42	1764	6.481	**92**	8464	9.592	**142**	20,164	11.916
43	1849	6.557	**93**	8649	9.644	**143**	20,449	11.958
44	1936	6.633	**94**	8836	9.695	**144**	20,736	12.000
45	2025	6.708	**95**	9025	9.747	**145**	21,025	12.042
46	2116	6.782	**96**	9216	9.798	**146**	21,316	12.083
47	2209	6.856	**97**	9409	9.849	**147**	21,609	12.124
48	2304	6.928	**98**	9604	9.899	**148**	21,904	12.166
49	2401	7.000	**99**	9801	9.950	**149**	22,201	12.207
50	2500	7.071	**100**	10,000	10.000	**150**	22,500	12.247

The following word rule and formula tell you how to find the length of the diagonal of a square.

Word Rule: The length of the diagonal equals the length of a side times the square root of 2.

Formula: $\mathbf{d = s\sqrt{2}}$

d = length of the diagonal
s = length of each side

EXAMPLE 4

The length of each side of a square is 15 centimeters. Approximate the length of a diagonal of the square to three decimal places.

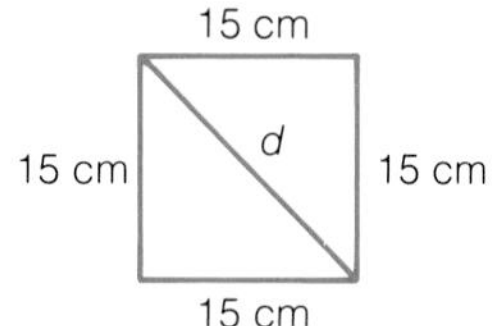

[1] Draw a figure. On the drawing show the known and unknown values.

[2] Write the formula. ⟶ $d = s\sqrt{2}$

[3] Substitute the known value in the formula. ⟶ $d = 15\sqrt{2}$

[4] Approximate $\sqrt{2}$ from the table. ⟶ $d \approx 15(1.414)$

$d \approx 21.210$ centimeters

CLASSROOM EXERCISES

Use the Table of Squares and Square Roots on page 458 to find each square.

1. 33^2 **2.** 37^2 **3.** 119^2 **4.** 95^2

5. 61^2 **6.** 142^2 **7.** 38^2 **8.** 104^2

Approximate each square root to three decimal places. Use the Table of Squares and Square Roots on page 458. (Example 1)

9. $\sqrt{84}$ **10.** $\sqrt{69}$ **11.** $\sqrt{136}$ **12.** $\sqrt{33}$

13. $\sqrt{44}$ **14.** $\sqrt{99}$ **15.** $\sqrt{131}$ **16.** $\sqrt{150}$

Use the Table of Squares and Square Roots on page 458 to write each square root. (Example 2)

17. $\sqrt{21{,}025}$ **18.** $\sqrt{8836}$ **19.** $\sqrt{4096}$ **20.** $\sqrt{1444}$

21. $\sqrt{17{,}956}$ **22.** $\sqrt{2401}$ **23.** $\sqrt{6084}$ **24.** $\sqrt{16{,}641}$

Write each of the following as a product of a positive integer and an approximate square root. (Steps 2, 3, and 4 of Example 3)

25. $\sqrt{36} \cdot \sqrt{23}$ **26.** $\sqrt{81} \cdot \sqrt{46}$ **27.** $\sqrt{49} \cdot \sqrt{94}$ **28.** $\sqrt{16} \cdot \sqrt{55}$

WRITTEN EXERCISES

Goal: To approximate the values of square roots by use of a Table of Squares and Square Roots

Sample Problem: Approximate $\sqrt{180}$ to three decimal places.

Answer: 13.416

Approximate each square root to three decimal places. Use the Table of Squares and Square Roots on page 458. (Example 1)

1. $\sqrt{62}$ **2.** $\sqrt{74}$ **3.** $\sqrt{37}$ **4.** $\sqrt{50}$

5. $\sqrt{93}$ **6.** $\sqrt{105}$ **7.** $\sqrt{113}$ **8.** $\sqrt{129}$

Use the Table of Squares and Square Roots on page 458 to write each square root. (Example 2)

9. $\sqrt{841}$ **10.** $\sqrt{1764}$ **11.** $\sqrt{7744}$ **12.** $\sqrt{4356}$

13. $\sqrt{10{,}609}$ **14.** $\sqrt{16{,}129}$ **15.** $\sqrt{324}$ **16.** $\sqrt{2304}$

Approximate each square root to three decimal places. (Example 3)

17. $\sqrt{172}$ **18.** $\sqrt{204}$ **19.** $\sqrt{153}$ **20.** $\sqrt{171}$

21. $\sqrt{176}$ **22.** $\sqrt{208}$ **23.** $\sqrt{275}$ **24.** $\sqrt{325}$

MIXED PRACTICE *Approximate to three decimal places or give the exact value if appropriate.*

25. $\sqrt{89}$ **26.** $\sqrt{7021}$ **27.** $\sqrt{4761}$ **28.** $\sqrt{66}$

29. $\sqrt{228}$ **30.** $\sqrt{272}$ **31.** $\sqrt{104}$ **32.** $\sqrt{12{,}544}$

33. $\sqrt{9409}$ **34.** $\sqrt{132}$ **35.** $\sqrt{304}$ **36.** $\sqrt{234}$

APPLICATIONS

37. On a baseball diamond the distance between bases is 90 feet. Find the distance from home plate to second base to the nearest tenth of a foot. HINT: Find the length of a diagonal to the nearest inch. (Example 4)

38. The size of a television screen is determined by the diagonal length. One model has a square screen with each side about $8\frac{1}{2}$ inches long. Compute the size (not the area) to the nearest inch. (Example 4)

A calculator without a square root key can be used to approximate the value of square roots by the "divide and average" method.

1. Choose a whole number as an estimate of the square root.
2. Divide the radicand by the estimated square root.
3. Find the average of the quotient and the estimated square root.
4. Divide the radicand by the average from step **3**.
5. Repeat steps **3** and **4** until two successive averages from step **3** are identical to one decimal place.

EXAMPLE Approximate to one decimal place: $\sqrt{647}$

SOLUTION 1 Use the table on page 458. ⟶ $\sqrt{647} \approx 25$

2 

3 [+] [2] [5] [÷] [2] [=] [M+]

The average is now stored in memory.

4 [6] [4] [7] [÷] [MR]

5 [+] [MR] [÷] [2] [=]

25.436194

Since the averages from steps 3 and 5 are identical to one decimal place, the answer is 25.4.

Approximate each square root to one decimal place.

39. $\sqrt{319}$ **40.** $\sqrt{813}$ **41.** $\sqrt{1027}$ **42.** $\sqrt{2433}$ **43.** $\sqrt{5029}$ **44.** $\sqrt{7105}$

45. $\sqrt{236}$ **46.** $\sqrt{418}$ **47.** $\sqrt{1240}$ **48.** $\sqrt{1876}$ **49.** $\sqrt{4608}$ **50.** $\sqrt{3156}$

REVIEW CAPSULE FOR SECTION 16.3

Factor. Use the Distributive Property. (Pages 37–39)

1. $3n^2 + 2n$ **2.** $15y^2 - 3y^3$ **3.** $6h^2 - h$ **4.** $7z^2 + 14z^3$

Combine like terms. (Pages 40–42)

5. $2p + 3p$ **6.** $3.2g + 8g$ **7.** $7u - 2u$ **8.** $9b - 4b$

9. $12w + w$ **10.** $4d + d$ **11.** $10.2q - q$ **12.** $9.7y - y$

Simplify. (Pages 452–455)

13. $\sqrt{12}$ **14.** $\sqrt{8}$ **15.** $\sqrt{30}$ **16.** $\sqrt{24}$

17. $\sqrt{18}$ **18.** $\sqrt{27}$ **19.** $\sqrt{45}$ **20.** $\sqrt{50}$

16.3 Sums and Differences

The radicals $2\sqrt{7}$ and $3\sqrt{7}$ have the same radicand, 7. They are called like radicals.

Definition

Square root radicals with equal radicands are ***like radicals.***	$2\sqrt{7}$ and $3\sqrt{7}$ $\sqrt{8}$, $5\sqrt{8}$ and $-2\sqrt{8}$

P–1 **Which of the following are like radicals?**

a. $2\sqrt{12}$ and $-\sqrt{12}$ **b.** $10\sqrt{7}$ and $7\sqrt{10}$ **c.** $-\sqrt{5}$, $\frac{1}{3}\sqrt{5}$, and $5\sqrt{5}$

The sum or difference of two like radicals can be simplified into a single radical.

EXAMPLE 1 Add: $2\sqrt{5} + 6\sqrt{5}$

Solution:

[1] Factor and add. ⟶ $\mathbf{2}\sqrt{5} + \mathbf{6}\sqrt{5} = (\mathbf{2} + \mathbf{6})\sqrt{5}$

[2] Simplest form. ⟶ $= \mathbf{8}\sqrt{5}$

Adding $2\sqrt{5}$ and $6\sqrt{5}$ is like adding 2x and 5x.

P–2 **Add.**

a. $3\sqrt{7} + 14\sqrt{7}$ **b.** $\frac{1}{5}\sqrt{13} + \frac{2}{5}\sqrt{13}$ **c.** $-2\sqrt{11} + 5\sqrt{11}$

EXAMPLE 2 Subtract: $5\sqrt{14} - 3\sqrt{14}$

Solution:

[1] Use the Distributive Property. ⟶ $5\sqrt{14} - 3\sqrt{14} = (5 - 3)\sqrt{14}$

[2] Subtract and simplify. ⟶ $= 2\sqrt{14}$

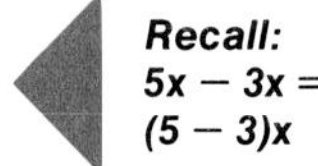

P–3

a. $9\sqrt{3} - 2\sqrt{3}$ **b.** $\frac{5}{3}\sqrt{11} - \frac{4}{3}\sqrt{11}$ **c.** $-6\sqrt{5} - 2\sqrt{5}$

When adding or subtracting, think of a radical such as $\sqrt{10}$ as $1\sqrt{10}$.

EXAMPLE 3 Add and subtract: $6\sqrt{10} - 3\sqrt{10} + \sqrt{10}$

Solution:

1 Write $\sqrt{10}$ as $1\sqrt{10}$. ⟶ $6\sqrt{10} - 3\sqrt{10} + \sqrt{10} = 6\sqrt{10} - 3\sqrt{10} + 1\sqrt{10}$

2 Use the Distributive Property. ⟶ $= (6 - 3 + 1)\sqrt{10}$

$= 4\sqrt{10}$

P–4 **Add and subtract.**

a. $3\sqrt{7} + \sqrt{7} - 2\sqrt{7}$ **b.** $2\sqrt{6} - 5\sqrt{6} + \sqrt{6}$ **c.** $9\sqrt{3} - \sqrt{3} + 9\sqrt{3}$

Radicals such as $5\sqrt{27}$ and $\sqrt{3}$ do not appear to be like radicals. However, simplifying $5\sqrt{27}$ shows that $5\sqrt{27} = 15\sqrt{3}$.

$$\begin{aligned} 5\sqrt{27} &= 5 \cdot \sqrt{9 \cdot 3} \\ &= 5 \cdot \sqrt{9} \cdot \sqrt{3} \\ &= 5 \cdot 3 \cdot \sqrt{3}, \text{ or } 15\sqrt{3} \end{aligned}$$

Thus, $5\sqrt{27}$ and $\sqrt{3}$ are like radicals.

EXAMPLE 4 Subtract: $5\sqrt{12} - \sqrt{3}$

Solution: 1 Simplify all radicals. ⟶ $5\sqrt{12} - \sqrt{3} = 5\sqrt{4 \cdot 3} - \sqrt{3}$

$= 5\sqrt{4} \cdot \sqrt{3} - \sqrt{3}$

$= 5 \cdot 2\sqrt{3} - \sqrt{3}$

2 Write $\sqrt{3}$ as $1\sqrt{3}$. Subtract. ⟶ $= 10\sqrt{3} - 1\sqrt{3}$

$= 9\sqrt{3}$

P–5 **Add or subtract.**

a. $3\sqrt{12} + 2\sqrt{3}$ **b.** $5\sqrt{2} - 2\sqrt{8}$ **c.** $4\sqrt{18} + \sqrt{2}$

Steps for Adding and Subtracting Radicals

1. Simplify all radicals.
2. Think of radicals such as $\sqrt{3}$ and $\sqrt{10}$ as $1\sqrt{3}$ and $1\sqrt{10}$.
3. Add and subtract like radicals.

CLASSROOM EXERCISES

Add. (Example 1)

1. $6\sqrt{2}+8\sqrt{2}$
2. $7\sqrt{3}+\sqrt{3}$
3. $3\sqrt{11}+12\sqrt{11}$
4. $4\sqrt{6}+13\sqrt{6}$
5. $\frac{1}{2}\sqrt{14}+\frac{1}{4}\sqrt{14}$
6. $\frac{5}{8}\sqrt{15}+\frac{7}{8}\sqrt{15}$

Subtract. (Example 2)

7. $9\sqrt{5}-6\sqrt{5}$
8. $10\sqrt{5}-\sqrt{5}$
9. $3\sqrt{7}-5\sqrt{7}$
10. $\sqrt{15}-3\sqrt{15}$
11. $\frac{2}{3}\sqrt{19}-\frac{5}{3}\sqrt{19}$
12. $\sqrt{23}-\frac{7}{6}\sqrt{23}$

Add and subtract. (Example 3)

13. $4\sqrt{2}+7\sqrt{2}+\sqrt{2}$
14. $6\sqrt{3}-2\sqrt{3}+5\sqrt{3}$
15. $8\sqrt{5}-\sqrt{5}-2\sqrt{5}$
16. $-12\sqrt{7}+\sqrt{7}-3\sqrt{7}$
17. $\frac{1}{4}\sqrt{10}+\frac{1}{2}\sqrt{10}-\frac{3}{8}\sqrt{10}$
18. $-\frac{2}{3}\sqrt{11}+\frac{4}{3}\sqrt{11}-\frac{1}{3}\sqrt{11}$

Add or subtract. (Example 4)

19. $\sqrt{12}+5\sqrt{3}$
20. $9\sqrt{2}+\sqrt{8}$
21. $5\sqrt{2}-\sqrt{8}$
22. $\sqrt{18}-\sqrt{2}$
23. $5\sqrt{5}-\sqrt{20}$
24. $\sqrt{18}-\sqrt{8}$

WRITTEN EXERCISES

Goal: To add or subtract with radicals
Sample Problem: $2\sqrt{3}+5\sqrt{12}$
Answer: $12\sqrt{3}$

Add. (Example 1)

1. $5\sqrt{2}+11\sqrt{2}$
2. $4\sqrt{3}+6\sqrt{3}$
3. $13\sqrt{5}+\sqrt{5}$
4. $\sqrt{7}+12\sqrt{7}$
5. $\frac{3}{4}\sqrt{6}+\frac{1}{8}\sqrt{6}$
6. $\frac{2}{3}\sqrt{10}+\frac{1}{6}\sqrt{10}$

Subtract. (Example 2)

7. $10\sqrt{6}-7\sqrt{6}$
8. $16\sqrt{10}-9\sqrt{10}$
9. $14\sqrt{11}-\sqrt{11}$
10. $21\sqrt{15}-\sqrt{15}$
11. $\frac{2}{3}\sqrt{3}-\frac{4}{3}\sqrt{3}$
12. $\frac{3}{4}\sqrt{5}-\frac{5}{4}\sqrt{5}$

For Exercises 13–42, perform the indicated operations. (Example 3)

13. $3\sqrt{6}-\sqrt{6}+4\sqrt{6}$
14. $5\sqrt{2}-\sqrt{2}+2\sqrt{2}$
15. $-7\sqrt{10}+3\sqrt{10}-5\sqrt{10}$
16. $-3\sqrt{11}+5\sqrt{11}-6\sqrt{11}$
17. $\frac{7}{2}\sqrt{6}+\frac{3}{2}\sqrt{6}-\frac{1}{2}\sqrt{6}$
18. $-\frac{1}{4}\sqrt{7}+\frac{7}{4}\sqrt{7}+\frac{3}{4}\sqrt{7}$

(Example 4)

19. $\sqrt{18} + 5\sqrt{2}$
20. $\sqrt{20} + 6\sqrt{5}$
21. $2\sqrt{18} + \sqrt{8}$
22. $\sqrt{27} + 3\sqrt{12}$
23. $3\sqrt{6} - \sqrt{24}$
24. $3\sqrt{7} - \sqrt{28}$

MIXED PRACTICE

25. $10\sqrt{6} - 7\sqrt{6}$
26. $13\sqrt{10} - 4\sqrt{10}$
27. $2\sqrt{3} + \sqrt{3} + 5\sqrt{3}$
28. $\frac{5}{6}\sqrt{13} - \frac{2}{3}\sqrt{13}$
29. $\frac{3}{8}\sqrt{17} - \frac{3}{4}\sqrt{17}$
30. $\sqrt{48} - \sqrt{75}$
31. $\sqrt{18} + \sqrt{50}$
32. $3\sqrt{5} - \sqrt{45} + 7\sqrt{5}$
33. $5\sqrt{3} - \sqrt{75} + 2\sqrt{3}$
34. $\sqrt{12} + \sqrt{27}$
35. $\sqrt{20} - \sqrt{45}$
36. $\sqrt{5} + 3\sqrt{5} + 4\sqrt{5}$
37. $\sqrt{2} + 5\sqrt{18}$
38. $4\sqrt{8} + \sqrt{18}$
39. $6\sqrt{12} - \sqrt{48}$
40. $7\sqrt{5} + 3\sqrt{5}$
41. $9\sqrt{15} + 6\sqrt{15}$
42. $2\sqrt{18} + \sqrt{32}$

MID-CHAPTER REVIEW

Simplify. (Section 16.1)

1. $\sqrt{81} \cdot \sqrt{37}$
2. $\sqrt{51} \cdot \sqrt{49}$
3. $\sqrt{144} \cdot \sqrt{43}$
4. $\sqrt{67} \cdot \sqrt{36}$
5. $\sqrt{104}$
6. $\sqrt{117}$
7. $\sqrt{120}$
8. $\sqrt{124}$
9. $\sqrt{140}$
10. $\sqrt{148}$
11. $\sqrt{207}$
12. $\sqrt{208}$

Approximate each square root to three decimal places. Use the Table of Squares and Square Roots on page 458. (Section 16.2)

13. $\sqrt{212}$
14. $\sqrt{232}$
15. $\sqrt{380}$
16. $\sqrt{425}$
17. $\sqrt{404}$
18. $\sqrt{480}$
19. $\sqrt{508}$
20. $\sqrt{909}$
21. $\sqrt{1552}$
22. $\sqrt{1251}$
23. $\sqrt{4067}$
24. $\sqrt{6592}$

25. A pipeline is built diagonally across a square field. Each side of the field measures 6 kilometers. What is the length of the pipeline?

26. John swims diagonally across a square swimming pool. Each side of the pool is 12 meters in length. Find the distance John swims.

Add or subtract. (Section 16.3)

27. $5\sqrt{17} + 4\sqrt{17}$
28. $8\sqrt{23} + 13\sqrt{23}$
29. $24\sqrt{17} - \sqrt{17}$
30. $19\sqrt{41} - \sqrt{41}$
31. $\sqrt{50} - \sqrt{18}$
32. $\sqrt{108} - \sqrt{75}$
33. $\sqrt{180} + \sqrt{80}$
34. $\sqrt{54} + \sqrt{150}$
35. $\sqrt{12} - \sqrt{3} + \sqrt{75}$
36. $\sqrt{128} - \sqrt{18} - \sqrt{72}$
37. $-\sqrt{40} + \sqrt{10} - \sqrt{250}$
38. $-\sqrt{15} + \sqrt{60} - \sqrt{135}$

16.4 Products

You multiply with radicals as with other real numbers.

EXAMPLE 1 Multiply: $-3(4\sqrt{2})$

Solution:

[1] Use the Associative Property of Multiplication. → $-3(4\sqrt{2}) = (-3 \cdot 4)\sqrt{2}$

[2] Multiply. → $= -12\sqrt{2}$

To multiply two or more radicals, you use the Product Property of Radicals ($\sqrt{a}\sqrt{b} = \sqrt{ab}$), as well as the Commutative and Associative Properties of Multiplication.

EXAMPLE 2 Multiply: $(-3\sqrt{5})(-7\sqrt{3})$

Solution:

[1] Use the Associative and Commutative Properties. → $(-3\sqrt{5})(-7\sqrt{3}) = (-3)(-7)(\sqrt{5})(\sqrt{3})$

[2] Use the Product Property of Radicals. → $= 21\sqrt{15}$

P–1 **Multiply.**

a. $2\sqrt{11}(4\sqrt{5})$ **b.** $(\frac{1}{2}\sqrt{3})(-\frac{2}{3}\sqrt{7})$ **c.** $(\sqrt{19})(-4\sqrt{2})$

When you multiply radicals, be sure to simplify the answer.

EXAMPLE 3 Multiply: $(-\sqrt{6})(\sqrt{10})$

Solution:

[1] The product of a positive number and a negative number is negative. → $(-\sqrt{6})(\sqrt{10}) = -(\sqrt{6} \cdot \sqrt{10})$

[2] Use the Product Property of Radicals. → $= -\sqrt{60}$

[3] Write 60 as a product of two factors, one of which is the greatest perfect square factor of 60. → $= -\sqrt{4 \cdot 15}$

[4] Simplify. → $= -\sqrt{4} \cdot \sqrt{15}$

$= -2\sqrt{15}$

P–2 **Multiply.**

a. $(\sqrt{8})(\sqrt{14})$ **b.** $(\sqrt{27})(\sqrt{3})$ **c.** $(3\sqrt{5})(-\sqrt{15})$

CLASSROOM EXERCISES

For Exercises 1–24, multiply. (Example 1)

1. $-5(4\sqrt{3})$ **2.** $-7(-8\sqrt{5})$ **3.** $10(-7\sqrt{6})$ **4.** $(9\sqrt{2})(-8)$

5. $-2(3\sqrt{10})$ **6.** $(2\sqrt{15})(-3)$ **7.** $-\frac{2}{3}(-12\sqrt{7})$ **8.** $(-7\sqrt{7})\frac{2}{5}$

(Example 2)

9. $(\sqrt{2})(\sqrt{7})$ **10.** $(-\sqrt{6})(\sqrt{5})$ **11.** $(3\sqrt{2})(-4\sqrt{13})$ **12.** $(-5\sqrt{5})(8\sqrt{6})$

13. $(-5\sqrt{5})(-2\sqrt{11})$ **14.** $(-4\sqrt{7})(8\sqrt{10})$ **15.** $(12\sqrt{7})(-\frac{3}{4}\sqrt{2})$ **16.** $(-\frac{1}{2}\sqrt{3})(14\sqrt{13})$

(Example 3)

17. $(-\sqrt{2})(-\sqrt{6})$ **18.** $(\sqrt{6})(-\sqrt{3})$ **19.** $(-3\sqrt{2})(2\sqrt{20})$ **20.** $(6\sqrt{6})(-4\sqrt{12})$

21. $(4\sqrt{3})(-3\sqrt{8})$ **22.** $(7\sqrt{2})(6\sqrt{14})$ **23.** $(-8\sqrt{6})(-\frac{1}{4}\sqrt{8})$ **24.** $(\frac{1}{3}\sqrt{10})(9\sqrt{6})$

WRITTEN EXERCISES

Goal: To multiply with radicals
Sample Problem: $(-4\sqrt{5})(5\sqrt{10})$
Answer: $-100\sqrt{2}$

For Exercises 1–32, multiply. (Example 1)

1. $-12(6\sqrt{5})$ **2.** $-8(9\sqrt{10})$ **3.** $15(-8\sqrt{7})$ **4.** $(-4\sqrt{3})12$

5. $14(-6\sqrt{13})$ **6.** $-3(-24\sqrt{2})$ **7.** $-\frac{3}{8}(-32\sqrt{3})$ **8.** $(18\sqrt{6})(-\frac{5}{9})$

(Example 2)

9. $(\sqrt{13})(\sqrt{5})$ **10.** $(\sqrt{13})(\sqrt{19})$ **11.** $(-\sqrt{6})(\sqrt{11})$ **12.** $(\sqrt{5})(-\sqrt{13})$

13. $(\sqrt{5})(-\sqrt{7})$ **14.** $(10\sqrt{10})(-\sqrt{3})$ **15.** $(6\sqrt{11})(-\frac{3}{4}\sqrt{2})$ **16.** $(\frac{2}{3}\sqrt{3})(-\sqrt{6})$

(Example 3)

17. $(\sqrt{12})(5\sqrt{2})$ **18.** $(\sqrt{20})(4\sqrt{3})$ **19.** $(-\sqrt{10})(\sqrt{2})$ **20.** $(\sqrt{8})(-\sqrt{6})$

21. $(\sqrt{2})(-\sqrt{14})$ **22.** $(-\sqrt{20})(-6\sqrt{6})$ **23.** $(-8\sqrt{27})(-\frac{3}{4}\sqrt{5})$ **24.** $(\frac{5}{8}\sqrt{12})(-32\sqrt{5})$

MIXED PRACTICE

25. $(-\sqrt{10})(-\sqrt{23})$ **26.** $(-\sqrt{45})(5\sqrt{3})$ **27.** $-\frac{3}{4}(12\sqrt{10})$ **28.** $(-\sqrt{63})(7\sqrt{3})$

29. $(-\sqrt{54})(-\sqrt{5})$ **30.** $(-3\sqrt{27})(5\sqrt{2})$ **31.** $(-3\sqrt{5})(4\sqrt{2})$ **32.** $(-\sqrt{17})(-\sqrt{11})$

REVIEW CAPSULE FOR SECTION 16.5

Evaluate. (Pages 225–228)

1. $\sqrt{25}$ 2. $\sqrt{49}$ 3. $\sqrt{\frac{25}{49}}$ 4. $\sqrt{16}$ 5. $\sqrt{81}$ 6. $\sqrt{\frac{16}{81}}$

16.5 Quotients

To simplify a fraction like $\frac{\sqrt{25}}{\sqrt{49}}$, evaluate the numerator and denominator separately.

$$\frac{\sqrt{25}}{\sqrt{49}} = \frac{5}{7}$$ ◀ **$\sqrt{25} = 5$; $\sqrt{49} = 7$**

Recall from Chapter 8 the meaning of a radical such as $\sqrt{\frac{25}{49}}$.

$$\sqrt{\frac{25}{49}} = \frac{5}{7} \quad \text{because} \quad \frac{5}{7} \cdot \frac{5}{7} = \frac{25}{49}.$$

Thus, $\frac{\sqrt{25}}{\sqrt{49}}$ and $\sqrt{\frac{25}{49}}$ both equal $\frac{5}{7}$.

Quotient Property of Radicals

For any nonnegative number a and any positive number b, $\frac{\sqrt{a}}{\sqrt{b}} = \sqrt{\frac{a}{b}}$.

$$\frac{\sqrt{4}}{\sqrt{9}} = \sqrt{\frac{4}{9}}$$

EXAMPLE 1 Write one radical for $\frac{\sqrt{20}}{\sqrt{5}}$. Then simplify.

Solution:

$$\frac{\sqrt{20}}{\sqrt{5}} = \sqrt{\frac{20}{5}}$$
$$= \sqrt{4} = 2$$

P–1 **Write one radical for each quotient. Then simplify.**

a. $\frac{\sqrt{8}}{\sqrt{2}}$ b. $\frac{\sqrt{63}}{\sqrt{7}}$ c. $\frac{\sqrt{99}}{\sqrt{11}}$ d. $\frac{\sqrt{75}}{\sqrt{3}}$

EXAMPLE 2

Write $\sqrt{\frac{3}{25}}$ as the quotient of two radicals. Then simplify.

Solution:

$$\sqrt{\frac{3}{25}} = \frac{\sqrt{3}}{\sqrt{25}}$$

$$= \frac{\sqrt{3}}{5} \quad \text{or} \quad \tfrac{1}{5}\sqrt{3}$$

Either answer is acceptable.

P–2 **Write each radical as the quotient of two radicals. Then simplify.**

a. $\sqrt{\frac{2}{49}}$ **b.** $\sqrt{\frac{7}{100}}$ **c.** $\sqrt{\frac{11}{144}}$ **d.** $\sqrt{\frac{5}{81}}$

A fraction with a radical in the denominator is not in simplest form. Eliminating the radical in the denominator is called ***rationalizing the denominator.***

EXAMPLE 3

Simplify: $\frac{\sqrt{2}}{\sqrt{5}}$

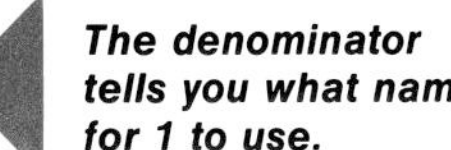

The denominator tells you what name for 1 to use.

Solution:

[1] Rationalize the denominator. ⟶ $\frac{\sqrt{2}}{\sqrt{5}} = \frac{\sqrt{2}}{\sqrt{5}} \cdot \frac{\sqrt{5}}{\sqrt{5}}$

[2] Use the Product Rule for Fractions. ⟶ $= \frac{\sqrt{2} \cdot \sqrt{5}}{\sqrt{5} \cdot \sqrt{5}}$

$$= \frac{\sqrt{10}}{5}$$

Simplest form

EXAMPLE 4

Simplify: $\sqrt{\frac{9}{33}}$

Solution: First, write the radicand in lowest terms.

$$\sqrt{\frac{9}{33}} = \sqrt{\frac{3}{11}}$$

$$= \frac{\sqrt{3}}{\sqrt{11}}$$

$$= \frac{\sqrt{3}}{\sqrt{11}} \cdot \frac{\sqrt{11}}{\sqrt{11}}$$

$$= \frac{\sqrt{33}}{11}$$

Simplest form

CLASSROOM EXERCISES

Write one radical for each quotient. Then simplify. (Example 1)

1. $\frac{\sqrt{32}}{\sqrt{8}}$ **2.** $\frac{\sqrt{27}}{\sqrt{3}}$ **3.** $\frac{\sqrt{50}}{\sqrt{2}}$ **4.** $\frac{\sqrt{32}}{\sqrt{2}}$ **5.** $\frac{\sqrt{72}}{\sqrt{2}}$

Write as the quotient of two radicals. Then simplify. (Example 2)

6. $\sqrt{\frac{5}{9}}$ **7.** $\sqrt{\frac{3}{64}}$ **8.** $\sqrt{\frac{17}{25}}$ **9.** $\sqrt{\frac{7}{36}}$ **10.** $\sqrt{\frac{19}{144}}$

For Exercises 11–30, write in simplest form.
(Example 3)

11. $\frac{\sqrt{3}}{\sqrt{7}}$ **12.** $\frac{\sqrt{2}}{\sqrt{11}}$ **13.** $\frac{\sqrt{5}}{\sqrt{3}}$ **14.** $\frac{\sqrt{15}}{\sqrt{2}}$ **15.** $\frac{3}{\sqrt{64}}$

16. $\frac{2}{\sqrt{15}}$ **17.** $\frac{3\sqrt{7}}{\sqrt{5}}$ **18.** $\frac{6\sqrt{2}}{\sqrt{3}}$ **19.** $\frac{9\sqrt{5}}{\sqrt{3}}$ **20.** $\frac{11\sqrt{2}}{\sqrt{13}}$

(Example 4)

21. $\sqrt{\frac{3}{6}}$ **22.** $\sqrt{\frac{4}{10}}$ **23.** $\sqrt{\frac{5}{35}}$ **24.** $\sqrt{\frac{16}{18}}$ **25.** $\sqrt{\frac{14}{49}}$

26. $\sqrt{\frac{18}{8}}$ **27.** $\sqrt{\frac{30}{9}}$ **28.** $\sqrt{\frac{8}{2}}$ **29.** $\sqrt{\frac{115}{23}}$ **30.** $\sqrt{\frac{72}{27}}$

WRITTEN EXERCISES

Goal: To apply the Quotient Property of Radicals

Sample Problems: a. $\frac{\sqrt{3}}{\sqrt{10}} = \sqrt{\frac{?}{?}}$ **b.** $\sqrt{\frac{5}{9}} = \frac{?}{?}$ **c.** $\frac{2}{\sqrt{5}} = \frac{?}{5}$

Answers: a. $\sqrt{\frac{3}{10}}$ **b.** $\frac{\sqrt{5}}{3}$ **c.** $\frac{2\sqrt{5}}{5}$

For Exercises 1–60, write in simplest form.
(Example 1)

1. $\frac{\sqrt{28}}{\sqrt{7}}$ **2.** $\frac{\sqrt{44}}{\sqrt{11}}$ **3.** $\frac{\sqrt{45}}{\sqrt{5}}$ **4.** $\frac{\sqrt{54}}{\sqrt{6}}$ **5.** $\frac{\sqrt{125}}{\sqrt{5}}$

6. $\frac{\sqrt{72}}{\sqrt{2}}$ **7.** $\frac{\sqrt{98}}{\sqrt{2}}$ **8.** $\frac{\sqrt{72}}{\sqrt{8}}$ **9.** $\frac{\sqrt{90}}{\sqrt{10}}$ **10.** $\frac{\sqrt{75}}{\sqrt{3}}$

(Example 2)

11. $\sqrt{\frac{2}{9}}$ 12. $\sqrt{\frac{4}{25}}$ 13. $\sqrt{\frac{13}{36}}$ 14. $\sqrt{\frac{7}{16}}$ 15. $\sqrt{\frac{29}{64}}$

16. $\sqrt{\frac{10}{49}}$ 17. $\sqrt{\frac{11}{64}}$ 18. $\sqrt{\frac{19}{81}}$ 19. $\sqrt{\frac{21}{100}}$ 20. $\sqrt{\frac{17}{121}}$

(Example 3)

21. $\frac{\sqrt{10}}{\sqrt{3}}$ 22. $\frac{\sqrt{6}}{\sqrt{5}}$ 23. $\frac{\sqrt{7}}{\sqrt{6}}$ 24. $\frac{\sqrt{3}}{\sqrt{11}}$ 25. $\frac{\sqrt{2}}{\sqrt{13}}$

26. $\frac{\sqrt{15}}{\sqrt{9}}$ 27. $\frac{1}{\sqrt{22}}$ 28. $\frac{1}{\sqrt{34}}$ 29. $\frac{3}{\sqrt{14}}$ 30. $\frac{5}{\sqrt{21}}$

(Example 4)

31. $\sqrt{\frac{6}{8}}$ 32. $\sqrt{\frac{9}{15}}$ 33. $\sqrt{\frac{9}{45}}$ 34. $\sqrt{\frac{3}{24}}$ 35. $\sqrt{\frac{5}{30}}$

36. $\sqrt{\frac{15}{3}}$ 37. $\sqrt{\frac{16}{12}}$ 38. $\sqrt{\frac{27}{15}}$ 39. $\sqrt{\frac{56}{35}}$ 40. $\sqrt{\frac{92}{69}}$

MIXED PRACTICE

41. $\sqrt{\frac{5}{81}}$ 42. $\frac{9}{\sqrt{10}}$ 43. $\sqrt{\frac{180}{5}}$ 44. $\frac{\sqrt{3}}{\sqrt{8}}$ 45. $\sqrt{\frac{12}{18}}$

46. $\frac{\sqrt{5}}{\sqrt{14}}$ 47. $\sqrt{\frac{405}{5}}$ 48. $\sqrt{\frac{19}{121}}$ 49. $\frac{\sqrt{3}}{\sqrt{17}}$ 50. $\frac{5}{\sqrt{2}}$

51. $\frac{5}{\sqrt{22}}$ 52. $\sqrt{\frac{10}{49}}$ 53. $\sqrt{\frac{112}{7}}$ 54. $\frac{4}{\sqrt{6}}$ 55. $\sqrt{\frac{32}{22}}$

56. $\frac{\sqrt{5}}{\sqrt{12}}$ 57. $\sqrt{\frac{192}{3}}$ 58. $\frac{6}{\sqrt{3}}$ 59. $\sqrt{\frac{23}{144}}$ 60. $\frac{\sqrt{10}}{\sqrt{24}}$

NON-ROUTINE PROBLEMS

61. In a traffic jam, twenty cars are lined up bumper to bumper. How many bumpers are touching one another?

62. Would it be cheaper for you to take one friend to the movies twice or two friends at the same time?

63. There are four children in a family. The sum of the squares of the ages of the three youngest children equals the square of the age of the oldest child. What are the ages of the four children?

64. The figure below shows how two squares can be formed by drawing just 7 lines. Show how two squares can be formed by drawing 6 lines.

Focal Length

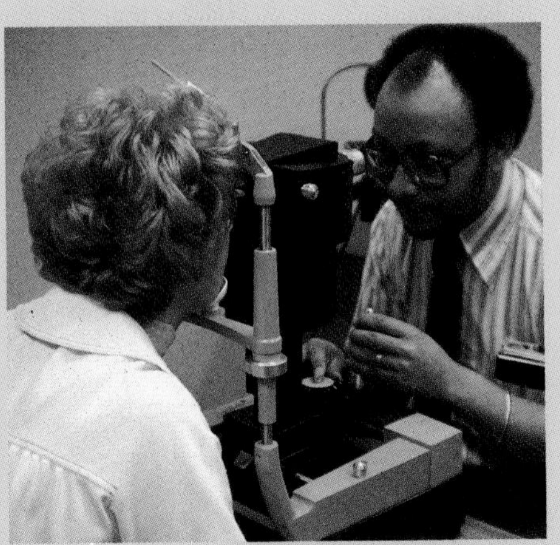

All measurements in the optical industry are made in metric units. One important unit of measure that **opticians** must know is the diopter. A lens having one **diopter** of power will bring parallel rays of light to a focus at a distance of one meter.

Lens power = 1 diopter (D)

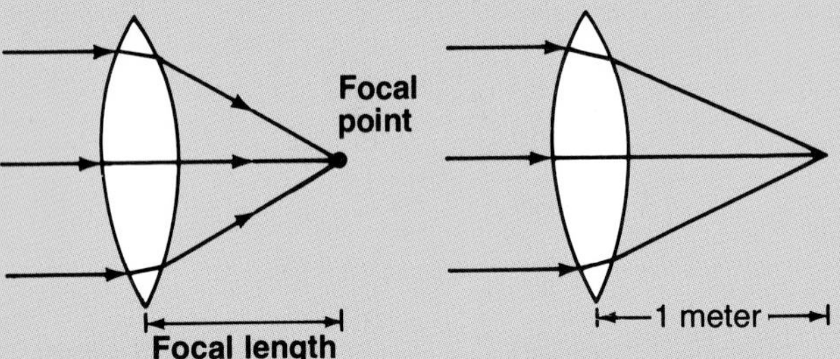

The **focal distance** of a lens is the distance between the focal point and the center of the lens. The formula below relates the power of a lens in diopters and its focal length in meters.

$$D = \frac{1}{f}$$

***D* = number of diopters**
***f* = focal length in meters**

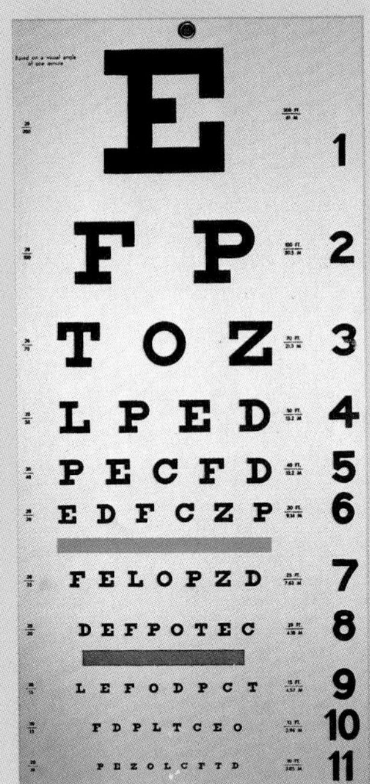

EXAMPLE: Find the focal length in centimeters of a 4.00 D lens.

SOLUTION:

$$D = \frac{1}{f}$$

$$4.00 = \frac{1}{f}$$

$$4.00f = 1$$

$$f = \frac{1}{4.00}$$

$f = 0.25$ meters, or **25 centimeters**

EXERCISES *Find the focal length, f, to the nearest whole centimeter.*

1. $D = 0.50$
2. $D = 6.50$
3. $D = 1.50$
4. $D = 3.00$
5. As the focal length, f, increases, does the number of diopters increase or decrease?
6. If the number of diopters is halved, what must happen to the focal length, f?

CHAPTER SUMMARY

IMPORTANT TERMS Like radicals *(p. 462)* Rationalizing the denominator *(p. 469)*

IMPORTANT

1. *Product Property of Radicals:* For any nonnegative numbers, a and b, $\sqrt{a} \cdot \sqrt{b} = \sqrt{ab}$.
2. A radical is in simplest form if the radicand has no perfect square factor other than 1.
3. Steps for Simplifying a Radical
   - **a.** Write the radicand as a product of a perfect square factor and another number.
   - **b.** Apply the Product Property of Radicals.
   - **c.** Write a square root of the perfect square radicand.
   - **d.** If necessary, repeat the above steps until no radicand has any perfect square factors other than 1.
4. Steps for Adding and Subtracting Radicals
   - **a.** Simplify all radicals.
   - **b.** Think of radicals such as $\sqrt{3}$ and $\sqrt{10}$ as $1\sqrt{3}$ and $1\sqrt{10}$.
   - **c.** Add and subtract like radicals.
5. *Quotient Property of Radicals:* For any nonnegative number a and any positive number b, $\frac{\sqrt{a}}{\sqrt{b}} = \sqrt{\frac{a}{b}}$.

CHAPTER REVIEW

SECTION 16.1

Simplify.

1. $\sqrt{3} \cdot \sqrt{13}$ **2.** $\sqrt{5} \cdot \sqrt{17}$ **3.** $\sqrt{64} \cdot \sqrt{11}$ **4.** $\sqrt{15} \cdot \sqrt{6}$

5. $\sqrt{81} \cdot \sqrt{23}$ **6.** $\sqrt{79} \cdot \sqrt{79}$ **7.** $\sqrt{93} \cdot \sqrt{93}$ **8.** $\sqrt{14} \cdot \sqrt{29}$

9. $\sqrt{25 \cdot 3}$ **10.** $\sqrt{36 \cdot 5}$ **11.** $\sqrt{50}$ **12.** $\sqrt{75}$ **13.** $\sqrt{63}$

14. $\sqrt{52}$ **15.** $\sqrt{108}$ **16.** $\sqrt{117}$ **17.** $\sqrt{12 \cdot 5}$ **18.** $\sqrt{9 \cdot 11}$

SECTION 16.2

Approximate each square root to three decimal places. Use the table on page 458.

19. $\sqrt{124}$ **20.** $\sqrt{117}$ **21.** $\sqrt{112}$ **22.** $\sqrt{150}$ **23.** $\sqrt{343}$

24. $\sqrt{356}$ **25.** $\sqrt{603}$ **26.** $\sqrt{596}$ **27.** $\sqrt{143}$ **28.** $\sqrt{600}$

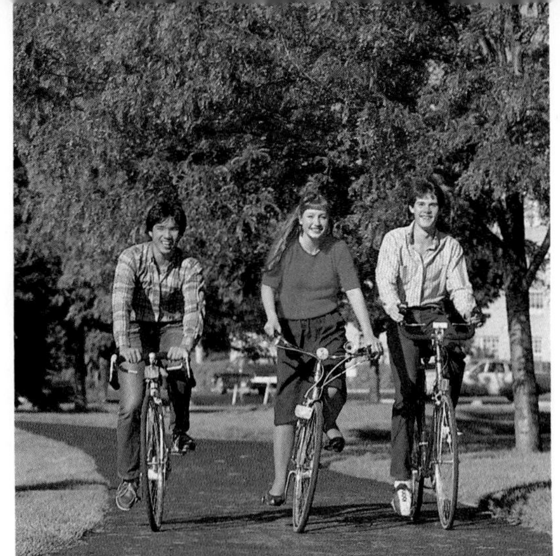

Approximate each square root to three decimal places. Use the Table of Squares and Square Roots on page 458.

29. A sidewalk is built diagonally across a square park. Each side of the park measures 1200 meters. What is the length of the sidewalk?

30. Anna runs diagonally across a square parking lot. Each side of the lot is 24 meters in length. Find the distance Anna runs.

SECTION 16.3

Add or subtract.

31. $7\sqrt{5} + 8\sqrt{5}$

32. $15\sqrt{11} + 14\sqrt{11}$

33. $18\sqrt{3} - \sqrt{3}$

34. $7\sqrt{10} - 13\sqrt{10}$

35. $3\sqrt{6} - 5\sqrt{6} - 10\sqrt{6}$

36. $5\sqrt{15} - \sqrt{15} - 3\sqrt{15}$

37. $\sqrt{24} - 5\sqrt{6}$

38. $-\sqrt{2} + \sqrt{18}$

39. $6\sqrt{8} + \sqrt{32} - 5\sqrt{8}$

SECTION 16.4

Multiply.

40. $7(-8\sqrt{5})$

41. $-11(6\sqrt{10})$

42. $-\frac{1}{3}(-12\sqrt{6})$

43. $-\frac{3}{4}(-24\sqrt{2})$

44. $(-12\sqrt{5})(4\sqrt{3})$

45. $(6\sqrt{7})(-9\sqrt{7})$

46. $(\sqrt{27})(2\sqrt{5})$

47. $(\sqrt{144})(3\sqrt{2})$

SECTION 16.5

Simplify.

48. $\frac{\sqrt{52}}{\sqrt{13}}$

49. $\frac{\sqrt{117}}{\sqrt{13}}$

50. $\frac{\sqrt{128}}{\sqrt{8}}$

51. $\frac{\sqrt{350}}{\sqrt{14}}$

52. $\frac{\sqrt{72}}{\sqrt{5}}$

53. $\sqrt{\frac{7}{36}}$

54. $\sqrt{\frac{13}{49}}$

55. $\sqrt{\frac{10}{81}}$

56. $\sqrt{\frac{23}{144}}$

57. $\sqrt{\frac{15}{76}}$

58. $\frac{\sqrt{5}}{\sqrt{13}}$

59. $\frac{\sqrt{15}}{\sqrt{2}}$

60. $\frac{4\sqrt{2}}{\sqrt{6}}$

61. $\frac{3\sqrt{6}}{\sqrt{3}}$

62. $\sqrt{\frac{9}{15}}$

More Algebraic Fractions

CHAPTER 17

The formula below can be used in a computer program to determine the monthly payment for a loan.

$$m = \frac{A\left(\frac{r}{12}\right)\left(1 + \frac{r}{12}\right)^n}{\left(1 + \frac{r}{12}\right)^n - 1}$$

m: monthly payment
A: amount borrowed
r: yearly rate of interest
n: number of months

17.1 Simplifying Algebraic Fractions

Recall from Chapter 4 how to write an algebraic fraction in lowest terms.

EXAMPLE 1 Write in lowest terms: $\frac{2x}{3x}$, $x \neq 0$

Solution: [1] Use the Product Rule for Fractions. $\longrightarrow$ $\frac{2x}{3x} = \frac{2}{3} \cdot \frac{\mathbf{x}}{\mathbf{x}}$

[2] Use the special property: $\frac{a}{a} = 1$. $\longrightarrow$ $= \frac{2}{3} \cdot \mathbf{1}$

[3] Use the Multiplication Property of One. $\longrightarrow$ $= \frac{2}{3}$

When you write an algebraic fraction in lowest terms, you are **simplifying** the fraction. The answer, such as $\frac{2}{3}$ in Example 1, is called the **simplest form.**

P–1 **Write in lowest terms.**

a. $\frac{-5y}{15y}$ **b.** $\frac{7r^2}{14r^2}$ **c.** $\frac{18b^2}{-24b}$ **d.** $\frac{-14k}{-7k^2}$

Example 2 illustrates a short method in which numerator and denominator are divided by their common factor (or common factors).

EXAMPLE 2 Simplify: $\frac{2x+4}{6x+2}$, $x \neq -\frac{1}{3}$

Solution:

[1] Use the Distributive Property. $\longrightarrow$ $\frac{2x+4}{6x+2} = \frac{\mathbf{2}(x+2)}{\mathbf{2}(3x+1)}$

[2] Divide the numerator and denominator by their common factor. $\longrightarrow$ $= \frac{\overset{1}{\cancel{2}}(x+2)}{\underset{1}{\cancel{2}}(3x+1)}$

[3] Write in lowest terms. $\longrightarrow$ $= \frac{\mathbf{1} \cdot (x+2)}{\mathbf{1} \cdot (3x+1)}$

$= \frac{x+2}{3x+1}$ ◀ ***Parentheses are not needed.***

P–2 **Simplify.**

a. $\frac{-4x^2}{9x^2}$ b. $\frac{3(x-4)}{3(x+4)}$ c. $\frac{4(y^2+1)}{10(y+1)}$ d. $\frac{5y-10}{15y+5}$

EXAMPLE 3

Simplify.

a. $\frac{-3x^2y}{x^2y+3xy}$ b. $\frac{2x+6}{9x^2+27x}$

Solutions: a. $\frac{-3x^2y}{x^2y+3xy} = \frac{-3x^2y}{xy(x+3)}$

$= \frac{-3\cancel{x}x\cancel{y}}{\cancel{x}\cancel{y}(x+3)}$

$= \frac{-3 \cdot 1 \cdot x \cdot 1}{1 \cdot 1 \cdot (x+3)}$

$= \frac{-3x}{(x+3)}$ or $-\frac{3x}{x+3}$

Simplest form

b. $\frac{2x+6}{9x^2+27x} = \frac{2(x+3)}{9x(x+3)}$

$= \frac{2\cancel{(x+3)}}{9x\cancel{(x+3)}}$

$= \frac{2 \cdot 1}{9x \cdot 1}$

$= \frac{2}{9x}$

P–3 **Simplify.**

a. $\frac{x(x+4)}{3(x+4)}$ b. $\frac{8(y-2)}{12(y-2)}$ c. $\frac{2x-6}{5x-15}$ d. $\frac{4x^2y^2}{xy+2x^2y}$

CLASSROOM EXERCISES

For Exercises 1–24 simplify each fraction. No divisor is zero. (Example 1)

1. $\frac{2a}{5a}$
2. $\frac{4t}{8t}$
3. $\frac{-5x}{10y}$
4. $\frac{6y}{-12x}$
5. $-\frac{2x}{4y}$
6. $\frac{12n}{-18n}$
7. $\frac{10x}{15x^2}$
8. $\frac{-8rt^2}{12r^2t}$

(Step 3, Examples 2 and 3)

9. $\frac{12(x-2)}{15(x+2)}$
10. $\frac{14(y+4)}{49(y+1)}$
11. $\frac{-3(r+1)}{2(r+1)}$
12. $\frac{-5(y+1)}{(y+3)5}$
13. $\frac{5t^2}{6st(s+3)}$
14. $\frac{3r^3s}{-6rs^2(r+s)}$
15. $\frac{-a^2b(a-b)}{-3ab(a-b)}$
16. $-\frac{xy(x+y)}{(x+y^2)x^2y}$

(Example 2)

17. $\frac{3x}{3x+3}$ 18. $\frac{2x+2}{2x-2}$ 19. $\frac{3x+3y}{5x+5y}$ 20. $\frac{7y-7x}{10y-10x}$

(Example 3)

21. $\frac{4m^2n}{mn^2+m^2n}$ 22. $\frac{-p^2qr}{pq-qr}$ 23. $\frac{a^2-ab}{a-ab^2}$ 24. $\frac{xz+xyz^2}{y+y^2z}$

WRITTEN EXERCISES

Goal: To write an algebraic fraction in simplest form

Sample Problem: Write in lowest terms: $\frac{2x+6}{4x+18}$

Answer: $\frac{x+3}{2x+9}$

For Exercises 1–32, simplify each fraction. No divisor is zero.
(Example 1)

1. $\frac{-5x}{5y}$ 2. $\frac{3m}{-3n}$ 3. $\frac{4rs}{8st}$ 4. $\frac{5pq}{15qr}$

5. $\frac{5x^2}{-3xy}$ 6. $\frac{-7ab}{12b^2}$ 7. $-\frac{12a^2b}{28ab}$ 8. $-\frac{18xy^2}{24x^2y^2}$

(Example 2)

9. $\frac{-3(x-7)}{6(x+7)}$ 10. $\frac{5(2x+1)}{-10(2x-1)}$ 11. $\frac{4y+8}{7y+14}$ 12. $\frac{12x+4}{8x+16}$

13. $-\frac{5x-5}{10x+10}$ 14. $-\frac{4x+4y}{8x+8y}$ 15. $\frac{12m+24}{24m+12}$ 16. $\frac{3x^2-9y}{4x^2-12y}$

(Example 3)

17. $\frac{4ab^2}{6ab-6b^2}$ 18. $\frac{5r^2s}{10r^2+10rs}$ 19. $\frac{2p^2q-2pq^2}{10pq^2-10q^3}$ 20. $\frac{3x^2-3xy}{6x^2y-6xy^2}$

21. $\frac{2x^2y}{xy-3xy^2}$ 22. $\frac{4rs-12r}{-8r^2}$ 23. $\frac{-2m^3n}{2m^3n-m^2n^2}$ 24. $\frac{9st^3}{3s^2t+6st^2}$

MIXED PRACTICE

25. $\frac{10y}{-15y}$ 26. $\frac{5ab^3}{-6a^2b}$ 27. $\frac{3x+6}{5x+10}$ 28. $\frac{-5r^2s^2}{r^2s^2-3rs^3}$

29. $\frac{4c^3d^2}{6c^2d-3cd^2}$ 30. $\frac{-6x}{8x}$ 31. $\frac{2xy+4y}{-6y}$ 32. $\frac{2x+10}{3x+15}$

APPLICATIONS

You can use a calculator to convert temperatures from degrees Celsius (°C) to degrees Fahrenheit (°F) by using the formula $\mathbf{F = 1.8C + 32}$.

EXAMPLE Convert 37°C to °F.

SOLUTION Substitute the known value in the formula $F = 1.8C + 32$.

[1] [.] [8] [×] [3] [7] [+] [3] [2] [=] 98.6

You can also convert temperatures from degrees Fahrenheit to degrees Celsius by using the formula $\mathbf{C = \frac{F - 32}{1.8}}$.

EXAMPLE Convert −40°F to °C.

SOLUTION [4] [0] [+/−] [−] [3] [2] [÷] [1] [.] [8] [=] -40.

Convert each temperature below to the alternate system of measurement.

33. 62.8°C **34.** 20.5°C **35.** −10.1°C **36.** 91.4°F **37.** 5.9°F **38.** 212°F

MORE CHALLENGING EXERCISES

Simplify. No divisor is zero.

39. $\frac{-1(a-b)}{a-b}$ **40.** $\frac{-1(-b+a)}{a-b}$ **41.** $\frac{b-a}{a-b}$ **42.** $\frac{a-b}{b-a}$

43. $\frac{3-y}{y-3}$ **44.** $\frac{x-2}{2-x}$ **45.** $\frac{3x-6}{12-6x}$ **46.** $\frac{15x^2y - x^3y}{2x-30}$

REVIEW CAPSULE FOR SECTION 17.2

Multiply. Write each answer in lowest terms. (Pages 86–88)

1. $\frac{2}{3} \cdot \frac{6}{5}$ **2.** $\frac{(5)13}{(4)(17)} \cdot \frac{34}{39}$ **3.** $\frac{18}{19} \cdot \frac{38}{99}$ **4.** $\frac{12}{49} \cdot \frac{77}{60}$

Name the factor or factors that the numerator and denominator have in common. (Pages 83–85)

5. $\frac{3x^2}{9y^2}$ **6.** $\frac{4y}{8x}$ **7.** $\frac{(5c)(8d)}{(2d)(15)}$ **8.** $\frac{(2m)(3p)}{(8p)(9m^2)}$

9. $\frac{(x-2)(3x)}{y(x-2)}$ **10.** $\frac{(8-4x)9}{(1)(6-12x)}$ **11.** $\frac{(a^2-a)(7)}{91(a-1)}$ **12.** $\frac{(y+1)(x-2)}{(2x-4)(1+y)}$

17.2 Products

Multiplying algebraic fractions is similar to multiplying arithmetic fractions.

Product Rule for Fractions

For any real numbers a, b, c, and d, with $b \neq 0$ and $d \neq 0$,

$$\frac{a}{b} \cdot \frac{c}{d} = \frac{ac}{bd}$$

$$\frac{-2}{5} \cdot \frac{7}{3} = \frac{-14}{15}$$

$$\frac{2x}{5} \cdot \frac{-x}{y} = \frac{-2x^2}{5y}$$

EXAMPLE 1 Multiply $\frac{2r}{3} \cdot \frac{-4}{r}$. Write the answer in simplest form.

Solution: [1] Multiply. ⟶ $\frac{2r}{3} \cdot \frac{-4}{r} = \frac{(2r)(-4)}{(3)(r)}$

Product Rule for Fractions

[2] Divide the numerator and denominator by their common factor. ⟶ $= \frac{(2\overset{1}{\cancel{r}})(-4)}{(3)(\underset{1}{\cancel{r}})}$

$$= \frac{2 \cdot 1 \cdot -4}{3 \cdot 1} = -\frac{8}{3}$$

P–1 **Multiply. Write each answer in simplest form.**

a. $\frac{6x}{5} \cdot \frac{10}{8x}$ **b.** $\frac{-y}{6} \cdot \frac{36}{y^2}$ **c.** $\frac{-3}{t} \cdot \frac{-t^2}{7}$ **d.** $\frac{-9}{15p^3} \cdot \frac{-5p}{2}$

The Product Rule for Fractions can be applied to fractions with binomial numerators and denominators. A ***binomial*** has two terms.

EXAMPLE 2 Multiply $\frac{4x}{x+5} \cdot \frac{x+5}{6xy}$. Write the answer in simplest form.

Solution:

$$\frac{4x}{x+5} \cdot \frac{x+5}{6xy} = \frac{4x(x+5)}{(x+5)6xy}$$

$$= \frac{\overset{2 \cdot 1}{\cancel{4x}}\overset{1}{\cancel{(x+5)}}}{\underset{1}{\cancel{(x+5)}}\underset{3 \cdot 1}{\cancel{6x}}y}$$

$$= \frac{2 \cdot 1 \cdot 1}{1 \cdot 3 \cdot 1 \cdot y} = \frac{2}{3y}$$

P–2 **Multiply. Write each answer in simplest form.**

a. $\frac{x+3}{x} \cdot \frac{2}{x+3}$ **b.** $\frac{3}{p-5} \cdot \frac{p-5}{p+1}$ **c.** $\frac{a-5}{a+2} \cdot \frac{a-6}{a-5}$

If possible, factor all binomials in the numerator and denominator before you apply the Product Rule for Fractions.

EXAMPLE 3 Multiply $\frac{2x-2}{3x^2} \cdot \frac{9x}{4x-4}$. Write the answer in simplest form.

Solution:

1 Use the Distributive Property. ⟶ $\frac{2x-2}{3x^2} \cdot \frac{9x}{4x-4} = \frac{2(x-1)}{3x^2} \cdot \frac{9x}{4(x-1)}$

2 Use the Product Rule for Fractions. ⟶ $= \frac{2 \cdot (x-1) \cdot 9x}{3x^2 \cdot 4(x-1)}$

3 Divide the numerator and denominator by their common factors. ⟶ $= \frac{\overset{1}{\cancel{2}} \cdot \overset{1}{\cancel{(x-1)}} \cdot \overset{3 \cdot 1}{\cancel{9x}}}{\underset{1 \cdot x}{\cancel{3x^2}} \cdot \underset{2}{\cancel{4}}\underset{1}{\cancel{(x-1)}}}$

$= \frac{1 \cdot 1 \cdot 3 \cdot 1}{1 \cdot x \cdot 2 \cdot 1}$

$= \frac{3}{2x}$

P–3 **Multiply. Write each answer in simplest form.**

a. $\frac{5}{3x-15} \cdot \frac{x-5}{25}$ **b.** $\frac{z-2}{4} \cdot \frac{-20}{10z-20}$ **c.** $\frac{y+6}{65y} \cdot \frac{5y^2}{12+2y}$

CLASSROOM EXERCISES

For Exercises 1–12, multiply. Write each answer in simplest form. (Example 1)

1. $\frac{x}{4} \cdot \frac{3}{x}$ **2.** $\frac{2}{t} \cdot \frac{-t}{4}$ **3.** $\frac{a}{5} \cdot \frac{-3}{x}$ **4.** $\frac{2x}{5} \cdot \frac{-3}{10}$

(Example 2)

5. $\frac{k-2}{5} \cdot \frac{3}{k-2}$ **6.** $\frac{f}{f+3} \cdot \frac{f+3}{-4}$ **7.** $\frac{2t}{v-8} \cdot \frac{v-8}{6t}$ **8.** $\frac{2r-1}{6st} \cdot \frac{9s}{2r-1}$

(Example 3)

9. $\frac{2s-6}{5s} \cdot \frac{10}{3s-9}$ **10.** $\frac{5s^2}{6r+2} \cdot \frac{12r+4}{3s}$ **11.** $\frac{2m+2n}{m+2} \cdot \frac{m-n}{m+n}$ **12.** $\frac{30y}{2z-8} \cdot \frac{z-4}{6y^2}$

WRITTEN EXERCISES

Goal: To multiply with algebraic fractions

Sample Problem: Multiply: $\frac{-6}{x-2y} \cdot \frac{x-2y}{-12}$

Write the answer in simplest form.

Answer: $\frac{1}{2}$

For Exercises 1–32, multiply. Write each answer in simplest form. (Example 1)

1. $\frac{3}{y} \cdot \frac{y}{2}$
2. $\frac{2a}{3} \cdot \left(-\frac{5}{a}\right)$
3. $\frac{-g}{5} \cdot \frac{2}{t}$
4. $\frac{a}{6} \cdot \frac{-3}{x}$
5. $\frac{7n}{4} \cdot \frac{10}{21n}$
6. $\frac{2}{-h} \cdot \frac{-5h}{3}$
7. $\frac{-3w}{7} \cdot \frac{5}{2w}$
8. $\frac{-3a}{2y^3} \cdot \frac{-5y}{a}$

(Example 2)

9. $\frac{2c}{c+4} \cdot \frac{c+4}{4c}$
10. $\frac{q+1}{5q} \cdot \frac{15q}{q+1}$
11. $\frac{3}{y+5} \cdot \frac{y+5}{12}$
12. $\frac{x-4}{4} \cdot \frac{2}{x-4}$
13. $\frac{n-7}{n+8} \cdot \frac{n+6}{n-7}$
14. $\frac{5}{x+3} \cdot \frac{x+3}{10x}$
15. $\frac{2f+3}{3f} \cdot \frac{-f}{2f+3}$
16. $\frac{3j+1}{-2j} \cdot \frac{j}{3j+1}$

(Example 3)

17. $\frac{4}{2d+2} \cdot \frac{d+1}{2d}$
18. $\frac{3m-3}{3m} \cdot \frac{2}{m-1}$
19. $\frac{2z-6}{5} \cdot \frac{10z}{3z-9}$
20. $\frac{3x}{2x-1} \cdot \frac{4x-2}{3x^2}$
21. $\frac{10u}{3u-12} \cdot \frac{5u-20}{4u}$
22. $\frac{14n^2}{15n-30} \cdot \frac{5n-10}{7n}$
23. $\frac{3t+15}{5t^2} \cdot \frac{15t}{4t+20}$
24. $\frac{5y+10}{3y-6} \cdot \frac{2y-4}{2y+4}$

MIXED PRACTICE

25. $\frac{-4}{g-5} \cdot \frac{g-5}{3}$
26. $\frac{2d}{d+4} \cdot \frac{d+4}{2d}$
27. $\frac{5x}{2a} \cdot \frac{-7a}{-3y}$
28. $\frac{4r}{2r+2} \cdot \frac{r+1}{2r^2}$
29. $\frac{b+2}{6b} \cdot \frac{4b^3}{4+2b}$
30. $\frac{-3a}{2y^3} \cdot \frac{-5y}{a^2}$
31. $\frac{2s-1}{5s} \cdot \frac{10s^2}{2s-1}$
32. $\frac{6}{3-3x} \cdot \frac{1-x}{4y}$

MID-CHAPTER REVIEW

Simplify. (Section 17.1)

1. $\frac{-3r}{2r}$
2. $\frac{-5ab}{-10ac}$
3. $\frac{12rs}{30rs^2}$
4. $\frac{4x-12}{4x+8}$
5. $-\frac{5f-10}{4f-8}$
6. $-\frac{6a+30}{ab+5b}$

Multiply. Write each answer in simplest form. (Section 17.2)

7. $-\frac{4a}{15c} \cdot \frac{5c}{12a}$
8. $\frac{-8t^2}{7r} \cdot \frac{21r^2}{-20t}$
9. $\frac{3k-1}{8k} \cdot \frac{-12}{3k-1}$
10. $\frac{t^2}{t+2} \cdot \frac{t+2}{-3t}$
11. $\frac{4w-4}{5w} \cdot \frac{-10}{3w-3}$
12. $\frac{2x+8}{10x} \cdot \frac{5y}{3x+12}$
13. $\frac{2}{2+t} \cdot \frac{t+2}{-4}$
14. $\frac{8+a}{b-3} \cdot \frac{b-3}{a+8}$

REVIEW CAPSULE FOR SECTION 17.3

Divide. (Pages 92–95)

1. $\frac{6}{35} \div \frac{1}{7}$
2. $\frac{12}{21} \div \frac{3}{7}$
3. $\frac{4}{5} \div \frac{8}{13}$
4. $\frac{24}{25} \div \frac{8}{15}$
5. $\frac{2}{11} \div \frac{9}{4}$
6. $\frac{6}{7} \div \frac{7}{8}$

Fill in the missing numeral or expression. (Pages 86–88 and 92–95)

7. $\frac{q}{2} \cdot \frac{?}{3} = \frac{q^2}{6}$
8. $\frac{5}{?} \cdot \frac{h-2}{-2} = \frac{5h-10}{-6}$
9. $\frac{3+x}{-4} \cdot \frac{?}{?} = \frac{9+3x}{8y}$
10. $\frac{8-3k}{15} \cdot \frac{?}{?} = \frac{3k-8}{30k}$
11. $\frac{2}{3} \div \frac{5}{7} = \frac{2}{3} \cdot \frac{?}{?}$
12. $\frac{3}{r} \div \frac{2}{r} = \frac{3}{r} \cdot \frac{?}{?}$

17.3 Quotients

You know that the equation

$$\frac{3}{4} \div \frac{7}{9} = \frac{3}{4} \cdot \frac{9}{7}$$

is true. Dividing by $\frac{7}{9}$ is the same as multiplying by $\frac{9}{7}$, the reciprocal of $\frac{7}{9}$. A similar rule holds true when you divide with algebraic fractions.

Quotient Rule for Fractions

For any numbers a, b, c, and d, with $b \neq 0$, $c \neq 0$, and $d \neq 0$,

$$\mathbf{\frac{a}{b} \div \frac{c}{d} = \frac{a}{b} \cdot \frac{d}{c}}.$$

$$\frac{2}{3} \div \left(-\frac{5}{7}\right) = \frac{2}{3} \cdot \left(-\frac{7}{5}\right) = -\frac{14}{15}$$

$$\frac{x}{6} \div \frac{3}{x} = \frac{x}{6} \cdot \frac{x}{3} = \frac{x^2}{18}$$

EXAMPLE 1 Divide: $\frac{5}{c} \div \frac{7}{c}$

Solution: [1] Use the Quotient Rule for Fractions. ⟶ $\frac{5}{c} \div \frac{7}{c} = \frac{5}{c} \cdot \frac{c}{7}$

[2] Use the Product Rule for Fractions. ⟶ $= \frac{5\overset{1}{\cancel{c}}}{7\underset{1}{\cancel{c}}}$

[3] Simplest form. ⟶ $= \frac{5}{7}$

P–1 **Divide.**

a. $\frac{j}{-2} \div \frac{j}{4}$ b. $\frac{3p}{7} \div \left(-\frac{14}{2p}\right)$ c. $\frac{2r}{s} \div \frac{2}{3}$ d. $\frac{3y}{4x} \div \frac{6x}{7y}$

EXAMPLE 2 Divide: $\frac{2}{x-1} \div \frac{-3}{x-1}$

Solution:

1 Use the Quotient Rule for Fractions. ⟶ $\frac{2}{x-1} \div \frac{-3}{x-1} = \frac{2}{x-1} \cdot \frac{x-1}{-3}$

2 Use the Product Rule for Fractions. ⟶ $= \frac{2\overset{1}{\cancel{(x-1)}}}{-3\underset{1}{\cancel{(x-1)}}}$

3 Simplest form. ⟶ $= -\frac{2}{3}$

P–2 **Divide.**

a. $\frac{1}{c+1} \div \frac{2}{c+1}$ b. $\frac{-3}{y+2} \div \frac{4}{3y+6}$ c. $\frac{-12}{7-z} \div \frac{6}{7+z}$

A fraction such as $\frac{\frac{2}{3}}{\frac{k}{3}}$ with a fractional numerator or denominator is called a ***complex fraction.***

Table

Complex Fraction	Quotient	Product	Simplest Form
$\frac{\frac{2}{3}}{\frac{k}{3}}$	$\frac{2}{3} \div \frac{k}{3}$	$\frac{2}{3} \cdot \frac{3}{k}$	$\frac{2}{\underset{1}{\cancel{3}}} \cdot \frac{\overset{1}{\cancel{3}}}{k} = \frac{2}{k}$
$\frac{\frac{9}{10x}}{\frac{3}{5x^2}}$	$\frac{9}{10x} \div \frac{3}{5x^2}$	$\frac{9}{10x} \cdot \frac{5x^2}{3}$	$\frac{\overset{3}{\cancel{9}}}{\underset{2 \cdot 1}{\cancel{10x}}} \cdot \frac{\overset{1 \cdot x}{\cancel{5x^2}}}{\underset{1}{\cancel{3}}} = \frac{3x}{2}$

P–3 **Divide. Write each answer in simplest form.**

a. $\frac{\frac{1}{2}}{\frac{3}{4}}$ b. $\frac{\frac{z}{5}}{\frac{3}{z}}$ c. $\frac{\frac{f}{3}}{\frac{f^2}{6}}$ d. $\frac{\frac{a^3}{4}}{\frac{2}{a}}$

CLASSROOM EXERCISES

For Exercises 1–12, divide. Write each answer in simplest form.
(Example 1)

1. $\frac{y}{4} \div \frac{3y}{4}$
2. $\frac{-z}{7} \div \frac{z}{14}$
3. $\frac{x}{3} \div \frac{5}{x}$
4. $\left(-\frac{r}{6}\right) \div \left(-\frac{s}{2}\right)$

(Example 2)

5. $\frac{10}{m+2} \div \frac{5}{m+2}$
6. $\frac{5t}{t-3} \div \frac{2t}{t-3}$
7. $\frac{7r}{g-1} \div \frac{g+1}{r}$
8. $\frac{2x+8}{y} \div \frac{x+4}{2y}$

(Table)

9. $\frac{\frac{x}{2}}{\frac{x}{5}}$
10. $\frac{\frac{6}{y}}{\frac{5}{y^2}}$
11. $\frac{\frac{y}{8}}{\frac{y}{16}}$
12. $\frac{-\frac{y}{3}}{\frac{y}{15}}$

WRITTEN EXERCISES

Goal: To divide with algebraic fractions

Sample Problem: Divide: $\frac{3x}{7} \div \frac{x^2}{14}$ Write the answer in simplest form.

Answer: $\frac{6}{x}$

For Exercises 1–20, divide. Write each answer in simplest form.
(Example 1)

1. $\frac{a}{3} \div \frac{2a}{9}$
2. $\frac{5}{8} \div \frac{2}{x}$
3. $\frac{5}{x} \div \frac{10}{7x}$
4. $\frac{3t}{5} \div \left(-\frac{7t}{10}\right)$
5. $\left(-\frac{4w}{7}\right) \div \frac{2w}{5}$
6. $\frac{10}{a} \div \frac{2}{3}$
7. $\left(-\frac{a}{5}\right) \div \left(-\frac{2a}{3}\right)$
8. $\left(-\frac{2x}{3}\right) \div \left(-\frac{x}{9}\right)$

(Example 2)

9. $\frac{5}{x-2} \div \frac{3}{x-2}$
10. $\frac{a+3}{2} \div \frac{a+3}{5}$
11. $\frac{y+3}{2y} \div \frac{y+3}{y}$
12. $\frac{3x}{2x-1} \div \frac{4x}{2x-1}$
13. $\frac{5y}{3x-2} \div \frac{10y}{2x+3}$
14. $\frac{b}{b-3} \div \frac{1}{b+3}$
15. $\frac{y+2}{y} \div \frac{3y+6}{5}$
16. $\frac{a+2}{b} \div \frac{4a+8}{4b}$

(Table)

17. $\frac{\frac{x}{5}}{\frac{2x}{5}}$
18. $\frac{\frac{a}{7}}{\frac{3a}{7}}$
19. $\frac{\frac{y}{4}}{\frac{5}{y}}$
20. $\frac{\frac{a}{3}}{\frac{6}{ab}}$

REVIEW CAPSULE FOR SECTION 17.4

Add or subtract. Pages 96–99)

1. $\frac{1}{5}+\frac{2}{5}$
2. $\frac{3}{8}-\frac{1}{8}$
3. $\frac{13}{27}-\frac{4}{27}$
4. $\frac{4}{57}+\frac{110}{57}$

Combine like terms. (Pages 43–45)

5. $2x + x$
6. $5t - t$
7. $3r + 2 - r$
8. $6w + 2 - 2w$
9. $7x + 13 + 4x - 7$
10. $5n + 6 - 2n - 5$

17.4 Sums and Differences

Adding or subtracting algebraic fractions is similar to adding or subtracting fractions of arithmetic.

Sum and Difference Rules for Fractions

For any numbers *a*, *b*, and *c*, with $c \neq 0$;

1. $\frac{a}{c}+\frac{b}{c}=\frac{a+b}{c}$ 2. $\frac{a}{c}-\frac{b}{c}=\frac{a-b}{c}$

$$\frac{3}{x}+\frac{5}{x}=\frac{8}{x}$$

$$\frac{2x}{y}-\frac{3}{y}=\frac{2x-3}{y}$$

EXAMPLE 1 Add: $\frac{5}{8x}+\frac{-3}{8x}$

Solution: [1] Use the Sum Rule for Fractions. ⟶ $\frac{5}{8x}+\frac{-3}{8x}=\frac{5+(-3)}{8x}$

[2] Add. ⟶ $=\frac{2}{8x}$

[3] Simplest form. ⟶ $=\frac{1}{4x}$

EXAMPLE 2 Subtract: $\frac{3x}{x+2}-\frac{-x}{x+2}$

Solution: [1] Use the Difference Rule for Fractions. ⟶ $\frac{3x}{x+2}-\frac{-x}{x+2}=\frac{3x-(-x)}{x+2}$

[2] Use the definition of subtraction. ⟶ $=\frac{3x+x}{x+2}$

[3] Simplest form. ⟶ $=\frac{4x}{x+2}$

P–1 **Add or subtract.**

a. $\frac{5}{n}+\frac{7}{n}$ **b.** $\frac{r}{2x}-\frac{7}{2x}$ **c.** $\frac{5y}{y+5}+\frac{5}{y+5}$

EXAMPLE 3 Add or subtract.

a. $\frac{2x}{x+1}+\frac{2}{x+1}$ **b.** $\frac{4}{3x-3}-\frac{5-x}{3x-3}$

Solutions:

a. $\frac{2x}{x+1}+\frac{2}{x+1}=\frac{2x+2}{x+1}$

$=\frac{2(x+1)}{x+1}$

$=\frac{2\overset{1}{\cancel{(x+1)}}}{\underset{1}{\cancel{(x+1)}}}$

$=2$

b. $\frac{4}{3x-3}-\frac{5-x}{3x-3}=\frac{4-(5-x)}{3x-3}$

Note the use of parentheses.

$=\frac{4-5+x}{3x-3}$

$=\frac{-1+x}{3(x-1)}$

$=\frac{\overset{1}{\cancel{(x-1)}}}{3\underset{1}{\cancel{(x-1)}}}=\frac{1}{3}$

P–2 **Add or subtract.**

a. $\frac{3x}{x+4}+\frac{12}{x+4}$ **b.** $\frac{3x}{x-2}-\frac{6}{x-2}$ **c.** $\frac{1}{6x-2}-\frac{4-9x}{6x-2}$

CLASSROOM EXERCISES

Add or subtract. Write each answer in simplest form. (Example 1)

1. $\frac{3}{y}+\frac{4}{y}$ **2.** $\frac{5}{t}-\frac{2}{t}$ **3.** $\frac{a}{x}+\frac{2a}{x}$ **4.** $\frac{5r}{3x}-\frac{r}{3x}$

(Example 2)

5. $\frac{3}{b-3}+\frac{7}{b-3}$ **6.** $\frac{3y}{w-2}-\frac{y}{w-2}$ **7.** $\frac{2y}{g+4}+\frac{3}{g+4}$ **8.** $\frac{3r}{s+1}-\frac{-r}{s+1}$

(Example 3)

9. $\frac{2n}{n-3}-\frac{6}{n-3}$ **10.** $\frac{9r}{3r+1}+\frac{3}{3r+1}$ **11.** $\frac{p}{p+2}+\frac{p-4}{p+2}$ **12.** $\frac{d}{2d-4}-\frac{2d+2}{2d-4}$

WRITTEN EXERCISES

Goal: To add or subtract with algebraic fractions that have like denominators

Sample Problem: Subtract: $\frac{5x-1}{2x-1} - \frac{3x}{2x-1}$

Write the answer in simplest form.

Answer: 1

For Exercises 1–36, add or subtract. Write each answer in simplest form. (Example 1)

1. $\frac{-12}{5x} + \frac{9}{5x}$ **2.** $\frac{13}{7t} + \frac{-9}{7t}$ **3.** $\frac{11}{12m} - \frac{5}{12m}$ **4.** $\frac{13}{8r} - \frac{7}{8r}$

5. $\frac{5s}{9t} + \frac{s}{9t}$ **6.** $\frac{a}{12b} + \frac{7a}{12b}$ **7.** $\frac{11x}{15y} - \frac{-x}{15y}$ **8.** $\frac{7m}{15n} - \frac{-2m}{15n}$

(Example 2)

9. $\frac{5f}{f+7} + \frac{f}{f+7}$ **10.** $\frac{7r}{2r-1} + \frac{r}{2r-1}$ **11.** $\frac{3p}{q+6} - \frac{2p}{q+6}$ **12.** $\frac{8t}{w+6} - \frac{3t}{w+6}$

13. $\frac{2w-3}{w-8} + \frac{-12}{w-8}$ **14.** $\frac{-b}{3b+1} + \frac{3b-5}{3b+1}$ **15.** $\frac{4z}{z+4} - \frac{7z+2}{z+4}$ **16.** $\frac{s-3}{5r+2} - \frac{s+3}{5r+2}$

(Example 3)

17. $\frac{3q+5}{q+1} + \frac{q-3}{q+1}$ **18.** $\frac{z-4}{z-3} + \frac{2z-5}{z-3}$ **19.** $\frac{2c}{c-3} - \frac{6}{c-3}$ **20.** $\frac{3m}{m+2} - \frac{2m-2}{m+2}$

21. $\frac{4}{1-3y} + \frac{2-18y}{1-3y}$ **22.** $\frac{5b}{2-3b} + \frac{b-4}{2-3b}$ **23.** $\frac{5h}{3h-3} - \frac{h+4}{3h-3}$ **24.** $\frac{7r}{5r+5} - \frac{3r-4}{5r+5}$

MIXED PRACTICE

25. $\frac{4y-1}{y+4} - \frac{3y-2}{y+4}$ **26.** $\frac{7n+1}{3n+1} - \frac{4n}{3n+1}$ **27.** $\frac{14p+4}{4p+1} - \frac{2p+1}{4p+1}$ **28.** $\frac{2}{x} - \frac{-5}{x}$

29. $\frac{a}{3a-9} - \frac{6-a}{3a-9}$ **30.** $\frac{1-v}{2v-1} - \frac{v}{2v-1}$ **31.** $\frac{5y-3}{y-3} - \frac{4y}{y-3}$ **32.** $\frac{2k+1}{k+2} - \frac{k-3}{k+2}$

33. $\frac{2-n}{3n-2} - \frac{2n}{3n-2}$ **34.** $\frac{6f}{3f-2} - \frac{4}{3f-2}$ **35.** $\frac{-3m}{2n} - \frac{m}{2n}$ **36.** $\frac{1}{2u-6} - \frac{4-u}{2u-6}$

NON-ROUTINE PROBLEM

37. This cube was painted on all sides and then cut into 27 equal smaller cubes.

a. How many smaller cubes have paint on only 2 sides?
b. How many smaller cubes have paint on only 1 side?

Using Guess and Check

Hidden Conditions

On first reading, some problems appear not to have enough information to be solved. Identifying a **"hidden" condition** in these problems can help you to solve them.

EXAMPLE: Sue bought 11 pounds of mixed nuts for gifts. She bought some two-pound boxes and some one-pound boxes, but not more than four boxes of each size. How many of each size did she buy?

SOLUTION:

[1] Represent the unknowns. ⟶ Let $x =$ the number of two-pound boxes.
Let $y =$ the number of one-pound boxes.

[2] Write an equation.

Think: Weight of two-pound boxes + Weight of one-pound boxes equals 11.

Translate: $2x + y = 11$

[3] Since there is one equation in two variables, use **guess and check.**

Think: x and y must be less than or equal to 4.
"Hidden" condition: x and y are whole numbers.

Guess 1: Let $x = 1$ and $y = 1$. **Check 1:** $2x + y = 2(1) + 1 = 3$
Guess 2: Let $x = 3$ and $y = 4$. **Check 2:** $2x + y = 2(3) + 4 = 10$
Guess 3: Let $x = 4$ and $y = 3$. **Check 3:** $2x + y = 2(4) + 3 = 11$

Too small! Try larger values.

So Sue bought **4 two-pound boxes** and **3 one-pound boxes.**

EXERCISES

1. Carlos bought some 22¢-stamps and some 17¢-stamps for \$1.44. There were fewer than 6 stamps of each kind. How many of each did he buy?

2. Amy bought used books for \$4.95. She paid 50¢ each for some books and 35¢ each for others. She bought fewer than 8 books at each price. How many did she buy for 50¢?

3. A basketball player scored 20 points in a game by making two-point goals and three-point goals. How many of each kind did the player score? (HINT: More than one answer is possible.)

4. Adult tickets to a school play sell for 75¢ each and student tickets for 50¢ each. Gary sold \$5.75 worth of tickets in all. How many of each kind did he sell? (HINT: More than one answer is possible.)

CHAPTER SUMMARY

IMPORTANT TERMS Complex fraction *(p. 484)*

IMPORTANT IDEAS

1. *Product Rule for Fractions:* For any numbers *a*, *b*, *c*, and *d*, with $b \neq 0$ and $d \neq 0$, $\frac{a}{b} \cdot \frac{c}{d} = \frac{ac}{bd}$.
2. *Quotient Rule for Fractions:* For any numbers, *a*, *b*, *c*, and *d*, with $b \neq 0$, $c \neq 0$, and $d \neq 0$, $\frac{a}{b} \div \frac{c}{d} = \frac{a}{b} \cdot \frac{d}{c}$.
3. *Sum and Difference Rules for Fractions:* For any numbers, *a*, *b*, and *c*, with $c \neq 0$, $\frac{a}{c} + \frac{b}{c} = \frac{a+b}{c}$ and $\frac{a}{c} - \frac{b}{c} = \frac{a-b}{c}$.

CHAPTER REVIEW

SECTION 17.1

Simplify.

1. $\frac{-8mn}{12np}$ **2.** $\frac{4x(x+5)}{6y(x+5)}$ **3.** $-\frac{3x-12}{4x-16}$ **4.** $-\frac{2a^2x+3a^2}{6x+9}$

SECTION 17.2

Multiply. Write each answer in simplest form.

5. $\frac{4r}{3s} \cdot \frac{9s^2}{-10r}$ **6.** $\frac{6k}{-5n} \cdot \frac{5n}{8k^2}$ **7.** $\frac{3t}{2t-3} \cdot \frac{2t-3}{-4t}$ **8.** $\frac{4r-1}{2s} \cdot \frac{-12s}{12r-3}$

SECTION 17.3

Divide. Write each answer in simplest form.

9. $-\frac{2a}{3} \div \frac{10a}{9}$ **10.** $\frac{5}{8x} \div \left(-\frac{25}{2x}\right)$ **11.** $\frac{4x-3}{3y} \div \frac{4x-3}{6}$ **12.** $\frac{3}{a-b} \div \frac{2}{a+b}$

13. $\frac{\frac{3}{x}}{\frac{1}{2x}}$ **14.** $\frac{\frac{3}{2y}}{\frac{2}{5y}}$ **15.** $\frac{\frac{a}{b}}{\frac{b}{a}}$ **16.** $\frac{\frac{x^2}{y}}{\frac{y^2}{x}}$

SECTION 17.4

Add or subtract. Write each answer in simplest form.

17. $\frac{7}{8x} + \frac{5}{8x}$ **18.** $\frac{19}{12t} - \frac{-10}{12t}$ **19.** $\frac{2r}{5s} + \frac{-r}{5s}$ **20.** $\frac{8q}{pt} - \frac{q}{pt}$

21. $\frac{b-6}{2b-3} + \frac{4b}{2b-3}$ **22.** $\frac{5m+12}{m+3} - \frac{3+2m}{m+3}$ **23.** $\frac{2q+7}{8q+32} + \frac{q+5}{8q+32}$ **24.** $\frac{12k-20}{5k-6} - \frac{2k-2}{5k-6}$

CHAPTER 18

Polynomials

To add a work island to a kitchen, an architect increases the length of the room by y feet. The remodeled and original kitchens will each have the same floor area, 400 ft^2. Knowing the area, A, of the island, you can use this formula to determine y.

$$(20 + y)20 - A = 400$$

18.1 Simplifying Polynomials

The following are examples of ***monomials.***

a. $2x^3$ **b.** $-p$ **c.** $\frac{1}{2}x$ **d.** -3.5 **e.** y^2

A ***binomial*** is the sum or difference of two monomials.

a. $x - 3$ **b.** $2c + 5$ **c.** $3x^2 - x$ **d.** $a + 1$

Monomials and binomials belong to a larger class of numerals or expressions called polynomials.

Definition

A ***polynomial*** is a monomial or the sum or difference of two or more monomials. $-2x^2$; $5u + 7$ $x^2 - 3x + 5$

P–1 **Which of the following are polynomials? binomials?**

a. $x^2 + 5x$ **b.** $-4n^5$ **c.** $-x^2 + 4x - 6$
d. -19 **e.** $x^3 - x^2 + x - 3$ **f.** $-5b + 1$

EXAMPLE 1 Simplify: $x - 3x^2 + 5x - 3 + 5x^2$

Solution:

1 Combine like terms. ⟶ $x - 3x^2 + 5x - 3 + 5x^2 = 6x + 2x^2 - 3$

2 Arrange terms in order with the largest exponent first. ⟶ $= 2x^2 + 6x - 3$

It is often helpful to begin by grouping like terms.

EXAMPLE 2 Simplify: $4x + x^3 + 12 - 2x^2 - 3x^3 + 2x^2 - x - 17$

Solution:

1 Group like terms. ⟶ $(4x - x) + (x^3 - 3x^3) + (12 - 17) + (-2x^2 + 2x^2)$

2 Combine like terms. ⟶ $3x - 2x^3 - 5 + 0$

3 Arrange terms in order. ⟶ $-2x^3 + 3x - 5$

P–2 **Simplify.**

a. $4x^2 - x + 2x^2$ **b.** $12 - 8z - 12z^3 + 4 - 10z^2 + 4z + z - 1$

CLASSROOM EXERCISES

Simplify. (Examples 1 and 2)

1. $3r - 5r + 2$
2. $2x^2 - 4 + 7$
3. $x^2 - x - 2x + 6$
4. $4x^2 + x - 5x - 3$
5. $y - 2y^2 + 3y - 7$
6. $2g + g^2 - 3g + 5$
7. $\frac{1}{2}x^2 + 2\frac{1}{2}x^2 - \frac{3}{4}$
8. $1.2x^2 + 1.8x^2 - x + 6$
9. $-3x + x^2 + x + 2x^2 - 3$
10. $5x + 8 - 2x^4 + 5x - 3x^2$
11. $2 + 3x^2 + 7x + 11 - 3x^2$
12. $9 - 21d^2 + 3d^4 + 21d^2$

WRITTEN EXERCISES

Goal: To simplify a polynomial
Sample Problem: Simplify: $x^2 + 5x - 2x^2 + 6 - 8x - 3x^3$
Answer: $-3x^3 - x^2 - 3x + 6$

Simplify. (Examples 1 and 2)

1. $-3x + 7x - 2$
2. $5t - 8t + 7$
3. $3x^2 - x^2 + x + 9$
4. $8x^2 - x^2 - 3x - 5$
5. $-5x^2 - x - x + 2 - 9$
6. $-x^2 + 3x + 4x - 11 + 8$
7. $10x^3 - x^2 + 3x^2 - x + 1$
8. $19x^3 + 4x^2 - 5x^2 + 4x - 6$
9. $3k - 10 + 4k^2 - 12k + 1$
10. $9 - 12x + x^2 - x - 5$
11. $12 - x + 4x^3 - 3x + 3$
12. $-2x^2 - 8 + 4x^3 - x^2 + 7$
13. $x^4 - 6x^2 - 10 + 2x^3 + 6x^2 - 1$
14. $-5x^2 - 11 + 8x - x^3 - 6 - 8x$
15. $-\frac{1}{2}x^2 + \frac{3}{4}x^3 - \frac{5}{2} + \frac{3}{2}x^2 + \frac{3}{2}$
16. $-\frac{5}{4}y + \frac{7}{2} - \frac{3}{4}y^2 + \frac{1}{4}y - \frac{5}{2}$
17. $2.7 - 1.3x^2 - 1.9 + 1.8x - 3.4x^2$
18. $-5.1x + 4.8 - 2.7x^2 - 0.8x - 3.7$

NON-ROUTINE PROBLEM

19. Find how many ways the word PROBLEM can be read from left to right in the diagram. Each letter in the word must be connected to the next letter by a marked line.

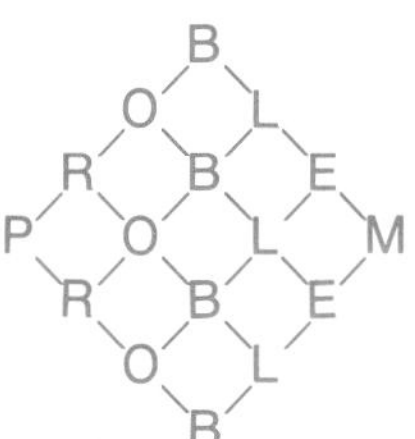

REVIEW CAPSULE FOR SECTION 18.2

Find the value of each monomial. (Pages 5–7)

1. $-5h$ when $h = -12$
2. $3x^2$ when $x = 2$
3. $4x^2$ when $x = 3$
4. $-x^2$ when $x = -2$
5. $(-b)^2$ when $b = -2$
6. $-2x^3$ when $x = -1$
7. $-x^3$ when $x = 2$
8. $-x^3$ when $x = -2$
9. s^4 when $s = -2$

18.2 Evaluating Polynomials

To evaluate a polynomial, evaluate each term first. Remember to raise a number to a power before performing any other operations.

EXAMPLE 1 Evaluate $x^2 - x - 2$ when $x = 3$.

Solution:

1 Replace each x by 3. → $x^2 - x - 2 = (3)^2 - 3 - 2$

2 Evaluate each term. → $= 9 - 3 - 2$

$= 4$

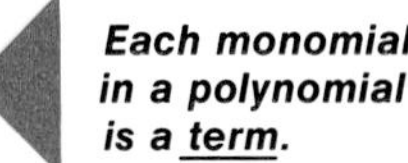

Each monomial in a polynomial is a term.

Remember that if you substitute a negative number for x, then x^2 is positive and x^3 is negative.

EXAMPLE 2 Evaluate $x^3 - x^2 + x - 1$ when $x = -1$.

Solution:

1 Replace each x by -1. → $x^3 - x^2 + x - 1 = (-1)^3 - (-1)^2 + (-1) - 1$

2 Evaluate each term. → $= -1 - 1 - 1 - 1$

$= -4$

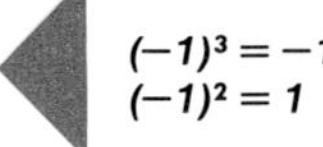

$(-1)^3 = -1$
$(-1)^2 = 1$

P–1 **Evaluate each polynomial when $a = -2$.**

a. $a^2 + 4$ **b.** $a^2 - a - 2$ **c.** $a^3 + 2a^2 - a - 7$

If possible you should simplify a polynomial before evaluating it.

EXAMPLE 3 Evaluate $-3 + x^2y - 3xy^2 + 4x^2y$ when $x = -1$ and $y = 2$.

Solution:

1 Combine like terms. → $-3 + x^2y - 3xy^2 + 4x^2y = -3 + 5x^2y - 3xy^2$

2 Rewrite the expression with the algebraic terms first. → $= 5x^2y - 3xy^2 - 3$

3 Replace x and y by their values. → $= 5(-1)^2(2) - 3(-1)(2)^2 - 3$

4 Evaluate each term. → $= 10 + 12 - 3$

$= 19$

CLASSROOM EXERCISES

Evaluate each polynomial when $x = -2$, $y = 1$ and $z = 3$. (Examples 1–3).

1. x^2
2. $-y^3$
3. $x^2 + y^2$
4. $x^2 - y^2$
5. $x + y - z$
6. $y^2 + z - 5$
7. $y^2z - yz$
8. $y^5 - 1$
9. $-3x^2 + 5x - 2$
10. $-y^3 + y^2 - y + 1$
11. $x^3 - x^2 + x - 1$
12. $-x^3 - x^2 - x - 1$

WRITTEN EXERCISES

Goal: To evaluate a polynomial
Sample Problem: $-2x^2 + 3xy - y^2$ when $x = 1$ and $y = -2$ **Answer:** -12

For Exercises 1–24, evaluate each polynomial.
(Example 1)

1. $m^2 - 2m + 3$ when $m = -2$
2. $-2x^2 + x - 1$ when $x = -2$
3. $3x^2 - 5x - 2$ when $x = 3$
4. $2v^2 + 3v - 10$ when $v = 1$
5. $-4x^2 + x - 2$ when $x = -3$
6. $-5x^2 + 2x - 3$ when $x = -3$

(Example 2)

7. $x^3 + 2x^2 - x - 3$ when $x = 1$
8. $w^3 - w^2 + 3w - 5$ when $w = 1$
9. $-x^3 + x^2 + 3x - 4$ when $x = -1$
10. $-x^3 + 2x^2 - 5x + 3$ when $x = -1$

(Example 3)

11. $d^2f - 2d^2f + 5$ when $d = -2$ and $f = 3$
12. $2x^2y - xy + y^2$ when $x = -1$ and $y = -2$
13. $-x^2y + 3xy^2 - 5x + 4y - 2$ when $x = 1$ and $y = -1$
14. $-p^3qr + 2p^2q^2r^2 - p^3qr$ when $p = -1$, $q = 1$, $r = -1$

MIXED PRACTICE

15. $t^4 - 5t^2 + 2$ when $t = 2$
16. $x^5 - x^3 + x + 3$ when $x = -2$
17. $-x^3 + x^2$ when $x = -4$
18. $n^3 - n^2 + n - 1$ when $n = -5$
19. $x^2 - 3x^2 - 3$ when $x = 2$
20. $-x^6 + x^3 - x$ when $x = -3$
21. $-x^3y + 2x^2y^2 - xy^3$ when $x = -1$ and $y = -3$
22. $-ab^2c + 2abc^2$ when $a = -2$, $b = -1$, and $c = 3$
23. $r^2st - 2rs^2t$ when $r = -1$, $s = 2$, and $t = -3$
24. $-3x^3y + 2x^2y - x^3y$ when $x = -2$ and $y = -1$

MORE CHALLENGING EXERCISES

Evaluate each polynomial.

25. $c^2 - 3c + \frac{1}{4}$ when $c = \frac{1}{4}$
26. $-x^2 + 2x - \frac{5}{2}$ when $x = -\frac{1}{2}$
27. $x^3 - x^2 + x - \frac{1}{2}$ when $x = \frac{1}{2}$
28. $-x^3 + x^2 - 2x + \frac{1}{3}$ when $x = -\frac{2}{3}$
29. $x^3 - x^2 + x - 1.2$ when $x = 0.4$
30. $-2u^3 + u^2 - u + 0.8$ when $u = -0.3$
31. $x^3 - x^2 - x + 2.4$ when $x = 0.2$
32. $-x^3 + 2x^2 - x - 0.4$ when $x = -0.1$

Let x be a positive integer. Find the value of x in each statement.

EXAMPLE $x^2 + x = 12$

SOLUTION Factor $x^2 + x$. ⟶ $x(x + 1) = 12$

Since x is a positive integer, $x + 1$ is also a positive integer. The two consecutive positive integers, x and $(x + 1)$, whose product is 12, are 3 and 4. Therefore, $x = 3$.

33. $x^2 - x = 12$
34. $x^2 + 2x = 8$
35. $x^2 - 2x = 24$
36. $3x^2 + 3x = 6$
37. $x^2 + x + 1 = 31$
38. $x^2 + 2x + 1 = 16$

NON-ROUTINE PROBLEMS

39. A train traveling at 60 miles per hour takes 3 seconds to enter a tunnel and an additional 30 seconds to pass completely through the tunnel. What is the length of the train? What is the length of the tunnel?

40. A customer at a shoe store bought a $15 pair of shoes and paid with a $50 bill. A day later, the store owner found that the $50 bill was counterfeit. How much did the store owner lose on the sale?

REVIEW CAPSULE FOR SECTION 18.3

Add. (Pages 247–250 and 267–269)

1. $(-5) + 12$
2. $8 + (-9)$
3. $(-3r) + (-4r)$
4. $(-5x^2) + 3x^2$
5. $(-8y) + (-7y)$
6. $4t^3 + (-5t^3)$
7. $(-xy^2) + (-\frac{3}{4}xy^2)$
8. $3\frac{3}{4}t + (-5\frac{1}{4}t)$
9. $(-n^2) - (-n^2) + (-n^2)$

18.3 Addition of Polynomials

EXAMPLE 1

Add: $(3x - 5x^2 + 3) + (x^2 - 2x - 8)$

Solution:

1. Write without parentheses, and combine like terms. → $3x - 5x^2 + 3 + x^2 - 2x - 8$

$x - 4x^2 - 5$

2. Arrange terms in order with the largest exponent first. → $-4x^2 + x - 5$

EXAMPLE 2

Add: $(x^3 - 3x) + (-3x^2 - 5x + 7) + (12 + 5x^2)$

Solution:

1. Write without parentheses, and combine like terms. → $x^3 - 3x - 3x^2 - 5x + 7 + 12 + 5x^2$

$x^3 - 8x + 2x^2 + 19$

2. Arrange terms in order. → $x^3 + 2x^2 - 8x + 19$

It is often helpful to arrange polynomials vertically for adding.

EXAMPLE 3

Add:

$$\begin{array}{r} 3x^2 - 5x + 6 \\ \underline{-2x^2 + 3x - 4} \end{array}$$

Like terms are in columns.

Solution:

$$\begin{array}{r} 3x^2 - 5x + 6 \\ \underline{-2x^2 + 3x - 4} \\ x^2 - 2x + 2 \end{array}$$

Find each sum: 6 and −4, −5x and 3x, 3x² and −2x²

EXAMPLE 4

Find the perimeter of the polygon.

Solution: Find the sum of the lengths of the sides.

$$\begin{array}{r} 4x^2 + x - 1 \\ 2x^2 \quad\quad + 3 \\ 5x - 2 \\ \underline{x + 1} \\ 6x^2 + 7x + 1 \end{array}$$

CLASSROOM EXERCISES

For Exercises 1–15, add. (Example 1)

1. $(x - 3) + (2x + 1)$
2. $(-3x^2 + 5) + (x^2 - 3)$
3. $(2c^2 - c) + (4c - 5)$
4. $(x^2 - 3x) + (4x - 2) + (3x^2 - 5)$
5. $(3x^3 + x - 5) + (x^2 - 5x + 3)$
6. $(2x^2y + xy - 5) + (3x^2y - 2xy + 7)$

(Examples 2, 3, and 4)

7. $\begin{array}{r} 3p - 2 \\ \underline{4p + 5} \end{array}$

8. $\begin{array}{r} -5x^2 + 1 \\ \underline{4x^2 - 3} \end{array}$

9. $\begin{array}{r} 2x^2 - x \\ \underline{x^2 + 5x} \end{array}$

10. $\begin{array}{r} x^2 - 3x + 2 \\ \underline{2x^2 + 5x - 6} \end{array}$

11. $\begin{array}{r} 4n^2 \qquad + 3 \\ \underline{n^2 - 3n + 4} \end{array}$

12. $\begin{array}{r} -5x^2 + 4x - 7 \\ \underline{-2x^2 \qquad + 3} \end{array}$

13. $\begin{array}{r} -s^2 - 3s + 1 \\ 2s^2 + s - 2 \\ \underline{s^2 - 2s + 6} \end{array}$

14. $\begin{array}{r} 2x^2 \qquad - 1 \\ 3x + 5 \\ \underline{-5x^2 - 2x \qquad} \end{array}$

15. $\begin{array}{r} 3x^2y \qquad + 4xy^2 - y^3 \\ -4x^2y + 6xy \qquad + 2y^3 \\ \underline{-3xy + 2xy^2 - 3y^3} \end{array}$

WRITTEN EXERCISES

Goal: To add polynomials

Sample Problem: Add: $(x^2y + 3xy - xy^2) + (2x^2y - xy - 4xy^2)$

Answer: $3x^2y + 2xy - 5xy^2$

For Exercises 1–14, add. (Example 1)

1. $(3x + 7) + (-5x - 2)$
2. $(-8j + 3) + (4j - 6)$
3. $(r^2 + 8) + (3r^2 - 11)$
4. $(7x^2 - 10) + (-x^2 + 8)$
5. $(5x^2 - x - 2) + (2x^2 - 2x + 7)$
6. $(-2z^2 + 3z - 4) + (5z^2 - z + 3)$

(Example 2)

7. $\begin{array}{r} 5n^2 - 2n + 3 \\ \underline{-2n^2 + 6n - 5} \end{array}$

8. $\begin{array}{r} 4x^2 - x + 5 \\ \underline{-x^2 + 6x - 3} \end{array}$

9. $\begin{array}{r} 4.3x^2 - 1.7x + 2.5 \\ \underline{3.9x^2 - 5.6x - 4.8} \end{array}$

10. $\begin{array}{r} 1.8x^2 - 3.7x + 2.9 \\ \underline{-2.4x^2 - 1.8x + 0.9} \end{array}$

11. $\begin{array}{r} x^2 - 5x + 3 \\ \underline{2x^2 + x - 7} \end{array}$

12. $\begin{array}{r} -6t^2 + 4t - 8 \\ \underline{3t^2 - 5t + 6} \end{array}$

13. $\begin{array}{r} 5x^2y - 4xy^2 + 2xy - 5 \\ \underline{-3x^2y + xy^2 - 5xy + 1} \end{array}$

14. $\begin{array}{r} -12x^2y + 4xy^2 - y^3 \\ \underline{5x^2y - 2xy^2 + 6y^3} \end{array}$

Arrange in columns of like terms. Then add. (Example 3)

15. $(3p^3 - 2p + 10) + (-2p^3 + 4p - 6)$

16. $(3x^2 - 17) + (5x^3 - x^2 + 9x)$

17. $(5x - 3x^3 + 9) + (x^2 + x^3 - 6x + 3)$

18. $(4w^4 - w^3 + 2w - 7) + (3w - 2w^2 + 5w^3 - w^4)$

19. $(13 - 2x^3) + (x^2 - 5x + 6) + (-3 + 4x^3 - x^2 + 2x)$

20. $(10x - 3x^2 + x^3) + (4x^2 - 5 - 6x) + (9 - 5x + 4x^3)$

Find the perimeter of each polygon. (Example 4)

21.

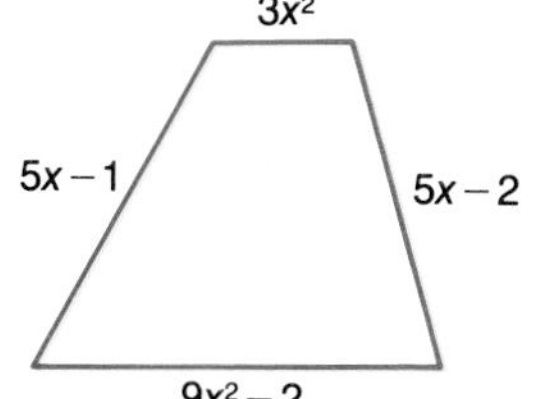

22.

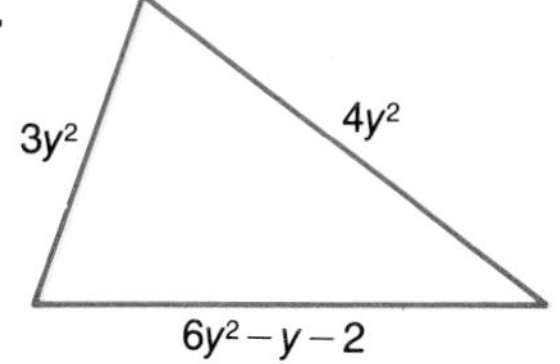

23.

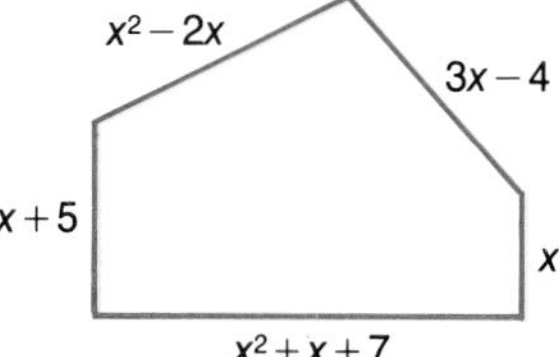

MIXED PRACTICE

24. $(4x - 3) + (3x + 1)$

25. $(-2y - 7) + (5y + 7)$

26. $(3m^2 + 2m) + (14m - 7)$

27. $(x^3 - 3x + 3) + (4x^3 - 10)$

28.
$$\begin{array}{r} 8x^3 \qquad\quad - x + 10 \\ \underline{-4x^3 + 9x^2 + 2x \qquad} \end{array}$$

29.
$$\begin{array}{r} x^2 - \ \ x + 9 \\ \underline{x^2 + 4x - 2} \end{array}$$

30.
$$\begin{array}{r} 2m^2 + 4m - \ \ 1 \\ 2m^2 + \ \ m + \ \ 4 \\ \underline{9m^2 - 6m - 11} \end{array}$$

31.
$$\begin{array}{r} 7s^2t + 9st \qquad\qquad - 9 \\ -2s^2t \qquad\quad + 3st^2 - 9 \\ \underline{8s^2t + 4st + 11st^2 + 18} \end{array}$$

32. Find the perimeter of a triangle with sides having lengths of $x - 10$, $3x^2 + 7x - 6$, and $-x^2 - 5x + 8$.

33. Find the perimeter of a rectangle having a length of $x^3 - 3x - 1$ and a width of $x^2 + 2x$.

MORE CHALLENGING EXERCISES

34. Find the polynomial that must be added to $4a - 3b$ to give a sum of $12a + 2b$.

35. Find the polynomial that must be added to $-3m^2 + 4m$ to give a sum of $m^2 - m$.

36. What polynomial must be added to $3xy - 4x^2 + y^2$ to give a sum of $-2x^2 + 4xy$?

37. What polynomial must be added to $-xy + 4x^2 - 3y^2$ to give a sum of $4x^2 + 5xy - 6$?

MID-CHAPTER REVIEW

Simplify. (Section 18.1)

1. $-t^2 + t - 3t + 4$
2. $-5k^2 - k - 6k + 12$
3. $1.2m - 2.7 + 1.8m^2 - 2.1m$
4. $4.8 + 7.2r - r^2 - 5.8r$
5. $12q - 28 + 9q^2 - q^3 + 2q^2$
6. $8w^2 - 9 - w + 3w^2 - 2w^3$
7. $-\frac{1}{4}a^2 + \frac{7}{4}a - \frac{3}{4}a + \frac{9}{4}$
8. $-\frac{3}{8}b^2 - \frac{7}{8}b - \frac{11}{8}b + \frac{9}{8}$
9. $9n^3 - 12n + 3n^4 - 24n^2 - 8n$
10. $49p - 28p^2 - 54 + 3p^4 - 4p^2$
11. $-y^5 + y^3 - y - 1 + 3y^5 - 2y^3 + y$
12. $2 - x^2 + 5x^4 - x^4 + 3x^2 - 17$

Evaluate each polynomial. (Section 18.2)

13. $2n^2 - 5n + 6$ when $n = -3$
14. $-3r^2 + 7r - 5$ when $r = -1$
15. $4t^3 - t^2 + 3t - 5$ when $t = 2$
16. $-p^3 + 4p^2 - 2p + 3$ when $p = 3$
17. $a^4 - a^2 - 1$ when $a = -2$
18. $y^5 - y^3 + y$ when $y = -2$
19. $0.4x^2 - 1.8x + 2.3$ when $x = 5$
20. $0.2m^2 - 0.8m - 1.5$ when $m = -4$
21. $-3r^2s + 2rs^2 - 4rs^2$ when $r = -1$ and $s = 4$
22. $2a^3b - 5ab^2 + 3a^2b^3$ when $a = -2$ and $b = 3$

Add. (Section 18.3)

23. $(2t^2 - 3t + 5) + (-t^2 + t - 12)$
24. $(-8r^2 + 6r - 3) + (3r^2 - r + 7)$
25. $\begin{array}{rrrr} 4t^3 & & -3t & +7 \\ \underline{-7t^3} & \underline{-8t^2} & & \underline{-6} \end{array}$
26. $\begin{array}{rrrr} 12q^3 & -5q^2 & & +12 \\ \underline{-5q^3} & & \underline{+9q} & \underline{-15} \end{array}$
27. $(42m^3 - 12m + 18) + (-21m^2 + 27m - 13) + (-18m^3 + 19m^2 - 30m)$
28. $(r^3s + 6rs^2) + (-3r^2s + 7rs) + (6r^2s^2 - 5rs^2 - 8rs)$
29. Find the perimeter of a triangle with sides having lengths of $2x^2 - 3x$, $3x^2 - 2x - 7$, and $5x - 4$.
30. Find the perimeter of a rectangle having a length of $4x + 3$ and a width of $x^2 - 4$.

REVIEW CAPSULE FOR SECTION 18.4

Subtract. (Pages 261–264 and 267–269)

1. $12 - 19$
2. $(-28) - (-17)$
3. $(-13) - (-21)$
4. $(-14) - 8$
5. $3r^2 - 12r^2$
6. $(-2m) - (-6m)$
7. $(-xy^2) - (-2xy^2)$
8. $\frac{7}{4}t^3 - (-\frac{5}{4}t^3)$
9. $(-10.9rs^2) - (-4.5rs^2)$

Using Patterns and Tables

Data Processing

High speed digital computers perform mathematical operations on numbers expressed in the **binary numeration system.** The binary system uses the digits 0 and 1 only.

$$\begin{aligned} 1101_{two} &= 1 \cdot 2^3 + 1 \cdot 2^2 + 0 \cdot 2^1 + 1 \cdot 2^0 \\ &= 8 + 4 + 0 + 1 \\ &= \mathbf{13} \end{aligned}$$

Expanded notation

Some computers use a **hexadecimal** (base 16) **numeration system.** The hexadecimal system uses the ten digits from 0 through 9 and the letters from A through F (see the table at the right).

$$\begin{aligned} 24C_{sixteen} &= 2 \cdot 16^2 + 4 \cdot 16^1 + 12 \cdot 16^0 \\ &= 512 + 64 + 12 \\ &= \mathbf{588} \end{aligned}$$

You can use the **patterns** in the table at the right to express numbers in base two and in base sixteen.

Base 10	Base 2	Base 16
1	1	1
2	10	2
3	11	3
4	100	4
5	101	5
6	110	6
7	111	7
8	1000	8
9	1001	9
10	1010	A
11	1011	B
12	1100	C
13	1101	D
14	1110	E
15	1111	F
16	10000	10
17	10001	11
18	10010	12
19	10011	13
20	10100	14
21	10101	15
22	10110	16
23	10111	17
24	11000	18
25	11001	19

EXAMPLE: Use the patterns in the table to write 26 and 27 in base two and in base sixteen.

SOLUTION: **Base 2** $26_{ten} = \mathbf{11010_{two}}$

$27_{ten} = \mathbf{11011_{two}}$

Base 16 $26_{ten} = \mathbf{1A_{sixteen}}$

$27_{ten} = \mathbf{1B_{sixteen}}$

EXERCISES

Write each number in base 10. Show the expanded notation.

1. 1011_{two} **2.** 110101_{two} **3.** 100101_{two} **4.** 10000011_{two}

Refer to the table for Exercises 5–7.

5. Write the next five counting numbers after 27 in base two.

6. Write the next five counting numbers after 27 in base sixteen.

7. Which statement is true?

a. $2E_{sixteen} < 35_{ten}$ **b.** $110111_{two} = 55_{ten}$ **c.** $11101_{two} > 3A_{sixteen}$

18.4 Subtraction of Polynomials

To subtract polynomials, subtract like monomials. Recall that when you subtract a number, you add its opposite.

EXAMPLE 1 Subtract:

$$\begin{array}{rr} & 5x^2 + 2x - 3 \\ (-) & \underline{2x^2 + 3x + 7} \end{array}$$

Add the opposite of $2x^2 + 3x + 7$.

Solution:

$$\begin{array}{rr} & 5x^2 + 2x - 3 \\ (-) & \underline{2x^2 + 3x + 7} \end{array} \longrightarrow \begin{array}{r} 5x^2 + 2x - 3 \\ \underline{-2x^2 - 3x - 7} \\ 3x^2 - x - 10 \end{array}$$

EXAMPLE 2 Subtract:

$$\begin{array}{rr} & -2x^3 \qquad\quad + 4x - 1 \\ (-) & \underline{x^3 + 4x^2 - 2x + 5} \end{array}$$

Solution:

$$\begin{array}{rr} & -2x^3 \qquad\quad + 4x - 1 \\ (-) & \underline{x^3 + 4x^2 - 2x + 5} \end{array} \longrightarrow \begin{array}{r} -2x^3 \qquad\quad + 4x - 1 \\ \underline{-x^3 - 4x^2 + 2x - 5} \\ -3x^3 - 4x^2 + 6x - 6 \end{array}$$

P–1 Subtract.

a. $\begin{array}{rr} & 3b^2 - 2b - 5 \\ (-) & \underline{b^2 + 4b + 2} \end{array}$ **b.** $\begin{array}{rr} & x^2 + 5x - 2 \\ (-) & \underline{3x^2 - 5x - 8} \end{array}$ **c.** $\begin{array}{rr} & 3.1q^2 \qquad\quad + 3.9 \\ (-) & \underline{1.7q^2 + 2.2q - 4.5} \end{array}$

You can also subtract polynomials horizontally.

EXAMPLE 3 Subtract: $(5x - 3x^3 + 17) - (15 - x^2 + 8x)$

Solution: 1 Write without parentheses, and combine like terms. ⟶ $5x - 3x^3 + 17 - 15 + x^2 - 8x$

$-3x - 3x^3 + 2 + x^2$

2 Arrange terms in order. ⟶ $-3x^3 + x^2 - 3x + 2$

P–2 Subtract.

a. $(y^2 - 2y - 8) - (2y^2 + 10y - 1)$

b. $(x^2 + x - 1) - (-7x + 2)$

CLASSROOM EXERCISES

For Exercises 1–7, subtract. (Examples 1 and 2)

1. $\begin{array}{rr} & 12x - 3 \\ (-) & \underline{8x + 4} \end{array}$

2. $\begin{array}{rr} & 4r^2 + 7 \\ (-) & \underline{r^2 - 1} \end{array}$

3. $\begin{array}{rr} & -x^2 + 2x - 3 \\ (-) & \underline{4x^2 + 6x - 5} \end{array}$

4. $\begin{array}{rr} & 3c^2 \qquad\quad - 8 \\ (-) & \underline{5c^2 + 2c + 3} \end{array}$

(Example 3)

5. $(5f - 2) - (3f - 7)$

6. $(-4x^2 + 3x) - (2x - 5x^2)$

7. $(3x^2 + 2x - 5) - (x^2 + x - 3)$

WRITTEN EXERCISES

Goal: To subtract polynomials
Sample Problem: Subtract: $(6x^2 + 10x - 3) - (-2x^2 - 5)$
Answer: $8x^2 + 10x + 2$

For Exercises 1–25, subtract. (Example 1)

1. $\begin{array}{rr} & 12x - 3 \\ (-) & \underline{8x + 2} \end{array}$

2. $\begin{array}{rr} & 8p + 6 \\ (-) & \underline{3p + 9} \end{array}$

3. $\begin{array}{rr} & 5x^2 + 2x \\ (-) & \underline{6x^2 - 5x} \end{array}$

4. $\begin{array}{rr} & 9w^3 - 4w \\ (-) & \underline{3w^3 + 3w} \end{array}$

5. $\begin{array}{rr} & 12x^2 - 2x + 3 \\ (-) & \underline{4x^2 + 3x - 7} \end{array}$

6. $\begin{array}{rr} & 10x^2 + 5x - 6 \\ (-) & \underline{8x^2 - 2x + 7} \end{array}$

(Example 2)

7. $\begin{array}{rr} & 8x^3 \qquad\quad - 5 \\ (-) & \underline{3x^3 - 4x^2 + 1} \end{array}$

8. $\begin{array}{rr} & 4m^3 + 9m^2 \qquad\quad - 2 \\ (-) & \underline{m^3 \qquad\qquad + 8m - 8} \end{array}$

9. $\begin{array}{rr} & 1.3x^2 \qquad\qquad + 2.7 \\ (-) & \underline{4.8x^2 - 3.7x - 1.9} \end{array}$

10. $\begin{array}{rr} & 4.1z^2 \qquad\qquad - 3.2 \\ (-) & \underline{5.9z^2 + 2.6z - 1.9} \end{array}$

11. $\begin{array}{rr} & 12x^2 - x \\ (-) & \underline{9x^2 \qquad - 6} \end{array}$

12. $\begin{array}{rr} & 5x^2 \qquad + 2 \\ (-) & \underline{x^2 - x \qquad} \end{array}$

(Example 3)

13. $(3x^2 - 2x + 5) - (x^2 + 5x - 6)$

14. $(-6v^2 + 4v - 1) - (2v^2 + 3v - 7)$

15. $(2a - 5a^3 + 7) - (15 + 2a^2 - 3a^3)$

16. $(12 + 3x^2 - x^3) - (-7x - 3x^2 + 19)$

17. $(-15x^3 - 2x + 6) - (-12x^3 + 3x^2 - 5x)$

18. $(13x^3 - 5x^2 + 8) - (9x^3 - 4x^2 - 5)$

MIXED PRACTICE

19. $\begin{array}{rr} & -3x^2 + x \\ (-) & \underline{5x^2 \qquad - 12} \end{array}$

20. $\begin{array}{rr} & 3x^3 \qquad\qquad - 4x \\ (-) & \underline{-x^3 + 2x^2 + \quad x - 19} \end{array}$

21. $\begin{array}{rr} & -j^2 + 8j - 9 \\ (-) & \underline{4j^3 \qquad - \quad j + 3} \end{array}$

22. $(1\frac{1}{4}b^2 - \frac{3}{8}b) - (\frac{3}{4}b^2 - \frac{5}{8})$

23. $(-5.2x + 4.7) - (-0.9x^2 + 2.8x)$

24. $(5g - 3g^4 + 2g^2 - 6) - (3g^3 - 6g + 15 - g^4)$

25. $(2x^2y^2 - 5xy^3 - 6xy + 1) - (5xy + 7xy^2 - x^2y^2 - 8)$

REVIEW CAPSULE FOR SECTION 18.5

Multiply. (Pages 278–280 and 285–288)

1. $(-7)(9)$	**2.** $(-12)(-5)$	**3.** $(-0.8)(-5)$	**4.** $(-2n)(3)$
5. $(-5d)(-3d)$	**6.** $(4x^2)(-xy)$	**7.** $(-\frac{3}{4}r)(12rs)$	**8.** $(-0.4t)(-0.8w)$
9. $(-3ab^2)(-abc)$	**10.** $3(x+2)$	**11.** $8(q-5)$	**12.** $a(2-x)$

18.5 Multiplication of Polynomials

EXAMPLE 1 Multiply: $(2x+3)(x-5)$

Solution:

1. Use the Distributive Property. ⟶ $(2x+3)\mathbf{(x-5)} = (2x+3)\mathbf{x} + (2x+3)\mathbf{(-5)}$
2. Use the Distributive Property. ⟶ $= 2x\mathbf{(x)} + 3\mathbf{(x)} + 2x\mathbf{(-5)} + 3\mathbf{(-5)}$
3. Simplify. ⟶ $= 2x^2 + 3x - 10x - 15$
4. Combine like terms. ⟶ $= 2x^2 - 7x - 15$

Example 2 shows a shorter method for multiplying two binomials. The method is called the FOIL method.

EXAMPLE 2 Multiply: $(3x-2)(x+4)$

Solution: **Multiply the first terms.** ⟶ **Multiply the outside terms.** ⟶ **Multiply the inside terms.** ⟶ **Multiply the last terms.**

$$(3x-2)(x+4) = \overset{F}{3x^2} + \overset{O}{12x} - \overset{I}{2x} - \overset{L}{8}$$
$$= 3x^2 + 10x - 8$$

***F** means First.*
***O** means Outside.*
***I** means Inside.*
***L** means Last.*

The problem can be solved step-by-step when arranged vertically.

1.
$$\begin{array}{r} \mathbf{x} + 4 \\ \underline{\mathbf{3x} - 2} \\ \mathbf{3x^2} \end{array}$$

2.
$$\begin{array}{r} x + \mathbf{4} \\ \underline{\mathbf{3x} - 2} \\ 3x^2 + \mathbf{12x} \end{array}$$

3.
$$\begin{array}{r} \mathbf{x} + 4 \\ \underline{3x - \mathbf{2}} \\ 3x^2 + 12x \\ \mathbf{-2x} \end{array}$$

Like terms are arranged vertically.

4.
$$\begin{array}{r} x + \mathbf{4} \\ \underline{3x - \mathbf{2}} \\ 3x^2 + 12x \\ \underline{- \quad 2x - \mathbf{8}} \end{array}$$

5.
$$\begin{array}{r} x + 4 \\ \underline{3x - 2} \\ 3x^2 + 12x \\ \underline{- \quad 2x - 8} \\ \mathbf{3x^2 + 10x - 8} \end{array}$$

P–1 **Multiply.**

a. $(r - 9)(r + 9)$ **b.** $(x + 5)(3x + 8)$ **c.** $(2y - 8)(y - 11)$

To multiply any two polynomials, multiply each term of one polynomial by each term of the other. Then combine like terms.

EXAMPLE 3 Multiply: $(3x - 1)(x^2 + 2x - 3)$

Solution: **Arrange vertically.** → **Multiply each term of $x^2 + 2x - 3$ by 3x.** → **Multiply each term of $x^2 + 2x - 3$ by −1.** → **Add like terms.**

$$\begin{array}{r} x^2 + 2x - 3 \\ \underline{3x - 1} \\ 3x^3 + 6x^2 - 9x \\ \underline{- x^2 - 2x + 3} \\ 3x^3 + 5x^2 - 11x + 3 \end{array}$$

First multiply each term by 3x. Then multiply each term by −1.

When you multiply two polynomials, arrange the terms in order. Leave space for the missing terms.

EXAMPLE 4 Multiply: $(2x - 3)(2x + 3 - x^3)$

Solution:

$$\begin{array}{rrrrr} -x^3 & & + 2x & + 3 & \\ & & 2x & - 3 & \\ \hline -2x^4 & & + 4x^2 & + 6x & \\ & 3x^3 & & - 6x & - 9 \\ \hline -2x^4 & + 3x^3 & + 4x^2 & & - 9 \end{array}$$

Arrange terms in order. Leave space for missing terms.

P–2 **Multiply.**

a. $(k - 4)(k^2 - k + 10)$ **b.** $(2x + 3)(x^2 - 7)$ **c.** $(5 + s)(s - 4s^3 + 9s^2)$

CLASSROOM EXERCISES

Multiply. (Examples 1 and 2)

1. $(x + 3)(x + 2)$ **2.** $(x + 4)(x - 1)$ **3.** $(x - 2)(x + 5)$ **4.** $(y + 5)(y + 5)$

5. $(2x + 1)(x - 3)$ **6.** $(3n - 1)(n + 3)$ **7.** $(u - 3)(u - 3)$ **8.** $(4x - 1)(x + 3)$

Write the terms missing from each row. (Examples 3 and 4)

9.
$$\begin{array}{r} 3p - 5 \\ \underline{2p + 7} \\ 6p^2 - 10p \\ \underline{21p - 35} \\ \boxed{} \end{array}$$

10.
$$\begin{array}{r} 2x^2 - 3x \\ \underline{2x + 4} \\ 4x^3 - 6x^2 \\ \underline{\boxed{}} \\ 4x^3 + 2x^2 - 12x \end{array}$$

11.
$$\begin{array}{r} 2x^2 - 3x + 4 \\ \underline{2x - 3} \\ 4x^3 - 6x^2 + 8x \\ \underline{\boxed{}} \\ 4x^3 - 12x^2 + 17x - 12 \end{array}$$

WRITTEN EXERCISES

Goal: To multiply polynomials
Sample Problem: Multiply: $(x + 1)(x^2 - x + 1)$
Answer: $x^3 + 1$

For Exercises 1–30, multiply. (Examples 1 and 2)

1. $(k - 2)(k + 5)$ **2.** $(x + 7)(x - 3)$ **3.** $(x - 8)(x - 9)$ **4.** $(x - 6)(x - 7)$

5. $(3q - 2)(q + 3)$ **6.** $(2x - 5)(x + 4)$ **7.** $(4x - 5)(4x - 5)$ **8.** $(6x + 7)(6x + 7)$

9. $(3n - 7)(3n + 7)$ **10.** $(2z + 5)(2z - 5)$ **11.** $(2x^2 + 3)(x - 1)$ **12.** $(x^2 - 3)(x^2 + 4)$

(Examples 3 and 4)

13. $\begin{array}{r} 2x^2 - x + 3 \\ \underline{x - 5} \end{array}$ **14.** $\begin{array}{r} -3x^2 + x - 2 \\ \underline{x + 4} \end{array}$ **15.** $\begin{array}{r} d^2 - 3d - 5 \\ \underline{d + 8} \end{array}$ **16.** $\begin{array}{r} 2x^2 + 4x - 3 \\ \underline{x - 6} \end{array}$

17. $\begin{array}{r} h^2 - 2h + 5 \\ \underline{h + 3} \end{array}$ **18.** $\begin{array}{r} x^2 + 5x - 3 \\ \underline{x + 2} \end{array}$ **19.** $\begin{array}{r} -3v^3 + 2v - 1 \\ \underline{v + 2} \end{array}$ **20.** $\begin{array}{r} x^3 - x + 3 \\ \underline{x - 5} \end{array}$

21. $\begin{array}{r} -2x^2 + 3x - 2 \\ \underline{x^2 + 3} \end{array}$ **22.** $\begin{array}{r} 4x^2 - x + 3 \\ \underline{x^2 - 4} \end{array}$ **23.** $\begin{array}{r} g^4 - g^2 + 1 \\ \underline{g - 1} \end{array}$ **24.** $\begin{array}{r} x^5 - x^3 + x \\ \underline{x + 2} \end{array}$

MIXED PRACTICE

25. $\begin{array}{r} -x^3 + 4x - 3 \\ \underline{x + 4} \end{array}$ **26.** $\begin{array}{r} b^2 + 2b - 3 \\ \underline{b - 5} \end{array}$ **27.** $\begin{array}{r} w^4 - 3w^2 + 2 \\ \underline{w - 5} \end{array}$ **28.** $\begin{array}{r} -3x^2 + 2 \\ \underline{x - 5} \end{array}$

29. $(f + 2)(f^2 - 4f - 1)$ **30.** $(2x - 1)(-x^2 + 2x - 5)$

NON-ROUTINE PROBLEM

31. The 5 short pieces of chain shown below must be linked together to form one long chain. Explain how the pieces can be linked together by cutting only three of the rings.

REVIEW CAPSULE FOR SECTION 18.6

Simplify. (Pages 476–479)

1. $\frac{2t^2}{t}$
2. $\frac{-6x^2}{2x}$
3. $\frac{10j^3}{-5j}$
4. $\frac{-5x^2}{-x^2}$

Multiply. (Pages 285–288)

5. $(x-5)(2x)$
6. $(3s-2)(-5)$
7. $(4p-3)(3p^2)$
8. $(-3m)(m^2+4)$

18.6 Division of Polynomials

Consider the product of these two binomials.

$$(2x+1)(x-2) = 2x^2 - 3x - 2$$

Two division problems can be written from this one multiplication problem.

1

$$(2x^2 - 3x - 2) \div (x - 2) = 2x + 1$$

2

$$(2x^2 - 3x - 2) \div (2x + 1) = x - 2$$

EXAMPLE Divide: $(3x^2 - 7x - 6) \div (x - 3)$

Solution: [1] Find $\frac{3x^2}{x}$ to get $3x$ as the first term of the quotient. ⟶

$$\begin{array}{r} \mathbf{3x} \\ x-3\,\overline{)\,\mathbf{3x^2} - 7x - 6} \end{array}$$

[2] Multiply $(x-3)$ by $3x$. ⟶

[3] Subtract the product from the dividend. ⟶

$$\begin{array}{r} \mathbf{3x} \\ \mathbf{x-3}\,\overline{)\,3x^2 - 7x - 6} \\ \underline{\mathbf{3x^2 - 9x}} \\ 2x - 6 \end{array}$$

[4] Find $\frac{2x}{x}$ to get 2 as the second term of the quotient. ⟶

$$\begin{array}{r} 3x + \mathbf{2} \\ \mathbf{x}-3\,\overline{)\,3x^2 - 7x - 6} \\ \underline{3x^2 - 9x} \\ \mathbf{2x} - 6 \end{array}$$

[5] Multiply $(x-3)$ by 2. ⟶

[6] Subtract the product from $2x - 6$. ⟶

$$\begin{array}{r} 3x + \mathbf{2} \\ \mathbf{x-3}\,\overline{)\,3x^2 - 7x - 6} \\ \underline{3x^2 - 9x} \\ 2x - 6 \\ \underline{\mathbf{2x - 6}} \\ 0 \end{array}$$

CLASSROOM EXERCISES

Find each quotient. (Steps 1 and 4 of Example)

1. $\frac{-5t}{t}$ 2. $\frac{3v^2}{-v}$ 3. $\frac{-4x^2}{-4x}$ 4. $\frac{10p^3}{5p}$ 5. $\frac{-2x^3}{x^2}$ 6. $\frac{-15x^3}{-5x^2}$

Subtract. (Steps 3 and 6 of Example)

7. $\begin{array}{l} x^2 - 3x + 2 \\ \underline{x^2 - x} \end{array}$

8. $\begin{array}{l} -f^2 - 5 \\ \underline{-f^2 + 2f + 3} \end{array}$

9. $\begin{array}{l} 6x^2 - x - 3 \\ \underline{6x^2 + 5} \end{array}$

Find each quotient. (Example)

10. $\begin{array}{r} 2r \\ r + 4\overline{)2r^2 + 5r - 12} \\ \underline{2r^2 + 8r} \\ -3r - 12 \end{array}$

11. $\begin{array}{r} x \\ x - 9\overline{)x^2 - 18x + 81} \\ \underline{x^2 - 9x} \\ -9x + 81 \end{array}$

12. $\begin{array}{r} 4x \\ x - 2\overline{)4x^2 - 7x - 2} \\ \underline{4x^2 - 8x} \\ x - 2 \end{array}$

WRITTEN EXERCISES

Goal: To divide polynomials

Sample Problem: Divide: $3x + 1\overline{)3x^2 - 14x - 5}$ **Answer:** $x - 5$

Divide. (Example)

1. $b - 4\overline{)2b^2 - 5b - 12}$
2. $j + 3\overline{)3j^2 + 5j - 12}$
3. $3x - 2\overline{)-6x^2 + 19x - 10}$
4. $3x - 6\overline{)-12x^2 + 27x - 6}$
5. $x - 7\overline{)x^2 - 14x + 49}$
6. $w + 12\overline{)w^2 + 24w + 144}$
7. $h + 3\overline{)h^3 + h^2 - 3h + 9}$
8. $x - 4\overline{)2x^3 - 3x^2 - 21x + 4}$
9. $2x - 2\overline{)2x^3 - 10x^2 + 20x - 12}$
10. $3s + 5\overline{)3s^3 - 4s^2 - 3s + 20}$

MORE CHALLENGING EXERCISES

Divide.

EXAMPLE: $x + 1\overline{)x^3 - 2x^2 + 3}$ **SOLUTION:**

$$\begin{array}{r} x^2 - 3x + 3 \\ x + 1\overline{)x^3 - 2x^2 + 3} \\ \underline{x^3 + x^2} \\ -3x^2 + 3 \\ \underline{-3x^2 - 3x} \\ 3x + 3 \\ \underline{3x + 3} \\ 0 \end{array}$$

Leave space for an x term.

11. $q - 1\overline{)q^3 - 6q^2 + 5}$
12. $2x - 6\overline{)2x^3 - 32x + 42}$

Focus on Reasoning: Key Words

Careful attention to **key words** is an aid to logical reasoning in drawing conclusions.

EXAMPLE Assume that the given statement is true. Write True, False, Possible, or Can't tell for each given conclusion. Give a reason for each answer.

Statement: Fairview Stadium has at least 5 gates.

Conclusions

a. Fairview Stadium has exactly 3 gates.

b. Fairview Stadium has exactly 5 gates.

c. Millard Stadium has at most 7 gates.

d. Fairview Stadium has at most 9 gates.

e. Fairview Stadium has 5 or more gates.

Analysis

a. **False.** "At least 5" means that the smallest possible number of gates is 5.

b. **Possible.** See the reason for conclusion **a.**

c. **Can't tell.** No information is given about Millard Stadium.

d. **Possible.** See the reason for conclusion **a.**

e. **True.** "At least 5 gates" means 5 gates, or 6 gates, or 7 gates, and so on.

EXERCISES

Assume that each given statement is true. Write True, False, Possible, or Can't tell for the conclusions that follow each statement. Explain each answer.

1. **Statement:** Fred spent $15 at most for the calculator.

   Conclusions:
   a. Fred spent exactly $13.75 for the calculator.
   b. Fred spent at least $20 for the calculator.
   c. Fred spent more than $8 for the calculator.
   d. Alice spent $4.50 more than Fred for a calculator.

2. **Statement:** Kristin saves at least $10 per week.

   Conclusions:
   a. In 4 weeks, Kristin has saved at least $40.
   b. Kristin saved less than $520 last year.
   c. Kristin saved at least $520 last year.
   d. In 4 weeks, Kristin saved exactly $50.

3. **Statement:** Between 30 and 40 students belong to the math club this year.

Conclusions:
- **a.** Exactly 35 students belong to the math club this year.
- **b.** More than 35 students belong to the math club this year.
- **c.** Last year, exactly 35 students belonged to the math club.
- **d.** Fewer than 45 students belong to the math club this year.

4. **Statement:** The Sandlot Hitters played more than 20 games last season.

Conclusions:
- **a.** The Sandlot Hitters had a winning season last year.
- **b.** The Sandlot Hitters played at most 20 games last season.
- **c.** The Sandlot Hitters played exactly 21 games last season.
- **d.** The Sandlot Hitters played between 21 and 28 games last season.

5. **Statement:** Exactly 5 out of every 7 students in Stevens Junior High received a grade of C or better on the test. There are 490 students at Stevens Junior High.

Conclusions:
- **a.** Of the 490 students, more than 350 received a grade of C on the test.
- **b.** Exactly 2 out of every 7 students received a grade lower than C on the test.
- **c.** Of the 490 students, exactly 140 received a grade lower than C on the test.
- **d.** Of the 490 students, more than 349 and fewer than 351 received a grade of C or better on the test.

6. **Statement:** Of 2500 people interviewed, fewer than 25% were under the age of 19.

Conclusions:
- **a.** Of the 2500 people interviewed, exactly 625 were 19 years old.
- **b.** Of the 2500 people interviewed, more than 75% were 19 years old or older.
- **c.** Of the 2500 people interviewed, 500 were between the ages of 12 and 18.
- **d.** Of the 2500 people interviewed, more than 1900 were over the age of 30.

CHAPTER SUMMARY

IMPORTANT TERMS	Monomial *(p. 492)* Binomial *(p. 492)*	Polynomial *(p. 492)* Term *(p. 494)*

IMPORTANT IDEAS

1. To simplify a polynomial:
   a. Combine like terms.
   b. Arrange terms in order with the largest exponent first, the next larger exponent second, and so on.
2. In evaluating a polynomial, raise a number to a power before performing any other operation.
3. Steps for Adding Two Polynomials
   a. Write without parentheses and combine like terms.
   b. Arrange the terms in order.
4. To subtract polynomials, subtract like monomials.
5. To multiply any two polynomials, multiply each term of one polynomial by each term of the other. Then combine like terms.

CHAPTER REVIEW

SECTION 18.1

Simplify.

1. $5t - 2t^2 + 3 + t^2$

2. $18x^2 + 11x - x^2 - 7$

3. $4x^2 - 2x^3 + 5 - x^2 + 3x^3$

4. $12r - 3r^2 + 6 - r - 2r^3 + r^2$

5. $7.3p - 12.6 + 1.9p^2 - 3.7p - 2.3p^2$

6. $2.9x^3 - 4.7 + 3.2x - 3.4x^3 - 1.9x$

SECTION 18.2

Evaluate each polynomial.

7. $-2w^2 + 5w - 3$ when $w = -4$

8. $3x^2 - 2x + 1$ when $x = -3$

9. $-x^3 + 2x^2 - x - 3$ when $x = -1$

10. $-q^5 + 3q^3 - q + 2$ when $q = -1$

11. $-2x^2y^2 - xy^2 + 3xy - 2$ when $x = -2$ and $y = 1$

12. $-x^3y + 2x^2y^2 - 3xy^3 + 1$ when $x = 3$ and $y = -2$

SECTION 18.3

Add.

13. $(4r - 7) + (-r + 3)$

14. $(-6x + 5) + (2x - 8)$

15. $(3b^2 - 2b + 3) + (-b^2 + 3b - 5) + (-4b^2 + 5b - 2)$

16. $(-5x^2 + 4x - 1) + (3x^2 - 8x + 5) + (x^2 - x - 2)$

17.
$$\begin{array}{rrrrr} -c^4 & & +2c^2 & -3c & \\ & 2c^3 & -c^2 & +7c & -3 \\ \underline{5c^4} & \underline{+c^3} & & \underline{-c} & \underline{-9} \end{array}$$

18.
$$\begin{array}{rrrrr} 5x^4 & -3x^3 & & +7x & -10 \\ & 4x^3 & -x^2 & -2x & +8 \\ \underline{-3x^4} & \underline{-2x^3} & \underline{+3x^2} & \underline{-x} & \end{array}$$

19. Find the perimeter of a rectangle having a length of $x^2 + x$ and a width of $-2x^2 + 4x - 1$.

20. Find the perimeter of a triangle having sides with lengths of $x + 3$, $2x^2 + 3x$, and $x^3 + x - 5$.

SECTION 18.4

Subtract.

21. $(4z^2 - 3z + 9) - (-z^2 + 2z - 6)$

22. $(-3x^2 + 7x - 5) - (4x^2 - x + 10)$

23. $(2.5g - 1.8 + 3.7g^2 - 0.9g^3) - (-2.7 + 1.9g^2 - 5.3g + 1.2g^3)$

24. $(5.1x^2 - 1.8x^4 + 3.7x - 1.5) - (5.3x + 0.6x^4 - 2.1x^3 - 2.3x^2)$

25.
$$\begin{array}{rrrrrr} & 19x^4 & & -12x^2 & +20x & \\ (-) & \underline{-x^4} & \underline{+24x^3} & & \underline{+3x} & \underline{-19} \end{array}$$

26.
$$\begin{array}{rrrrr} & -2h^3 & +5h^2 & & -35 \\ (-) & \underline{9h^3} & \underline{-h^2} & \underline{-27h} & \underline{+18} \end{array}$$

SECTION 18.5

Multiply.

27. $\begin{array}{r} 2j^2 - 3j + 5 \\ \underline{2j - 3} \end{array}$

28. $\begin{array}{r} -3x^2 + 4x - 6 \\ \underline{5x - 2} \end{array}$

29. $\begin{array}{r} -3d^3 - d^2 + 3 \\ \underline{d^2 + 2} \end{array}$

30. $\begin{array}{r} 5x^3 - 3x + 7 \\ \underline{x^2 - 3} \end{array}$

31. $\begin{array}{r} 4t^2 - t + 2 \\ \underline{t^2 + 3} \end{array}$

32. $\begin{array}{r} x^2 - 2x + 3 \\ \underline{-2x^2 + 5} \end{array}$

SECTION 18.6

Divide.

33. $4n - 3 \overline{) 20n^2 - 7n - 6}$

34. $3x - 7 \overline{) 6x^2 - 23x + 21}$

35. $x + 3 \overline{) -3x^3 - 8x^2 + 5x + 6}$

36. $2w + 1 \overline{) 8w^3 - 2w^2 + 3w + 3}$

CUMULATIVE REVIEW: CHAPTERS 1–18

Evaluate each expression. (Section 1.2)

1. $9p + 8q$ when $p = 5$ and $q = 3$
2. $15x - 3y$ when $x = 8$ and $y = 6$
3. $18 - 3 \cdot d$ when $d = 4$
4. $s \div 3 + t \cdot 4$ when $s = 24$ and $t = 5$

Solve each problem. (Sections 4.7 and 6.5)

5. The base of a triangular sail is $12\frac{1}{4}$ feet. The height of the sail is 18 feet. Find the area of the sail.
6. The length of a rectangular sign is 3 meters more than the width. The perimeter of the sign is 26 meters. Find the length and width.

Multiply. (Sections 10.1 through 10.3)

7. $(-4.3)12$
8. $-5(-\frac{3}{4})(-12)$
9. $(8 - 3a)a$
10. $2c^2(5c - 4)$

Solve and check. (Section 11.4)

11. $22 = \frac{m}{4} - 6$
12. $2m + 13.8 = -24.8$
13. $3(2b + 5) - 8b = 10$

Solve and check. (Section 12.5)

14. $-3b - 12 = b - 24$
15. $-4(3x - 2) = 27 - 9x - 34$

Solve and graph. (Sections 13.2 through 13.4)

16. $x + 12 < 10$
17. $9 > -2a$
18. $2r - 7 < 6r + 2 - r$

Write the domain and range of each function. (Section 14.3)

19. $\{(0,2), (-2,-4), (2,8), (-1,-1)\}$
20. $\{(-1,0), (-3,2), (4,1), (0,1)\}$

Graph each function. (Section 14.4)

21. $y = x - 3$
22. $y = 2x + 1$
23. $y = \frac{1}{2}x + 3$

Use Condition 1 and two variables to represent the unknowns. Use Conditions 1 and 2 to write a system of equations for the problem. Solve. (Sections 15.4 and 15.7)

24. The length of a rectangle increased by twice the width is 77 units. The perimeter of the rectangle is 120 units. Find the length and width.
25. The sum of two numbers is 84. The difference of the two numbers is 22. Find the unknown numbers.

Simplify. (Section 16.1)

26. $\sqrt{72}$ **27.** $\sqrt{80}$ **28.** $\sqrt{112}$ **29.** $\sqrt{275}$

Approximate each square root to three decimal places or give the exact value if appropriate. Refer to the table on page 458. (Section 16.2)

30. $\sqrt{65}$ **31.** $\sqrt{5625}$ **32.** $\sqrt{200}$ **33.** $\sqrt{164}$

Perform the indicated operation. (Sections 16.3 and 16.4)

34. $3\sqrt{5} + 4\sqrt{5}$ **35.** $4\sqrt{3} - \sqrt{12}$ **36.** $(3\sqrt{15})(4\sqrt{2})$ **37.** $(-5\sqrt{6})(3\sqrt{2})$

Simplify. (Section 16.5)

38. $\frac{\sqrt{12}}{\sqrt{3}}$ **39.** $\sqrt{\frac{5}{9}}$ **40.** $\frac{\sqrt{3}}{\sqrt{7}}$ **41.** $\sqrt{\frac{10}{15}}$

Simplify each fraction. No divisor is zero. (Section 17.1)

42. $-\frac{10x^2y}{15xy}$ **43.** $\frac{3k-6}{4k-8}$ **44.** $\frac{6xy^2}{2y^2-4xy}$ **45.** $\frac{3x+6}{6x^2+12x}$

Perform the indicated operations. Write each answer in simplest form. (Sections 17.2 through 17.4)

46. $\frac{-2a}{3} \cdot \frac{-5}{6a}$ **47.** $\frac{3x-3}{x} \cdot \frac{4x^2}{9x-9}$ **48.** $\frac{a+5}{4a} \div \frac{a+5}{a}$

49. $\frac{2r}{r-2} \div \frac{10r}{3r-6}$ **50.** $\frac{13}{12b} + \frac{5}{12b}$ **51.** $\frac{3}{x-3} - \frac{x+6}{x-3}$

Evaluate each polynomial. (Section 18.2)

52. $2x^2 + 7x - 3$ when $x = 3$ **53.** $x^3 - x^2 + 4x + 9$ when $x = -2$

Add or subtract. (Sections 18.3 and 18.4)

54.
$$\begin{array}{r} r^2 - 5r + 3 \\ (+)\ \underline{3r^2 + 7r + 10} \end{array}$$

55.
$$\begin{array}{r} 4d^2 + d - 7 \\ (-)\ \underline{d^2 \qquad - 10} \end{array}$$

56.
$$\begin{array}{r} y^2 - 3y - 9 \\ (-)\ \underline{7y^2 + 6y - 2} \end{array}$$

Multiply. (Section 18.5)

57. $(g + 9)(g + 4)$ **58.** $(4f + 5)(f - 3)$ **59.** $(c - 6)(3c^2 - c + 2)$

Divide. (Section 18.6)

60. $a + 3\overline{)4a^2 + 10a - 6}$ **61.** $3p - 5\overline{)6p^3 - p^2 - 18p + 5}$

Using Manipulatives

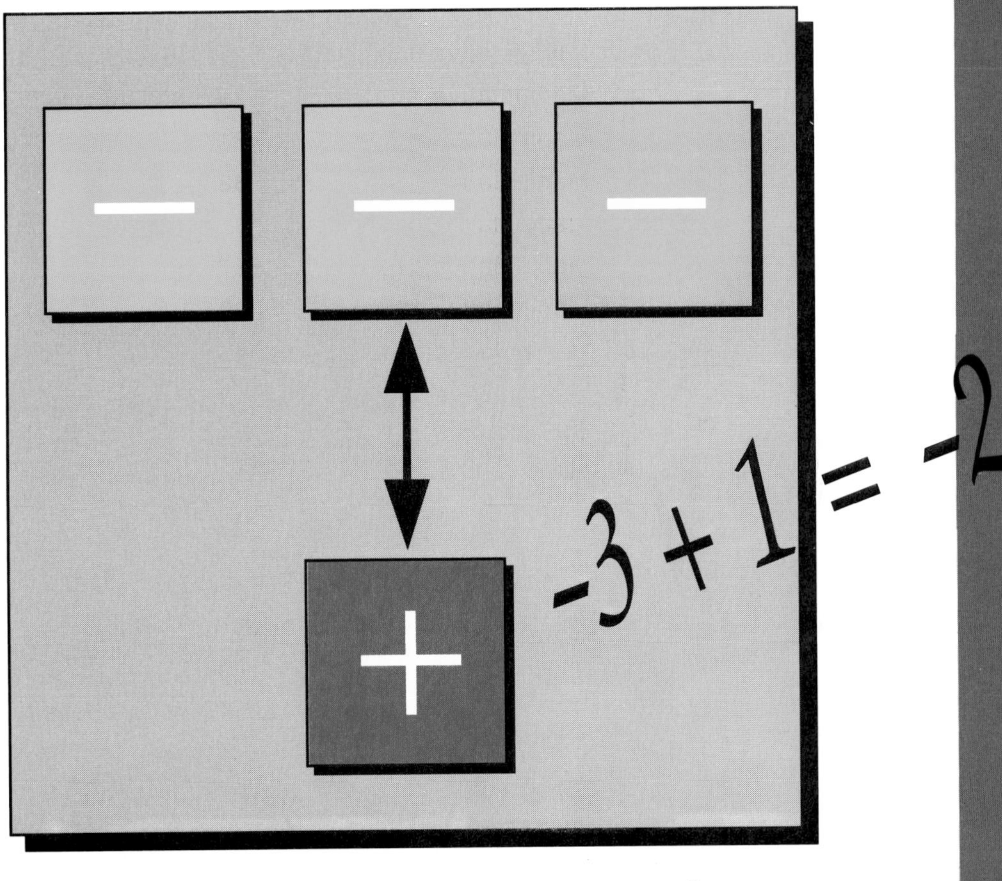

$$-3 + 1 = -2$$

CONTENTS

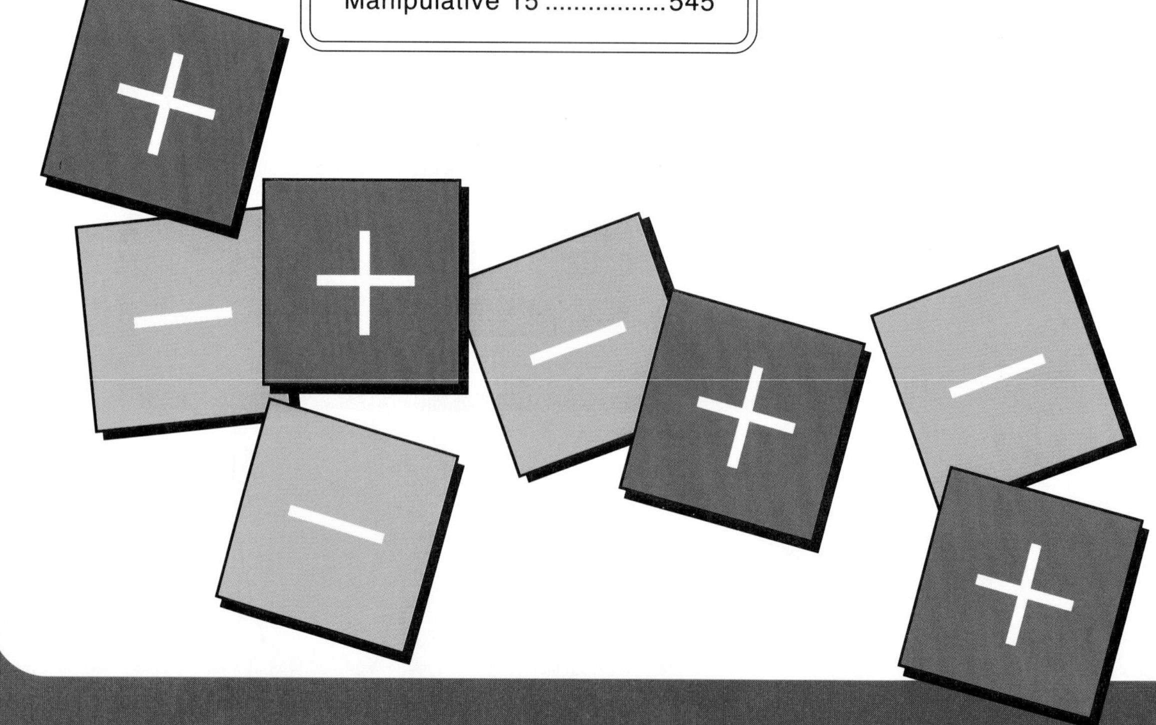

Manipulative 1 *(Use with Lesson 5.1.)*

You can model equations with tiles such as these.

x-tile **1-tile**

EXAMPLE 1

Use algebra tiles to model each equation.

Equation **Model**

a. $x + 2 = 5$

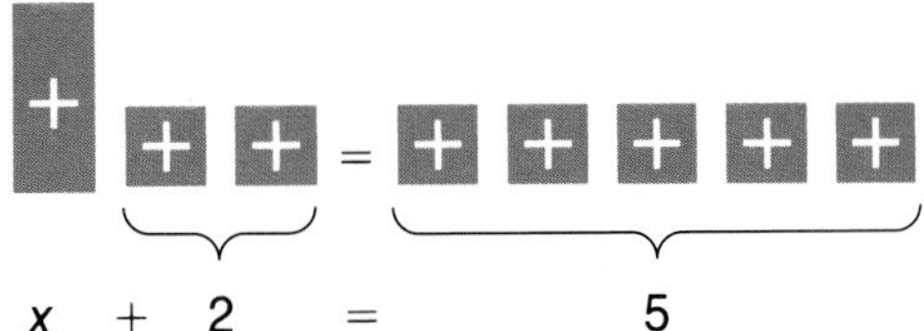

$x + 2 = 5$

b. $2x + 1 = 3$

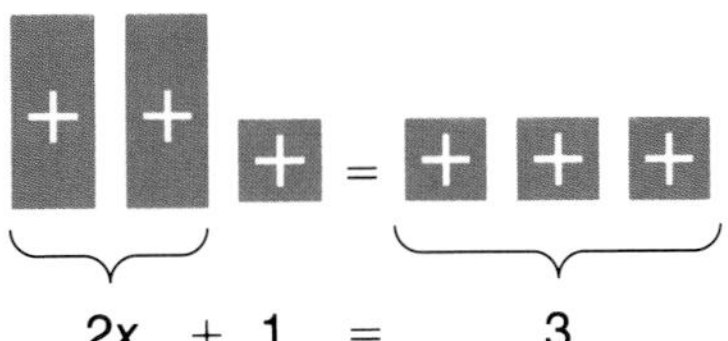

$2x + 1 = 3$

EXAMPLE 2

Solve and check: $x + 3 = 5$

Solution: **A.** Model the equation.

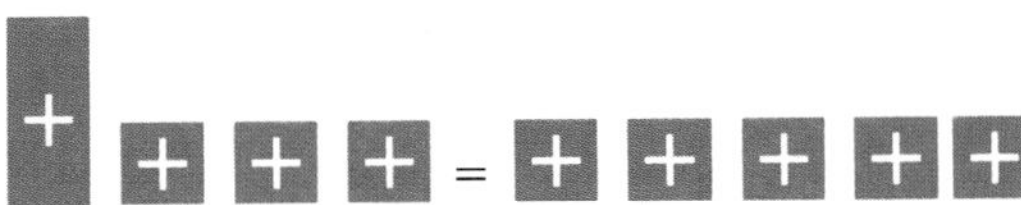

$x + 3 = 5$

B. **Think:** To get x alone on the left side of the equation, remove three (1)–tiles from each side.

$x + 3 - 3 = 5 - 3$

C. Show the result.

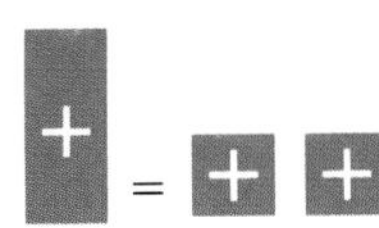

$x = 2$

Check: To check your answer, replace the x-tile in the original equation with the two (1)-tiles.

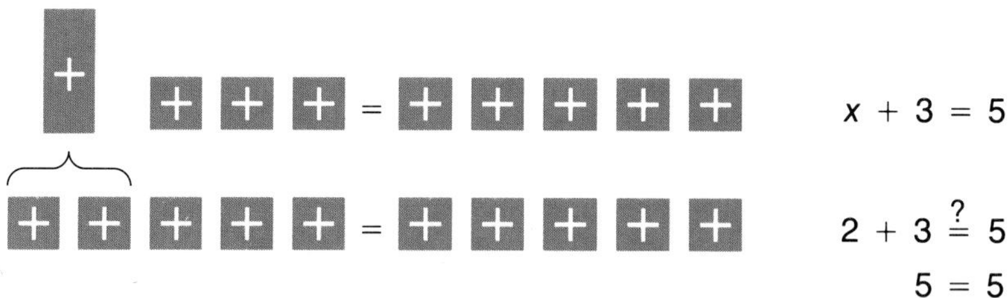

$x + 3 = 5$

$2 + 3 \stackrel{?}{=} 5$

$5 = 5$

EXERCISES

Goal: To use algebra tiles to solve equations of the form $x + a = b$ or $b = x + a$

Write the equation shown by each model.

1. 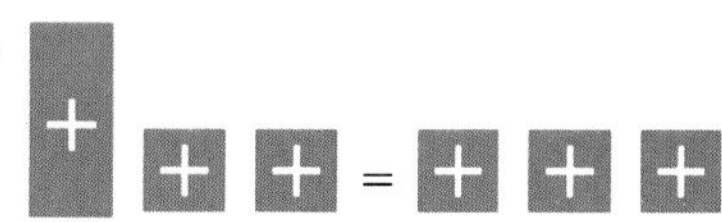

2. 

3. 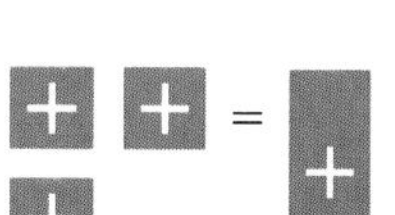

4. 

5. Use algebra tiles to show that the solution of $x + 7 = 9$ is 2.
6. Use algebra tiles to show that the solution of $x + 5 = 6$ is 1.

Use algebra tiles to solve and check each equation.

7. $x + 1 = 5$	**8.** $x + 6 = 9$	**9.** $x + 5 = 9$
10. $x + 2 = 8$	**11.** $x + 3 = 5$	**12.** $x + 4 = 5$
13. $6 = x + 4$	**14.** $4 = x + 1$	**15.** $3 = x + 3$

Use algebra tiles to model an equation that has the given solution.

16. 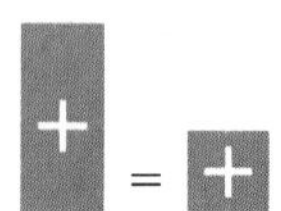

17. 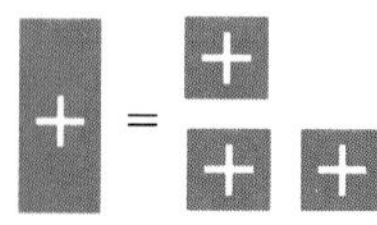

18. 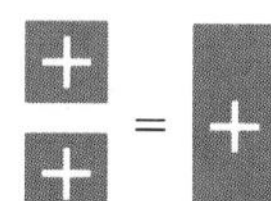

Manipulative 2 *(Use with Lesson 5.2.)*

To represent an equation such as $2x - 3 = 1$, you will need two x-tiles, a (1)-tile, and tiles to represent -3. Since [+] represents adding 1, let [−] represent subtracting 1.

Equation | **Model**

$2x - 3 = 1$

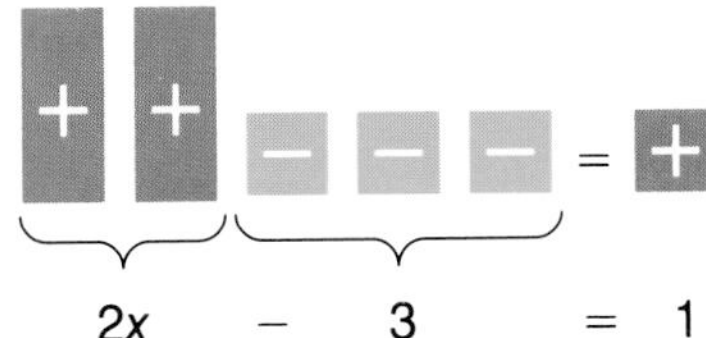

You already know that $1 - 1 = 0$. So a pair of [+]↔[−] tiles equals $1 - 1$, or 0.

This example shows you how to use algebra tiles to solve $x - 2 = 4$.

EXAMPLE Solve and check: $x - 2 = 4$

Solution: **A.** Model the equation.

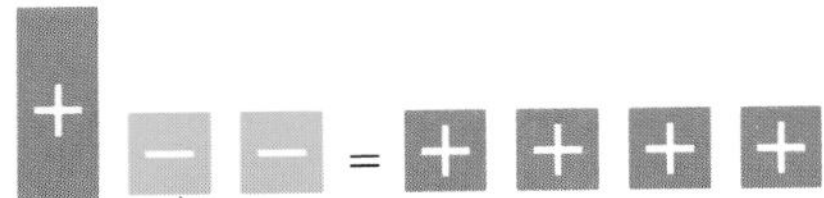

$x - 2 = 4$

B. **Think:** There are two (-1)-tiles on the left side of the equation. To get x alone, I need to add two [+]-tiles to make two [+]↔[−] pairs with a value of 0. So add two [+]-tiles to each side.

$x - 2 + \mathbf{2} = 4 + \mathbf{2}$

or $x + \mathbf{2} - 2 = 4 + \mathbf{2}$

C. Identify and remove any [+]↔[−] pairs from each side of the equation.

$x + 0 = 6$

D. Show the result.

$x = 6$

Check: To check your answer, replace x in the original equation with six (1)-tiles.

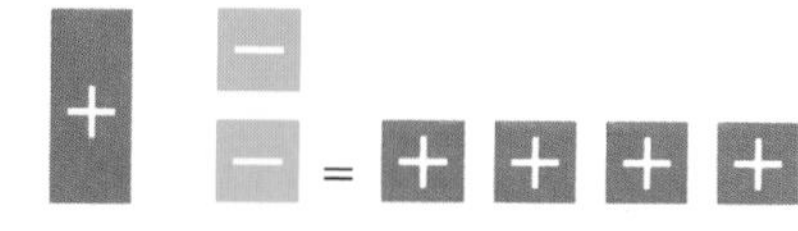

$x - 2 = 4$

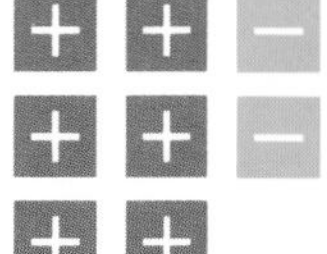

$6 - 2 \stackrel{?}{=} 4$

Identify and remove the  pairs.

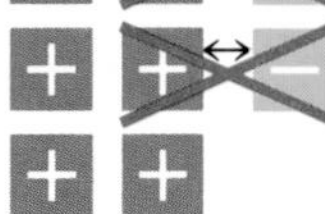

$4 = 4$

EXERCISES

Goal: To use algebra tiles to solve equations of the form $x - a = b$

Write the equation shown by each model.

1. 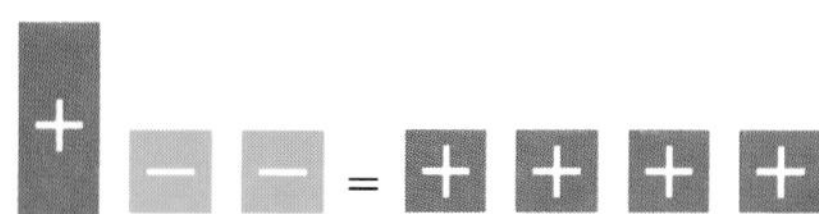

2. 

3. Use algebra tiles to show that the solution of $x - 3 = 5$ is 8.

4. Use algebra tiles to show that the solution of $x - 3 = 2$ is 5.

Use algebra tiles to solve and check each equation.

5. $x - 1 = 2$

6. $x - 2 = 2$

7. $x - 3 = 5$

8. $x - 3 = 1$

9. $x - 4 = 2$

10. $x - 3 = 6$

11. $x - 3 = 8$

12. $x - 2 = 7$

13. $x - 4 = 4$

Manipulative 3 *(Use with Lesson 5.3.)*

You can use algebra tiles to solve equations like $2x = 6$.

EXAMPLE

Solve and check: $2x = 6$

Solution:

A. Model the equation.

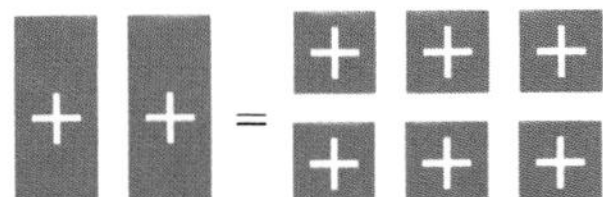

$2x = 6$

B. **Think:** There are two x-tiles. I need to find how many (1)-tiles there are for each x-tile. So divide each side of the equation into two equal groups.

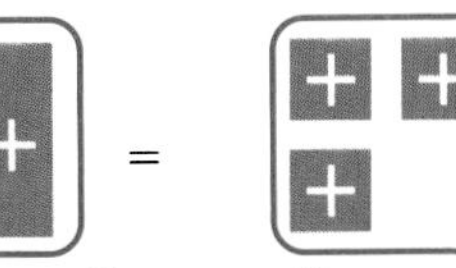

Group 1 Group 2 = Group 1 Group 2

$$\frac{2x}{2} = \frac{6}{2}$$

C. Show the value of one x-tile.

$x = 3$

Check: To check your answer, replace each x-tile in the original equation with three (1)-tiles.

$2x = 6$

$(2 \cdot 3) \stackrel{?}{=} 6$

$6 = 6$

Since the value, 3, that you found checks in the original equation, it is the solution of $2x = 6$.

EXERCISES

Goal: To use algebra tiles to solve equations of the form $ax = b$

Write the equation shown by each model.

1.

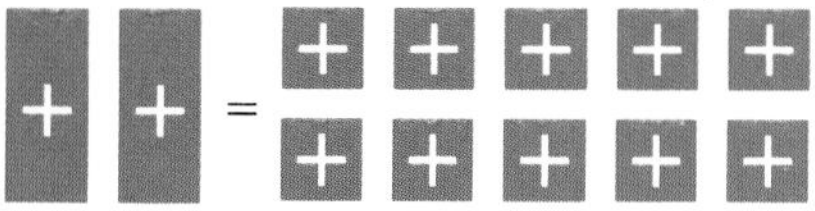

2.

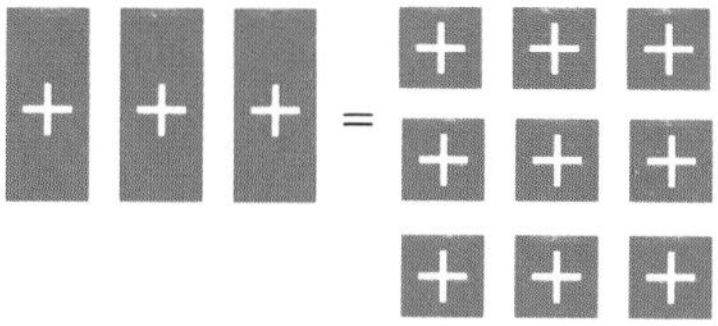

When you solve the equation for x, how many equal groups will there be on each side of the equation?

3.

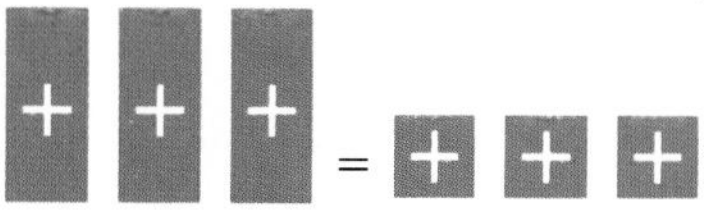

4.

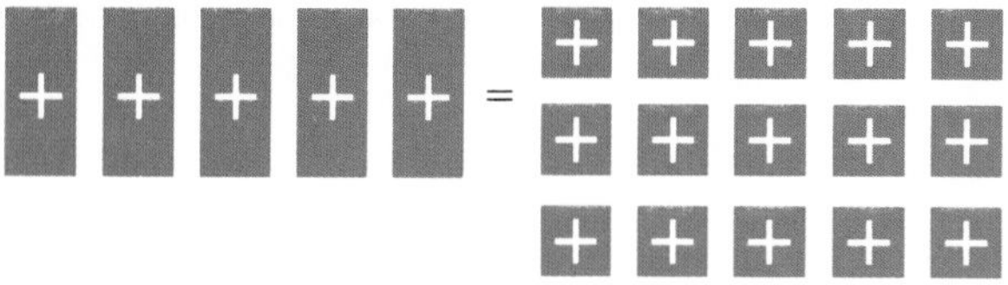

Use algebra tiles to solve and check each equation.

5. $2x = 10$	**6.** $3x = 6$	**7.** $2x = 4$	**8.** $2x = 8$
9. $5x = 10$	**10.** $5x = 5$	**11.** $6x = 12$	**12.** $3x = 12$
13. $3x = 15$	**14.** $4x = 8$	**15.** $6x = 6$	**16.** $4x = 0$

Write an equation that satisfies the description.

17. There are 8 equal groups on each side of the equation.

18. There are 15 equal groups on each side of the equation.

19. There are three equal groups on each side of the equation and the solution is $x = 9$.

20. There are seven equal groups on each side of the equation and the solution is $x = 5$.

Manipulative 4 *(Use with Lesson 6.1.)*

To solve equations such as

$2x - 1 = 3$, $\quad 3x + 5 = 8$, $\quad$ and $\quad 5 = 3x + 2$,

the first step is to get the term with the variable alone on one side of the equation.

EXAMPLE

Solve and check: $2x - 1 = 3$

Solution:

A. Model the equation.

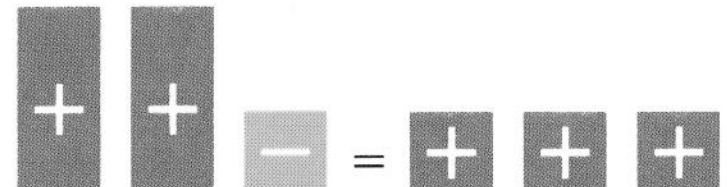

$2x - 1 = 3$

B. To get the x-tiles alone on the left side of the equation, add a (1)-tile to each side.

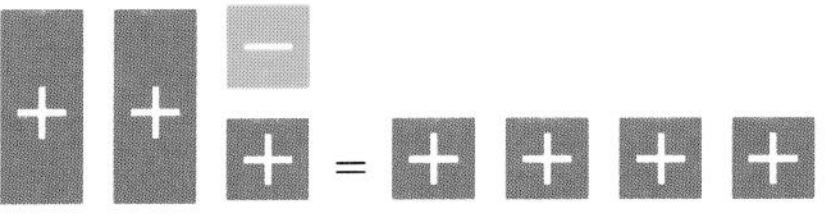

$2x - 1 + \mathbf{1} = 3 + \mathbf{1}$

or $2x + 1 - 1 = 3 + 1$

C. Identify and remove any + ↔ − pairs.

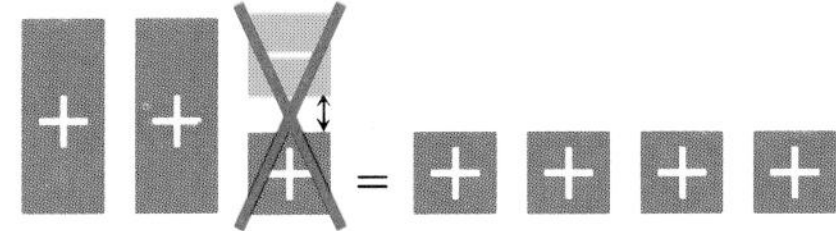

$2x = 4$

D. **Think:** The left side of the equation has two x-tiles. Divide each side into two equal groups.

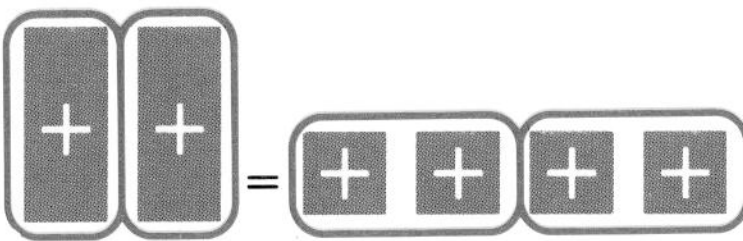

$$\frac{2x}{2} = \frac{4}{2}$$

E. Show the value of one x-tile.

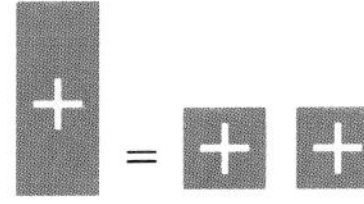

$x = 2$

Check: Replace each x-tile in the original equation with two (1)-tiles.

$$2x - 1 = 3$$

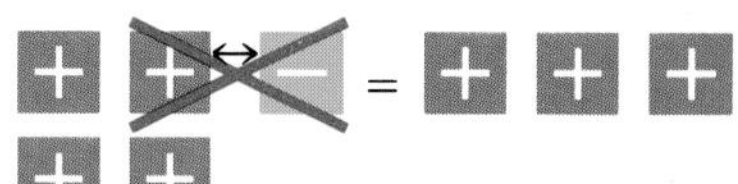

$$2(2) - 1 \stackrel{?}{=} 3$$

$$4 - 1 \stackrel{?}{=} 3$$

Identify and remove any + ↔ − pairs.

$$3 = 3$$

EXERCISES

Goal: To use algebra tiles to solve equations involving more than one operation

1. Write the equation shown by the model in each step of the solution of $3x + 5 = 8$.

	Model	Equation
a.		?
b.		?
c.		?
d.		?

Use algebra tiles to solve and check each equation.

2. $2x + 3 = 5$ **3.** $3x - 1 = 5$ **4.** $2x + 2 = 6$

5. $4x - 1 = 3$ **6.** $5x + 1 = 6$ **7.** $2x - 3 = 5$

8. $8 = 2x + 8$ **9.** $5 = 4x + 1$ **10.** $5 = 6x - 13$

Manipulative 5 *(Use with Lesson 6.2.)*

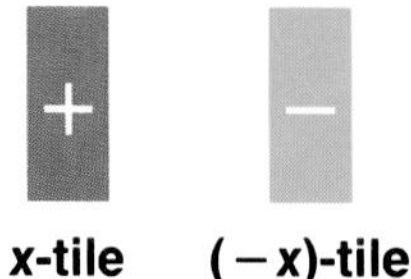

You have used a ▭ -tile to represent subtracting 1. In this lesson, you will use a $(-x)$-tile to represent subtracting x. You know that a pair of ▭↔▭ tiles represents $1 - 1 = 0$. So, the pair

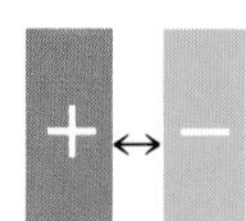

represents $x - x = 0$.

Pairs that equal 0 are called **neutral pairs**.

You can use algebra tiles to model solving equations with like terms. Algebra tiles that model **like terms** will have the same size and shape. The tiles may have the same sign, or they may have opposite signs. These pairs of tiles show like terms.

  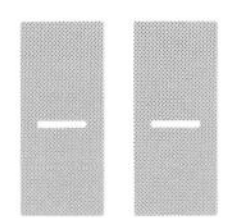 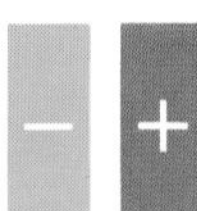 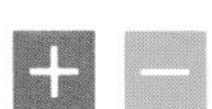

To solve an equation having like terms, you must first combine (add or subtract) the like terms. The Example below shows you how to do this.

EXAMPLE

Solve: $3x + 2 - x = 4$

Solution:

A. Model the equation.

$3x + 2 - x = 4$

B. Group like terms together on each side of the equation.

$3x - x + 2 = 4$

C. Identify and remove any neutral pairs.

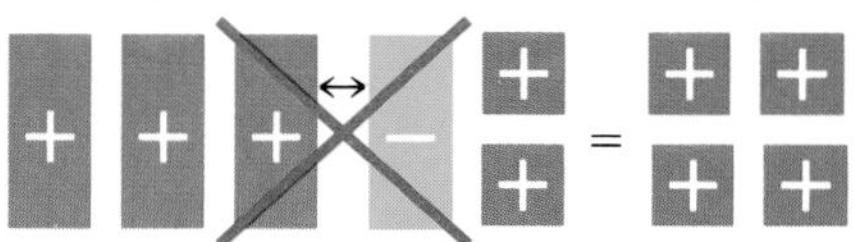

$2x + 2 = 4$

D. **Think:** To get $2x$ alone, remove two + -tiles from each side of the equation.

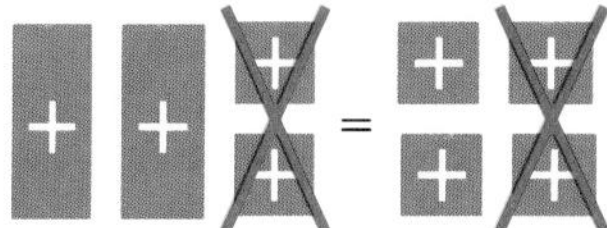

$$2x + 2 - \mathbf{2} = 4 - \mathbf{2}$$
$$2x = 2$$

E. Form two equal groups on each side of the equation.

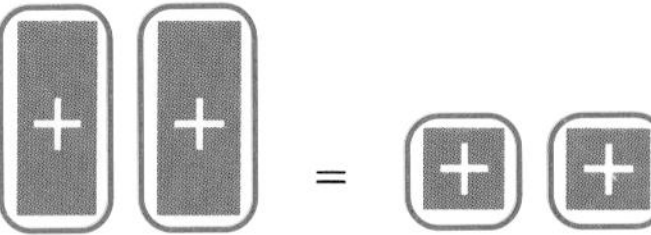

$$\frac{2x}{2} = \frac{2}{2}$$

F. Show the value of one x-tile.

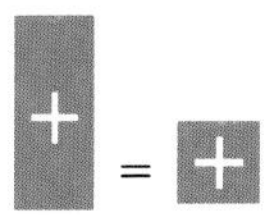

$$x = 1$$

EXERCISES

Goal: To use algebra tiles to solve equations involving like terms

Which equations can be used to represent the model?

1.

a. $3x - 2 = 2$
b. $3x - 2 = 2x$
c. $3x - 2x = 2$
d. $2x + x - 2x = 2$

2.

a. $3x = 7$
b. $x + 2x + 1 = 7$
c. $3x + 1 = 7$
d. $x + 2x = 7$

Use algebra tiles to solve each equation.

3. $2x + x + 1 = 4$

4. $5x - x + 1 = 5$

5. $2x - x - 1 = 4$

6. $x + x - 2 = 4$

7. $3x - 2x + 1 = 5$

8. $3x + 2x - 1 = 4$

9. $3x - x = 4$

10. $2x + x = 0$

Manipulative 6 *(Use with Lesson 7.2.)*

For this lesson, you will need pencils, paper clips, scissors, and Teaching Aid 5. Use the materials to construct a spinner.

STEP 1 Cut out Circle A and Circle B on Teaching Aid 5. Bend one end of a paper clip to form a pointer.

STEP 2 To make spinner B, place the loop of the paper clip over the center of Circle B as shown at the right.

STEP 3 Hold a pencil in a vertical position with the point facing downward. Place the point through the loop of the paper clip so that it touches the center of the circle.

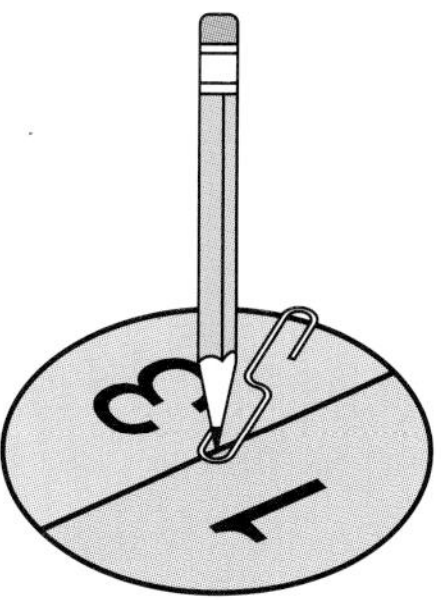

STEP 4 Holding the pencil firmly against the center of the circle, flick the paper clip with your finger so that it spins smoothly around the circle.

STEP 5 Make Spinner A. Follow the same steps as for spinner B.

EXERCISES

Goal: To use a spinner to determine the probability of an event

Work with a partner. On a piece of paper, make a table like the one below. Your table should have 40 rows. Number the rows 1 through 40.

Turn	Spinner A	Spinner B	Sum of Spinner A and Spinner B
1			
2			

1. For each of the 40 turns, spin spinner A and record the result in your table. Then spin spinner B 40 times and record the result. Compute the sum of the two spins and write the sum in your table.

Use the results recorded in your table to complete these exercises. Write the probability ratio in the fourth column in lowest terms.

	Sum of	Number of Times	Total Number of Turns	$P = \dfrac{\text{Number of Times}}{\text{Total Number of Turns}}$
2.	2	?	40	$P = \dfrac{?}{40} = ?$
3.	3	?	40	$P = \dfrac{?}{40} = ?$
4.	4	?	40	$P = \dfrac{?}{40} = ?$
5.	5	?	40	$P = \dfrac{?}{40} = ?$
6.	6	?	40	$P = \dfrac{?}{40} = ?$
7.	7	?	40	$P = \dfrac{?}{40} = ?$

8. For which sums were the probability ratios about the same?
9. Which two sums had the highest probability ratios?
10. Examine Spinner A and Spinner B. Explain why, on a given turn, you are more likely to spin a sum of 4 or a sum of 5 than a sum of 2, 3, 6, or 7.

The table below gives the mathematical probabilities of spinning sums of 2, 3, 4, 5, 6, and 7. Compare your results from Exercise 1 with these probabilities.

Sum	2	3	4	5	6	7
Probability	$\frac{1}{8}$	$\frac{1}{8}$	$\frac{1}{4}$	$\frac{1}{4}$	$\frac{1}{8}$	$\frac{1}{8}$

11. Is your probability for each sum about the same as the mathematical probability shown in the table for that sum?

Manipulative 7 *(Use with Lesson 9.2.)*

You can use algebra tiles to model addition and subtraction of integers. Think of a + -tile as a positive 1-tile. Think of a − -tile as a (−1)-tile.

Since a + -tile and a − -tile have opposite signs, they form a **neutral pair**. A neutral pair has a value of 0.

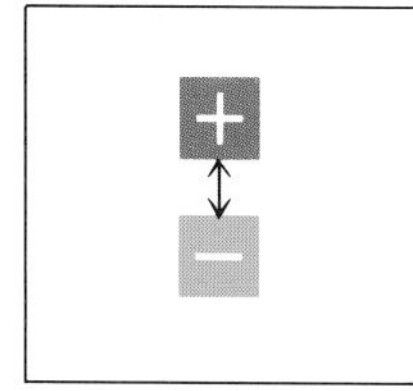

Value: 0

EXAMPLE 1 Find the value of each square.

a.

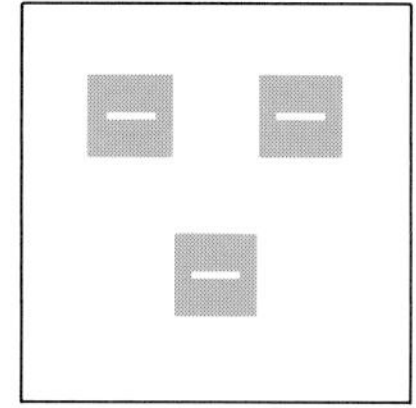

b.

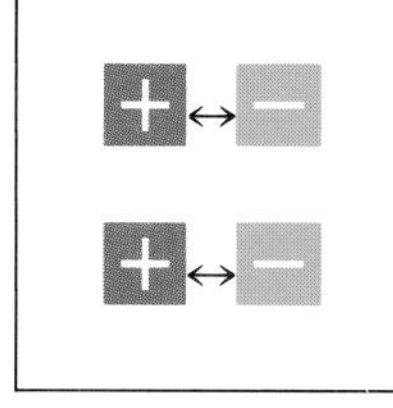

c.

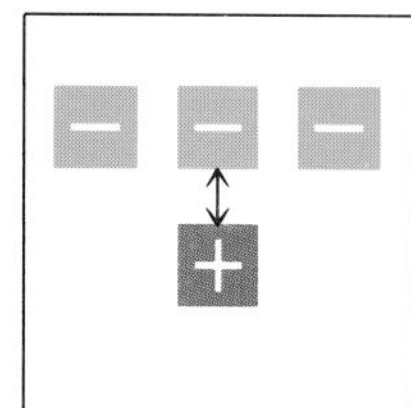

Solutions: **Value: −3** **Value: 0** **Value: −2**

EXAMPLE 2 Use algebra tiles to find each sum.

a. $-1 + (-3) = \underline{\ ?\ }$

Solution: **A.** Put one − -tile in the square.

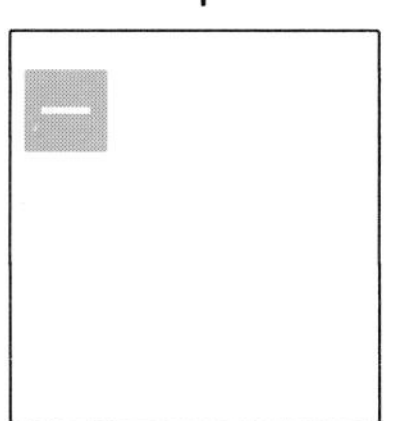

B. Add three

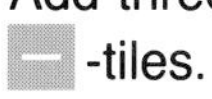

− -tiles.

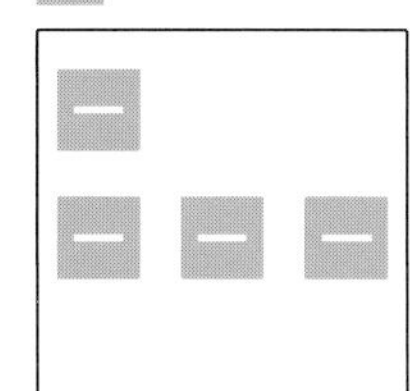

C. Record the sum.

$-1 + (-3) = \mathbf{-4}$

b. $-4 + 3 = \underline{\ ?\ }$

Solution: **A.** Put in four − -tiles.

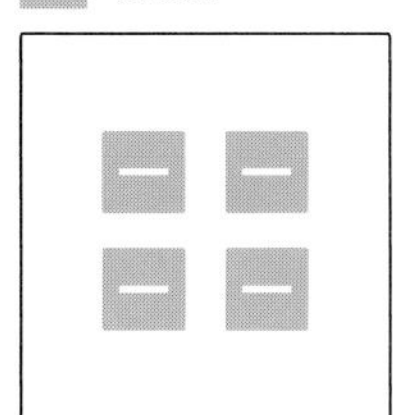

B. Add three + -tiles.

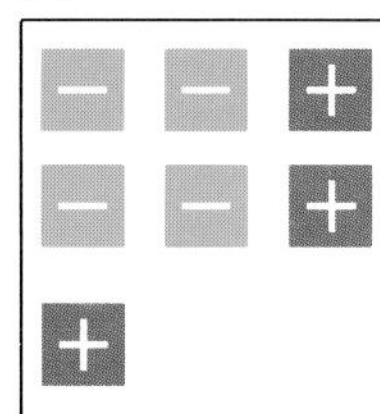

C. Remove neutral pairs. Record the sum.

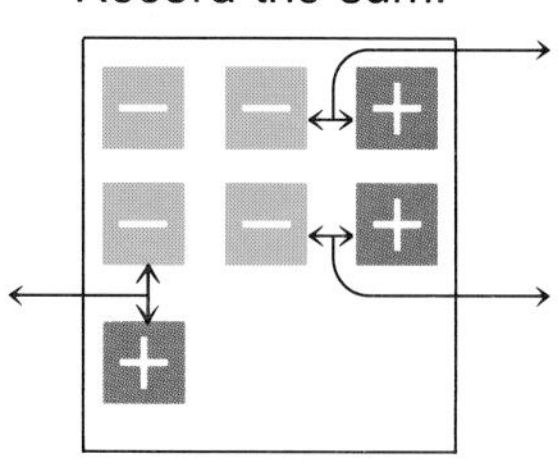

$-4 + 3 = \mathbf{-1}$

c. $5 + (-3) = \underline{\ ?\ }$

Solution:

A. Put in five + -tiles.

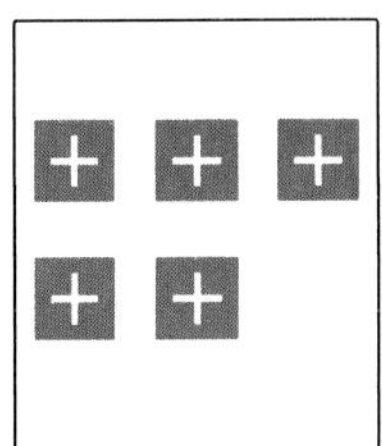

B. Add three − -tiles.

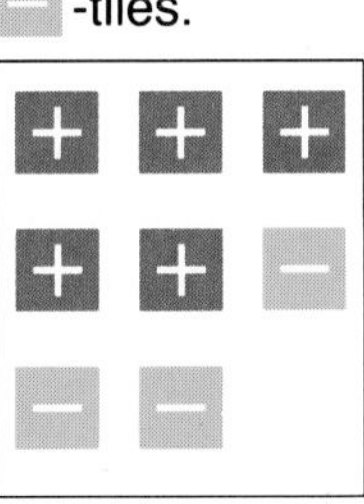

C. Remove neutral pairs. Record the sum.

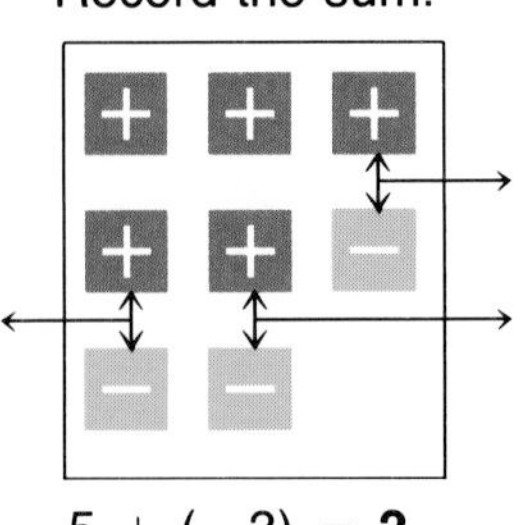

$5 + (-3) = \mathbf{2}$

EXERCISES

Goal: To use algebra tiles to add integers

Give the value of each square.

1.

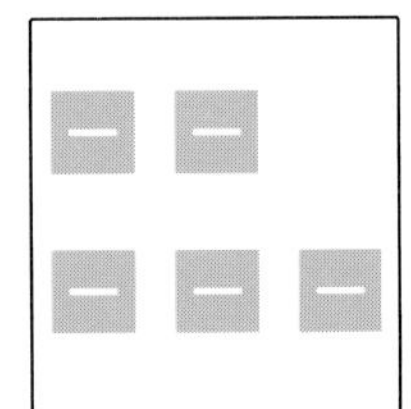

2.

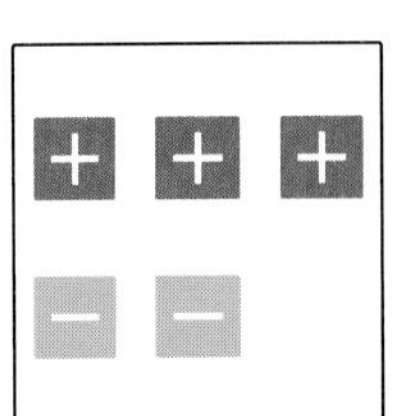

3.

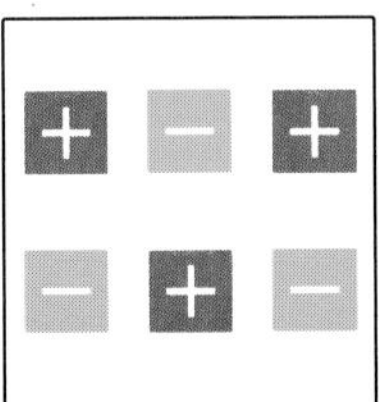

Use algebra tiles to model each sum.

4. $5 + 3$
5. $-3 + 7$
6. $-7 + (-1)$
7. $4 + (-6)$
8. $-2 + 2$
9. $6 + (-7)$
10. $-3 + (-3)$
11. $8 + (-3)$
12. $0 + (-4)$
13. Use algebra tiles to show that $-2 + 4 = 4 + (-2)$. Check students' models.

Use algebra tiles to find each answer.

14. $-3 + \underline{\ ?\ } = -1$
15. $6 + \underline{\ ?\ } = -1$
16. $3 + \underline{\ ?\ } = 0$
17. $-2 + \underline{\ ?\ } = 0$
18. $-8 + \underline{\ ?\ } = -5$
19. $-2 + \underline{\ ?\ } = 4$
20. $-1 + \underline{\ ?\ } = 5$
21. $-5 + \underline{\ ?\ } = -2$
22. $-10 + \underline{\ ?\ } = -10$

Manipulative 8 *(Use with Lesson 9.3.)*

Using algebra tiles to model the addition of two or more numbers is similar to modeling the addition of two numbers.

EXAMPLE Use algebra tiles to find the sum: $1 + (-3) + 2$

Solution: **Method 1:** Place all the tiles that represent the addends in the square. Then remove neutral pairs and find the sum.

A. Place a (1)-tile in the square.

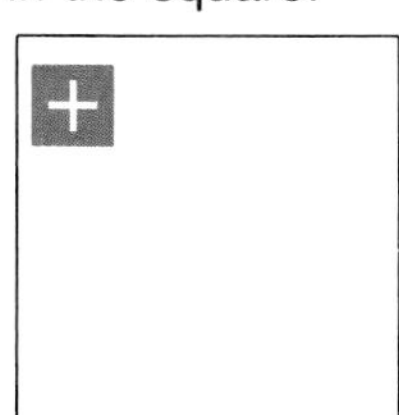

B. Add three (−1)-tiles.

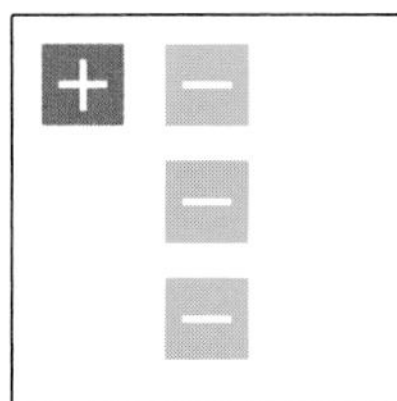

C. Add two (1)-tiles.

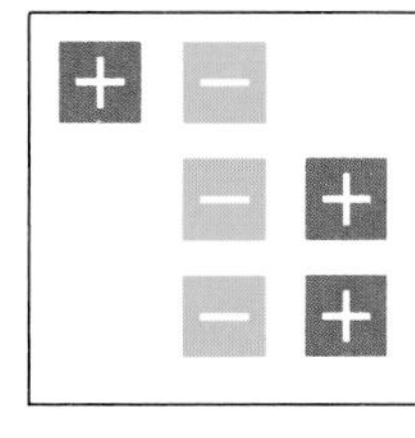

D. Remove neutral pairs.

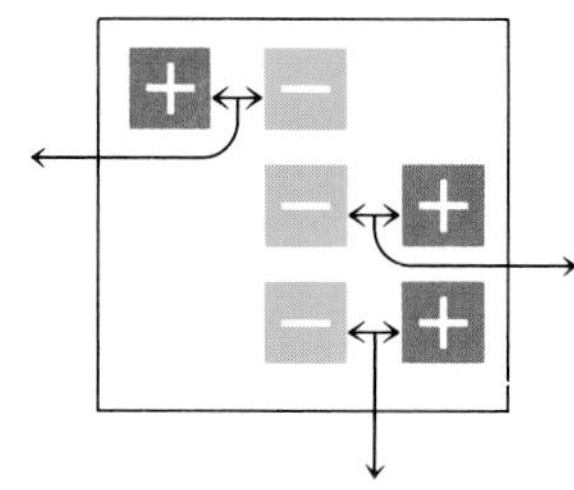

E. Record the sum: $1 + (-3) + 2 = \mathbf{0}$

Method 2: Use the tiles to find the sum of 1 and (−3). Then use tiles to add 2 to this sum.

A. Put a (1)-tile in the square.

B. Add three (−1)-tiles.

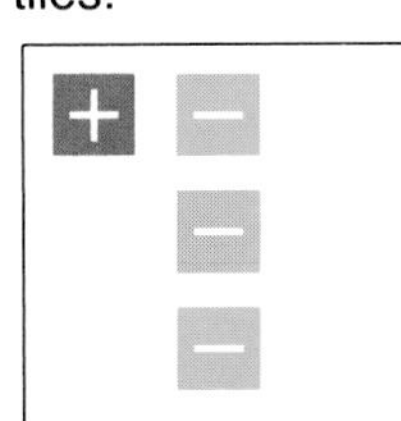

C. Remove neutral pairs.

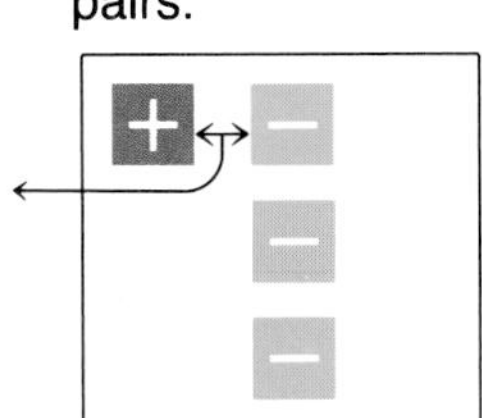

D. Add two (1)-tiles.

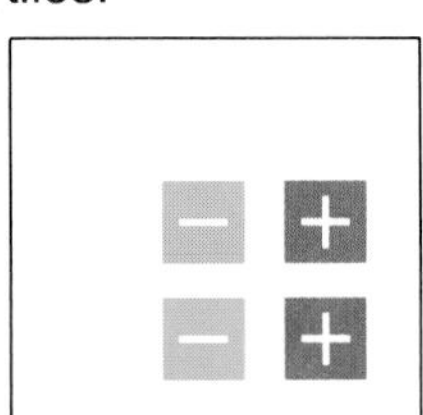

E. Remove neutral pairs.

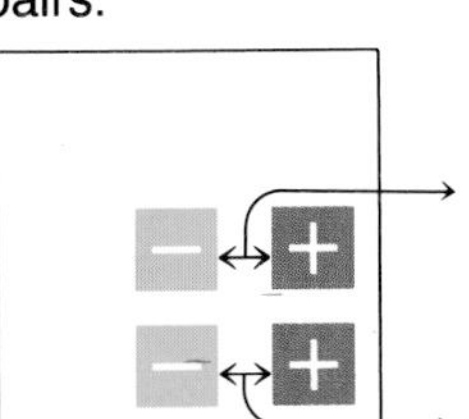

F. Record the sum.

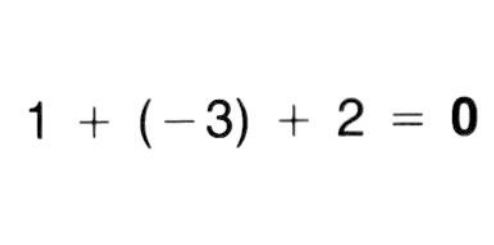

$1 + (-3) + 2 = \mathbf{0}$

EXERCISES

Goal: To use algebra tiles to add more than two integers

Use algebra tiles to find each sum.

1. $4 + 2 + (-6)$
2. $3 + (-2) + 6$
3. $-3 + 4 + (-5)$
4. $6 + (-3) + (-2)$
5. $-3 + (-4) + 5$
6. $-6 + 3 + (-4)$
7. $-6 + 3 + (-1)$
8. $1 + (-2) + (-5)$
9. $7 + (-8) + 1$

10. Use algebra tiles to show that $(-2 + 3) + (-5) = -2 + [3 + (-5)]$. What property have you demonstrated?

11. Use algebra tiles to show that $(-6 + 4) + (-2) = -2 + [-6 + 4]$. What property have you demonstrated?

12. Use algebra tiles to show that $[(-7 + (-5)] + (7 + 5) = 0$. What property have you demonstrated?

13. If you add two positive numbers and a negative number, will the sum always be a positive number? Give an example to explain your answer.

14. If you add two negative numbers and a positive number, can the sum equal 0? Give an example to explain your answer.

15. If the sum of three numbers is -35 and you add the opposite of the opposite of each number to this, what is the new sum? Explain.

Use algebra tiles to find the missing addend.

16. $-4 + (-3) + \underline{\ ?\ } = 2$
17. $-3 + \underline{\ ?\ } + 2 = -4$
18. $5 + (-6) + \underline{\ ?\ } = -1$
19. $4 + \underline{\ ?\ } + (-6) = -4$
20. $-5 + \underline{\ ?\ } + 2 = 0$
21. $-2 + 3 + \underline{\ ?\ } = 4$

Manipulative 9 *(Use with Lesson 9.5.)*

You can use algebra tiles to show that the answers to these problems are the same.

$$4 - 3 \quad \text{and} \quad 4 + (-3)$$

EXAMPLE 1

Use algebra tiles to show that $4 - 3 = 4 + (-3)$.

a. $4 - 3 = \underline{\ ?\ }$

Solution: **A.** Place four (1)-tiles in the square **B.** Remove three (1)-tiles. **C.** Record the result.

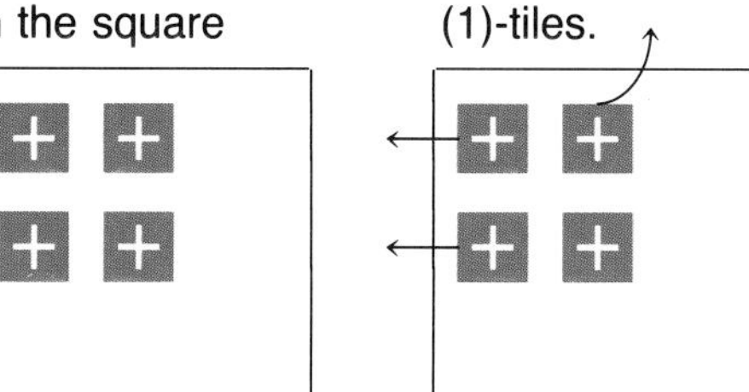

$4 - 3 = \mathbf{1}$

b. $4 + (-3) = \underline{\ ?\ }$

A. Place four (1)-tiles in the square.

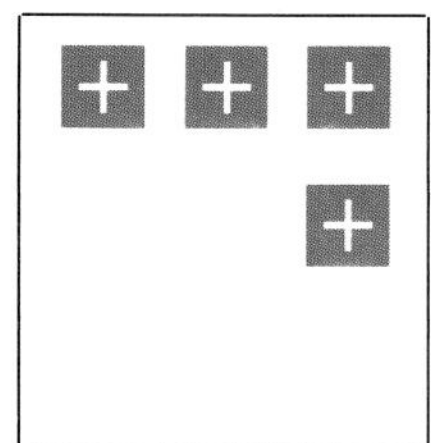

B. Add three (−1)-tiles.

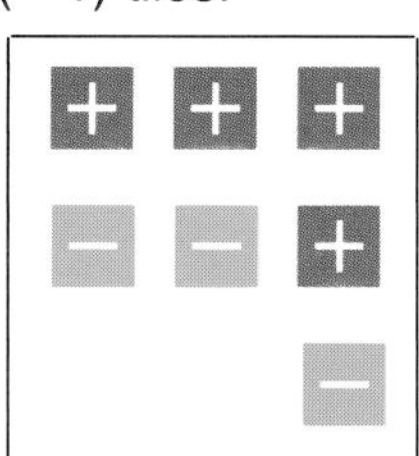

C. Remove the

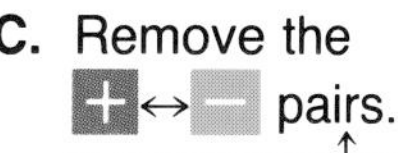

pairs.

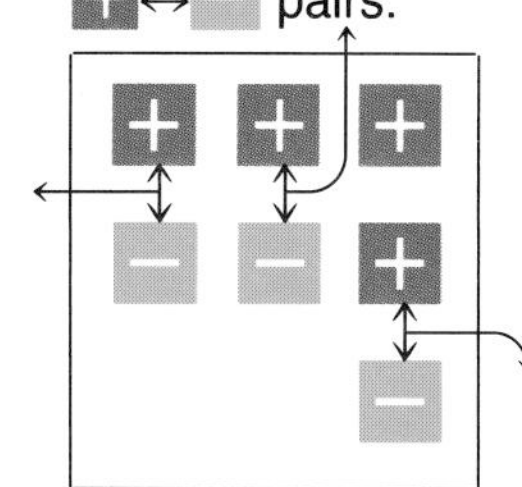

D. Record the result.

$4 + (-3) = \mathbf{1}$

Example 1 shows that $4 - 3 = 4 + (-3)$. That is, subtracting 3 from 4 gives the same result as adding the opposite of 3 to 4. This suggests the following rule.

> **To subtract a number *b* from a number *a*, add the opposite of *b* to *a*.**

EXAMPLE 2

Use algebra tiles to subtract: $-4 - (-3)$

Solution: Rewrite $-4 - (-3)$ as $-4 + 3$.

A. Place four (-1)-tiles in the square.

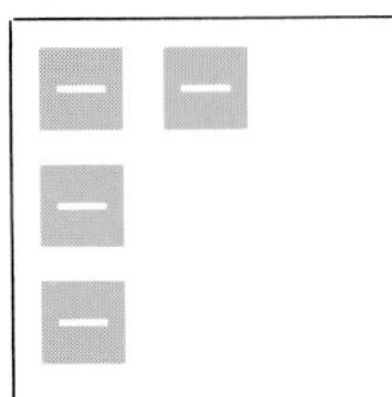

B. Add three (1)-tiles.

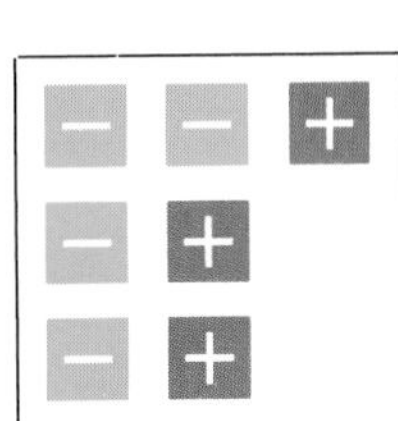

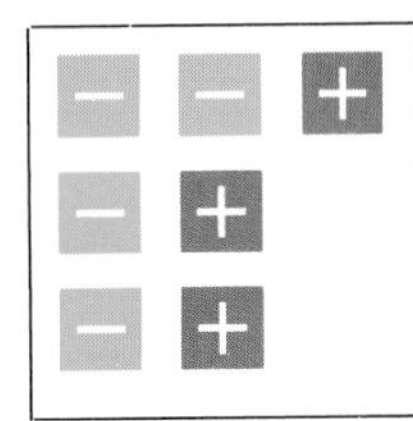

C. Remove neutral pairs.

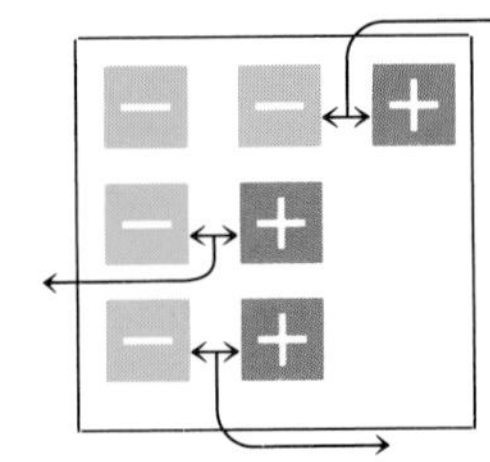

D. Record the result.

$-4 - (-3) = \mathbf{-1}$

EXERCISES

Goal: To use algebra tiles to subtract integers

Write an addition problem that will give the same answer as the subtraction problem.

1. $-5 - 3$ **2.** $4 - 6$ **3.** $4 - (-2)$ **4.** $-4 - (-6)$

5. $0 - 9$ **6.** $0 - (-9)$ **7.** $-1 - (-1)$ **8.** $-1 - 4$

Use algebra tiles to subtract.

9. $3 - 5$ **10.** $-1 - 4$ **11.** $-2 - (-3)$ **12.** $6 - (-1)$

13. $4 - 8$ **14.** $-5 - 3$ **15.** $-2 - (-1)$ **16.** $-6 - (-5)$

17. Use algebra tiles to show that $-3 - (-2) \neq -2 - (-3)$.

18. Does $-4 - 0 = 0 - (-4)$? Use algebra tiles to explain your answer.

Use algebra tiles to solve each problem.

19. $2 - 4 - 6 + 4$ **20.** $4 - 1 + 3 - 5$

21. $-1 + 4 - (-2) - (-3)$ **22.** $2 - (-3) + 6 - (-3)$

23. Ellen said that when she was solving Exercise 19, she added 2 and 4 before subtracting 4 and 6. Amos said that he did the addition and subtraction in order from left to right. Are both methods correct? Explain.

24. Leighton said that when she solved Exercise 19, she thought: "$-4 + 4 = 4 - 4 = 0$. So all I have to do is find $2 - 6$." Was Leighton's thinking correct? Explain.

Manipulative 10 *(Use with Lesson 9.7.)*

You can model algebraic expressions with tiles that look like this.

1-tile **x-tile** **x^2-tile**

For each positive tile, there is a corresponding negative tile.

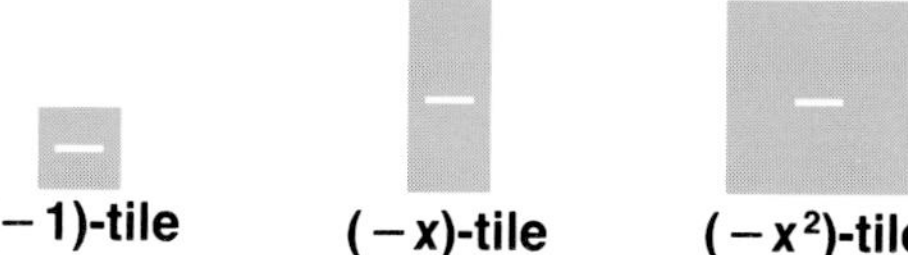

(−1)-tile **(−x)-tile** **($-x^2$)-tile**

Just as with the [+]↔[−] pairs you used to model a neutral pair of integers, neutral pairs can also be formed as shown at the right. The value of each neutral pair is 0.

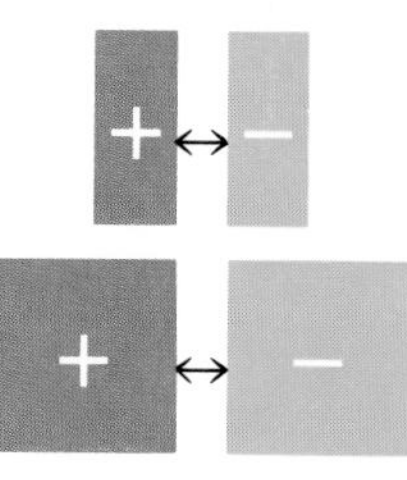

Neutral Pairs

Recall that algebra tiles that represent like terms have the same size and shape.

So neutral pairs of tiles show like terms.

EXAMPLE 1 Simplify: $2x + 3 - x - 5$

Solution: Think of $2x + 3 - x - 5$ as $2x + 3 + (-x) + (-5)$.

A. Model the expression.

$2x + 3 + (-x) + (-5)$

B. Group like terms. Remove neutral pairs.

$[2x + (-x)] + [3 + (-5)]$

C. Record the result.

$x - 2$

EXAMPLE 2 Simplify: $2x - 2x^2 + x^2 - x$

Solution: Rewrite the expression as $2x + (-2x^2) + x^2 + (-x)$.

A. Model the expression.

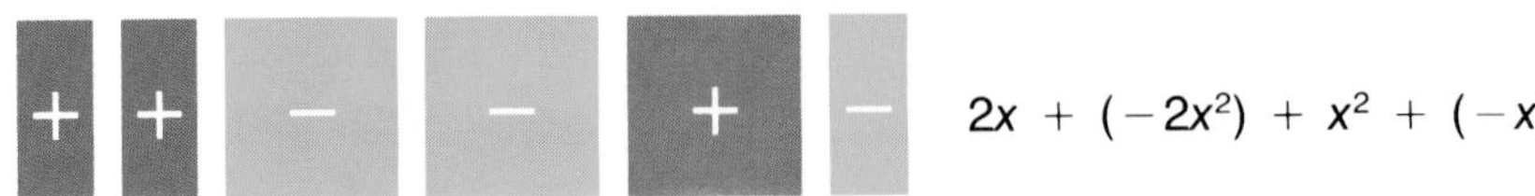

$2x + (-2x^2) + x^2 + (-x)$

B. Group like terms. Then remove neutral pairs and record the result.

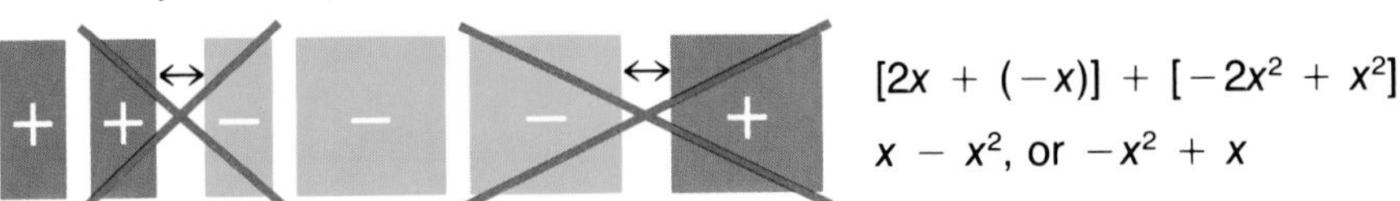

$[2x + (-x)] + [-2x^2 + x^2]$

$x - x^2$, or $-x^2 + x$

EXERCISES

Goal: To use algebra tiles to simplify algebraic expressions

Use algebra tiles to simplify each expression.

1. $2x^2 + 3x + x^2 - 4x$

2. $2x + 4 - 4x - 1$

3. $-2x^2 + x^2 - 3 + 2$

4. $3x^2 + 2x - x^2 + x$

5. $2x - (-2x^2) + 6 - (-3x)$

6. $10 - 4x + x^2 - (-2x)$

7. $4x - 3 - x^2 + 2x^2$

8. $-x^2 + 2 - 3 + x$

9. $4x - 5 - 2x + 6$

10. $3x - 4 - x^2 + x$

Use algebra tiles to determine whether each statement is true (T) or false (F).

11. $x^2 - 2x + (-x^2 + 2x) = 0$

12. $-x^2 + 2x - (-x^2) = x - (-x)$

Manipulative 11 *(Use with Lesson 10.3.)*

You can use algebra tiles to model the product of a monomial and a binomial.

EXAMPLE 1 Use algebra tiles to find $x(2x + 1)$.

Solution:

A. Draw a grid as shown at the right. Place one positive x-tile at the left of the grid. Place two positive x-tiles and one (1)-tile at the top of the grid.

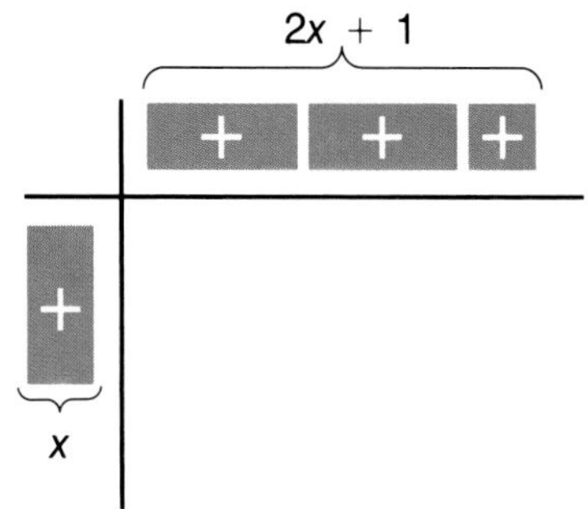

B. Build a rectangle using x^2-tiles, x-tiles, and 1-tiles. Use tiles that have the same length as the tiles to the left of the vertical line and the same width as the tile above the horizontal line.

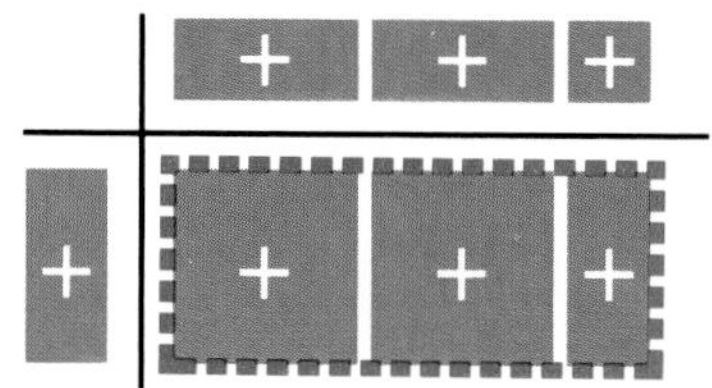

C. Count the tiles in the completed rectangle. Write the product.

$$x(2x + 1) = 2x^2 + x$$

To use a model to find the product $2x(x - 2)$, think of $x - 2$ as $x + (-2)$.

EXAMPLE 2 Use algebra tiles to find $2x(x - 2)$.

Solution:

A. Start with two x-tiles on the left side of the grid. Place an x-tile and two (-1)–tiles at the top of the grid.

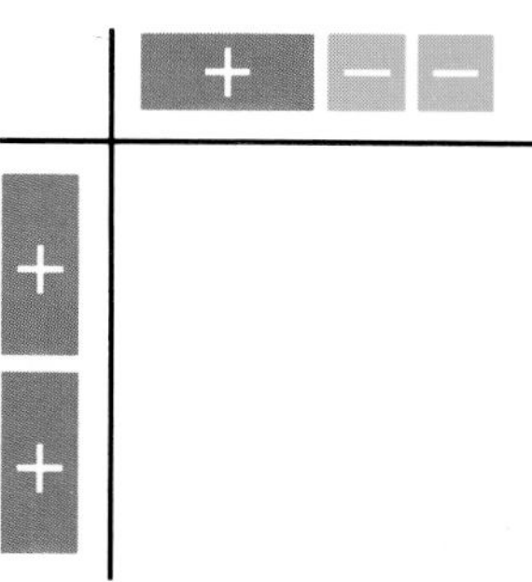

B. Build a rectangle. Notice that the x-tiles in the product are negative. This happens because the x-tiles (factors) at the left of the vertical line and the (1)-tiles (factors) above the horizontal line have different signs.

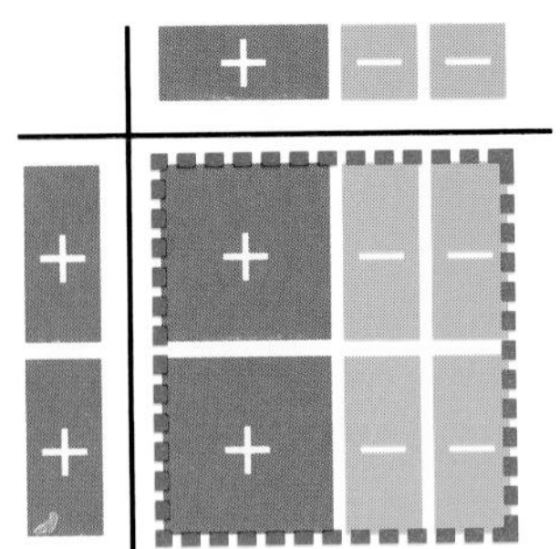

C. Write the product.

$2x(x - 2) = 2x^2 - 4x$

EXERCISES

Goal: To use algebra tiles to multiply a monomial and a binomial

Write the multiplication problem shown by each model.

1.

2.

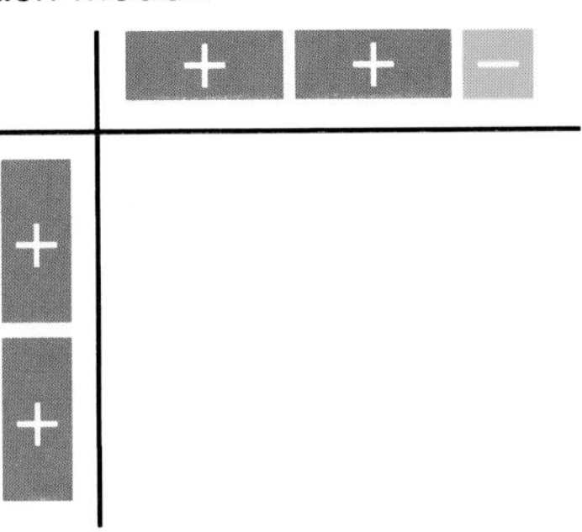

3.

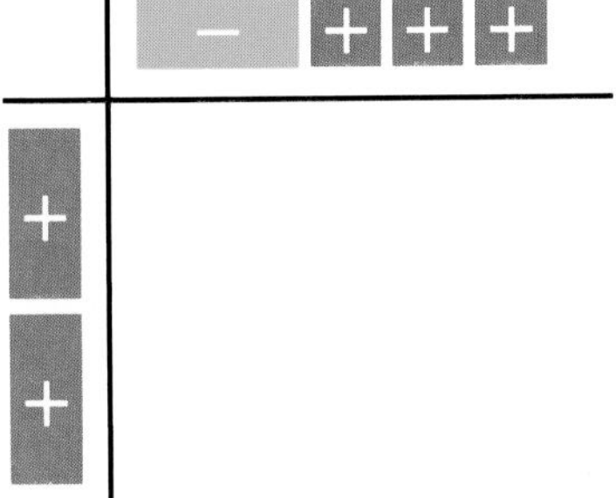

4.

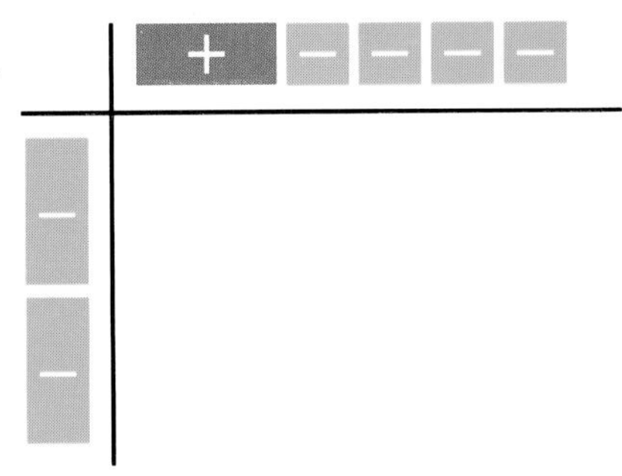

Use algebra tiles to find each product.

5. $x(2x - 1)$

6. $3(x + 4)$

7. $2x(x - 3)$

8. $-x(2x + 3)$

9. $-x(x - 5)$

10. $x(3x + 6)$

11. $4(2x + 1)$

12. $-2x(x + 4)$

13. $3x(-x + 1)$

14. Use algebra tiles to model $2x(x - 5)$. Then model $2x(x) + 2x(-5)$.

a. What conclusion can you draw?

b. What property have you illustrated?

Manipulative 12 *(Use with Lesson 18.1.)*

You can also use algebra tiles to simplify polynomial expressions.

Recall that algebra tiles that represent like terms will have the same size and shape. They may have the same sign, or they may have different signs.

Each group of 3 tiles below shows like terms.

EXAMPLE 1

Use algebra tiles to simplify: $-x^2 + 3x - x + 1$

Solution: **A.** Use algebra tiles to model the polynomial.

B. Group like terms. Remove neutral pairs.

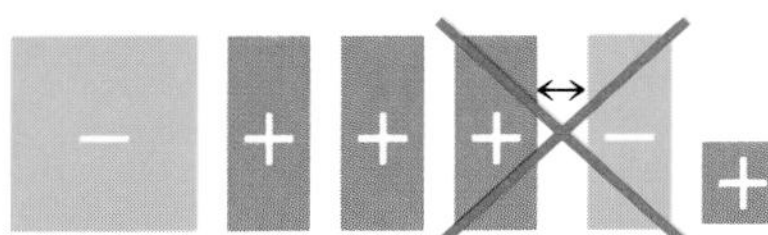

C. Record the result.

$-x^2 + 3x - x + 1 = -x^2 + 2x + 1$

EXAMPLE 2

Use algebra tiles to simplify: $x - 3x^2 + 2 + x^2 + 4x - 6$

Solution: Think of $x - 3x^2$ as $x + (-3x^2)$ and think of $4x - 6$ as $4x + (-6)$.

A. Use algebra tiles to model the polynomial.

B. Group like terms.

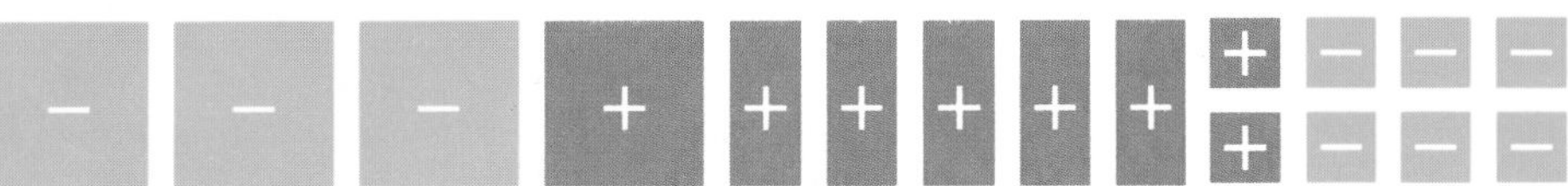

C. Remove any neutral pairs.

D. Record the result.

$x - 3x^2 + 2 + x^2 + 4x - 6 = -2x^2 + 5x - 4$

EXERCISES

Goal: To use algebra tiles to simplify polynomials

Tell if each pair shows like terms. Answer Yes or No.

1. 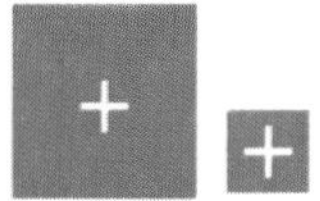

2. 

3. 

Use algebra tiles to simplify each polynomial.

4. $x^2 + 2x - 4 + 3x + 5$

5. $x^2 - 3 + 4x - 5x + 6$

6. $5 + x^2 + 7x - 8 + 2x^2$

7. $1 - x^2 + 2x + x^2 - 1$

8. $2x + x^2 - 5x - 3$

9. $7x - 8 - 3x + 6 + x^2$

10. $-2x^2 + 6x - 5 + x^2 - 7x$

11. $-3x^2 + x + x^2 - 8x$

12. $3 - 5x + x^2 + 8x - 7$

13. $2x + 3 + x^2 - 4x - 1$

14. Use algebra tiles to represent a polynomial whose simplified form is $x^2 - 2x + 3$.

15. Use algebra tiles to represent a polynomial whose simplified form is $-3x^2 + 9x$.

Manipulative 13 *(Use with Lesson 18.3.)*

You have already used algebra tiles to model polynomials. In this lesson you will use algebra tiles to find the sum of polynomials.

Example 1 shows how to model the sum of two trinomials.

EXAMPLE 1

Use algebra tiles to add:

$$\begin{array}{r} 2x^2 + 3x - 4 \\ \underline{-x^2 + x + 2} \end{array}$$

Solution: **A.** Model each polynomial. Align tiles representing like terms.

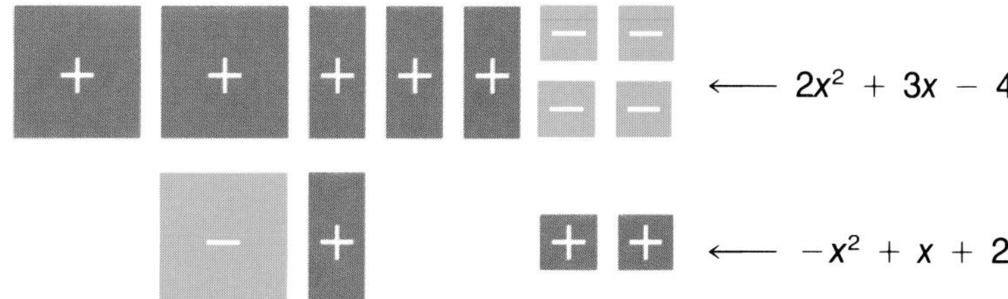

B. Remove neutral pairs and count the remaining tiles.

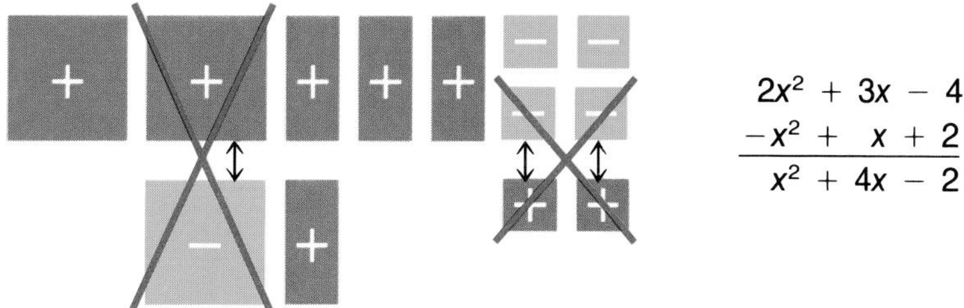

Example 2 shows that the sum of two binomials can be a trinomial.

EXAMPLE 2

Use algebra tiles to add: $(x^2 - 3x) + (5x - 1)$

Solution: **A.** Model each polynomial. Align tiles representing like terms.

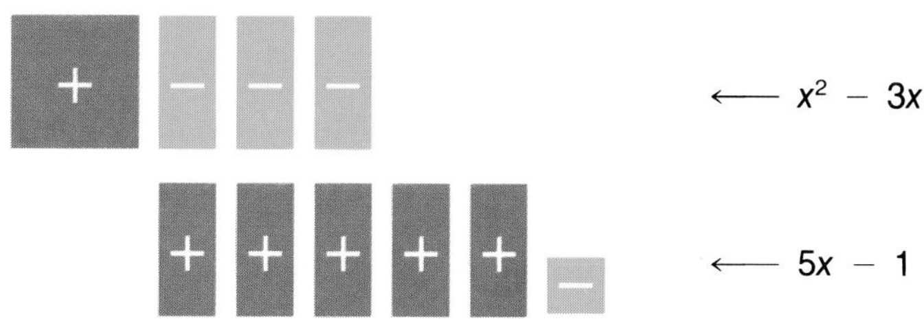

B. Remove neutral pairs and count the remaining tiles.

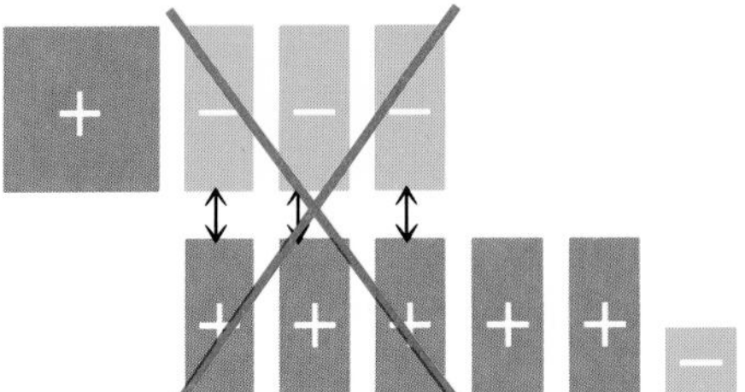

$$\begin{array}{r} x^2 - 3x \\ 5x - 1 \\ \hline x^2 + 2x - 1 \end{array}$$

EXERCISES

Goal: To use algebra tiles to add polynomials

Write the polynomial shown by each model.

1.   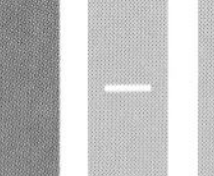  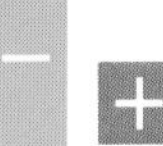  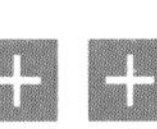

2.   

Use algebra tiles to find each sum.

3. $$\begin{array}{r} 2x^2 + 5x - 1 \\ + \; -x^2 - 3x + 4 \\ \hline \end{array}$$

4. $$\begin{array}{r} x^2 + 2x + 3 \\ + \; x^2 + 4x - 8 \\ \hline \end{array}$$

5. $$\begin{array}{r} x^2 + 5x - 6 \\ + \; -2x^2 + 3x + 4 \\ \hline \end{array}$$

6. $$\begin{array}{r} -x^2 + 5x - 6 \\ + \; -x^2 - 2x + 5 \\ \hline \end{array}$$

7. $(x^2 + 3x - 9) + (x^2 - 4x + 5)$
8. $(-x^2 + 2x - 6) + (x^2 - 4x + 5)$
9. $(x^2 - 4x + 3) + (-x^2 + 4x - 3)$
10. $(2x^2 + 6x - 1) + (x^2 - 4x + 8)$
11. Use algebra tiles to show that the sum of two binomials can equal 0.
12. Use algebra tiles to show that the sum of two trinomials can be a binomial.
13. Use algebra tiles to show that the sum of two trinomials can be a monomial.
14. Use algebra tiles to show that the sum of two trinomials can equal zero.
15. Use algebra tiles to find the binomial that must be added to $6x^2 - 5x$ to give a sum of $x^2 + 6$.

Manipulative 14 *(Use with Lesson 18.4.)*

In this lesson, you will use algebra tiles to model subtraction of polynomials.

EXAMPLE 1

Use algebra tiles to subtract:

$$\begin{array}{r} 2x^2 + 3x + 5 \\ (-)\ \underline{x^2 + 2x + 1} \end{array}$$

Solution:

A. Make a model to show $2x^2 + 3x + 5$. Then remove one x^2-tile, two x-tiles, and one (1)-tile.

B. Record the result.
There are one x^2-tile, one x-tile, and four (1)-tiles remaining.

$$\begin{array}{r} 2x^2 + 3x + 5 \\ (-)\ \underline{x^2 + 2x + 1} \\ x^2 + x + 4 \end{array}$$

Recall that when you subtract a number, you add its opposite.

EXAMPLE 2

Subtract by using opposites:

$$\begin{array}{r} 2x^2 - x - 3 \\ (-)\ \underline{x^2 - x + 1} \end{array}$$

Solution:

A. Add the opposite of $(x^2 - x + 1)$ to $2x^2 - x - 3$.

$$\begin{array}{r} 2x^2 - x - 3 \\ (-)\ \underline{x^2 - x + 1} \end{array} \longrightarrow \begin{array}{r} 2x^2 - x - 3 \\ (+)\ \underline{-x^2 + x - 1} \end{array}$$

B. Model the polynomials. Align like terms.

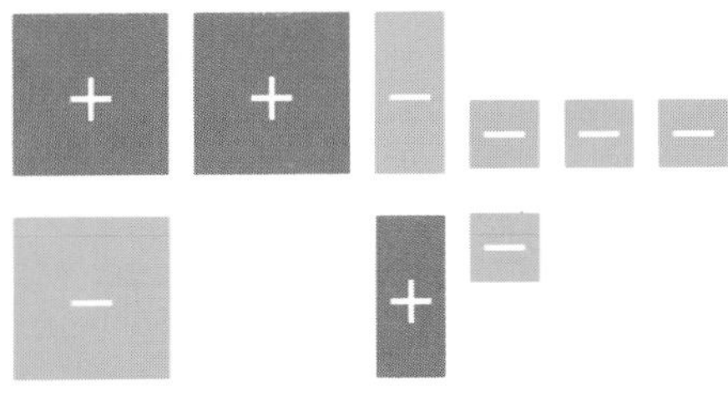

C. Remove neutral pairs and count the remaining tiles.

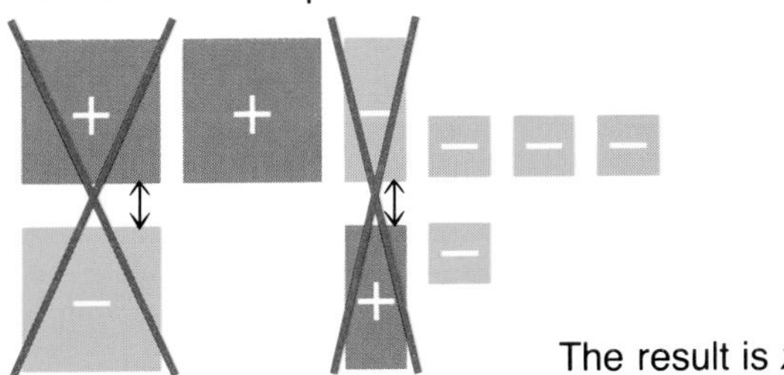

The result is $x^2 - 4$.

EXERCISES

Goal: To use algebra tiles to subtract polynomials

Which expression shows the opposite of the polynomial modeled by the algebra tiles?

1. 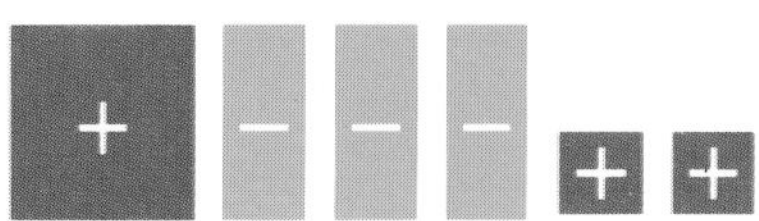

a. $x^2 - 3x + 2$
b. $x^2 + 3x - 2$
c. $-x^2 + 3x - 2$

2. 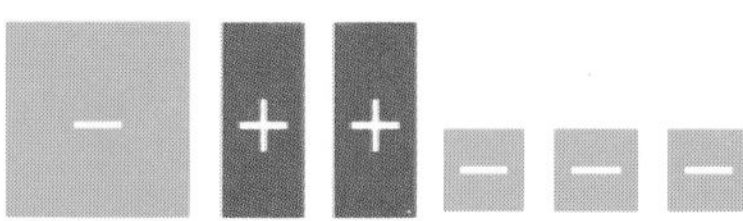

a. $x^2 - 2x - 3$
b. $-x^2 + 2x - 3$
c. $x^2 - 2x + 3$

3. 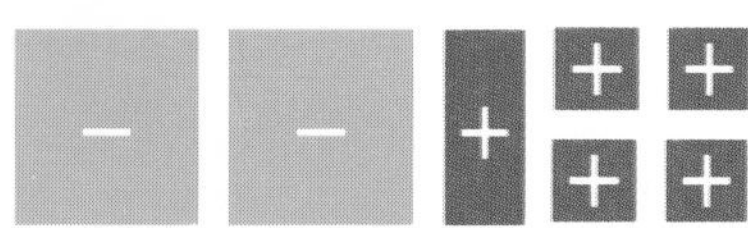

a. $4 - x + 2x^2$
b. $2x^2 - x - 4$
c. $2x^2 - x + 4$

Subtract by using opposites and algebra tiles.

4. $\begin{array}{rr} & 3x^2 + 4x - 2 \\ (-) & \underline{x^2 + x + 4} \end{array}$

5. $\begin{array}{rr} & x^2 - 5x + 2 \\ (-) & \underline{-x^2 + 3x - 1} \end{array}$

6. $(x^2 + 6x - 2) - (x^2 + 5x - 3)$

7. $(-2x^2 + 4x - 5) - (-x^2 - 2x + 4)$

8. $(2x^2 + 7x - 3) - (x^2 + 6x + 5)$

9. $(2x^2 - 5x + 7) - (x^2 - 5x + 4)$

10. $(2x^2 - 8x + 6) - (x^2 - 4x + 4)$

11. $(2x^2 + 3x - 1) - (-x^2 + 2x - 1)$

12. $(2x^2 - 5x + 6) - (x^2 + 5x - 8)$

13. $(3x^2 + 5x - 9) - (3x^2 + 7x + 6)$

14. Use algebra tiles to show that the difference of two trinomials can be a binomial.

15. Use algebra tiles to show that the difference of two trinomials can be a monomial.

Manipulative 15 *(Use with Lesson 18.5.)*

You have already used algebra tiles to model the product of a monomial and a binomial. Now you will use the tiles to model other polynomial products.

EXAMPLE Use algebra tiles to find the product: $(2x + 1)(x + 3)$

Solution: **A.** Model the factors as shown.

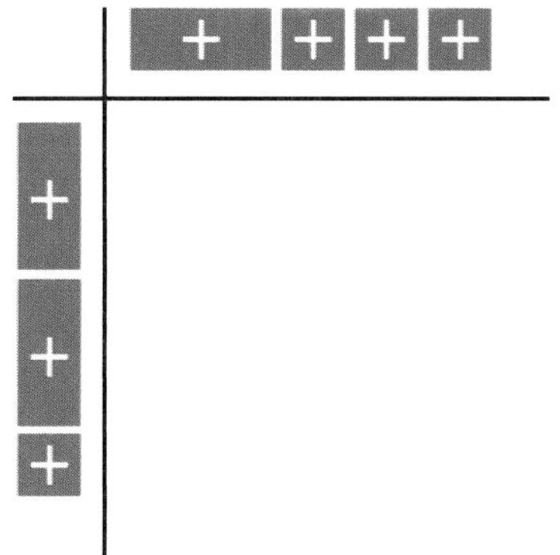

B. Build a rectangle that has the dimensions $(2x + 1)$ and $(x + 3)$.

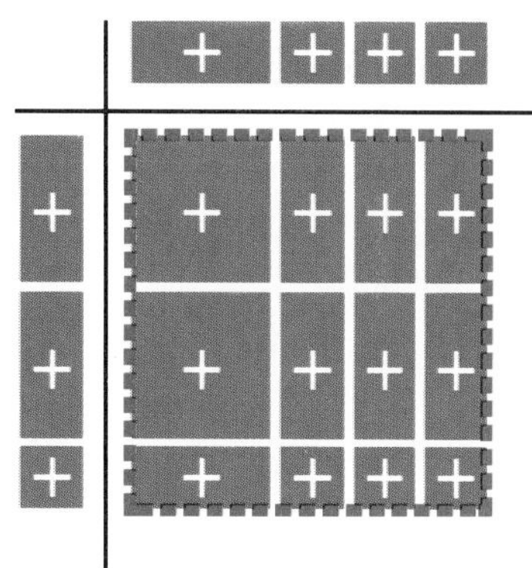

C. Count the tiles in the rectangle and write the product.

$(2x + 1)(x + 3) = 2x^2 + 6x + x + 3 = 2x^2 + 7x + 3$

The product $2x^2 + 7x + 3$ represents the area of the rectangle. The area is the sum of the areas of the tiles in the rectangle.

EXERCISES

Goal: To use algebra tiles to model multiplication of binomials

For each model, name the factors and state the product.

1. 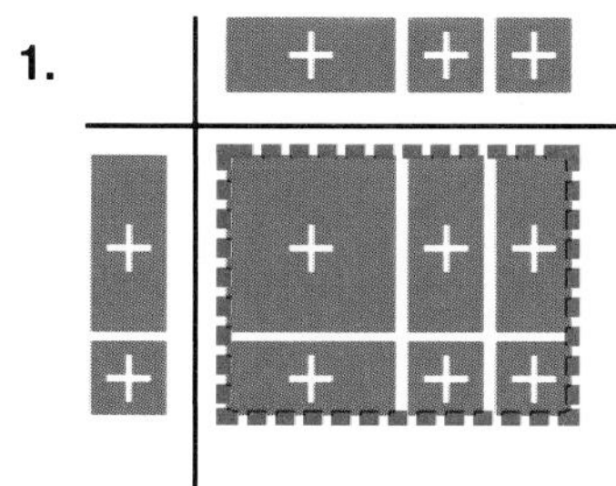

2. 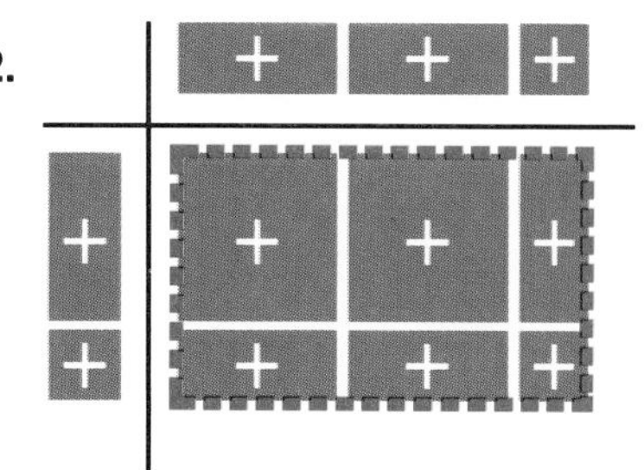

3. 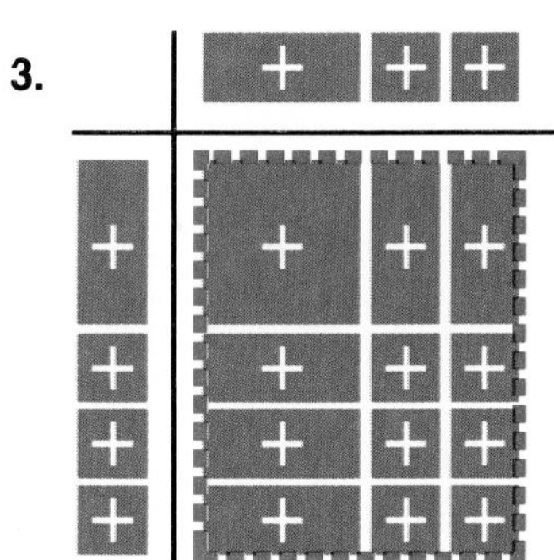

4.

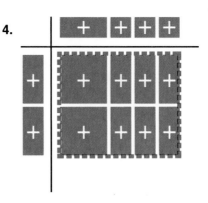

Use algebra tiles to find the product.

5. $(x + 3)(x + 4)$

6. $(x + 6)(x + 2)$

7. $(x + 3)(x + 5)$

8. $(x + 2)(x + 2)$

9. $(x + 1)(x + 1)$

10. $(2x + 1)(x + 2)$

11. $(x + 3)(2x + 1)$

12. $(3x + 1)(3x + 1)$

13. $(2x + 5)(x + 2)$

14. $(x + 3)(2x + 1)$

15. $(3x + 1)(x + 2)$

16. $(2x + 1)(x + 4)$

Use algebra tiles to find each answer.

17. $(x + 2)(x + ?) = x^2 + 5x + 6$

18. $(x + 5)(x + ?) = x^2 + 6x + 5$

19. $(x + 3)(x + ?) = x^2 + 7x + 12$

20. $(x + 6)(x + ?) = x^2 + 9x + 18$

21. Use algebra tiles to show that $(x + 5)(x + 4) = x(x + 4) + 5(x + 4)$.

22. What property is illustrated in Exercise 21?

GLOSSARY

The following definitions and statements reflect the usage of terms in this textbook.

Absolute Value To indicate distance, but not direction, from zero, you use *absolute value.* (Page 237)

Algebraic Expression An *algebraic expression* contains at least one variable. (Page 5)

Algebraic Fraction In an *algebraic fraction*, a variable appears in either the numerator or denominator or both. (Page 83)

Arithmetic Mean The *arithmetic mean* of a collection of numbers, scores, measures, and so on, is the sum of the items in the collection divided by the number of items. (Page 296)

Binomial A *binomial* is the sum or difference of two monomials. (Page 492)

Combining Like Terms When you add or subtract like terms, you are *combining like terms.* (Page 41)

Complex Fraction A fraction such as $\frac{\frac{2}{3}}{\frac{k}{3}}$ with a fractional numerator or denominator is called a *complex fraction.* (Page 484)

Composite Number A *composite number* is a counting number greater than 1 that is not a prime number. (Page 58)

Coordinate Plane The *coordinate plane* has four regions, or quadrants, which are separated by the x and y axes. (Page 393)

Discount The difference between the list and sale prices is called the *discount.* (Page 204)

Divisible A number is *divisible* by another number if the remainder is 0. (Page 50)

Domain of a Function The *domain of a function* is the set of x values of its ordered pairs (x,y). (Page 397)

Equation An *equation* is a sentence that contains the equality symbol "=." (Page 110)

Equivalent Fractions *Equivalent fractions* name the same number. (Page 80)

Equivalent Inequalities Inequalities that have the same solution set are called *equivalent inequalities.* (Page 368)

Evaluate *Evaluate* means to find the value. (Page 5)

Even Number A number that is divisible by 2 is an *even number.* (Page 54)

Exponent The second power of 3 is written as 3^2. The raised two is called an *exponent*. The exponent indicates how many times a number is multiplied by itself. (Page 28)

Factor Since 24 is divisible by 8, 8 is a *factor* of 24. (Page 50)

Factoring The Distributive Property can be used to write sums or differences as products. If the terms of the sum or difference share a common factor, a product can be written with the common factor as one of the factors. This is called *factoring*. (Page 35)

Formula A *formula* is a shorthand way of writing a word rule. In a formula, variables and symbols are used to represent words. (Page 8)

Function A *function* is a relation in which no two ordered pairs have the same x value. (Page 397)

Hypotenuse The *hypotenuse* of a right triangle is the side opposite the right angle. It is the longest side of the right triangle. (Page 229)

Integers The set of *integers* is made up of the positive integers, zero, and the negative integers. (Page 214)

Interest *Interest* is the amount of money that is charged for the use of borrowed money. (Page 200)

Interest Rate Interest is usually expressed as a per cent of the principal. This per cent is called the *interest rate.* (Page 200)

Intersection of Two Lines The *intersection of two lines* is the set of points common to both lines. (Page 408)

Least Common Multiple The *least common multiple* (LCM) of two or more counting numbers is the smallest counting number that is divisible by the given numbers. (Page 67)

Like Fractions *Like fractions* have a common denominator. (Page 96)

Like Radicals Square root radicals with equal radicands are *like radicals.* (Page 462)

Like Terms Two or more *like terms* have the same variables, and the same powers of these variables. (Page 40)

Linear Function A *linear function* is a function in which the points of its graph lie in a straight line. (Page 402)

Mean See arithmetic mean.

Median The *median* of a collection of data is the middle item when the items are arranged in order of size. When there is an even number of items, the median is the mean of the two middle items. (Page 297)

Mode The *mode* is the item in a collection of data that occurs most often. (Page 297)

Multiple A *multiple* of a given number is a product that has the given number as one of its factors. (Page 50)

Net Price The price of an article after a discount is deducted is the *net price* or *sale price.* (Page 204)

Numerical Expression A *numerical expression* includes at least one of the operations of addition, subtraction, multiplication, or division. (Page 2)

Odd Number A number that is not divisible by 2 is an *odd number.* (Page 54)

Odds The ratio of the number of ways that an event can occur to the number of ways an event cannot occur is called the *odds.* (Page 80)

Ordered Pair An *ordered pair* is a pair of numbers one of which is designated as the first and the other is the second. (Page 390)

Origin In the coordinate plane, the point with coordinates (0,0) is called the *origin.* (Page 390)

Per Cent A *per cent* is a ratio in which the second term is 100. (Page 185)

Perfect Square A rational number whose square root is also a rational number is a *perfect square.* (Page 234)

Perimeter The *perimeter* of a figure is the distance around it. To find perimeter, add the lengths of the sides. (Page 37)

Polynomial A *polynomial* is a monomial or the sum or difference of two or more monomials. (Page 492)

Power of a Number When a number is multiplied by itself n times, the number is raised to nth *power.* (Page 28)

Prime Factorization Every composite number can be expressed as a product of prime factors. Such a product is called the *prime factorization* of the number. (Page 64)

Prime Number A *prime number* is a counting number greater than 1 that has exactly two counting-number factors, 1 and the number itself. (Page 58)

Principal The amount of money borrowed on which interest is paid is called the *principal.* (Page 200)

Probability The *probability* of an event is the ratio of the number of favorable outcomes of the event to the total number of possible outcomes of the event. (Page 177)

Proportion A *proportion* is an equation that states that two ratios are equal. (Page 181)

Range of a Function The *range of a function* is the set of y values of its ordered pairs (x,y). (Page 397)

Rate of Discount Discounts are often expressed as a certain per cent of the list price. This per cent is called the *rate of discount.* (Page 204)

Ratio A *ratio* is a quotient that compares the two numbers. (Page 174)

Rational Numbers The set of *rational numbers* consists of the negative rational numbers, zero, and the positive rational numbers. (Page 219)

Rationalizing the Denominator Eliminating the radical in the denominator of a fraction is called *rationalizing the denominator.* (Page 469)

Reaction Distance *Reaction distance* is the distance through which an automobile travels while the driver's foot is moving from the accelerator to the brake pedal. (Page 411)

Real Numbers The set of *real numbers* contains all the rational numbers and all the irrational numbers. (Page 235)

Reciprocals Two numbers having 1 as their product are called *reciprocals.* (Page 92)

Relation A *relation* is a set of ordered pairs. (Page 393)

Replacement Set The *replacement set* is the set of numbers from which replacements for the variable are chosen. (Page 364)

Retail Price The public pays the *retail price* for items sold in stores. (Page 194)

Simplified Radical A *radical is simplified* if the radicand has no perfect square factor other than 1. (Page 453)

Solve an Equation To *solve an equation* means to find the number or numbers that will make the equation true. (Page 110)

Systems of Equations Two equations in the same two variables form a *system of equations.* (Page 420)

Unit Price *Unit price* is the price per gram, per ounce, and so on. (Page 8)

Value of Numerical Expression To find the *value of a numerical expression*, perform the indicated operation or operations. (Page 2)

Variable A *variable* is a letter representing one or more numbers. (Page 5)

Wholesale Price *Wholesale price* is the price that a store pays for items it will sell. (Page 194)

X Intercept The *x intercept* of a line is the x value of its point on the x axis. (Page 406)

Y Intercept The *y intercept* of a line is the y value of its point on the y axis. (Page 405)

INDEX

ANSWERS TO SELECTED EXERCISES

The answers to the odd-numbered problems in the *Classroom Exercises, Written Exercises, Mid-Chapter Reviews, Chapter Reviews, Cumulative Reviews, Special Topic Applications,* and *Focus on Reasoning* sections are given on the following pages.

The answers are provided for all of the problems in the *Review Capsules* and *Pivotal Exercises* (P-1, P-2, etc.).

CHAPTER 1 EXPRESSIONS AND FORMULAS

Page 3 **P-1** a. 9 b. 8 c. 14 d. 16 e. 1 f. 6

Pages 3-4 **Classroom Exercises** 1. A 3. S 5. M 7. D 9. A 11. M 13. D 15. M 17. M 19. M 21. M 23. D 25. 9; 5 27. 9; 17 29. 4; 4; 15; 19

Page 4 **Written Exercises** 1. 8 3. 27 5. 28 7. 8 9. 24 11. 54 13. 100 15. 4 17. 4 19. 47 21. 60 23. 27 25. 28 27. 15 29. 45 31. 50 33. 35 35. 3 37. 19 39. 50 41. 18 43. 56 45. 4 47. 90 49. 21 51. 1

Pages 5-6 **P-1** a. algebraic expression b. not an algebraic expression c. not an algebraic expression d. algebraic expression **P-2** a. 10 b. 60 c. 11 **P-3** a. 33 b. 81 c. 15

Page 6 **Classroom Exercises** 1. 15(5) + 32 3. 24(4) + 15(7) 5. 24(19) + 18(15) 7. 2(3)(4) − 4(4) 9. 2 11. 4 13. 8 15. 4 17. 24 19. 0 21. 12 23. 100

Pages 6-7 **Written Exercises** 1. 5 3. 8 5. 4 7. 4 9. 36 11. 51 13. 30 15. 46 17. 29 19. 100 21. 42 23. 10 25. 42 27. 7 29. 23 31. 31 33. 625 35. 29 37. 71

Page 7 **Review Capsule for Section 1.3** 1. 2.1 2. 4.6 3. 6.0 4. 3.0 5. 1.1 6. 3.3 7. 9.3 8. 7.1 9. 5.9 10. 1.5 11. 4.0 12. 2.6

Page 8 **P-1** a. U = 89 ÷ 12 b. 65 ÷ 100

Page 9 **Classroom Exercises** 1. 17.8 3. 0.735 5. 13.495 7. 4.09 9. 34.319 11. 0.6 13. 20.555556 15. 4.8 17. 24,733.333 19. U = 65 ÷ 8 21. U = 98 ÷ 3.5 23. U = 119 ÷ 10 25. U = 295 ÷ 3

Pages 10-11 **Written Exercises** 1. 3.8¢ per ounce 3. 4.5¢ per fluid ounce 5. 0.5¢ per gram 7. 6.6¢ per fluid ounce 9. 748.8¢ or $7.488 per fluid ounce 11. 0.6¢ per ounce; 0.5¢ per ounce; the 450-ounce size is the better buy. 13. 4.9¢ per ounce; 5.0¢ per ounce; the 16-ounce size is the better buy. 15. 38.3¢ per pound; 41.8¢ per pound; the 3-pound size is the better buy. 17. The 2000-milliliter size is the better buy. 19. The 24-ounce box is the better buy.

Page 11 **Mid-Chapter Review** 1. 6 3. 10 5. 43 7. 42 9. 4 11. 13 13. 0 15. 6.5¢ per ounce

Page 11 **Review Capsule for Section 1.4** 1. $5 2. $8 3. $7 4. $10 5. $18 6. $44 7. $53 8. $61 9. $37 10. $50 11. $38 12. $87 13. $24 14. $33 15. $75 16. $18 17. $54 18. $100

Page 12 **P-1** a. 0; 0.00 b. 3; 0.03 c. 3; 0.03

Page 14 **Classroom Exercises** 1. a. 1 b. 0.05 3. a. 3 b. 0.03 5. a. 3 b. 0.03 7. S = (0.025)(725)(2) 9. S = (0.05)(585)(4) 11. S = (0.02)(230)(2) 13. S = (0.02)(550)(3.5)

Pages 14-15 **Written Exercises** 1. $51 3. $95 5. $121 7. $57 9. $78 11. $6 13. $26 15. $24 17. a. S = 6; S = 12 b. When k changes from 0.02 to 0.04, S doubles. c. If k is doubled while c and t remain the same, then S doubles. 19. 12 dimes 21. 5 persons

Page 17 **Classroom Exercises** 1. a. Subtraction b. $300; $350; $400; $450; $500 3. a. Multiplication b. $20; $22.50; $25; $27.50; $30 5. a. Multiplication b. 5; 7; 9; 11; 14

Pages 18-19 **Written Exercises** 1. a 3. c 5. d 7. g 9. 19; 21; 23; 25; 27; Word rule: Dexter's age, d, equals 7 subtracted from Reggie's age, r.; Formula: $d = r - 7$ 11. 84; 88; 92; 96; 100; Word rule: The average test score, a, equals the total score, T, divided by 5.; Formula: $a = T \div 5$

Page 20 1. 15 ways 3. 36 choices 5. 3 choices

Pages 21-22 **Chapter Review** 1. 12 3. 20 5. 62 7. 18 9. 31 11. 21 13. 1 15. 250 milliliters 17. $31 19. $69 21. 30; 35; 40; 45; 50; Word rule: The wages, w, equal 5 times the hours worked, h.

CHAPTER 2 PROPERTIES OF NUMBERS

Page 25 **P-1** a. r = 0 b. y = 1 c. b = 0 d. q = 3

Page 25 **Classroom Exercises** 1. n = 0 3. v = 18 5. k = 0 7. c = 0.75 9. m = 24 11. s = 7 13. k = 0.3 15. f = 2.5

Pages 25-26 **Written Exercises** 1. n = 0 3. x = 0.45 5. y = 0 7. m = 0 9. n = 1 11. k = 1 13. x = 13 15. z = 19 17. n = 9 19. k = 23 21. c = 3 23. m = 4 25. n = 1 27. r = 0 29. x = 0 31. q = 8 33. w = 25 35. p = 10.83 37. x = 0 39. q = 4 41. Any number 43. None 45. None 47. Any number 49. n = 6 51. x = 4 53. x = 12 55. a = 5

Page 26 **Review Capsule for Section 2.2** 1. a. 42 b. 42 2. a. 27 b. 27 3. a. 22 b. 22 4. a. 80 b. 80 5. a. 24 b. 24 6. a. 378 b. 378 7. a. 960 b. 960 8. a. 175 b. 175 9. a. 550 b. 550 10. a. 144 b. 144 11. a. 180 b. 180 12. a. 360 b. 360

Pages 27-28 **P-1** a. c = 8; Associative Property of Multiplication b. p = 6; Commutative Property of Multiplication c. h = 5; Commutative and Associative Properties of Multiplication **P-2** a. 30y b. 48km c. 36st d. 33cz **P-3** a. $40c^2d$ b. $64m^2p^2$ c. $30x^3yz$

Page 29 **Classroom Exercises** 1. g = 5 3. m = 20 5. y = 30 7. s = 4 9. 50y 11. 200t 13. 28k 15. 18xy 17. 7m 19. 30rs 21. $56a^2$ 23. $48m^3n^2$ 25. $24b^3c^2$ 27. $6r^2s^3t^2$

Pages 29-30 **Written Exercises** 1. a = 7 3. n = 2 5. k = 4 7. x = 2 9. $b = \frac{1}{2}$ 11. 35x 13. 42a 15. 56t 17. 8.4n 19. 63ab 21. 39xy 23. 100rs 25. 11.2pqr 27. $39x^2$ 29. $36mn^2$ 31. $36x^2y$

33. $2np^2$ 35. $24x^3y$ 37. $120gk^2t^3$ 39. 18xy 41. 84t 43. $9x^3y$ 45. 7.5w 47. 30rst 49. $6r^2s^2t^3$
51. 12 53. 75 55. No, because 3 * 5 = 45 is not the same as 5 * 3 = 75. 57. 2304 59. 200 61. 576

Page 30 Review Capsule for Section 2.3 1. 18 + 15 2. 28 − 14 3. 45 + 63 4. 40 − 10 5. 6p − 6q 6. 5a + 10 7. $2r^2 + 10r$ 8. $12d - 3d^2$ 9. 8c + 8d 10. 12f − 12g 11. km − jm 12. yz + xz

Page 32 P-1 a. 44 + 11k b. 3a − 12 c. 80f + 16g d. 6r − 10s **P-2** a. $18c^2 - 24cd$ b. $2a^2 + 28a$ c. $6fg + 3f^2$ d. $30rs^2 - 20r^3s$ **P-3** a. 560 b. 192 c. 1144

Page 33 Classroom Exercises 1. 35 + 7c 3. 12j − 96 5. 20 − 4d 7. 81 − 45w 9. 28 + 8a 11. 190m − 90 13. $x^2 + xy$ 15. $2m^2 + 4m$ 17. $12pq + 6q^2$ 19. $a^2 - 1.8a$ 21. $10mn - 2n^2$ 23. $6k^2 - 2k^3$ 25. $6mv + 12v^2$ 27. $35s^2 + 5k^2s$ 29. 315 31. 350 33. 2964 35. 5440

Pages 33-34 Written Exercises 1. 16 − 4d 3. 9g + 45 5. 5r + 30 7. 35w − 21 9. 10r + 30 11. 84 − 18m 13. $15m^2 + 15mt$ 15. $rx + 2x^2$ 17. $2n^2 + 32n$ 19. $6p^2 + 3pq$ 21. $2x^2 + 2xy$ 23. $12r^2 + 15rv$ 25. $28k^2 + 20gk$ 27. 12t − 14s 29. 8a + 4b 31. 192 + 112s 33. $5m^2 - 2mn$ 35. $24y^2z - 6yz^2$ 37. 8q − 40 39. 13.5m + 15q 41. $12a^2 + 36a + 28$ 43. $10x^4 + 50x^2 + 35x$ 45. $12m^3n + 48m^2n^2 + 36mn$ 47. $54r^3 + 54r^2 + 54r$ 49. $24y^3z^2 + 16y^3z + 18yz^2$ 51. 9 seconds

Page 34 Mid-Chapter Review 1. t = 0 3. m = 0 5. a = 275 7. b = 14.3 9. y = 5 11. 63c 13. 54r 15. 51cd 17. $140y^2$ 19. $36a^2b^2c$ 21. $126p^3q^3r^2$ 23. 45c + 135 25. $54a^2 + 42a$ 27. $40nt + 8n^2$ 29. $2rs^2 + 6s^2$

Page 35 P-1 a. 4(t − 1) b. 12(m + 1) c. 8(q + p) d. j(17 − 1) or 16j **P-2** a. m(m + 7) b. r(r − 12) c. 6(k − 6) d. 9(t − 5)

Page 36 Classroom Exercises 1. 3(p + q) 3. 12(m − q) 5. 2(r − 1) 7. x(r − 3) 9. n(n + 2) 11. t(t − 10) 13. w(5 + w) 15. c(12 − c) 17. 3(x − 2) 19. 2(y − 3) 21. 6(b + 2) 23. 3(4 + r)

Page 36 Written Exercises 1. 7(k + n) 3. 19(p − q) 5. 5(h − 7) 7. y(4 − 1) or 3y 9. y(y + 5) 11. k(k − 4) 13. t(13 + t) 15. $a(\frac{1}{4} - a)$ 17. 2(n + 2) 19. 2(r − 3) 21. 3(y + 5) 23. 7(3 + c) 25. $\frac{1}{2}(y + 1)$ 27. 7(n + 4) 29. t(n + q) 31. a(n − t)

Page 36 Review Capsule for Section 2.5 1. a 2. c 3. c 4. b

Page 38 Classroom Exercises 1. 30 cm 3. 8 cm 5. 46 ft 7. 12 yd 9. 88 ft 11. 144 km^2 13. 1.44 m^2 15. 144 yd^2 17. 64 mi^2

Pages 38-39 Written Exercises 1. 11.4 meters 3. 91.3 meters 5. 900 square centimeters 7. 196 square inches 9. 272.25 square meters 11. 52 square meters 13. 4536 square yards

Page 39 Review Capsule for Section 2.6 1. 2x + 3x 2. 6mp + 8mp 3. $20r^2s + 5r^2s$ 4. $1.8ab^2c + 4.7ab^2c$ 5. $19t^2 - 2t^2$ 6. $23pq^2 - 14pq^2$ 7. $7.8xy^2 - 5.3xy^2$ 8. $13.6a^3d - 12.6a^3d$ 9. 18rs + 5rs + 3rs 10. $4.7mn^2p + 1.3mn^2p + 36mn^2p$ 11. $5\frac{3}{4}t + 3\frac{1}{8}t + 1\frac{1}{2}t$ 12. $56c^3d^2 + 39c^3d^2 + 11c^3d^2$

Pages 40-41 P-1 a. $\frac{1}{2}$ b. 0.8 c. 1 d. 12 **P-2** a. 4n b. 14cd c. 2a d. $22v^2$ **P-3** a. 3b b. a c. r^2 d. 0

Page 41 **Classroom Exercises** 1. 3mn and mn; $7m^2n$ and $\frac{1}{2}m^2n$; 6x and 1.8x 3. 22 5. 13 7. 1 9. 0.5 11. 11x 13. 34cd 15. $21r^2$ 17. 0.9ad 19. 5y 21. 25ab 23. 3.5mn 25. $20r^3st$

Page 42 **Written Exercises** 1. G 3. F 5. M 7. B, I 9. D 11. 16a 13. 13ab 15. 39y 17. 3.9x 19. 12mn 21. $14x^2$ 23. 7s 25. $5a^2$ 27. 6.8rt 29. ab 31. $1.3pq^2$ 33. $3t^3$ 35. 2np 37. 45stv 39. 27abc 41. $159m^5n^2$ 43. $146.4p^2q^2$

Page 42 **Review Capsule for Section 2.7** 1. a. 21 b. 21 2. a. 14.4 b. 14.4 3. a. 27 b. 27 4. a. 25.9 b. 25.9 5. a. 79 b. 79 6. a. 39 b. 39

Pages 43-44 **P-1** a. g = 42; Commutative Property of Addition b. z = 18; Commutative and Associative Properties of Addition c. c = 3; Associative Property of Addition d. y = 25; Commutative Property of Addition **P-2** a. $33r^2 + 9$ b. 25pq + 3 c. 5kr + 16 **P-3** a. 11a b. $7c^2d + 4cd$ c. $14p + 21q^3$

Pages 44-45 **Classroom Exercises** 1. a = 15 3. r = 11 5. b = 15 7. 11x + 3 9. 34y + 23 11. 52r + 10 13. $8.7n^2 + 9$ 15. 18x 17. $4ab + 6a^2b$ 19. 21rs + 18

Page 45 **Written Exercises** 1. r = 27 3. x = 4 5. c = 6 7. $b = \frac{1}{4}$ 9. 29a + 15 11. 4.3m + 4.2 13. 55k + 7 15. 34t 17. 1.2p + 2.8r 19. $6q^3 + 15q + 7$ 21. $20xy^2$ 23. $25r^2s + 47rs^2 + 29rs$ 25. $60a^2 + 18b^2$ 27. 4.2ma 29. $29.00

Page 46 1. a. 10,300 watts b. 7 circuits 3. increase

Pages 47-48 **Chapter Review** 1. n = 0 3. t = 1 5. x = 0 7. q = 0 9. y = 1 11. 96r 13. 30mn 15. $117k^2g$ 17. $8.8p^2q^2$ 19. $56d^3f^2$ 21. $15m^3p^2$ 23. 12k − 48 25. 25c + 25d 27. 18p − 54 29. $54g^2 + 18gh$ 31. $48s - 27s^2$ 33. $12p^2q + 16pq^2$ 35. 24(m − w) 37. $\frac{1}{3}(r + 5)$ 39. q(23 + p) 41. 39(1 − z) 43. p(2.2b + a) 45. $c(c^2 - 2)$ 47. 144 inches 49. 28t 51. 4.2n 53. $25x^3y$ 55. $3.6a^2bc$ 57. 79ab 59. $15n^3 + 13n^2$ 61. $18.2ab + 3.8a^2b^2 + a^2b$ 63. $8p^3 + 13p^2 + 14p + 15$

CHAPTER 3 BASIC NUMBER CONCEPTS

Page 51 **P-1** a. 1, 2, 11, 22 b. 1, 29 c. 1, 2, 17, 34 d. 1, 5, 7, 35 e. 1, 41 **P-2** a. 2 · 5 b. 2 · 11 c. 3 · 13 d. 3 · 11 e. 6 · 5 or 15 · 2 or 10 · 3

Pages 51-52 **Classroom Exercises** 1. 35 is divisible by 7. 3. 10 is divisible by 10. 5. 30 is divisible by 5. 7. 18 is divisible by 1. 9. 2 is a factor of 10. 11. 7 is a factor of 35. 13. 1 is a factor of 19. 15. 17 is a factor of 34. 17. 28 is a multiple of 4. 19. 36 is a multiple of 9. 21. 27 is a multiple of 27. 23. 45 is a multiple of 5. 25. F; 6 is not divisible by 18. 27. T; 17 ÷ 17 = 1 29. F; 23 ÷ 7 = 3R2 31. T; 2 · 13 = 26 33. F; 33 ÷ 13 = 2R7 35. T; 19 ÷ 19 = 1 37. 1, 2, 4 39. 1, 5 41. 1, 2, 4, 8 43. 1, 2, 5, 10 45. 1, 17 47. 1, 2, 3, 4, 6, 8, 12, 24 49. 2 · 3 51. 3 · 5 53. 3 · 3 55. 3 · 7 57. 2 · 19 59. 4 · 11 or 2 · 22

Pages 52-53 **Written Exercises** 1. T; 21 ÷ 7 = 3 3. F; 43 is not divisible by 6. 5. T; 5 · 11 = 55 7. F; 258 ÷ 13 = 19 R11 9. T; 126 · 1 = 126 11. F; 237 ÷ 13 = 18 R3 13. 1, 2, 3, 4, 6, 12 15. 1, 11 17. 1, 19 19. 1, 2, 4, 5, 10, 20 21. 1, 23 23. 1, 5, 25 25. 1, 2, 4, 7, 14, 28 27. 1, 2, 3, 4, 6, 8, 12, 16, 24, 48 29. 2 · 6 or 3 · 4 31. 2 · 9 or 3 · 6 33. 3 · 9 35. 2 · 18 or 3 · 12 or 4 · 9 or 6 · 6 37. 3 · 11

39. $7 \cdot 11$ 41. $7 \cdot 7$ 43. $3 \cdot 31$ 45. a. No b. Yes c. No d. 1, 2, 4, 7, 8, 14, 28, 56 47. a. No b. Yes c. Yes d. 1, 7, 11, 77 49. a. Yes b. Yes c. No d. 1, 2, 3, 4, 6, 8, 9, 12, 18, 24, 36, 72 51. a. Yes b. No c. No d. 1, 2, 3, 4, 5, 6, 8, 10, 12, 15, 20, 24, 30, 40, 60, 120 53. The number is 13425.

Pages 54-55 **P-1** a. None b. 2 and 4 c. 2, 4, 5, and 10 d. 5 only e. 2, 4, 5, and 10 **P-2** a **P-3** a, b, c

Page 56 **Classroom Exercises** 1. T; 958 ends in 8. 3. T; 706 ends in 6. 5. F; 756 does not end in 0 or 5. 7. T; 12,690 ends in 0. 9. T; $9 + 6 = 15$, which is divisible by 3. 11. T; $2 + 7 + 6 = 15$, which is divisible by 3. 13. T; $8 + 1 + 2 + 3 + 4 = 18$, which is divisible by 9.

Pages 56-57 **Written Exercises** 1. T; 546 ends in 6. 3. F; 811 does not end in 0, 2, 4, 6, or 8. 5. F; 14 is not a multiple of 4. 7. T; 28 is a multiple of 4. 9. F; 56 does not end in 0 or 5. 11. T; $8 + 1 + 3 = 12$, which is divisible by 3. 13. F; sum of digits, 10, is not divisible by 3. 15. F; sum of digits, 22, is not divisible by 3. 17. T; sum of digits, 27, is divisible by 9. 19. F; sum of digits, 30, is not divisible by 9. 21. F; 95 is not a multiple of 4. 23. T; 2160 ends in 0. 25. T; sum of digits, 36, is divisible by 3. 27. T; 222 ends in 2. 29. F; 829 does not end in 0 or 5. 31. T; sum of digits, 18, is divisible by 9. 33. a. No b. 2 apples 35. a. No b. \$3 37. a. No b. 4 toys

Page 57 **Review Capsule for Section 3.3** 1. 4 2. 9 3. 27 4. 8 5. 100 6. 225 7. 882 8. 9075

Pages 58-59 **P-1** a. 2 b. 2, 3 c. 2, 4 d. 3, 5 e. 2, 5, 10, 25 f. 2, 4, 7, 14 **P-2** a. 2 b. 3 c. 2 d. 5 e. 11

Page 59 **Classroom Exercises** 1. prime 3. composite 5. composite 7. composite 9. 2 11. 3 13. 2 15. 2 17. 5 19. 3 21. $2^2 \cdot 3 \cdot 5$ 23. $2^3 \cdot 3^2 \cdot 5 \cdot 11$ 25. $5^2 \cdot 11 \cdot 17^3$ 27. $2^3 \cdot 3^4 \cdot 7^2$ 29. $5^3 \cdot 7^2 \cdot 11^3 \cdot 17^2$

Pages 60-61 **Written Exercises** 1. 2 3. 2 5. 5 7. 3 9. 5 11. 7 13. 7 15. 11 17. 3 19. $2 \cdot 3^2 \cdot 5$ 21. $2 \cdot 3 \cdot 5^2 \cdot 7^2$ 23. $2 \cdot 3^3 \cdot 5 \cdot 7^2$ 25. $2 \cdot 3^2 \cdot 5 \cdot 7^3 \cdot 13$ 27. $2^2 \cdot 3^2 \cdot 5^2 \cdot 7 \cdot 11^2 \cdot 17^2$ 29. 3 31. 127 33. 131,071 35. 2, 3, 5, 7, 11, 13, 17, 19, 23, 29, 31, 37, 41, 43, 47, 53, 59, 61, 67, 71, 73

Page 61 **Mid-Chapter Review** 1. 1, 2, 11, 22; $2 \cdot 11$ 3. 1, 2, 19, 38; $2 \cdot 19$ 5. 1, 2, 4, 5, 8, 10, 20, 40; $2 \cdot 20$, $4 \cdot 10$, $8 \cdot 5$ 7. F; sum of digits, 16, is not divisibel by 3. 9. T; 5610 ends in 0. 11. F; 4 is a factor of 1008. 13. a. No b. \$1 15. 5 17. 11 19. $2 \cdot 3 \cdot 5^3 \cdot 7^2$ 21. $3 \cdot 5 \cdot 7 \cdot 11^2 \cdot 13^3$

Page 63 1. \$23.97 3. 1 gallon plus 2 quarts; 1 gallon plus 1 quart; 2 gallons plus 3 quarts 5. \$32.91 7. Both amounts of paint would cost more than 3 gallons.

Page 64 **Review Capsule for Section 3.4** 1. a. 126 b. 126 2. a. 126 b. 126 3. a. 42 b. 42 4. a. 68 b. 68 5. a. 150 b. 150 6. a. 315 b. 315

Pages 64-65 **P-1** $3 \cdot 5 \cdot 7$ **P-2** $2^3 \cdot 3 \cdot 5$

Pages 65-66 **Classroom Exercises** 1. 2 3. 3 5. 3 7. $5 \cdot 7 \cdot 13$ 9. $2 \cdot 3 \cdot 5^2$ 11. $2^2 \cdot 3 \cdot 41$ 13. $2 \cdot 3^2 \cdot 11$ 15. $3^2 \cdot 29$ 17. $2 \cdot 3^2 \cdot 5^2$ 19. $2 \cdot 3$ 21. $3 \cdot 5$ 23. $5 \cdot 7$ 25. $3 \cdot 13$ 27. 2^3 29. 2^4 31. $2 \cdot 3^2$ 33. 3^3 35. $2^2 \cdot 3^2$ 37. $2^2 \cdot 5^2$

Page 66 Written Exercises 1. $2 \cdot 11$ 3. 5^2 5. $3 \cdot 7$ 7. $3 \cdot 13$ 9. $3 \cdot 17$ 11. $7 \cdot 11$ 13. $5 \cdot 17$ 15. $11 \cdot 13$ 17. $2^4 \cdot 3$ 19. $2 \cdot 3^3$ 21. $2 \cdot 3 \cdot 13$ 23. $2^2 \cdot 3 \cdot 7$ 25. 5^3 27. $2^3 \cdot 3 \cdot 7$ 29. $2 \cdot 3^3 \cdot 7$ 31. $2^2 \cdot 3^4$ 33. 110 miles

Page 67 Review Capsule for Section 3.5 1. 2, 4, 6, 8, 10, 12 2. 4, 8, 12, 16, 20, 24 3. 3, 6, 9, 12, 15, 18 4. 5, 10, 15, 20, 25, 30 5. 9, 18, 27, 36, 45, 54 6. 6, 12, 18, 24, 30, 36 7. 4, 8, 12 8. 9, 18 9. 12, 24 10. 10

Page 69 Classroom Exercises 1. $2^2 \cdot 3 \cdot 5$ 3. $11 \cdot 13 \cdot 17$ 5. $2^2 \cdot 3^2 \cdot 5$ 7. $2 \cdot 3^2 \cdot 5^2$ 9. 12 11. 6 13. 15 15. 24 17. 14 19. 18 21. 30 23. 35 25. 15 27. 24 29. 8 31. 18 33. 30 35. 100 37. 36 39. 30

Pages 69-70 Written Exercises 1. 12 3. 140 5. 65 7. 270 9. 120 11. 56 13. 150 15. 672 17. 120 19. 240 21. 168 23. 720 25. 840 27. 945 29. 36 31. 27 33. 140 35. 72 37. 247 39. 81 41. 60 43. 1260 45. 144 47. 210 49. 306 51. 15,470 53. 12 days 55. 60 days 57. 173 59. 461 61. 1231

Pages 71-72 1. a. False. The circle for "Those who work hard" does not completely fill the rectangle for "students." b. True. The rectangle for "students" contains some students "who work hard." c. False. The rectangle for "students" contains some students "who like holidays." d. False. The overlap of the two circles represents those students who work hard and like holidays. 3. a. False. The circle for "Those who are healthy" is not entirely contained in the circle for "Those who eat yogurt." b. False. The circle for "Those who are healthy" is not entirely contained in the circle for "Those who eat yogurt." c. True. The circle for "Those who eat yogurt" is not entirely contained in the circle for "Those who are healthy." Thus, there are some people who eat yogurt but are not healthy. d. True. The overlap of the two circles represents those people who are healthy and who eat yogurt. 5. a. False. The circle for "Divisible by 5" does not completely fill the rectangle for "Whole Numbers." b. False. The circle for "Divisible by 10" does not completely fill the circle for "Divisible by 5." c. False. The circle for "Divisible by 10" is entirely contained in the circle for "Divisible by 5." d. True. The circle for "Divisible by 10" does not completely fill the circle for "Divisible by 5." Thus, there are some numbers that are divisible by 5 but are not divisible by 10.

Pages 73-74 Chapter Review 1. 1, 3, 9, 27; $3 \cdot 9$ 3. 1, 2, 5, 10, 25, 50; $2 \cdot 25$, $5 \cdot 10$ 5. 1, 2, 4, 11, 22, 44; $2 \cdot 22$, $4 \cdot 11$ 7. 1, 2, 41, 82; $2 \cdot 41$ 9. 1, 2, 3, 5, 6, 9, 10, 15, 18, 30, 45, 90; $2 \cdot 45$, $3 \cdot 30$, $5 \cdot 18$, $6 \cdot 15$, $9 \cdot 10$ 11. T; sum of digits, 18, is divisible by 3 13. F; 2 is not a multiple of 4 15. T; sum of digits, 18, is divisible by 9 17. F; does not end in 0 19. F; does not end in 0, 2, 4, 6, or 8 21. a. No b. 3425 pounds 23. $2^3 \cdot 3 \cdot 5^2 \cdot 7^2 \cdot 19$ 25. $2^3 \cdot 3^4 \cdot 5^4$ 27. $2^2 \cdot 3$ 29. $2^2 \cdot 7$ 31. $2^2 \cdot 17$ 33. $2^2 \cdot 23$ 35. $2 \cdot 3^3 \cdot 5^2 \cdot 11 \cdot 13^2$ 37. $2^3 \cdot 3^4 \cdot 5^2 \cdot 7 \cdot 11$ 39. 40 41. 36 43. 385 45. 120 47. 525 49. 360 51. 15 days

Pages 75-76 Cumulative Review: Chapters 1-3 1. 59 3. 9 5. 13 7. 17 9. 25-pound bag: 38.8¢; 10-pound bag: 49.5¢; The 25-pound bag is the better buy. 11. $38.75 13. The sale price, s, is equal to $5 subtracted from the original price, p. 15. $x = 0$ 17. $y = 17$ 19. $4hk$ 21. $12g^3t^3$ 23. $12 - 48p$ 25. $18x^2 + 3xy$ 27. $2(x + 6)$ 29. $x(6x + 1)$ 31. $25m^2$ 33. $23ab$ 35. $1 + 12y$ 37. $7x^2y + 3xy^2$ 39. 1, 2, 17, 34 41. 1, 2, 4, 5, 10, 20, 25, 50, 100 43. No; sum of digits, 24, is not divisible by 9 45. No; 2 is not divisible by 436. 47. $2^2 \cdot 3 \cdot 5 \cdot 7^3$ 49. $2^3 \cdot 3^2 \cdot 5^3 \cdot 17$ 51. $2 \cdot 3^2 \cdot 5$ 53. $2^3 \cdot 5$ 55. 60 57. 90 59. 35 days

CHAPTER 4 WRITING FRACTIONS IN LOWEST TERMS

Pages 78-81 **P-1** a. 10 b. 0 c. 1 d. 1 e. 1 **P-2** a. 3 b. 2 c. none d. 5 e. 3 **P-3** a. $\frac{1}{4}$ b. $\frac{1}{3}$ c. $\frac{1}{3}$ d. $\frac{1}{11}$ e. $\frac{1}{3}$ **P-4** a. $\frac{5}{9}$ b. $\frac{3}{5}$ c. $\frac{5}{7}$ d. $\frac{4}{9}$ e. $\frac{3}{7}$ **P-5** a and d; b and e

Page 81 **Classroom Exercises** 1. $\frac{1}{5}$ 3. $\frac{1}{5}$ 5. $\frac{2}{3}$ 7. 1 9. $\frac{7}{9}$ 11. $\frac{2}{5}$ 13. $\frac{5}{6}$ 15. $\frac{1}{12}$ 17. $\frac{3}{4}$ 19. $\frac{1}{8}$ 21. $\frac{1}{5}$ 23. $\frac{2}{7}$ 25. $\frac{2}{3}$ 27. $\frac{5}{7}$ 29. $\frac{5}{9}$

Pages 81-82 **Written Exercises** 1. $\frac{1}{3}$ 3. $\frac{1}{9}$ 5. $\frac{1}{3}$ 7. $\frac{1}{2}$ 9. $\frac{1}{2}$ 11. $\frac{1}{3}$ 13. $\frac{2}{3}$ 15. $\frac{4}{5}$ 17. $\frac{3}{16}$ 19. $\frac{3}{4}$ 21. $\frac{5}{6}$ 23. $\frac{3}{8}$ 25. $\frac{1}{2}$ 27. $\frac{5}{7}$ 29. $\frac{1}{14}$ 31. $\frac{13}{16}$ 33. $\frac{1}{6}$ 35. $\frac{10}{21}$ 37. $\frac{20}{77}$ 39. $\frac{10}{11}$ 41. $\frac{3}{4}$ 43. $\frac{13}{16}$ 45. $\frac{1}{6}$ 47. $\frac{13}{19}$ 49. $\frac{7}{8}$ 51. $\frac{6}{7}$ 53. $\frac{5}{41}$ 55. $\frac{3}{4}$ 57. $\frac{1}{2}$

Page 82 **Review Capsule for Section 4.2** 1. $x \cdot x \cdot x$ 2. $m \cdot m$ 3. $t \cdot t \cdot t \cdot t \cdot t \cdot t$ 4. $a \cdot a \cdot a \cdot a \cdot a$ 5. $r \cdot r \cdot s \cdot s \cdot s$ 6. $m \cdot m \cdot m \cdot n$ 7. $p \cdot p \cdot q \cdot r \cdot r$ 8. $a \cdot a \cdot a \cdot b \cdot b \cdot c$ 9. $3 \cdot n \cdot n \cdot n$ 10. $5 \cdot b \cdot b \cdot b \cdot b$ 11. $7 \cdot c \cdot c \cdot c \cdot d \cdot d \cdot d \cdot d \cdot d$ 12. $13 \cdot r \cdot r \cdot r \cdot r \cdot t \cdot t \cdot t$

Pages 83-84 **P-1** a, d, e **P-2** a. $\frac{3}{4}$ b. $\frac{1}{3}$ c. $\frac{3}{5}$ d. $\frac{1}{3}$ e. $\frac{8}{9}$ **P-3** a. $\frac{5}{8}$ b. $\frac{2}{5t}$ c. $\frac{3x}{5y}$ d. $\frac{3j^2}{5k}$ e. $\frac{b^2}{2}$

Pages 84-85 **Classroom Exercises** 1. $\frac{2 \cdot 2 \cdot n}{2 \cdot 3 \cdot m}$ 3. $\frac{3 \cdot 3 \cdot r \cdot t}{3 \cdot 5 \cdot s}$ 5. $\frac{2 \cdot 5 \cdot z \cdot x}{2 \cdot 3 \cdot 3 \cdot t}$ 7. $\frac{2 \cdot x \cdot x \cdot y}{2 \cdot 3 \cdot 3 \cdot x \cdot y \cdot y \cdot y}$ 9. $\frac{2 \cdot 2 \cdot 2 \cdot 3 \cdot a \cdot b \cdot c}{2 \cdot 13 \cdot b \cdot b \cdot c \cdot c \cdot c}$ 11. $\frac{5 \cdot x \cdot x \cdot y \cdot y}{2 \cdot 2 \cdot 5 \cdot x \cdot y \cdot y \cdot y}$ 13. $\frac{2}{3}$ 15. $\frac{r}{t}$ 17. $\frac{bc}{5a}$ 19. $\frac{m}{7n}$ 21. $\frac{2}{3}$ 23. $\frac{1}{4}$ 25. $\frac{w}{7y}$ 27. $\frac{y}{3}$ 29. $\frac{2s}{7r}$ 31. $\frac{8a^2}{9b}$

Page 85 **Written Exercises** 1. $\frac{5}{11}$ 3. $\frac{3}{x}$ 5. $\frac{3}{4}$ 7. $\frac{3n}{5p}$ 9. $\frac{2}{3y}$ 11. $\frac{1}{n}$ 13. $\frac{4m}{9p}$ 15. $\frac{5ra}{8b}$ 17. $\frac{5}{6yt}$ 19. $\frac{2}{15}$ 21. $\frac{3ad}{2bc}$ 23. $\frac{10a}{21c}$ 25. $\frac{r}{s}$ 27. $\frac{1}{5qr^2}$ 29. $\frac{3}{2x^2yz}$ 31. $\frac{5x^3}{21yz^2}$ 33. $\frac{3r}{5tp}$ 35. $\frac{9y^2}{20xz^2}$ 37. Because only when the value of x is greater than 5 does the numerator become larger than the denominator 39. n = 4

Pages 86-87 **P-1** $3 \cdot 3$; $2 \cdot 7$; $2 \cdot 3 \cdot 5$ **P-2** a. $\frac{4}{5}$ b. $\frac{6}{7}$ c. $\frac{6}{7}$ d. $\frac{2}{5}$ **P-3** a. 8n b. $\frac{14p}{s}$ c. $\frac{6ry}{5}$ d. $\frac{56p^2t}{3q}$ **P-4** a. $\frac{3}{5c}$ b. $\frac{3g}{7f}$ c. $\frac{1}{12}$ d. $\frac{8g}{15fh^2}$

Pages 87-88 **Classroom Exercises** 1. $\frac{8}{3}$ 3. $\frac{5}{2}$ 5. $\frac{1}{14}$ 7. $\frac{3}{5}$ 9. $\frac{3}{4}$ 11. $\frac{1}{4}$ 13. $\frac{2x}{3}$ 15. 2x 17. 10 19. $\frac{6}{xy}$ 21. $\frac{6r}{st}$ 23. $\frac{3a}{5}$

Pages 88-89 **Written Exercises** 1. $\frac{7}{2}$ 3. $\frac{10}{3}$ 5. $\frac{40}{3}$ 7. $\frac{44}{5}$ 9. $\frac{5}{3}$ 11. $\frac{7}{9}$ 13. $\frac{3x}{4}$ 15. $\frac{2r}{3}$ 17. $\frac{21}{b}$ 19. $\frac{4a}{7b}$ 21. $\frac{3}{7x}$ 23. $\frac{3st}{2}$ 25. $\frac{1}{2}$ 27. $\frac{66}{7n}$ 29. $\frac{c}{2ab}$ 31. $\frac{3}{10}$ 33. $\frac{51}{65b}$ 35. $\frac{72}{5t}$ 37. 814 feet 39. 2156 inches

Page 89 Mid-Chapter Review 1. $\frac{5}{8}$ 3. $\frac{9}{16}$ 5. $\frac{15}{16}$ 7. $\frac{5}{6}$ 9. $\frac{1}{2}$ 11. $\frac{7}{10}$ 13. $\frac{5}{3}$ 15. $\frac{1}{4m}$ 17. $\frac{2k}{9}$ 19. $\frac{3m}{7p}$ 21. $\frac{3x}{5y^2}$ 23. $\frac{8q^2}{3rp}$ 25. $\frac{5m^2}{6n^2}$ 27. $\frac{15}{2}$ 29. $\frac{5t}{4}$ 31. $\frac{20sz}{9yw}$ 33. $\frac{px}{15qy^2}$

Pages 90-91 1. a. An estimate will be close enough. Reasons will vary. b. Answers will vary. c. His estimate was 26¢ higher than the actual cost. d. Yes 3. a. No; The amounts in a recipe should be exact. b. Answers will vary. c. flour: $11\frac{1}{4}$ cups; sugar: 6 cups; eggs: 6; molasses: $1\frac{1}{2}$ cups; oleo: $2\frac{1}{4}$ cups; soda: 3 teaspoons; cinnamon: $1\frac{1}{2}$ teaspoons; cloves: $\frac{3}{4}$ teaspoon

Pages 92-94 **P-1** a. $\frac{13}{8}$ b. $\frac{1}{9}$ c. $\frac{y}{x}$ d. $\frac{1}{6g}$ e. $\frac{10}{2}$ or $\frac{5}{1}$ or 5 **P-2** a. $\frac{k}{5}$ b. $\frac{2}{t}$ c. t d. $\frac{7}{3w}$ **P-3** a. $\frac{1}{18p}$ b. $\frac{1}{6n}$ c. $\frac{2}{7}$ d. $\frac{1}{15gh}$ **P-4** a. $\frac{20y}{7}$ b. 4 c. $\frac{8b^2}{27c}$ d. $\frac{12nk}{m}$

Page 94 Classroom Exercises 1. $\frac{3}{4} \cdot \frac{3}{y}$ 3. $\frac{4y}{3} \cdot \frac{8}{1}$ 5. $\frac{3}{8} \cdot \frac{1}{y}$ 7. $\frac{4a^2}{2b} \cdot \frac{1}{5b^2}$ 9. $\frac{7a^2}{8} \cdot \frac{3ab}{4}$ 11. $\frac{r}{s} \cdot \frac{t}{1}$ 13. $\frac{2}{n}$ 15. $\frac{3x}{2}$ 17. $\frac{3}{8y}$ 19. $\frac{2}{11}$ 21. $\frac{5}{4}$ 23. 2x 25. $\frac{3g}{8}$ 27. $\frac{16z^3}{5xy^3}$

Pages 94-95 Written Exercises 1. $\frac{3x}{10}$ 3. $\frac{2}{k}$ 5. $\frac{5m}{6}$ 7. $\frac{15}{2p}$ 9. $\frac{2}{3}$ 11. 8c 13. $\frac{5x}{91}$ 15. $\frac{38c}{5e}$ 17. $\frac{s}{t}$ 19. $\frac{6p}{q^2}$ 21. $\frac{s}{t^2}$ 23. $\frac{27kp}{20dt}$ 25. $\frac{3m}{16}$ 27. $\frac{2r}{3q}$ 29. $\frac{28y}{45}$ 31. $\frac{66ad}{7b^2}$ 33. $\frac{3y}{7z}$ 35. a^2bcrst^2 37. 47 seconds

Page 95 Review Capsule for Section 4.5 1. 3t 2. 12r 3. 14m 4. 18k 5. $7a^2b$ 6. $6rs^3$ 7. 2.6pq 8. $3.3x^2y$ 9. 9x + 2y 10. 8w − 4r 11. 14a − 5 12. 4t + 5v 13. $31x^2$ 14. $17rs^2 - r^2s$

Pages 96-97 **P-1** a. 10 b. 13 c. 4 d. 16 **P-2** a. 8 b. 4 c. 6 d. 6 **P-3** a. $\frac{7k}{11}$ b. $\frac{b}{9}$ c. $\frac{9c}{7r}$ d. $\frac{5r}{4s}$ **P-4** a. $\frac{t+u}{6}$ b. $\frac{3m+4n}{5}$ c. $\frac{7p-5q}{8r}$ d. $\frac{19j-11k}{20z}$ **P-5** a. $\frac{3g}{5}$ b. $\frac{10s}{3t}$ c. $\frac{1}{y^2}$ d. $\frac{2a^2b}{5c}$

Page 98 Classroom Exercises 1. $\frac{2}{3}$ 3. $\frac{5}{x}$ 5. $\frac{3}{n}$ 7. $\frac{9s}{17}$ 9. $\frac{3a}{4}$ 11. $\frac{x+y}{3}$ 13. $\frac{4-2r}{9}$ 15. $\frac{12-m}{n}$ 17. $\frac{6p-3q}{5}$ 19. $\frac{5+5a}{6z}$ 21. $\frac{11y}{2x}$ 23. $\frac{3y}{8d}$ 25. $\frac{11f}{w}$ 27. $\frac{17m}{5jk}$ 29. $\frac{14c}{s^2}$

Pages 98-99 Written Exercises 1. $\frac{8}{11}$ 3. $\frac{19}{y}$ 5. $\frac{2r}{9}$ 7. $\frac{13}{16}$ 9. $\frac{5}{3x}$ 11. $\frac{a+b}{5}$ 13. $\frac{2y-x}{3}$ 15. $\frac{4b-k}{15}$ 17. $\frac{4p+5q}{n}$ 19. $\frac{6s-5t}{7w}$ 21. $\frac{b}{4}$ 23. $\frac{4x}{5}$ 25. $\frac{1}{6x}$ 27. $\frac{3r}{4s}$ 29. $\frac{2k}{3m^2}$ 31. $\frac{8h}{11}$ 33. $\frac{3m+2n}{5}$ 35. $\frac{5r^2}{g}$ 37. $\frac{8w}{13}$ 39. $\frac{5m}{9n}$ 41. $\frac{k+7}{9}$ 43. $\frac{5t}{8r^2}$ 45. $\frac{3mn^2}{p}$ 47. $\frac{a}{n}$ 49. $\frac{3r^2}{s}$ 51. If the value of the fraction is less than 1, the numerator must be smaller than the denominator. Thus the value of a + b must be less than 7. 53. $\frac{2(3)}{3(2)} = 1$; $\frac{2(6)}{3(4)} = 1$; $\frac{2(9)}{3(6)} = 1$; An infinite number of values for a and b are possible. 55. r = 3

Page 99 Review Capsule for Section 4.6 1. 18 2. 63 3. 20 4. 15 5. 560 6. 180 7. 60 8. 56 9. 12 10. 60 11. 180 12. 63 13. 70 14. 24 15. 60 16. 90

Pages 100-102 **P-1** a. $\frac{3}{2}$ b. $\frac{1}{8}$ c. $\frac{1}{3}$ d. $\frac{33}{20}$ **P-2** a. $\frac{4}{4}$ b. $\frac{3}{3}$ c. $\frac{2}{2}$ d. $\frac{6}{6}$ **P-3** a. $\frac{p}{10}$ b. $\frac{5b}{6}$ c. $\frac{h}{2}$ d. $\frac{m}{40}$ **P-4** a. $\frac{1}{2y}$ b. $\frac{11}{6j}$ c. $\frac{3}{10x}$ d. $\frac{11}{12v}$

Page 102 **Classroom Exercises** 1. 8 3. 6 5. 18 7. 24 9. 30r 11. $\frac{2}{2}$ 13. $\frac{4}{4}$ 15. $\frac{4}{4}$ 17. $\frac{3}{3}$ 19. $\frac{3}{4}$ 21. $\frac{1}{6}$ 23. $\frac{1}{9}$ 25. $\frac{x}{8}$ 27. $\frac{11y}{12}$ 29. $\frac{7}{3x}$ 31. $\frac{9}{8y}$ 33. $\frac{25}{14a}$

Pages 102-103 **Written Exercises** 1. 5; 3 3. 10; 3 5. $\frac{7}{12}$ 7. $\frac{7}{16}$ 9. $\frac{8x}{15}$ 11. $\frac{13d}{10}$ 13. $\frac{18m}{35}$ 15. $\frac{5t}{12}$ 17. $\frac{13n}{18}$ 19. $\frac{5}{2n}$ 21. $\frac{7}{5x}$ 23. $\frac{1}{4s}$ 25. $\frac{3}{4y}$ 27. $\frac{7}{10q}$ 29. $\frac{31}{12}$ 31. $\frac{11}{6r}$ 33. $\frac{2z}{63}$ 35. $\frac{8}{3w}$ 37. $\frac{11n}{20}$ 39. $\frac{7}{15q}$ 41. $\frac{9p}{10}$ 43. $\frac{53t}{45}$ 45. $\frac{59}{10s}$ 47. $\frac{1}{20b}$ 49. 59 minutes

Page 103 **Review Capsule for Section 4.7** 1. $\frac{11}{4}$ 2. $\frac{15}{2}$ 3. $\frac{37}{16}$ 4. $\frac{47}{12}$ 5. $\frac{61}{5}$ 6. 80 7. 50 8. 306 9. 195 10. 1225

Page 105 **Classroom Exercises** 1. $18\frac{1}{4}$ in 3. 14 yd 5. $20\frac{1}{4}$ in^2 7. 183 yd^2

Page 106 **Written Exercises** 1. $16\frac{1}{4}$ inches 3. $8\frac{7}{12}$ feet 5. 28 square feet 7. $108\frac{1}{3}$ square feet 9. $9\frac{4}{5}$ miles 11. $52\frac{1}{2}$ square inches

Pages 107-108 **Chapter Review** 1. $\frac{6}{5}$ 3. $\frac{4}{7}$ 5. $\frac{7}{9}$ 7. $\frac{1}{14}$ 9. $\frac{5}{4s}$ 11. $\frac{1}{5x}$ 13. $\frac{7a}{12b}$ 15. $\frac{7}{15}$ 17. 9r 19. $\frac{10m}{9a}$ 21. $\frac{5}{3b}$ 23. $\frac{1}{4}$ 25. $\frac{3}{20b}$ 27. $\frac{2a^2c}{15}$ 29. $\frac{3rt^2}{2}$ 31. $\frac{15}{17}$ 33. $\frac{9x}{11}$ 35. $\frac{3}{4y}$ 37. $\frac{x}{2}$ 39. $\frac{11}{3t}$ 41. $\frac{1}{15n}$ 43. $5505\frac{1}{2}$ square yards

CHAPTER 5 SOLVING EQUATIONS

Pages 110-111 **P-1** a, b, and c **P-2** x = 9 **P-3** n = 8.2

Page 112 **Classroom Exercises** 1. x 3. b 5. n 7. k 9. 12 11. 112 13. x + 5 − 5 = 19 − 5 15. 105 − 92 = m + 92 − 92 17. t + 1.8 − 1.8 = 2.3 − 1.8 19. 184 − 102 = c + 102 − 102 21. x + 12.5 − 12.5 = 21.3 − 12.5 23. $b + \frac{1}{2} - \frac{1}{2} = 19 - \frac{1}{2}$

Pages 112-113 **Written Exercises** 1. n = 7 3. y = 17 5. t = 24 7. a = 17 9. q = 27 11. h = 49 13. k = 1.2 15. y = 3.5 17. t = 3.1 19. r = 11.5 21. y = 0.38 23. m = 21.74 25. a = 13 27. y = 2131 29. x = 6 31. x = 1.07 33. b = 8.15 35. $m = \frac{1}{2}$ 37. $5 39. $2.86 41. $5.45 43. $0.88 45. $32

Page 113 **Review Capsule for Section 5.2** 1. 23 2. 8.3 3. 98 4. 107 5. 10 6. 45 7. 18 8. 42 9. 102

Pages 114-115 **P-1** 5 **P-2** a, b, and c **P-3** $x = 18$ **P-4** $x = 7.8$

Pages 115-116 **Classroom Exercises** 1. r 3. x 5. n 7. b 9. $3\frac{1}{4}$ 11. $a - 12 + 12 = 5 + 12$ 13. $47 + 83 = t - 83 + 83$ 15. $12.9 + 5.6 = b - 5.6 + 5.6$ 17. $w - 3 + 3 = 21 + 3$ 19. $8.3 + 15.9 = t - 15.9 + 15.9$ 21. $q - \frac{3}{4} + \frac{3}{4} = 6\frac{1}{2} + \frac{3}{4}$ 23. Yes

Pages 116-117 **Written Exercises** 1. $x = 58$ 3. $t = 101$ 5. $n = 203$ 7. $a = 392$ 9. $m = 338$ 11. $y = 4107$ 13. $s = 10.1$ 15. $w = 1.13$ 17. $g = 48$ 19. $r = 24.37$ 21. $c = 20.5$ 23. $p = 110.18$ 25. $r = 127$ 27. $n = 43$ 29. $k = 373$ 31. $t = 583$ 33. $y = 3.05$ 35. $b = 1110$ 37. $q = 1795$ 39. $h = 5\frac{3}{8}$ 41. \$54 43. \$485 45. $p - c = m$ 47. $p - a - c = b$

Page 117 **Review Capsule for Section 5.3** 1. 8 2. 15 3. 17 4. 27 5. 4 6. 1 7. x 8. y 9. a 10. n 11. r 12. p

Pages 118-119 **P-1** a, b, and c **P-2** 8 **P-3** a. $n = 4\frac{4}{5}$ or $n = 4.8$ b. $n = 0.3$ c. $n = 4$

Pages 119-120 **Classroom Exercises** 1. x 3. t 5. k 7. 35 9. $\frac{4n}{4} = \frac{32}{4}$ 11. $\frac{44}{11} = \frac{11y}{11}$ 13. $\frac{25a}{25} = \frac{75}{25}$ 15. $\frac{1.8}{2.4} = \frac{2.4x}{2.4}$

Pages 120-121 **Written Exercises** 1. $x = 9$ 3. $n = 5$ 5. $y = 32$ 7. $q = 5$ 9. $r = 8$ 11. $k = 3\frac{3}{4}$ 13. $m = 9$ 15. $s = 1.2$ 17. $x = 12$ 19. $t = 24$ 21. $n = 125$ 23. $k = 13$ 25. $s = 2\frac{1}{3}$ 27. $a = 6\frac{1}{2}$ 29. $h = 65$ 31. $x = 1\frac{1}{2}$ 33. 120.4 miles per hour 35. 254.5 hours

Page 121 **Mid-Chapter Review** 1. $r = 47$ 3. $k = 36$ 5. $t = 0.47$ 7. \$0.81 9. $y = 66$ 11. $p = 58$ 13. $a = 142.2$ 15. \$98 17. $q = 19$ 19. $x = 540$ 21. $s = 5\frac{1}{3}$ 23. 230 miles per hour

Pages 122-123 **P-1** a, b, and c **P-2** a. $t = 54$ b. $m = 200$ c. $s = 45$ d. $r = 9.6$

Page 124 **Classroom Exercises** 1. a 3. r 5. h 7. $7(23) = \frac{s}{23}(23)$ 9. $\frac{w}{14}(14) = 19(14)$ 11. $x = 10$ 13. $n = 16$ 15. $a = 20$ 17. $m = 99$ 19. $p = 5$ 21. $c = 10$

Pages 124-125 **Written Exercises** 1. $t = 136$ 3. $x = 276$ 5. $w = 882$ 7. $p = 54$ 9. $a = 72$ 11. $s = 747$ 13. $w = 29.6$ 15. $x = 94$ 17. $r = 33.6$ 19. $n = 234$ 21. $q = 270$ 23. $a = 896$ 25. $m = 48$ 27. $d = 207$ 29. $t = 768$ 31. $w = 18\frac{2}{3}$, or $w = 18.\overline{6}$ 33. 408 miles 35. 750 kilometers 37. a. $d = 400$; $d = 800$ b. d doubles. c. d also doubles.

Pages 127-128 **Classroom Exercises** 1. c 3. e 5. b 7. $n - 20$ 9. $n - 5$ 11. $n + 5$ 13. $n - 15$ 15. $3n$ 17. $15 - n$ 19. $\frac{n}{42}$ 21. a

Pages 128-129 Written Exercises 1. $n + 12$ 3. $n - 2$ 5. $n - 15$ 7. $n - 120$ 9. $n - 3$ 11. $2n$ 13. $\frac{n}{500}$ 15. $5 - n$ 17. $2n + 7$ 19. $16n - 8$ 21. $\frac{n}{100} - 15$ 23. $800 - n + 425$ 25. $n + 36$ 27. $n - 25$ 29. $25 - n$ 31. $2n - 850$ 33. $7(y - x)$ 35. $\frac{y}{7} - x$

Pages 131-132 Classroom Exercises 1. h 3. d 5. c 7. g 9. $n + 75 = 108$ 11. $m + 13 = 20$ 13. $t - 7 = 76$ 15. $\ell - 5 = 14$ 17. $22c = 220$ 19. $6h = 72$ 21. $\frac{d}{16} = 24$

Pages 132-133 Written Exercises 1. a. Let s = the number of sailboats. b. $s + 4 = 15$ 3. a. Let c = the cost of a ticket. b. $c + 2 = 6$ 5. a. Let a = the number of animals. b. $a - 14 = 57$ 7. a. Let m = the number of millimeters of rain. b. $m - 3.2 = 24.5$ 9. a. Let n = the number. b. $6n = 42$ 11. a. Let k = the number of kilometers walked. b. $2k = 12.4$ 13. a. Let a = the amount of the loan. b. $\frac{a}{12} = 72$ 15. a. Let c = the number of cars. b. $\frac{c}{18} = 5$ 17. $6p = 54$ 19. $\frac{m}{12} = 18$ 21. $w - 16 = 33$ 23. 10 boxes can be shaded. Designs may vary. A sample design is shown at the right.

23.

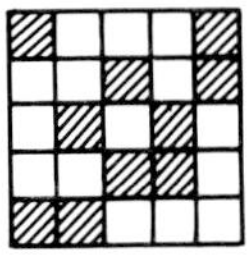

Page 133 Review Capsule for Section 5.7 1. $g = 35$ 2. $r = 126$ 3. $f = 26.66$ 4. $n = 5$ 5. $y = 97$ 6. $k = 21$ 7. $v = 75$ 8. $h = 6\frac{6}{7}$

Page 136 Classroom Exercises 1. d 3. c 5. $x + 7 = 5$ 7. $t + 7 = 18$ 9. $28 - n = 13$ 11. $\frac{10}{n} = 42$

Pages 136-137 Written Exercises 1. \$496 3. 134 5. 75 tickets 7. 84 inches 9. 9 miles 11. 97 students 13. 28 days

Pages 138-139 1. WILL ARRIVE NEW YORK FRIDAY 3. DEPOSIT \$250,000 IN ACCOUNT 10043

Pages 140-142 Chapter Review 1. $t = 45$ 3. $x = 11$ 5. $y = 12.8$ 7. \$22 9. $a = 82$ 11. $k = 96.4$ 13. $m = 612$ 15. \$8.95 17. $x = 9$ 19. $t = 4\frac{1}{2}$, or $t = 4.5$ 21. $a = 430$ 23. 789 seconds 25. $x = 133$ 27. $t = 216$ 29. $r = 35.1$ 31. 384 miles 33. $n + 25$ 35. $2n - 5$ 37. $\frac{n}{45}$ 39. Let p = the number of passengers. $p + 7 = 146$ 41. Let y = the number of years. $y - 15 = 18$ 43. 15 cats 45. 161 students

CHAPTER 6 EQUATIONS AND PROBLEM SOLVING

Pages 144-145 P-1 a. $x = 4$ b. $x = 6$ c. $x = 12$ **P-2** a. $x = 7$ b. $x = 2$ c. $x = 4$ **P-3** a. $x = 90$ b. $x = 114$ c. $x = 760$

Pages 145-146 Classroom Exercises 1. addition 3. subtraction 5. addition 7. $4n - 1 + 1 = 5 + 1$ 9. $13 + 5 = 12s - 5 + 5$ 11. $21.8 - 5.6 = 8a + 5.6 - 5.6$ 13. $19 + 2\frac{1}{2} = 3b - 2\frac{1}{2} + 2\frac{1}{2}$ 15. $3.1 - 0.9 = 0.7x + 0.9 - 0.9$ 17. $\frac{x}{5} - 2 + 2 = 13 + 2$ 19. Yes 21. No 23. No

Pages 146-147 Written Exercises 1. $x = 8$ 3. $x = 8\frac{1}{2}$ 5. $x = 8$ 7. $x = 14$ 9. $x = 12$ 11. $x = 3.9$ 13. $x = 105$ 15. $x = 216$ 17. $x = 97.6$ 19. $x = 9$ 21. $x = 60$ 23. $x = 2$ 25. $x = 12$ 27. $x = 13$

29. $x = 1\frac{5}{6}$ 31. $x = 1$ 33. $x = 6\frac{1}{11}$ 35. $x = 20$ 37. 82 points 39. \$34 41. $\frac{6}{5}$, or 1.2

Page 147 **Review Capsule for Section 6.2** 1. $5n$ 2. $8t$ 3. $1.3r$ 4. $16a - 11$ 5. $3 + 7y$ 6. $7m - 6$ 7. $3y + 12$ 8. $2x + 6$ 9. $2a + 1$ 10. 13 11. 6 12. 6 13. 7 14. 28 15. 11 16. 19 17. 8 18. 32

Pages 148-149 **P-1** a. $y = 2$ b. $t = 4$ **P-2** a. $x = 3$ b. $p = 5$ **P-3** a. $r = 31$ b. $y = 3$

Pages 149-150 **Classroom Exercises** 1. $20r + 3 = 10$ 3. $7a + 11 = 27$ 5. $56 = 12w - 4$ 7. $2n - 6 = 28$ 9. $3b + 6 = 19$ 11. $6t - 3 = 15$ 13. $108 = 17t + 3$ 15. $33 = 15m - 12$ 17. $4p + 3.5 = 24.5$ 19. $6t + 1.7 = 49.7$ 21. $14 = 8r - 2$ 23. $1.3p = 5.2$

Page 150 **Written Exercises** 1. $x = 7$ 3. $x = 9$ 5. $x = 9$ 7. $x = 14$ 9. $x = 27$ 11. $t = 15$ 13. $q = 9$ 15. $r = 17$ 17. $t = 6$ 19. $d = 2$ 21. $x = 15$ 23. $t = 5$ 25. $x = 5$ 27. $y = 6$ 29. $t = 4.5$ 31. $n = 27$

Page 151 **Review Capsule for Section 6.3** 1. $m = 12$ 2. $x = 11$ 3. $n = 30\frac{1}{3}$, or $n = 30.\overline{3}$ 4. $p = 22$ 5. $f = 101$ 6. $b = 42$

Pages 152-153 **Classroom Exercises** 1. t = smallest number; t + 1 = second number; t + 2 = largest number 3. t = smallest number; t + 2 = second number; t + 4 = third number; t + 6 = largest number 5. $h + (h + 1) = 5$ 7. $h + (h + 2) = 28$

Page 153 **Written Exercises** 1. 12 and 13 3. 33 and 34 5. 42, 43, and 44 7. 48 and 50 9. 33, 35, and 37 11. 19, 21, and 23 13. 20 and 21 15. 120 and 122 17. 102, 103, and 104

Page 155 1. 75 feet 3. 119 feet 5. About 98 feet 7. The breaking distance increases.

Page 156 **Mid-Chapter Review** 1. $m = 5$ 3. $r = 12$ 5. $t = 6$ 7. 16 9. $n = 7$ 11. $r = 2\frac{4}{9}$, or $r = 2.\overline{4}$ 13. $d = 14$ 15. 36 and 37 17. 85, 87, and 89

Page 156 **Review Capsule for Section 6.4** 1. $9 - 3t$ 2. $6b + 24$ 3. $10c - 15$ 4. $72 + 18s$ 5. $54 + 9y$ 6. $60x - 144$ 7. t 8. $11s$ 9. $16b + 6$ 10. $4g - 9$ 11. $4d - 5$ 12. $4n + 4$ 13. 32 14. 78 15. 38

Page 157 **P-1** $x = 5$ **P-2** $b = 4$

Page 158 **Classroom Exercises** 1. $7n + 14 = 56$ 3. $20n - 45 = 144$ 5. $4a + 3 = 42$ 7. $40 = 9r - 6$ 9. $2x - 6 + x = 9$ 11. $19 = 20x - 4 - 3x$ 13. $7x + 3x - 15 = 42$ 15. $24 = 7t - 4 + 3.5t$

Pages 158-159 **Written Exercises** 1. $x = 4\frac{1}{2}$ 3. $x = 14$ 5. $x = 8$ 7. $x = 6\frac{2}{3}$ 9. $x = 7$ 11. $x = 9$ 13. $x = 48$ 15. $x = 54$ 17. $x = 10\frac{1}{3}$ 19. $x = 8$ 21. $s = 3\frac{1}{4}$ 23. $p = 16\frac{7}{8}$ 25. $x = 7$ 27. $x = 5$ 29. 20 31. The value of $(d + e)$ is twice the value of $(b + c)$.

Page 159 **Review Capsule for Section 6.5** 1. $p = 4 + 5 + 7$; $p = 16$ 2. $p = 9 + 15 + 13$; $p = 37$ 3. $p = 12.1 + 13.5 + 15.6$; $p = 41.2$ 4. $p = 3.2 + 6.5 + 9.4$; $p = 19.1$ 5. $p = 9 + 2x + x + 5$; $p = 3x + 14$ 6. $p = s + 3s + s - 8$; $p = 5s - 8$ 7. $p = 2(8 + 12)$; $p = 40$ 8. $p = 2(16 + 6)$; $p = 44$ 9. $p = 2(5.8 + 0.9)$; $p = 13.4$ 10. $p = 2(12.5 + 13.2)$; $p = 51.4$ 11. $p = 2(q + q + 3.2)$; $p = 4q + 6.4$ 12. $p = 2(2d + 7.8 + d)$; $p = 6d + 15.6$

Pages 161-162 Classroom Exercises 1. x = shortest side; $x + 8$ = longest side; $x + 4$ = third side 3. x = shortest side; $x + 9$ = longest side; $\frac{1}{2}(x + 9)$ = third side 5. $60 = d + d + 1 + 2d + 5$ 7. $72 = 2m - 1 + m + 3m$ 9. w = width; $w + 4$ = length 11. w = width; $10w - 7$ = length 13. $2(w + 8 + w) = 49$ 15. $2(\ell + \ell - 5) = 35$

Pages 162-164 Written Exercises 1. 19 units, 19 units, and 22 units 3. 798 kilometers 5. width: 8 units; length: 14 units 7. length: 457 meters; width: 341 meters 9. distance from home plate to first base = distance from first base to second base = 90 feet; distance from home plate to second = 127.3 feet 11. distance from A to C = distance from B to C = 781.6 meters; distance from A to B = 366.6 meters 13. length of each sloping side = 15 feet

Page 166 Written Exercises 1. You do not need to know that the unknown number is even; the number is 224. 3. You cannot solve the problem because you need to know the relationship of the legs. 5. You cannot solve the problem because you need to know the distance from the sun to Venus. 7. You do not need to know that each number is two more than the preceding number; the numbers are 21, 23, and 25. 9. You do not need to know that one of the numbers is even and the other is odd; the numbers are 27 and 28. 11. You cannot solve the problem because you need to know the distance the skater traveled.

Pages 167-168 1. Ann: banana; Beth: apple; Carol: orange

3.

	Teacher	Detective	Mayor
Jim	✓	x	x
Sarah	x	x	✓
Jane	x	✓	x

Jim: Teacher

Sarah: Mayor

Jane: Detective

Pages 169-170 Chapter Review 1. $x = 19$ 3. $w = 40$ 5. $a = 518$ 7. 36 hours 9. $x = 7\frac{3}{5}$ 11. $t = 17\frac{1}{2}$ 13. $w = 6$ 15. 83, 84, and 85 17. $x = 5$ 19. $r = 5\frac{1}{4}$ 21. first leg: 5.0 kilometers; second leg: 2.6 kilometers; third leg: 5.2 kilometers 23. You cannot solve the problem because you need to know the average annual rainfall. You do not need to know the average windspeed.

Pages 171-172 Cumulative Review: Chapters 1-6 1. 31 3. $48t$ 5. $132a^2b$ 7. $27r + 36$ 9. $12p^3 - 10p^2$ 11. $13t^2$ 13. $v^2 + 3v + 2$ 15. $2^3 \cdot 11$ 17. $3^2 \cdot 5^2$ 19. 84 21. 90 23. $\frac{2}{11}$ 25. $\frac{2a}{5b^2}$ 27. $\frac{3x}{2}$ 29. $\frac{2}{3ab^2}$ 31. $\frac{13a - 9h}{4c}$ 33. $\frac{25x}{24}$ 35. $22\frac{5}{8}$ inches 37. $n = 9$ 39. $x = 38$ 41. $b = 2.4$ 43. $x = 49$ 45. $9n$ 47. Let m = the number of miles; $2m = 92$ 49. \$338 51. $w = 29$ 53. $m = 80$ 55. $n = 23$ 57. 73, 75, and 77 59. You do not need to know the car's fuel economy. The trip took $6\frac{1}{2}$ hours.

CHAPTER 7 RATIO, PROPORTION, AND PER CENT

Pages 174-175 P-1 a. $\frac{2}{3}$ b. $\frac{3}{4}$ c. $\frac{1}{4}$ d. $\frac{7}{8}$ e. $\frac{2}{9}$ f. $\frac{2}{7}$ **P-2** a. $\frac{1}{4}$ b. $\frac{2}{3}$ c. $\frac{2}{5}$ d. $\frac{5}{3}$ **P-3** a. $\frac{2}{5}$ b. $\frac{2}{7}$ c. $\frac{6}{5}$ d. $\frac{2}{3}$

Page 175 **Classroom Exercises** 1. $\frac{3}{4}$ 3. $\frac{10}{7}$ 5. $\frac{2}{5}$ 7. $\frac{5}{1}$ 9. $\frac{1}{10}$ 11. $\frac{12}{5}$ 13. $\frac{3}{5}$ 15. $\frac{7p}{2}$ 17. $\frac{2}{5y}$ 19. $\frac{4}{1}$

Pages 175-176 **Written Exercises** 1. $\frac{5}{9}$ 3. $\frac{4}{9}$ 5. $\frac{1}{7}$ 7. $\frac{4}{21}$ 9. $\frac{7}{2}$ 11. $\frac{5}{6}$ 13. $\frac{5r}{2}$ 15. $\frac{10}{7}$ 17. $\frac{3}{7}$ 19. $\frac{5}{12}$ 21. $\frac{1}{2}$ 23. $\frac{6g}{5}$ 25. $\frac{20}{11d}$ 27. $\frac{3}{10}$ 29. $\frac{3}{2}$ 31. $\frac{8}{17b}$

Pages 177-178 **P-1** a. $\frac{2}{10}$, or $\frac{1}{5}$ b. $\frac{4}{10}$, or $\frac{2}{5}$ **P-2** a. $\frac{0}{6}$, or 0 b. $\frac{6}{6}$, or 1

Pages 178-179 **Classroom Exercises** 1. 3 3. 7 5. 10 7. $\frac{3}{15}$, or $\frac{1}{5}$ 9. $\frac{7}{15}$ 11. $\frac{10}{15}$, or $\frac{2}{3}$ 13. 1 15. 3 17. $\frac{1}{6}$ 19. $\frac{3}{6}$, or $\frac{1}{2}$

Pages 179-180 **Written Exercises** 1. $\frac{3}{12}$, or $\frac{1}{4}$ 3. $\frac{5}{12}$ 5. $\frac{7}{12}$ 7. 0 9. $\frac{4}{20}$, or $\frac{1}{5}$ 11. $\frac{8}{20}$, or $\frac{2}{5}$ 13. $\frac{8}{20}$, or $\frac{2}{5}$ 15. $\frac{5}{20}$, or $\frac{1}{4}$ 17. 0 19. $\frac{6}{20}$, or $\frac{3}{10}$ 21. $\frac{5}{20}$, or $\frac{1}{4}$ 23. $\frac{7}{19}$ 25. $1 - n$

Page 180 **Review Capsule for Section 7.3** 1. $2\frac{2}{3}$ 2. $2\frac{3}{4}$ 3. $8\frac{1}{4}$ 4. $6\frac{5}{12}$ 5. $7\frac{2}{3}$ 6. $9\frac{5}{8}$ 7. $z = 64$ 8. $b = 18$ 9. $w = 11\frac{1}{4}$ 10. $q = 5\frac{3}{4}$

Page 182 **P-1** a. T b. F c. T d. F **P-2** a. $m = 42$ b. $y = 9\frac{3}{5}$ c. $x = 18\frac{2}{3}$ d. $w = 2\frac{2}{5}$

Page 183 **Classroom Exercises** 1. T 3. T 5. T 7. F 9. F 11. F 13. $9r = 20$ 15. $7x = 45$ 17. $4m = 27$ 19. $x = 3$ 21. $a = 1$ 23. $w = 2$ 25. $p = 4\frac{1}{3}$ 27. $m = 3\frac{3}{4}$ 29. $b = 2$

Page 184 **Written Exercises** 1. T 3. F 5. T 7. F 9. F 11. F 13. $x = 18$ 15. $n = 18$ 17. $x = 15$ 19. $p = 5$ 21. $m = 7$ 23. $k = 11\frac{1}{4}$ 25. $q = 10$ 27. $a = 6\frac{2}{5}$ 29. $m = 5$ 31. $n = 4\frac{2}{7}$ 33. $d = 10$ 35. $y = 13\frac{1}{8}$ 37. 24 games 39. 800 girls

Page 184 **Review Capsule for Section 7.4** 1. 0.08 2. 0.35 3. 0.02 4. 0.002 5. 0.005 6. 2.5 7. 0 8. 0.61 9. 0.00003 10. 0.98

Pages 185-187 **P-1** a. 12% b. 2.8% c. $12\frac{1}{2}$% d. $33\frac{1}{3}$% **P-2** a. 25% b. $37\frac{1}{2}$% c. $41\frac{2}{3}$% d. $233\frac{1}{3}$% e. 300% f. 0% **P-3** a. 4% b. 52% c. 30.1% d. 559% e. 630% **P-4** a. 0.02 b. 0.83 c. 0.169 d. 1.01 e. 0.007

Page 187 **Classroom Exercises** 1. $\frac{3}{8} = \frac{t}{100}$ 3. $\frac{2}{3} = \frac{t}{100}$ 5. $\frac{13}{15} = \frac{t}{100}$ 7. $\frac{5}{4} = \frac{t}{100}$ 9. $\frac{20}{19} = \frac{t}{100}$ 11. $\frac{20}{7} = \frac{t}{100}$ 13. $\frac{43}{100}$ 15. $\frac{87}{100}$ 17. $\frac{270}{100}$ 19. $\frac{26}{100}$ 21. $\frac{4.7}{100}$ 23. $\frac{0.13}{100}$

Page 188 Written Exercises 1. 25% 3. $56\frac{1}{4}$% 5. $83\frac{1}{3}$% 7. 120% 9. 150% 11. $131\frac{1}{4}$% 13. 40% 15. 940% 17. 20.3% 19. 0.15 21. 0.661 23. 1.01 25. 17%; $\frac{17}{100}$ 27. 0.22; $\frac{11}{50}$ 29. 95%; 0.95 31. 0.4; $\frac{2}{5}$. 33. 70%; 0.7 35. Multiply the per cent by 3.; $3 \times 12\frac{1}{2}\% = 37\frac{1}{2}\%$; $37\frac{1}{2}\% = \frac{375}{1000}$, or $\frac{3}{8}$. 37. $\frac{g}{25} = 4g\%$

Pages 190-191 Classroom Exercises 1. 25% of 36 is x. 3. 8% of 90 is g. 5. t% of 156 is 23. 7. 15% of a is 10. 9. 15% of n is 26. 11. 100% of 86 is 86. $\frac{300}{100} = \frac{k}{86}$ 13. 100% of 25.2 is 25.2. $\frac{v}{100} = \frac{70.8}{25.2}$ 15. 100% of 1 is 1. $\frac{w}{100} = \frac{\frac{1}{4}}{1}$ 17. 100% of z is z. $\frac{5.2}{100} = \frac{14}{z}$

Page 191 Written Exercises 1. 25% of 28 is 7. 3. 12% of 13 is 1.56. 5. $66\frac{2}{3}$% of 126 = 84. 7. 0.736 is 0.8% of 92. 9. 75% of 92 is 69. 11. 150 is 150% of 100. 13. 15 is $37\frac{1}{2}$% of 40. 15. 58.24 is 112% of 52. 17. 15 is 20% of 75. 19. 14 = $12\frac{1}{2}$% of 112. 21. 20 is 0.8% of 2500. 23. 84 is $83\frac{1}{3}$% of 100.8. 25. 42 is 96% of 43.75. 27. 200% of $1\frac{1}{2}$ is 3. 29. 28 = 115% of 24.35. 31. 12% of $20.8\overline{3}$ is 2.5.

Page 192 Mid-Chapter Review 1. $\frac{4}{5}$ 3. $\frac{8}{5}$ 5. $\frac{4x}{15}$ 7. $\frac{1}{6}$ 9. $\frac{5}{10}$, or $\frac{1}{2}$ 11. d = 15 13. $\frac{1}{2}$ 15. 95% 17. $87\frac{1}{2}$% 19. 192% 21. 0.25 23. 0.04 25. 3.04 27. 42% of 14 is 5.88. 29. 102 is 120% of 85.

Page 192 Review Capsule for Section 7.6 1. 3900 2. 6800 3. 1080 4. 9 5. 4550 6. 13 7. 13 8. 28 9. 109 10. 1 11. \$18 12. \$24 13. \$60 14. \$119 15. \$247

Page 194 Classroom Exercises 1. 75% of 120 is y. 100% of 120 is 120. $\frac{75}{100} = \frac{y}{120}$ 3. y% of 65 is 45.50. 100% of 65 is 65. $\frac{y}{100} = \frac{45.50}{65}$ 5. 5% of y is 2.03. 100% of y is y. $\frac{5}{100} = \frac{2.03}{y}$

Page 195 Written Exercises 1. 3 seniors 3. 5% increase 5. \$960 7. $83\frac{1}{3}$% 9. 504 miles

Page 195 Review Capsule for Section 7.7 1. 0.15d 2. 0.354z 3. 4.12m 4. 32.45s

Page 196 P-1 a. 0.09 b. 0.152 c. $0.33\frac{1}{3}$ d. 0.01q e. 0.01r **P-2** a. $0.18 \cdot 250 = c$ b. $0.01c \cdot 96 = 32$

Page 198 Classroom Exercises 1. 14% of 21 is some number. $0.14 \cdot 21 = n$ 3. 120% of 83 is some number. $1.20 \cdot 83 = n$ 5. 5% of 42 is some number. $0.05 \cdot 42 = b$ 7. 70% of 6400 is some number. $0.70 \cdot 6400 = g$ 9. Some % of 10 is 14. $0.01x \cdot 10 = 14$ 11. Some % of 15.8 is 10. $0.01r \cdot 15.8 = 10$ 13. Some % of 2600 is 500. $0.01p \cdot 2600 = 500$ 15. Some % of 6500 is 10,000. $0.01z \cdot 6500 = 10{,}000$ 17. 15% of some number is 9. $0.15n = 9$ 19. 74% of some number is 18. $0.74w = 18$

Pages 198-199 Written Exercises 1. \$3900 3. $4\frac{1}{2}$% 5. 20 students 7. 80% 9. \$306 11. \$16.00

Page 199 **Review Capsule for Section 7.8** 1. $\frac{1}{6}$ 2. $\frac{3}{4}$ 3. 1 4. $\frac{5}{4}$ 5. $\frac{13}{6}$ 6. $\frac{7}{2}$ 7. $\frac{1}{6}$ 8. $\frac{7}{12}$ 9. $\frac{1}{2}$ 10. $\frac{3}{2}$ 11. $\frac{7}{3}$ 12. $\frac{13}{4}$ 13. 0.05 14. 0.12 15. 0.10 16. 0.095 17. 0.215 18. 0.1225

Page 200 **P-1** $120

Pages 201-202 **Classroom Exercises** 1. $i = (850)(0.08)(1)$ 3. $i = (1956)(0.085)(\frac{1}{2})$ 5. $i = (2575)(0.095)(2)$ 7. $5 9. $12 11. $5.50 13. $3 15. $20 17. $50 19. $15

Pages 202-203 **Written Exercises** 1. $40 3. $8.50 5. $11.20 7. $15 9. $80 11. $39 13. $80 15. $1912.50 17. $2945 19. $412.50 21. $3420 23. 6% 25. Answers may vary. Sample answer is shown at the right. 27. 25 blocks

25. 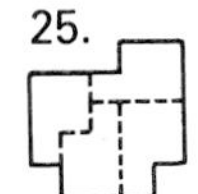

Page 203 **Review Capsule for Section 7.9** 1. 0.60 2. 0.063 3. 1.01 4. 0.09 5. 0.416 6. 2.50 7. 0.04 8. 0.25 9. 0.085 10. 0.006 11. 0.01x 12. 0.01k

Page 204 **P-1** $35.00

Page 206 **Classroom Exercises** 1. $2.00; $1.63 3. $12.00; $12.53 5. $45.00 7. $24.00 9. 40% 11. 25%

Page 207 **Written Exercises** 1. $120; $117.80 3. $300; $298.86 5. $56,000 7. $504 9. 28% 11. 14% 13. $85; 15% 15. $400; $7\frac{1}{2}$% 17. $29.25; $165.75 19. $207.20; 20%

Page 209 1. 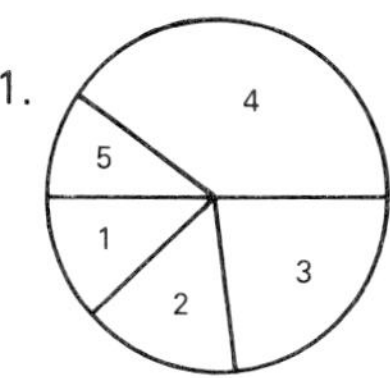

1	Warm-up:	12% ⟶	43°
2	Team Drills:	15% ⟶	54°
3	Special Drills:	23% ⟶	83°
4	Scrimmage:	40% ⟶	144°
5	Team Meeting:	10% ⟶	36°

3. About 33% 5. About 30%

Pages 211-212 **Chapter Review** 1. $\frac{1}{8}$ 3. $\frac{3}{7c}$ 5. $\frac{7y}{12}$ 7. $\frac{3}{6}$, or $\frac{1}{2}$ 9. $p = 10$ 11. $m = 20$ 13. $c = 7\frac{1}{2}$ 15. 130 people 17. 83% 19. $57\frac{1}{7}$% 21. 309% 23. 0.17 25. 0.01 27. 1.5 29. 33% of 28 is 9.24. 31. $12 = 33\frac{1}{3}$% of 36. 33. 3 is 15% of 20. 35. 60% 37. $0.75 39. $11.13

CHAPTER 8 REAL NUMBERS

Page 215 **P-1** a. $^-9, ^-7, ^-6, ^-1, 2$ b. $^-12, ^-6, 0, 9, 11$ **P-2** a. 175 b. $^-500$ c. $^-16$ d. 19 **P-3** a. $^-17 < ^-7$; $^-7 > ^-17$ b. $^-1 < 8$; $8 > ^-1$ c. $^-5 < 0$; $0 > ^-5$ d. $^-6 < ^-5$; $^-5 > ^-6$

Page 216 **Classroom Exercises** 1.-9. See below. 11. $^-7, 0, 3, 4, 8$ 13. $^-15, 0, 3, 12$ 15. $^-14, ^-9, 7, 8, 11$ 17. $^-9, ^-6, 0, 10, 12$ 19. 10 21. $^-25$ 23. 37 25. $^-6 < ^-1$; $^-1 > ^-6$ 27. $^-4 < 8$; $8 > ^-4$ 29. $^-10 < 10$; $10 > ^-10$ 31. $^-21 < ^-12$; $^-12 > ^-21$

1. 3. 5.

7.

−12 −8 −4 0 4 8 12

9.

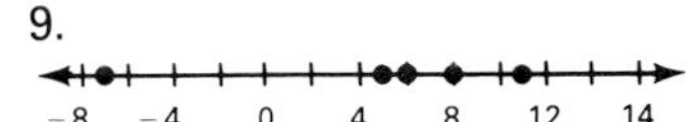

Pages 216-217 Written Exercises For the graphs for Ex. 1-17 odd, see below. 1. $^{-}3, ^{-}1, 2, 4, 6$ 3. $^{-}9, ^{-}5, 3, 5, 7$ 5. $^{-}4, ^{-}3, 0, 2, 7$ 7. $^{-}6, ^{-}4, 2, 6, 8$ 9. $^{-}7, ^{-}4, ^{-}3, 2, 7$ 11. $^{-}8, ^{-}6, 5, 7, 9$ 13. $^{-}6, ^{-}5, ^{-}3, 1, 7$ 15. $^{-}10, ^{-}7, ^{-}3, ^{-}2, 0$ 17. $^{-}7, ^{-}1, 4, 7, 8, 11$ 19. 2 21. 1000 23. $^{-}105$ 25. 17 27. $^{-}7$ 29. 17 31. $^{-}11 < 4; 4 > ^{-}11$ 33. $^{-}12 < 2; 2 > ^{-}12$ 35. $^{-}7 < ^{-}3; ^{-}3 > ^{-}7$ 37. $^{-}9 < ^{-}6; ^{-}6 > ^{-}9$ 39. $0 < 6; 6 > 0$ 41. $^{-}14 < 0; 0 > ^{-}14$ 43. $^{-}7 < ^{-}2; ^{-}2 > ^{-}7$ 45. $^{-}13 < 23; 23 > ^{-}13$ 47. $^{-}21 < 21; 21 > ^{-}21$ 49. $^{-}72 < 27; 27 > ^{-}72$ 51. 1 half-dollar, 1 quarter, 4 dimes, and 4 pennies

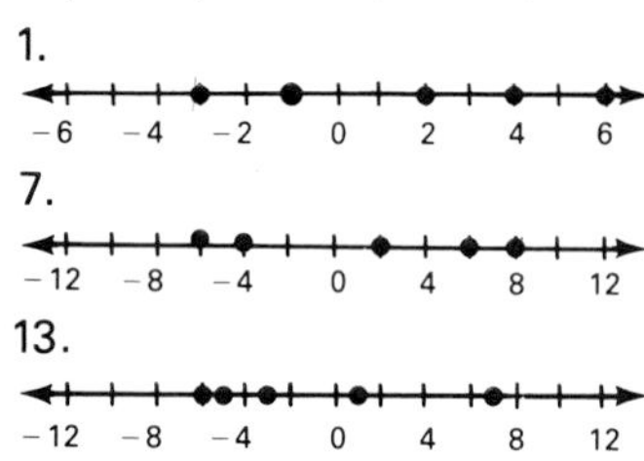

3.

−12 −8 −4 0 4 8 12

9.

−12 −8 −4 0 4 8 12

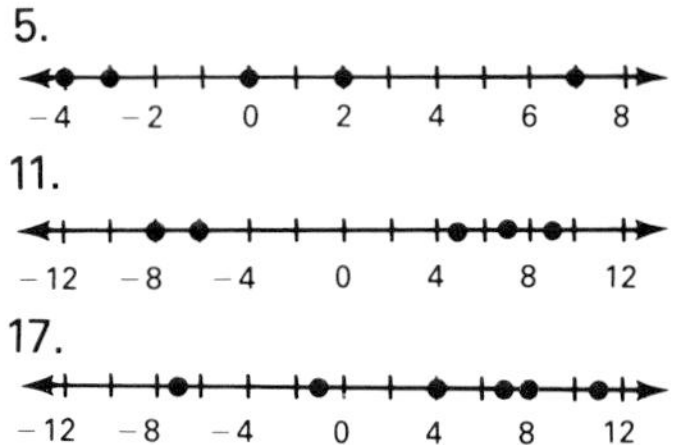

13.

−12 −8 −4 0 4 8 12

15.

−10 −8 −6 −4 −2 0 2

17.

−12 −8 −4 0 4 8 12

Page 217 Review Capsule for Section 8.2 1. $3\frac{2}{3}$ 2. $6\frac{3}{5}$ 3. $10\frac{1}{2}$ 4. $2\frac{1}{9}$ 5. $3\frac{7}{8}$ 6. $2\frac{2}{7}$ 7. $3\frac{9}{10}$ 8. $9\frac{1}{6}$ 9. $2\frac{1}{2}$ 10. $2\frac{17}{100}$ 11. $1\frac{7}{10}$ 12. $2\frac{3}{10}$ 13. $3\frac{9}{10}$ 14. $2\frac{1}{2}$ 15. $8\frac{1}{10}$ 16. $4\frac{1}{10}$ 17. $6\frac{2}{5}$ 18. $10\frac{4}{5}$ 19. $13\frac{1}{50}$ 20. $4\frac{3}{50}$

Pages 218-219 P-1 a. −11 b. 0 c. 24 d. 35 **P-2** a. $\frac{7}{4}$ b. $\frac{5}{2}$ c. $\frac{4}{1}$ d. $\frac{0}{1}$ **P-3** a. $-4 = -(4)$ b. $-(2.6) = -2.6$

Pages 219-220 Classroom Exercises 1.-9. See below. 11. −12 13. 0 15. $-\frac{3}{4}$ 17. 34 19. 9 21. $36.55 = -(-36.55)$ 23. $-8 = -(8)$ 1.

−8 −4 0 4 8

3. $-3\frac{1}{3}$ $3\frac{1}{3}$

−8 −4 0 4 8

5. −2.3 2.3

−4 −2 0 2 4

7. −2.6 2.6

−4 −2 0 2 4

9. $-\frac{3}{4}$ $\frac{3}{4}$

−4 −2 0 2 4

Page 220 Written Exercises 1. −9 3. $\frac{1}{4}$ 5. 5.3 7. 10 9. $3\frac{1}{5}$ 11. 0 13. $-2\frac{3}{8}$ 15. $\frac{7}{4}$ 17. $-1\frac{1}{4} = -(1\frac{1}{4})$ 19. $-(5) = -5$ 21. $-(500) = -500$

Page 220 Review Capsule for Section 8.3 1. 0.625 2. 0.6 3. 2.0 4. 1.25 5. 0.4375 6. $0.91\overline{6}$ 7. 3.6875 8. 6.75 9. 0.45 10. 3.5 11. 0.3 12. 0.08

Pages 221-222 P-1 a. −0.2 b. −0.125 c. −0.6 d. −0.3125 **P-2** a. −0.04 b. −0.15 c. −0.2875 d. −0.35 **P-3** a. −2.4 b. 5.1875 c. 8.0833 · · · d. −6.625 **P-4** a. $0.\overline{3}$ b. $0.\overline{39}$ c. $0.\overline{285714}$ **P-5** a. $0.1\overline{5}$ b. $-0.8\overline{3}$ c. $0.\overline{4}$ d. $0.\overline{857142}$ e. $-0.\overline{09}$

Page 223 Classroom Exercises 1. 0.0625 3. 0.875 5. 0.375 7. 0.2 9. 0.4 11. 0.25 13. −0.15 15. −0.9 17. −0.4375 19. −0.68 21. −0.38 23. −0.175 25. −4.125 27. −8.25 29. 2.8 31. 7.025 33. −10.7 35. 11.16 37. $0.\overline{428571}$ 39. $0.3\overline{571428}$ 41. $0.\overline{4}$ 43. $-0.0\overline{6}$ 45. $-0.\overline{857142}$ 47. $0.\overline{39}$

Pages 223-224 Written Exercises 1. -0.32 3. 0.225 5. 0.42 7. -0.5875 9. -0.3 11. 0.98 13. -3.25 15. -9.125 17. 1.9375 19. 6.36 21. 3.175 23. -9.65 25. $0.1\overline{6}$ 27. $-0.7\overline{3}$ 29. $-0.\overline{428571}$ 31. $0.\overline{2}$ 33. $1.\overline{5}$ 35. $0.5\overline{3}$ 37. -0.6875 39. -5.5625 41. $-0.\overline{90}$ 43. 0.12 45. 7.65 47. $2.\overline{571428}$

Page 224 Mid-Chapter Review For the graphs for Ex. 1-5 odd, see below. 1. $-8, -4, 0, 4, 8$ 3. $-8, -6, -4, 0, 2$ 5. $-14, -12, -10, -6, -2$ 7. $-3 < 3; 3 > -3$ 9. $-4 < 0; 0 > -4$ 11. $-88 < -86; -86 > -88$ 13. $-12 < 21; 21 > -12$ 15. -1700 17. 5 19. 10 21. $\frac{3}{5}$ 23. $\frac{9}{2}$ 25. 1 27. $12\frac{9}{10}$ 29. -0.6 31. $0.6\overline{81}$ 33. 17.25

Page 224 Review Capsule for Section 8.4 1. $\frac{3}{100}$ 2. $\frac{7}{100}$ 3. $\frac{13}{20}$ 4. $\frac{18}{25}$ 5. $1\frac{7}{50}$ 6. $3\frac{2}{25}$ 7. $\frac{1}{100}$ 8. $\frac{1}{25}$ 9. $\frac{9}{100}$ 10. $\frac{4}{25}$ 11. $\frac{1}{4}$ 12. $\frac{9}{25}$

Pages 225-227 P-1 a. 1 b. 144 c. 0.09 d. $7\frac{9}{16}$ **P-2** a. 196 b. 529 c. 22 d. 13 **P-3** a. $\frac{8}{9}$ b. $\frac{12}{6} = 2$ c. $\frac{2}{14} = \frac{1}{7}$ d. $\frac{0}{23} = 0$ **P-4** a. -3 b. -16 c. $-\frac{2}{5}$ d. $-\frac{15}{13}$, or $-1\frac{2}{13}$ **P-5** a. -0.4 b. -0.7 c. -1.2 d. -2.4

Page 227 Classroom Exercises 1. 27.04 3. 256 5. 0.01 7. 0.0225 9. $2\frac{1}{4}$ 11. $\frac{1}{4}$ 13. 256 15. 0 17. 441 19. 324 21. 361 23. 49 25. $(\frac{2}{3})(\frac{2}{3})$ 27. $(\frac{1}{5})(\frac{1}{5})$ 29. $(\frac{6}{7})(\frac{6}{7})$ 31. $(\frac{4}{11})(\frac{4}{11})$ 33. $(\frac{2}{15})(\frac{2}{15})$ 35. $(\frac{5}{9})(\frac{5}{9})$ 37. 12 39. -9 41. 24 43. -13 45. 20 47. $\frac{4}{5}$ 49. $(\frac{6}{10})(\frac{6}{10})$ 51. $(\frac{3}{10})(\frac{3}{10})$ 53. $(\frac{4}{10})(\frac{4}{10})$ 55. $(\frac{12}{10})(\frac{12}{10})$ 57. $(\frac{24}{10})(\frac{24}{10})$ 59. $(\frac{8}{10})(\frac{8}{10})$

Page 228 Written Exercises 1. $\frac{9}{64}$ 3. 144 5. $\frac{1}{49}$ 7. 3.24 9. 0.0144 11. 0.0064 13. $\frac{4}{3}$ 15. $\frac{5}{6}$ 17. $\frac{7}{10}$ 19. $\frac{2}{11}$ 21. $\frac{13}{20}$ 23. $\frac{1}{2}$ 25. -13 27. $-\frac{4}{5}$ 29. -24 31. $-\frac{5}{9}$ 33. $-\frac{12}{13}$ 35. $-\frac{6}{19}$ 37. -0.7 39. -0.8 41. -0.2 43. -0.3 45. -2.5 47. -1.1 49. $-\frac{3}{4}$ 51. $\frac{4}{25}$ 53. $\frac{6}{15}$, or $\frac{2}{5}$ 55. -1.4 57. $\frac{2}{23}$ 59. -2.1

Page 228 Review Capsule for Section 8.5 1. 25 2. 61 3. 265 4. 250 5. 549 6. 13 7. 4.25 8. $\frac{26}{16}$, or $\frac{13}{8}$, or $1\frac{5}{8}$

Pages 230-231 P-1 a. Yes b. No **P-2** a. 3 b. 5 **P-3** a. 4 b. 15

Pages 231-232 Classroom Exercises 1. Yes 3. Yes 5. 15 7. $b = 16$ 9. $a = 24$

Pages 232-233 Written Exercises 1. No 3. Yes 5. 17 7. 24 9. 3 11. 25 feet 13. 10 kilometers

Page 233 Review Capsule for Section 8.6 1. $0.625\overline{0}$ 2. $1.8\overline{0}$ 3. $2.\overline{0}$ 4. $0.\overline{39}$ 5. $1.4\overline{0}$ 6. $4.1\overline{6}$

7. $0.\overline{0}$ 8. $7.75\overline{0}$ 9. $0.1875\overline{0}$ 10. $0.\overline{3}$ 11. $7.5\overline{0}$ 12. $0.08\overline{3}$ 13. 225 14. 2.25 15. 16 16. 0.16 17. $\frac{4}{9}$ 18. $\frac{9}{16}$

Pages 234-235 **P-1** a. rational b. irrational c. rational d. irrational **P-2** a. rational b. irrational c. rational d. rational

Page 235 **Classroom Exercises** 1. irrational 3. irrational 5. irrational 7. rational 9. irrational 11. irrational 13. rational 15. rational 17. rational 19. irrational 21. rational 23. irrational

Page 236 **Written Exercises** 1. rational; 25 is a perfect square. 3. rational; 100 is a perfect square. 5. irrational; 18 is not a perfect square. 7. irrational; 96 is not a perfect square. 9. irrational; 7 is not a perfect square. 11. rational; 529 is a perfect square. 13. irrational; $\frac{1}{10}$ is not a perfect square. 15. rational; 2.56 is a perfect square. 17. irrational; $4\frac{1}{6}$ is not a perfect square. 19. irrational; $\frac{7}{12}$ is not a perfect square. 21. rational; $\frac{9}{36}$ is a perfect square. 23. rational; 1.69 is a perfect square. 25. irrational; 5 is not a perfect square. 27. rational; $\frac{81}{25}$ is a perfect square. 29. rational; 3.61 is a perfect square. 31. irrational; 112 is not a perfect square. 33. rational; $\frac{1}{36}$ is a perfect square. 35. irrational; $3\frac{1}{5}$ is not a perfect square. 37. B, C, E 39. B, C, E 41. A, B, C, E 43. B, C, E; -17 45. C, E; $-\frac{10}{9}$ 47. C, E

Page 236 **Review Capsule for Section 8.7** 1. -9 2. -1.5 3. $\frac{1}{3}$ 4. 80 5. 4.9 6. $-6\frac{1}{4}$ 7. $12\frac{1}{8}$ 8. 0 9. $\frac{10}{3}$ 10. -0.05 11. 14.1 12. 2.7 13. $\frac{1}{7}$ 14. -0.1 15. $\frac{2}{3}$

Pages 237-238 **P-1** a. 14 b. $2\frac{3}{4}$ c. 0.8 d. 12.04 **P-2** a. 9.5 b. $5\frac{3}{4}$ c. 17

Page 238 **Classroom Exercises** 1. 8 3. 1.5 5. $\frac{3}{8}$ 7. 14.7 9. $2\frac{1}{2}$ 11. $\sqrt{21}$ 13. 3 15. 5 17. 12 19. 18.2

Page 238 **Written Exercises** 1. 0.8 3. 6 5. $3\frac{1}{2}$ 7. 16 9. 18 11. $\sqrt{11}$ 13. 9 15. 13 17. 13 19. 7 21. False; If $x < 0$, then $|x| > x$. 23. False; If the values of both x and y are negative and $x < y$, then $|x| > |y|$.

Page 239 1. 1123 cubic centimeters 3. 1584 cubic centimeters 5. The bore must be increased.

Pages 240-242 **Chapter Review** For the graphs for Ex. 1-5 odd, see p. 543. 1. $-5, -1, 0, 3, 7$ 3. $-4, -3, 0, 5, 6$ 5. $-11, -4, 2, 5, 12$ 7. $-5 < 3$; $3 > -5$ 9. $-8 < 0$; $0 > -8$ 11. $-7 < 9$; $9 > -7$ 13. $-15 < -12$; $-12 > -15$ 15. 12 17. -7.1 19. $-2\frac{1}{5}$ 21. 2.3 23. $9\frac{1}{2}$ 25. $-9.1 = -(9.1)$ 27. $-(-82.3) = 82.3$ 29. 0.45 31. 1.4375 33. $3.\overline{2}$ 35. -0.6875 37. -3.75 39. $0.\overline{12}$ 41. 196 43. 3.61 45. 289 47. 0.09 49. 121 51. 0.0225 53. $\frac{4}{9}$ 55. $-\frac{6}{11}$ 57. 1.4 59. -20 61. -0.5 63. $\frac{4}{23}$ 65. Yes 67. No 69. 20 71. 15 73. 20 feet 75. irrational; 95 is not a perfect square. 77. irrational; $2\frac{1}{3}$ is not a perfect square. 79. rational; $\frac{100}{169}$ is a perfect square. 81. irrational;

5 is not a perfect square. 83. rational; 1 is a perfect square. 85. irrational; 6.5 is not a perfect square. 87. 8 89. $3\frac{1}{5}$ 91. $11\frac{1}{9}$ 93. 9 95. 52

1.
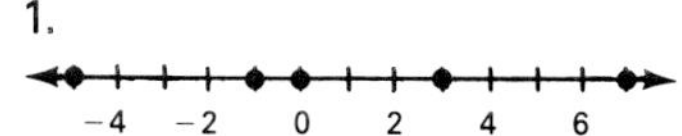

3.
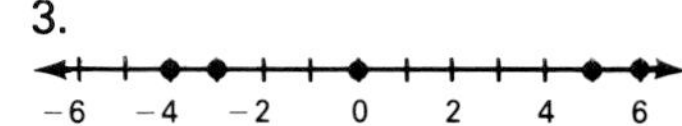

5.

CHAPTER 9 REAL NUMBERS: ADDITION AND SUBTRACTION

Pages 244-245

P-1 c. e. b. d. a.

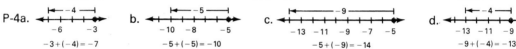

P-2a. $3+(-6)=-3$

b. $2+(-9)=-7$

c. $5+(-5)=0$

d. $12+(-6)=6$

P-3a. $-4+8=4$

b. $-1+15=14$

c. $-5+3=-2$

d. $-10+9=-1$

P-4a. $-3+(-4)=-7$

b. $-5+(-5)=-10$

c.
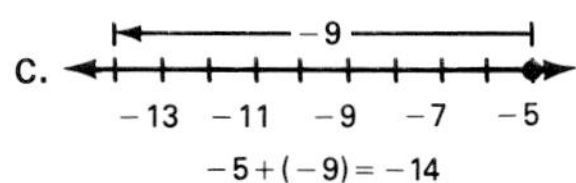

$-5+(-9)=-14$

d. $-9+(-4)=-13$

Pages 245-246 Classroom Exercises

1.
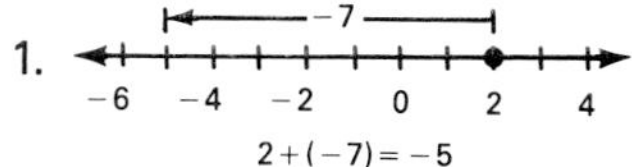

$2+(-7)=-5$

3.
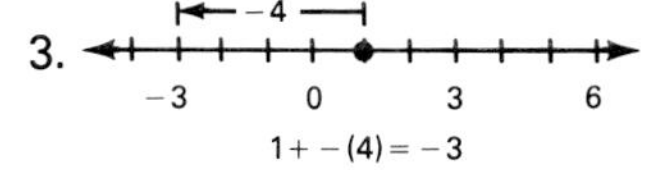

$1+-(4)=-3$

5.
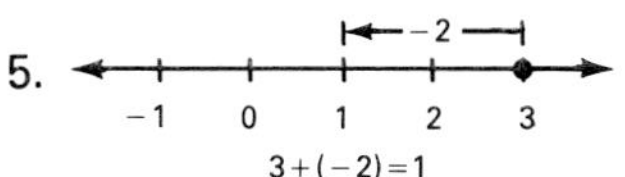

$3+(-2)=1$

7.
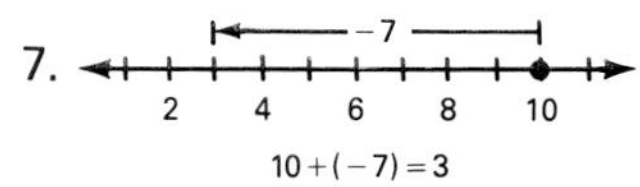

$10+(-7)=3$

9.
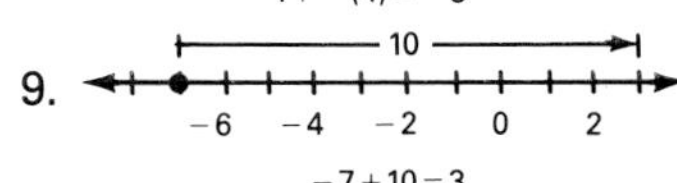

$-7+10=3$

11. $-1+3=2$

13. $-9+6=-3$

15. $-12+9=-3$

17. $-2+(-4)=-6$

19. $-3+(-3)=-6$

21.
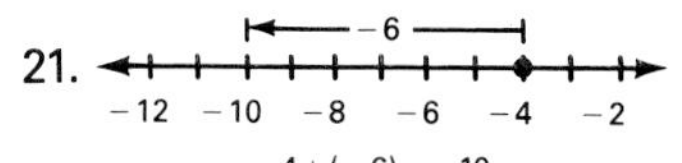

$-4+(-6)=-10$

23. $-5+(-2)=-7$

Page 246 Written Exercises

1. $3+(-5)=-2$

3.
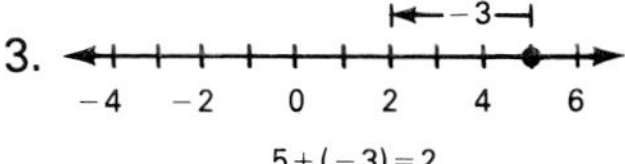

$5+(-3)=2$

5.
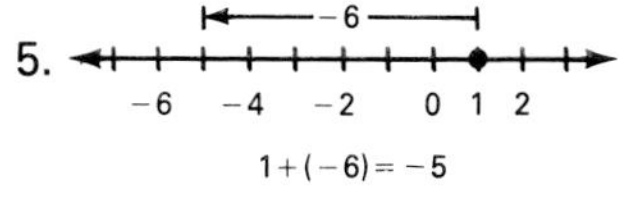

$1+(-6)=-5$

7. $7+(-2)=5$

9. $-4+7=3$

11. $-5+2=-3$

13. $-2+8=6$

15. $-7+3=-4$

17. $-5+(-4)=-9$

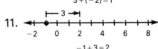

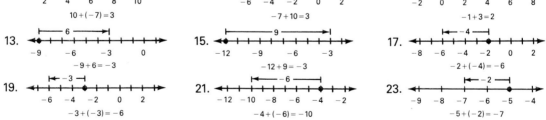

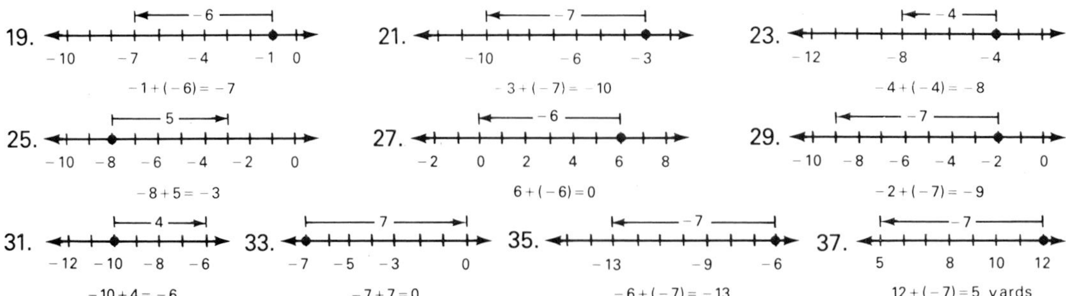

Page 247 **Review Capsule for Section 9.2** 1. −20 2. −6 3. 2 4. −75 5. −12 6. −9.6 7. $3\frac{1}{8}$ 8. 0.6 9. $-\frac{3}{10}$ 10. 15 11. 28 12. $14\frac{1}{4}$ 13. 8.5 14. $\frac{7}{2}$

Page 248 **P-1** a. −15 b. −18 c. −11 d. −14 **P-2** a. 5 b. 4 c. 2 d. 11 **P-3** a. −7 b. −2 c. −3 d. −8

Page 249 **Classroom Exercises** 1. −18 3. −21 5. −60 7. −10 9. 2 11. 11 13. 4 15. 3.6 17. −8 19. −10 21. −60 23. −3.6

Pages 249-250 **Written Exercises** 1. −16 3. −28 5. −56 7. −16.7 9. 5 11. 8 13. 1 15. 3.9 17. −5 19. −12 21. −31 23. −1.1 25. 5 27. −125 29. −5 31. 15 33. −108 35. −30 37. −11 39. 48 41. −5.7 43. $4\frac{1}{2}$ 45. −11 47. $2\frac{1}{2}$ 49. dropped 4° 51. fell $3.00

Page 250 **Review Capsule for Section 9.3** 1. b 2. c 3. a

Page 252 **P-1** a. −10 b. 13 c. 6 d. 4

Page 252 **Classroom Exercises** 1. 11 3. −10 5. 21 7. 1 9. 5 11. 1.5 13. 0 15. −3 17. −34

Pages 252-253 **Written Exercises** 1. −5 3. 2 5. −29 7. −64 9. −11 11. $-\frac{11}{16}$ 13. −10 15. 33 17. −13 19. −13 21. −4 23. −25 25. −25 27. −6.2 29. −16 31. −10.7 33. $-7\frac{1}{3}$ 35. −97 37. −0.67 39. −86 41. See the figure at the right.

41. 3 7 8 6 4 1 2 5 9

Page 255 1. Less than 10 centimeters 3. Larger 5. 6 centimeters 7. 236%

Page 257 **Classroom Exercises** 1. 1.7 centimeters excess 3. 0.3 centimeters deficiency 5. $33.50 7. $12.25

Pages 258-260 **Written Exercises** 1. 0.5 centimeters excess 3. 3.5 kilograms over 5. $44.88 7. 19 feet $11\frac{11}{16}$ inches 9. 2345 students 11. 24-car decrease 13.
15. The three persons were grandmother, mother, and daughter.

Page 260 **Mid-Chapter Review** 1.-7. See p. 545. 9. 6 11. −110 13. 56 15. −6 17. fell 9° 19. −18 21. 30 23. 11 25. 1.4 centimeters excess

1. $2+(-9)=-7$

3. $-2+4=2$

5. $-5+(-5)=-10$

7. $-3+(-7)=-10$

Pages 261-262 **P-1** a. $6+(-14)$ b. $24+(-37)$ c. $0.5+(-2.1)$ **P-2** a. -1 b. -106 c. -0.7
P-3 a. 26 b. 94 c. 20.9 **P-4** a. -4 b. -183 c. -3

Pages 262-263 **Classroom Exercises** 1. $14+(-5)$ 3. $5+14$ 5. $16+7$ 7. $0+(-17)$ 9. $13+(-19)$ 11. $2.6+14.9$ 13. -3 15. -15 17. -12 19. -2.4 21. 9 23. -17 25. 12 27. 11.6 29. -3 31. 0 33. -5 35. 0.1 37. 32 39. -9 41. 15

Pages 263-264 **Written Exercises** 1. -4 3. -7 5. -24 7. -25 9. -0.69 11. $-1\frac{3}{4}$ 13. 25 15. 53 17. 36 19. 57 21. 23.4 23. $5\frac{1}{2}$ 25. -8 27. 25 29. -1 31. 14 33. 0.8 35. $\frac{1}{8}$ 37. 0 39. 7 41. -171 43. 135 45. 134 47. 3 49. -5 51. 112 53. 1.9 55. -85 57. $-2\frac{7}{8}$ 59. 8.3 61. 4.3°C less 63. 43 yards

Page 264 **Review Capsule for Section 9.6** 1. -5; 3 2. -10; -4 3. -3; -4 4. -19; -13 5. -40; 19 6. -8; $\frac{1}{2}$ 7. $3\frac{11}{15}$; $5\frac{7}{15}$

Page 265 **P-1** a. -7 b. -37 c. -7.5

Pages 265-266 **Classroom Exercises** 1. -10 3. -25 5. -5 7. 30 9. 3.6 11. 9 13. -11 15. -2

Page 266 **Written Exercises** 1. -23 3. -8 5. -20 7. -23 9. 112 11. 6.8 13. -9 15. -31 17. -70 19. 62 21. -1.6 23. $-10\frac{1}{4}$ 25. -96 27. -42.2 29. -9 31. $-7\frac{1}{2}$

Page 266 **Review Capsule for Section 9.7** 1. c 2. c 3. a 4. a

Pages 267-268 **P-1** $-7+y$ **P-2** $10p-21$ **P-3** a. $-8n-26$ b. $-3z$ c. $-6.5-10.7c$
P-4 a. $-3t^2-3t-12$ b. $4.6-1.3f-2f^2$

Page 268 **Classroom Exercises** 1. $14+y$ 3. $u+27$ 5. $4+a$ 7. $-4.2+p$ 9. $y-4$ 11. $-x-3$ 13. $-2a-1$ 15. $1.7-3.8t$ 17. $m+1$ 19. $-3s-4$ 21. $2x-2$ 23. $4.2g-5.8$ 25. $-y^2-4$ 27. $5s-6s^2-11$ 29. -3 31. $-1.8w+3.6-8.4w^2$

Page 269 **Written Exercises** 1. $52+k$ 3. $m+4$ 5. $r+86$ 7. $-4\frac{1}{2}+t$ 9. $k-7$ 11. $-15s+7$ 13. $-7x-6$ 15. $-1.6n-3.4$ 17. $-3a-11$ 19. $-13y-43$ 21. $-15-51p$ 23. $-9\frac{3}{4}w-1\frac{1}{4}$ 25. $-13x^2-4x$ 27. $-13s-12s^2-13$ 29. $-16r^2-4r-2$ 31. $10.4n^2-13.2n-9.2$ 33. $4x-20$ 35. $-8.3t-0.5$ 37. $5w$ 39. $-8.8y-3.7$ 41. $-2a-23$ 43. $16+2t$ 45. $\frac{5}{6}w-4$ 47. $-\frac{3}{8}c+19$

Pages 270-271 1. C 3. B 5. B 7. A 9. C 11. A

Pages 272-274 **Chapter Review** 1.-11. See below. 13. −17 15. −24 17. 19 19. 39 21. −6.9 23. $-1\frac{5}{6}$ 25. 6° rise 27. −3 29. −58 31. −11 33. −47.3 35. $-1\frac{1}{4}$ 37. 1.9 centimeters deficiency 39. debit of $145.39 41. 7 43. 65 45. 4 47. 23 49. 18.1 51. $1\frac{3}{4}$ 53. 14,776 feet 55. −18 57. −74 59. −9 61. $-11\frac{2}{3}$ 63. 16 + 2s 65. $4a^2 - 16a + 6$ 67. −0.9w − 12 69. $-\frac{3}{10}x + \frac{8}{15}$

1.
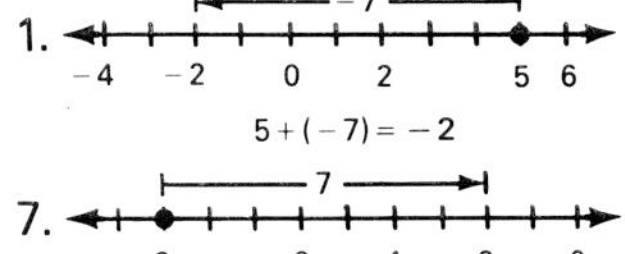

3.
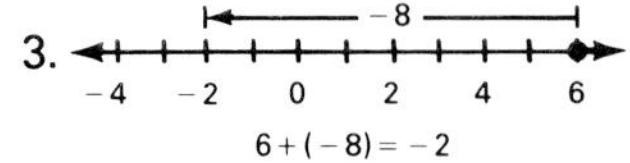

5.
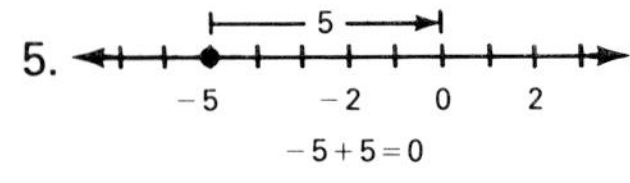

7.

9.
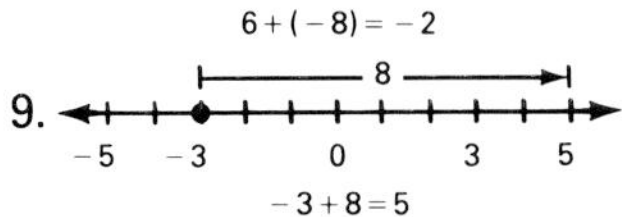

11.
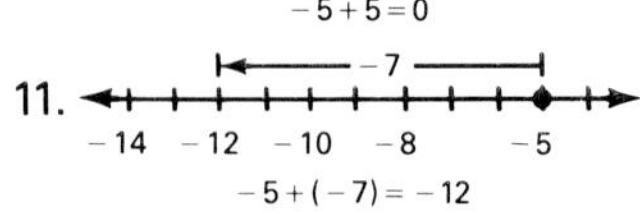

Pages 275-276 **Cumulative Review: Chapters 1-9** 1. 15 3. 56w 5. $36a^2b$ 7. 72 − 24n 9. $15x^2 + 20x$ 11. $\frac{2}{3}$ 13. $\frac{m}{9k^2}$ 15. $\frac{2}{7e}$ 17. $\frac{13y}{15}$ 19. $152\frac{1}{3}$ miles or $152.\overline{3}$ miles 21. w = 48 23. $x = 7\frac{1}{2}$ or x = 7.5 25. $\frac{6}{11}$ 27. x = 5 29. $c = 3\frac{1}{5}$ or c = 3.2 31. 160% 33. 17% 35. 120 37. 65% 39. $240 41. 8 43. −24 45. 9 feet 47. −13 49. 10 51. −118 53. $211.40 55. −47 57. −26 59. −12 61. $41a - 25a^2$

CHAPTER 10 REAL NUMBERS: MULTIPLICATION AND DIVISION

Pages 278-279 **P-1** a. −3; −6; −9 b. −6; −12; −18 c. −5; −10; −15 **P-2** a. −36 b. −84 c. −7 d. −121.6 **P-3** a. 0 b. −4 c. 0 d. −16.8

Pages 279-280 **Classroom Exercises** 1. −6 3. −20 5. −28 7. −42 9. −32 11. −27 13. −48 15. −44 17. −18 19. −36 21. 0 23. −1200 25. −16 27. −32 29. −40 31. −66 33. $-\frac{1}{3}$ 35. 0 37. −9 39. −30

Page 280 **Written Exercises** 1. −12 3. −35 5. −30 7. −63 9. −54 11. −96 13. −64 15. −54 17. −3000 19. −56 21. −96 23. −96 25. −156 27. −602 29. −5 31. −0.39 33. $-\frac{1}{10}$ 35. 0 37. −420 39. −638 41. −7 43. 0

Page 280 **Review Capsule for Section 10.2** 1. 192 2. 700 3. 1 4. 855 5. $\frac{1}{4}$ 6. 5 7. 2 8. 247 9. 8 10. 12 11. 1.7 12. 0.3 13. $2\frac{1}{4}$ 14. $7\frac{1}{8}$ 15. $\frac{5}{4}$ 16. $\frac{17}{9}$

Pages 281-282 **P-1** a. 4; 8; 12 b. 5; 10; 15 c. 5; 10; 15 **P-2** a. 56 b. 10 c. 10 d. 4

Pages 282-283 **Classroom Exercises** 1. 35 3. 9 5. 45 7. 80 9. 36 11. 18 13. 66 15. 27 17. 7 19. 14.4 21. $\frac{3}{8}$ 23. 0.54 25. 12 27. −6 29. 30 31. 40 33. 120 35. −24 37. 4 39. −2

Pages 283-284 **Written Exercises** 1. 24 3. 24 5. 84 7. 72 9. 128 11. 126 13. 108 15. 100 17. 110 19. 135 21. 10.8 23. 0.14 25. 16 27. $3\frac{17}{20}$ 29. −42 31. −48 33. −90 35. 180 37. 38 39. −30 41. −10 43. −110 45. −14.4 47. 12,600 49. 67 51. −216 53. 192 55. −90 57. $-\frac{3}{8}$

59. -2325 61. -90 63. 18,833 65. 45 67. -80 69. 7 71. 33 73. 41 75. positive 77. cannot be determined 79. zero 81. negative 83. negative 85. positive

Page 284 Review Capsule for Section 10.3 1. $48 + 6x$ 2. $10z + 5$ 3. $46 + 16y$ 4. $8t + st$ 5. $4 + \frac{1}{4}s$ 6. $2r + 4r^2$ 7. $x + 8x^2$ 8. $30x^2 + 30$ 9. -40 10. -72 11. -150 12. -84 13. -0.56 14. -20 15. -7 16. $-\frac{3}{8}$

Pages 285-286 P-1 a. Associative b. Distributive c. Commutative **P-2** a. $48k$ b. $-7cd$ c. $-27cd$ d. $44m^2n$ **P-3** a. $18n + 27$ b. $40t - 60$ c. $-2p - 8$ d. $-16k - 12$ **P-4** a. $-12p^2 - 21p$ b. $\frac{3}{4}r^2 + 9r$ c. $8x^2 - 6x$ d. $4s^2 - 6s$ **P-5** a. $24x - 18$ b. $-x^2 + 12x$ c. $6x - 2x^3$ d. $3x^3 - x^2$

Page 287 Classroom Exercises 1. $-72y$ 3. $-15rs$ 5. $-72s^3$ 7. $-2x - 10$ 9. $-n - 3$ 11. $-10x - 15$ 13. $-3r^2 - 6r$ 15. $-2k^2 + 5kt$ 17. $10w^2 - w$ 19. $5y - 30$ 21. $-3a + 6$ 23. $8x - x^2$

Pages 287-288 Written Exercises 1. $24g$ 3. $-88nt$ 5. $2rt^2w$ 7. $-16a - 10$ 9. $15r + 35$ 11. $-35 - 21x$ 13. $5x + 2x^2$ 15. $-6a^2 - 4a$ 17. $-6n^2 + 15n$ 19. $12x - 84$ 21. $-24 + 8x$ 23. $9x^2 - \frac{1}{3}x$ 25. $2a^2 + a$ 27. $-4r^2 + 12r$ 29. $45y + 18$ 31. $-x^3y^2 + 4x$ 33. $-16x^2 - 30x$ 35. $-45t^2 - 35rt$ 37. $x^2 + 3x + 2$ 39. $r^2 + 13r + 30$ 41. $x^2 + x - 2$ 43. 9 minutes

Page 289 Mid-Chapter Review 1. -2000 3. 0 5. -348 7. -78 9. -5.4 11. -36 13. 140 15. 84 17. 29 19. 84 21. 1.6 23. 832 25. $42y$ 27. $3ab$ 29. $-72x - 56$ 31. $-32m^2 + 28m$ 33. $6c^2 + 4cd$ 35. $35t - 5$ 37. $-108p + 36p^2$ 39. $9g^2 - 2\frac{1}{2}g$

Page 289 Review Capsule for Section 10.4 1. Sum 2. Sum 3. Product 4. Product 5. Sum 6. Sum 7. Product 8. Sum 9. Sum 10. Product 11. $3(x + 6)$ 12. $6(x + 2)$ 13. $4(3 + x)$ 14. $2(3 + x)$ 15. $2(2x + 1)$ 16. $12(2y + 3x)$ 17. $(3y + 2x)6$ 18. $2x(y + 6)$ 19. $2(xy + 3)$ 20. $6(2x + y)$

Pages 290-291 P-1 a. $-4(p + q)$ b. $5(h + 3g)$ c. $-q(1 + t)$ d. $-7(n + 1)$ **P-2** a. $-3p(1 + q)$ b. $12p(1 + p^2)$ c. $6h(-h + 1)$ d. $a(b - 2)$ **P-3** a. $4x(2y - 1)$ b. $3(3f^2 - g^2)$ c. $-2(1 + 2a)$ d. $-6y(2y + z)$

Pages 291-292 Classroom Exercises 1. 2 3. d 5. -7 7. m 9. $3(y + x)$ 11. $-10(r + 7)$ 13. $-4(t + 1)$ 15. $x(5a + 3)$ 17. $x(m - 2n)$ 19. $13r(t - s)$ 21. $-12(m + 1)$ 23. $15(k - 1)$ 25. $-r(4 + 1)$ or $-5r$

Page 292 Written Exercises 1. $7(x + y)$ 3. $-8(t + s)$ 5. $12(r + 1)$ 7. $5r(w + x)$ 9. $-9k(n + t)$ 11. $t(-13r + 5s)$ 13. $9x(2x - 1)$ 15. $-3b(a^2 + 3)$ 17. $s(5t - 1)$ 19. $-8(r + 1)$ 21. $-5(y + 1)$ 23. $3p(q - 3p)$ 25. $13(x - y)$ 27. $s(-34 + 1)$, or $-33s$ 29. $5n(5m + 2n)$ 31. $8(13 - 4g)$ 33. $m(1 - 12a)$ 35. $-x(1 + x)$

Page 292 Review Capsule for Section 10.5 1. 62 2. 25 3. 8 4. 34 5. 270 6. 145 7. 48 8. 24

Pages 293-294 P-1 a. -9 b. -1 c. 13 d. 0 **P-2** a. $\frac{1}{3}$ b. 5 c. $-\frac{3}{4}$ d. $-\frac{3}{10}$ e. 0 has no reciprocal. **P-3** a. -63 b. -15 c. 40 d. $\frac{4}{5}$

Pages 294-295 **Classroom Exercises** 1. −6 3. −9 5. 10 7. −13 9. 9 11. −11 13. $-\frac{1}{5}$ 15. $-\frac{8}{7}$ 17. $-\frac{2}{3}$ 19. $-\frac{2}{9}$ 21. $-\frac{4}{13}$ 23. $-\frac{10}{9}$ 25. $-16 \cdot 3$ 27. $25 \cdot (-\frac{5}{3})$ 29. $10 \cdot (-\frac{2}{5})$ 31. $-6 \cdot \frac{3}{4}$ 33. −12 35. −10 37. $-3\frac{1}{5}$ 39. $4\frac{1}{2}$ 41. 28 43. −60

Page 295 **Written Exercises** 1. −4 3. −10 5. 49 7. −17 9. −4 11. 17 13. −10 15. $-2\frac{4}{5}$ 17. $-4\frac{1}{6}$ 19. −6 21. $-\frac{1}{5}$ 23. $-\frac{2}{3}$ 25. 10 27. $\frac{1}{4}$ 29. −25 31. 19 33. $\frac{3}{4}$ 35. −30 37. −5 39. −3 41. −37 43. −14

Page 296 **Review Capsule for Section 10.6** 1. 748 2. 607 3. 581 4. 7.5 5. 7.9 6. 6.2 7. 155 8. 171 9. 0.62 10. 0.86

Page 296 **P-1** a. 0.4°C b. −0.1°C

Pages 298-299 **Classroom Exercises** 1. −0.2°C 3. −0.1°C 5. median: −2; mode: −2 7. median: 3; modes: 5 and 2 9. mean: −2.3 centimeters; median: −1.6 centimeters; The median would best represent the data because five of the six amounts of rainfall are closer to the median than to the mean.

Pages 299-301 **Written Exercises** 1. −0.1°C 3. 0.3 feet per second per second 5. median: 80.5 inches; mode: 79 inches 7. mean: 27.3; median: 27; Since the mean and the median are close, either can be used to represent the data. 9. mean: −1.2 centimeters; median: −0.6 centimeters; modes 1.2 centimeters and −0.3 centimeter 11. mean: $8545; median: $2240.50; mode: none

Page 302 1. 5 milliliters 3. 3 tablets

Pages 303-304 **Chapter Review** 1. −108 3. −208 5. −6 7. −3740 9. 120 11. 27 13. 108 15. −36 17. $-36r - 24$ 19. $-32s - 60$ 21. $-24x^2 + 72x$ 23. $19(r + s)$ 25. $c(c + 3)$ 27. $-12n(a + k)$ 29. −5 31. 9 33. −64 35. 24 37. mean: 169; median: 167.5; mode: none 39. mean: 140; median: 144; mode: 144 41. mean: 189; median: 186; mode: 180 43. 1.7 cm

CHAPTER 11 SOLVING EQUATIONS: REAL NUMBERS

Pages 306-307 **P-1** a. 3.2 b. 8 c. $2\frac{1}{4}$ **P-2** a. $x = -6\frac{1}{3}$ b. $x = -3$ c. $x = 11.8$ **P-3** a. 14 b. 3.2 c. $5\frac{3}{4}$ **P-4** a. $y = -4$ b. $x = -5$ c. $z = 0$

Page 308 **Classroom Exercises** 1. $y + 18 - 18 = 7 - 18$ 3. $a + 5 - 5 = -22 - 5$ 5. $k - \frac{3}{4} + \frac{3}{4} = -4\frac{1}{2} + \frac{3}{4}$ 7. $28 + 17 = -17 + m + 17$ 9. $-12.8 + 3.9 = y - 3.9 + 3.9$ 11. $15.6 + 0.9 = -0.9 + w + 0.9$ 13. $43 + f - 43 = 24 - 43$ 15. $-56 - 27 = 27 + z - 27$ 17. $y = 84$ 19. $c = -5$ 21. $d = -1$ 23. $k = -10$ 25. $n = -111.8$ 27. $w = -3\frac{1}{5}$

Pages 308-309 **Written Exercises** 1. $x = 39$ 3. $d = 9$ 5. $p = -3$ 7. $q = -1.8$ 9. $y = 2\frac{1}{2}$ 11. $m = -3\frac{2}{5}$ 13. $p = -13$ 15. $t = -27$ 17. $x = -30$ 19. $c = -4.3$ 21. $t = -11.1$ 23. $y = -19$ 25. $j = -25$ 27. $x = 18$ 29. $f = 0.9$ 31. $c = -28.2$ 33. $y = 2\frac{1}{2}$ 35. $y = 64$ 37. $m = 7.3$ 39. $n = -10.1$ 41. $x = -\frac{2}{3}$ 43. −3°F 45. 7°C

Page 309 **Review Capsule for Section 11.2** 1. $18 + x$ 2. $2p - 12$ 3. $-0.8n + 1.9$ 4. $8.5x - 2$ 5. $-3p + 2$ 6. $5q - 8$ 7. $2x + \frac{1}{4}$ 8. $3\frac{1}{2} - 10x$ 9. $x - 7\frac{1}{2}$ 10. $-24d + 32$ 11. $-72 + 12g$ 12. $9h - 6$ 13. $-10a - 2\frac{1}{2}$ 14. $25w + 8\frac{1}{3}$ 15. $9p + 3$ 16. $\frac{1}{2}h + \frac{3}{4}$ 17. $-1\frac{2}{5} + 7c$ 18. $-0.28q + 0.5$ 19. $24x - 12$ 20. $-15m - 10$ 21. $6g + 4$

Page 310 **P-1** a. $x = -38$ b. $x = 7.8$

Page 311 **Classroom Exercises** 1. $p - 10 = -5$ 3. $z + 6 = -11$ 5. $m + 5.1 = -5.3$ 7. $y + 6 = 19$ 9. $x = 2$ 11. $x = -1$ 13. $t = -13$ 15. $x = 0$

Page 311 **Written Exercises** 1. $q = 16$ 3. $d = -5.3$ 5. $y = 4$ 7. $h = 2.2$ 9. $s = -3$ 11. $y = 19$ 13. $n = 58$ 15. $r = 22$ 17. $k = 3.9$ 19. $r = 1\frac{1}{2}$

Pages 312-313 **P-1** a. $x = -42$ b. $y = -112$ c. $d = -1.92$ d. $f = -11$ **P-2** a. $q = -16$ b. $n = \frac{3}{4}$ c. $b = -2.2$ d. $y = -193$ **P-3** a. $y = -8\frac{1}{2}$ b. $d = -62$ c. $p = -3$ d. $t = -4.7$ **P-4** a. $n = -27$ b. $p = -105$ c. $t = -126$ d. $x = -16$

Pages 314-315 **Classroom Exercises** 1. $\frac{x}{4}(4) = -7(4)$ 3. $\frac{r}{-7}(-7) = 10(-7)$ 5. $\frac{12y}{12} = \frac{-48}{12}$ 7. $\frac{-6p}{-6} = \frac{42}{-6}$ 9. $p = -96$ 11. $s = -60$ 13. $x = -16.8$ 15. $a = 32.4$ 17. $c = -8$ 19. $x = -11$ 21. $z = -2.7$ 23. $w = -\frac{1}{21}$ 25. $r = -34$ 27. $x = -45$ 29. $n = 32$ 31. $v = 96$

Pages 315-316 **Written Exercises** 1. $a = -114$ 3. $d = -51.6$ 5. $x = -4$ 7. $b = -9\frac{4}{5}$ 9. $z = 4$ 11. $d = -7$ 13. $h = -\frac{1}{10}$ 15. $x = -1\frac{1}{2}$ 17. $y = -18$ 19. $d = -12$ 21. $m = 28$ 23. $d = -80$ 25. $a = -8$ 27. $y = -\frac{1}{3}$ 29. $m = -32$ 31. $s = -7$ 33. $c = -\frac{1}{2}$ 35. $r = 6$ 37. $j = -28$ 39. $r = \frac{1}{18}$ 41. $c = -100$ 43. $m = -0.72$ 45. $-12°F$

Page 316 **Mid-Chapter Review** 1. $c = -32$ 3. $n = -4.5$ 5. $p = 227$ 7. $t = -6\frac{1}{2}$ 9. $h = -11.6$ 11. $-56°F$ 13. $r = -228$ 15. $m = -169$ 17. $t = -219$ 19. $z = -6$ 21. $r = 77$ 23. $m = -\frac{9}{40}$ 25. $b = 2\frac{5}{6}$ 27. $\frac{3}{8}$ point

Page 317 **Review Capsule for Section 11.4** 1. $g = 4$ 2. $u = 5$ 3. $h = 12$ 4. $x = 4$ 5. $k = 36$ 6. $c = 32$

Pages 317-318 **P-1** a. $n = -1$ b. $x = -2$ c. $y = -15$ **P-2** a. $s = -24$ b. $t = -30$ c. $n = 36$

Page 318 **Classroom Exercises** 1. $3k + 2 - 2 = 5 - 2$ 3. $4n - 6 + 6 = -19 + 6$ 5. $12 - 8 = 8 - 3w - 8$ 7. $5.8 - 12.4 = 12.4 - 2z - 12.4$ 9. $\frac{m}{3} - 5 + 5 = -22 + 5$ 11. $-3.5 - 1.9 = 6x + 1.9 - 1.9$ 13. $-\frac{3}{5}v - 1\frac{2}{3} + 1\frac{2}{3} = 4\frac{1}{3} + 1\frac{2}{3}$ 15. $-3\frac{1}{2} + 8\frac{1}{4} = \frac{g}{-2} - 8\frac{1}{4} + 8\frac{1}{4}$ 17. $0.6r + 1.3 - 1.3 = -5.3 - 1.3$ 19. $\frac{1}{2}n - 3 + 3 = -6 + 3$ 21. $\frac{d}{0.2} + 9 - 9 = 3 - 9$

Page 319 **Written Exercises** 1. $m = -5$ 3. $t = -2$ 5. $a = -4.5$ 7. $b = -12$ 9. $c = 42$ 11. $d = 36$ 13. $p = 3.3$ 15. $q = 35$ 17. $x = -1\frac{1}{2}$, or $x = -1.5$ 19. $r = -8.25$ 21. $b = -145.6$ 23. $a = -1980$ 25. 3 miles 27. gained 386 yards

Page 320 **P-1** $\frac{1}{2}w + 4$

Pages 321-322 **Classroom Exercises** 1. $n - 5$ 3. $n + 60$ 5. $1.8n + 32$ 7. f 9. a 11. d 13. b

Pages 322-323 **Written Exercises** For Ex. 1-25, variables may vary. 1. Let t = Tuesday's low temperature; $t - 7$ 3. Let p = the number of players on the team; $4.50p$ 5. Let q = the number of questions; $\frac{100}{q}$ 7. Let m = the number of millimeters of rain; $2m + 20.4$ 9. Let p = the profit of a business; $2.5p - 2685$ 11. Let p = the number of payments; $\frac{570}{p} = 47.50$ 13. Let y = the number of yards gained; $11y = 619.3$ 15. Let t = the number of tugboats; $4t - (-10) = 26$, or $4t + 10 = 26$ 17. Let t = the average daily temperature in June; $t - 35 = 30$ 19. $t - 29 = -29$ 21. $d - 1200$ 23. $35 - t$ 25. $2h - 15 = -3$ 27. See the figure at the right.

27.

Page 326 **Classroom Exercises** 1. c 3. b 5. a. Let x = the cost of the slacks. Then $x + 46$ = the cost of the jacket. b. $x + x + 46 = 155$ 7. a. Let x = the distance driven on Monday. Then $x + 120$ = the distance driven on Tuesday. b. $x + x + 120 = 960$ 9. a. Let x = the amount of interest for last year. Then $x - 10.11$ = the amount of interest for this year. b. $x + x - 10.11 = 58.32$

Pages 326-328 **Written Exercises** For Ex. 1-19, variables may vary. 1. a. Let x = the number of runners last year. Then $x + 31$ = the number of runners this year. b. $x + x + 31 = 279$ c. There were 124 runners last year. 3. a. Let x = the amount of rain. Then $x + 1.8$ = the amount of snow. b. $x + x + 1.8 = 159.2$ c. 78.7 centimeters of rain and 80.5 centimeters of snow fell in Detroit that year. 5. a. Let x = the amount of money in Max's checking account. Then $x + 238.57$ = the amount of money in Max's savings account. b. $x + x + 238.57 = 1526.95$ c. Max has $644.19 in his checking account and $882.76 in his savings account. 7. a. Let d = the distance from New Orleans to Miami. Then $d - 366$ = the distance from Fort Worth to New Orleans. b. $d + d - 366 = 1788$ c. The distance from Fort Worth to New Orleans is 711 kilometers. 9. a. Let x = the number of girls. Then $x - 9$ = the number of boys. b. $x + x - 9 = 37$ c. There are 23 girls and 14 boys in the class. 11. a. Let x = the cost of the flight on Saturday. Then $x - 37.50$ = the cost of the flight during the week. b. $x + x - 37.50 = 246.90$ c. The return flight costs $142.20. 13. a. Let x = the number of gallons of fuel in the station wagon. Then $x - 13.6$ = the number of gallons of fuel in the sedan. b. $x + x - 13.6 = 36.8$ c. The station wagon holds 25.2 gallons of fuel and the sedan holds 11.6 gallons of fuel. 15. a. Let x = the cost of father's ticket. Then $x - 2$ = the cost of Myra's ticket. b. $x + x - 2 = 7$. c. Myra's ticket cost $2.50 and her father's ticket cost $4.50. 17. a. Let x = the amount earned by the assistant. Then $x + 14.75$ = the amount earned by the mechanic. b. $x + x + 14.75 = 89.25$ c. The assistant earned $37.25 and the mechanic earned $52.00. 19. a. Let x = the number of perfect checkups in the group using Toothpaste B. Then $x - 176$ = the number of perfect checkups in the group using Toothpaste A. b. $x + x - 176 = 1418$ c. There were 797 perfect checkups for the group using Toothpaste B and 621 perfect checkups for the group using Toothpaste A.

Page 329 1. 4%; 160 3. 3%; 24 5. 1300

Pages 330-332 **Chapter Review** 1. $p = 14$ 3. $a = -44$ 5. $g = -5.0$ 7. 77°F 9. $m = -6$ 11. $d = -2$

13. $x = 19$ 15. $w = 55$ 17. $n = -5\frac{3}{4}$ 19. $c = 32.2$ 21. $y = -14$ 23. $d = -1\frac{3}{125}$ 25. $-20°F$
27. $m = -4$ 29. $r = -24$ 31. $z = 5$ 33. \$3250 For Ex. 35-41, variables may vary.
35. Let r = the deficiency in rainfall; $\frac{r}{6} = -1.2$ 37. Let y = the number of net yards running; $20y - 4 = 156$ 39. a. Let x = the time it took to come down the mountain. Then x + 95 = the time it took to climb up the mountain. b. $x + x + 95 = 245$ c. It took 170 minutes to climb to the top. 41. a. Let x = the number of persons who entered on the second day. Then x − 235 = the number of persons who entered on the first day. b. $x + x - 235 = 885$ c. 325 persons entered the store the first day and 560 persons entered the store the second day.

CHAPTER 12 EQUATIONS AND PROBLEM SOLVING: REAL NUMBERS

Pages 335-336 Classroom Exercises 1. Let 2n = the amount of sand; n = the amount of cement. 3. Let 3n = the number of games won; 2n = the number of games lost. For Exercises 5-7, variables may vary. 5. a. Let 7n = the amount of alcohol. Then n = the amount of iodine. b. $7n + n = 720$ 7. a. Let 4n = the amount spent on schools. Then 2n = the amount spent on the library and 2n = the amount spent on the hospital. b. $4n + 2n + 2n = 555{,}000$

Page 336 Written Exercises For Exercises 1-7, variables may vary. 1. a. Let 11x = the number of kilograms of copper. Then 9x = the number of kilograms of zinc. b. $11x + 9x = 220$ c. There are 220 kilograms of copper and 180 kilograms of zinc. 3. a. Let 5x = the amount of brown sugar. Then 3x = the amount of flour. b. $5x + 3x = 1$ c. There is $\frac{3}{8}$ cup of flour. 5. a. Let 6x = the amount of the first investment. Then 5x = the amount of the second investment; 4x = the amount of the third investment. b. $6x + 5x + 4x = 390{,}000$ c. The amounts invested were \$156,000, \$130,000, and \$104,000. 7. a. Let 2x = the number of kilograms of Brand A. Then 5x = the number of kilograms of Brand B; 11x = the number of kilograms of Brand C. b. $2x + 5x + 11x = 54$ c. He used 6 kilograms of Brand A, 15 kilograms of Brand B, and 33 kilograms of Brand C.

Page 338 Classroom Exercises 1. $x + x + 17 + 25 = 180$ 3. $t + 3t - 1 + 25 = 180$

Pages 338-340 Written Exercises For Exercises 1-13, variables may vary. 1. a. Let x = the measure of angle B. Then 5x = the measure of angle A. b. $x + 5x + 150 = 180$ c. The measure of angle A is 25°. 3. a. Let x = the measure of angle C. Then 2x + 3 = the measure of angle B. b. $x + 2x + 3 + 27 = 180$ c. The measure of angle B is 103°. 5. a. Let x = the measure of angle B. Then 3x + 14 = the measure of angle A. b. $x + 3x + 14 + 106 = 180$ c. The measure of angle A is 59°. 7. a. Let x = the measure of each of the equal angles. b. $x + x + 70 = 180$ c. Each of the equal angles measures 55°. 9. a. Let x = the measure of angle X. Then 25x − 8 = the measure of angle Z. b. $x + 25x - 8 + 90 = 180$ c. The measure of angle X is approximately 3.77°. 11. a. Let x = the measure of angle A. Then 3x = the measure of angle C. b. $x + 3x + 90 = 180$ c. The measure of angle A is $22\frac{1}{2}°$, or 22.5°. 13. a. Let x = the measure of angle C. Then 4x + 10 = the measure of angle A. b. $x + 4x + 10 + 90 = 180$ c. The measure of angle A is 74°.

Page 340 Review Capsule for Section 12.3 1. $t = 9$ 2. $r = 4$ 3. $q = 12$ 4. $p = 8$ 5. $b = 2$ 6. $z = 1$ 7. $m = 6.5$, or $m = 6\frac{1}{2}$ 8. $f = 13\frac{1}{3}$, or $f = 13.\overline{3}$ 9. $a = 11$ 10. $g = 1$ 11. $p = 17$ 12. $d = 15$

Pages 342-343 Classroom Exercises 1. d 3. c 5. d 7. a

Pages 343-344 Written Exercises 1. a. 50¢; 50h; 5¢; 5(h + 2) b. 50h + 5(h + 2) = 340 c. Juan gave the driver 6 half-dollars and 8 nickels. 3. a. 25¢; 25x; x – 3; 6¢; 6(x – 3) b. 25x + 6(x – 3) = 106 c. There are four 25-cent stamps and one 6-cent stamp. For Ex. 5 and 7, variables may vary. 5. Let x = the number of nickels; total value: 5x; Let x – 13 = the number of quarters; total value: 25(x – 13); 5x + 25(x – 13) = 425; There are 12 quarters. 7. Let x = the number of 3-cent stamps; total value: 3x; Let x + 5 = the number of 1-cent stamps; total value: x + 5; Let x + 2 = the number of 10-cent stamps; total value: 10(x + 2); 3x + x + 5 + 10(x + 2) = 53; There are four ten-cent stamps.

Page 345 Mid-Chapter Review For Exercises 1-11, variables may vary. 1. Let 4x = the number of grams of copper. Then x = the number of grams of tin. 4x + x = 832; There are 665.6 grams of copper and 166.4 grams of tin. 3. Let 2x = the amount one partner receives. Then 5x = the amount the other partner receives. 2x + 5x = 21,315; One partner receives $6090 and the other receives $15,225. 5. Let x = the measure of angle B. Then 2x + 8 = the measure of angle A. x + 2x + 8 + 34 = 180; The measure of angle A is 100°. 7. Let x = the measure of each of the equal angles. x + x + 48 = 180; Each of the equal angles measures 66°. 9. Let x = the number of quarters. Then x + 3 = the number of nickels and 1 = the number of dimes. 25x + 5(x + 3) + 10 = 115; She has 6 nickels. 11. Let x = the number of 2-cent stamps. Then 3x + 2 = the number of 10-cent stamps. 2x + 10(3x + 2) = 116; There were three 2-cent stamps and eleven 10-cent stamps.

Page 346 1. –2034 Btu's per hour 3. Concrete wall 6 inches thick; the value of U for concrete (0.58) is higher than the value of U for brick (0.41).

Page 347 Review Capsule for Section 12.4 1. 3n + 13 2. –6m – 11 3. 3.8 – 3.7w 4. –3y – 13 5. 17 6. –9 7. x 8. 9a 9. 5b 10. q 11. 2.1p 12. 5.5j 13. $\frac{2}{3}z$ 14. $\frac{3}{4}t$ 15. r

Pages 347-348 P-1 a. x = –24 b. x = 34 c. x = –4 **P-2** n = –6

Page 348 Classroom Exercises 1. n = –4 3. a = –9 5. x = –4 7. y = 9 9. x = 1 11. Yes 13. Yes 15. No

Page 349 Written Exercises 1. a = –12 3. g = –20 5. m = –17 7. b = 6.6 9. q = 16 11. k = –10 13. p = –15 15. d = –2 17. n = 10 19. f = –84 21. m = 16 23. w = –1 25. a = –8 27. v = –6.7 29. b = 31 31. c = 20 33. m = 71 35. x = b + a; positive 37. x = b – a; negative 39. x = b – a; negative 41. x = –a – b; negative

Page 350 Review Capsule for Section 12.5 1. 2x 2. 3y 3. 11z 4. $8\frac{3}{4}p$ 5. $4\frac{4}{5}r$ 6. $4\frac{5}{6}t$ 7. 1.2f 8. 4.8m 9. 0.6s 10. –8x + 30 11. –4x – 24 12. –30n + 24 13. –5t – 10 14. –3r + 8 15. 5 – w 16. –6.1m – 7.8 17. 1.8y + 4 18. $6\frac{3}{8}p - 9$

Pages 350-351 P-1 a. m = –2 b. x = –2 c. y = –2 **P-2** x = –5

Page 351 Classroom Exercises For Exercises 1-9, answers may vary. 1. 4 – m + 5m = 6 – 5m + 5m 3. 12 + 6y – 9y = 9y + 15 – 9y 5. 20n – 6 – 18n = 18n + 4 – 18n 7. 2.8 + 0.5f + 4.5f = 3.7 – 4.5f + 4.5f 9. $\frac{1}{2}a + 3 - \frac{1}{4}a = \frac{1}{4}a - 2 - \frac{1}{4}a$ 11. –3q = –4q – 12 13. –10r + 15 + 3r = 3r

Page 352 Written Exercises 1. m = –5 3. n = 3 5. w = 15 7. $x = -5\frac{1}{2}$ 9. $c = \frac{1}{3}$ 11. $k = -\frac{1}{2}$

13. $q = 3$ 15. $n = -1\frac{3}{4}$ 17. $b = 3$ 19. $p = 3$ 21. $g = 2\frac{1}{4}$ 23. checks 25. checks 27. checks

Page 353 **Review Capsule for Section 12.6** 1. $t = 15$ 2. $m = 8$ 3. $r = 29$ 4. $x = 42$ 5. $y = 16$ 6. $w = 50$

Page 355 **Classroom Exercises** 1. $r + 6$ 3. $3m - 5$ 5. $n + 4$ 7. $2n - 2$

Pages 355-356 **Written Exercises** 1. a. $x + 1$; $6x + 1$ b. $6x + 1 = 5(x + 1)$ c. Joe is 24 years old and Lisa is 4 years old. 3. a. $3y$; $y - 3$; $3y - 3$ b. $3y - 3 = 4(y - 3)$ c. Pete is 9 years old and Carl is 27 years old. 5. $2x - 12 = 3(x - 12)$; Mrs. Cooke is 24 years old and Mrs. Montez is 48 years old. 7. $4x + 20 = 2(x + 20)$; Mr. Adams is 40 years old and Luann is 10 years old. 9. 40; 8; $40 + n$; $8 + n$; $40 + n = 3(8 + n)$; Mr. Conner's age will be exactly 3 times Robert's age in 8 years.

Pages 357-358 1. 74 3. 7 5. 4.8 7. 72 9. 20 11. 2 13. 0 15. 18 17. 15 19. 16 21. 3 23. 12 25. 6 27. 61 29. 7.5

Pages 359-360 **Chapter Review** For Exercises 1-6, variables may vary. 1. Let $3x$ = the number of pounds of propane. Then $2x$ = the number of pounds of butane. $3x + 2x = 25$; There are 15 pounds of propane and 10 pounds of butane. 3. Let x = the measure of angle C. Then $x - 18$ = the measure of angle A. $x + x - 18 + 72 = 180$; The measure of angle C is 63°. 5. Let x = the number of 15-cent stamps. Then $x + 2$ = the number of 25-cent stamps. $15x + 25(x + 2) = 290$; There are eight 25-cent stamps. 7. $p = 9$ 9. $w = 9.3$ 11. $r = -15$ 13. $f = -5$ 15. $p = 13\frac{1}{2}$ 17. $x = 1\frac{4}{11}$ For Exercise 19, the variable may vary. 19. $4x + 6 = 2(x + 6)$; Maria is 12 years old and Ann is 3 years old.

Pages 361-362 **Cumulative Review: Chapters 1-12** 1. 18 3. $15c$ 5. $18hp^2$ 7. $\frac{2}{3}$ 9. $9c^2$ 11. $\frac{5x - 2y}{3}$ 13. $\frac{3}{10f}$ 15. Nancy sold 375 tickets. 17. $\frac{5}{12}$ 19. $\frac{9}{12}$, or $\frac{3}{4}$ 21. 13% 23. 20 25. $0.8k + 6$ 27. $6y + 16$ 29. $-2a^2 - 10a$ 31. 14 33. $12a^2 - 20a^3$ 35. $-6x(1 + x)$ 37. $2a(4b - 3a)$ 39. 5 41. -6 43. mean: $2.\overline{8}$ or $2\frac{8}{9}$; median: 7; mode: 9 45. $x = -4$ 47. $x = 128$ 49. $r = -15$ For Ex. 51-53, variables may vary. 51. Let $8x$ = the number of gallons of red paint. Then $3x$ = the number of gallons of white paint. $8x + 3x = 88$; 64 gallons of red paint and 24 gallons of white paint are needed. 53. Let x = the number of quarters. Then $x + 7$ = the number of dimes. $25x + 10(x + 7) = 910$; Kirk has 31 dimes. 55. $x = -1$ For Ex. 57, the variable may vary. 57. Let x = Carol's age now. Then $3x$ = Beth's age now. $9(x - 6) = 3x - 6$; Beth is 24 years old and Carol is 8 years old.

CHAPTER 13 INEQUALITIES

Pages 364-365 **P-1** a, b, and d **P-2** a

Page 366 **Classroom Exercises** 1. Not a solution 3. Not a solution 5. Not a solution 7. Solution 9. Not a solution 11. Not a solution 13. Solution 15. Not a solution 17. Solution 19. $f > 1$; replacement set: {real numbers} 21. $f > -4$, or $f \geq -3$; replacement set: {integers} 23. $f \geq -1.5$; replacement set: {real numbers} 25. $f > -1$, or $f \geq 0$; replacement set: {integers}

Pages 366-367 Written Exercises

1. etc. −1 0 1 2 3 4 5 6 3. etc. −6 −5 −4 −3 −2 −1 0 1 5. etc. −1 0 1 2 3 4 5 6 7. etc. −1 0 1 2 3 4 5 6

9. −2 0 2 4 11. −6 −5 −3 0 13. −4 −2 0 2 15. $-1\frac{1}{3}$ −2 0 2 4

17. $q > -3$, or $q \geq -2$; replacement set: {integers} 19. $q \leq 4$; replacement set: {real numbers} 21. $q > 1.5$; replacement set: {real numbers} 23. $q < \frac{13}{4}$; replacement set: {real numbers} 25. c 27. b 29. a 31. $y > w$

Pages 368-369 P-1 a. 5 b. 6 c. 6.5 **P-2** a. $h > 13$ b. $g < -2$ c. $m > 2.5$ **P-3** a. $x > 4$ b. $c > 34$ c. $r < 3\frac{1}{2}$

Page 370 Classroom Exercises 1. No 3. Yes 5. No 7. $r + 5 - 5 < 3 - 5$ 9. $w - 1 + 1 > 2 + 1$ 11. $-13 - 7 > a + 7 - 7$ 13. $k - 234 + 234 > -189 + 234$ 15. $-3\frac{1}{4} + 5\frac{1}{8} > t - 5\frac{1}{8} + 5\frac{1}{8}$ 17. $-5.6 - 1.8 < 1.8 + c - 1.8$

Pages 370-371 Written Exercises For the graphs of Ex. 1-33 odd, see below. 1. $t < -7$ 3. $p > 0$ 5. $b > -3.3$ 7. $w > 2\frac{3}{8}$ 9. $s > -9$ 11. $r > -22$ 13. $d < -152$ 15. $m > -14.6$ 17. $t > -3\frac{5}{8}$ 19. $t > 27$ 21. $n > -8$ 23. $y > -177$ 25. $f > 111$ 27. $t > 27.2$ 29. $x < -10.8$ 31. $n > -8\frac{1}{4}$ 33. $m > 125$ 35. 8°F 37. 139 points

1.

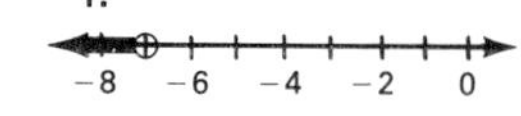

3. −4 0 4 8

5. and 7.

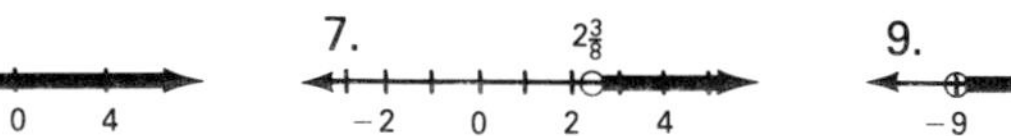

9.

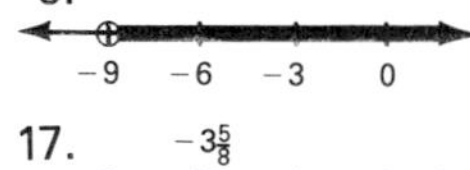

11.

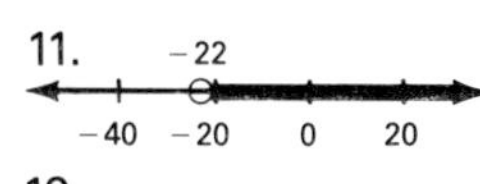

13. and 15.

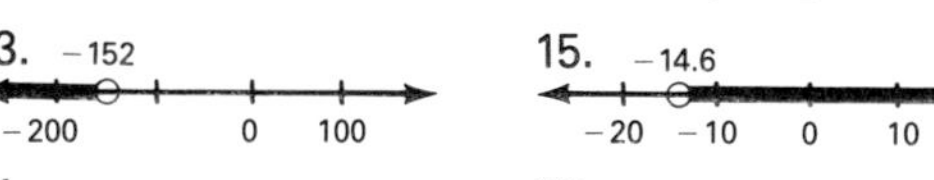

17. $-3\frac{5}{8}$ −6 −4 −2 0

19.

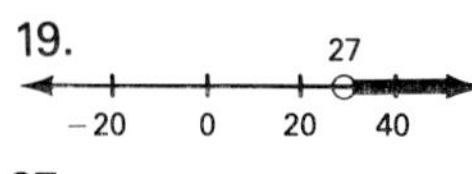

21. −8 −4 0 4 23. −177 −200 0 25. 111 0 100

27. 27.2 −20 0 20 40 29. −10.8 −15 −10 −5 0 31. $-8\frac{1}{4}$ −12 −8 −4 0 33. 125 0 50 100 150

Page 371 Review Capsule for Section 13.3 1. $-20 < 40$ 2. $-30 < -10$ 3. $15 > 10$ 4. $25 > -25$ 5. $-6 < 13$ 6. $-29 > -37$ 7. $-20 < 80$ 8. $-50 < -10$ 9. $40 < 100$ 10. $40 > -30$ 11. $18 < 32$ 12. $69 > 27$

Pages 372-374 P-1 a. $-15 > -18$ b. $-1 < 2$ c. $0 > -3.6$ d. $36 > -36$ **P-2** a. $-30 > -54$ b. $42 > 12$ c. $0 < 30$ d. $2 < 36$ **P-3** a. $n < 64$ b. $t < 36$ c. $r < -175$ d. $d < -144$ **P-4** a. $p < 16$ b. $b < -6$ c. $a < -4$ d. $f > -6$

Page 375 Classroom Exercises 1. $-10 < 35$ 3. $15 < 25$ 5. $-1 < 10$ 7. $-20 < 4$ 9. $24 < 32$ 11. $2\frac{1}{2} < 12$ 13. $3(\frac{1}{3}x) < 3(-2)$ 15. $\frac{3t}{3} < \frac{-12}{3}$ 17. $\frac{4m}{4} < \frac{-21}{4}$ 19. $-3(-\frac{1}{3}b) > -3(2\frac{1}{4})$

Pages 375-376 Written Exercises For the graphs of Ex. 1-43 odd, see p. 555. 1. $y < 45$ 3. $w > -72$ 5. $d > -2$ 7. $r > 4$ 9. $y < 6$ 11. $t > 3$ 13. $p < -7$ 15. $p > -13\frac{1}{3}$ 17. $s > -3$

19. $t < 2$ 21. $a < 2\frac{2}{5}$ 23. $w < 5\frac{1}{4}$ 25. $w < -6$ 27. $y < -8$ 29. $r < -\frac{1}{2}$ 31. $w > 9$ 33. $g > 5\frac{3}{8}$ 35. $r < -28$ 37. $b > 32$ 39. $g < 20$ 41. $t < -4$ 43. $q > -1\frac{1}{2}$ For Ex. 45 and 47, variables may vary. 45. Since $e < 30{,}000$, it is not possible that the yearly auto expenses total more than $31,500. 47. Since $s < 160$, it is possible that 148 sophomores were polled.

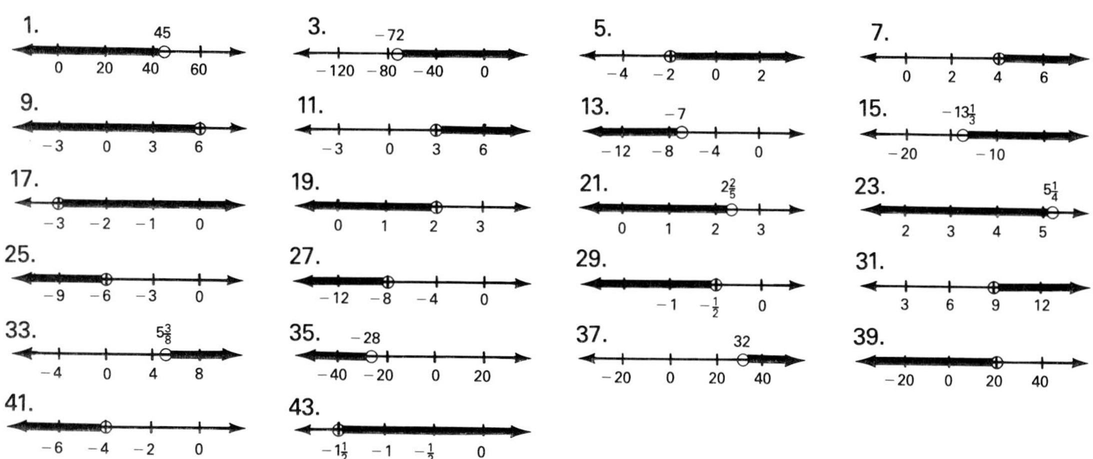

Page 377 Mid-Chapter Review 1.-3. See below. For Ex. 5 and 7, variables may vary. 5. $x > -2$, or $x \geq -1$; replacement set: {integers} 7. $x > -2\frac{1}{2}$; replacement set: {real numbers} 9. c. For the graphs of Ex. 11-25 odd, see below. 11. $w < -110$ 13. $p > -2$ 15. $r < 24$ 17. $x < 1$ 19. $r < -6$ 21. $w < 3\frac{3}{4}$ 23. $n < -8$ 25. $g > -5$ 27. 19 strikes

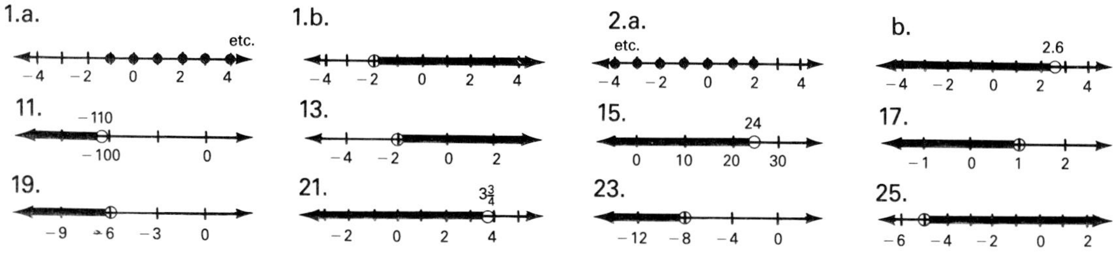

Page 379 1. See table and graph below. 3. $\overline{x} = 18$ 5. $\overline{x} = 76$

1. a.

Interval	Midpoint	Tally	Frequency
1	1	II	2
2	2	I	1
3	3	IIII	4
4	4	III	3
5	5	~~IIII~~ I	6
6	6	~~IIII~~	5
7	7	II	2
8	8	IIII	4
9	9	II	2
10	10	I	1

b.

Quiz Scores of Thirty Students

Number of Students: 0 1 2 3 4 5 6

Score: 1 2 3 4 5 6 7 8 9 10

Page 380 **Review Capsule for Section 13.4** 1. $g = -8$ 2. $k = 4$ 3. $c = -13$ 4. $b = 21$ 5. $z = -\frac{3}{8}$ 6. $m = 1\frac{2}{3}$

Pages 380-381 **P-1** a. 6 b. 5 c. 2 **P-2** a. $v < 7$ b. $t < -8$ **P-3** a. $h < -5$ b. $g > -6$

Pages 381-382 **Classroom Exercises** 1. $3n - 13 < -17$ 3. $2r - 6 < -4$ 5. $-4y - 23 > -17$ 7. $-28 > -23m + 17$ 9. $-b - 42 < -56$ 11. $8 - 3y + 4y > -4y - 2 + 4y$ 13. $4 - 6d + 6d < 5d - 3 + 6d$ 15. $6p + 4 - 6p > 7p - 12 - 6p$ 17. $12 - t + 5t < -8 - 5t + 5t$ 19. $x < -5$ 21. $g < 3$ 23. $z > 0$

Page 382 **Written Exercises** For the graphs of Ex. 1-23 odd, see below. 1. $n > -5$ 3. $a < -5$ 5. $y > 4$ 7. $w > -18$ 9. $x < 1$ 11. $m < 5$ 13. $t > -1\frac{1}{2}$ 15. $k < 4$ 17. $w > -15$ 19. $p < -9$ 21. $d < 7.1$ 23. $n < 1$ 25. 3 socks

1. −6 −4 −2 0 2
3. −8 −6 −4 −2 0
5. −4 0 4 8
7. −18; −20 −10 0 10
9. −1 0 1 2
11. −2 0 2 4 6
13. −3 −2 −1 0 1
15. −2 0 2 4 6
17. −20 −10 0 10
19. −18 −9 0 9
21. 7.1; −4 0 4 8
23. −1 0 1 2

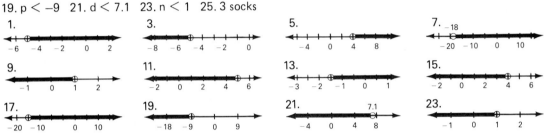

Page 384 **P-1** a. \$4.99 b. \$15.49 c. \$19.94 **P-2** a. 82 b. 84 c. $\frac{79 + 81 + x}{3}$

Page 383 **Classroom Exercises** 1. $b > 45$ 3. $b + 275 < 1050$ 5. $\frac{87 + 92 + z}{3} < 91$ 7. $\frac{65 + 70 + 68 + y}{4} > 66$

Pages 385-386 **Written Exercises** 1. The greatest amount they can pay each month for food is \$179.99 3. The greatest amount they can spend for clothing is \$1599.99. 5. The lowest score Ava can get on the fourth quiz is 89. 7. The lowest score Rocco can bowl in the next game is 101. 9. 24 ways 11. 3 majorettes

Pages 387-388 **Chapter Review** 1. See below. 3. See below. 5. $p \leq 3.3$; replacement set: {real numbers} For the graphs of Ex. 7-11 odd and Ex. 15-19 odd, see below. 7. $n < -5$ 9. $p > 2$ 11. $t > -4.9$ 13. 68°F 15. $r < 5\frac{3}{4}$ 17. $n < 3\frac{1}{2}$ 19. $x < 7$ For Ex. 21, the variable may vary. 21. Since $w < 67$, the building could not be 70 meters wide. For the graphs of Ex. 23-31 odd, see below. 23. $d > -2$ 25. $m > 6\frac{1}{2}$ 27. $p < 2\frac{2}{3}$ 29. $c < 5$ 31. $g > -1\frac{1}{3}$ 33. The most Gail can pay each month is \$201.38.

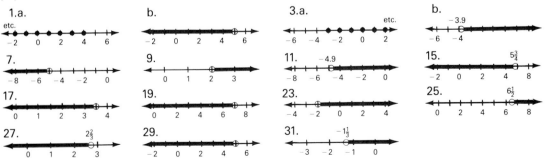

1.a. etc.; −2 0 2 4 6
b. −2 0 2 4 6
3.a. etc.; −6 −4 −2 0 2
b. −3.9; −6 −4
7. −8 −6 −4 −2 0
9. 0 1 2 3
11. −4.9; −8 −6 −4 −2 0
15. $5\frac{3}{4}$; −2 0 2 4 8
17. 0 1 2 3 4
19. 0 2 4 6 8
23. −4 −2 0 2 4
25. $6\frac{1}{2}$; 0 2 4 6 8
27. $2\frac{2}{3}$; 0 1 2 3
29. −2 0 2 4 6
31. $-1\frac{1}{3}$; −3 −2 −1 0

CHAPTER 14 RELATIONS AND FUNCTIONS

Pages 390-391 **P-1** 0, 6, −5, 7; 3, −2, 8, 0 **P-2** a. (3,1) b. (−2,2) c. (−1,−2)

Page 391 **Classroom Exercises** For Ex. 1-15 odd, see the graph at the right. 17. (−3,1) 19. (2,−2) 21. (−2,2) 23. (5,3) 25. (0,3) 27. (−5,−3) 29. (2,2) 31. (4,0)

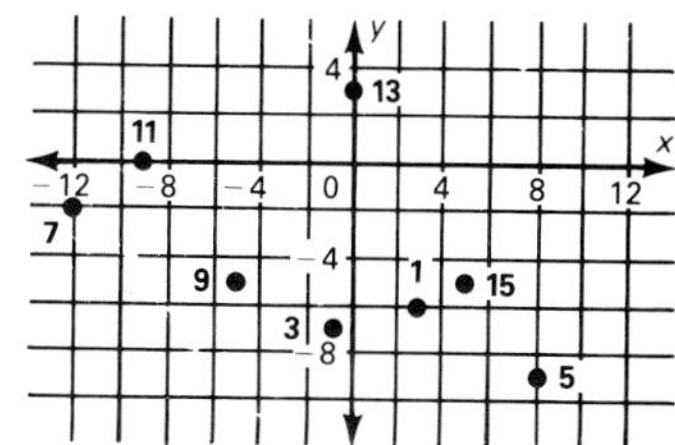

Page 392 **Written Exercises** 1.-11. 13. (−4,2) 15. (−3,−2) 17. (0,4) 19. (4,−2) 21. (2,0) 23. (2,3) For Ex. 25 and 27, see the graph at the right. 29. (0,−2) 31. (−1,2)

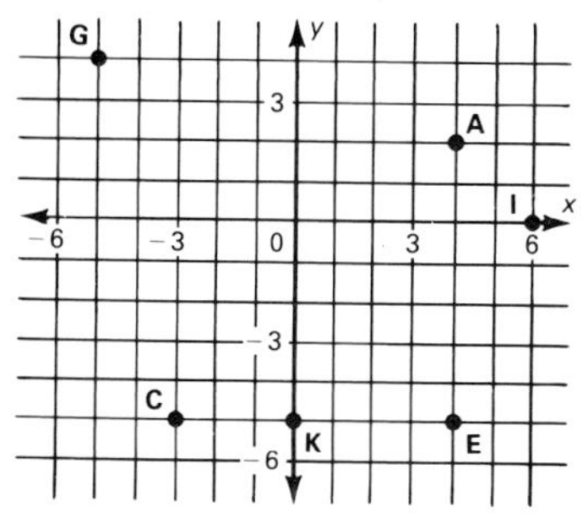

25.-27.

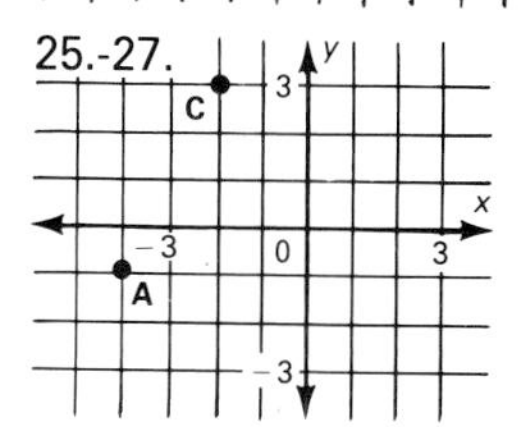

Page 392 **Review Capsule for Section 14.2** 1. 11 2. 12 3. $-7\frac{1}{2}$ 4. 11 5. $3\frac{1}{2}$ 6. −15 7. 7 8. $-\frac{1}{2}$ 9. $-4\frac{1}{7}$

Pages 393-394 **P-1** a. II b. IV c. I **P-2** c

Pages 394-395 **Classroom Exercises**

1.

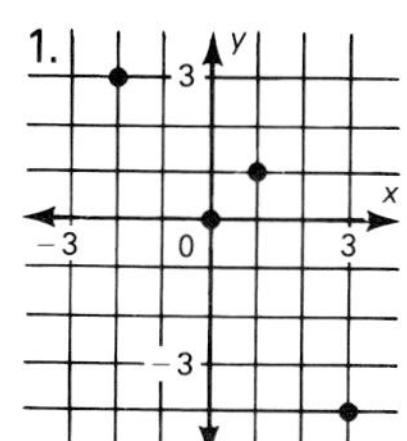

3.

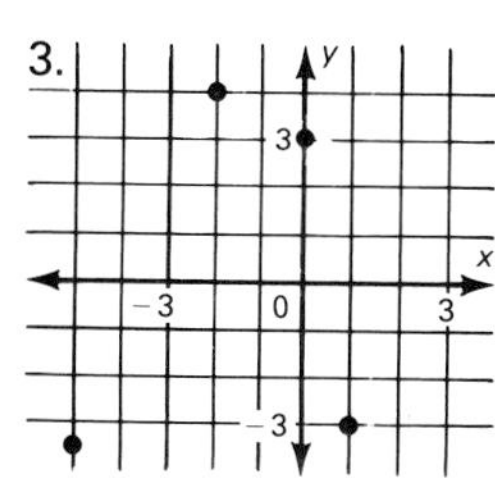

5.

x	−5	−1	0	4
y	−3	1	2	6

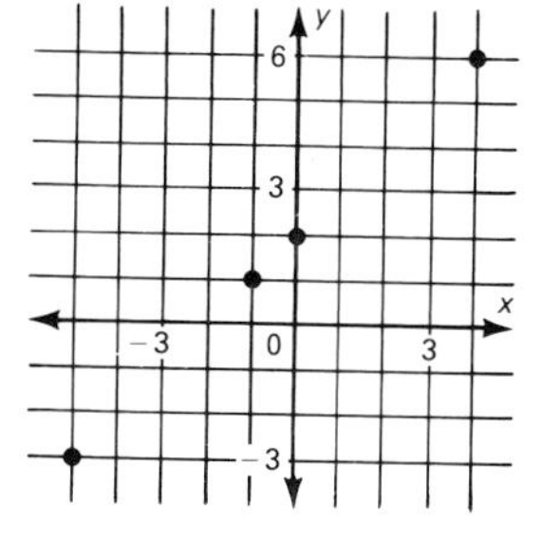

7.

x	−3	−1	0	$1\frac{1}{2}$
y	−1	3	5	8

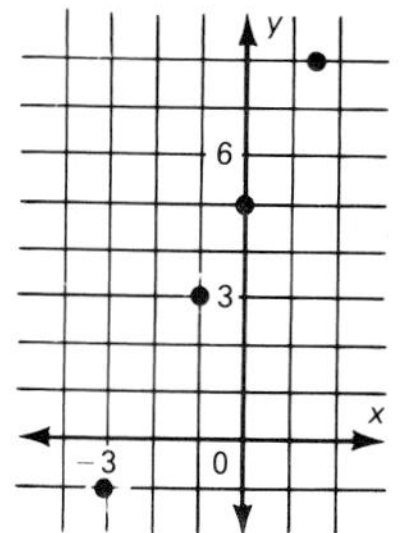

9. Multiply each x value by 2.
11. Add 3 to each x value.

Pages 395-396 **Written Exercises** For 1, 3, 5, and 7, see below. For 9 and 11, see page 558.

7.

x	−4	0	1	3
y	8	0	−2	−6

9.

x	−4	0	1	3
y	−10	−2	0	4

11.

x	−4	0	1	3
y	$-3\frac{1}{2}$	$-1\frac{1}{2}$	−1	0

13. Subtract 2 from each x value or add −2 to each x value. 15. Take the opposite of each x value.

1.

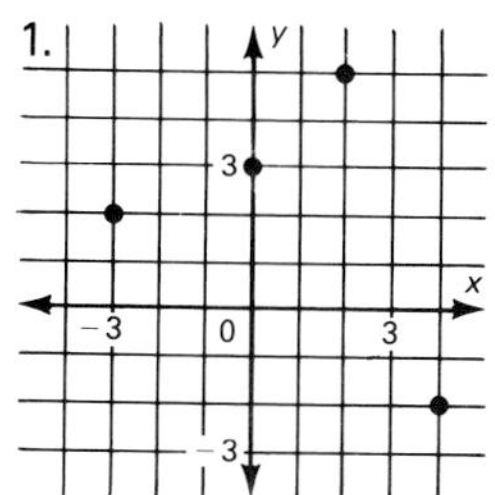

3.

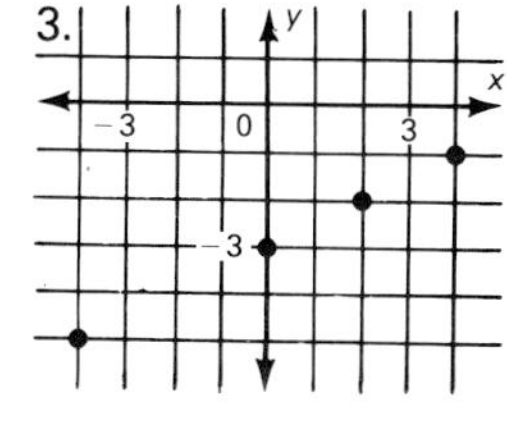

5.

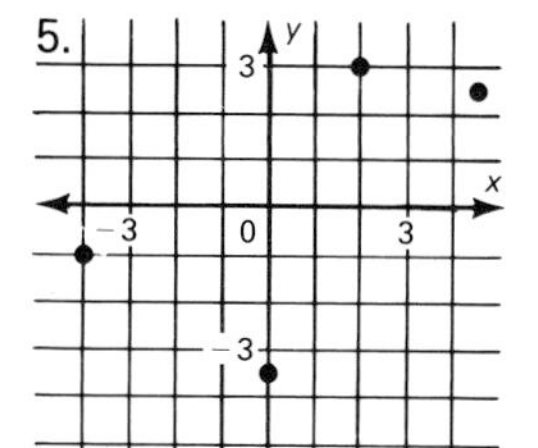

7.

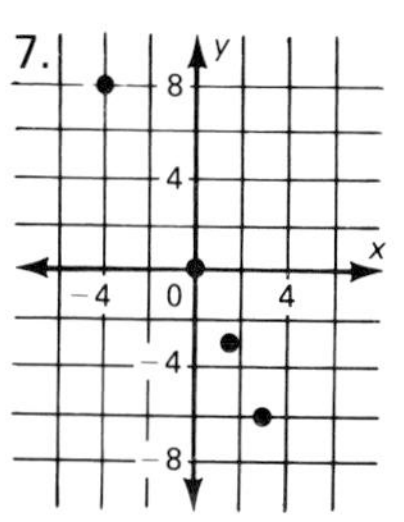

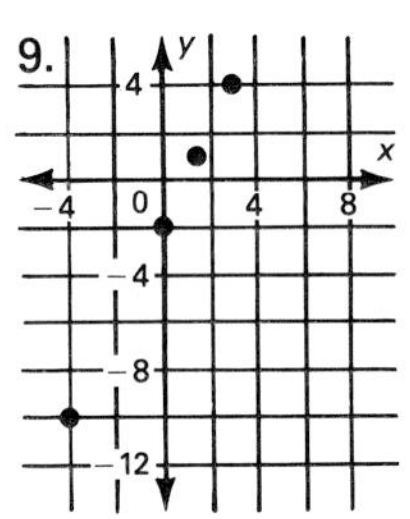

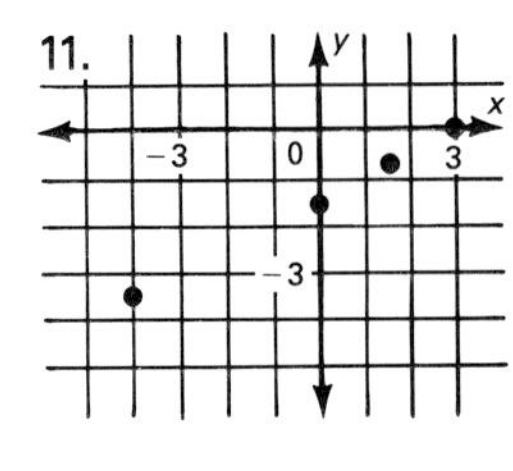

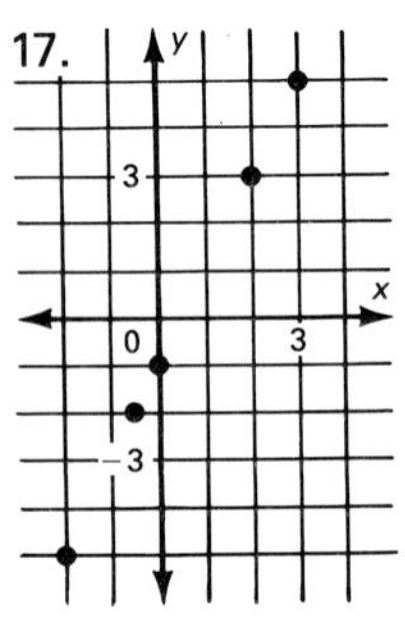

21. y: 3, $-1\frac{1}{2}$, −1, 0, $1\frac{1}{2}$

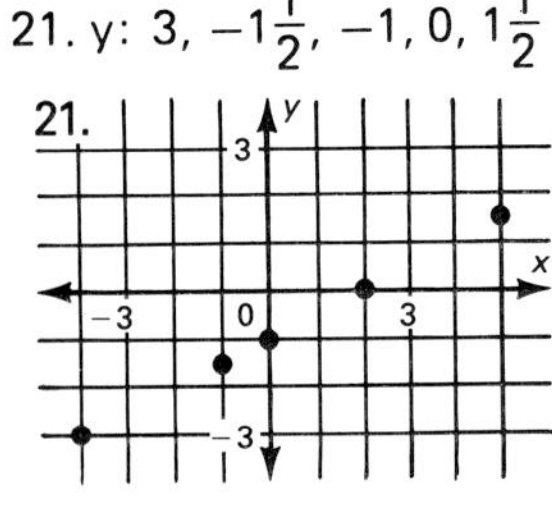

19. Multiply each x value by 3.
23. Total cost: $8.70

Page 396 **Review Capsule for Section 14.3** 1. No 2. Yes 3. No 4. No 5. No 6. Yes

Pages 397-398 **P-1** a. No, because (−4,2) and (−4,3) have the same x value. b. Yes **P-2** a. Domain: {−1, 3, 2}; Range: {1, 6, 0} b. Domain: {0, 1, 3, 2}; Range: {0, $-\frac{1}{2}$}

Pages 398-399 **Classroom Exercises** 1. Yes 3. No, all pairs have the same x value. 5. Yes 7. Domain: {1, −3, −1}; Range: {4, 5} 9. Domain: {$\frac{1}{2}$, $-\frac{1}{4}$, 7}; Range: {3, 2, $\frac{3}{4}$} 11. Domain: {0, −1, 3, 5}; Range: {5, 6, −2, −1} 13. {−5, 0, 5, 10} 15. {2, 0, −2, −4} 17. {−4, −1, 2, 5} 19. {−1, 0, 1, 2}

Page 399 **Written Exercises** 1. Yes 3. No, because the pairs (0,−1) and (0,1) have the same x value. 5. No, because the pairs ($\frac{1}{2}$,1) and (0.5,2) have the same x value. 7. Domain: {5, 3, 8}; Range: {0, −2, 3} 9. Domain: {−1, 5, 0, 2}; Range: {8, 2, −3} 11. Domain: {1, −2, −3, −4}; Range: {2, 3, 4, 0} 13. Range: {−11, 3, 1, 9} 15. Range: {7, 3, −1, −3} 17. Range: {3, −2, $-2\frac{1}{2}$, −4}

Page 400 **Mid-Chapter Review** 1.-7. See the graph at the right. 9. (−3,2) 11. (1,−1) 13. (−2,−1) 15. (−4,−1) 17. (3,$-1\frac{1}{2}$) 19. A 21. {10, 5, 2, 7} 23. {−5, 2, −3, $-3\frac{1}{2}$} 25. {−3, 0, −2, −1}

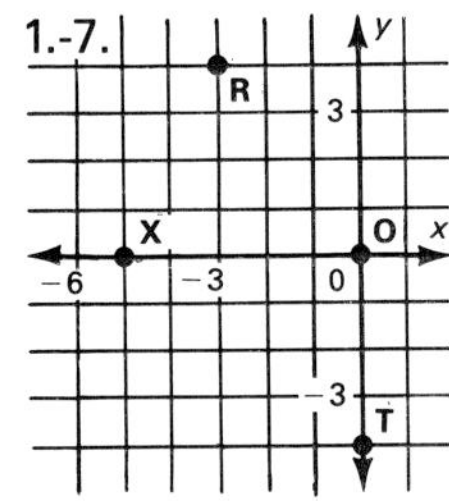

Page 400 **Review Capsule for Section 14.4**

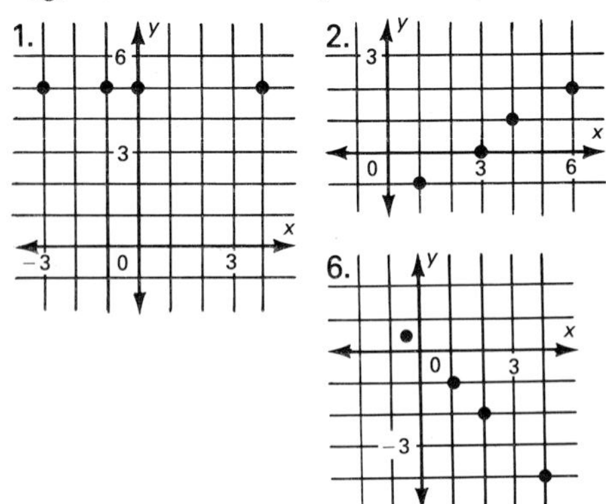

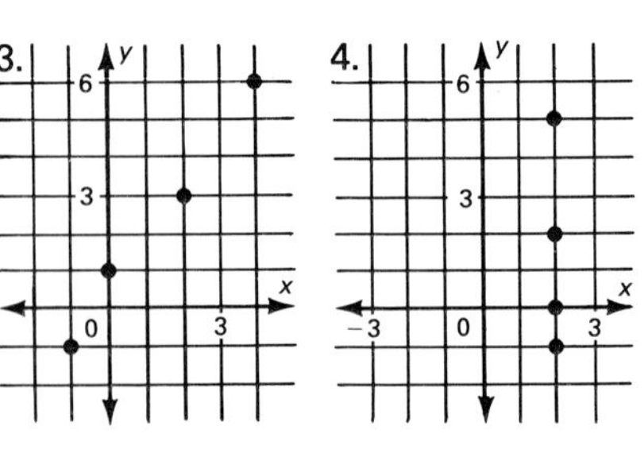

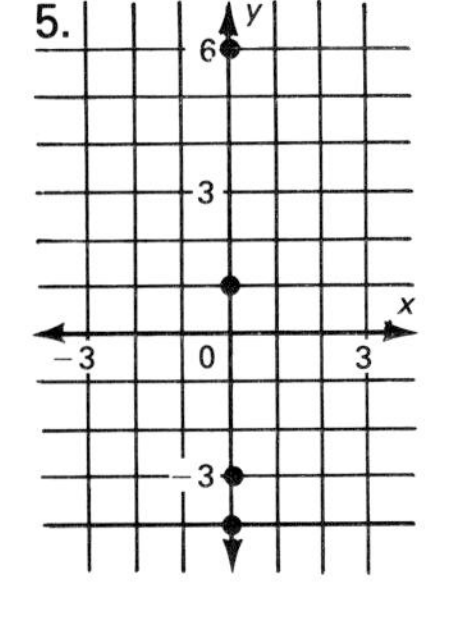

Page 402 **P-1** a. 2 + 1 = 3; 4 + 1 = 5 b. 2(−3) = −6; 2(7) = 14 c. 2(−1) + 1 = −1; 2(3) + 1 = 7

Pages 402-403 **Classroom Exercises** 1. −2 3. −1 5. −1 7. 1

9.

x	−2	−1	0	1	2
y	0	1	2	3	4

11.

x	−2	−1	0	1	2
y	3	2	1	0	−1

13.

x	−2	−1	0	1	2
y	−5	−2	1	4	7

Pages 403-404 **Written Exercises** For the graphs for Ex. 1-23 odd, see below.

1. 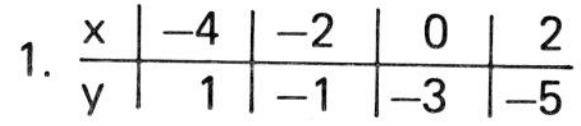

x	−4	−2	0	2
y	1	−1	−3	−5

3.

x	−2	0	2	3
y	5	1	−3	−5

5.

x	−4	−2	0	2
y	−2	0	2	4

7.

x	−4	−2	0	2
y	5	4	3	2

21.

x	−4	−2	0	2
y	1	1	1	1

23.

x	−2	−1	0	1
y	2	2	2	2

25. $R = \left\{-\frac{1}{2}, -3\frac{1}{2}, -11\frac{1}{2}, -29\frac{1}{2}\right\}$

27. $R = \{-40, -30, -12, 5\}$

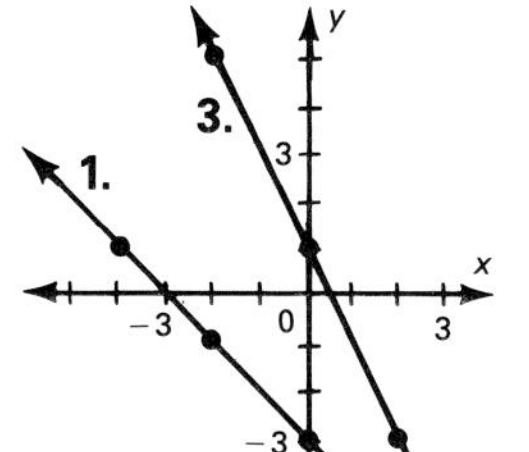

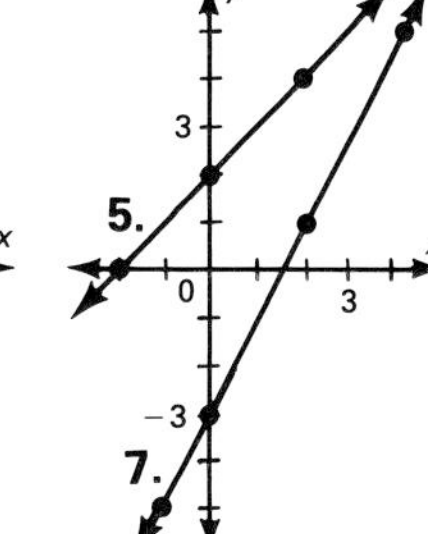

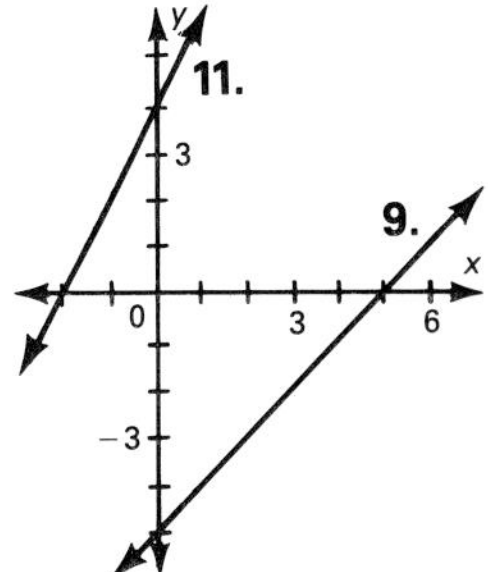

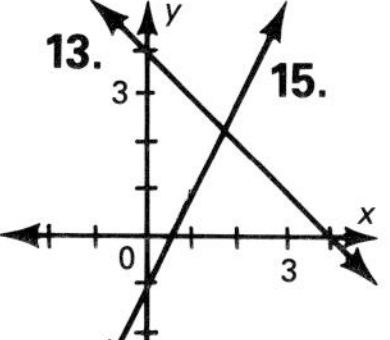

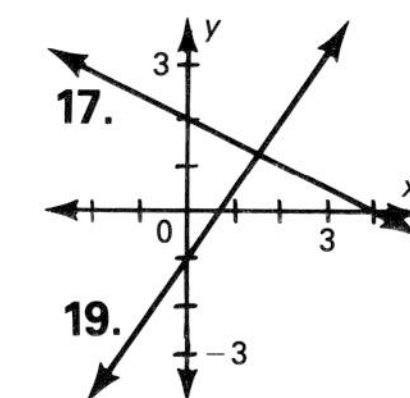

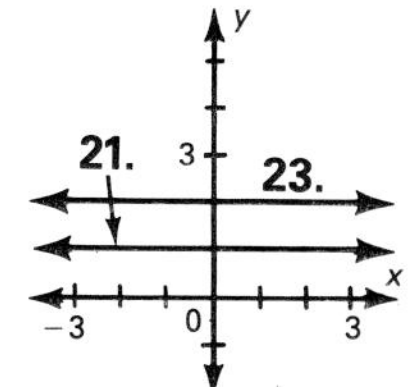

Page 404 **Review Capsule for Section 14.5** 1. y = −4 2. y = 5 3. y = 1 4. y = 0 5. x = −1 6. x = 1 7. x = 1 8. x = 6

Pages 405-406 **P-1** a. 1 b. −1 **P-2** a. 2 b. 3 c. 6

Page 407 **Classroom Exercises** 1. −2 3. 10 5. $-\frac{1}{2}$ 7. 0.7 9. 1 11. 5 13. 2 15. $\frac{1}{2}$

Page 407 **Written Exercises** 1. 5 3. −4 5. −1 7. $-\frac{1}{2}$ 9. −2.1 11. 6.9 13. 3 15. −10 17. −2 19. $\frac{1}{3}$ 21. $-\frac{5}{2}$ 23. $-\frac{7}{3}$

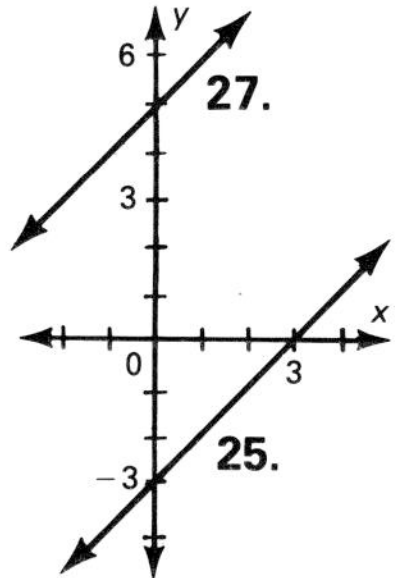

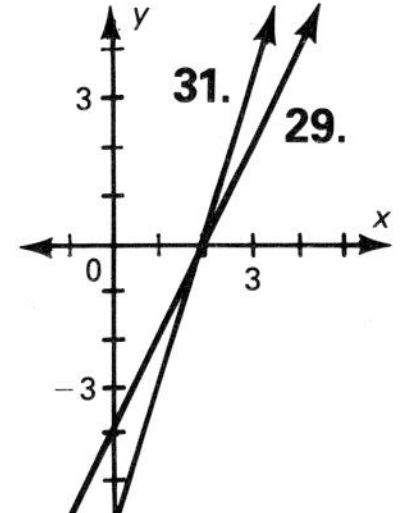

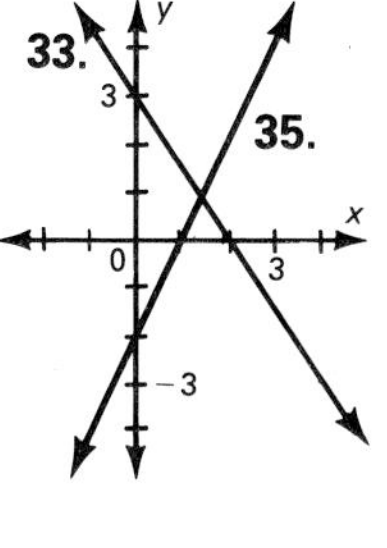

Page 407 **Review Capsule for Section 14.6** 1. m = 3; b = 7 2. m = −5; b = −2 3. m = 1; b = −3 4. m = −1; b = 6 5. m = $\frac{1}{2}$; b = 3 6. m = $-\frac{1}{3}$; b = $-\frac{4}{3}$ 7. m = 13; b = 13 8. m = −2; b = $\frac{7}{5}$

Page 409 Classroom Exercises 1. (–2,2) 3. (0,0) 5. No 7. Yes 9. No

Pages 409-410 Written Exercises

1.

x	–2	0	4
y	–4	–2	2

x	–3	0	4
y	4	1	–3

Intersection: $(1\frac{1}{2}, -\frac{1}{2})$

3.

x	–2	0	2
y	–5	–1	3

x	–4	–2	3
y	–3	–1	4

Intersection: (2,3)

5.

x	–1	1	3
y	0	4	8

x	–3	–1	3
y	–9	–5	3

Intersection: ϕ

7.

x	–2	0	2
y	5	–1	–7

x	–1	1	3
y	4	–2	–8

Intersection: ϕ

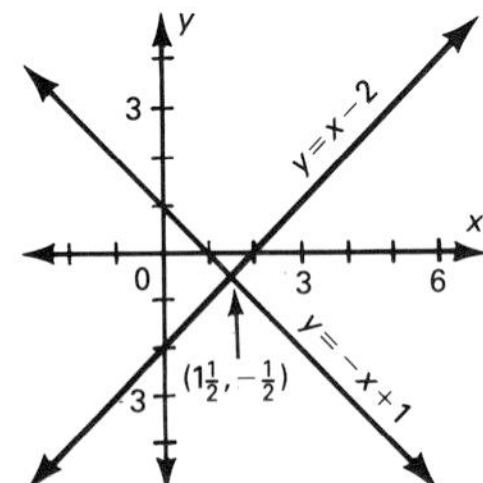

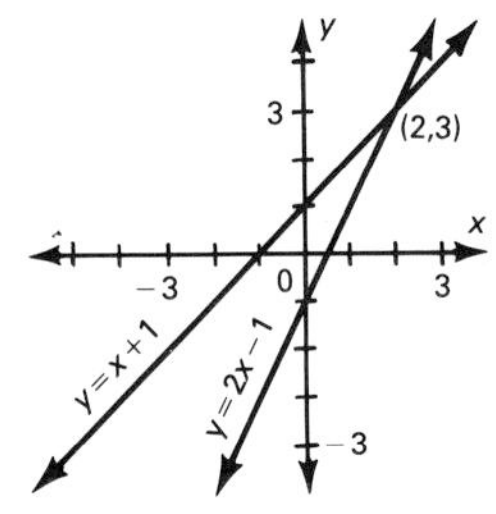

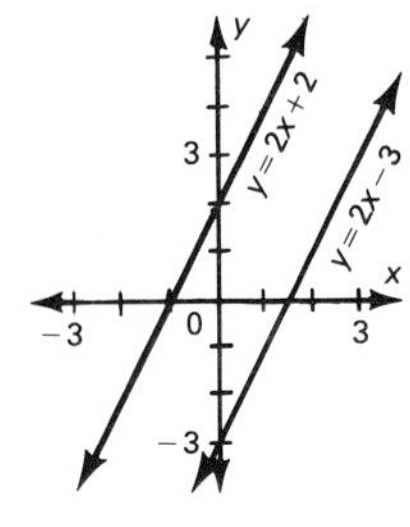

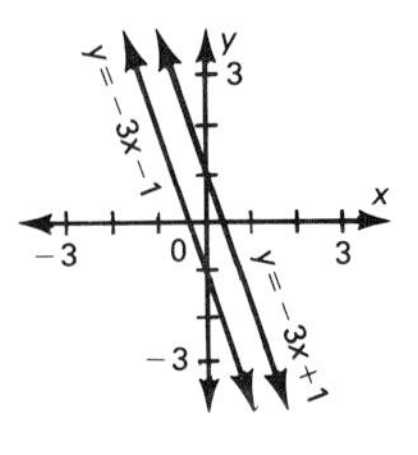

9. Intersection: (1,3)

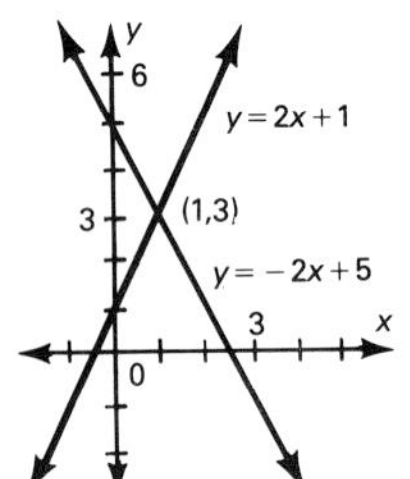

11. Intersection: ϕ

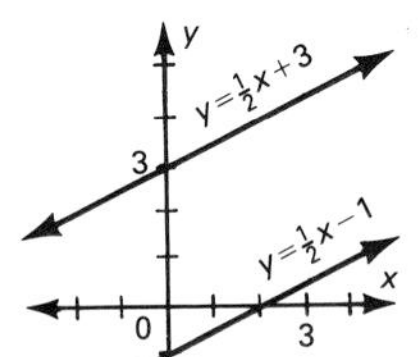

13. The lines intersect.

15. The lines intersect.

17.

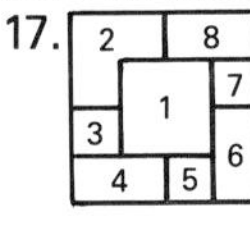

Page 412 Classroom Exercises For Ex. 1-11 odd, answers may vary slightly. 1. 55 feet 3. 22 feet 5. 149 miles 7. 75 miles 9. 6 feet 11. 8 feet

Page 413 Written Exercises For Ex. 1-15 odd, answers may vary slightly. 1. 90 miles 3. 56 miles 5. 22 miles 7. 123 million 9. 132 million 11. 270 million 13. 9 kilometers 15. 15 kilometers

Pages 414-415 1. 40°F 3. b 5. About 20°F 7. See the graph at the right.

7. Wind Chill Temperature (in °F); Thermometer Reading (in °F); 30 mph; 40 mph

Pages 416-418 Chapter Review 1. (–3,2) 3. (3,–1) 5. (–3,0)

For the graphs of Ex. 7-21, odd, see page 561.

19.

x	–4	0	1	3
y	–9	–5	–4	–2

21.

x	–4	0	1	3
y	–7	–3	–2	0

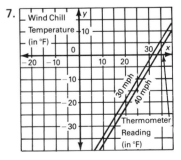

23. Subtract 5 from each x value, or add –5 to each x value. 25. Yes
27. No, because the pairs (3,1) and (3,–1) have the same x value.

29. Domain: $\{7, 4, -6\}$; Range: $\{-2, 0, 1\}$ 31. Domain: $\{-\frac{1}{2}, 3, 0, 2\}$; Range: $\{1, -4, -1\}$ 33. Range: $\{-1, 2, 3, 4\}$ 35. Range: $\{-10, -4, 1, -\frac{1}{4}\}$ 37. See p. 561.

39. See p. 561. 41. –6 43. 15 45. 12 47. –6 49. Intersection: (–2,–5) See p. 561 for the graph.
For Ex. 51-55 odd, answers may vary slightly. 51. 50 miles 53. 150 miles 55. 300 miles

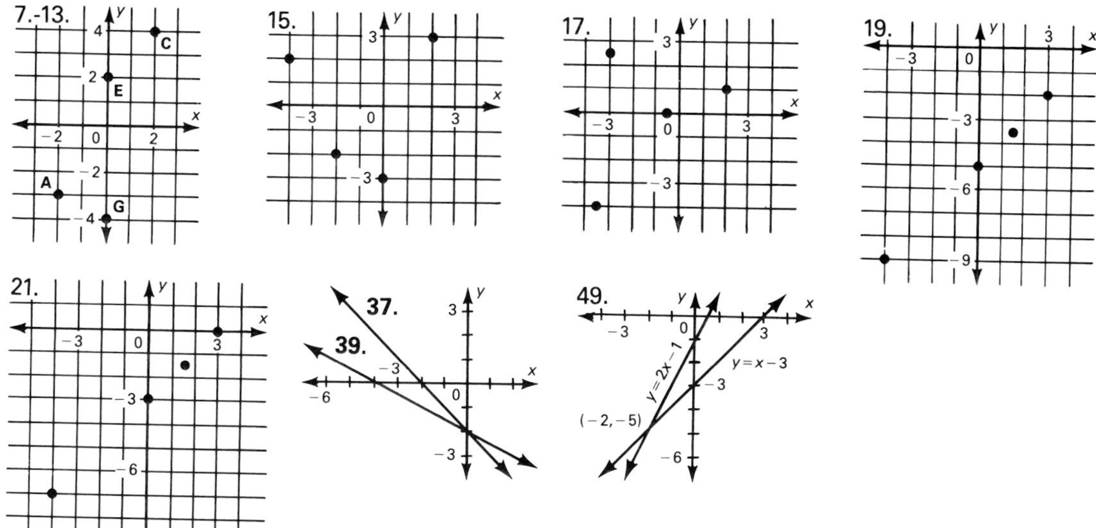

CHAPTER 15 EQUATIONS WITH TWO VARIABLES

Page 420 **P-1** A = 1, B = 1, C= −2

Page 421 **Classroom Exercises** 1. $y = x - 1$ 3. $y = 2x + 5$ 5. $y = -x + 6$ 7. $y = -4x + 3$ 9. Yes 11. Yes 13. No

Page 421 **Written Exercises**

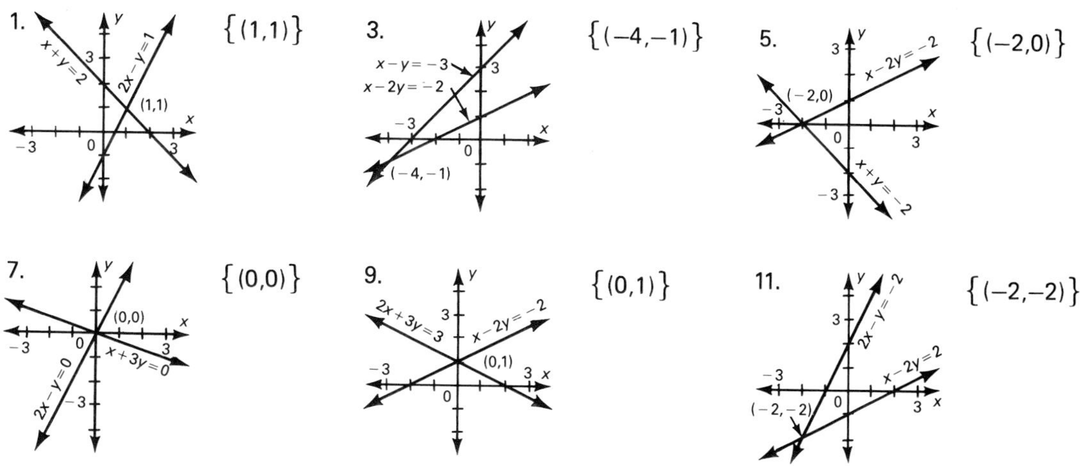

Pages 422-423 **P-1** a. $2x = 6$ b. $2y = -12$ c. $2y = 12$ **P-2** a. $2y = 4$ b. $3y = 6$ c. $2x = 0$

Page 423 **Classroom Exercises** 1. $3y = 2$ 3. $5x = 6$ 5. $-2y = 4$ 7. $-3y = -1$ 9. $8x = 16$; $x = 2$ 11. $2y = -2$; $y = -1$

Page 424 **Written Exercises** 1. $\{(-1,2)\}$ 3. $\{(3,6)\}$ 5. $\{(0,1)\}$ 7. $\{(5,1)\}$ 9. $\{(0,2)\}$ 11. $\{(3,-\frac{2}{3})\}$ 13. $\{(-2,-1)\}$ 15. $\{(7,2)\}$ 17. $\{(-3,-2)\}$ 19. $\{(-1,6)\}$ 21. $\{(4,\frac{3}{2})\}$ 23. Fill the 5-liter jar. Pour 3 liters into 3-liter jar. Empty the 3-liter jar. Pour the remaining 2 liters from 5-liter jar into 3-liter jar. Fill 5-liter jar again. Fill 3-liter jar with 1 liter from 5-liter jar. The 5-liter jar will now contain exactly 4 liters of water. 25. 56 apples

Page 425 **Review Capsule for Section 15.3** 1. 7 2. $\frac{1}{4}$ 3. 14 4. 1 5. $4x - 8y$ 6. $6x - 8y$ 7. $6x - 3y$ 8. $6x - 18y$ 9. $-6x + 3y + 12$ 10. $2x - 6y + 2$

Page 425 **P-1** a. -1 b. 3 c. $-\frac{1}{5}$

Pages 426-427 **Classroom Exercises** 1. Multiply each side of the first equation by 2. 3. Multiply each side of the second equation by 3. 5. Multiply each side of the first equation by 5; then multiply each side of the second equation by -2. 7. B 9. C

Page 427 **Written Exercises** 1. $\{(5,4)\}$ 3. $\{(-4,7)\}$ 5. $\{(0,-1)\}$ 7. $\{(3,-1)\}$ 9. $\{(6,-1)\}$ 11. $\{(-2,2)\}$ 13. $\{(-1,-2)\}$ 15. $\{(2,2)\}$ 17. $\{(3,0)\}$

Page 428 **Review Capsule for Section 15.4** 1. $x + y = -10$ 2. $y - x = 5$ 3. $x = y - 4$ 4. $y = x + 12$

Page 428 **P-1** $x + y$; $x - y$

Page 430 **Classroom Exercises** For Exercises 1-9, x = the smaller number and y = the greater number. (Variables may vary.) 1. $x = y - 5$ 3. $y - x = -3$ 5. $y = 2x + 3$ 7. $y = x + 16$ 9. $2x = y + 2$

Pages 430-431 **Written Exercises** For Exercises 1-17 variables may vary. 1. a. Let x = the smaller number. Let y = the greater number. b. $x + y = 31$; $y - x = 3$ c. $x = 14$, $y = 17$; The numbers are 14 and 17. 3. a. Let x = the smaller number. Let y = the greater number. b. $x + y = 5$; $x - y = -2$ c. $x = 1\frac{1}{2}$, $y = 3\frac{1}{2}$; The numbers are $1\frac{1}{2}$ and $3\frac{1}{2}$. 5. a. Let x = the smaller number. Let y = the greater number. b. $x + y = 40$; $y - x = 2$ c. $x = 19$, $y = 21$; The numbers are 19 and 21. 7. a. Let x = the record low temperature. Let y = the record high temperature. b. $y - x = 82$; $y + x = 138$ c. $x = 28$, $y = 110$; The record low temperature is 28°F and the record high temperature is 110°F. 9. a. Let x = the low temperature in Dallas. Let y = the low temperature in Seattle. b. $x = 2y - 2$; $y - x = 13$ c. $x = -24$, $y = -11$; The record low temperature is -24°C in Dallas and -11°C in Seattle. 11. a. Let = the temperature at an altitude of 12 kilometers. Let y = the temperature at 1 kilometer. b. $x = y - 63$; $4y = -x - 3$ c. $x = -51$, $y = 12$; The temperature is 12°C at an altitude of 1 kilometer and -51°C at an altitude of 12 kilometers. 13. a. Let x = the smaller number. Let y = the greater number. b. $y = x + 13$; $x + y = 38$ c. $x = 12\frac{1}{2}$, $y = 25\frac{1}{2}$; The numbers are $12\frac{1}{2}$ and $25\frac{1}{2}$. 15. a. Let x = the smaller number. Let y = the greater number. b. $y = x + 8$; $2x = y + 4$ c. $x = 12$, $y = 20$; The numbers are 12 and 20. 17. a. Let x = the smaller number. Let y = the greater number. b. $x + y = -20$; $3y = 2x + 10$ c. $x = -14$, $y = -6$; The numbers are -14 and -6.

Page 432 **Mid-Chapter Review** For the graphs for Ex. 1-5 odd, see p. 563. 1. $\{(1,0)\}$ 3. $\{(-2,-2)\}$ 5. $\{(0,1)\}$ 7. $\{(-2,-1)\}$ 9. $\{(-4,1)\}$ 11. $\{(-3,10)\}$ 13. $\{(3,3)\}$ 15. $\{(-22,-9)\}$ 17. $\{(-\frac{1}{7}, -\frac{5}{7})\}$ For Exercises 19-21 variables may vary. 19. Let x = the smaller number; let y = the greater number; $x + y = 13$; $y - x = 7$; $x = 3$, $y = 10$; The numbers are 3 and 10. 21. Let x = the low termperature in Miami; let y = the low temperature in Milwaukee; $y = x - 54$; $x + y = 42$; $x = 48$, $y = -6$; The low temperature is 48°F in Miami and −6°F in Milwaukee.

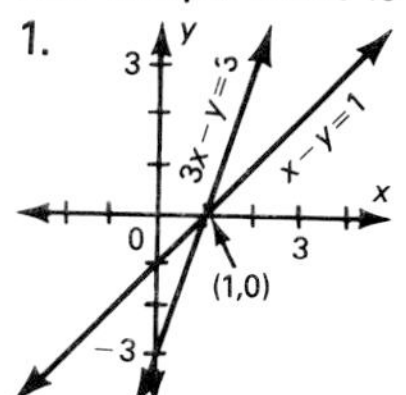

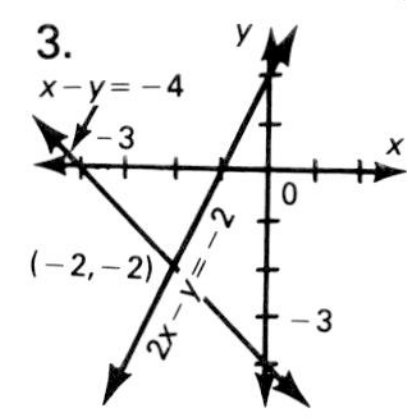

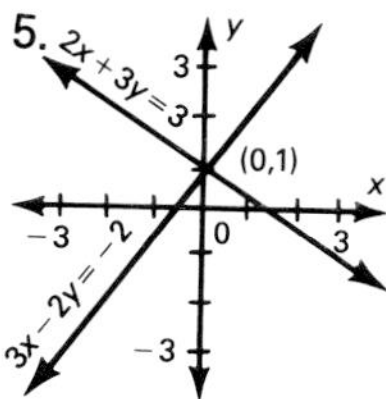

Page 433 1. 37 miles per hour 3. 52 miles per hour

Page 434 **Review Capsule for Section 15.5** 1. $x = -4$ 2. $x = 10$ 3. $x = -1$ 4. $x = 2\frac{3}{4}$ 5. $x = 23$ 6. $x = -\frac{1}{3}$

Pages 434-435 **P-1** $y = -2x + 1$; $x = -\frac{1}{2}y + \frac{1}{2}$ **P-2** a. $x + (x + 2) = 6$ b. $(y + 1) + 2y = -2$ c. $2x + (x + 4) = 7$

Page 436 **Classroom Exercises** 1. $x = -y + 10$ or $y = -x + 10$ 3. $y = -3x + 1$ or $x = -\frac{1}{3}y + \frac{1}{3}$ 5. $x = 2y - 7$ or $y = \frac{1}{2}x + \frac{7}{2}$ 7. $y = 3x - 9$ or $x = \frac{1}{3}y + 3$ 9. $3x - 5x - 10 = 3$ 11. $6y - 6 - 4y = -5$ 13. $-3x - 4x + 2 = 6$

Page 436 **Written Exercises** 1. $\{(2,-3)\}$ 3. $\{(-5,-2)\}$ 5. $\{(7,-3)\}$ 7. $\{(-7,-18)\}$ 9. $\{(1\frac{1}{2},3)\}$ 11. $\{(14,10)\}$ 13. $\{(-3,-\frac{1}{2})\}$ 15. $\{(-2,1\frac{1}{3})\}$

Page 436 **Review Capsule for Section 15.6** 1. $6x + 5y$ 2. $7x + 9y$ 3. $1.25x + 0.85y$ 4. $1.95x + 2.25y$

Pages 438-439 **Classroom Exercises** 1. \$0.18; $0.25x$; $0.18y$; $x + y = 8$; $0.25x + 0.18y = 1.65$ 3. \$5; \$2; $5x$; $2y$; $x + y = 48$; $5x + 2y = 216$ 5. $0.12x$; $0.15y$; $x + y = 12{,}000$; $0.12x + 0.15y = 1560$

Page 440 **Written Exercises** For Ex. 1-5, variables may vary. 1. a. Let x = the number of \$22 shirts; total cost 22x; Let y = the number of \$28 shirts; total cost: 28y b. $x + y = 8$; $22x + 22y = 206$ c. There are 3 shirts that cost \$22 each and 5 shirts that cost \$28 each. 3. a. Let x = the amount invested at 8%; interest: 0.08y; Let y = the amount invested at 10%; interest: 0.10y b. $x + y = 5000$; $0.08x + 0.10y = 456$ c. \$2200 was invested at 8% and \$2800 was invested at 10% 5. a. Let x = the amount borrowed at 12%; total interest: 0.12x; Let y = the amount borrowed at 15%; total interest: 0.15y b. $x + y = 25{,}000$; $0.12x + 0.15y = 3510$ c. She borrowed \$8000 at 12% and \$17,000 at 15%.

Page 442 **Classroom Exercises** For Ex. 1, variables may vary. 1. $2x + 2(x + 12) = 36$ 3. $x + y = 37$ 5. $x + y = 27$ 7. $92x + 85y = 1225$

Page 443 **Written Exercises** 1. $x = y - 5$; $2x + 2y = 54$; $x = 11$, $y = 16$; The length of the rectangle is 16 millimeters and the width is 11 millimeters. For Ex. 3-9, variables may vary. 3. $x = y + 5$; $2x + y = 35$; $x = 13\frac{1}{3}$, $y = 8\frac{1}{3}$; The lengths of the sides are $13\frac{1}{3}$ inches, $13\frac{1}{3}$ inches, and $8\frac{1}{3}$ inches.

5. $y = 2x$; $x + y = 60$; $x = 20$, $y = 40$; Mrs. Johner is 40 and her son is 20. 7. $x = 2y + 2$; $x - y = 6$; $x = 10$, $y = 4$; John is 10 and his brother is 4. 9. $x = y + 111$; $x + y = 697$; $x = 404$, $y = 293$; The distance from St. Louis to Chicago is 404 miles. The distance from Chicago to Minneapolis is 293 miles.

Page 444 1. 21, 25, 29, 33 3. $7\frac{1}{8}, 7, 6\frac{7}{8}, 6\frac{3}{4}$ 5. −64, −128, −256, −512 7. 4.52, 5.02, 5.52, 6.02 9. 20, 26, 33, 41 11. 36, 36, 45, 45 13. 25, 16, 9, 4 15. $\frac{1}{36}, \frac{1}{49}, \frac{1}{64}, \frac{1}{81}$

Page 445 1.  3. 5. 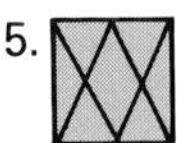 7. 9. 11.

Pages 446-448 Chapter Review For the graphs for Ex. 1-5 odd, see below. 1. $\{(-2,-1)\}$ 3. $\{(-1,1)\}$ 5. $\{(0,1)\}$ 7. $\{(-2,0)\}$ 9. $\{(\frac{1}{2},-1)\}$ 11. $\{(2,\frac{5}{3})\}$ 13. $\{(-3,4)\}$ 15. $\{(1,-1)\}$ 17. $\{(-6,-1)\}$ For Exercises 19-21, variables may vary. 19. a. Let x = the smaller number. Let y = the greater number. b. $x + y = 32$; $y - x = 7$ c. $x = 12\frac{1}{2}$, $y = 19\frac{1}{2}$; The numbers are $12\frac{1}{2}$ and $19\frac{1}{2}$. 21. a. Let x = the higher temperature. Let y = the lower temperature. b. $2x = 9y - 1$; $x + y = 104$ c. $x = 85$, $y = 19$; The higher temperature is 85°F. The lower temperature is 19°F. 23. $\{(-1,-1)\}$ 25. $\{(\frac{5}{2},\frac{1}{2})\}$ 27. $\{(14,10)\}$ For Exercises 29-37, variables may vary. 29. a. Let x = the number of student tickets. Let y = the number of adult tickets. b. $x + y = 205$; $2x + 3y = 485$ c. $x = 130$, $y = 75$; 130 student tickets and 75 adult tickets were sold. 31. a. Let x = the number of glasses of iced tea. Let y = the number of glasses of lemonade. b. $x + y = 30$; $0.15x + 0.20y = 5.10$ c. $x = 18$, $y = 12$; They sold 18 glasses of iced tea and 12 glasses of lemonade. 33. $x = 2y - 10$; $2x + 2y = 28$; $x = 6$, $y = 8$; The length is 8 centimeters and the width is 6 centimeters. 35. $y = x + 29$; $x + y = 61$; $x = 16$, $y = 45$; Pat is 16 and her teacher is 45. 37. $x = 2y + 105$; $x + y = 685$; $x = 491\frac{2}{3}$, $y = 193\frac{1}{3}$; The distance from San Francisco to Portland is $491\frac{2}{3}$ miles. The distance from Portland to Seattle is $193\frac{1}{3}$ miles.

1. $x - y = -1$; $x + 2y = -4$; $(-2,-1)$
3. $x - 2y = -3$; $x - y = -2$; $(-1,1)$
5. $2x + 3y = 3$; $x - 2y = -2$; $(0,1)$

Pages 449-450 Cumulative Review: Chapters 1-15 1. $\frac{ax}{3}$ 3. $2t$ 5. 16% 7. $-13k$ 9. $a(-5a - 8)$ 11. $-4y(y + 1)$ 13. Let $15x$ = the amount of milk; $2x$ = the amount of cream; $15x + 2x = 102$; There are 90 gallons of milk and 12 gallons of cream. For the graphs for Ex. 15-21 odd, see page 565.

19. $a < 2$ 21. $x > 5$ 23. Variable may vary. Let x = the lowest score she can get on the fourth test; $\frac{78 + 93 + 81 + x}{4} > 85$; The lowest score Hilda can get on the fourth test is 89. 25. See page 565.

27. See page 565. 29. Domain: $\{-3,5,7\}$; Range: $\{0,2,11\}$ 31. See page 565. 33. x intercept = 3; y intercept = −3 35. x intercept = −8; y intercept = 2 37. See page 565. $\{(-3,-1)\}$ 39. $\{(2,-1)\}$ 41. $\{(4,2)\}$ 43. $\{(6,2)\}$ For Exercises 45-47, variables may vary. 45. Let x = the smaller number; let y = the greater number; $x + y = 52$; $y - x = 20$; $x = 16$, $y = 36$; The numbers are 16 and 36. 47. Let x = the number of miles Emily walked; let y = the number of miles Janet walked; $x + y = 42$; $0.30x + 0.20y = 10.40$; $x = 20$, $y = 22$; Emily walked 20 miles and Janet walked 22 miles.

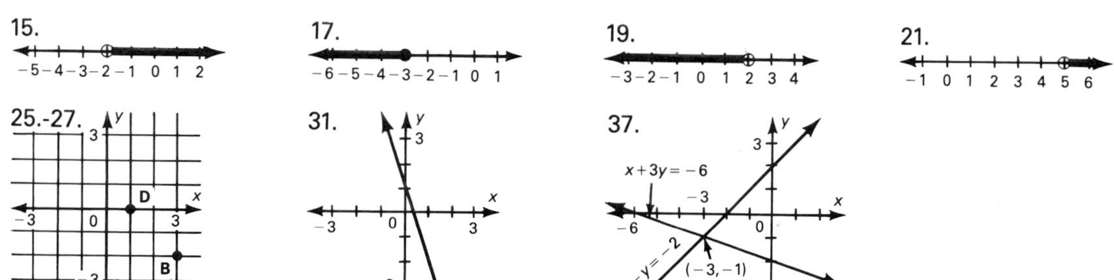

CHAPTER 16 RADICALS

Pages 452-454 **P-1** a. $\sqrt{144}$ b. $\sqrt{150}$ c. $\sqrt{35}$ **P-2** a. $\sqrt{3}\cdot\sqrt{11}$ b. $\sqrt{2}\cdot\sqrt{5}$ c. $\sqrt{2}\cdot\sqrt{21}$ or $\sqrt{6}\cdot\sqrt{7}$ or $\sqrt{3}\cdot\sqrt{14}$ d. $\sqrt{6}\cdot\sqrt{9}$ or $\sqrt{3}\cdot\sqrt{18}$ or $\sqrt{2}\cdot\sqrt{27}$ **P-3** a. $2\sqrt{11}$ b. $\sqrt{21}$ c. $\sqrt{71}$ d. $3\sqrt{7}$ **P-4** a. $6\sqrt{2}$ b. $9\sqrt{2}$ c. $4\sqrt{11}$ d. $5\sqrt{3}$

Pages 454-455 **Classroom Exercises** 1. $\sqrt{12}$ 3. $\sqrt{35}$ 5. $\sqrt{28}$ 7. $\sqrt{90}$ 9. $\sqrt{3}\cdot\sqrt{5}$ 11. $\sqrt{6}\cdot\sqrt{11}$ or $\sqrt{2}\cdot\sqrt{33}$ or $\sqrt{3}\cdot\sqrt{22}$ 13. $\sqrt{2}\cdot\sqrt{36}$ or $\sqrt{8}\cdot\sqrt{9}$ or $\sqrt{4}\cdot\sqrt{18}$ or $\sqrt{6}\cdot\sqrt{12}$ or $\sqrt{3}\cdot\sqrt{24}$ 15. $\sqrt{4}\cdot\sqrt{13}$ or $\sqrt{2}\cdot\sqrt{26}$ 17. $2\sqrt{3}$ 19. $2\sqrt{5}$ 21. $3\sqrt{3}$ 23. $3\sqrt{6}$

Page 455 **Written Exercises** 1. $\sqrt{14}$ 3. $\sqrt{45}$ 5. $\sqrt{56}$ 7. $\sqrt{36}$ 9. $\sqrt{3}\cdot\sqrt{2}$ 11. $\sqrt{8}\cdot\sqrt{8}$ or $\sqrt{2}\cdot\sqrt{32}$ or $\sqrt{16}\cdot\sqrt{4}$ 13. $\sqrt{9}\cdot\sqrt{4}$ or $\sqrt{6}\cdot\sqrt{6}$ or $\sqrt{3}\cdot\sqrt{12}$ or $\sqrt{2}\cdot\sqrt{18}$ 15. $\sqrt{11}\cdot\sqrt{15}$ or $\sqrt{3}\cdot\sqrt{55}$ or $\sqrt{5}\cdot\sqrt{33}$ 17. $2\sqrt{6}$ 19. $3\sqrt{5}$ 21. $4\sqrt{5}$ 23. $4\sqrt{7}$ 25. $\sqrt{199}$ 27. $6\sqrt{5}$ 29. $\sqrt{30}$ 31. $\sqrt{143}$ 33. $10\sqrt{2}$ 35. $2\sqrt{23}$ 37. $3\sqrt{15}$ 39. $5\sqrt{38}$ 41. The triangles have the same area. A = 48 square meters

Pages 456-457 **P-1** a. 9 b. 16 c. 1225 d. 1296 **P-2** a. 5.745 b. 8.718 c. 10.863 d. 11.358 **P-3** a. 21 b. 139 c. 95 d. 52 **P-4** a. 14.140 b. 36.249 c. 15.652 d. 14.388

Pages 459-460 **Classroom Exercises** 1. 1089 3. 14,161 5. 3721 7. 1444 9. 9.165 11. 11.662 13. 6.633 15. 11.446 17. 145 19. 64 21. 134 23. 78 25. 6(4.796) 27. 7(9.695)

Page 460 **Written Exercises** 1. 7.874 3. 6.083 5. 9.644 7. 10.630 9. 29 11. 88 13. 103 15. 18 17. 13.114 19. 12.369 21. 13.268 23. 16.585 25. 9.434 27. 69 29. 15.100 31. 10.198 33. 97 35. 17.436 37. 127.3 feet 39. 17.9 41. 32.0 43. 70.9 45. 15.4 47. 35.2 49. 67.9

Page 461 **Review Capsule for Section 16.3** 1. $n(3n+2)$ 2. $3y^2(5-y)$ 3. $h(6h-1)$ 4. $7z^2(1+2z)$ 5. 5p 6. 11.2g 7. 5u 8. 5b 9. 13w 10. 5d 11. 9.2q 12. 8.7y 13. $2\sqrt{3}$ 14. $2\sqrt{2}$ 15. $\sqrt{30}$ 16. $2\sqrt{6}$ 17. $3\sqrt{2}$ 18. $3\sqrt{3}$ 19. $3\sqrt{5}$ 20. $5\sqrt{2}$

Pages 462-463 **P-1** a, c **P-2** a. $17\sqrt{7}$ b. $\frac{3}{5}\sqrt{13}$ c. $3\sqrt{11}$ **P-3** a. $7\sqrt{3}$ b. $\frac{1}{3}\sqrt{11}$ c. $-8\sqrt{5}$ **P-4** a. $2\sqrt{7}$ b. $-2\sqrt{6}$ c. $17\sqrt{3}$ **P-5** a. $8\sqrt{3}$ b. $\sqrt{2}$ c. $13\sqrt{2}$

Page 464 **Classroom Exercises** 1. $14\sqrt{2}$ 3. $15\sqrt{11}$ 5. $\frac{3}{4}\sqrt{14}$ 7. $3\sqrt{5}$ 9. $-2\sqrt{7}$ 11. $-\sqrt{19}$ 13. $12\sqrt{2}$ 15. $5\sqrt{5}$ 17. $\frac{3}{8}\sqrt{10}$ 19. $7\sqrt{3}$ 21. $3\sqrt{2}$ 23. $3\sqrt{5}$

Pages 464-465 **Written Exercises** 1. $16\sqrt{2}$ 3. $14\sqrt{5}$ 5. $\frac{7}{8}\sqrt{6}$ 7. $3\sqrt{6}$ 9. $13\sqrt{11}$ 11. $-\frac{2}{3}\sqrt{3}$ 13. $6\sqrt{6}$ 15. $-9\sqrt{10}$ 17. $\frac{9}{2}\sqrt{6}$ 19. $8\sqrt{2}$ 21. $8\sqrt{2}$ 23. $\sqrt{6}$ 25. $3\sqrt{6}$ 27. $8\sqrt{3}$ 29. $-\frac{3}{8}\sqrt{17}$ 31. $8\sqrt{2}$ 33. $2\sqrt{3}$ 35. $-\sqrt{5}$ 37. $16\sqrt{2}$ 39. $8\sqrt{3}$ 41. $15\sqrt{15}$

Page 465 **Mid-Chapter Review** 1. $9\sqrt{37}$ 3. $12\sqrt{43}$ 5. $2\sqrt{26}$ 7. $2\sqrt{30}$ 9. $2\sqrt{35}$ 11. $3\sqrt{23}$ 13. 14.560 15. 19.494 17. 20.100 19. 22.539 21. 39.395 23. 63.773 25. 8.484 kilometers 27. $9\sqrt{17}$ 29. $23\sqrt{17}$ 31. $2\sqrt{2}$ 33. $10\sqrt{5}$ 35. $6\sqrt{3}$ 37. $-6\sqrt{10}$

Pages 466-467 **P-1** a. $8\sqrt{55}$ b. $-\frac{1}{3}\sqrt{21}$ c. $-4\sqrt{38}$ **P-2** a. $4\sqrt{7}$ b. 9 c. $-15\sqrt{3}$

Page 467 **Classroom Exercises** 1. $-20\sqrt{3}$ 3. $-70\sqrt{6}$ 5. $-6\sqrt{10}$ 7. $8\sqrt{7}$ 9. $\sqrt{14}$ 11. $-12\sqrt{26}$ 13. $10\sqrt{55}$ 15. $-9\sqrt{14}$ 17. $2\sqrt{3}$ 19. $-12\sqrt{10}$ 21. $-24\sqrt{6}$ 23. $8\sqrt{3}$

Page 467 **Written Exercises** 1. $-72\sqrt{5}$ 3. $-120\sqrt{7}$ 5. $-84\sqrt{13}$ 7. $12\sqrt{3}$ 9. $\sqrt{65}$ 11. $-\sqrt{66}$ 13. $-\sqrt{35}$ 15. $-\frac{9}{2}\sqrt{22}$ 17. $10\sqrt{6}$ 19. $-2\sqrt{5}$ 21. $-2\sqrt{7}$ 23. $18\sqrt{15}$ 25. $\sqrt{230}$ 27. $-9\sqrt{10}$ 29. $3\sqrt{30}$ 31. $-12\sqrt{10}$

Page 468 **Review Capsule for Section 16.5** 1. 5 2. 7 3. $\frac{5}{7}$ 4. 4 5. 9 6. $\frac{4}{9}$

Pages 468-469 **P-1** a. 2 b. 3 c. 3 d. 5 **P-2** a. $\frac{\sqrt{2}}{7}$ b. $\frac{\sqrt{7}}{10}$ c. $\frac{\sqrt{11}}{12}$ d. $\frac{\sqrt{5}}{9}$

Page 470 **Classroom Exercises** 1. 2 3. 5 5. 6 7. $\frac{\sqrt{3}}{8}$ 9. $\frac{\sqrt{7}}{6}$ 11. $\frac{\sqrt{21}}{7}$ 13. $\frac{\sqrt{15}}{3}$ 15. $\frac{3}{8}$ 17. $\frac{3\sqrt{35}}{5}$ 19. $3\sqrt{15}$ 21. $\frac{\sqrt{2}}{2}$ 23. $\frac{\sqrt{7}}{7}$ 25. $\frac{\sqrt{14}}{7}$ 27. $\frac{\sqrt{30}}{3}$ 29. $\sqrt{5}$

Pages 470-471 **Written Exercises** 1. 2 3. 3 5. 5 7. 7 9. 3 11. $\frac{\sqrt{2}}{3}$ 13. $\frac{\sqrt{13}}{6}$ 15. $\frac{\sqrt{29}}{8}$ 17. $\frac{\sqrt{11}}{8}$ 19. $\frac{\sqrt{21}}{10}$ 21. $\frac{\sqrt{30}}{3}$ 23. $\frac{\sqrt{42}}{6}$ 25. $\frac{\sqrt{26}}{13}$ 27. $\frac{\sqrt{22}}{22}$ 29. $\frac{3\sqrt{14}}{14}$ 31. $\frac{\sqrt{3}}{2}$ 33. $\frac{\sqrt{5}}{5}$ 35. $\frac{\sqrt{6}}{6}$ 37. $\frac{2\sqrt{3}}{3}$ 39. $\frac{2\sqrt{10}}{5}$ 41. $\frac{\sqrt{5}}{9}$ 43. 6 45. $\frac{\sqrt{6}}{3}$ 47. 9 49. $\frac{\sqrt{51}}{17}$ 51. $\frac{5\sqrt{22}}{22}$ 53. 4 55. $\frac{4\sqrt{11}}{11}$ 57. 8 59. $\frac{\sqrt{23}}{12}$ 61. 38 bumpers 63. 1, 4, 8, and 9 years old

Page 472 1. 200 centimeters 3. 67 centimeters 5. decrease

Pages 473-474 **Chapter Review** 1. $\sqrt{39}$ 3. $8\sqrt{11}$ 5. $9\sqrt{23}$ 7. 93 9. $5\sqrt{3}$ 11. $5\sqrt{2}$ 13. $3\sqrt{7}$ 15. $6\sqrt{3}$ 17. $2\sqrt{15}$ 19. 11.136 21. 10.583 23. 18.520 25. 24.556 27. 11.958 29. 1696.8 meters 31. $15\sqrt{5}$ 33. $17\sqrt{3}$ 35. $-12\sqrt{6}$ 37. $-3\sqrt{6}$ 39. $6\sqrt{2}$ 41. $-66\sqrt{10}$ 43. $18\sqrt{2}$ 45. 378 47. $36\sqrt{2}$ 49. 3 51. 5 53. $\frac{\sqrt{7}}{6}$ 55. $\frac{\sqrt{10}}{9}$ 57. $\frac{\sqrt{285}}{38}$ 59. $\frac{\sqrt{30}}{2}$ 61. $3\sqrt{2}$

CHAPTER 17 MORE ALGEBRAIC FRACTIONS

Pages 476-477 **P-1** a. $-\frac{1}{3}$ b. $\frac{1}{2}$ c. $-\frac{3b}{4}$ d. $\frac{2}{k}$ **P-2** a. $-\frac{4}{9}$ b. $\frac{x-4}{x+4}$ c. $\frac{2(y^2+1)}{5(y+1)}$ d. $\frac{y-2}{5(3y+1)}$ **P-3** a. $\frac{x}{3}$ b. $\frac{2}{3}$ c. $\frac{2}{5}$ d. $\frac{4xy}{1+2x}$

Pages 477-478 **Classroom Exercises** 1. $\frac{2}{5}$ 3. $-\frac{x}{2y}$ 5. $-\frac{x}{2y}$ 7. $\frac{2}{3x}$ 9. $\frac{4(x-2)}{5(x+2)}$ 11. $-\frac{3}{2}$ 13. $\frac{5t}{6s(s+3)}$ 15. $\frac{a}{3}$ 17. $\frac{x}{x+1}$ 19. $\frac{3}{5}$ 21. $\frac{4m}{n+m}$ 23. $\frac{a-b}{1-b^2}$

Pages 478-479 Written Exercises 1. $-\frac{x}{y}$ 3. $\frac{r}{2t}$ 5. $-\frac{5x}{3y}$ 7. $-\frac{3a}{7}$ 9. $-\frac{x-7}{2(x+7)}$ 11. $\frac{4}{7}$ 13. $-\frac{x-1}{2(x+1)}$ 15. $\frac{m+2}{2m+1}$ 17. $\frac{2ab}{3(a-b)}$ 19. $\frac{p}{5q}$ 21. $\frac{2x}{1-3y}$ 23. $-\frac{2m}{2m-n}$ 25. $-\frac{2}{3}$ 27. $\frac{3}{5}$ 29. $\frac{4c^2d}{3(2c-d)}$ 31. $-\frac{x+2}{3}$ 33. 145.04°F 35. 13.82°F 37. −14.5°C 39. −1 41. −1 43. 1 45. $-\frac{1}{2}$

Page 479 Review Capsule for Section 17.2 1. $\frac{4}{5}$ 2. $\frac{5}{6}$ 3. $\frac{4}{11}$ 4. $\frac{11}{35}$ 5. 3 6. 4 7. 10d 8. 6mp 9. (x − 2) 10. 6 11. 7; (a − 1) 12. (y + 1); (x − 2)

Pages 480-481 P-1 a. $\frac{3}{2}$ b. $-\frac{6}{y}$ c. $\frac{3t}{7}$ d. $\frac{3}{2p^2}$ **P-2** a. $\frac{2}{x}$ b. $\frac{3}{p+1}$ c. $\frac{a-6}{a+2}$ **P-3** a. $\frac{1}{15}$ b. $-\frac{1}{2}$ c. $\frac{y}{26}$

Page 481 Classroom Exercises 1. $\frac{3}{4}$ 3. $-\frac{3a}{5x}$ 5. $\frac{3}{5}$ 7. $\frac{1}{3}$ 9. $\frac{4}{3s}$ 11. $\frac{2(m-n)}{m+2}$

Page 482 Written Exercises 1. $\frac{3}{2}$ 3. $-\frac{2g}{5t}$ 5. $\frac{5}{6}$ 7. $-\frac{15}{14}$ 9. $\frac{1}{2}$ 11. $\frac{1}{4}$ 13. $\frac{n+6}{n+8}$ 15. $-\frac{1}{3}$ 17. $\frac{1}{d}$ 19. $\frac{4z}{3}$ 21. $\frac{25}{6}$ 23. $\frac{9}{4t}$ 25. $-\frac{4}{3}$ 27. $\frac{35x}{6y}$ 29. $\frac{b^2}{3}$ 31. 2s

Page 482 Mid-Chapter Review 1. $-\frac{3}{2}$ 3. $\frac{2}{5s}$ 5. $-\frac{5}{4}$ 7. $-\frac{1}{9}$ 9. $-\frac{3}{2k}$ 11. $-\frac{8}{3w}$ 13. $-\frac{1}{2}$

Page 483 Review Capsule for Section 17.3 1. $\frac{6}{5}$ 2. $\frac{4}{3}$ 3. $\frac{13}{10}$ 4. $\frac{9}{5}$ 5. $\frac{8}{99}$ 6. $\frac{48}{49}$ 7. q 8. 3 9. $\frac{3}{2y}$ 10. $\frac{-1}{2k}$ 11. $\frac{7}{5}$ 12. $\frac{r}{2}$

Page 484 P-1 a. −2 b. $-\frac{3p^2}{49}$ c. $\frac{3r}{s}$ d. $\frac{7y^2}{8x^2}$ **P-2** a. $\frac{1}{2}$ b. $-\frac{9}{4}$ c. $\frac{-2(7+z)}{7-z}$ **P-3** a. $\frac{2}{3}$ b. $\frac{z^2}{15}$ c. $\frac{2}{f}$ d. $\frac{a^4}{8}$

Page 485 Classroom Exercises 1. $\frac{1}{3}$ 3. $\frac{x^2}{15}$ 5. 2 7. $\frac{7r^2}{(g-1)(g+1)}$ 9. $\frac{5}{2}$ 11. 2

Page 485 Written Exercises 1. $\frac{3}{2}$ 3. $\frac{7}{2}$ 5. $-\frac{10}{7}$ 7. $\frac{3}{10}$ 9. $\frac{5}{3}$ 11. $\frac{1}{2}$ 13. $\frac{2x+3}{2(3x-2)}$ 15. $\frac{5}{3y}$ 17. $\frac{1}{2}$ 19. $\frac{y^2}{20}$

Page 486 Review Capsule for Section 17.4 1. $\frac{3}{5}$ 2. $\frac{1}{4}$ 3. $\frac{1}{3}$ 4. 2 5. 3x 6. 4t 7. 2r + 2 8. 4w + 2 9. 11x + 6 10. 3n + 1

Page 487 P-1 a. $\frac{12}{n}$ b. $\frac{r-7}{2x}$ c. $\frac{5y+5}{y+5}$ **P-2** a. 3 b. 3 c. $\frac{3}{2}$

Page 487 Classroom Exercises 1. $\frac{7}{y}$ 3. $\frac{3a}{x}$ 5. $\frac{10}{b-3}$ 7. $\frac{2y+3}{g+4}$ 9. 2 11. $\frac{2(p-2)}{p+2}$

Page 488 Written Exercises 1. $-\frac{3}{5x}$ 3. $\frac{1}{2m}$ 5. $\frac{2s}{3t}$ 7. $\frac{4x}{5y}$ 9. $\frac{6f}{f+7}$ 11. $\frac{p}{q+6}$ 13. $\frac{2w-15}{w-8}$ 15. $\frac{-(3z+2)}{z+4}$ 17. $\frac{4q+2}{q+1}$ 19. 2 21. 6 23. $\frac{4}{3}$ 25. $\frac{y+1}{y+4}$ 27. 3 29. $\frac{2}{3}$ 31. 1 33. −1 35. $-\frac{2m}{n}$ 37. a. 12 cubes b. 6 cubes

Page 489 1. Five 22¢ stamps and two 17¢ stamps 3. The following combinations are possible: 10 two-point goals and 0 three-point goals, 7 two-point goals and 2 three-point goals, 4 two-point goals and 4 three-point goals, or 1 two-point goal and 6 three-point goals.

Page 490 **Chapter Review** 1. $-\frac{2m}{3p}$ 3. $-\frac{3}{4}$ 5. $-\frac{6s}{5}$ 7. $-\frac{3}{4}$ 9. $-\frac{3}{5}$ 11. $\frac{2}{y}$ 13. 6 15. $\frac{a^2}{b^2}$ 17. $\frac{3}{2x}$ 19. $\frac{r}{5s}$ 21. $\frac{5b-6}{2b-3}$ 23. $\frac{3}{8}$

CHAPTER 18 POLYNOMIALS

Page 492 **P-1** a. polynomial; binomial b. polynomial c. polynomial d. polynomial e. polynomial f. polynomial; binomial **P-2** a. $6x^2 - x$ b. $-12z^3 - 10z^2 - 3z + 15$

Page 493 **Classroom Exercises** 1. $-2r + 2$ 3. $x^2 - 3x + 6$ 5. $-2y^2 + 4y - 7$ 7. $3x^2 - \frac{3}{4}$ 9. $3x^2 - 2x - 3$ 11. $7x + 13$

Page 493 **Written Exercises** 1. $4x - 2$ 3. $2x^2 + x + 9$ 6. $-5x^2 - 2x - 7$ 7. $10x^3 + 2x^2 - x + 1$ 9. $4k^2 - 9k - 9$ 11. $4x^3 - 4x + 15$ 13. $x^4 + 2x^3 - 11$ 15. $\frac{3}{4}x^3 + x^2 - 1$ 17. $-4.7x^2 + 1.8x + 0.8$ 19. 20 ways

Page 493 **Review Capsule for Section 18.2** 1. 60 2. 12 3. 36 4. −4 5. 4 6. 2 7. −8 8. 8 9. 16

Page 494 **P-1** a. 8 b. 4 c. −5

Page 495 **Classroom Exercises** 1. 4 3. 5 5. −4 7. 0 9. −24 11. −15

Pages 495-496 **Written Exercises** 1. 11 3. 10 5. −41 7. −1 9. −5 11. −41 13. −7 15. −2 17. 80 19. −11 21. −12 23. −30 25. $-\frac{7}{16}$ 27. $-\frac{1}{8}$ 29. −0.896 31. 2.168 33. $x = 4$ 35. $x = 6$ 37. $x = 5$ 39. The train is $\frac{1}{20}$ of a mile long. The length of the tunnel is $\frac{1}{2}$ mile.

Page 496 **Review Capsule for Section 18.3** 1. 7 2. −1 3. $-7r$ 4. $-2x^2$ 5. $-15y$ 6. $-t^3$ 7. $-1\frac{3}{4}xy^2$ 8. $-1\frac{1}{2}t$ 9. $-n^2$

Page 498 **Classroom Exercises** 1. $3x - 2$ 3. $2c^2 + 3c - 5$ 5. $3x^3 + x^2 - 4x - 2$ 7. $7p + 3$ 9. $3x^2 + 4x$ 11. $5n^2 - 3n + 7$ 13. $2s^2 - 4s + 5$ 15. $-x^2y + 3xy + 6xy^2 - 2y^3$

Pages 498-499 **Written Exercises** 1. $-2x + 5$ 3. $4r^2 - 3$ 5. $7x^2 - 3x + 5$ 7. $3n^2 + 4n - 2$ 9. $8.2x^2 - 7.3x - 2.3$ 11. $3x^2 - 4x - 4$ 13. $2x^2y - 3xy^2 - 3xy - 4$ 15. $p^3 + 2p + 4$ 17. $-2x^3 + x^2 - x + 12$ 19. $2x^3 - 3x + 16$ 21. $12x^2 + 10x - 5$ 23. $2x^2 + 4x + 8$ 25. $3y$ 27. $5x^3 - 3x - 7$ 29. $2x^2 + 3x + 7$ 31. $13s^2t + 13st + 14st^2$ 33. $2x^3 + 2x^2 - 2x - 2$ 35. $4m^2 - 5m$ 37. $6xy + 3y^2 - 6$

Page 500 **Mid-Chapter Review** 1. $-t^2 - 2t + 4$ 3. $1.8m^2 - 0.9m - 2.7$ 5. $-q^3 + 11q^2 + 12q - 28$ 7. $-\frac{1}{4}a^2 + a + \frac{9}{4}$ 9. $3n^4 - 15n^2 - 20n$ 11. $2y^5 - y^3 - 1$ 13. 39 15. 29 17. 11 19. 3.3 21. 20 23. $t^2 - 2t - 7$ 25. $-3t^3 - 8t^2 - 3t + 1$ 27. $24m^3 - 2m^2 - 15m + 5$ 29. $5x^2 - 11$

Page 500 **Review Capsule for Section 18.4** 1. -7 2. -11 3. 8 4. -22 5. $-9r^2$ 6. $4m$ 7. xy^2 8. $3t^3$ 9. $-6.4rs^2$

Page 501 1. $1 \cdot 2^3 + 0 \cdot 2^2 + 1 \cdot 2^1 + 1 \cdot 2^0 = 11$ 3. $1 \cdot 2^5 + 0 \cdot 2^4 + 0 \cdot 2^3 + 1 \cdot 2^2 + 0 \cdot 2^1 + 1 \cdot 2^0 = 37$ 5. 11100_{two}, 11101_{two}, 11110_{two}, 11111_{two} 7. b

Page 502 **P-1** a. $2b^2 - 6b - 7$ b. $-2x^2 + 10x + 6$ c. $1.4q^2 - 2.2q + 8.4$ **P-2** a. $-y^2 - 12y - 7$ b. $x^2 + 8x - 3$

Page 503 **Classroom Exercises** 1. $4x - 7$ 3. $-5x^2 - 4x + 2$ 5. $2f + 5$ 7. $2x^2 + x - 2$

Page 503 **Written Exercises** 1. $4x - 5$ 3. $-x^2 + 7x$ 5. $8x^2 - 5x + 10$ 7. $5x^3 + 4x^2 - 6$ 9. $-3.5x^2 + 3.7x + 4.6$ 11. $3x^2 - x + 6$ 13. $2x^2 - 7x + 11$ 15. $-2a^3 - 2a^2 + 2a - 8$ 17. $-3x^3 - 3x^2 + 3x + 6$ 19. $-8x^2 + x - 12$ 21. $-4j^3 - j^2 + 9j - 12$ 23. $0.9x^2 - 8x + 4.7$ 25. $3x^2y^2 - 7xy^2 - 5xy^3 - 11xy + 9$

Page 504 **Review Capsule for Section 18.5** 1. -63 2. 60 3. 4 4. $-6n$ 5. $15d^2$ 6. $-4x^3y$ 7. $-9r^2s$ 8. $0.32tw$ 9. $3a^2b^3c$ 10. $3x + 6$ 11. $8q - 40$ 12. $2a - ax$

Page 505 **P-1** a. $r^2 - 81$ b. $3x^2 + 23x + 40$ c. $2y^2 - 30y + 88$ **P-2** a. $k^3 - 5k^2 + 14k - 40$ b. $2x^3 + 3x^2 - 14x - 21$ c. $-4s^4 - 11s^3 + 46s^2 + 5s$

Pages 505-506 **Classroom Exercises** 1. $x^2 + 5x + 6$ 3. $x^2 + 3x - 10$ 5. $2x^2 - 5x - 3$ 7. $u^2 - 6u + 9$ 9. $6p^2 + 11p - 35$ 11. $-6x^2 + 9x - 12$

Page 506 **Written Exercises** 1. $k^2 + 3k - 10$ 3. $x^2 - 17x + 72$ 5. $3q^2 + 7q - 6$ 7. $16x^2 - 40x + 25$ 9. $9n^2 - 49$ 11. $2x^3 - 2x^2 + 3x - 3$ 13. $2x^3 - 11x^2 + 8x - 15$ 15. $d^3 + 5d^2 - 29d - 40$ 17. $h^3 + h^2 - h + 15$ 19. $-3v^4 - 6v^3 + 2v^2 + 3v - 2$ 21. $-2x^4 + 3x^3 - 8x^2 + 9x - 6$ 23. $g^5 - g^4 - g^3 + g^2 - g - 1$ 25. $-x^4 - 4x^3 + 4x^2 + 13x - 12$ 27. $w^5 - 5w^4 - 3w^3 + 15w^2 + 2w - 10$ 29. $f^3 - 2f^2 - 9f - 2$ 31. Cut each ring of the first chain. Separate the rings and insert one ring between each of the other four chains.

Page 507 **Review Capsule for Section 18.6** 1. $2t$ 2. $-3x$ 3. $-2j^2$ 4. 5 5. $2x^2 - 10x$ 6. $-15s + 30$ 7. $12p^3 - 9p^2$ 8. $-3m^3 - 12m$

Page 508 **Classroom Exercises** 1. -5 3. x 5. $-2x$ 7. $-2x + 2$ 9. $-x - 8$ 11. $x - 9$

Page 508 **Written Exercises** 1. $2b + 3$ 3. $-2x + 5$ 5. $x - 7$ 7. $h^2 - 2h + 3$ 9. $x^2 - 4x + 6$ 11. $q^2 - 5q - 5$

Pages 509-510 1. a. Possible. "$15 at most" means he spent $15 or less. b. False. "$15 at most" means the largest possible amount was $15. c. Possible. See conclusion a. d. Can't tell. No information is given about how much Alice spent. 3. a. Possible. "Between 30 and 40" means 31, 32, 33, 34, 35, 36, 37, 38, or 39 students belong to the club. b. Possible. See conclusion a. c. Can't tell. No information is given about last year's membership. d. True. All of the possible numbers given in conclusion a are below 45. 5. a. False. "Exactly 5 out of every 7 students" means exactly 350 of the 490 students. b. True. If "exactly 5 out of every 7 students" received a grade of C or better, then the remaining 2 out of every 7 students received a grade lower than C. c. True. See conclusion b. "Exactly 2 out of every 7 students" means exactly 140 of the 490 students. d. True. See conclusion a. 350 is "more than 349 and fewer than 351."

Pages 511-512 Chapter Review 1. $-t^2 + 5t + 3$ 3. $x^3 + 3x^2 + 5$ 5. $-0.4p^2 + 3.6p - 12.6$ 7. -55 9. 1 11. -14 13. $3r - 4$ 15. $-2b^2 + 6b - 4$ 17. $4c^4 + 3c^3 + c^2 + 3c - 12$ 19. $-2x^2 + 10x - 2$ 21. $5z^2 - 5z + 15$ 23. $-2.1g^3 + 1.8g^2 + 7.8g + 0.9$ 25. $20x^4 - 24x^3 - 12x^2 + 17x + 19$ 27. $4j^3 - 12j^2 + 19j - 15$ 29. $-3d^5 - d^4 - 6d^3 + d^2 + 6$ 31. $4t^4 - t^3 + 14t^2 - 3t + 6$ 33. $5n + 2$ 35. $-3x^2 + x + 2$

Pages 513-514 Cumulative Review: Chapters 1-18 1. 69 3. 6 5. $110\frac{1}{4}$ ft^2 7. -51.6 9. $8a - 3a^2$ 11. $m = 112$ 13. $b = 2\frac{1}{2}$ 15. $x = 5$ 17. $a > -4\frac{1}{2}$

$-4\frac{1}{2}$ −6 −5 −4 −3 −2 −1 0 1

19. Domain: $\{0, -2, 2, -1\}$; Range: $\{2, -4, 8, -1\}$

For 21 and 23, see the graph at the right.

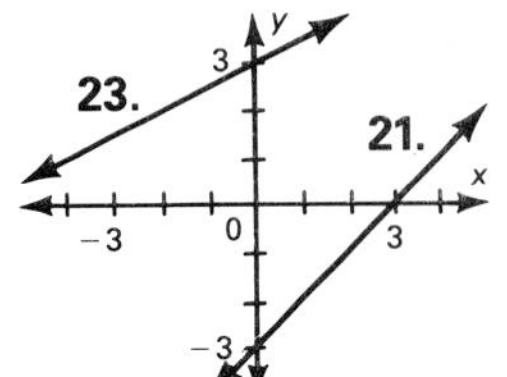

25. Let x = the greater number; y = smaller number; $x + y = 84$; $x - y = 22$; The numbers are 53 and 31. 27. $4\sqrt{5}$ 29. $5\sqrt{11}$ 31. 75 33. 12.806 35. $2\sqrt{3}$ 37. $-30\sqrt{3}$ 39. $\frac{\sqrt{5}}{3}$ 41. $\frac{\sqrt{6}}{3}$ 43. $\frac{3}{4}$ 45. $\frac{1}{2x}$ 47. $\frac{4x}{3}$ 49. $\frac{3}{5}$ 51. $\frac{-x-3}{x-3}$ 53. -11 55. $3d^2 + d + 3$ 57. $g^2 + 13g + 36$ 59. $3c^3 - 19c^2 + 8c - 12$ 61. $2p^2 + 3p - 1$

PICTURE CREDITS

Key: *(t)* top; *(b)* bottom; *(l)* left; *(r)* right; *(c)* center.

CHAPTER ONE: Page 1 *(left)*, James Israel/Shostal Associates, *(tr)*, Bill Bailey/Shostal Associates, *(br)*, Joseph Nettis/Photo Researchers, Inc. CHAPTER TWO: 23 *(left)*, Andy Howarth/Woodfin Camp and Associates, *(tr)*, John Serafin/Peter Arnold, Inc.; 45, Robert R. McElroy/Woodfin Camp and Associates; 46, Renate Hiller/Monkmeyer. CHAPTER THREE: 49 *(left)*, Richard Weiss/Peter Arnold, Inc., *(tr)*, The Granger Collection, *(br)*, Ancient Art and Architecture Collection/Robert Sheridan Photo Library; 55, HBJ Photo/Blaise Zito Associates; 57, HBJ Photo/Blaise Zito Associates; 62 *(t)*, HBJ Photo/Rodney Jones; 63, HBJ Photo/Jerry White. CHAPTER FOUR: 771, National Bureau of Standards/U. S. Dept. of Commerce, *(tr)*, HBJ Photo/Blaise Zito Associates, *(br)* Joan Menschenfreund/The Stock Market; 90, HBJ Photo/Blaise Zito Associates; 91, Stan Pantovic/Photo Researchers, Inc. CHAPTER FIVE: 109 *(left)*, The Granger Collection, *(tr)*, P. Wallich/Alpha-FPG, *(br)*, *(c)* Mark Greenberg/Visions; 113, HBJ Photo/Blaise Zito Associates; 120, *(c)* Jeffrey Vock/Visions; 125, HBJ Photo/Blaise Zito Associates; 137, Burton McNeely/The Image Bank; 142, Frederic Lewis. CHAPTER SIX: 143 *(left)*, Langley Research Center/NASA, *(tr)*, *(c)* Chris Sorensen, *(br)*, Lockheed-Georgia Co.; 154, Galen Rowell/Peter Arnold, Inc. CHAPTER SEVEN: 173 *(left)*, The Granger Collection, *(tr)*, HBJ Photo/Jerry White, *(br)*, Don Renner; 178, HBJ Photo/Blaise Zito Associates; 194, HBJ Photo/Blaise Zito Associates; 197, H. Armstrong Roberts; 205 both, HBJ Photo/Blaise Zito Associates; 208 *(l)*, George Dodge/DPI, *(c)*, HBJ Photo/Blaise Zito Associates, *(r)*, Shostal Associates; 212 Focus on Sports. CHAPTER EIGHT: 213, *(l)*, Dr. E. R. Degginger, *(tr)*, *(c)* Art Attack, *(br)*, National Park Service/Death Valley Monument; 239, Michal Heron/Monkmeyer. CHAPTER NINE: 242, *(all)*, The Granger Collection; 254, HBJ Photo/Blaise Zito Associates; 255, Don Renner; 258, HBJ Photo/Blaise Zito Associates. CHAPTER TEN: 277 *(l)*, Ken Karp/Omni Photo Communications, *(br)*, Jon Riley/The Stock Shop; 288, Alan Carey/The Image Works; 298, HBJ Photo/Blaise Zito Associates; 302, Steven Young/Taurus Photos. CHAPTER ELEVEN: 305 *(l)*, Jose Maria Rosario/Shostal Associates, *(tr)*, H. Armstrong Roberts, *(br)*, Dr. E. R. Degginger/Bruce Coleman, Inc.; 328, 329, HBJ Photo/Blaise Zito Associates; 332, David Madison/Focus on Sports. CHAPTER TWELVE: 333 *(l)*, David Pollack/The Stock Market, *(tr)*, Peter Arnold, Inc., *(br)*, HBJ Photo/Jerry White; 334, 341, 344, HBJ Photo/Blaise Zito Associates; 346, Joe Barnell/Shostal Associates; 360, HBJ Photo/Blaise Zito Associates. CHAPTER THIRTEEN: 363 *(l)*, Phil Degginger/Bruce Coleman, Inc., *(tr)*, J. R. Simon/Bruce Coleman, Inc., *(br)*, J. Fennell/Bruce Coleman, Inc.; 371, Mark Stevenson/Nawrocki Stock Photo; 378, Ewing Galloway; 383, 386, HBJ Photo/Blaise Zito Associates. CHAPTER FOURTEEN: 389 *(l)*, Shostal Associates, *(tr)*, Peter Arnold, Inc.; 412, HBJ Photo/Blaise Zito Associates; 414, John Lei/Omni Photo Communications. CHAPTER FIFTEEN: 419, *(l)* and *(tr)*, NASA, *(br)*, Omni Photo Communications; 433, Photri; 437, 438, HBJ Photo/Blaise Zito Associates; 439, AP/Wide World Photos. CHAPTER SIXTEEN: 451, *(l)* United Nations Photo, *(tr and br)*, NASA; 472 *(t)*, Photri, *(b)*, L. L. T. Rhodes/Taurus Photos; 474, *(c)* Jim Whitmer. CHAPTER SEVENTEEN: 475 *(l)*, Shostal Associates, *(tr)*, Bill Anderson/Shostal Associates, *(br)*, Gabe Palmer/Palmer-Kane. CHAPTER EIGHTEEN: 491 *(l)*, Browning/Shostal Associates, *(tr)*, Shostal Associates, *(br)*, Gabe Palmer/Palmer-Kane; 496, Dykstra/Index Stock International; 510, HBJ Photo/Blaise Zito Associates.